QSAR IN SAFETY EVALUATIC AND RISK ASSESSMENT

QSAR IN SAFETY EVALUATION AND RISK ASSESSMENT

Edited by

HUIXIAO HONG

SBRBPAS Expert, Chief of Bioinformatics Branch — Division of Bioinformatics and Biostatistics, National Central for Toxicological Research (NCTR) at US Food and Drug Administration (FDA), USA

ELSEVIER

ACADEMIC PRESS

An imprint of Elsevier

Academic Press is an imprint of Elsevier
125 London Wall, London EC2Y 5AS, United Kingdom
525 B Street, Suite 1650, San Diego, CA 92101, United States
50 Hampshire Street, 5th Floor, Cambridge, MA 02139, United States
The Boulevard, Langford Lane, Kidlington, Oxford OX5 1GB, United Kingdom

Notices

Knowledge and best practice in this field are constantly changing. As new research and experience broaden our understanding, changes in research methods, professional practices, or medical treatment may become necessary.

Practitioners and researchers must always rely on their own experience and knowledge in evaluating and using any information, methods, compounds, or experiments described herein. In using such information or methods they should be mindful of their own safety and the safety of others, including parties for whom they have a professional responsibility.

To the fullest extent of the law, neither the Publisher nor the authors, contributors, or editors, assume any liability for any injury and/or damage to persons or property as a matter of products liability, negligence or otherwise, or from any use or operation of any methods, products, instructions, or ideas contained in the material herein.

ISBN: 978-0-443-15339-6

For information on all Academic Press publications visit our website at https://www.elsevier.com/books-and-journals

Publisher: Matthew Deans
Acquisitions Editor: Kattie Washington
Editorial Project Manager: Kathy Padilla
Production Project Manager: Fahmida Sultana
Cover Designer: Christian J. Bilbow

Typeset by TNQ Technologies

Contents

IV QSAR models for risk assessment of chemicals

List of contributors

Eric Akortia Ghana Atomic Energy Commission (GAEC), Legon-Accra, Ghana

Sai Radha Mani Alla MultiCASE Inc., Mayfield Heights, OH, United States

Vinicius M. Alves Laboratory for Molecular Modeling, University of North Carolina at Chapel Hill, Chapel Hill, NC, United States

Carolina H. Andrade Laboratory for Molecular Modeling and Design, Faculty of Pharmacy, Universidade Federal de Goiás, Goiania, GO, Brazil

Speranta Avram Department of Anatomy, Animal Physiology and Biophysics, Faculty of Biology, University of Bucharest, Bucharest, Romania

Tong Bao State Key Laboratory of Pollution Control and Resource Reuse, School of the Environment, Nanjing University, Nanjing, China

Emilio Benfenati Istituto di Ricerche Farmacologiche Mario Negri IRCCS, Department of Environmental Health Sciences, Milano, Italy

Rodolpho C. Braga InsilicAll Ltda, São Paulo, SP, Brazil

Catalin Buiu Department of Automatic Control and Systems Engineering, Politehnica University of Bucharest, Bucharest, Romania

Suman K. Chakravarti MultiCASE Inc., Mayfield Heights, OH, United States

Jingwen Chen Key Laboratory of Industrial Ecology and Environmental Engineering (Ministry of Education), Dalian Key Laboratory on Chemicals Risk Control and Pollution Prevention Technology, School of Environmental Science and Technology, Dalian University of Technology, Dalian, China

Minjun Chen Division of Bioinformatics and Biostatistics, National Center for Toxicological Research, United States Food and Drug Administration, Jefferson, AR, United States

Krzesimir Ciura QSAR Lab Ltd, Gdańsk, Poland; Department of Physical Chemistry, Faculty of Pharmacy, Medical University of Gdansk, Gdansk, Poland

Erika Colombo Istituto di Ricerche Farmacologiche Mario Negri IRCCS, Department of Environmental Health Sciences, Milano, Italy

Kevin P. Cross Instem, Columbus, OH, United States

Shixuan Cui Key Laboratory of Environment Remediation and Ecological Health, Ministry of Education, College of Environmental and Resource Sciences, Zhejiang University, Hangzhou, China

Giusy del Giudice Faculty of Medicine and Health Technology, Tampere University, Tampere, Finland; Finnish Hub for Development and Validation of Integrated Approaches (FHAIVE), Tampere, Finland

Fan Dong Division of Bioinformatics and Biostatistics, National Center for Toxicological Research, United States Food and Drug Administration, Jefferson, AR, United States

Pablo R. Duchowicz Instituto de Investigaciones Fisicoquímicas Teóricas y Aplicadas (INIFTA), La Plata, Argentina

Courage Egbi Ghana Atomic Energy Commission (GAEC), Legon-Accra, Ghana

Prasanna Egodawatta Queensland University of Technology (QUT), Brisbane, QLD, Australia

Elango Elakkiya Department of Computer Applications, Alagappa University, Karaikudi, Tamil Nadu, India

Antonio Federico Faculty of Medicine and Health Technology, Tampere University, Tampere, Finland; Finnish Hub for Development and Validation of Integrated Approaches (FHAIVE), Tampere, Finland; Tampere Institute for Advanced Study, Tampere, Finland

Michele Fratello Faculty of Medicine and Health Technology, Tampere University, Tampere, Finland; Finnish Hub for Development and Validation of Integrated Approaches (FHAIVE), Tampere, Finland

Zhiqiang Fu Key Laboratory of Industrial Ecology and Environmental Engineering (Ministry of Education), Dalian Key Laboratory on Chemicals Risk Control and Pollution Prevention Technology, School of Environmental Science and Technology, Dalian University of Technology, Dalian, China

Ayako Furuhama Division of Genetics and Mutagenesis, National Institute of Health Sciences, Kawasaki, Kanagawa, Japan

Yuchen Gao Key Laboratory of Environment Remediation and Ecological Health, Ministry of Education, College of Environmental and Resource Sciences, Zhejiang University, Hangzhou, China; Women's Reproductive Health Key Laboratory of Zhejiang Province, Women's Hospital, School of Medicine, Zhejiang University, Hangzhou, China

Juan C. Garro Martinez Centro Científico Tecnológico San Luis (CCT-CONICET), Facultad de Química, Bioquímica y Farmacia, Universidad Nacional de San Luis, San Luis, Argentina

Gustav Gbeddy Ghana Atomic Energy Commission (GAEC), Legon-Accra, Ghana

Beatriz Giner Facultad de Ciencias de la Salud, Universidad San Jorge, Campus Universitario, Villanueva de Gállego, Zaragoza, Spain

Eric Glover Ghana Atomic Energy Commission (GAEC), Legon-Accra, Ghana

Ashantha Goonetilleke Queensland University of Technology (QUT), Brisbane, QLD, Australia

Dario Greco Faculty of Medicine and Health Technology, Tampere University, Tampere, Finland; Finnish Hub for Development and Validation of Integrated Approaches (FHAIVE), Tampere, Finland; Institute of Biotechnology, University of Helsinki, Helsinki, Finland; Division of Pharmaceutical Biosciences, Faculty of Pharmacy, University of Helsinki, Helsinki, Finland

Maciej Gromelski QSAR Lab Ltd, Gdańsk, Poland

Yaxin Gu Laboratory of Molecular Modeling and Design, School of Pharmacy, East China University of Science and Technology, Shanghai, China

Bingxin Gui Key Laboratory for Wetland Ecology and Vegetation Restoration of National Environmental Protection, School of Environment, Northeast Normal University, Changchun, Jilin, PR China

Jing Guo State Key Laboratory of Pollution Control and Resource Reuse, School of the Environment, Nanjing University, Nanjing, China

Wenjing Guo Division of Bioinformatics and Biostatistics, National Center for Toxicological Research, United States Food and Drug Administration, Jefferson, AR, United States

Jiale He Key Laboratory of Industrial Ecology and Environmental Engineering (MOE), School of Environmental Science and Technology, Dalian University of Technology, Dalian, China

Minghui He Key Laboratory for Wetland Ecology and Vegetation Restoration of National Environmental Protection, School of Environment, Northeast Normal University, Changchun, Jilin, PR China

Huixiao Hong Division of Bioinformatics and Biostatistics, National Center for Toxicological Research, United States Food and Drug Administration, Jefferson, AR, United States

Masamitsu Honma Division of General Affairs, National Institute of Health Sciences, Kawasaki, Kanagawa, Japan

Shuheng Huang Key Laboratory of Biorheological Science and Technology (Ministry of Education), College of Bioengineering, Chongqing University, Chongqing, China; Key Laboratory of Tropical Biological Resources of Ministry of Education and One Health Institute, School of Pharmaceutical Sciences, Hainan University, Haikou, China

Ruili Huang Division of Preclinical Innovation (DPI), National Center for Advancing Translational Sciences (NCATS), National Institutes of Health (NIH), Rockville, MD, United States

Meiling Huang Key Laboratory of Environment Remediation and Ecological Health, Ministry of Education, College of Environmental and Resource Sciences, Zhejiang University, Hangzhou, China

Martin Jaeger Niederrhein University of Applied Sciences, Department of Chemistry and ILOC, Krefeld, Germany

Gopala Krishna Jillella National Institute of Pharmaceutical Education and Research-Kolkata, Kolkata, India

Candice Johnson Instem, Columbus, OH, United States

Toshio Kasamatsu Division of Genetics and Mutagenesis, National Institute of Health Sciences, Kawasaki, Kanagawa, Japan

Pathan Mohsin Khan National Institute of Pharmaceutical Education and Research-Kolkata, Kolkata, India

Sander Koster Nestlé Research, Lausanne, Switzerland

Rebecca Kusko Immuneering Corp, Cambridge, MA, United States

Myeongsang Lee Division of Bioinformatics and Biostatistics, National Center for Toxicological Research, United States Food and Drug Administration, Jefferson, AR, United States

Chao Li State Environmental Protection Key Laboratory of Wetland Ecology and Vegetation Restoration, Engineering Lab for Water Pollution Control and Resources Recovery, School of Environment, Northeast Normal University, Changchun, China

Fei Li CAS Key Laboratory of Coastal Environmental Processes and Ecological Remediation, Yantai Institute of Coastal Zone Research (YIC), Chinese Academy of Sciences (CAS), Shandong Key Laboratory of Coastal Environmental Processes, YICCAS, Yantai, PR China

Bohao Li College of Geography and Environmental Sciences, Zhejiang Normal University, Jinhua, China

Menglong Li College of Chemistry, Sichuan University, Chengdu, China

Dongying Li Division of Bioinformatics and Biostatistics, National Center for Toxicological Research, United States Food and Drug Administration, Jefferson, AR, United States

Jinjie Li Key Laboratory for Wetland Ecology and Vegetation Restoration of National Environmental Protection, School of Environment, Northeast Normal University, Changchun, Jilin, PR China; College of Marine Ecology and Environment, Shanghai Ocean University, Shanghai, PR China

Xuehua Li Key Laboratory of Industrial Ecology and Environmental Engineering (Ministry of Education), Dalian Key Laboratory on Chemicals Risk Control and Pollution Prevention Technology, School of Environmental Science and Technology, Dalian University of Technology, Dalian, China

Zoe Li Division of Bioinformatics and Biostatistics, National Center for Toxicological Research, United States Food and Drug Administration, Jefferson, AR, United States

Tsung-Jen Liao Division of Bioinformatics and Biostatistics, National Center for Toxicological Research, United States Food and Drug Administration, Jefferson, AR, United States

Yuan Liu College of Chemistry, Sichuan University, Chengdu, China; University of California, Department of Bioengineering, San Diego, CA, United States

Huihui Liu Jiangsu Key Laboratory of Chemical Pollution Control and Resources Reuse, School of Environmental and Biological Engineering, Nanjing University of Science and Technology, Nanjing, China

Jie Liu Division of Bioinformatics and Biostatistics, National Center for Toxicological Research, United States Food and Drug Administration, Jefferson, AR, United States

Wenjia Liu Key Laboratory of Industrial Ecology and Environmental Engineering (Ministry of Education), Dalian Key Laboratory on Chemicals Risk Control and Pollution Prevention Technology, School of Environmental Science and Technology, Dalian University of Technology, Dalian, China

Laura Lomba Facultad de Ciencias de la Salud, Universidad San Jorge, Campus Universitario, Villanueva de Gállego, Zaragoza, Spain

Chaofeng Lou Laboratory of Molecular Modeling and Design, School of Pharmacy, East China University of Science and Technology, Shanghai, China

Guangcai Ma College of Geography and Environmental Sciences, Zhejiang Normal University, Jinhua, China

Serena Manganelli Nestlé Research, Lausanne, Switzerland

Hu Mei Key Laboratory of Biorheological Science and Technology (Ministry of Education), College of Bioengineering, Chongqing University, Chongqing, China

Sabrina S. Mendonca Laboratory for Molecular Modeling and Design, Faculty of Pharmacy, Universidade Federal de Goiás, Goiania, GO, Brazil

Tadahaya Mizuno Laboratory of Molecular Pharmacokinetics, Graduate School of Pharmaceutical Sciences, The University of Tokyo, Tokyo, Japan

Glenn J. Myatt Instem, Columbus, OH, United States

Deborah K. Ngan Division of Preclinical Innovation (DPI), National Center for Advancing Translational Sciences (NCATS), National Institutes of Health (NIH), Rockville, MD, United States

Baitang Ning Division of Bioinformatics and Biostatistics, National Center for Toxicological Research, United States Food and Drug Administration, Jefferson, AR, United States

Shraddha Pandit CSIR- Indian Institute of Toxicology Research, Lucknow, Uttar Pradesh, India

Ramakrishnan Parthasarathi CSIR- Indian Institute of Toxicology Research, Lucknow, Uttar Pradesh, India; Academy of Scientific and Innovative Research (AcSIR), Ghaziabad, Uttar Pradesh, India

Tucker A. Patterson Office of Research, National Center for Toxicological Research, United States Food and Drug Administration, Jefferson, AR, United States; Division of Bioinformatics and Biostatistics, National Center for Toxicological Research, United States Food and Drug Administration, Jefferson, AR, United States

Alisa Pavel Faculty of Medicine and Health Technology, Tampere University, Tampere, Finland; Finnish Hub for Development and Validation of Integrated Approaches (FHAIVE), Tampere, Finland

Paulraj Prabhu Department of Computer Applications, Alagappa University, Karaikudi, Tamil Nadu, India

Tomasz Puzyn QSAR Lab Ltd, Gdańsk, Poland; Laboratory of Environmental Chemoinformatics, Faculty of Chemistry, University of Gdańsk, Gdańsk, Poland

Jiao Qu Key Laboratory for Wetland Ecology and Vegetation Restoration of National Environmental Protection, School of Environment, Northeast Normal University, Changchun, Jilin, PR China

Alessandra Roncaglioni Istituto di Ricerche Farmacologiche Mario Negri IRCCS, Department of Environmental Health Sciences, Milano, Italy

Kunal Roy Drug Theoretics and Cheminformatics Laboratory, Department of Pharmaceutical Technology, Jadavpur University, Kolkata, India

Laura Aliisa Saarimäki Faculty of Medicine and Health Technology, Tampere University, Tampere, Finland; Finnish Hub for Development and Validation of Integrated Approaches (FHAIVE), Tampere, Finland

Laura M. Saavedra Instituto de Investigaciones Fisicoquímicas Teóricas y Aplicadas (INIFTA), La Plata, Argentina

Igor H. Sanches Laboratory for Molecular Modeling and Design, Faculty of Pharmacy, Universidade Federal de Goiás, Goiania, GO, Brazil

Benoit Schilter Nestlé Research, Lausanne, Switzerland

Chandrabose Selvaraj Center for Transdisciplinary research, Department of Pharmacology, Saveetha Dental College and Hospital, Saveetha Institute of Medical and Technical Sciences (SIMATS), Saveetha University, Chennai, Tamil Nadu, India

Angela Serra Faculty of Medicine and Health Technology, Tampere University, Tampere, Finland; Finnish Hub for Development and Validation of Integrated Approaches (FHAIVE), Tampere, Finland; Tampere Institute for Advanced Study, Tampere, Finland

Wei Shi State Key Laboratory of Pollution Control and Resource Reuse, School of the Environment, Nanjing University, Nanjing, China

Prakrity Singh CSIR- Indian Institute of Toxicology Research, Lucknow, Uttar Pradesh, India; Academy of Scientific and Innovative Research (AcSIR), Ghaziabad, Uttar Pradesh, India

Sanjeev Kumar Singh Department of Data Sciences, Centre of Biomedical Research, SGPGIMS Campus, Lucknow, Uttar Pradesh, India

Catalina Sogor Department of Anatomy, Animal Physiology and Biophysics, Faculty of Biology, University of Bucharest, Bucharest, Romania

Meng Song Division of Bioinformatics and Biostatistics, National Center for Toxicological Research, United States Food and Drug Administration, Jefferson, AR, United States

Anita Sosnowska QSAR Lab Ltd, Gdańsk, Poland

Limin Su Key Laboratory for Wetland Ecology and Vegetation Restoration of National Environmental Protection, School of Environment, Northeast Normal University, Changchun, Jilin, PR China

Kei-ichi Sugiyama Division of Genetics and Mutagenesis, National Institute of Health Sciences, Kawasaki, Kanagawa, Japan

Marta Swirog QSAR Lab Ltd, Gdańsk, Poland

Yun Tang Laboratory of Molecular Modeling and Design, School of Pharmacy, East China University of Science and Technology, Shanghai, China

Yuefa Teng CAS Key Laboratory of Coastal Environmental Processes and Ecological Remediation, Yantai Institute of Coastal Zone Research (YIC), Chinese Academy of Sciences (CAS), Shandong Key Laboratory of Coastal Environmental Processes, YICCAS, Yantai, PR China; University of Chinese Academy of Sciences, Beijing, PR China

Weida Tong Division of Bioinformatics and Biostatistics, National Center for Toxicological Research, United States Food and Drug Administration, Jefferson, AR, United States

Alexander Tropsha Laboratory for Molecular Modeling, Division of Chemical Biology and Medicinal Chemistry, UNC Eshelman School of Pharmacy, University of North Carolina, Chapel Hill, NC, United States

Ana-Maria Udrea Laser Department, National Institute for Laser, Plasma and Radiation Physics, Magurele, Romania

Devadasan Velmurugan Department of Biotechnology, School of Bioengineering, SRM Institute of Science and Technology, Kattankulathur, Tamil Nadu, India

Melanie Voigt Niederrhein University of Applied Sciences, Department of Chemistry and ILOC, Krefeld, Germany

Haobo Wang Key Laboratory of Industrial Ecology and Environmental Engineering (Ministry of Education), Dalian Key Laboratory on Chemicals Risk Control and Pollution Prevention Technology, School of Environmental Science and Technology, Dalian University of Technology, Dalian, China

Xiaoqing Wang CAS Key Laboratory of Coastal Environmental Processes and Ecological Remediation, Yantai Institute of Coastal Zone Research (YIC), Chinese Academy of Sciences (CAS), Shandong Key Laboratory of Coastal Environmental Processes, YICCAS, Yantai, PR China; University of Chinese Academy of Sciences, Beijing, PR China

Zhongyu Wang Key Laboratory of Industrial Ecology and Environmental Engineering (MOE), School of Environmental Science and Technology, Dalian University of Technology, Dalian, China

Xueyu Wang College of Geography and Environmental Sciences, Zhejiang Normal University, Jinhua, China

Xiaoxuan Wei College of Geography and Environmental Sciences, Zhejiang Normal University, Jinhua, China

Zhining Wen College of Chemistry, Sichuan University, Chengdu, China

Yiqu Wu Key Laboratory of Environment Remediation and Ecological Health, Ministry of Education, College of Environmental and Resource Sciences, Zhejiang University, Hangzhou, China

Ewelina Wyrzykowska QSAR Lab Ltd, Gdańsk, Poland

Fang Xiao College of Geography and Environmental Sciences, Zhejiang Normal University, Jinhua, China

Hong-Bin Xie Key Laboratory of Industrial Ecology and Environmental Engineering (Ministry of Education) Dalian Key Laboratory on Chemicals Risk Control and Pollution Prevention Technology, School of Environmental Science and Technology, Dalian University of Technology, Dalian, China

Tuan Xu Division of Preclinical Innovation (DPI), National Center for Advancing Translational Sciences (NCATS), National Institutes of Health (NIH), Rockville, MD, United States

Liang Xu Division of Bioinformatics and Biostatistics, National Center for Toxicological Research, United States Food and Drug Administration, Jefferson, AR, United States

Xianhai Yang Jiangsu Key Laboratory of Chemical Pollution Control and Resources Reuse, School of Environmental and Biological Engineering, Nanjing University of Science and Technology, Nanjing, China

Haiying Yu College of Geography and Environmental Sciences, Zhejiang Normal University, Jinhua, China

Yuxuan Zhang Key Laboratory of Industrial Ecology and Environmental Engineering (Ministry of Education), Dalian Key Laboratory on Chemicals Risk Control and Pollution Prevention Technology, School of Environmental Science and Technology, Dalian University of Technology, Dalian, China

Jiachen Zhang Key Laboratory of Environment Remediation and Ecological Health, Ministry of Education, College of Environmental and Resource Sciences, Zhejiang University, Hangzhou, China

Jingwen Zhao Bioinformatics Program, Department of Information Science, University of Arkansas at Little Rock, Little Rock, AR, United States

Yuanhui Zhao Key Laboratory for Wetland Ecology and Vegetation Restoration of National Environmental Protection, School of Environment, Northeast Normal University, Changchun, Jilin, PR China

Minghua Zhu Key Laboratory of Industrial Ecology and Environmental Engineering (Ministry of Education), Dalian Key Laboratory on Chemicals Risk Control and Pollution Prevention Technology, School of Environmental Science and Technology, Dalian University of Technology, Dalian, China

Shulin Zhuang Key Laboratory of Environment Remediation and Ecological Health, Ministry of Education, College of Environmental and Resource Sciences, Zhejiang University, Hangzhou, China; Women's Reproductive Health Key Laboratory of Zhejiang Province, Women's Hospital, School of Medicine, Zhejiang University, Hangzhou, China

Wen Zou Division of Bioinformatics and Biostatistics, National Center for Toxicological Research, United States Food and Drug Administration, Jefferson, AR, United States

Preface

The assumption of quantitative structure–activity relationship (QSAR) is that the structure of a compound determines its physical, chemical, and biological properties. QSAR has been realized for a long time. Early in 1868, Crum-Brown and Fraser found that the physiological action of a compound in a biological system is correlated with the composition and constitution of the compound. Examples of early time QSAR include the dependency of narcotic action of compounds on their oil/water partition coefficients separately found by Meyer and Overton in 1899 and 1901, and the seminal finding of Hammett in 1935 that described substituent effects on organic reactions. The first QSAR revival started in 1960s which includes just a few well-known QSAR works, the dependency of plant growth regulators on Hammett constants and hydrophobicity published by Hansch et al. in 1962, the QSAR models in leucomycin and lincomycin antibiotics developed by Martin and Lynn in 1971, and the comparative molecular field analysis method proposed by Cramer et al. in 1988. With many years of advances in computational tools, hardware, available databases, and especially machine learning and deep learning algorithms, QSAR modeling is equipped enough for applications to all scientific fields including drug safety evaluation and chemical risk assessment.

More than 350,000 chemicals are registered for production and use, and many have already been used in products and entered the environment. Safety evaluation of the chemical containing products and risk assessment of the environmental chemicals are a huge challenge for the scientific community and regulatory agencies. Experimentally evaluating safety and assessing risk of such many chemicals is very challenging, if not impossible, in practice. Therefore, the scientific community and regulatory agencies have been looking for alternative in silico methods to facilitate safety evaluation and risk assessment of chemicals. QSAR is a promising alternative in silico method that has been extensively investigated in the scientific community and is attracting attentions of regulatory agencies. This book is compiled with ambition to capturing the latest advancements of QSAR for safety evaluation and risk assessment of chemicals.

The motivation in producing this book was twofold. First, more and more new chemical entities are being developed and used by the ever growing biotechnology, pharmaceutical, and food industry. Absolutely all these chemicals must be screened for safety at various stages of product development. Second, the ecosystem is increasingly buried in a shroud of chemicals, many of which safety and risk to humans and animals have never been assessed before. The surge of new chemicals entering the environment and the human body brings an insurmountable challenge to the scientific community and regulatory agencies. Impressively, the advancements in applications of QSAR to safety evaluation and risk assessment described in this book are helping to meet the challenge.

This book consists of a background introduction and five parts to cover various aspects of QSAR methods and applications to safety evaluation and risk assessment. The first part collected chapters describing QSAR methods and recent advances, focusing on emerging methods for QSAR modeling such as machine learning and deep learning. The QSAR field had many efforts on implementing QSAR tools and creating data sources to facilitate QSAR modeling. Therefore, the second part comprises chapters introducing popular QSAR tools and frequently used data sources. The third part presented chapters on applications of QSAR models for safety evaluation of drugs and consumer products. The fourth part gathered various applications of QSAR models for risk assessment of chemicals. The last part compiled chapters demonstrating applications of QSAR models in material science and other areas such as cancer risk assessment and marine ecotoxicology of chemicals.

This book is anticipated as a text for established computational chemists looking to enter the QSAR field, experienced QSAR scientists seeking to enhance QSAR research, or students and trainees stepping into the QSAR field. Therefore, each chapter not only includes an introductory section to help readers to understand new concepts but also provides details actionable methods and exampled models which can be deployed by the reader. Furthermore, each chapter can stand alone as a summary on a specific topic of interest. Available as another possibility, this textbook could be read in sections as the sections are organized by topics. For a graduate course on QSAR, this book would provide students excellent material for reading.

While this book does provide readers a comprehensive coverage of methods and advances of applications of QSAR to safety evaluation and risk assessment, it does not include any learning exercises and quizzes for readers to evaluate and test the knowledge and techniques they learn. Thus, for the use of this book in a classroom setting, the course instructor needs to create these materials. Any corrections, comments, criticism, and feedback on improvement for a future body of work from readers are appreciative.

Inspired by how QSAR modeling is rising to meet the challenges in safety evaluation and risk assessment, this book was compiled for the scientific community. This book would not be born without the arduous efforts of and gigantic cooperation from the authors, for which I am greatly appreciative. This book also would not be possible without the help and support of Tim Eslava, Kattie Washington, and Andre Wolff from Elsevier, whom I acknowledge for their visionary understanding of the importance of QSAR in safety evaluation and risk assessment.

This preface reflects the views of the author and should not be construed to represent the FDA's views or policies.

Huixiao Hong, PhD
SBRBPAS Expert of US Department of Health and Human Services
Chief, Bioinformatics Branch
Division of Bioinformatics and Biostatistics
National Center for Toxicological Research
U.S. Food and Drug Administration
Jefferson, AR, USA

CHAPTER

1

QSAR facilitating safety evaluation and risk assessment

Rebecca Kusko[1] *and Huixiao Hong*[2]

[1]Immuneering Corp, Cambridge, MA, United States [2]Division of Bioinformatics and Biostatistics, National Center for Toxicological Research, United States Food and Drug Administration, Jefferson, AR, United States

Abbreviations

CoMFA comparative molecular field analysis
CoMSIA comparative molecular similarity indices analysis
CV cross-validation
DILI drug-induced liver injury
EPA Environmental Protection Agency
FDA Food and Drug Administration
FN false negative
FP false positive
LD50 lethal dose 50%
LOO leave one out
MCC Matthews correlation coefficient
NCATS National Center for Advanced Translational Sciences
NICEATM NTP interagency center for the evaluation of alternative toxicological methods
NOAEL NO adverse effect level
QSAR quantitative structure-activity relationship
R&D research and development
TN true negative
TP true positive

Disclaimer: This chapter reflects the views of the authors and does not necessarily reflect those of the U.S. Food and Drug Administration.

Introduction

The relationship between the structure of a chemical and the potential for toxicity has been undergoing modeling efforts for a very long time, as this information is relevant to agriculture, drug development, ecology, industry, and more. The concept that the shape of a molecule is related to its physiochemical properties is thought to have first emerged from an 1863 thesis from the University of Strasbourg where a relationship was noted between water solubility and toxicity of certain alcohols [1]. Later in 1869, two scientists bravely noted "... There cannot be any reasonable objection against a fact that a relationship between physiological effect of a compound and its chemical constitution exists ....," positing that structure is a major driver of the biological effects of chemicals [2]. The era of using mathematical models to link physiochemical properties to biological effect began during the 1960s [3], and the term "QSAR" (quantitative structure activity relationship) was first reported at a conference in the early 1970s [4]. In the modern era, QSAR involves statistics to query for relationships between a) the molecular properties or structure of a group of chemicals and b) the biological activity of those chemicals [5].

After the successful development of QSAR, in the modern era, it has been applied to a number of fields, including product safety evaluation and environmental chemical risk assessment. To evaluate drug and consumer products, QSAR models are used to replace animal methods, support prioritization, and add to limited experimental data [6]. In earlier stages, QSAR models were used to evaluate leads and optimized leads before conducting costly and often inaccurate animal testing. Moreover, QSAR methods have multiple applications to protect life on earth from environmental toxicants. In this space, QSAR can be leveraged to assess the bioavailability, health effects, exposure, and pharmacokinetics of a potential toxicant [7].

QSAR in Safety Evaluation and Risk Assessment
https://doi.org/10.1016/B978-0-443-15339-6.00036-9

QSAR has the ability to overcome a panoply of limitations that are inherent to experimental toxicology. The number of environmental toxicants, new drug products, new consumer products, and their interactions are increasing exponentially as various industries expand. While there have been promising wet lab and artificial intelligence hybrid approaches, any in vitro or in vivo system will be fundamentally limited by throughput and unable to scale up to meet the current needs of the environmental, pharmaceutical, and regulatory industries. Moreover, experimental systems utilize time in a highly inefficient manner. Animal models must be established, bred, and housed; in vivo cultures must be split and maintained. Lastly, these experimental systems often do not clinically translate. Humans and higher organisms represent a complex interplay of immune cells, hepatocytes, epithelial cells, blood cells, gametes, and more, which are organized into organ systems, containing a complex mixture of nonbiological elements coming from the environment. In vitro and in vivo systems do not accurately capture this complex dynamic and, therefore, often fail to accurately predict toxicity. Animal testing also raises obvious ethical concerns. Lastly, experimental toxicology is very costly, creating challenges for academics working with limited grant funding as well as the pharmaceutical industry trying to develop safer medicines in a cost-efficient manner. Taken together, the challenges of experimental toxicology demonstrate a clear need for lower cost, more accurate, ethical, faster, and higher throughput approaches. As a fully computational method which can be performed in a time and cost-efficient manner without any ethical concerns, QSAR is an improvement or supplement to experimental toxicology.

The advent of new data types, expanded datasets, increased public data, synergy with artificial intelligence and machine learning approaches, and improved statistics, as well as improved measurements of physiochemical properties, have led to great progress in the use of QSAR for the evaluation of safety and risk assessment. Recent progress in the field of QSAR will be the subject of this text.

Data sources for QSAR

Any computational predictive method will succeed or fail based on the quality and quantity of input data. A QSAR dataset must have two elements: (a) biological activity data and (b) chemical structure information such as molecular descriptors. Biological data must come from a rigorous assay, as well as an assay appropriate for predicting the effect of interest. For toxicology purposes, the most ideal biological subject is the human being. Human toxicity is the ultimate endpoint and thus is the ultimate input data but often comes with other issues including smaller dataset size and confounding factors [8]. Concerningly, a reproducibility crisis of published biological experiments has been reported by many independent studies across multiple fields [9–11]. In addition to leveraging multiple independent orthogonal biological datasets, it is recommended to initially treat biological datasets with suspicion until they have proven their worth or in other words "trust but verify" [12]. In addition to inaccurate biological activity information, public datasets can contain structural errors or other incorrect physiochemical property information [13]. Here, the "trust but verify" approach has also been encouraged; any public dataset should be carefully spot-checked for erroneous structures [14] and cleaned of structures that will not be handled by the method and duplicates should be dropped [15]. After carefully checking both the chemical and biological data, the size of a dataset must be assessed for appropriateness before proceeding. The mandatory size of the dataset will depend not only on the scope of the question but also the statistical approach. More advanced algorithms such as machine learning and deep learning are quite data hungry and require massive datasets [16]. For example, often it is recommended to split a dataset into a training, test, and validation set. Simpler approaches such as linear modeling can tolerate smaller but robust datasets. Lastly, experimental campaigns often focus on linear sets of models, thus the resulting dataset can fundamentally constrain the predictive space of a QSAR model. A chemically and biologically diverse dataset is optimal for achieving maximum impact with QSAR [8].

A number of public databases exist, which could prove excellent starting points to begin dataset assessment and quality control. The US Environmental Protection Agency (EPA) and the National Toxicology Program (NTP) Interagency center for the Evaluation of Alternative Toxicological Methods (NICEATM) assembled a dataset of rat LD50 (lethal dose 50%) values spanning 12000 chemicals available to the research community [17]. Drugs were classified by toxicity level according to multiple different schemes, and rodent LD50 was estimated [18]. The National Center for Advanced Translational Sciences (NCATS) generated rat liver microsome data on 24000 compounds by aggregating more than 250 projects spanning nearly a decade [19]. Competitions also make datasets available; for example, a Kaggle QSAR competition included a manually curated dataset of 779 chemicals including number of Heavy atoMs (nHM), molecular multiple path count (piPC09), difference between multiple path count and path count (PCD), average valence connectivity (X2AV), Moriguchi octanol-water partition coefficient (MLOGP), overall modified Zagreb index by valence vertex degrees (ON1V), frequency of RCO-N</>-X = X fragments (N.072), bioconcentration factor, and bioaccumulation [20]. Additionally, in 2009, the Tox21 dataset generated more than 100 million data points spanning 8000 chemicals in over 70 biological assays [21]. The number of available datasets potentially appropriate for QSAR is ever expanding. When initiating a QSAR study, it is always recommended to check for newly released datasets.

QSAR methods

A QSAR model can be generally represented using Eq. (1.1).

$$y = f(X) + \delta \tag{1.1}$$

Where y is the predicted biological activity of chemical; X is the numerical description of structure of the chemical; f is the mathematical function, the QSAR model, used to transform the structure description to the activity; and δ is the difference between the actual biological activity obtained from experimentally testing the chemical and the predicted activity. The overall process to develop a QSAR model is illustrated in Fig. 1.1. In short, a training dataset of structures and biological activity of chemicals is generated in a lab, collected from public databases, curated from the literature, or a combination of these approaches. Different internal validation methods could be applied to the training dataset for QSAR modeling. The internal validation results are employed to interrogate the goodness of QSAR modeling on the training dataset. Occasionally, the same biological activity for a set of chemicals that are not included in the training dataset is generated in a different lab, collected from databases, or curated from the literature as an external validation dataset. A QSAR model is first developed using the whole training dataset. The

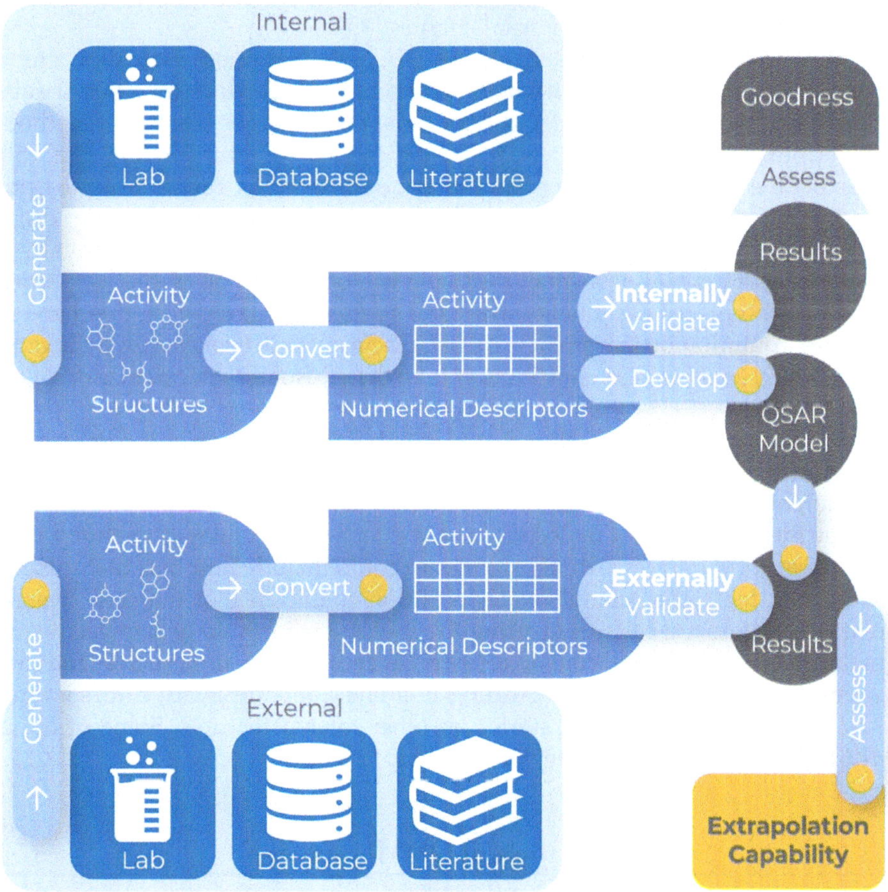

FIGURE 1.1 Overview of the quantitative structure activity relationship (QSAR) modeling process.

external validation is then used to test the predictivity of the developed QSAR model. The external validation results are used to assess the extrapolation capability of the developed QSAR model.

A cornucopia of approaches to QSAR exist, focusing either on the biological molecule of concern (DNA or protein, "macromolecule") or the small molecule in question itself. Macromolecule-based methods focus on pinpointing the interaction between a potential toxicant and a biological target, which could be a receptor, a transporter, genetic material (RNA/DNA), enzymes, hormones, and more. One such approach, molecular docking, simulates the binding of a toxicant to a macromolecule target of interest by calculating the energy of a variety of binding poses [22–24]. High scoring toxicants are more likely to have an undesired interaction with the macromolecule and should be selected for follow-up experimentation. Molecular dynamics simulations take into account the time evolution of the macromolecule-toxicant system and consider all elements inside of the simulation to be flexible [25–28]. Atomic movements are computed using classical Newtonian mechanics rules, with time steps being partitioned [29]. Leveraging shape complementarity and pharmacophore modeling represents the active shape of the ligand imputed from the binding site of the macromolecule and then interrogates potential toxicants for the binding shape [30]. Potential toxicants with the correct shape to bind to the site on the macromolecule would be flagged for potential toxicity and follow-up investigation. The pharmacophore approach can have advantages when few ligands are known for the macromolecule [31].

As an alternative to macromolecule-based approaches, small molecule-based approaches focus on the potential toxicant itself rather than attempting to model an interaction between the toxicant and a macromolecule. comparative molecular field analysis (CoMFA) begins with a set of small molecules that all have the same interaction with the same macromolecule. Unlike above methods, CoMFA is alignment based and places ligands in an energy grid with a probe at every lattice point [32,33]. Model development descriptors come from this energy calculation, which is correlated with biological output. Comparative molecular similarity indices analysis (CoMSIA) is a variant of CoMFA, which takes grid nodes falling within the molecular volume into account [34]. Moreover, the CoMFA steric and electrostatic grid calculation method is replaced with Steric and Electrostatic Alignment (SEAL), which performs additional optimization based on atomic partial charges [35].

While other QSAR approaches involve sophisticated models, certain phenomenon allows for a more straightforward solution. Structural alerts represent a streamlined method to tag hazardous moieties or substructures into categories, which enable read-across to other similar compounds. Unlike other black box methods, structural alerts represent the opposite approach where the structure causing the alert is immediately clear. Structural alerts can be used not only to predict the toxicity of a potential toxicant but moreover can be used to optimize research and development (R&D) stage molecules away from moieties triggering alerts. For example, during small molecule drug R&D, it is common practice to search discovery candidates against various structural alert databases to avoid toxicity issues from an earlier stage [36]. If a small molecule has otherwise desirable properties, the questionable moiety can be partially or fully replaced. Originally, structural alerts were manually curated and added to databases; however, in the current era, computational approaches exist. Chemicals are broken down into addressable substructures, and each substructure is analyzed for contribution to toxicity, usually via calculation of a substructure accuracy or *P* value [37].

For computational modeling purposes, it is often useful to extract information about the structure and features of a molecule into a numerical or otherwise digital representation to facilitate further work. Molecular fingerprints are a numerical representation of a molecular structure, and molecular descriptors represent the features of said molecule. Together, they allow the structure of a molecule to be connected to a biological activity through computation.

Molecular descriptors were originally one-dimensional and could include numbers including but not limited to: a steric parameter, atom count, logP (the octanol–water partition coefficient), or molecule weight [38]. Often, one-dimensional molecular descriptors are calculated using the molecular formula of a molecule [39]. By comparison, a two-dimensional molecular descriptor will contain an extra layer of information, such as the electronic distribution of the molecule or number of stereocenters [40]. Adding in a layer of complexity, a three-dimensional molecular descriptor could be related to volume, surface, sterics, or other related features. Some of the methods described above (such as CoMFA) can generate four-dimensional descriptors.

Evaluation of QSAR models

A QSAR model must prove its utility via thoughtful evaluation before it is applied to predict toxicity. The utility of a QSAR model will be fundamentally limited by its true predictive ability and accuracy. The task of predicting toxicity is paramount to human health, and thus, an improperly evaluated model could potentially have serious consequences. Various performance metrics are used to evaluate QSAR models. The performance of a qualitative QSAR model (classification) is assessed by comparing its predictions on a set of chemicals with their actual activity data. The comparison divides the chemicals into four classes: true positives (TP) which are positive in both actual data and predictions, true negatives (TN) which are negative in both actual data and predictions, false positives (FN) which are actual positives but predicted as negatives, and false positives (FP) which are actual negatives but predicted as positives. Below are the seven metrics that are commonly used for evaluation of QSAR models; they are calculated using Eqs. (1.2)-(1.8).

$$Sensitivity = \frac{TP}{TP + FN} \tag{1.2}$$

$$Specificity = \frac{TN}{TN + FP} \tag{1.3}$$

$$Positive - Rate = \frac{TP}{TP + FP} \tag{1.4}$$

$$Negative - Rate = \frac{TN}{TN + FN} \tag{1.5}$$

$$Accuracy = \frac{TP + TN}{TP + FP + TN + FN} \tag{1.6}$$

$$Balanced - Accuracy = \frac{Sensitivity + Specificity}{2} \tag{1.7}$$

$$MCC = \frac{TP * TN - FP * FN}{\sqrt{(TP + FP)(TP + FN)(TN + FP)(TN + FN)}} \tag{1.8}$$

Where MCC represents the Matthews correlation coefficient. Sensitivity, specificity, positive rate, and negative rate are used to measure individual aspects of model performance, while accuracy, balanced accuracy, and MCC are used to assess the overall model. When a dataset is not imbalanced (positives are much more or less than negatives), accuracy can be used. However, when the prevalence of positives is too low or too high, accuracy is biased to the majority, and thus balanced accuracy and MCC are preferred for assessing model performance. For a quantitative model (regression), the performance of the model is usually measured using the coefficient of determination (Q^2) which can be calculated using Eq. (1.9).

$$Q^2 = 1 - \frac{\sum \left(y_{actual} - y_{predicted}\right)^2}{\sum \left(y_{actual} - y_{average}\right)^2} \tag{1.9}$$

Where y_{actual} and $y_{predicted}$ are the actual and predicted activity values, respectively, and $y_{average}$ indicates the average actual activity value of the training chemicals.

Evaluation methods for QSAR include but are not limited to: leave one out (LOO) cross-validation, k-fold cross-, holdout validation, external validation, and experimental validation. All cross-validation (CV) strategies involve repeating the analysis on a subset of data, leaving out data points, molecules, groups of data points, or other aspects of the data. The CV procedure not only evaluates the predictive ability of a model but importantly assesses for over-fitting. An over-fit model will not predict new unseen data after training and is best avoided. LOO cross-validation involves removing an individual data point or molecule iteratively for the entire training set. An average is computed of all resulting values to assess model performance. Many consider this LOO method to have critical limitations and inconsistent performance [41], but others posit it has superior performance [42]. In comparison, k-fold cross-validation splits the dataset in question into k equal parts. Each part is iteratively held out, while the model is run on the remaining data points, generating model evaluation scores [43]. As an example, in 10-fold cross-validation, a dataset will be split into 10 groups, the model will be trained and tested 10 times (once with each group removed) and would produce 10 scores. Choosing the right "size" k will depend not only on the dataset size but also dataset diversity and type of model [44].

Not all validation strategies involve iteratively "hiding" data from the model and quizzing it with a different test set. For example, holdout validation involves splitting the original dataset into a "training" and "test" dataset. The model is run on the "training" dataset and evaluated on the "test" dataset. Here, the subtlety is selecting the correct split between training/test, for example, 80% training and 20% test or 90% training and 10% test [45].

The ultimate gold standards of validation are both external validation and experimental validation, especially when used together. External validation involves testing the model on a completely independent set of data, ideally derived by a different group, study, or during a different period of time. Examples of an external validation dataset could include clinical trial toxicity data from a different country, drug-induced liver injury (DILI) in vitro screening from a different lab, or rodent screen in a different mouse model. A second example could be training with ChEMBL data and validating with PubChem. External validation is an excellent method to test not only for model overfitting but model generalizability. Experimental validation has the advantage of providing true biological validation but at the cost of being lower throughput. This means that a QSAR model will be deployed to make predictions, and a selection of data points will be chosen for experimental wet lab confirmation. This form of biological validation is important for ensuring the translational relevance of a QSAR model. The ultimate goal of any toxicity QSAR model is to determine the safety of a toxicant in a biological system; thus, testing the predictions of a model in a true biological system is the ultimate validation. However, experimental validation is limited by throughput and thus will not be the best tool for determining model overfitting.

Machine learning and deep learning accelerate QSAR development

Machine learning

New advanced algorithms have ushered in a new era of QSAR modeling. Now that larger training datasets are available, there are numerous cases of machine learning being successfully applied to QSAR [46–56]. At a high level, machine learning seeks to pinpoint groupings within data via a predictive model. For small datasets with clear patterns, a human expert can quickly and easily pick out patterns. However, as datasets grow larger and larger and patterns grow increasingly abstract, the capacity of the individual toxicologist to pick out relevant molecular descriptors or structures becomes limited [57]. Unlike classical modeling techniques such as linear modeling, machine learning aims to not only build a single model but also train and improve it.

Both supervised and unsupervised machine learning approaches can be applied to QSAR. Supervised approaches are superior in cases when data labeling is clear (for example, the readout is a very clear binary "toxic" or "not toxic"). Unsupervised learning could be superior in cases where the goal is to find a pattern of toxicity in the first place. This could include identifying a subset of enantiomers more likely to cause toxicity. In the drug safety and toxicity space, machine learning is mostly applied to solve a classification problem. Machine learning algorithms tuned to perform this task are often termed "classifiers." However, there are cases when the desired result is a continuous number; there machine learning regression is more appropriate. Examples of machine learning which have been applied to QSAR include support vector machine [58], random forest [59], LASSO [60], gradient boosting [61], and naive Bayes [62].

Recently, machine learning has been deployed to predict whether certain environmental toxicants cross the placenta and could potentially damage a fetus. A group of researchers at the Universite de Montreal retrieved cord:maternal exposure ratios for 105 toxicants and then obtained or calculated 214 molecular descriptors for each of these toxicants [63]. The authors parameterized their QSAR using a variety of machine learning approaches, including random forest, support vector regression, and LASSO. In their study, random forest was superior at predicting crossing of the placenta in external validation.

Deep learning

Deep learning is evolved from a machine learning technique called artificial neural network. Deep learning approaches such as deep neural networks are not particularly new, but the computation power for processing and requisite dataset size have evolved. A simplified architecture of deep neural network is illustrated in Fig. 1.2. The key difference between an artificial neural network and a deep neural network is the hidden layers of neurons: an artificial neural network has one hidden layer, while a deep neural network has multiple hidden layers. Owing

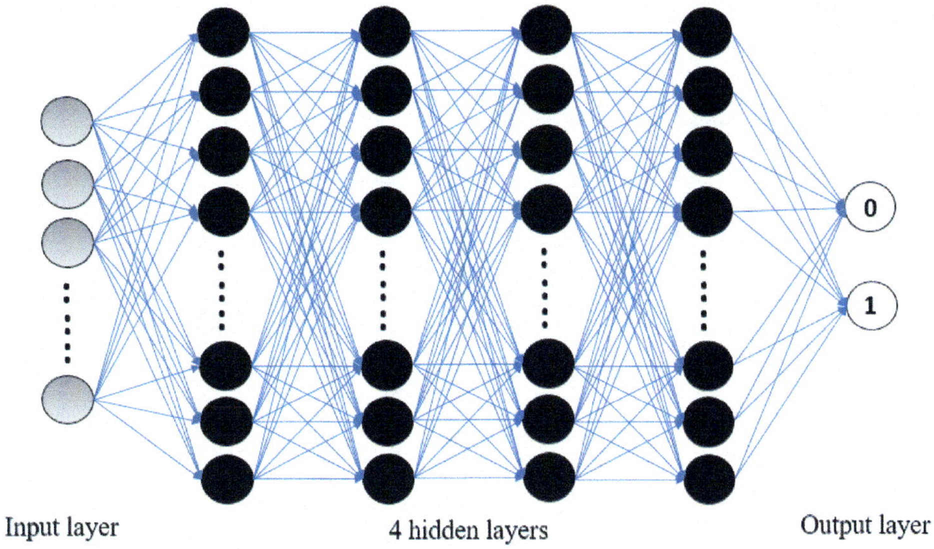

FIGURE 1.2 Illustrative deep neural network architecture.

to the large number of neurons in hidden layers of deep learning, more algorithmic parameters need to be determined. Therefore, more data and more intensive computation are needed to train a deep learning model.

Compared to machine learning, deep learning provides more adaptability around predicting complex interactions (such as concomitant medicine interactions) and can be more flexible with molecular representations. The major limitation of deep learning is interpretability. Put simply, the aspects of the model causing it to arrive at a conclusion are not always immediately obvious. In contrast, a machine learning approach usually produces weights which can show the user the key features underlying a prediction. Deep learning models usually take a molecule itself as input rather than a long list of descriptors typical of machine learning. The deep learning model determines or calculates the important properties, rather than relying on them being input. Importantly, this removes operator bias where a scientist might approach a question with a preconceived notion about which descriptors are important for predicting a certain event, outcome, or other toxicity. Similar to machine learning, deep learning can be divided into supervised, unsupervised, and hybrid. Supervised approaches include convolutional neural networks and recurrent neural networks. Unsupervised approaches include general adversarial networks and self-organizing maps, while hybrid approaches include deep transferred learning and deep reinforcement learning.

A variety of QSAR models have been developed using deep learning techniques for toxicity prediction [64–66]. Here, the deep learning model that was developed for predicting DILI [64] is taken to illustrate the advantage of deep learning in computational toxicology. DILI is a major safety concern for pharmaceutical industry and regulatory agencies. Many efforts have been made by scientists to assess DILI risk in humans for drugs using different approaches such as machine learning and deep learning techniques. Based on Mold2 descriptors [38] and PaDEL descriptors [67], Xu et al. developed traditional neural network models and deep neural network models by integrating three datasets with different DILI annotations obtained from the literature [64]. The performance of the machine learning and deep learning models was evaluated using 10-fold cross-validations. The 10-fold cross-validation results were evaluated using five performance metrics which are summarized in Fig. 1.3. As it can be seen from Fig. 1.3, for both Mold2 descriptors and PaDEL descriptors, the QSAR models developed using deep neural network outperformed the QSAR models developed using traditional neural network, demonstrating the advantage of deep learning over machine learning for development of QSAR models for predicting DILI in humans.

Perspectives

Owing to the rapid growth of the pharmaceutical, agricultural, and other industries, the number of species exposed to potential toxicants zoo is rising. Humankind as well as other species are being exposed to new chemicals through medicines, food products, supplements, and the environment. To keep up with this pace of ever increasing chemicals to evaluate, high speed and accurate safety evaluation and risk assessment are critical. Existing wet lab-

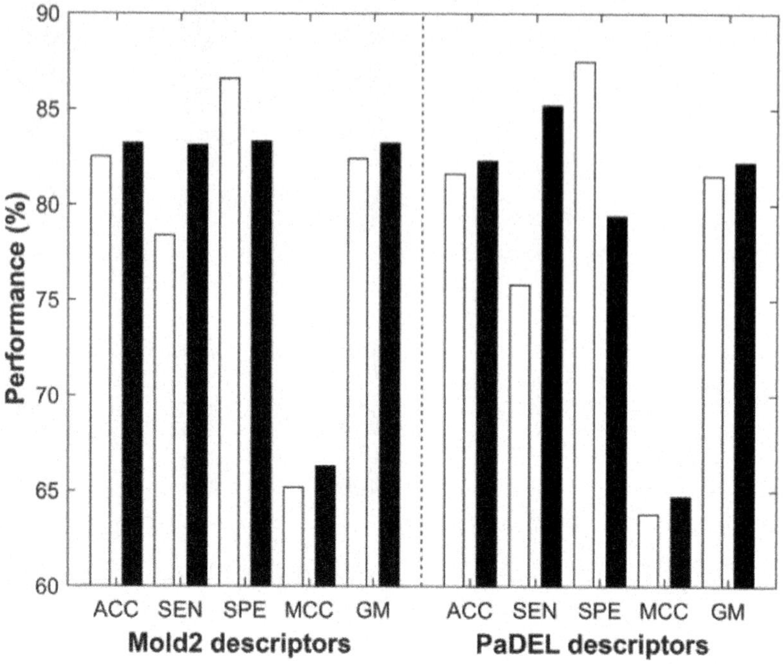

FIGURE 1.3 Performance of the neural network models (empty bars) and deep neural network models (solid bars). Performance parameters are marked on the x-axis ticks, and results from Mold2 descriptors and PaDEL descriptors are separated by the dashed vertical line and indicated by the x-axis labels. *ACC*, accuracy; *GM*, geometric mean of sensitivity and specificity; *MCC*, Matthews correlation coefficient; *SEN*, sensitivity; *SPE*, specificity.

based methods are too costly from a finance, time, and ethical point of view. Computational toxicology approaches including QSAR can be performed at high throughput. The ever increasing size of datasets relevant to QSAR has enabled the use of advanced algorithms including machine learning and deep learning. Machine learning and deep learning are able to model more complex interactions and work with datasets lacking customary molecular descriptors.

Machine learning and deep learning are only as smart as the input datasets, so it is critical to "trust but verify" any dataset before applying one of these approaches to computational toxicology. Moreover, datasets must be scrutinized for appropriate size, diversity of chemical structure, and other key descriptors before advanced algorithms blended with QSAR are employed. As dataset size and quality continue to improve, this will enable more robust deployment of machine learning and deep learning into the safety evaluation and risk assessment regulatory space. As the validation of these models becomes more trustworthy, they will inevitably be adopted as the mainstay of toxicology studies.

References

[1] Cros AFA. Action de l'alcool amylique sur l'organisme. Faculté de médecine de Strasbourg; 1863.

[2] Brown AC, Fraser TR. On the connection between chemical constitution and physiological action; with special reference to the physiological action of the salts of the ammonium bases derived from Strychnia, Brucia, Thebaia, Codeia, Morphia, and Nicotia. J Anat Physiol 1868;2:224–42.

[3] Hansch C, Maloney PP, Fujita T, Muir RM. Correlation of biological activity of phenoxyacetic acids with hammett substituent constants and partition coefficients. Nature 1962;194:178–80.

[4] Tichý M, Hanzlíková I, Rucki M, et al. Acute toxicity of binary mixtures: alternative methods, QSAR and mechanisms. Interdiscipl Toxicol 2008;1:15–7.

[5] Hong H, editor. Advances in computational toxicology: methodologies and applications in regulatory science. Cham, ZG, Switzerland: Springer International Publishing; 2019.

[6] Cronin MT, Jaworska JS, Walker JD, et al. Use of QSARs in international decision-making frameworks to predict health effects of chemical substances. Environ Health Perspect 2003;111:1391–401.

[7] Demchuk E, Ruiz P, Chou S, Fowler BA. SAR/QSAR methods in public health practice. Toxicol Appl Pharmacol 2011;254:192–7.

[8] Cronin MT, Schultz TW. Pitfalls in QSAR. J Mol Struct: THEOCHEM 2003;622:39–51.

[9] Prinz F, Schlange T, Asadullah K. Believe it or not: how much can we rely on published data on potential drug targets? Nat Rev Drug Discov 2011;10. 712–712.

[10] Ioannidis JP, Allison DB, Ball CA, et al. Repeatability of published microarray gene expression analyses. Nat Genet 2009;41:149–55.

[11] Bell AW, Deutsch EW, Au CE, et al. A HUPO test sample study reveals common problems in mass spectrometry–based proteomics. Nat Methods 2009;6:423–30.

[12] Brothers 2nd JF, Ung M, Escalante-Chong R, et al. Integrity, standards, and QC-related issues with big data in pre-clinical drug discovery. Biochem Pharmacol 2018;152:84–93.

[13] Zhao L, Wang W, Sedykh A, Zhu H. Experimental errors in QSAR modeling sets: what we can do and what we cannot do. ACS Omega 2017;2: 2805–12.

[14] Young D, Martin T, Venkatapathy R, Harten P. Are the chemical structures in your QSAR correct? QSAR Comb Sci 2008;27:1337–45.

[15] Fourches D, Muratov E, Tropsha A. Trust, but verify: on the importance of chemical structure curation in cheminformatics and QSAR modeling research. J Chem Inf Model 2010;50:1189–204.

[16] Kim J, Kusko R, Zeskind B, Zhang J, Escalante-Chong R. A primer on applying AI synergistically with domain expertise to oncology. Biochim Biophys Acta Rev Cancer 2021;1876:188548.

[17] Gadaleta D, Vuković K, Toma C, et al. SAR and QSAR modeling of a large collection of LD50 rat acute oral toxicity data. J Cheminf 2019;11:58.

[18] Kleinstreuer NC, Karmaus A, Mansouri K, et al. Predictive models for acute oral systemic toxicity: a workshop to bridge the gap from research to regulation. Comput Toxicol 2018;8:21–4.

[19] Siramshetty VB, Shah P, Kerns E, et al. Retrospective assessment of rat liver microsomal stability at NCATS: data and QSAR models. Sci Rep 2020;10:20713.

[20] Grisoni F, Consonni V, Vighi M, Villa S, Todeschini R. Investigating the mechanisms of bioconcentration through QSAR classification trees. Environ Int 2016;88:198–205.

[21] Richard AM, Huang R, Waidyanatha S, et al. The Tox21 10K compound library: collaborative chemistry advancing toxicology. Chem Res Toxicol 2021;34:189–216.

[22] Yang X, Liu H, Kusko R. Molecular modeling method applications: probing the mechanism of endocrine disruptor action. In: Hong H, editor. Advances in computational toxicology: methodologies and applications in regulatory science. Springer International Publishing; 2019. p. 315–35. https://doi.org/10.1007/978-3-030-16443-0_16.

[23] Sakkiah S, Leggett C, Pan B, et al. Development of a nicotinic acetylcholine receptor nAChR α7 binding activity prediction model. J Chem Inf Model 2020;60(4):2396–404. https://doi.org/10.1021/acs.jcim.0c00139.

[24] Ng HW, Zhang W, Shu M, et al. Competitive molecular docking approach for predicting estrogen receptor subtype α agonists and antagonists. BMC Bioinf 2014;15:S4. https://doi.org/10.1186/1471-2105-15-S11-S4. Suppl 11(Suppl 11).

[25] Sakkiah S, Kusko R, Tong W, Hong H. Applications of molecular dynamics simulations in computational toxicology. In: Hong H, editor. Advances in computational toxicology: methodologies and applications in regulatory science. Springer International Publishing; 2019. p. 181–212. https://doi.org/10.1007/978-3-030-16443-0_10.

[26] Selvaraj C, Sakkiah S, Tong W, Hong H. Molecular dynamics simulations and applications in computational toxicology and nanotoxicology. Food Chem Toxicol 2018;112:495–506. https://doi.org/10.1016/j.fct.2017.08.028.

[27] Sakkiah S, Guo W, Pan B, et al. Elucidating interactions between SARS-CoV-2 trimeric spike protein and ACE2 using homology modeling and molecular dynamics simulations. Front Chem 2021;8:622632. https://doi.org/10.3389/fchem.2020.622632.

[28] Sakkiah S, Selvaraj C, Guo W, et al. Elucidation of agonist and antagonist dynamic binding patterns in ER-α by integration of molecular docking, molecular dynamics simulations and quantum mechanical calculations. Int J Mol Sci 2021;22(17):9371. https://doi.org/10.3390/ijms22179371.

[29] Alonso H, Bliznyuk AA, Gready JE. Combining docking and molecular dynamic simulations in drug design. Med Res Rev 2006;26:531–68.

[30] Hong H, Neamati N, Winslow H, et al. Identification of HIV-1 integrase inhibitors based on a four-point pharmacophore. Antiviral Chem Chemother 1998;9:461–72. https://doi.org/10.1177/095632029800900602.

[31] Ebalunode J, Zheng W, Tropsha A. Application of QSAR and shape pharmacophore modeling approaches for targeted chemical library design. Methods in molecular biology. vol 685; 2011. p. 111–33. Clifton, N.J.

[32] Cramer RD, Patterson DE, Bunce JD. Comparative molecular field analysis (CoMFA). 1. Effect of shape on binding of steroids to carrier proteins. J Am Chem Soc 1988;110:5959–67.

[33] Shi L, Tong W, Fang H, et al. An integrated "4-phase" approach for setting endocrine disruption screening priorities–phase I and II predictions of estrogen receptor binding affinity. SAR QSAR Environ Res 2002;13:69–88.

[34] Klebe G, Abraham U, Mietzner T. Molecular similarity indices in a comparative analysis (CoMSIA) of drug molecules to correlate and predict their biological activity. J Med Chem 1994;37:4130–46.

[35] Korhonen S-P, Tuppurainen K, Laatikainen R, Peräkylä M. Comparing the performance of FLUFF-BALL to SEAL-CoMFA with a large diverse estrogen data set: from relevant superpositions to solid predictions. J Chem Inf Model 2005;45:1874–83.

[36] Yang H, Lou C, Li W, Liu G, Tang Y. Computational approaches to identify structural alerts and their applications in environmental toxicology and drug discovery. Chem Res Toxicol 2020;33:1312–22.

[37] Ahlberg E, Carlsson L, Boyer S. Computational derivation of structural alerts from large toxicology data sets. J Chem Inf Model 2014;54(10): 2945–52. https://doi.org/10.1021/ci500314a.

[38] Hong H, Xie Q, Ge W, et al. Mold2, molecular descriptors from 2D structures for chemoinformatics and toxicoinformatics. J Chem Inf Model 2008;48:1337–44.

[39] Chandrasekaran B, Abed SN, Al-Attraqchi O, Kuche K, Tekade RK. Chapter 21—computer-aided prediction of pharmacokinetic (ADMET) properties. In: Tekade RK, editor. Dosage form design parameters. Academic Press; 2018. p. 731–55. https://doi.org/10.1016/B978-0-12-814421-3.00021-X.

[40] Hong H, Fang H, Xie Q, et al. Comparative molecular field analysis (CoMFA) model using a large diverse set of natural, synthetic and environmental chemicals for binding to the androgen receptor. SAR QSAR Environ Res 2003;14:373–88. https://doi.org/10.1080/1062936031000l623962.

[41] Wong T-T. Performance evaluation of classification algorithms by k-fold and leave-one-out cross validation. Pattern Recogn 2015;48:2839–46.

[42] Majumdar S, Basak SC. Beware of external validation! - a comparative study of several validation techniques used in QSAR modelling. Curr Comput Aided Drug Des 2018;14:284–91.

[43] Rácz A, Bajusz D, Héberger K. Modelling methods and cross-validation variants in QSAR: a multi-level analysis. SAR QSAR Environ Res 2018;29:661–74.

[44] Idakwo G, Luttrell J, Chen M, et al. A review on machine learning methods for in silico toxicity prediction. J Environ Sci Health C Environ Carcinog Ecotoxicol Rev 2018;36(4):169–91. https://doi.org/10.1080/10590501.2018.1537118.
[45] Pal K, Patel Biraj V. Data classification with k-fold cross validation and holdout accuracy estimation methods with 5 different machine learning techniques. In: 2020 fourth international conference on computing methodologies and communication. ICCMC; 2020. p. 83–7. https://doi.org/10.1109/ICCMC48092.2020.ICCMC-00016.
[46] Ng HW, Shu M, Luo H, et al. Estrogenic activity data extraction and in silico prediction show the endocrine disruption potential of bisphenol A replacement compounds. Chem Res Toxicol 2015;28(9):1784–95. https://doi.org/10.1021/acs.chemrestox.5b00243.
[47] Shen J, Xu L, Fang H, et al. EADB: an estrogenic activity database for assessing potential endocrine activity. Toxicol Sci 2013;135(2):277–91. https://doi.org/10.1093/toxsci/kft164.
[48] Hong H, Tong W, Xie Q, Fang H, Perkins R. An in silico ensemble method for lead discovery: decision forest. SAR QSAR Environ Res 2005;16: 339–47. https://doi.org/10.1080/10659360500203022.
[49] Hong H, Thakkar S, Chen M, Tong W. Development of decision forest models for prediction of drug-induced liver injury in humans using A large set of FDA-approved drugs. Sci Rep 2017;27(1):17311. https://doi.org/10.1038/s41598-017-17701-7.
[50] Luo H, Ye H, Ng HW, et al. Machine learning methods for predicting HLA-peptide binding activity. Bioinf Biol Insights 2015;9(Suppl. 3):21–9. https://doi.org/10.4137/BBI.S29466.
[51] Ji Z, Guo W, Wood EL, et al. Machine learning models for predicting cytotoxicity of nanomaterials. Chem Res Toxicol 2022;35(2):125–39. https://doi.org/10.1021/acs.chemrestox.1c00310.
[52] Liu J, Guo W, Sakkiah S, et al. Machine learning models for predicting liver toxicity. Methods Mol Biol 2022;2425:393–415. https://doi.org/10.1007/978-1-0716-1960-5_15.
[53] Tang W, Liu W, Wang Z, Hong H, Chen J. Machine learning models on chemical inhibitors of mitochondrial electron transport chain. J Hazard Mater 2022;426:128067. https://doi.org/10.1016/j.jhazmat.2021.128067.
[54] Sakamuru S, Zhao J, Xia M, et al. Predictive models to identify small molecule activators and inhibitors of opioid receptors. J Chem Inf Model 2021;61(6):2675–85. https://doi.org/10.1021/acs.jcim.1c00439.
[55] Wang Z, Chen J, Hong H. Developing QSAR models with defined applicability domains on PPARγ binding affinity using large data sets and machine learning algorithms. Environ Sci Technol 2021;55(10):6857–66. https://doi.org/10.1021/acs.est.0c07040.
[56] Tang W, Liu W, Wang Z, Hong H, Chen J. Machine learning models on chemical inhibitors of mitochondrial electron transport chain. J Hazard Mater 2021;426:128067. https://doi.org/10.1016/j.jhazmat.2021.128067.
[57] Greener JG, Kandathil SM, Moffat L, Jones DT. A guide to machine learning for biologists. Nat Rev Mol Cell Biol 2022;23:40–55.
[58] Rahman F, Lhaksmana KM, Kurniawan I. Implementation of simulated annealing-support vector machine on QSAR study of fusidic acid derivatives as anti-malarial agent. In: 2020 6th international conference on interactive digital media (ICIDM); 2020. p. 1–4. https://doi.org/10.1109/ICIDM51048.2020.9339632.
[59] Motamedi F, Pérez-Sánchez H, Mehridehnavi A, Fassihi A, Ghasemi F. Accelerating big data analysis through LASSO-random forest algorithm in QSAR studies. Bioinformatics 2022;38:469–75.
[60] Mozafari Z, Chamjangali MA, Arashi M, Goudarzi N. Application of the LAD-LASSO as a dimensional reduction technique in the ANN-based QSAR study: discovery of potent inhibitors using molecular docking simulation. Chemometr Intell Lab Syst 2022;222:104510.
[61] Wu Z, Zhu M, Kang Y, et al. Do we need different machine learning algorithms for QSAR modeling? A comprehensive assessment of 16 machine learning algorithms on 14 QSAR data sets. Briefings Bioinf 2021;22. bbaa321.
[62] Keyvanpour MR, Shirzad MB. An analysis of QSAR research based on machine learning concepts. Curr Drug Discov Technol 2021;18:17–30.
[63] Lévêque L, Tahiri N, Goldsmith M-R, Verner M-A. Quantitative Structure-Activity Relationship (QSAR) modeling to predict the transfer of environmental chemicals across the placenta. Comput Toxicol 2022;21:100211.
[64] Xu Y, Dai Z, Chen F, Gao S, Pei J, Lai L. Deep learning for drug-induced liver injury. J Chem Inf Model 2015;55(10):2085–93.
[65] Tang W, Chen J, Wang Z, Xie H, Hong H. Deep learning for predicting toxicity of chem-icals: a mini review. J Environ Sci Health C Environ Carcinog Ecotoxicol Rev 2018;36(4):252–71.
[66] Pan X, Zuallaert J, Wang X. ToxDL: deep learning using primary structure and domain embeddings for assessing protein toxicity. Bioinformatics 2021;36(21):5159–68. https://doi.org/10.1093/bioinformatics/btaa656.
[67] Yap CW. PaDEL-descriptor: an open source software to calculate molecular descriptors and fingerprints. J Comput Chem 2011;32(7):1466–74.

PART I

Methods and advances of QSAR

CHAPTER

2

Development of QSAR models as reliable computational tools for regulatory assessment of chemicals for acute toxicity

Alexander Tropsha

Laboratory for Molecular Modeling, Division of Chemical Biology and Medicinal Chemistry, UNC Eshelman School of Pharmacy, University of North Carolina, Chapel Hill, NC, United States

Abbreviations

3R's reduce, refine, and replace
AOP adverse outcome pathway
NAMs new approach methods
QSAR quantitative structure activity relationships

Introduction: growing regulatory pressure to develop alternative computational methods for chemical toxicity assessment

Historically, regulatory agencies have required animal testing for hazard categorization and labeling [1]. However, there have been multiple calls, especially in the last 2 decades, to reduce, refine, and replace (3Rs) animal tests [2,3]. *Interagency Coordinating Committee on the Validation of Alternative Methods* (ICCVAM), a permanent committee of the National Institute of Environmental Health Sciences (NIEHS), created a U.S. Strategic Roadmap in 2018 for establishing new methods to evaluate the chemical safety of medical products [4]. This Roadmap pushed to expedite the development of New Approach Methods (NAMs), that is, chemical hazard/risk assessment technologies that combine a series of in vitro toxicity assays and computational approaches, which do not involve animal use. In 2017, FDA published its Predictive Toxicology Roadmap [5] emphasizing the importance of computational toxicology and, specifically, "new quantitative structure-activity relationship (QSAR) programs." In 2021, the FDA published *"2021: Advancing Regulatory Science at FDA: Focus Areas of Regulatory Science (FARS)"* report [6]. One of the critical areas was listed as "Advancing Novel Technologies to Improve Predictivity of Non-clinical Studies and Replace, Reduce, and Refine Reliance on Animal Testing." Finally, recent 2022 FDA Modernization Act 2.0 [7] called to restrict and, eventually, eliminate animal testing of medical and cosmetic products [8] and integrate alternative NAMs including computational tools into regulatory safety assessment programs. These recent documents highlight the continuing emphasis within the FDA to promote and support the use of NAMs robust in vitro and computational NAMs tools for accurate and reliable hazard identification in cosmetics products. To respond to these directives, novel reliable computational tools address these regulatory demands and help manufacturers of chemical products across multiple industries reduce costs and expedite safety assessment of their products.

This chapter focuses on approaches and tools for predicting acute toxicity. The most common type of animal assays for acute toxicity assessment of chemicals used as cosmetic or pharmaceutical products or medical devices includes topical skin sensitization, skin irritation, and skin corrosion end points, as well as eye irritation and corrosion [9]. The challenge of developing alternative computational tools effectively implies that model-based predictions should be at least as accurate as the assays themselves. This premise can be tested by developing respective end

https://doi.org/10.1016/B978-0-443-15339-6.00003-5

point models using carefully curated data and rigorous model building and validation workflows, and we consider examples of such studies in peer-reviewed literature [10–12]. We also reflect on the importance of meaningful model interpretation and elucidation of mechanistic pathways underlying chemical effects on biological systems (often referred to as adverse outcome pathways, AOPs) using emerging approaches of knowledge graph mining. We summarize the current challenges and describe in silico solution to the problem of acute toxicity prediction in Fig. 2.1. We posit that the successful development of reliable computational tools for acute toxicity prediction will replace current animal texting protocols, shorten the regulatory review time, and ultimately bring safer products to the market faster.

Comparison of computational approaches for chemical toxicity prediction

Computational models have earned recognition as reliable, fast, and inexpensive alternatives for the toxicity evaluation of chemicals during the early stages of drug discovery or environmental safety assessment [13]. This assessment often relies on structural alerts, chemical grouping, and read-across. Structural alerts, otherwise known as "expert rules," are molecular substructures that are associated with a particular adverse outcome [14]. The popularity of alerts dates back to a series of studies by Ashby and Tenant [15,16] in late 80s and early 90s on chemical carcinogenicity and mutagenicity. As the name suggests, expert rules are based on human expertise and are intended to reflect the chemical basis of the mechanism of action or, at least, the molecular initiating event in the case of more complex endpoints [17].

Alerts are used to flag potential hazards and to group compounds into categories for read-across [18]. Chemical read-across is a data gap filling procedure to assess certain end point effect of a chemical by using data for the same end point from another chemical (or a group of chemicals), which is (are) considered structurally similar [19]. These methods have earned acceptance among toxicologists due to their simplicity, transparency, and ease of interpretation [20]. However, although chemical read across has been accepted by regulatory agencies, there have been

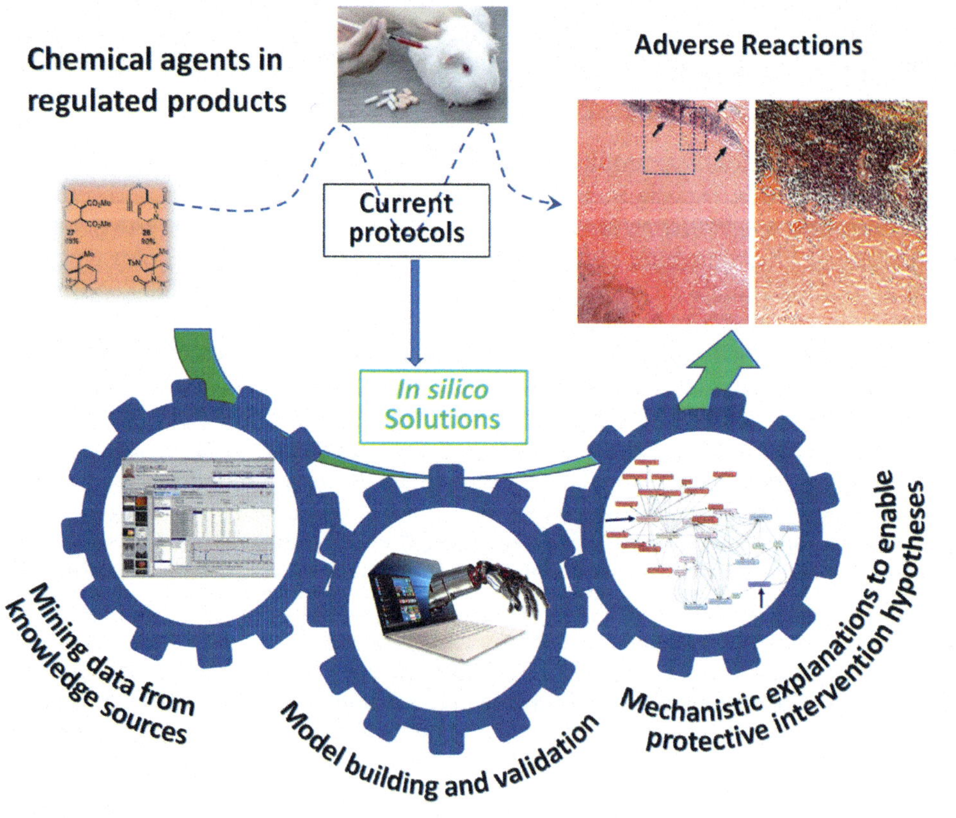

FIGURE 2.1 The key challenge of in silico alternative methods is to develop predictive computational toxicology assessment tools and provide mechanistic explanation of chemical toxicity.

observations that this approach is prone to bias [21–24]. On the other hand, though supported by rigorously evaluated statistical significance, quantitative structure-activity relationships (QSARs) modeling is usually considered as a "black box" [25], referencing the perceived lack of interpretability.

The use of structural alerts and QSAR models (often collectively referred to as (Q)SAR) has become a major concept in chemical safety assessment and regulatory decision support since the acceptance of Registration, Evaluation, Authorization, and Restriction of Chemicals (REACH) legislation in 2006 b y the European Union [26]. This law compels manufacturers to provide detailed information on chemicals that are manufactured, marketed, or imported. In the United States, the Environmental Protection Agency (EPA) has been using (Q)SAR methods to support a weight-of-evidence in the hazard assessment of chemicals under the US Toxic Substances Control Act (TSCA) [27].

Qualitative approaches, such as alerts, chemical grouping, and read-across, have been incorporated into several computerized expert systems, often employing additional layers of secondary modulating effects around each generic expert alert (e.g., OECD QSAR Toolbox [28], OCHEM ToxAlerts [29], Lhasa's Derek [30], etc.). Also, because analyzing and processing large volumes of experimental data by human experts can be very slow, current expert-based systems often use statistical approaches to discover candidate substructures strongly associated with the target activity. These candidate substructures are then reviewed and curated by experts in the field to provide mechanistic interpretation.

Albeit less commonly appreciated, statistically based QSAR models can indeed afford mechanistic interpretation [31]. However, the difficulty of interpreting QSAR models, especially by nonexperts in the field, has led regulatory agencies to prefer the use of simple structural alerts for the prediction of various properties. On the other hand, the ability of QSAR to provide a reliable quantitative assessment of the toxicity potential of a chemical is highly advantageous. Hence, there is a strong need for the development of a method that would combine the predictive power of QSAR models while maintaining transparency for expert evaluation. Recently, integrative approaches of QSAR and read-across have been proposed by our [32] and other groups [33–35]. Specifically, our team proposed chemical-biological read-across (CBRA), which combines quantitative toxicity prediction with a visualization methodology, ensuring both high predictivity and interpretability of models [32].

Contrasting alerts and QSAR-based predictions of acute toxicity

Skin sensitization, being a major environmental and occupational health hazard [36], represents a highly illustrative opportunity to contrast the results of QSAR modeling and alert-based approaches. Previously, our group developed and published skin sensitization QSAR models [21] that can be compared with the outcome of the local lymph node assay (LLNA) node of OECD QSAR Toolbox [28] and OCHEM ToxAlerts [29] web server. To make the comparison fair, we used compounds (i) not present in databases and (ii) inside the applicability domain of our models. Therefore, 90 compounds (38 sensitizers and 52 nonsensitizers) were used for comparison with QSAR Toolbox, and 246 compounds (160 sensitizers and 86 nonsensitizers) were used for comparison with ToxAlerts [37]. The resulting QSAR models showed a significantly higher prediction accuracy for the same sets of external compounds when compared to both alerts-based tools as evaluated by positive predictive value (PPV), specificity, negative predictive value (NPV), and correct classification rate (CCR) (Table 2.1). Although predictions made with QSAR Toolbox and ToxAlerts showed higher sensitivity than our models when evaluating the same set of structures, the developed

TABLE 2.1 Comparison of the prediction accuracies of QSAR models, skin sensitization module of the OECD QSAR Toolbox, and OCHEM ToxAlerts.

QSAR vs. SAR rule tool	CCR	Sensitivity	PPV	Specificity	NPV
QSAR toolbox					
QSAR	0.74	0.50	0.94	0.98	0.75
QSAR toolbox	0.46	0.53	0.38	0.38	0.53
OCHEM ToxAlerts					
QSAR	0.80	0.74	0.90	0.85	0.64
ToxAlerts	0.60	0.84	0.71	0.36	0.55

models featured a much higher PPV. These results indicate that the probability of correctly classifying sensitizers is much higher using QSAR models and that alert-based predictions have a bias toward sensitizers.

In Table 2.2, we show three examples of structural alerts flagged by QSAR Toolbox or by ToxAlerts: activated pyridine/pyrimidine, formyl group (aldehydes), and aromatic amines. In addition, we compare experimental toxicity assessment with predictions obtained by our QSAR models for compounds containing the alert substructures obtained by QSAR Toolbox or ToxAlerts. As one can see, all compounds with an activated pyridine/pyrimidine subgroup are nonsensitizers in LLNA assays. These compounds were flagged as sensitizers by QSAR Toolbox but correctly predicted as nonsensitizers by the QSAR models. Molecules containing two other alerts flagged by ToxAlerts (formyl group and aromatic amines) are not always the sensitizers: examples are provided by 3-hydroxy-4-methoxybenzaldehyde (correctly predicted by QSAR as a nonsensitizer) and (2E)-2-methyl-2-butenal (incorrectly predicted by QSAR as a sensitizer) that contain the formyl group. We also identified the following 11 nonsensitizers containing five or more ToxAlerts each: 3-hydroxy-4-methoxybenzaldehyde, vanillin (10 alerts both), and ethyl vanillin (nine alerts) (see Supplementary Information). These results highlight that even the presence of multiple structural alerts for skin sensitization does not automatically suggest that the respective compounds are toxic and as such should be avoided in commercial products. At the same time, our results show that externally validated QSAR models afford higher accuracy of assessing chemical toxicity than structural alerts.

Integration of interpretative QSAR models and chemical alerts

Despite the aforementioned deficiencies, especially, exceeding specificity of chemical alerts, it may be beneficial to combine mechanism-based alerts with QSAR predictions especially taking into account the often overlooked ability to provide structural interpretation of "black box" QSAR models. Indeed, especially, when best practices [38,39] of QSAR model development are strictly followed, statistical models produce more accurate predictions of specific endpoints. However, both researchers and regulators are interested not only in quantitative prediction but also in understanding both structural determinants of toxicity and underlying mechanisms of action [40]. Despite common beliefs, statistical QSAR models can be interpreted in terms of relative contributions of the underlying chemical fragments [41]. For instance, one can use the predicted probability maps [42] available in RDKit [43], which provide a graphical visualization of the contribution of structural fragments to the activity that can be derived from QSAR models [43]. The atoms are highlighted according to their predicted activity contribution. Atoms and molecular fragments highlighted in green are responsible for an increase in toxicity, whereas pink fragments lead to toxicity decrease. Activities of the target molecule and its virtual analog derived from this initial molecule by the elimination of a molecular fragment with a predefined structure are estimated using the respected QSAR model. The difference of calculated activities of a target molecule and its virtual analog is considered as the influence (contribution) of the eliminated fragment on activity. As we demonstrated earlier [37], fragment/atom contributions are not constant and depend on the structural environment. Fig. 2.2 illustrates how one can explain the predictions generated for N-phenyl-*p*-phenylenediamine, a known skin sensitizer that is usually added to temporary black henna tattoos leading to many cases of contact allergy [44]. The figure shows the predicted probability maps, providing a graphical visualization of structural fragments predicted by the QSAR models to contribute positively (green) or negatively (pink) to toxicity. Such capabilities have been illustrated in several studies [24,45], and our group offered an approach for integrating common chemical alerts and QSAR model predictions [37].

Indeed, as described above, structural alerts, while transparent, are extremely promiscuous if used alone to predict biological activity, and their use may be harmful in the drug discovery pipeline as well as for the safety assessment. Certainly, structural alerts can drive the design of compounds with a desired toxicity profile. However, the global effect of such modification can only be predicted by a QSAR model and not by isolated alerts alone. These considerations advance the integration of structural alerts and QSAR models for designing novel compounds and predicting their toxicity. To this point, we previously proposed an integrative approach for safety assessment of new chemicals using both structural alerts and QSAR models (Fig. 2.3). In this approach, the alerts serve as transparent hypotheses (which could be derived by traditional SAR analysis or model interpretation) that can be either reinforced or refuted by statistically significant and externally validated QSAR models. Any alert should be viewed as a *structural hypothesis* of chemical action, but its predictive power should be buttressed by QSAR predictions and, if possible, by experimental validation.

TABLE 2.2 Examples of structural alerts and comparison of experimental and predicted skin sensitization effects for alert-containing compounds.

Alerts	Compounds containing the alert substructure		
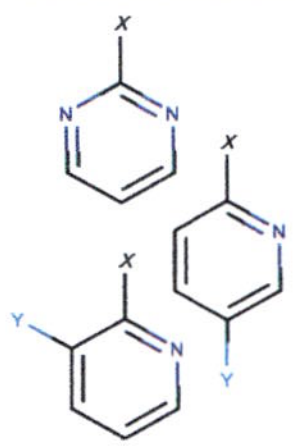 **Activated pyridine/pyrimidine** QSAR toolbox	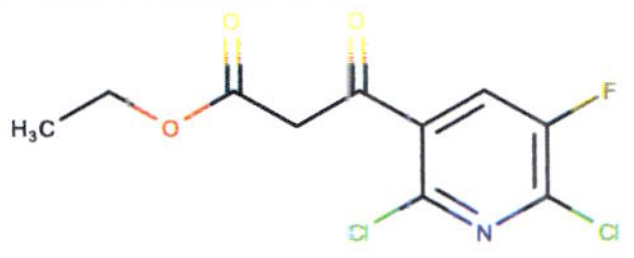 **Ethyl 2,6-dichloro-5-fluoro-β-oxo-3-pyridinepropanoate** LLNA: Nonsensitizer QSAR: Nonsensitizer QSAR toolbox: Sensitizer	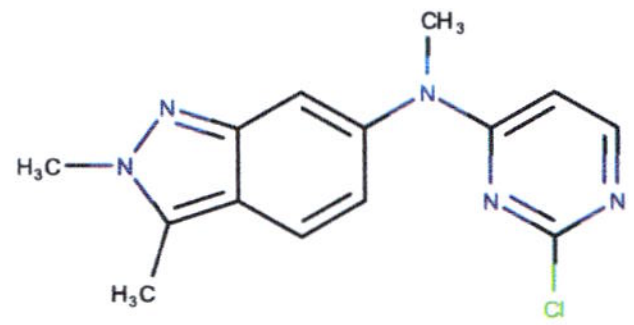 **N-(2-chloro-4-pyrimidinyl)-N-2,3-trimethyl-2H-indazol-6-amine** LLNA: Nonsensitizer QSAR: Nonsensitizer QSAR toolbox: Sensitizer	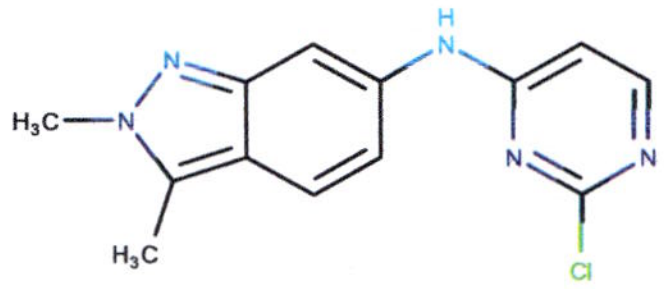 **N-(2-chloro-4-pyrimidinyl)-2 ,3-dimethyl-2H-indazol-6-amine** LLNA: Nonsensitizer QSAR: Nonsensitizer QSAR toolbox: Sensitizer
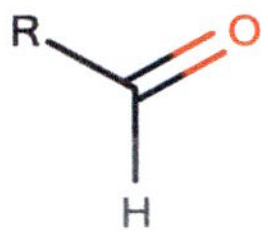 **Formyl group** OCHEM ID: TA264	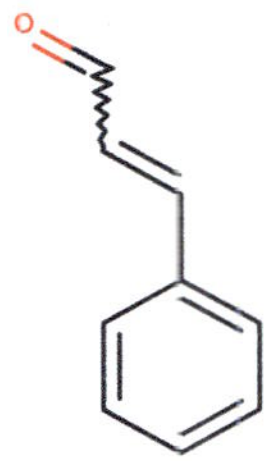 **Cinnamic aldehyde** LLNA: Sensitizer ToxAlerts: Sensitizer QSAR: Sensitizer	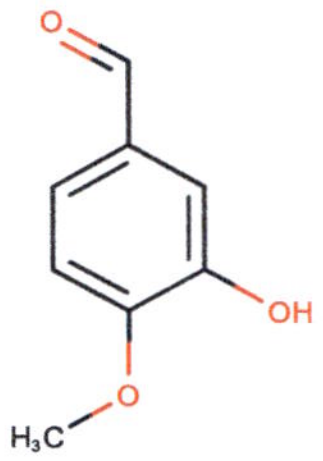 **3-Hydroxy-4-methoxybenzaldehyde** LLNA: Nonsensitizer ToxAlerts: Sensitizer QSAR: Nonsensitizer	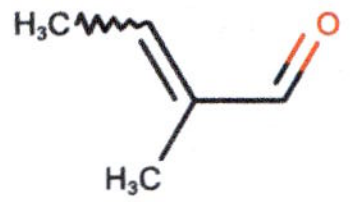 **(2E)-2-methyl-2-butenal** LLNA: Nonsensitizer ToxAlerts: Sensitizer QSAR: Sensitizer
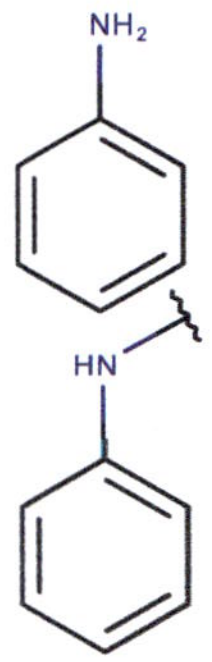 **Aromatic amines** OCHEM ID: TA311	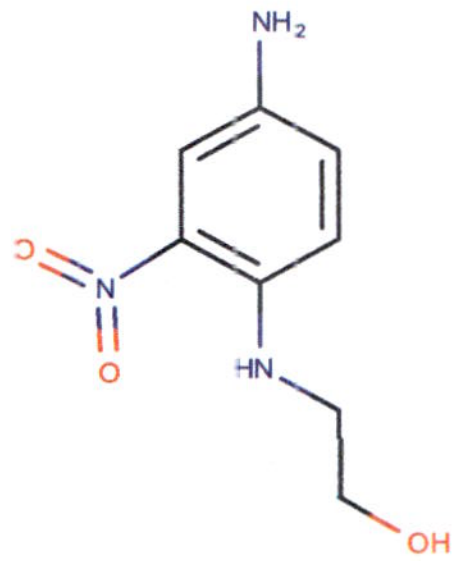 **2-(4-Amino-2nitro-phenylamino)-ethanol** LLNA: Sensitizer ToxAlerts: Sensitizer QSAR: Sensitizer	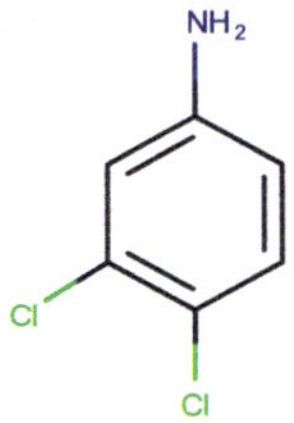 **3,4-Dichloroaniline hydrochloride** LLNA: Sensitizer ToxAlerts: Sensitizer QSAR: Nonsensitizer	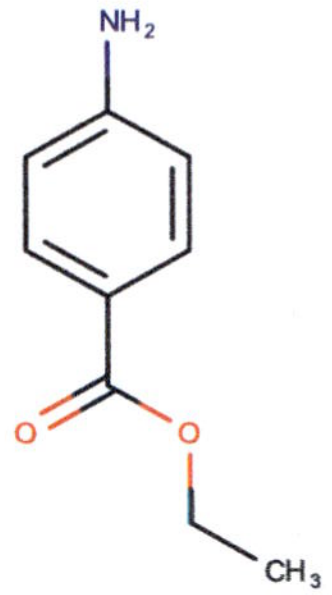 **Benzocaine** LLNA: Nonsensitizer ToxAlerts: Sensitizer QSAR: Nonsensitizer

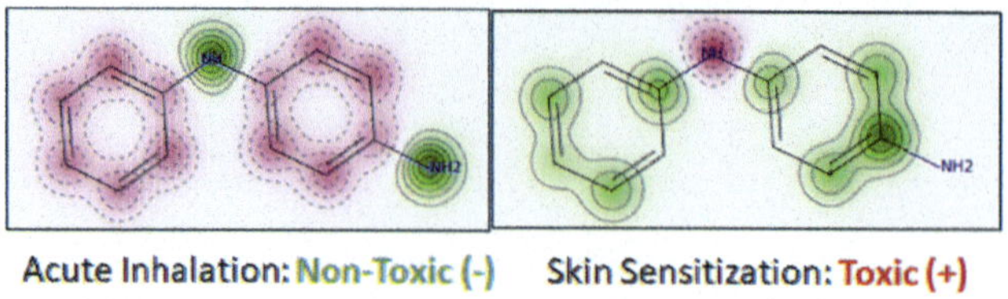

FIGURE 2.2 QSAR model interpretation. *Green* contours indicate that respective groups are responsible for compound toxicity.

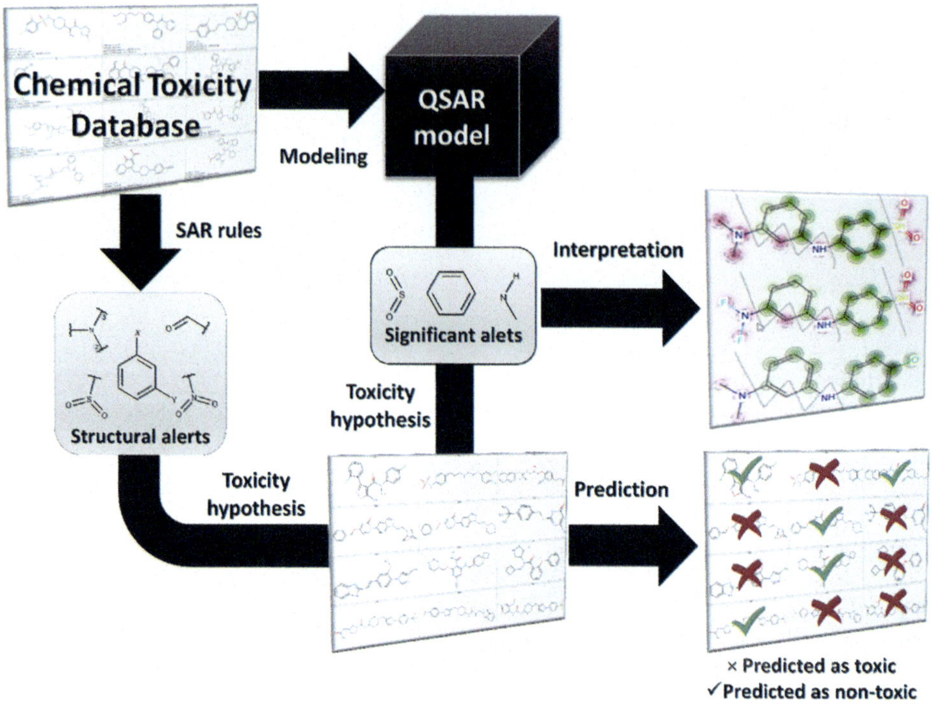

FIGURE 2.3 Integrative approach for chemical safety assessment of new chemicals by combining structural alerts and QSAR models.

The continuing importance of data quality and curation in the age of big data and AI

The recent emergence of big data and AI methods has generated optimistic expectations in chemical toxicology as the field strived to reduce or replace animal testing by NAMs [46,47]. Despite this excitement and current progress, the main limitation of AI models remains the quality of the toxicology and chemistry data used to train the models. Indeed, a key consideration universally applied to any computational modeling approach is the need for careful data curation before initiating any model development [48]. While data preparation has been repeatedly emphasized as a critical element to ensure rigor and reproducibility of experimental and computational research [12,49], many publications seem to overlook this critical step [50–54]. Thus, it is still important to highlight challenges with experimental data reproducibility to reinforce the need for high-quality data curation as an essential initial step when developing predictive and robust AI models that could eventually reduce or replace animal testing.

It has been shown repeatedly in the field of QSAR modeling that data curation strongly affects the predictive accuracy of the models [38,39,49,55]. This experience suggests that data curation and rigorous external model validation should be made mandatory for employing computational models in regulatory assessment to account for data inconsistency or relevance and avoid overly optimistic evaluation of a computational model's power. Chemical and biological data curation is not the only factor that affects the accuracy and utility of in silico approaches as alternatives to animal testing. For instance, improper statistical analysis has been heavily discussed as a factor as well [56–58], but inaccurate chemical and biological data are still the central issue in modeling chemogenomic data [59].

For instance, it was recently reported [60] that some of the data contained in the European Chemicals Agency (ECHA) Registration, Evaluation, Authorization and Restriction of Chemicals (REACH) database are not from the guideline studies but were predicted using QSAR models or read-across. Therefore, those data points are not suitable for building new models, since models that predict compound categories that have been predicted by similar models are likely to suffer from inflated accuracy. Also, much of the data in this database have been marked as "not reliable" by ECHA. As part of a more robust data ecosystem, experimental and predicted data employed by regulatory agencies need to be better structured and properly annotated to be easily integrated within informatics systems and needs not to be mined from free text present in reports. Another critical issue is the variability or errors in reported units of measurement, especially the use of concentrations or doses in weight rather than in molar units, since the biological effect of a chemical is due to the number of molecules present and not their weight (even though the weight can affect pharmacokinetic properties or dispersion in the environment) [49]. To provide an example of data harmonization, the ChEMBL [61] database has standardized all its data to nanomolar units.

In vivo test reproducibility is not the only factor that affects the accuracy and utility of in silico approaches as alternatives to animal testing. Recently [62], we have reported the collection, curation, and integration of the most extensive publicly available datasets for acute toxicity tests collectively known as the "six-pack" (acute oral toxicity, acute dermal toxicity, acute inhalation toxicity, skin irritation and corrosion, eye irritation and corrosion, and skin sensitization). The data used in that study were collected from multiple sources, including the scientific publications, the National Toxicology Program Interagency Center for the Evaluation of Alternative Toxicological Methods on behalf of ICCVAM, and the REACH study results database (https://iuclid6.echa.europa.eu/reach-study-results). As a result of curation, the datasets' size was considerably diminished before QSAR model development. However, upon inspection, we found many data points initially processed as experimental records to be, in fact, predictions made by QSAR models, read-across, or expert-based systems. In addition, many experimental data points have been flagged as "not reliable." The "shrinkage" was substantial for most of the datasets we examined. For skin sensitization end point, we reduced the database from 10,861 records to about 1000; for skin irritation/corrosion, from 5274 to 1012; for eye irritation and corrosion, from 7322 to 3547; for acute dermal toxicity from 29,824 to 2622; and for acute inhalation from 8176 to 681. Lastly, because acute oral toxicity data were already extensively curated, we reduced the dataset from 8994 records to 8495. More details about the data curation and how many compounds were reduced at each step can be found elsewhere [62]. The models were implemented on a freely available web application termed Systemic and Topical Toxicity (STopTox, https://stoptox.mml.unc.edu/).

The rapid growth of publicly available toxicology data [63] does not prevent the need for careful data curation before model development. Addressing and ensuring data quality and assay reproducibility is the first step in curating chemogenomics data. Several studies have pointed out errors in chemical structures deposited in public databases [55,64]. Even a relatively small number of erroneous data points can significantly affect modeling outcomes, especially when assessing the external model accuracy [11]. Many investigations have shown that the model's predictivity is affected by the accuracy of chemical structure, choice of descriptors, modeling algorithms, and data curation [65–67]. To help to address this issue, Mansouri et al. [68] developed a KNIME workflow to curate and correct errors in the structure and identify chemicals using environmental fate datasets. Gadaleta et al. [69] developed a workflow for retrieving chemical structures from several web-based databases. The workflow automatically compares these data and performs some structural cleaning. The developers of many publicly accessible databases have made substantial efforts to ensure that chemical structures in their databases are correct. For example, ChemSpider has promoted a crowdsourced collaboration to curate chemical structures derived from public compound databases [70]. More recently, the US EPA CompTox Chemistry Dashboard [63] performed curation of the substances linked to chemical structures and integrated with other diverse types of publicly available data from multiple sources.

Biomedical knowledge mining to identify mechanistic pathways underlying chemical toxicity effects

Scientists and regulators are interested in understanding biological mechanisms underlying toxic effects of chemicals or AOPs [71]. The latter are typically developed through laborious tests and thorough manual mining of the research literature [72]. Tools to facilitate AOPs elucidation based on biomedical knowledge mining are emerging, such as Chemotext [73,74] and ROBOKOP [75,76]. Chemotext extracts specific assertions linking any specific annotations, for example, chemicals (C), targets (T), and effect (E), from the over 30 million papers annotated in PubMed. These assertions reflect the possible mechanistic relationships in chemical toxicity, that is, chemical–protein (target), protein–effects, or chemical–effect relationships forming CTE interaction triangles.

ROBOKOP is a biomedical knowledge graph that consolidates data from more than 30 biomedical databases including Comparative Genomics Database (CTD) [77] and Pharos [78]. ROBOKOP can be queried to identify mechanistic connections (paths) such as *Clinical Outcome Pathways* [79] between query terms. "Queries" are defined as paths through the knowledge graph that meets some type and length constraints. Queries that will be investigated in this project could be formulated as: "why the exposure to a certain chemical causes an adverse effect?" For instance, previously, we reported that our skin sensitization models based on human data identified styrene as a potential skin sensitizer [80]. By running Chemotext, we found epidemiological evidence that this compound does cause skin sensitization [80]. By running ROBOKOP, we could reconfirm this link as illustrated in Fig. 2.4. As such methods and tools developed further, for every new compound of interest predicted toxic by QSAR models, one may identify the most chemically similar known toxicant in any of the assays for sensitization, irritation, or corrosion and use it to run Chemotext and ROBOKOP to elucidate respective AOPs.

Conclusions and perspectives

To employ computational models as a reliable means to reduce animal testing, we need to ensure model transparency and reproducibility. To this end, all the curated data employed for model development need to be made publicly available following the FAIR (findability, accessibility, interoperability, and reusability) data principles [81]. In the field of molecular modeling, this means that the chemical structures and associated data should be made available in a machine-readable format. The data must be well organized and structured in databases (e.g., Integrated Chemical Environment [ICE] [82], Tox21 [83], ToxCast [84], ChEMBL [85], PubChem [86]) and made publicly available [87]. In addition, all the scripts used to process or model the data should also be available. In the context of regulatory use of QSAR models, the harmonized template for summarizing and reporting key information on QSAR models, named QSAR Model Reporting Format (QMRF), is recommended, especially when validation studies are available.

We envision further adoption of the knowledge-driven approaches for chemical toxicity prediction that can be used to reduce or replace animal testing (Fig. 2.5). As more robust in vitro and organ on a chip data becomes available, after validation and acceptance for regulatory use, these assays need to be scaled for high-throughput screening. All the data resulting from such screening campaigns, and all the scripts used to process or model the data, need to be stored and managed, following the FAIR data principles [81], and curated based on standard approaches described above. Subsequent cheminformatics analysis and generation and validation of AI models should be done following the best practices promoted by OECD [39] and universally accepted by the community. Only then should these models be used to predict biological response and, potentially, be employed to design safer chemicals.

We hope that, as high-quality data accumulates and mechanistic understanding informs new experimental approaches, nonanimal, especially computational models, will gradually reduce or replace animal testing. However, chemogenomic data curation would still be necessary before embarking on model development. We strongly caution against overinterpreting results from models built on noncurated datasets. Following the editorial requirements implemented by the ACS *Journal of Chemical Information and Modeling* [88], we also support the idea that all scientific journals should require a section on data curation in any manuscript submitted for publication. We expect that careful collection, curation, model development, and data/model sharing practices adopted by computational toxicologists will create an unprecedented ability to analyze historical data and employ modern data analytics to derive knowledge and generate safer products smarter and faster.

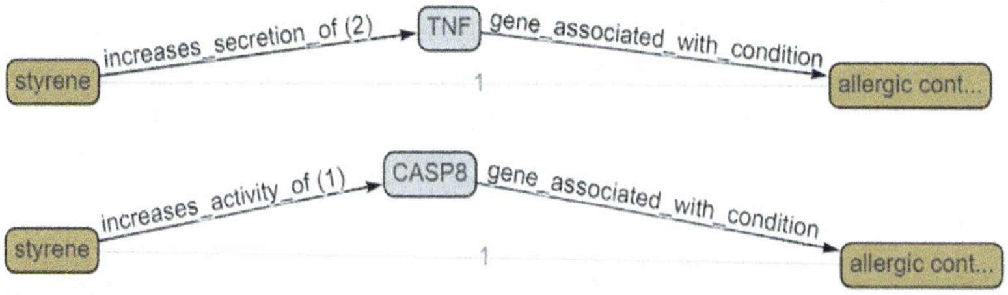

FIGURE 2.4 Results of running a ROBOKOP query for styrene—contact allergic dermatitis pair. The number of papers supporting the edge are shown in parentheses.

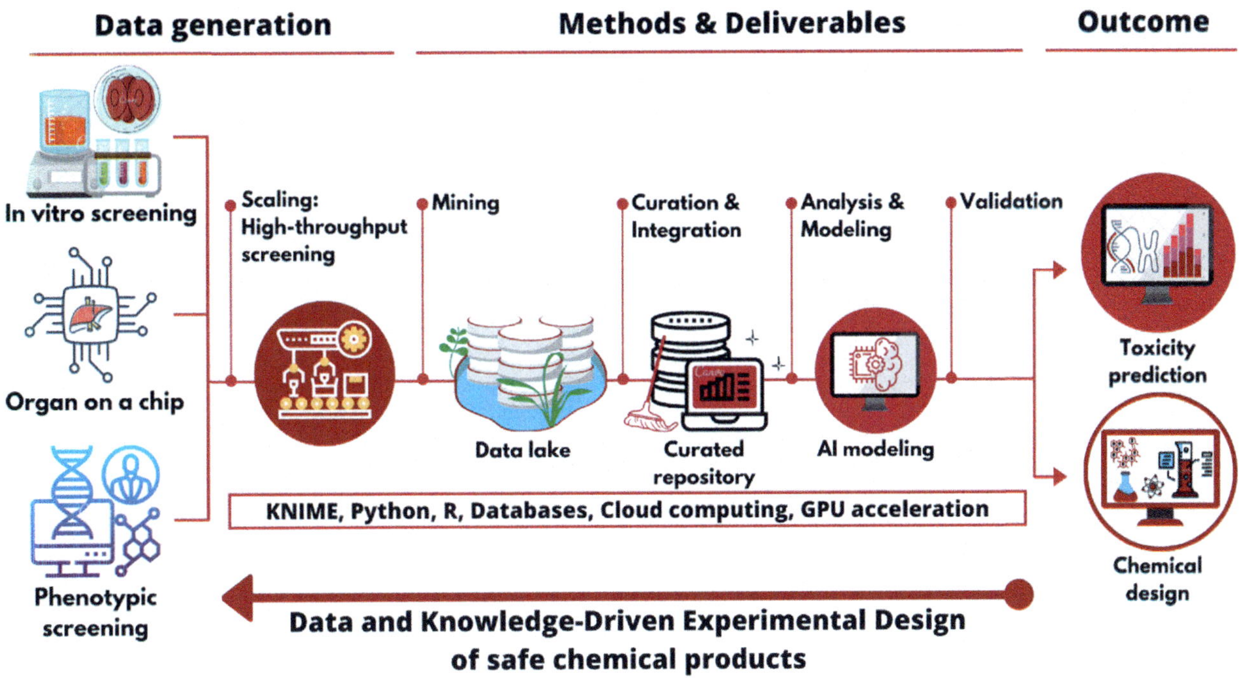

FIGURE 2.5 An in silico workflow for data and knowledge-driven discovery of safer chemicals that can reduce or replace animal testing.

Acknowledgments

The author thanks many colleagues and members of his laboratory for multiple methodological contributions and application studies that supported the analysis discussed in this chapter. The author is especially grateful to Drs. Vinicius Alves, Eugene Muratov, Denis Fourches, Nicole Kleinstreuer, and Carolina Andrade for their friendship and contributions that enabled this writing.

References

[1] National Research Council (US). Commitee on animals as monitors of environmental hazards. In: Animals as sentinels of environmental health hazards. Washington, D.C.: National Academies Press; 1991. https://doi.org/10.17226/1351.
[2] Flecknell P. Replacement, reduction and refinement. ALTEX 2002;19:73–8.
[3] Patlewicz G, Fitzpatrick JM. Current and future perspectives on the development, evaluation, and application of in silico approaches for predicting toxicity. Chem Res Toxicol 2016;29:438–51. https://doi.org/10.1021/acs.chemrestox.5b00388.
[4] ICCVAM A. Strategic roadmap for establishing new approaches to evaluate the safety of chemicals and medical products in the United States. 2018. https://doi.org/10.22427/NTP-ICCVAM-ROADMAP2018.
[5] FDA. FDA's predictive toxicology roadmap. https://www.fda.gov/media/109634/download (accessed 10 Aug 2022).
[6] FDA. 2021 advancing regulatory science at FDA: focus areas of regulatory science (FARS). 2021. https://www.fda.gov/media/145001/download. [Accessed 1 January 2022].
[7] S.5002–117th Congress. FDA modernization act 2.0 | Congress.gov | library of congress. 2021–2022. https://www.congress.gov/bill/117th-congress/senate-bill/5002. [Accessed 16 December 2022].
[8] FDA. Use of International Standard ISO 10993-1, 'Biological evaluation of medical devices—Part 1: evaluation and testing within a risk management process'. 2020. https://www.fda.gov/regulatory-information/search-fda-guidance-documents/use-international-standard-iso-10993-1-biological-evaluation-medical-devices-part-1-evaluation-and. [Accessed 4 January 2021].
[9] Data Requirements for Pesticide Registration | Pesticide Registration | US EPA. https://www.epa.gov/pesticide-registration/data-requirements-pesticide-registration (accessed 5 Jan 2021).
[10] Fourches D, Muratov E, Tropsha A, et al. Trust, but verify II: a practical guide to chemogenomics data curation. J Chem Inf Model 2016;56:1243–52. https://doi.org/10.1021/acs.jcim.6b00129.
[11] Fourches D, Muratov E, Tropsha A. Trust, but verify: on the importance of chemical structure curation in cheminformatics and QSAR modeling research. J Chem Inf Model 2010;50:1189–204. https://doi.org/10.1021/ci100176x.
[12] Fourches D, Muratov E, Tropsha A. Curation of chemogenomics data. Nat Chem Biol 2015;11:535.
[13] Naven R, Louise-May S. Computational toxicology: its essential role in reducing drug attrition. Hum Exp Toxicol 2015;34:1304–9. https://doi.org/10.1177/0960327115605440.
[14] Blagg J. Structural alerts for toxicity. In: Burger's medicinal chemistry and drug discovery. Hoboken, NJ, USA: John Wiley and Sons, Inc.; 2010. p. 301–34. https://doi.org/10.1002/0471266949.bmc128.
[15] Ashby J, Tennant RW. Definitive relationships among chemical structure, carcinogenicity and mutagenicity for 301 chemicals tested by the U.S. NTP. Mutat Res 1991;257:229–306.

[16] Ashby J, Tennant RW. Chemical structure, Salmonella mutagenicity and extent of carcinogenicity as indicators of genotoxic carcinogenesis among 222 chemicals tested in rodents by the U.S. NCI/NTP. Mutat Res 1988;204:17–115.
[17] Allen TEH, Goodman JM, Gutsell S, et al. Defining molecular initiating events in the adverse outcome pathway framework for risk assessment. Chem Res Toxicol 2014;27:2100–12. https://doi.org/10.1021/tx500345j.
[18] Enoch SJ, Roberts DW. Approaches for grouping chemicals into categories. In: Cronin M, Madden J, Enoch S, et al., editors. Chemical toxicity prediction: category formation and read-across. Royal Society of Chemistry; 2013. p. 30–43. https://doi.org/10.1039/9781849734400-00030.
[19] Cronin MTD. An introduction to chemical grouping, categories and read-across to predict toxicity. In: Chemical toxicity prediction: category formation and read-across; 2013. p. 1–29. https://doi.org/10.1039/9781849734400-00001.
[20] Cronin MTD. Evaluation of categories and read-across for toxicity prediction allowing for regulatory acceptance. In: Cronin M, Madden J, Enoch S, et al., editors. Chemical toxicity prediction: category formation and read-across. Royal Society of Chemistry; 2013. p. 155–67. https://doi.org/10.1039/9781849734400-00155.
[21] Alves VM, Muratov E, Fourches D, et al. Predicting chemically-induced skin reactions. Part I: QSAR models of skin sensitization and their application to identify potentially hazardous compounds. Toxicol Appl Pharmacol 2015;284:262–72. https://doi.org/10.1016/j.taap.2014.12.014.
[22] Stepan AF, Walker DP, Bauman J, et al. Structural alert/reactive metabolite concept as applied in medicinal chemistry to mitigate the risk of idiosyncratic drug toxicity: a perspective based on the critical examination of trends in the top 200 drugs marketed in the United States. Chem Res Toxicol 2011;24:1345–410. https://doi.org/10.1021/tx200168d.
[23] Liu R, Yu X, Wallqvist A. Data-driven identification of structural alerts for mitigating the risk of drug-induced human liver injuries. J Cheminf 2015;7:4. https://doi.org/10.1186/s13321-015-0053-y.
[24] Low YS, Caster O, Bergvall T, et al. Cheminformatics-aided pharmacovigilance: application to Stevens–Johnson syndrome. J Am Med Inf Assoc 2016;23:968–78. https://doi.org/10.1093/jamia/ocv127.
[25] Raunio H. In silico toxicology—non-testing methods. Front Pharmacol 2011;2:33. https://doi.org/10.3389/fphar.2011.00033.
[26] European Union. Regulation (EC) No 1907/2006. Off J Eur Union; 2007. p. 3–280.
[27] ECHA. The use of alternatives to testing on animals for the REACH regulation. 2014. p. 131. https://doi.org/10.2823/22471. Second Report under Article 117(3) of the REACH Regulation.
[28] OECD. QSAR Toolbox v. 3.2. 2014.
[29] Sushko I, Salmina E, Potemkin V a, et al. ToxAlerts: a Web server of structural alerts for toxic chemicals and compounds with potential adverse reactions. J Chem Inf Model 2012;52:2310–6. https://doi.org/10.1021/ci300245q.
[30] Ridings JE, Barratt MD, Cary R, et al. Computer prediction of possible toxic action from chemical structure: an update on the DEREK system. Toxicology 1996;106:267–79.
[31] Polishchuk P, Kuz'min V, Artemenko A, et al. Universal approach for structural interpretation of QSAR/QSPR models. Mol Inform 2013;32: 843–53.
[32] Low Y, Sedykh AY, Fourches D, et al. Integrative chemical-biological read-across approach for chemical hazard classification. Chem Res Toxicol 2013;26:1199–208. https://doi.org/10.1021/tx400110f.
[33] Benfenati E, Roncaglioni A, Petoumenou MII, et al. Integrating QSAR and read-across for environmental assessment. SAR QSAR Environ Res 2015;26:605–18. https://doi.org/10.1080/1062936X.2015.1078408.
[34] Price N, Chaudhry Q. Application of in silico modelling to estimate toxicity of migrating substances from food packaging. Food Chem Toxicol 2014;71:136–41. https://doi.org/10.1016/j.fct.2014.05.022.
[35] Lozano S, Poezevara G, Halm-Lemeille MP, et al. Introduction of jumping fragments in combination with QSARs for the assessment of classification in ecotoxicology. J Chem Inf Model 2010;50:1330–9. https://doi.org/10.1021/ci100092x.
[36] Diepgen TL, Coenraads PJ. Occupational contact dermatitis. In: Kanerva's occupational dermatology. Berlin, Heidelberg: Springer Berlin Heidelberg; 2012. p. 51–8. https://doi.org/10.1007/978-3-642-02035-3_8.
[37] Alves V, Muratov E, Capuzzi S, et al. Alarms about structural alerts. Green Chem 2016;18:4348–60. https://doi.org/10.1039/C6GC01492E.
[38] Tropsha A. Best practices for QSAR model development, validation, and exploitation. Mol Inform 2010;29:476–88. https://doi.org/10.1002/minf.201000061.
[39] OECD. Guidance document on the validation of (quantitative) structure-activity relationship [(Q)sar] models. Transport 2007;2:1–154.
[40] Kleinstreuer NC, Karmaus AL, Mansouri K, et al. Predictive models for acute oral systemic toxicity: a workshop to bridge the gap from research to regulation. Comput Toxicol 2018;8:21–4. https://doi.org/10.1016/j.comtox.2018.08.002.
[41] Polishchuk P. Interpretation of quantitative structure-activity relationship models: past, present, and future. J Chem Inf Model 2017. https://doi.org/10.1021/acs.jcim.7b00274.
[42] Riniker S, Landrum G a. Similarity maps—a visualization strategy for molecular fingerprints and machine-learning methods. J Cheminf 2013; 5:43. https://doi.org/10.1186/1758-2946-5-43.
[43] Landrum G. RDKit: open-source cheminformatics software. 2021. http://rdkit.org/. [Accessed 24 March 2021].
[44] Panfili E, Esposito S, Di Cara G. Temporary black henna tattoos and sensitization to para-phenylenediamine (PPD): two paediatric case reports and a review of the literature. Int J Environ Res Publ Health 2017;14:421. https://doi.org/10.3390/ijerph14040421.
[45] Low YS, Alves VM, Fourches D, et al. Chemistry-wide association studies (CWAS): a novel framework for identifying and interpreting structure–activity relationships. J Chem Inf Model 2018;58:2203–13. https://doi.org/10.1021/acs.jcim.8b00450.
[46] Ciallella HL, Zhu H. Advancing computational toxicology in the big data era by artificial intelligence: data-driven and mechanism-driven modeling for chemical toxicity. Chem Res Toxicol 2019;32:536–47. https://doi.org/10.1021/acs.chemrestox.8b00393.
[47] Luechtefeld T, Rowlands C, Hartung T. Big-data and machine learning to revamp computational toxicology and its use in risk assessment. Toxicol Res 2018;7:732–44. https://doi.org/10.1039/C8TX00051D.
[48] Chu X, Ilyas IF, Krishnan S, et al. Data cleaning: overview and emerging challenges. In: Proceedings of the 2016 international conference on management of data. New York, NY, USA: ACM; 2016. p. 2201–6. https://doi.org/10.1145/2882903.2912574.
[49] Dearden JC, Cronin MTD, Kaiser KLE. How not to develop a quantitative structure–activity or structure–property relationship (QSAR/QSPR). SAR QSAR Environ Res 2009;20:241–66. https://doi.org/10.1080/10629360902949567.

[50] Pogodin PV, Lagunin AA, Rudik AV, et al. AntiBac-pred: a web application for predicting antibacterial activity of chemical compounds. J Chem Inf Model 2019;59:4513–8. https://doi.org/10.1021/acs.jcim.9b00436.
[51] Gao H, Struble TJ, Coley CW, et al. Using machine learning to predict suitable conditions for organic reactions. ACS Cent Sci 2018;4:1465–76. https://doi.org/10.1021/ACSCENTSCI.8B00357/ASSET/IMAGES/LARGE/OC-2018-00357G_0010.JPEG.
[52] Luechtefeld T, Marsh D, Rowlands C, et al. Machine learning of toxicological big data enables read-across structure activity relationships (RASAR) outperforming animal test reproducibility. Toxicol Sci 2018;165:198–212. https://doi.org/10.1093/toxsci/kfy152.
[53] Matsuzaka Y, Uesawa Y. Optimization of a deep-learning method based on the classification of images generated by parameterized deep snap a novel molecular-image-input technique for quantitative structure–activity relationship (QSAR) analysis. Front Bioeng Biotechnol 2019;7. https://doi.org/10.3389/fbioe.2019.00065.
[54] Alves VM, Borba J, Capuzzi SJ, et al. Oy Vey! A comment on "machine learning of toxicological big data enables read-across structure activity relationships outperforming animal test reproducibility". Toxicol Sci 2019;167:3–4. https://doi.org/10.1093/toxsci/kfy286.
[55] Young D, Martin D, Venkatapathy R, et al. Are the chemical structures in your QSAR correct? QSAR Comb Sci 2008;27:1337–45.
[56] Golbraikh A, Tropsha A. Beware of q2. J Mol Graph Model 2002;20:269–76. https://doi.org/10.1016/S1093-3263(01)00123-1.
[57] Todeschini R, Ballabio D, Grisoni F. Beware of unreliable Q^2! A comparative study of regression metrics for predictivity assessment of QSAR models. J Chem Inf Model 2016;56:1905–13. https://doi.org/10.1021/acs.jcim.6b00277.
[58] Alexander DLJ, Tropsha A, Winkler DA. Beware of R2: simple, unambiguous assessment of the prediction accuracy of QSAR and QSPR models. J Chem Inf Model 2015;55:1316–22. https://doi.org/10.1021/ACS.JCIM.5B00206.
[59] Muratov EN, Bajorath J, Sheridan RP, et al. QSAR without borders. Chem Soc Rev 2020;49:3525–64. https://pubs.rsc.org/en/content/articlehtml/2020/cs/d0cs00098a. [Accessed 9 July 2020].
[60] Borba JVVB, Alves V, Braga R, et al. STopTox: an in-silico alternative to animal testing for acute systemic and TOPical TOXicity. ChemRxiv; 2020. https://doi.org/10.26434/chemrxiv.13283930.v1. chemrxiv.13283930.vol. 1.
[61] Gaulton A, Hersey A, Nowotka M, et al. The ChEMBL database in 2017. Nucleic Acids Res 2017;45:D945–54. https://doi.org/10.1093/nar/gkw1074.
[62] Borba JVVB, Alves V, Braga R, et al. STopTox: an in-silico alternative to animal testing for acute systemic and topical toxicity. Environ Health Perspect 2022;130:27012.
[63] Williams AJ, Grulke CM, Edwards J, et al. The CompTox Chemistry Dashboard: a community data resource for environmental chemistry. J Cheminf 2017;9:61. https://doi.org/10.1186/s13321-017-0247-6.
[64] Olah M, Mracec M, Ostopovici L, et al. WOMBAT: world of molecular bioactivity. In: Oprea TI, editor. Chemoinformatics in drug discovery. New York: Wiley-VCH; 2005. p. 221–39. https://doi.org/10.1002/3527603743.ch9.
[65] Mansouri K, Kleinstreuer N, Abdelaziz AM, et al. Compara: collaborative modeling project for androgen receptor activity. Environ Health Perspect 2020;128. https://doi.org/10.1289/EHP5580.
[66] Mansouri K, Abdelaziz A, Rybacka A, et al. CERAPP: collaborative estrogen receptor activity prediction project. Environ Health Perspect 2016;124:1023–33. https://doi.org/10.1289/ehp.1510267.
[67] Zhu H, Tropsha A, Fourches D, et al. Combinatorial QSAR modeling of chemical toxicants tested against Tetrahymena pyriformis. J Chem Inf Model 2008;48:766–84. https://doi.org/10.1021/ci700443v.
[68] Mansouri K, Grulke CM, Richard AM, et al. An automated curation procedure for addressing chemical errors and inconsistencies in public datasets used in QSAR modelling. SAR QSAR Environ Res 2016;27:939–65. https://doi.org/10.1080/1062936X.2016.1253611.
[69] Gadaleta D, Lombardo A, Toma C, et al. A new semi-automated workflow for chemical data retrieval and quality checking for modeling applications. J Cheminf 2018;10:60. https://doi.org/10.1186/s13321-018-0315-6.
[70] Williams AJ. Chemspider: a platform for crowdsourced collaboration to curate data derived from public compound databases. In: Ekins S, Hupcey MAZ, Williams AJ, editors. Collaborative computational technologies for biomedical research. Hoboken, NJ, USA: John Wiley and Sons, Inc.; 2011. p. 363–86. https://doi.org/10.1002/9781118026038.ch22.
[71] Ankley GT, Bennett RS, Erickson RJ, et al. Adverse outcome pathways: a conceptual framework to support ecotoxicology research and risk assessment. Environ Toxicol Chem 2010;29:730–41. https://doi.org/10.1002/etc.34
[72] Pittman ME, Edwards SW, Ives C, et al. AOP-DB: a database resource for the exploration of adverse outcome pathways through integrated association networks. Toxicol Appl Pharmacol 2018;343:71–83. https://doi.org/10.1016/j.taap.2018.02.006.
[73] Capuzzi SJ, Thornton TE, Liu K, et al. Chemotext: a publicly available web server for mining drug–target–disease relationships in PubMed. J Chem Inf Model 2018;58:212–8. https://doi.org/10.1021/acs.jcim.7b00589.
[74] Chemotext | Home. http://chemotext.mml.unc.edu/(accessed 3 Oct 2020).
[75] Bizon C, Cox S, Balhoff J, et al. ROBOKOP KG and KGB: integrated knowledge graphs from federated sources. J Chem Inf Model 2019;59:4968–73. https://doi.org/10.1021/acs.jcim.9b00683.
[76] ROBOKOP. https://robokop.renci.org/(accessed 8 Jul 2022).
[77] Davis AP, Grondin CJ, Johnson RJ, et al. The comparative toxicogenomics database: update 2019. Nucleic Acids Res 2019;47:D948–54. https://doi.org/10.1093/nar/gky868.
[78] Nguyen D-T, Mathias S, Bologa C, et al. Pharos: collating protein information to shed light on the druggable genome. Nucleic Acids Res 2017;45:D995–1002. https://doi.org/10.1093/nar/gkw1072.
[79] Korn D, Thieme AJ, Alves VM, et al. Defining clinical outcome pathways. Drug Discov Today 2022;27:1671–8. https://doi.org/10.1016/J.DRUDIS.2022.02.008.
[80] Alves VM, Capuzzi SJ, Muratov EN, et al. QSAR models of human data can enrich or replace LLNA testing for human skin sensitization. Green Chem 2016;18:6501–15. https://doi.org/10.1039/C6GC01836J.
[81] Wilkinson MD, Dumontier M, Aalbersberg IJ, et al. The FAIR Guiding Principles for scientific data management and stewardship. Sci Data 2016;3:160018. https://doi.org/10.1038/sdata.2016.18.
[82] Bell S, Abedini J, Ceger P, et al. An integrated chemical environment with tools for chemical safety testing. Toxicol Vitro 2020;67:104916. https://doi.org/10.1016/j.tiv.2020.104916.
[83] Richard AM, Huang R, Waidyanatha S, et al. The Tox21 10K compound library: collaborative chemistry advancing toxicology. Chem Res Toxicol 2020:0c00264. https://doi.org/10.1021/acs.chemrestox.0c00264. **Just accep**:acs.chemrestox.

[84] Richard AM, Judson RS, Houck KA, et al. ToxCast chemical landscape: paving the road to 21st century toxicology. Chem Res Toxicol 2016;29: 1225–51. https://doi.org/10.1021/acs.chemrestox.6b00135.

[85] Mendez D, Gaulton A, Bento AP, et al. ChEMBL: towards direct deposition of bioassay data. Nucleic Acids Res 2019. https://doi.org/10.1093/nar/gky1075. Published Online First.

[86] Wang Y, Suzek T, Zhang J, et al. PubChem BioAssay: 2014 update. Nucleic Acids Res 2014;42:D1075–82. https://doi.org/10.1093/nar/gkt978.

[87] Merz KM, Amaro R, Cournia Z, et al. Editorial: method and data sharing and reproducibility of scientific results. J Chem Inf Model 2020;60: 5868–9. https://doi.org/10.1021/acs.jcim.0c01389.

[88] Merz KM, Rarey M, Tropsha A, et al. Letter from the editors. J Chem Inf Model 2015;55:719–20. https://doi.org/10.1021/acs.jcim.5b00180.

C H A P T E R

3

Neural network-based descriptors as input for QSAR

Tadahaya Mizuno

Laboratory of Molecular Pharmacokinetics, Graduate School of Pharmaceutical Sciences, The University of Tokyo, Tokyo, Japan

Abbreviations

2D Two-dimensional
3D Three-dimensional
ECFP Extended-connectivity fingerprints
ED Encoder—decoder
GNN Graph neural network
GRU Gated recurrent unit
InChI International chemical identifier
LSTM Long short-term memory
QSAR Quantitative structure—activity relationship
seq2seq Sequence to sequence
SMILES Simplified molecular input line-entry system

Introduction

It is estimated that the total number of chemicals that could exist with a molecular weight less than 500 is about 10^{60}, with the total being about 10^{32} when restricted to druggable compounds [1,3—5]. In this huge chemical space, chemicals with structures that have yet to be imagined are expected to exist and would have unknown effects on living organisms. On the other hand, the number of existing chemicals registered in PubChem in October 2022 was only about 10^{10}. The difference is enormous, as can be clearly seen from the fact that the largest decimal unit prefix in the International System of Units is *yotta*, indicating $\times 10^{24}$. In Japanese, there are also *goku* ($\times 10^{48}$) and *kogasha* ($\times 10^{52}$). In other words, it is unrealistic to search the huge chemical space of 10^{60} by synthesizing and evaluating all the chemicals in an experimental way. What alternative methods exist? One of the solutions is a chemoinformatics approach such as the quantitative structure—activity relationship (QSAR), which utilizes computers and data science to address problems in various chemical fields [6].

When we humans (among those who have some knowledge of chemistry) solve various problems in chemistry, we recognize chemical structures in various ways. For example, all the representations in Fig. 3.1 involve D-glucose, and we can recognize them as such, including their steric properties. However, a computer cannot recognize chemical structures in these forms without appropriate conversion and will interpret simplified molecular-input line-entry system (SMILES) or in international chemical identifier (InChI) simply as strings, and two-(2D) and three-dimensional (3D) images as pixel information [7]. For computers, it is necessary to encode these low-level representations into numerical information that reflects the chemical structure in some way for a chemical structure to be recognized.

A multivariate entity that can be recognized by a computer and that represents a chemical structure is called a "descriptor" in this chapter. Various synonyms of "descriptor" exist, including "fingerprint" and "molecular representation," although they have slightly different nuances and a particular tendency in the context in which they are used. For example, "fingerprint" is often used when the representation is binary information within the field of chemoinformatics, whereas "molecular representation" is often employed in the machine learning context [8,9].

https://doi.org/10.1016/B978-0-443-15339-6.00007-2

SMILES

O[C@@H]1[C@H](O)[C@@H](O)[C@H](O)[C@@]([H])(CO)O1

InChI

1/C6H12O6/c7-1-2-3(8)4(9)5(10)6(11)12-2/h2-11H,1H2/t2-,3-,4+,5-,6+/m1/s1

2D-image

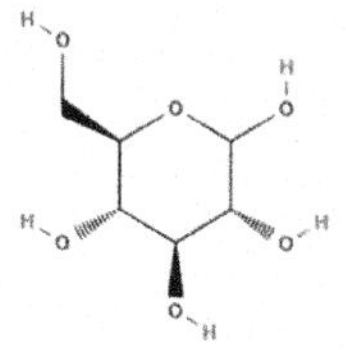

3D-image

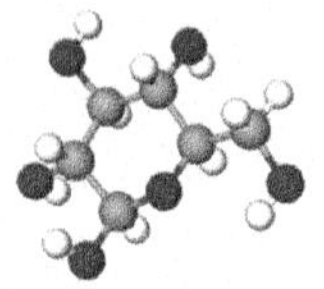

FIGURE 3.1 Various representations of D-glucose.

	Binary representation	Continuous representation
Structure representability?	Low	High
Feature interpretability?	High	Low
Decodable?	No	Yes*

* Dependent on models

FIGURE 3.2 Characteristic differences between descriptors of binary and continuous representations.

Although the use of different designations is not so significant, it is important to note that the binary information type of descriptor, often used for fingerprints, and the continuous-quantity type of descriptor, mainly derived from machine learning, are different in nature (see Fig. 3.2). In many cases involving the binary information type of descriptor, each bit represents a substructure of chemicals, enabling its meaning in structural terms to be quite clear. This is a strong advantage for the descriptor because we can easily interpret the results after data analyses such as toxicity prediction using machine learning, using the descriptor as input. On the other hand, the binary information type has a limited representation capability in terms of chemical structures (structure representability), and once encoded, it cannot be decoded (restored) back to the original structure [10]. Descriptors derived from machine learning are often continuous quantities with rich structure representability, and some of them have the advantage that they can be decoded back to the original structure (structure restorability). On the other hand, it is not possible to know directly what kind of structure each variable came from, which calls for various kinds of ingenuity [11–13]. In addition, particularly in the case of methods derived from deep learning, the descriptors themselves are highly dependent on the training used to build the method's model. It is therefore likely that much further research and evaluation will be necessary before these types of descriptors can be used in regulatory science, such as in the safety assessment of pharmaceuticals and other chemical substances.

In this chapter, we introduce methods for generating descriptors that are mainly derived from deep learning. After a brief review of recent methods, the "black box" approach to chemical structure recognition by deep learning models is discussed and its future prospects are discussed [14].

Deep learning–based methods for generating descriptors

First of all, it is important to note that we regard the numerical representation of fixed-length multivariate models used to encode a wide range of chemicals as "descriptors" in this chapter. In other words, features acquired in the learning process of end-to-end models oriented to a specific task (e.g., searching for chemicals that bind to protein X or making toxicity predictions) are not regarded as descriptors here because they are specific to a particular task and are not used in general-purpose applications. This recognition is important because feature generation and data modeling are performed simultaneously during learning in most of the deep learning models [15]. The deep learning paradigm that focuses on feature generation is called "representation learning" [16]; therefore, we focus on the representation learning of chemical structures in this chapter.

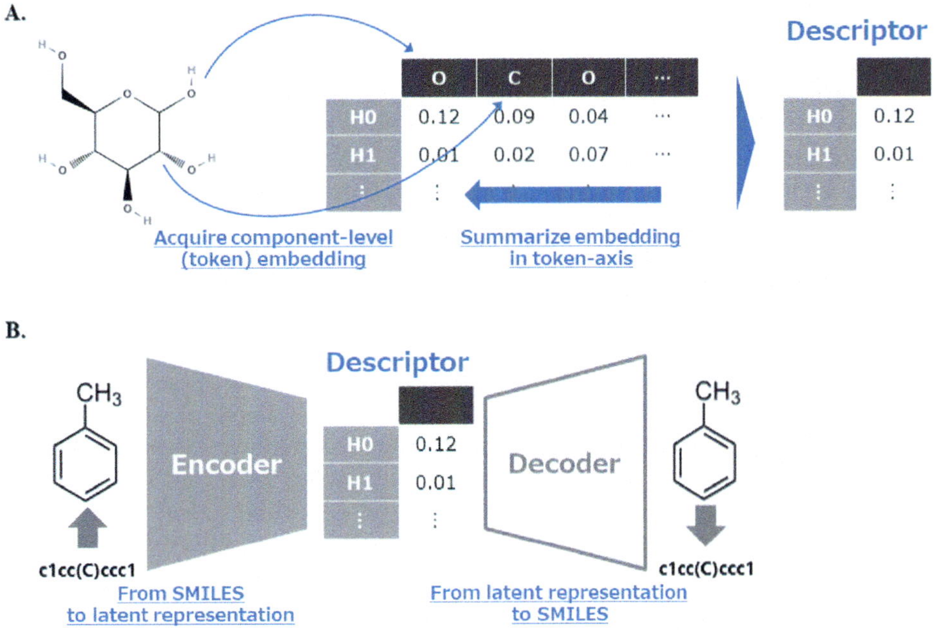

FIGURE 3.3 Two types of deep learning–based methods for generating descriptors: (A) The element expression and summarization type; (B) the latent representation of the whole structure type.

Deep learning–based methods for generating descriptors can be classified broadly into two types (Fig. 3.3). One type includes methods that obtain hidden representations of elements constituting a chemical such as its atoms and then transforms them into a fixed-length vector by operations such as *sum* and *max*. The second type includes methods that obtain a latent representation in an encoder–decoder (ED) structure to reproduce the input after the input has been reduced to a lower dimension, in which case the latent representations can be used as descriptors.

Element expression and summarization

The first application of neural network methods in generating descriptors was the neural fingerprint by Duvenaud et al. [17], which is a natural expansion of the well-known extended-connectivity fingerprint (ECFP) [17,18]. In this method, atoms are considered as nodes and bonds between atoms considered as edges, with the chemical structure being represented as a 2D graph. Because the hidden representation is the matrix whose size is the number of atoms multiplied by the number of features, which varies from chemical to chemical, we should summarize the matrix in terms of atoms to obtain a fixed-length multivariate model that captures the chemical structure by collapsing (e.g., summing up or taking the maximum value), which is kind of "message passing" [19] (see Fig. 3.3A).

An approach such as neural fingerprint, which represents the chemical structure as a 2D graph and handles the chemical structure by aggregating the information of surrounding atoms based on the graph structure, is a representative architecture for neural network models called graph neural networks (GNNs). As the name indicates, GNNs are neural network models that handle graphs as inputs. Many GNN models have been devised, and their details are summarized in excellent reviews of the field [20–22]. As in the neural fingerprint approach, all of these models can be used as fixed-length descriptor generation methods that obtain a representation of each component and then summarize the structure in terms of its components. From the viewpoint of descriptor generation, the difference between the methods is "*how to aggregate information on surrounding atoms.*" Methods that use 3D graphs as input also belong to this category.

It is important to note that these methodologies require learning some external information for the model to be able to learn chemical-like properties. On the one hand, most of these methodologies are focused on solving downstream tasks such as predicting the actions and properties of chemicals. They do not aim at descriptor generation. On the other hand, it is possible to learn a wide range of chemical structures by training multiple tasks on multiple heads. In this respect, these methods can be applied to descriptor generation.

Latent representation of the whole structure

In their 2016 preprint, which was followed by publication in *ACS Central Science* in 2018, [23] presented an innovative proposal for compound structure representation. The ED structure is one of the best-known neural network architectures derived from the AutoEncoder introduced by Hinton et al. [24] and has become firmly established as an unsupervised learning and generative model in recent years, following the achievements of Hinton and Salakhutdinov [24], Kingma and Welling [25]. The idea is to encode the input data and squeeze it into a low-dimensional space by learning to reproduce the original input. This is based on the idea that low-dimensional representations capture the essential information of the original data if they can be decoded back to the original data. Therefore, by feeding SMILES into the ED structure, it is possible to obtain the essential information in the string, that is, the chemical structure (see Fig. 3.3B). In the field of natural language processing, this latent representation is called a "sentence embedding" or "sentence vector" [26].

The aforementioned approach by Gómez-Bombarelli et al. [23] used a simple ED structure that matched input and output and learning processes to reduce errors at each character level. Therefore, although syntactic information such as letter sequences and repetitions can be learned, the notion of the meaning of the string as a whole cannot be learned (i.e., the entire chemical structure) in applications involving SMILES. Winter et al. [27] proposed the use of neural machine translation and developed continuous and data-driven descriptors. In the ED structure of strings used in translation, called sequence to sequence (seq2seq), learning different languages at the input and output enables the acquisition of latent expressions that have learned the grammars of both languages [28]. Although standardized forms of SMILES (called "canonical SMILES") are frequently used, it is possible to obtain multiple SMILES by changing the starting point (called "random SMILES") [29]. By building a model that translates random SMILES into canonical SMILES, the model can learn the nature of the chemical structure and obtain a latent expression that reflects it. Although many methods position the approach of translating random SMILES to canonical SMILES as an augmentation of the original Bjerrum (2017) work, which is used to improve the robustness of the model, the viewpoint of Winter et al. [27] regarding the approach as neural machine translation is important in interpreting the meaning attached to the latent expressions. In fact, it has been shown that it is possible to convert not only between random SMILES and canonical SMILES but also between SMILES and other string representations, such as InChI [27].

Regarding architectures of models [23], Winter et al., who introduced SMILES2Vec, employed the gated recurrent unit (GRU) architecture, which is a type of long short-term memory (LSTM) architecture [27,30,31]. In seq2seq with LSTM and GRU, only the last hidden state of an LSTM/GRU block, which is not a matrix but a vector, is often employed for the latent representation although all hidden states can be employed for the latent representation. Thus, the hidden state of LSTM and GRU is directly utilized as a descriptor. Unlike the applicability of LSTM and GRU systems, however, a transformer–decoder model uses all hidden representations. To obtain a fixed-length latent representation [32], it would have to summarize the latent representation in some way in terms of the elements, called "tokens" in the field of natural language processing, as was required for the methods introduced in the Element expression and summarization section. There are various methods for extracting sentence embeddings. One approach involves using of the representation of the token in terms of the token's first character, maximum value, or sum. Another involves the output from fully connected layers fed with the concatenated vector of all the aforementioned summarizations. However, there is currently no de facto standard for the extraction of sentence embeddings. Note that transformed descriptors are not easy to return to the original structure by themselves because the summarization operation used to extract sentence embeddings loses much information for decoding, whereas GRU-derived descriptors can be easily restorable as long as the latent representations comprise only the last hidden layer of the GRU. Combination with latent variable modeling could be one of the solutions to decode transformer-derived embedding to the original structure although such technique has not been employed for descriptor generation [33,34].

Other methods for generating descriptors

In recent times, the aforementioned type of descriptor generation using SMILES as input seems to have become the mainstream approach in deep learning–based methods for generating descriptors. On the other hand, most of the examples of using an ED structure are for the purposes of de novo drug design [35]. In these cases, the obtained expression is not a descriptor that can be applied to a variety of chemicals, but the idea itself can be used for descriptor generation in some cases, so care should be taken when surveying the literature. In fact, the work of Gómez-Bombarelli et al. [23], which was the origin of the ED model using the SMILES approach, is also oriented toward de novo drug design and was structured in that way. However, the possibility that it could be used as a descriptor was described previously.

Another important method in the field involves mol2vec, which is not a deep learning model, but a shallow neural network composed of two layers that provides a continuous representation as a descriptor [36]. This method was also inspired by natural language processing and applies that field's word2vec, a method for inference-based acquisition of word representations and distributed representations, to chemical structures [37,38]. The word2vec method solves the task of masking some words in a sentence and making predictions from other words in the same sentence. In mol2vec, chemical structures are treated as sentences, and substructures obtained by the Morgan algorithm used in ECFP are treated as words (RDKit). The resulting word-by-word representations are collapsed as described in the Element expression and summarization section to obtain final fixed-length representations for chemical structures [18]. Because of the limitations on the substructures that can be detected by the Morgan algorithm, the ability to represent compound structures may be lower than previous deep learning–based methods. However, it is relatively well accepted personally by end users who actually perform toxicity evaluations, rather than by data scientists from other fields. This might be because the application package (mol2vec GitHub) comprises a lightweight shallow layer model and is easy to handle. For deep learning–based methods to take root in the field of chemoinformatics, it may be necessary to devise a more user-friendly interface.

Black box approach in deep learning–based descriptors

It has been a long time since neural nets were introduced to chemoinformatics. More than 10 years have passed [39,40] and even with respect to deep learning, nearly 10 years have passed [41]. During this period, one of the big topics has been the "black box" problem of deep learning [42–44]. By development of sophisticated techniques such as integrated gradients and attention, the end-to-end tasks related to chemicals have been largely solved using a variety of techniques [13,32]. On the other hand, it remains unclear how deep learning models recognize the various chemical structures required for descriptor generation, rather than the structures associated with specific tasks.

We have been investigating how the descriptor generation–oriented ED model (a GRU model similar to Winter et al.) recognizes chemical structures in the learning process. In general, a performance evaluation of descriptor generation methods is mainly based on the performance of downstream tasks, implying that "*good representation leads to good downstream results*" [27,45]. We therefore prepared ED models trained with various numbers of steps and evaluated downstream tasks using their latent representations as descriptors. We found it interesting that the performance of the downstream tasks quickly reached a plateau after the early stages of the learning progress and deviated substantially from the performance of the ED models themselves (see Fig. 3.4, which shows the correct SMILES rate when recovering the input SMILES). When the similarity between the input and output was evaluated using the MACCS key fingerprint (as defined in RDKit), which gives binary information based on the presence or absence of predefined substructures, the relationship between substructure similarity and the learning process was similar to that between the performances of downstream tasks and the learning process (RDKit) (see Fig. 3.4). These results suggest that the ED model recognizes substructures of chemical structures that work for downstream tasks in the early stage of the learning process. On the other hand, the ED model would be expected to take longer to recognize the entire chemical structure because more learning would be required to recover the correct SMILES that corresponded to the entire chemical structure. In cases where the model failed to recover the input appropriately, the model tended to output a larger chemical structure than the input in the middle of the learning process and was able to output an appropriately sized structure as the learning proceeded. Please refer to recent our work for details of this issue [46]. Although research into how an ED model recognizes structures is still in its infancy, we expect that understanding of this phenomenon will be useful for the development of new methods for generating descriptors.

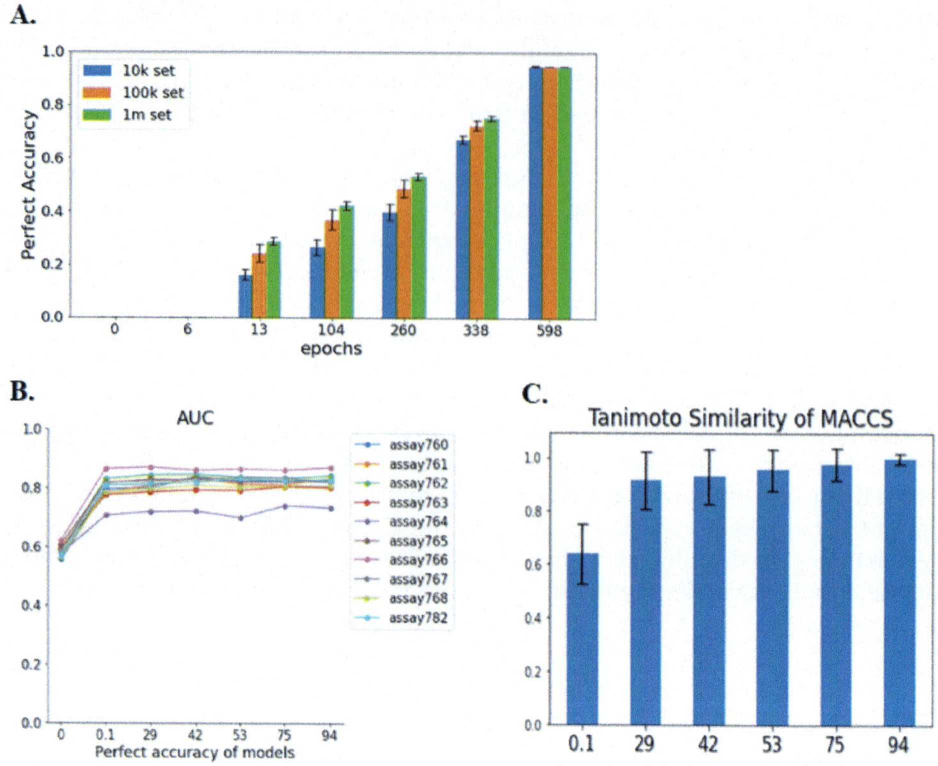

FIGURE 3.4 The relationship between the learning progress and descriptor properties of a descriptor generation method based on the encoder–decoder model: (A) the learning progress of the encoder–decoder model; (B) the relationship between the learning progress and the performances of downstream tasks; (C) the relationship between the learning progress and similarities in input–output substructure. Note that these figures are modified versions of those in our recent work [46].

Summary

Combination with different modalities: phenotype-based descriptors

As shown by the success of the connectivity map, the library of integrated network–based cellular signatures, the DrugMatrix, and TG-GATEs, phenotype-based methods are widely known as effective tools in applications other than describing chemical properties [47–53]. Although most of these methods are based on transcriptome data, there are a wide variety of approaches ranging from cell morphology and proteome analysis based on 2D electrophoresis to the integration of various high-throughput screening of assay results [54–60]. These approaches, which consider the cell as providing some kind of *function*, are a modality that provides a different layer of information than just the structure, and their combination has been shown to be useful in areas such as toxicity prediction [58]. There are various existing methods for integrating multimodal data, including statistical and network methods, but deep learning is also a strong player [61–63]. In particular, methods for multimodal embedding, such as in the task of annotating images, have recently become popular applications in a deep learning field such as vision–language pretraining [64,65]. Of particular importance is that they can offer complementary performance for missing layers, as in zero-shot learning [66]. That is, even though the training phase requires dense information from two different layers, there are often cases where the inference phase can be complemented from one of the layers. By combining these informatics developments with developments in the life science field such as omics analysis technology and high-throughput screening systems, it is expected that it will be possible to create descriptors that can be calculated from chemical structures with appropriate biological information attached to them.

Future perspectives

Machines cannot take as input the various representations that we humans usually work with in chemistry. Therefore, it is important to convert chemical structure information into descriptors that are easy for machines to recognize, and it is therefore an important task to extract as much information as possible from chemical structures for the effective performance of any subsequent chemoinformatics. With the development of deep learning models that excel in simultaneous feature extraction and modeling, it is becoming possible to solve specific tasks with sufficient information being extracted, and their interpretability is also improving [13,42–44]. On the other hand, we should keep in mind that the training of a deep learning model for a specific task often requires a large amount of data. Despite chemical-related databases being well developed, the range of tasks that can be solved remains limited and the cost of data preparation remains high in many cases (Environmental Protection Agency [EPA]; European Molecular Biology Laboratory [EMBL]; PubChem). Although there are methods that can compensate for the lack of data in some cases, a personal estimate of more than 10,000 items of data may be required to obtain reliable results [67]. In this respect, it is useful to separate feature generation from modeling, enabling feature generation to be performed with ample data that does not require labeling for a specific task and modeling to be performed with a lesser amount of labeled data, by using sophisticated machine learning algorithms that demand relatively small amounts of data (perhaps some hundreds) such as the support vector machine, the random forest algorithm, and XGBoost [68–70]. The first task (feature generation) is exemplified exactly by descriptor generation in the field of chemoinformatics, and the second (modeling) is a chemoinformatics method using descriptors such as QSAR.

The recent development of sophisticated deep learning–based methods, as described before in the Deep learning–based methods for generating descriptors section, is quite advantageous for our field. In particular, interpretability is expected to be enhanced by using the transformer's attention mechanism, and the method using SMILES as input, which started with Gómez-Bombarelli et al. is now considered to be in a mature stage. Since SMILES itself is a one-dimensional graph representation, the next development is expected to be descriptor generation from lower-level inputs than SMILES, such as an approach that uses the 3D coordinates of atom arrangements, which has recently been adopted for end-to-end tasks [35,71]. Such improvements in descriptor representability and interpretability are indispensable for chemoinformatics to be able to penetrate into the world of 10^{60} possible chemicals [1,3,4]. I hope that the field will continue to develop more expressive, interpretable, and usable methods for generating descriptors in close collaboration with informatics, including deep learning, and I would like to make a personal effort to contribute to the field.

Acknowledgments

This research was supported by AMED under Grant Number 22mk0101250h0001 and the JSPS KAKENHI Grant-in-Aid for Scientific Research (C) (grant number 21K06663) from the Japan Society for the Promotion of Science.

Conflicts of Interest

The author declares that he has no conflicts of interest.

References

[2] Bohacek RS, Mcmartin C, Guida WC. The art and practice of structure-based drug design: a molecular modeling perspective. Med Res Rev 1996;16:3–50.
[3] Lowe D. Chemical space is big. Really big. Med chem comm 2015;6. 12–12.
[4] Polishchuk PG, Madzhidov TI, Varnek A. Estimation of the size of drug-like chemical space based on GDB-17 data. J Comput Aided Mol Des 2013;27:675–9.
[5] Dobson CM. Chemical space and biology. Nature 2004;432:824–8.
[6] Hong H, Chen M, Ng HW, Tong W. QSAR models at the US FDA/NCTR. 2016. p. 431–59.
[7] Hähnke VD, Kim S, Bolton EE. PubChem chemical structure standardization. J Cheminf 2018;10:36.
[8] Sawada R, Kotera M, Yamanishi Y. Benchmarking a wide range of chemical descriptors for drug-target interaction prediction using a chemogenomic approach. Mol Inform 2014;33:719–31.
[9] Medina-Franco JL, Chávez-Hernández AL, López-López E, Saldívar-González FI. Chemical multiverse: an expanded view of chemical space. Mol Inform 2022:2200116.
[10] Le T, Winter R, Noé F, Clevert D-A. Neuraldecipher–reverse-engineering extended-connectivity fingerprints (ECFPs) to their molecular structures. Chem Sci 2020;11:10378–89.
[11] Lundberg S, Lee S-I. A unified approach to interpreting model predictions. ArXiv 2017.
[12] Molnar C, Freiesleben T, König G, Casalicchio G, Wright MN, Bischl B. Relating The Partial Dependence Plot and Permutation Feature Importance To The Data Generating Process. 2021.
[13] Sundararajan M, Taly A, Yan Q. Axiomatic Attribution for Deep Networks. 2017.
[14] Zhang D, Song J, Dharmarajan S, Jung TH, Lee H, Ma Y, et al. The use of machine learning in regulatory drug safety evaluation. Stat Biopharm Res 2022:1–5.
[15] Bodapati JD, Veeranjaneyulu N. Feature extraction and classification UsingDeep convolutional neural networks. Journal of Cyber Security and Mobility 2019;8:261–76.

[16] Bengio Y, Courville A, Vincent P. Representation Learning: A Review and New Perspectives. 2012.
[17] Duvenaud D, Maclaurin D, Aguilera-Iparraguirre J, Gómez-Bombarelli R, Hirzel T, Aspuru-Guzik A, et al. Convolutional Networks on Graphs for Learning Molecular Fingerprints. 2015.
[18] Rogers D, Hahn M. Extended-connectivity fingerprints. J Chem Inf Model 2010;50:742–54.
[19] Gilmer J, Schoenholz SS, Riley PF, Vinyals O, Dahl GE. Neural Message Passing for Quantum Chemistry. 2017.
[20] Wu Z, Pan S, Chen F, Long G, Zhang C, Yu PS. A Comprehensive Survey on Graph Neural Networks. 2019. https://doi.org/10.1109/TNNLS.2020.2978386.
[21] Zhang Z, Cui P, Zhu W. Deep Learning on Graphs: A Survey. 2018.
[22] Zhou J, Cui G, Hu S, Zhang Z, Yang C, Liu Z, et al. Graph Neural Networks: A Review of Methods and Applications. 2018.
[23] Gómez-Bombarelli R, Wei JN, Duvenaud D, Hernández-Lobato JM, Sánchez-Lengeling B, Sheberla D, et al. Automatic chemical design using a data-driven continuous representation of molecules. ACS Cent Sci 2018;4:268–76.
[24] Hinton GE, Salakhutdinov RR. Reducing the dimensionality of data with neural networks. 1979 Science 2006;313:504–7.
[25] Kingma DP, Welling M. Auto-encoding Variational Bayes. 2013.
[26] Li R, Zhao X, Moens M-F. A brief overview of universal sentence representation methods: a linguistic view. ACM Comput Surv 2023;55:1–42.
[27] Winter R, Montanari F, Noé F, Clevert D-A. Learning continuous and data-driven molecular descriptors by translating equivalent chemical representations. Chem Sci 2019;10:1692–701.
[28] Yousuf H, Lahzi M, Salloum SA, Shaalan K. A systematic review on sequence-to-sequence learning with neural network and its models. Int J Electr Comput Eng 2021;11:2315.
[29] Bjerrum EJ. SMILES Enumeration as Data Augmentation for Neural Network Modeling of Molecules. 2017.
[30] Hochreiter S, Schmidhuber J. Long short-term memory. Neural Comput 1997;9:1735–80.
[31] Cho K, van Merrienboer B, Gulcehre C, Bahdanau D, Bougares F, Schwenk H, et al. Learning Phrase Representations Using RNN Encoder-Decoder for Statistical Machine Translation. 2014.
[32] Vaswani A, Shazeer N, Parmar N, Uszkoreit J, Jones L, Gomez AN, et al. Attention is all You Need. 2017.
[33] Liu D, Liu G. A transformer-based variational autoencoder for sentence generation. In: 2019 international joint conference on neural networks (IJCNN). IEEE; 2019. p. 1–7.
[34] Fang L, Zeng T, Liu C, Bo L, Dong W, Chen C. Transformer-Based Conditional Variational Autoencoder for Controllable Story Generation. 2021.
[35] Martinelli DD. Generative machine learning for de novo drug discovery: a systematic review. Comput Biol Med 2022;145:105403.
[36] Jaeger S, Fulle S, Turk S. Mol2vec: unsupervised machine learning approach with chemical intuition. J Chem Inf Model 2018;58:27–35.
[37] Mikolov T, Chen K, Corrado G, Dean J. Efficient Estimation of Word Representations in Vector Space. 2013.
[38] Mikolov T, Sutskever I, Chen K, Corrado G, Dean J. Distributed Representations of Words and Phrases and Their Compositionality. 2013.
[39] Merkwirth C, Lengauer T. Automatic generation of complementary descriptors with molecular graph networks. J Chem Inf Model 2005;45:1159–68.
[40] Micheli A. Neural network for graphs: a contextual constructive approach. IEEE Trans Neural Network 2009;20:498–511.
[41] Lusci A, Pollastri G, Baldi P. Deep architectures and deep learning in chemoinformatics: the prediction of aqueous solubility for drug-like molecules. J Chem Inf Model 2013;53:1563–75.
[42] Vilone G, Longo L. Explainable Artificial Intelligence: A Systematic Review. 2020.
[43] Preece A. Asking 'Why' in AI: explainability of intelligent systems – perspectives and challenges. Intell Syst Account Finance Manag 2018;25:63–72.
[44] Tjoa E, Guan C. A Survey on Explainable Artificial Intelligence (XAI): Towards Medical XAI. 2019. https://doi.org/10.1109/TNNLS.2020.3027314.
[45] Goh GB, Hodas NO, Siegel C, Vishnu A. SMILES2Vec: An Interpretable General-Purpose Deep Neural Network for Predicting Chemical Properties. 2017.
[46] Nemoto S, Mizuno T, Kusuhara H. Investigation of Chemical Structure Recognition By Encoder-Decoder Models in Learning Progress. 2022.
[47] Morita K, Mizuno T, Kusuhara H. Decomposition profile data analysis of multiple drug effects identifies endoplasmic reticulum stress-inducing ability as an unrecognized factor. Sci Rep 2020;10:13139.
[48] Igarashi Y, Nakatsu N, Yamashita T, Ono A, Ohno Y, Urushidani T, et al. Open TG-GATEs: a large-scale toxicogenomics database. Nucleic Acids Res 2015;43:D921–7.
[49] Ganter B, Snyder RD, Halbert DN, Lee MD. Toxicogenomics in drug discovery and development: mechanistic analysis of compound/class-dependent effects using the DrugMatrix ® database. Pharmacogenomics 2006;7:1025–44.
[50] Lamb J, Crawford ED, Peck D, et al. The connectivity map: using gene-expression signatures to connect small molecules, genes, and disease. 1979 Science 2006;313:1929–35.
[51] Subramanian A, Narayan R, Corsello SM, et al. A next generation connectivity map: L1000 platform and the first 1,000,000 profiles. Cell 2017;171:1437–1452.e17.
[52] Mizuno T, Kinoshita S, Ito T, Maedera S, Kusuhara H. Development of orthogonal linear separation analysis (OLSA) to decompose drug effects into basic components. Sci Rep 2019;9:1824.
[53] Nemoto S, Morita K, Mizuno T, Kusuhara H. Decomposition profile data analysis for deep understanding of multiple effects of natural products. J Nat Prod 2021;84:1283–93.
[54] Creech AL, Taylor JE, Maier VK, et al. Building the Connectivity Map of epigenetics: chromatin profiling by quantitative targeted mass spectrometry. Methods 2015;72:57–64.
[55] Abelin JG, Patel J, Lu X, et al. Reduced-representation phosphosignatures measured by quantitative targeted MS capture cellular states and enable large-scale comparison of drug-induced phenotypes. Mol Cell Proteomics 2016;15:1622–41.
[56] Kinoshita S, Mizuno T, Hori M, Kohno M, Kusuhara H. Development of a novel platform of proteome profiling based on an easy-to-handle and informative 2D-DIGE system. Biol Pharm Bull 2019;42:2069–75.
[57] Young DW, Bender A, Hoyt J, et al. Integrating high-content screening and ligand-target prediction to identify mechanism of action. Nat Chem Biol 2008;4:59–68.

[58] Vo AH, van Vleet TR, Gupta RR, Liguori MJ, Rao MS. An overview of machine learning and big data for drug toxicity evaluation. Chem Res Toxicol 2020;33:20–37.
[59] Muroi M, Kazami S, Noda K, Kondo H, Takayama H, Kawatani M, et al. Application of proteomic profiling based on 2D-DIGE for classification of compounds according to the mechanism of action. Chem Biol 2010;17:460–70.
[60] Bray M-A, Singh S, Han H, Davis CT, Borgeson B, Hartland C, et al. Cell Painting, a high-content image-based assay for morphological profiling using multiplexed fluorescent dyes. Nat Protoc 2016;11:1757–74.
[61] Wang B, Mezlini AM, Demir F, Fiume M, Tu Z, Brudno M, et al. Similarity network fusion for aggregating data types on a genomic scale. Nat Methods 2014;11:333–7.
[62] Argelaguet R, Velten B, Arnol D, Dietrich S, Zenz T, Marioni JC, et al. Multi-Omics Factor Analysis—a framework for unsupervised integration of multi-omics data sets. Mol Syst Biol 2018. https://doi.org/10.15252/msb.20178124.
[63] Gao J, Li P, Chen Z, Zhang J. A survey on deep learning for multimodal data fusion. Neural Comput 2020;32:829–64.
[64] Li LH, Yatskar M, Yin D, Hsieh C-J, Chang K-W. Visualbert: A Simple and Performant Baseline for Vision and Language. 2019.
[65] Lu J, Batra D, Parikh D, Lee S. Vilbert: Pretraining Task-Agnostic Visiolinguistic Representations for Vision-and-Language Tasks. 2019.
[66] Socher R, Ganjoo M, Sridhar H, Bastani O, Manning CD, Ng AY. Zero-Shot Learning Through Cross-Modal Transfer. 2013.
[67] Li X, Deng S, Wang S, Lv Z, Wu L. Review of small data learning methods. In: 2018 IEEE 42nd annual computer software and applications conference (COMPSAC). IEEE; 2018. p. 106–9.
[68] Cristianini N, Ricci E. Support vector machines. In: Encyclopedia of algorithms. Boston, MA: Springer US; 2008. p. 928–32.
[69] Breiman L. Random forests. Mach Learn 2001;45:5–32.
[70] Chen T, Guestrin C. XGBoost: A Scalable Tree Boosting System. 2016. https://doi.org/10.1145/2939672.2939785.
[71] Wang M, Hsieh C-Y, Wang J, et al. RELATION: a deep generative model for structure-based de novo drug design. J Med Chem 2022;65: 9478–92.

Further reading

[1] PubChem. In: https://pubchem.ncbi.nlm.nih.gov/.
[2] RDKit. In: https://www.rdkit.org/.
[3] Turk S mol2vec github repository. In: https://github.com/samoturk/mol2vec.
[4] Environmental Protection Agency (EPA) ToxCast. In: https://www.epa.gov/chemical-research/exploring-toxcast-data-downloadable-data.
[5] European Molecular Biology Laboratory (EMBL) ChEMBL. In: https://www.ebi.ac.uk/chembl/.

CHAPTER

4

Decision forest—a machine learning algorithm for QSAR modeling

Huixiao Hong[1], Jie Liu[1], Wenjing Guo[1], Fan Dong[1], Myeongsang Lee[1], Liang Xu[1], Zoe Li[1], Meng Song[1], Minjun Chen[1], Wen Zou[1], Weida Tong[1] and Tucker A. Patterson[2]

[1]Division of Bioinformatics and Biostatistics, National Center for Toxicological Research, United States Food and Drug Administration, Jefferson, AR, United States [2]Office of Research, National Center for Toxicological Research, United States Food and Drug Administration, Jefferson, AR, United States

Abbreviations

3Rs replacement, reduction, and refinement
ANN artificial neural networks
AR androgen receptor
CERAPP Collaborative Estrogen Receptor Activity Prediction Project
CID PubChem chemical identification
CoMFA comparative molecular field analysis
CoMPARA Collaborative Modeling Project for Androgen Receptor Activity
DILI drug-induced liver injury
DT decision tree
EPA Environmental Protection Agency
ER estrogen receptor
FDA Food and Drug Administration
kNN k-nearest neighbors
QSAR quantitative structure–activity relationship
RF random forest
SNP single nucleotide polymorphism
SVM support vector machine
TSCA Toxic Substances Control Act

Introduction

Large numbers of chemicals are used in drugs, consumer products, and in the environment. About 62,000 chemicals were reported by manufacturers and processors to the US Environmental Protection Agency (EPA) from 1978 to 1982, with the Toxic Substances Control Act (TSCA) of 1976 to create an inventory of chemicals [1]. The inventory was increased to about 84,000 chemicals in 2012 by adding 22,000 new chemicals [2]. Chemical toxicity evaluation is needed for the safety evaluation of drugs and consumer products and for hazard identification in risk assessment of environmental chemicals. Experimental evaluation of toxicity by animal testing for such large numbers of chemicals is very challenging because the time and cost needed are practically incalculable. In addition, the principles of the 3Rs (replacement, reduction, and refinement) on the use of animals in research discourage animal testing for so many chemicals [3]. Therefore, alternative approaches to animal testing are needed for the safety evaluation of drugs and consumer products and for risk assessment of environmental chemicals.

Various computational methods have been developed and used as potential alternative methods for toxicity assessment of chemicals. Quantitative structure–activity relationship (QSAR) modeling is an attractive computational approach for predicting toxicity of chemicals. The hypothesis of QSAR modeling is that chemicals with similar structures have similar biological activities. QSAR models can be divided into two types based on their modeling activity: regression and classification models. QSAR models can also be put into three categories based on how

QSAR in Safety Evaluation and Risk Assessment
https://doi.org/10.1016/B978-0-443-15339-6.00029-1

structure similarity between chemicals is measured: group- or fragment-based QSAR, 3D-QSAR, and chemical descriptor–based QSAR. Group- or fragment-based QSAR modeling explores the relationship between various molecular fragments of chemicals and their biological activity. One example is pharmacophore QSAR modeling [4,5]. 3D-QSAR modeling establishes relationships between biological activity of a set of chemicals and the force fields that are calculated from the superimposed 3D structures of the chemicals. The first 3D-QSAR method is comparative molecular field analysis (CoMFA) that has been widely applied [6,7]. Chemical descriptor-based QSAR modeling creates models for predicting biological activity of chemicals based on descriptors that quantify various geometric, electronic, and steric properties of the chemicals. Various descriptors have been developed for QSAR modeling, including DRAGON [8], Mold2 [9], Mordred [9], and PaDEL-descriptor [10].

Mathematically, a QSAR model can be represented using Eq. (4.1).

$$\mathbf{A} = f(\mathbf{S}) + \varepsilon \tag{4.1}$$

Where **A** represents the activity vector of chemicals; **S** denotes the matrix describing the structures such as the molecular fragments, force fields, and chemical descriptors; f is the function (relationship) that predicts activity of a chemical based on its structure descriptors; and ε depicts prediction error that is the difference between actual activity and predicted activity. QSAR modeling is to determine the function f using a training set by minimizing the prediction error ε. A variety of machine learning algorithms have been developed for QSAR modeling, including decision tree (DT) [11], k-nearest neighbors (kNN) [12], Naïve Bayes [13], support vector machine (SVM) [14,15], artificial neural networks (ANN) [16], random forest (RF) [17,18], logistic regression [19], and others.

Each machine learning algorithm has its mathematical principle or statistical hypothesis. Therefore, each algorithm fits the dataset in which the relationship among the data matches or is similar as its mathematical principle or statistical hypothesis. One algorithm that fits all datasets is not available in practice. Therefore, selection of machine learning algorithms for a specific learning data set is very challenging. One solution explored in the field is to adopt consensus learning such as random forest [17]. Decision forest is a consensus machine learning algorithm that was developed based on consensus modeling of decision tree models [20–22]. Decision forest has been widely applied to QSAR modeling [23–28]. This chapter will introduce the principle of this machine learning algorithm, discuss its features, and showcase its applications in QSAR modeling.

Decision forest algorithm

Decision forest is a machine learning algorithm that takes consensus results from multiple decision trees. Different versions of decision forest have been developed for two-class classification [20] and multiclass classification [21] problems.

Two-class classification

In two-class decision forest, multiple two-class decision trees are first constructed from a training dataset. Determining sample class is conducted by the consensus of the results from the member decision trees as illustrated in Fig. 4.1. Algorithmically, the whole training dataset is used to build the first decision tree. Next, the variables (e.g., molecular descriptors in QSAR modeling) used in the first decision tree are removed from the training dataset, and the resulting reduced training dataset is used to build the second decision tree. Training performance is assessed by combining the training results from all decision trees. This process is repeated to generate N decision trees so that an additional decision tree would not improve the training performance by combining the N+1 decision trees. These N decision trees form a decision forest.

Different from random forest which generates a large number of shallow trees using part of the samples and variables that are randomly selected from the training data set, the decision trees in a decision forest are constructed using the whole training dataset, all samples, and all available variables. The decision trees in decision forest are deep trees. Therefore, a decision forest model usually has a small number of trees. Another feature of decision forest algorithm is the diversity of member decision trees. The hypothesis behind this is that combining different trees increases prediction performance more than combining similar trees. Therefore, decision forest ensures a member decision tree is very different from other member decision trees by using different variables.

Decision forest adopts two strategies to combine decision trees to make a consensus decision. The first strategy is majority voting. In this strategy, the prediction class for a sample from a decision forest is made by counting its decision trees, which predict the sample as the two individual classes: the class predicted from more decision trees. A prediction probability to the winning class is further calculated as the ratio of number of decision trees predicting

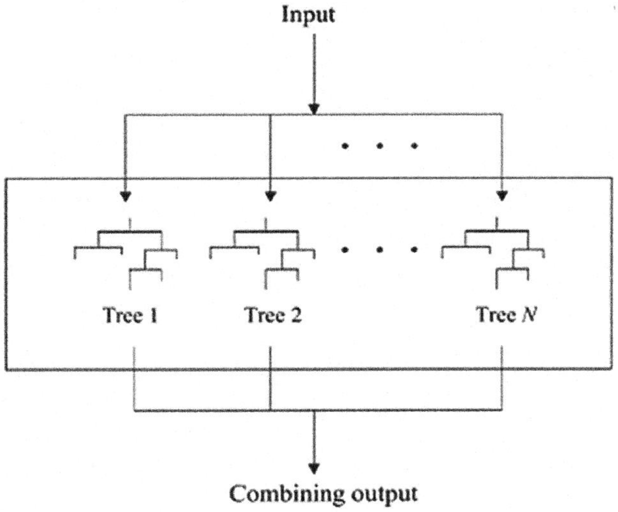

FIGURE 4.1 Overview of two-class decision forest algorithm. Two-class decision trees are constructed from a training dataset. Sample prediction takes the consensus of all predictions for the same sample from the N member decision trees.

the sample to the winning class to the total of decision trees in the decision forest model. The second strategy uses the average probability predicting a sample to the target class (such as positive) from the decision trees in a decision forest model. The sample is predicted as the target class (usually positive) if the average probability is greater than 0.5; otherwise, it is predicted as the other class (negative).

The probability to predict a sample to a class calculated by decision forest is a measurement of likelihood of the sample to be in the class. It can be used to indicate the confidence on a prediction from the decision forest model. Analysis on prediction confidence provides an additional aspect for evaluating a decision forest model. A good decision forest model should have high prediction confidence for predictions with high confidence. Furthermore, a good decision forest model should predict most samples with high confidence. Prediction confidence is a useful metric when utilizing models constructed with decision forest algorithm. Predictions with higher prediction confidence are more reliable than predictions with lower prediction confidence. Low prediction confidence warns that caution is needed to use the prediction.

In addition to prediction confidence, the distance of a chemical in prediction to the chemical space of training chemicals (applicability domain) is also important in a decision forest model. The farther away from the training chemicals, the chemical in prediction is less similar to the training chemicals, and, thus, the prediction from the decision forest model is less reliable. Various applicability domains have been proposed in QSAR studies. Our two-class version of decision forest algorithm provides two methods to calculate applicability domains. The descriptors used in a decision forest model are checked for the chemical in prediction. If all descriptors are within the ranges of training chemicals, it is within the training domain, and the applicability domain value is set as zero. If any descriptors are out of the ranges of training chemicals, the first method counts the number of descriptors with values out of the training chemicals and uses the ratio of this number to the total number of descriptors used in the decision forest model to represent the distance to the applicability domain. In the second method, the centroid of training chemicals in the space represented by scaled descriptors used in the decision forest model is first calculated. A Euclidean distance to the centroid is then calculated for the chemical with descriptor values out of training chemicals as the distance to the applicability domain.

The two-class decision forest algorithm works on variables that can be sorted so that the decision trees can be constructed by values to split samples into two nodes. It does not work for categorical variables such as genotypes of patients for a single nucleotide polymorphism (SNP) because their values are not sortable. We modified the two-class decision forest algorithm to work on such categorical variables [29]. Using genotype data of SNPs as an example to illustrate how to work on categorical variables, when splitting patients into two groups such as cancer patients and healthy individuals based on their genotypes of an SNP, patients with a single genotype of the SNP (homozygous common, heterozygous, homozygous variant) are put into one node and the patients with the remaining two genotypes are placed into the other node. The impurity (the percentage of majority samples, cancer patients, or healthy individuals) is calculated for the two nodes and their parent node. A deviance is calculated by comparing total impurity of the nodes and their parent node to measure how good is the splitting by this SNP.

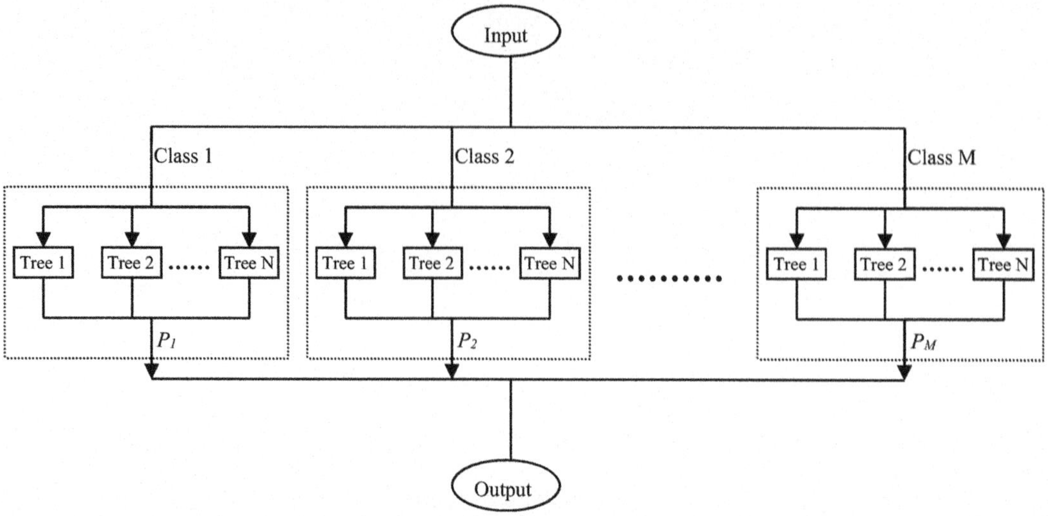

FIGURE 4.2 Overview of the multiclass decision forest algorithm. For each class in the training dataset with M classes of samples, a two-class decision forest model is developed for each of the M classes by aggregating all other classes of samples as the second class. When predicting a sample, the M two-class decision forest models are used to predict the sample separately, obtaining M probabilities ($P_1, P_2, \ldots, P_M$) from the M two-class decision forest models as the output to indicate the likelihood values of the sample can be predicted to the M classes. A sample is predicted to the class with the highest probability.

Multiclass classification

The two-class decision forest algorithm was extended to applications of multiclass classification, deriving the multiclass decision forest algorithm. The overview of the multiclass decision forest algorithm is illustrated in Fig. 4.2.

In brief, the multiclass decision forest algorithm consists of the following steps for a dataset containing M classes of samples: (1) for each class i ($i = 1, 2, \ldots, M$), the samples in the dataset are divided into two classes with each class: class i and class non-i that contains all the samples in the dataset except those in class i; (2) develop a qualified decision tree; where a qualified decision tree is the decision tree having a misclassification rate smaller than a predefined value; (3) remove variables such as molecular descriptors that have been used in previous trees and develop the next qualified decision tree based on the remaining variables; (4) repeat steps 2 and 3 until no qualified decision trees can be generated; (5) the number of decision trees included in the two-class decision forest model is determined to be the minimum number of decision trees that result in the fewest misclassified samples; (6) each terminal node of a qualified decision tree contains a specific percentage of class i samples, and this percentage is calculated as the probability for those samples of being class i; (7) the probability value (P_{ij}) for sample j to be predicted to class i is calculated using the average value of the probabilities from all decision trees in the two-class decision forest model; (8) repeat steps one to seven for each class i, and at the end of the process, probability values for sample j are generated from the M two-class decision forest models; (9) the final class to predict sample j is determined by a *winner-take-all* strategy, where the two-class decision forest model for class i that has the largest probability P_{ij} wins; thus, sample j is predicted as class i.

QSAR models developed using decision forest

QSAR modeling is to discover the relationship between chemical structures and their toxicological activity hiding in a training dataset. Machine learning algorithms are the necessary techniques that are used to discover the relationship between structures and their toxicological activity by learning from a training data set. Decision forest has been used in the development of various models, including QSAR models. Here, some examples are given to showcase the successful applications of decision forest in QSAR modeling.

QSAR models for predicting estrogen receptor activity

The estrogen receptor (ER) is a member of the steroid hormone receptor family in the nuclear receptor superfamily [30]. ER regulates endothelial functions, lipid profile, hemostasis, the growth of different tissues, inflammatory markers, bone integrity, and both prenatal and postnatal development [31]. Estrogenic chemicals can impact human health, both male and female, and the impact could be both adverse and beneficial for males and females. Testing the ability of binding ER is necessary for assessing the risk to humans of environmental chemicals in terms of endocrine disrupting potential. Because there are large numbers of chemicals in the environment, experimentally testing estrogenic activity for all environmental chemicals is very difficult, if not impossible. Thus, computationally predicting estrogenic activity of chemicals in the environment is needed for prioritizing chemicals for experimental testing. The US EPA led a large-scale project called the Collaborative Estrogen Receptor Activity Prediction Project (CERAPP) to develop computational models using various machine learning algorithms to predict estrogenic activity of large numbers of environmental chemicals [32]. The results from CERAPP are expected to help prioritize chemicals in the environment for experimentally testing ER-related activity. Seventeen modeling teams from US and European institutes participated in CERAPP to develop computational models for predicting ER-related activity using a variety of machine learning algorithms.

The ER binding, ER agonistic, and ER antagonistic activity data of 1677 chemicals that were generated with ToxCast high-throughput screening assays were used as the training dataset. The estrogenic activity data and structures of the training chemicals were provided to the 17 modeling teams to develop computational models using their preferred machine learning algorithms and molecular descriptors for predicting ER binding, ER agonistic, and ER antagonistic activity. An evaluation dataset of 7522 chemicals with ER binding, ER agonistic, and ER antagonistic activity data were curated from publicly accessible databases and the literature. The estrogenic activity data were withheld by US EPA and only structures of the 7522 chemicals were provided to the 17 modeling teams. ER binding, ER agonistic, and ER antagonistic activity prediction results for the 7522 chemicals from the 17 modeling teams were submitted to US EPA for evaluating performance of their developed models. US EPA also provided 32,464 environmental chemicals to the 17 modeling teams for predicting ER binding, ER agonistic, and ER antagonistic activity using their developed models. Two scores were calculated and used by US EPA for evaluating performance of the models from the 17 modeling teams based the prediction results of the 7522 chemicals. The first score was calculated using Eq. (4.2).

$$\text{score}-1=\frac{1}{3}\left(\frac{BA_{\text{ToxCast}}N\text{pred}_{\text{ToxCast}}}{N_{\text{ToxCast}}}+\frac{N\text{pred}}{N\text{total}}+\frac{1}{N\text{filter}}\sum_{i=1}^{N\text{filter}}\frac{BA_i N\text{pred}_\text{i}}{N\text{total}_\text{i}}\right) \tag{4.2}$$

where BA_{ToxCast} represents balanced accuracy from training model, $N\text{pred}_{\text{ToxCast}}$ is the number of chemicals used in training model, N_{ToxCast} indicates the total number of chemicals in the training dataset (1677), N_{pred} is the number of chemicals predicted, N_{total} indicates the total number of chemicals needing to be predicted, N_{filter} depicts the number of filters applied in selection of chemicals for the evaluation of model performance, i indicates ith filter, BA_i is balanced accuracy with ith filter, $N\text{pred}_\text{i}$ indicates the number of chemicals predicted with ith filter, and $N\text{total}_\text{i}$ is the number of chemicals with ith filter. This score considers goodness of training, chemical space that can be applied to, and accuracy for predicting various unknown chemicals. The second score is calculated using Eq. (4.3).

$$\text{score}-2=\frac{1}{2}\left(BA_{\text{ToxCast}}+BA_{\text{all-filters}}\right) \tag{4.3}$$

where $BA_{\text{all-filters}}$ represents the balanced accuracy of predictions on the various groups of chemicals such as chemicals that were curated from multiple literature sources, chemicals that displayed very week estrogenic activity, and chemicals that are not in the applicability domain of a model, if provided. This score focuses more on prediction accuracy than the chemical space a model can be applied.

The 17 modeling teams provided results of 21 models (some teams developed more than one model) for predicting ER binding activity. The machine learning methods and types of molecular descriptors used in the 21 models are given in Table 4.1. Some teams developed models using the same algorithms for predicting all three ER-related activity, while some teams did not develop models for predicting ER agonistic and ER antagonistic activity. The machine learning models from the 17 teams were compared using the two scores calculated by Eqs. (4.1) and (4.2), as well as the percentage of the 32,464 environmental chemicals that were predicted by the models. The results of comparing the performance of the prediction models from the 17 modeling teams are shown in Fig. 4.3. The models abbreviated as FDA1 were developed by the team from the Division of Bioinformatics and Biostatistics at the National Center for Toxicological Research of the US Food and Drug Administration using the machine learning algorithm decision forest based on Mold2 descriptors. The prediction models developed using machine learning decision forest (FDA1 in Fig. 4.3) successfully predicted all 32,464 environmental chemicals. The FDA1 models and the team NIH1 models from the National Cancer Institute at the National Institutes of Health performed similarly in

TABLE 4.1 Methods used in the estrogenic activity prediction models.

Model[a]	Modeling method[b]	Descriptors software/type
DTU	PLS/fragments	Leadscope
EPA	GA + PLSDA	PADEL
FDA1	Decision forest	Mold2
FDA2	PLS	3D-SDAR
ILS	SVM + RF	Qikprop
IRC1	CART-VEGA	2D descriptors
IRC2	Ruleset	SMATS
JRC	PASS	MNA
LM1	kNN	Fingerprints
LM2	kNN	Fingerprints
NIH1	RBF-SCR	MNA, QNA
NIH2	PASS	MNA
NIH3	Docking	AutoDock score
OCHE	*Consensus*	11 descriptor types
RIFM	SVM	Fingerprints
Umea	ASNN	DRAGON
UNC	SVM + RF	DRAGON
UBA	Docking	GLIDE score
UMB1	kNN	DRAGON + fingerprints
UMB2	kNN	DRAGON + fingerprints
USTR	SVM	ISIDA

[a]*Institute submitting model: DTU (Technical University of Denmark), EPA (U.S. Environmental Protection Agency), FDA1 (Division of Bioinformatics and Biostatistics, National Center for Toxicological Research, U.S. Food and Drug Administration), FDA2(Division of Systems Biology, National Center for Toxicological Research, U.S. Food and Drug Administration), ILS (ILS Inc and U.S. EPA), IRC (Istituto di Ricerche Farmacologiche "Mario Negri," Italy), JRC (Joint Research Center of the European Commission in Ispra, Italy), LM (Lockheed Martin and U.S. EPA), NIH1 and NIH2 (National Cancer Institute, National Institutes of Health), NIH3 (National Center for Advancing Translational Sciences, National Institutes of Health), OCHE (Institute of Structural Biology Helmholtz Zentrum Muenchen, Germany), RIFMs (Research Institute for Fragrance Materials, Inc), Umea (Umeå University, Sweden), UNC (University of North Carolina), UBA (University of Bari, Italy), UMB (University of Milano-Bicocca, Italy), USTR (University of Strasbourg, France).*

[b]*Abbreviations for modeling methods: PLS (partial least-squares), PLS-DAs (partial least-squares discriminant analysis), SVMs (support vector machines), RF (random forest), kNN (k-nearest neighbors), ASNNs (associative artificial neural networks), PASS (algorithm derived from Naïve Bayes classifier), RBF-SCR (self-consistent regression with radial basis function interpolation).*

terms of both score-1 and score-2 for all the three ER-related activities (ER binding, ER agonistic, and ER antagonistic activity) while outmatching the models from other teams.

QSAR models for predicting androgen receptor activity

The androgen receptor (AR) is another member of the steroid hormone receptor family in the nuclear receptor superfamily [33]. AR is a ligand dependent transcriptional factor in the endocrine system and is expressed in different tissues of many species. It regulates multiple physiological functions such as bone density, cognition, muscle hypertrophy, and prostate growth and differentiation [34]. AR plays a critical role in male sexual development and is another important and well-investigated receptor for evaluation of potential endocrine activity of chemicals.

Following CERAPP, the US EPA led the Collaborative Modeling Project for Androgen Receptor Activity (CoMPARA) project [35]. Twenty-five modeling teams from multiple countries participated in CoMPARA. The data generated using 11 AR-related ToxCast assays were integrated to form the training datasets of AR binding activity (1662 chemicals), AR agonistic activity (1659 chemicals), and AR antagonistic activity (1525 chemicals). US EPA curated

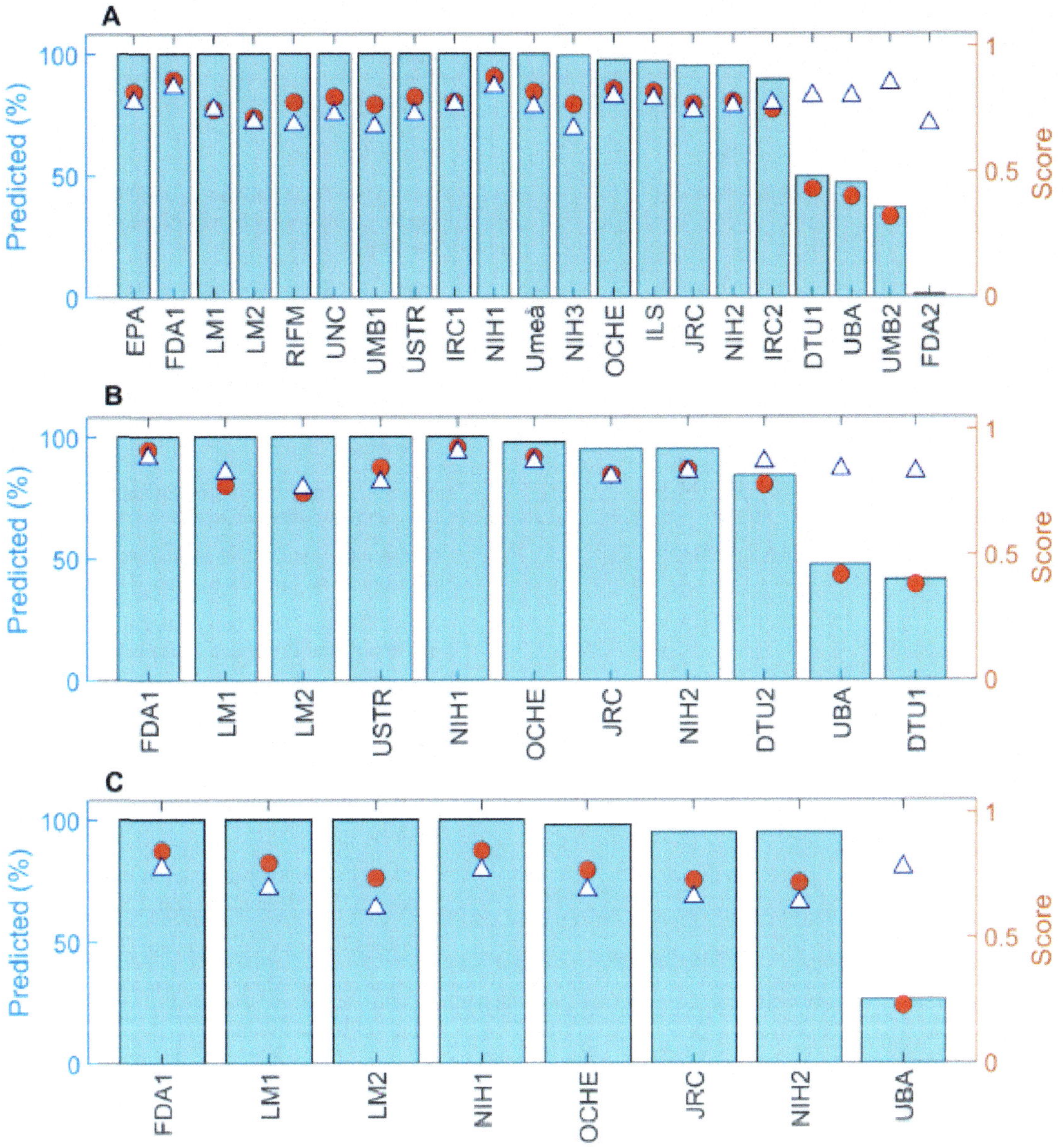

FIGURE 4.3 Performance of machine learning models for predicting estrogen receptor binding activity (A), agonistic activity (B), and antagonistic activity (C). The models are marked on the x-axis. The percentages of the environmental chemicals predicted by the models are plotted as the cyan bars (*the left y axis*). The overall performance score 1 and score 2 were plotted as solid red circles and empty blue triangles (*the right y-axis*), respectively. The models developed using decision forest algorithm are labeled as FDA1.

chemicals with experimental androgenic activity data from the PubChem database and literature using established procedures [36]. After excluding the chemicals which are included in the training dataset, 3882 chemicals with AR binding activity data, 4839 chemicals with AR agonistic activity data, and 4040 chemicals with AR antagonistic activity data were used as the external validation data sets. In CoMPARA, 55,450 chemicals were collected as the prediction data set, including the chemicals used as the prediction dataset in CERAPP and the metabolites generated from the ToxCast parent chemicals. US EPA provided the structures and AR binding, AR agonistic, and AR antagonistic activity data of the training chemicals to the 25 modeling teams to develop prediction models. The structures of chemicals in the external validation datasets and the prediction dataset were also provided to the modeling teams. In total, 91 predictive models, including 78 qualitative and 13 quantitative models, were developed by the modeling teams using various machine learning algorithms for predicting AR binding, AR agonistic, and AR antagonistic activity of chemicals in the external validation datasets and the prediction dataset. Table 4.2 lists the qualitative models for predicting AR binders. The prediction results on the external validation datasets were submitted to US EPA by

TABLE 4.2 AR binding activity prediction models.

Model	Modeling method[a]	Institute
ATSDR1	ANN+SVM+DT	Agency for Toxic Substances and Disease Registry, Centers for Disease Control and Prevention, USA
ATSDR2	ANN+SVM+DT	Agency for Toxic Substances and Disease Registry, Centers for Disease Control and Prevention, USA
ATSDR3	ANN+SVM+DT	Agency for Toxic Substances and Disease Registry, Centers for Disease Control and Prevention, USA
DTU	PLR	Technical University of Denmark, Denmark
ECUST	SVM	East China University of Science and Technology, China
EPA1	kNN+PLSDA	National Center for Computational Toxicology, Environmental Protection Agency, USA
EPA2	kNN+PLSDA	National Center for Computational Toxicology, Environmental Protection Agency, USA
EPA3	kNN+PLSDA	National Center for Computational Toxicology, Environmental Protection Agency, USA
EPA4	SVM+HC	National Risk Management Research Laboratory, Environmental Protection Agency, USA
EPA5	SVM+HC	National Risk Management Research Laboratory, Environmental Protection Agency, USA
FDA	Decision Forest	National Center for Toxicological Research, Food and Drug Administration, USA
IBMC1	SRC	Institute of Biomedical Chemistry, Russia
IBMC2	SRC	Institute of Biomedical Chemistry, Russia
IDEA	RF	Ideaconsult, Ltd., Bulgaria
INSLA	kNN+LDA+SVM+ RF+DT	University of Insubria, Italy
CMPLI	RF+kNN+NB+SVM MLR+PLS+ANN+RF	Computational Modelling of Protein-Ligand Interactions, France
LM		Lockheed Martin IS&GS, USA
IRFMN	Structuralalerts	Istituto Di Ricerche Farmacologiche "Mario Negri", Italy
NIH1	Docking+	National Center for Advancing Translational Sciences, NIH, USA
NIH2		National Institute of Environmental Health Sciences, Nihs, USA
NCS	RF	North Carolina State University, USA
SWET1	RF	Swedish Toxicology Sciences Research Center, Sweden

TABLE 4.2 AR binding activity prediction models.—cont'd

Model	Modeling method[a]	Institute
SWET2	RF	Swedish Toxicology Sciences Research Center, Sweden
TART1	SVM	University of Tartu, Estonia
TART2	SVM	University of Tartu, Estonia
TUM	MLR+	Technical University Munich. Germany
UFG	GBM	Federal University of Goiás, Brazil
UMEA	ASSN	Umeå University, Sweden
UNC	RF+RBF	University of North Carolina at Chapel Hill, USA
UBARI	Docking	University of Bari, Italy
UMIB1	kNN+NB+RF	University of Milano-Bicocca, Italy
UMIB2	kNN+NB+RF	University of Milano-Bicocca, Italy
USTRA	SVM	University of Strasbourg, France
VCCL	ASNN+	German Research Center for Environmental Health, Germany

[a]*ANN: artificial neural networks, ASNN: associative artificial neural networks, DT: decision tree, GBM: gradient boosting method, HC: hierarchical clustering, kNN: k-nearest neighbors, LDA: linear discriminant analysis, MLR: multilinear regression, NB: Naïve Bayes, PLS: partial least squares, PLSDA: partial least squares discriminative approach, RBF: radial basis function, RF: random forest, SCR: self-consistent regression, SVM support vector machine.*

the 25 modeling teams. A score was then calculated by the US EPA using Eqs. (4.3)–(4.6) to measure the performance of each model on each external validation dataset.

$$\text{score} = 0.3 * \text{Goodness} + 0.45 * \text{Predictivity} + 0.25 * \text{Robustness} \tag{4.4}$$

$$\text{Goodness} = 0.7 * BA_{\text{Train}} + 0.3 * (1 - |\text{SENS}_{\text{Train}} - \text{SPEC}_{\text{Train}}|) \tag{4.5}$$

$$\text{Predictivity} = 0.7 * BA_{\text{test}} + 0.3 * (1 - |\text{SENS}_{\text{test}} - \text{SPEC}_{\text{Test}}|) \tag{4.6}$$

$$\text{Robustness} = 1 - |BA_{\text{Train}} - BA_{\text{Test}}| \tag{4.7}$$

Where Goodness represents the training performance of a model and is calculated using Eq. (4.5), Predictivity is the prediction performance on an external validation dataset and is calculated using Eq. (4.6), and robustness is the balance between training and evaluation performance and is calculated using Eq. (4.7). BA_{Train}, $\text{SENS}_{\text{Train}}$, and $\text{SPEC}_{\text{Train}}$ indicate balance accuracy, sensitivity, and specificity in training the model, while BA_{Test}, $\text{SENS}_{\text{Test}}$, and $\text{SPEC}_{\text{Test}}$ are balance accuracy, sensitivity, and specificity of the predictions on the external validation datasets.

The performances of the qualitative prediction models were compared with results shown in Fig. 4.4. The models abbreviated as FDA were developed by the team from the Division of Bioinformatics and Biostatistics at the National Center for Toxicological Research of the US Food and Drug Administration using the machine learning algorithm decision forest based on Mold2 descriptors. As shown in Fig. 4.4, the FDA models predicted all 55,450 chemicals in the prediction dataset. The FDA models also outperformed the other models in terms of *score* values, except the models UMB1 and UMEA for AR binding activity and UMEA for AR agonistic and AR antagonistic activity. However, those four models were not able to predict a large number of chemicals in the prediction dataset. The results from CoMPARA support the notion that the machine learning algorithm decision forest is good for QSAR modeling.

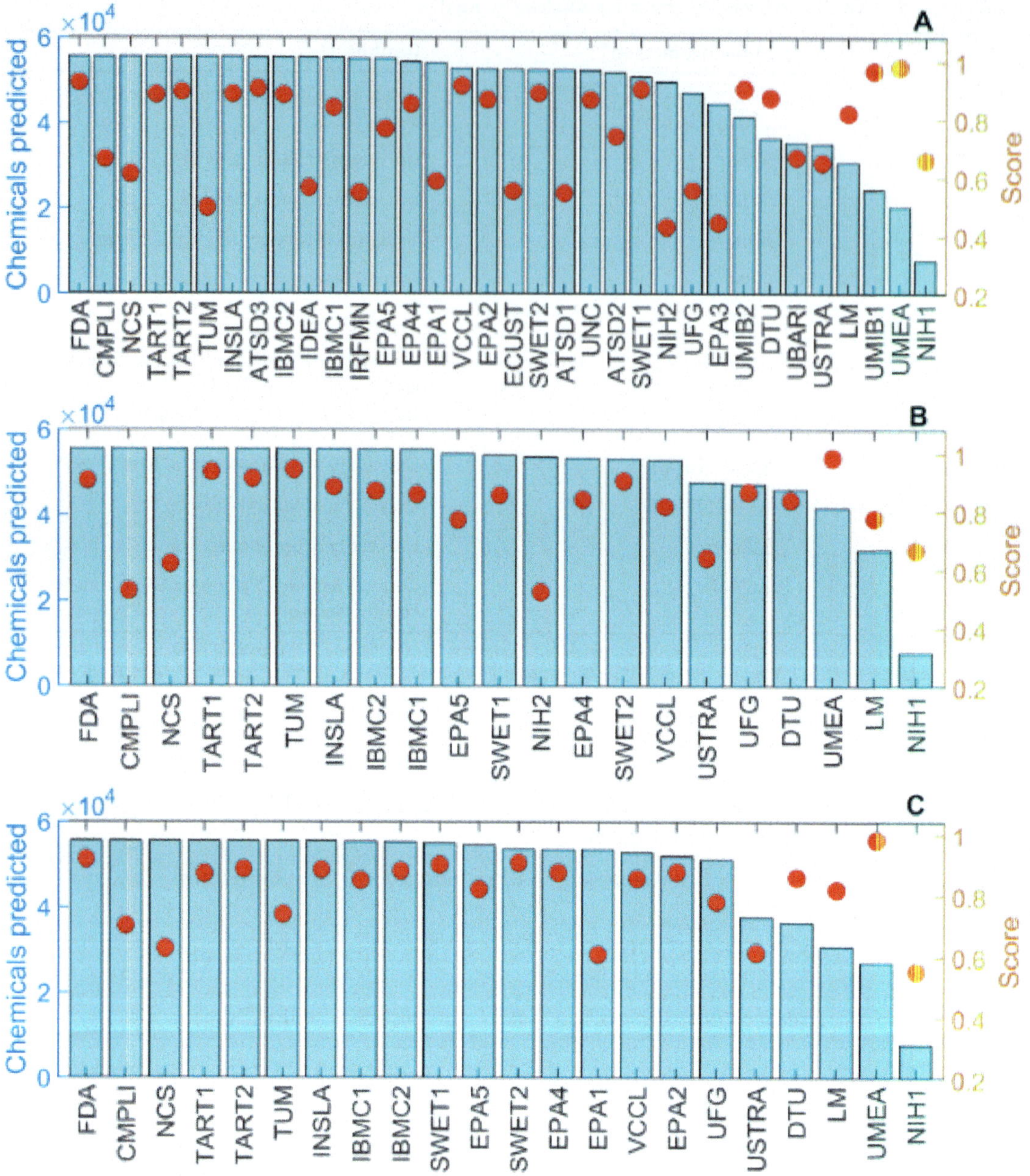

FIGURE 4.4 Performance of machine learning models for predicting androgen receptor binding activity (A), agonistic activity (B), and antagonistic activity (C). The models are marked at the x-axis. The numbers of environmental chemicals predicted by the models are plotted as bars (*the left y-axis*). The overall performance scores are plotted as solid red circles (*the right y-axis*). The models constructed using decision forest algorithm are labeled as FDA.

QSAR models for predicting drug-induced liver injury

Drug-induced liver injury (DILI) is the adverse reaction that a drug product causes to the liver, including damage to liver cells and the bile duct. It is one of the main causes for discontinuing the drug development process and has led to over 50 drugs being withdrawn from the market after approval [37,38]. Although many efforts have been made to remove candidate drugs with DILI potential in drug development, some drugs with DILI potential are not identified during preclinical toxicity testing but in later stages of the drug development pipeline and sometimes, after they are already on the market [39]. Therefore, predicting the DILI potential of drugs is needed but very challenging for the pharmaceutical industry and regulatory agencies [40]. Many approaches have been explored in prediction of human DILI potential, including machine learning–based QSAR modeling [25,27,41–46].

Determining if a drug has the potential to cause DILI in humans is complicated, and different methods may be used for this purpose. Therefore, reliable annotation of a drug's DILI potential in humans is critical for the development of DILI prediction models. A drug is annotated as a DILI-concern drug (positive) if it was withdrawn/discontinued from the market, or a Boxed Warning is included in its labeling document because of clinical cases of DILI, or a high/moderate DILI risk is indicated in the Warnings and Precautions sections in its labeling document. A drug is considered as no-DILI-concern drug (negatives) if no, DILI risk can be found in the Boxed Warning, Warnings and Precautions, and Adverse Reaction sections in its drug labeling document. Using this approach, we annotated 176 drugs as positives and 211 drugs as negatives. Based on their PubChem chemical identification (CID) numbers, the 387 drugs were divided into two groups. The group of 197 drugs (81 positives and 116 negatives) that have even CID numbers were used as the training dataset, while the group of 190 drugs (95 positive and 95 negative drugs) having odd CID numbers were used as the external validation dataset [25]. In addition, the 214 drugs with evidence for hepatotoxicity in humans (positives) and the 114 drugs with no evidence for hepatotoxicity in any species (negatives) reported by Greene et al. [41] were used as an additional external validation dataset (termed as Greene). Xu et al. annotated DILI risk of drugs according to human hepatotoxicity [46]. From this data source, 132 positives and 109 negatives were used as another external validation dataset (termed as Xu).

Mold2 descriptors [47] were calculated for the drugs in all datasets. The training dataset was used to construct a prediction model using the two-class decision forest algorithm. The developed decision forest model was then used to predict DILI risk in humans for the drugs of the external validation dataset, Greene dataset, and Xu dataset. The predictions were compared with the DILI annotations in the three external validation datasets to evaluate the performance of the developed model, which are shown as the red, cyan, and magenta bars in Fig. 4.5. In addition, 2000 iterations of 10-fold cross validations were conducted on the training dataset using the two-class decision forest algorithm. The performance metrics values of the models in the cross validations are summarized as the blue bars and the sticks above the bars in Fig. 4.5. As can be seen in Fig. 4.5, the performance of the decision forest models in the cross validations is similar to the external validation results, indicating the DILI prediction model is robust to predicting DILI risk of unseen drugs. The developed decision forest model could be useful in identification of putative DILI liability of drug candidates in the early stages of drug development.

Recently, we have further enhanced the drug-labeling based method and annotated 1036 drugs for their DILI risk in humans by combining their FDA drug labeling information and the DILI causality reported in the literature. This enhanced DILI annotation method classified the drugs into four classes: 192 most-DILI, 312 less-DILI, 278 no-DILI, and 254 ambiguous DILI [48]. Drugs that have DILI-related information in FDA labeling documents but do not have evidenced causality reported in the literature were annotated as ambiguous DILI drugs because the evidence from the two sources are not consistent and, thus, were not included in the development of multiclass decision forest

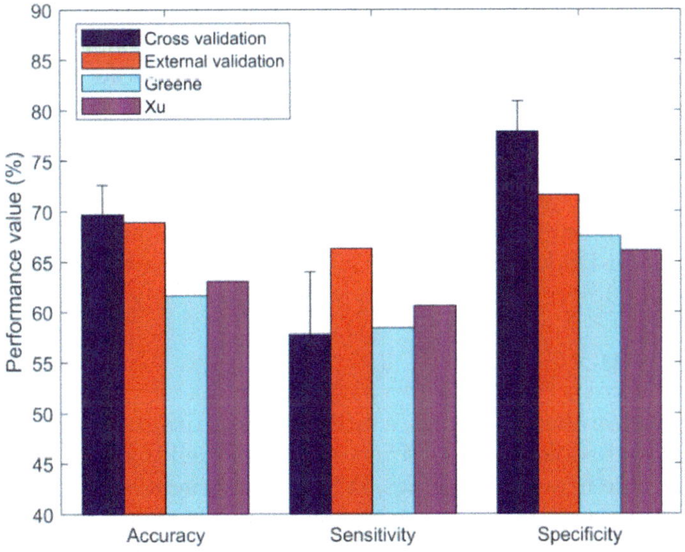

FIGURE 4.5 Performance of machine learning models for predicting DILI potential of drugs. Performance metrics (accuracy, sensitivity, and specificity) are grouped and marked at the x-axis. The blue bars give the average performance metrics values in the cross validations, and the stick above the blue bars represent the standard deviations in the cross validations. The red, cyan, and magenta bars depict prediction results on the external validation, Greene, and Xu datasets, respectively.

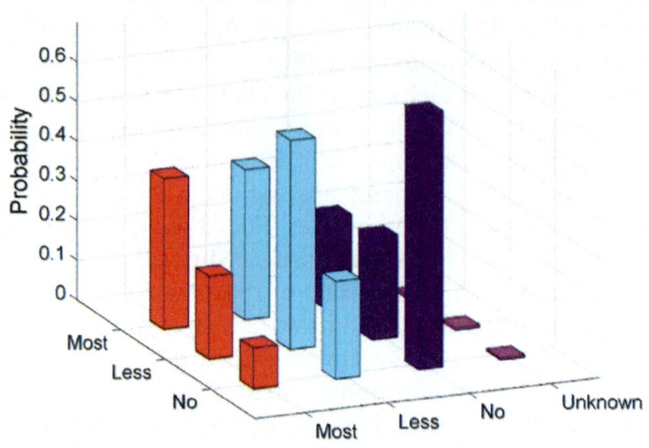

FIGURE 4.6 DILI risk prediction performance of the three-class decision forest models in the 5-fold cross validations. The predictions of most-DILI, less-DILI, no-DILI, and unknown (indicated at the y-axis) were given in red, cyan, blue, and magenta bars, respectively. The actual DILI classes (most-DILI, less-DILI, no-DILI) of the drugs were given at the x-axis. The z-axis represents the rate of predicting drugs to a type of drugs.

models. After removing drugs of inorganic compounds, mixtures of compounds, and compounds with ions, 183 most-DILI, 270 less-DILI, and 268 no-DILI risk drugs remained. multiclass decision forest was used to develop three-class prediction models based on the 721 drugs for classifying drugs into most-DILI, less-DILI, and no-DILI.

Mold2 descriptors [47] were first calculated for the 721 drugs. The DILI annotation (most-DILI, less-DILI, and no-DILI) and Mold2 descriptors formed the dataset, which was used in 1000 iterations of 5-fold cross-validations for measuring the performance of three-class DILI prediction decision forest models. The performances of three-class DILI prediction models from the 1000 iterations of 5-fold cross-validations are summarized in Fig. 4.6. The hypothetical random recall rate for each of the three classes of drugs is 33.3% if the prevalence of drugs in the three classes is balanced (37.2%, 25.4%, and 37.4% for no-, most-, and less-DILI, respectively, for this dataset), while the hypothetical random accuracy is 50% for a dataset with two classes of drugs. The average recall rates for most-DILI, less-DILI, no-DILI, and all drugs were 37.9%, 52.7%, 64.6%, and 53.4%, respectively, indicating the three-class DILI prediction models had moderate prediction performances. Examination of the detailed predictions for the three classes of drugs from the cross validations given in Fig. 4.6 found that 37.9% most-DILI drugs were correctly predicted as most-DILI drugs, which are slightly higher than the hypothetical random recall rate (25.4%). In addition, 37.8% most-DILI drugs were predicted as less-DILI drugs and could be considered as reasonable predictions. Only 23.4% most-DILI drugs were incorrectly predicted as no-DILI drugs, indicating a lower error rate for the three-class decision forest models than the two-class DILI prediction models that incorrectly predicted 42.2% most-DILI drugs as no-DILI drugs in the cross validations (Fig. 4.5). For the less-DILI drugs, 52.7% were correctly predicted as less-DILI drugs, and 20.8% were predicted as most-DILI drugs. Only 25.9% were predicted as no-DILI drugs which is also lower than the error rate for DILI drugs from the two-class DILI prediction models. For the no-DILI drugs, 64.6% were correctly predicted as no-DILI drugs and 24.6% were predicted as less-DILI drugs. Only 10.3% no-DILI drugs were predicted as most-DILI drugs that should be considered as the incorrect predictions. The results demonstrated the DILI risk prediction models developed using our multiclass decision forest algorithm not only had a higher resolution in prediction of DILI in humans (most-DILI, less-DILI, and no-DILI) than the two-class models reported previously, including our two-class decision forest model, but also had a better ability to differentiate most-DILI drugs from no-DILI drugs.

It is particularly noteworthy that 0.8% most-DILI, 0.6% less-DILI, and 0.5% no-DILI drugs were not predicted as any of the three classes in the training dataset but predicted as an unknown class of drugs. Different from models developed using other machine learning algorithms which predict all drugs belong to one of the classes in training, the multiclass decision forest methodology allows drugs to be predicted to none of the classes of drugs in training and to be determined as a possible new class of drugs. Predicting a drug to none of the drug classes in training implies that the drug is different from all drugs in training, which reflects the real biological phenomena. Therefore, prediction models developed using our multiclass decision forest could identify new classes of drugs for their potential DILI risk.

Conclusion remarks

Machine learning algorithms play an important role in QSAR modeling. A variety of machine learning algorithms have been developed based on various mathematical and statistical theories and have been applied in QSAR modeling. Consensus machine learning algorithms such as RF are attractive because they can overcome some algorithmically intrinsic limitations of individual single models that are used to make a consensus decision. Both RF and decision forest use decision trees as the member models for consensus decision-making. However, RF usually uses a large number of shallow decision trees while decision forest only needs a small number of deep decision trees. Therefore, decision forest is suitable for modeling on datasets with a limited number of independent variables such as most datasets used in QSAR modeling. Decision forest ensures consensus benefit by warranting diversity in its member decision trees. It also offers the ability for prediction confidence analysis and applicability domain assessment. A decision forest algorithm has been used in applications of developing prediction models in various fields including QSAR modeling. This chapter described its application in QSAR modeling in both two-class and multiclass classifications. The examples given in this chapter demonstrate the success of decision forest algorithm in QSAR modeling.

Acknowledgments

This research was supported in part by an appointment to the Research Participation Program at the National Center for Toxicological Research (Fan Dong, Myeongsang Lee, Zoe Li, Meng Song), administered by the Oak Ridge Institute for Science and Education through an interagency agreement between the US Department of Energy and the US Food and Drug Administration.

Disclaimer: This chapter reflects the views of the authors and does not necessarily reflect those of the U.S. Food and Drug Administration.

References

[1] Toxic Substances Control Act of 1976, Public Law 94-469, 94th Congress (https://www.congress.gov/94/statute/STATUTE-90/STATUTE-90-Pg2003.pdf, accessed on July 15, 2022).
[2] GAO (U.S. Government Accountability Office). Toxic substances: EPA has increased efforts to assess and control chemicals but could strengthen its approach. Washington, DC: GAO; 2013. https://www.gao.gov/products/gao-13-249. [Accessed 15 July 2022].
[3] Hubrecht RC, Carter E. The 3Rs and humane experimental technique: implementing change. Animals 2019;9(10):754. https://doi.org/10.3390/ani9100754.
[4] Prasanth Kumar S, Jasrai YT, Pandya HA, Rawal RM. Pharmacophore-similarity-based QSAR (PS-QSAR) for group-specific biological activity predictions. J Biomol Struct Dynam 2013;33(1):56–69. https://doi.org/10.1080/07391102.2013.849618.
[5] Hong H, Neamati N, Winslow HE, Christensen JL, Orr A, Pommier Y, et al. Identification of HIV-1 integrase inhibitors based on a four-point pharmacophore. Antiviral Chem Chemother 1998;9:461–72.
[6] Hong H, Fang H, Xie Q, Perkins R, Sheehan DM, Tong W. Comparative molecular field analysis (CoMFA) model using a large diverse set of natural, synthetic and environmental chemicals for binding to the androgen receptor. SAR QSAR Environ Res 2003;14(5–6):373–88. https://doi.org/10.1080/10629360310001623962.
[7] Cramer RD, Wendt B. Template CoMFA: the 3D-QSAR grail? J Chem Inf Model 2014;54(2):660–71. https://doi.org/10.1021/ci400696v.
[8] Helguera AM, Combes RD, González MP, Cordeiro MN. Applications of 2D descriptors in drug design: a DRAGON tale. Curr Top Med Chem 2008;8(18):1628–55. https://doi.org/10.2174/156802608786786598.
[9] Moriwaki H, Tian YS, Kawashita N, Takagi T. Mordred: a molecular descriptor calculator. J Cheminf 2018;10(1):4. https://doi.org/10.1186/s13321-018-0258-y.
[10] Yap CW. PaDEL-descriptor: an open source software to calculate molecular descriptors and fingerprints. J Comput Chem 2011;32(7):1466–74. https://doi.org/10.1002/jcc.21707.
[11] Enoch SJ, Schultz TW, Popova IG, Vasilev KG, Mekenyan OG. Development of a decision tree for mitochondrial dysfunction: uncoupling of oxidative phosphorylation. Chem Res Toxicol 2018;31(8):814–20. https://doi.org/10.1021/acs.chemrestox.8b00132.
[12] Hsieh JH, Wang XS, Teotico D, Golbraikh A, Tropsha A. Differentiation of AmpC beta-lactamase binders vs. decoys using classification kNN QSAR modeling and application of the QSAR classifier to virtual screening. J Comput Aided Mol Des 2008;22(9):593–609. https://doi.org/10.1007/s10822-008-9199-2.
[13] Chen B, Sheridan RP, Hornak V, Voigt JH. Comparison of random forest and pipeline pilot Naïve Bayes in prospective QSAR predictions. J Chem Inf Model 2012;52(3):792–803. https://doi.org/10.1021/ci200615h.
[14] Cortes C, Vapnik V. Support-vector networks. Mach Learn 1995;20(3):273–97. https://doi.org/10.1007/BF00994018.
[15] Hong H, Hong Q, Perkins R, Shi L, Fang H, Su Z, et al. The accurate prediction of protein family from amino acid sequence by measuring features of sequence fragments. J Comput Biol 2009;16(12):1671–88. https://doi.org/10.1089/cmb.2008.0115.
[16] Barzegar A, Zamani-Gharehchamani E, Kadkhodaie-Ilkhchi A. ANN QSAR workflow for predicting the inhibition of HIV-1 reverse transcriptase by pyridinone non-nucleoside derivatives. Future Med Chem 2017;9(11):1175–91. https://doi.org/10.4155/fmc-2017-0040.
[17] Breiman L. Random forests. Mach Learn 2001;45:5–32. https://doi.org/10.1023/A:1010933404324.
[18] Lee K, Lee M, Kim D. Utilizing random Forest QSAR models with optimized parameters for target identification and its application to target-fishing server. BMC Bioinf 2017;18(Suppl. 16):567. https://doi.org/10.1186/s12859-017-1960-x.

[19] Li Y, Pan D, Liu J, Kern PS, Gerberick GF, Hopfinger AJ, et al. Categorical QSAR Models for skin sensitization based upon local lymph node assay classification measures part 2: 4D-fingerprint three-state and two-2-state logistic regression models. Toxicol Sci 2007;99(2):532–44. https://doi.org/10.1093/toxsci/kfm185.
[20] Tong W, Hong H, Fang H, Xie Q, Perkins R. Decision Forest: combining the predictions of multiple independent decision tree models. J Chem Inf Comput Sci 2003;43(2):525–31. https://doi.org/10.1021/ci020058s.
[21] Hong H, Tong W, Perkins R, Fang H, Xie Q, Shi L. Multiclass Decision Forest–a novel pattern recognition method for multiclass classification in microarray data analysis. DNA Cell Biol 2004;23(10):685–94. https://doi.org/10.1089/dna.2004.23.685.
[22] Hong H, Tong W, Xie Q, Fang H, Perkins R. An in silico ensemble method for lead discovery: decision Forest. SAR QSAR Environ Res 2005;16: 339–47. https://doi.org/10.1080/10659360500203022.
[23] Sakkiah S, Leggett C, Pan B, Guo W, Valerio Jr LG, Hong H. Development of a nicotinic acetylcholine receptor nAChR α7 binding activity prediction model. J Chem Inf Model 2020;60(4):2396–404. https://doi.org/10.1021/acs.jcim.0c00139.
[24] Shen J, Xu L, Fang H, Richard AM, Bray JD, Judson RS, et al. EADB: an estrogenic activity database for assessing potential endocrine activity. Toxicol Sci 2013;135(2):277–91. https://doi.org/10.1093/toxsci/kft164.
[25] Chen M, Hong H, Fang H, Kelly R, Zhou G, Borlak J, et al. Quantitative structure-activity relationship models for predicting drug-induced liver injury based on FDA-approved drug labeling annotation and using a large collection of drugs. Toxicol Sci 2013;136(1):242–9. https://doi.org/10.1093/toxsci/kft189.
[26] Sakkiah S, Selvaraj C, Gong P, Zhang C, Tong W, Hong H. Development of estrogen receptor beta binding prediction model using large sets of chemicals. Oncotarget 2017;8(54):92989–3000. https://doi.org/10.18632/oncotarget.21723.
[27] Hong H, Thakkar S, Chen M, Tong W. Development of decision forest models for prediction of drug-induced liver injury in humans using A large set of FDA-approved drugs. Sci Rep 2017;7(1):17311. https://doi.org/10.1038/s41598-017-17701-7.
[28] Ng HW, Doughty SW, Luo H, Ye H, Ge W, Tong W, et al. Development and validation of decision forest model for estrogen receptor binding prediction of chemicals using large data sets. Chem Res Toxicol 2015;28(12):2343–51. https://doi.org/10.1021/acs.chemrestox.5b00358.
[29] Xie Q, Ratnasinghe LD, Hong H, Perkins R, Tang ZZ, Hu N, et al. Decision Forest analysis of 61 single nucleotide polymorphisms in a case-control study of esophageal cancer; a novel method. BMC Bioinf 2005;6(Suppl. 2):S4. https://doi.org/10.1186/1471-2105-6-S2-S4. Suppl 2.
[30] Kampa M, Pelekanou V, Notas G, Stathopoulos EN, Castanas E. The estrogen receptor: two or more molecules, multiple variants, diverse localizations, signaling and functions. Are we undergoing a paradigm-shift as regards their significance in breast cancer? Hormones (Basel) 2013;12(1):69–85. https://doi.org/10.1007/BF03401288.
[31] Ng HW, Perkins R, Tong W, Hong H. Versatility or promiscuity: the estrogen receptors, control of ligand selectivity and an update on subtype selective ligands. Int J Environ Res Publ Health 2014;11(9):8709–42. https://doi.org/10.3390/ijerph110908709.
[32] Mansouri K, Abdelaziz A, Rybacka A, Roncaglioni A, Tropsha A, Varnek A, et al. CERAPP: collaborative estrogen receptor activity prediction project. Environ Health Perspect 2016;124(7):1023–33. https://doi.org/10.1289/ehp.1510267.
[33] Sakkiah S, Ng HW, Tong W, Hong H. Structures of androgen receptor bound with ligands: advancing understanding of biological functions and drug discovery. Expert Opin Ther Targets 2016;20(10):1267–82. https://doi.org/10.1080/14728222.2016.1192131.
[34] Gelmann EP. Molecular biology of the androgen receptor. J Clin Oncol 2002;20:3001–15. https://doi.org/10.1200/JCO.2002.10.018.
[35] Mansouri K, Kleinstreuer N, Abdelaziz AM, Alberga D, Alves VM, Andersson PL, et al. CoMPARA: collaborative modeling project for androgen receptor activity. Environ Health Perspect 2020;128(2):27002. https://doi.org/10.1289/EHP5580.
[36] Williams AJ, EkinsS. A quality alert and call for improved curation of public chemistry databases. Drug Discov Today 2011;16(17–18):747–50. https://doi.org/10.1016/j.drudis.2011.07.007.
[37] Mosedale M, Watkins PB. Understanding idiosyncratic toxicity: lessons learned from drug-induced liver injury. J Med Chem 2020;63(12): 6436–61. https://doi.org/10.1021/acs.jmedchem.9b01297.
[38] Chen M, Vijay V, Shi Q, Liu Z, Fang H, Tong W. FDA-approved drug labeling for the study of drug-induced liver injury. Drug Discov Today 2011;16(15–16):697–703. https://doi.org/10.1016/j.drudis.2011.05.007.
[39] Sarges P, Steinberg JM, Lewis JH. Drug-induced liver injury: highlights from a review of the 2015 literature. Drug Saf 2016;39:801–21. https://doi.org/10.1007/s40264-016-0427-8.
[40] Mosedale Mand Watkins PB. Drug-induced liver injury: advances in mechanistic understanding that will inform risk management. Clin Pharmacol Ther 2016;101(4):469–80. https://doi.org/10.1002/cpt.564.
[41] Greene N, Fisk L, Naven RT, Note RR, Patel ML, Pelletier DJ. Developing structure-activity relationships for the prediction of hepatotoxicity. Chem Res Toxicol 2010;23(7):1215–22. https://doi.org/10.1021/tx1000865.
[42] Ekins S, Williams AJ, Xu JJ. A predictive ligand-based Bayesian model for human drug-induced liver injury. Drug Metab Dispos 2010;38(12): 2302–8. https://doi.org/10.1124/dmd.110.035113.
[43] Chen M, Bisgin H, Tong L, Hong H, Fang H, Borlak J, et al. Toward predictive models for drug-induced liver injury in humans: are we there yet? Biomarkers Med 2014;8(2):201–13. https://doi.org/10.2217/bmm.13.146.
[44] Liu J, Guo W, Sakkiah S, Ji Z, Yavas G, Zou W, et al. Machine learning models for predicting liver toxicity. Methods Mol Biol 2022;2425: 393–415. https://doi.org/10.1007/978-1-0716-1960-5_15.
[45] Hong H, Chen M, Ng HW, Tong W. QSAR models at the US FDA/NCTR. Methods Mol Biol 2016;1425:431–59. https://doi.org/10.1007/978-1-4939-3609-0_18.
[46] Xu JJ, Henstock PV, Dunn MC, Smith AR, Chabot JR, de Graaf D. Cellular imaging predictions of clinical drug-induced liver injury. Toxicol Sci 2008;105(1):97–105. https://doi.org/10.1093/toxsci/kfn109.
[47] Hong H, Xie Q, Ge W, Qian F, Fang H, Shi L, et al. Mold2, molecular descriptors from 2D structures for chemoinformatics and toxicoinformatics. J Chem Inf Model 2008;48(7):1337–44. https://doi.org/10.1021/ci800038f.
[48] Chen M, Suzuki A, Thakkar S, Yu K, Hu C, Tong W. DILIrank: the largest reference drug list ranked by the risk for developing drug-induced liver injury in humans. Drug Discov Today 2016;21(4):648–53. https://doi.org/10.1016/j.drudis.2016.02.015.

CHAPTER

5

Integrated modeling for compound efficacy and safety assessment

Angela Serra[1,2,3], Michele Fratello[1,2], Giusy del Giudice[1,2], Laura Aliisa Saarimäki[1,2], Alisa Pavel[1,2], Antonio Federico[1,2,3] and Dario Greco[1,2,4,5]

[1]Faculty of Medicine and Health Technology, Tampere University, Tampere, Finland [2]Finnish Hub for Development and Validation of Integrated Approaches (FHAIVE), Tampere, Finland [3]Tampere Institute for Advanced Study, Tampere, Finland [4]Institute of Biotechnology, University of Helsinki, Helsinki, Finland [5]Division of Pharmaceutical Biosciences, Faculty of Pharmacy, University of Helsinki, Helsinki, Finland

Abbreviations

AOPs Adverse outcome pathways
BAL Bronchoalveolar lavage
BMD Benchmark dose
BMDL Denchmark dose lower-bound
BMDU Benchmark dose upper-bound
CTD Comparative Toxicogenomics Database
EBI European Bioinformatics Institute
ECFP Extended-connectivity fingerprint
EMBL European Molecular Biology Laboratory
ENM/s Engineered nanomaterial/s
EPA Environmental Protection Agency
FDA Food and Drug Administration
GANs Generative adversarial networks
GEO Gene expression omnibus
GO Gene ontology
HTS High-throughput screening
InChI International chemical identifier
KGs Knowledge graph/s
MACCS Molecular ACCess System
MIAMI Minimum Information About a Microarray Experiment
MINSEQE Minimum Information About a Next-generation Sequencing Experiment
MOA Mechanism of action
NAMs New approach methodologies
NCBI National Center for Biotechnology Information
NIH National Institutes of Health
PDD Phenotypic drug discovery
POD Point of departure
QSAR Quantitative structure–activity relationship
RNASeq RNA sequencing
RRNs Recurrent neural networks
RT-PCR Real-time polymerase chain reaction
SMILES Simplified molecular-input line-entry system
SVM Support vector machine
TDD Target-based drug discovery
TGP Japanese Toxicogenomics Project consortium
TGx Toxicogenomic
ToxCast Toxicity Forecasting

Introduction

The research and development process behind new drugs, chemicals, and engineered nanomaterials (ENMs) is a long and resource-intensive endeavor. Despite rapid data generation, careful engineering, and advances in target identification and prediction, high failure rates regarding safety and efficacy form a bottleneck that prevents new compounds from entering the market.

For drugs, the failure rate is suggested to be up to 96%, with unexpected drug toxicity during clinical trials being one of the core reasons [1]. Ineffective trials are now considered an economic and social burden, which should be avoided with thorough assessment in early phases of efficacy testing and safety assessment [2]. Classical approaches

https://doi.org/10.1016/B978-0-443-15339-6.00011-4

to drug discovery rely on the selection of compounds with the best affinity to the intended target, best specificity, low toxicity, and desirable physicochemical characteristics (e.g., stability or the ease of administration) [3]. This results in screening large compound libraries, which is lengthy, expensive, and potentially ineffective. However, the drug design process is not simply a problem of optimizing the binding to a single target, but the prediction of effects derived from the interaction of the chemical with the exposed system [4].

Chemical safety assessment aims to uncover the risks of chemical exposures and to define limits for safe or effective doses for the intended use of the compound. The risks for human health and the environment are evaluated throughout the life cycle of the compound, from manufacturing to waste disposal using a variety of methods. Although the process of regulatory safety assessment still largely relies on animal experimentation, the role of in vitro and in silico approaches is becoming increasingly prominent. The expanding toolkit of computational models for chemical safety assessment provides the means for compound prioritization and hazard prediction through quantitative structure–activity relationship (QSAR) modeling and read across.

While these approaches can be accepted as part of the regulatory safety assessment procedure, they can be challenging to implement for compounds with distinct physicochemical properties and unknown biological interactions, such as ENMs [5]. Unlike small molecules, ENMs have not been consistently shown to have direct interactions with surface receptors, and the adverse effects related to ENM exposures are still largely unknown [5]. This, together with the growing interest toward ENMs in industry and medicine, sets ENMs to the center of the efforts to advance the methods of safety assessment.

However, current data-driven and computational methods aimed at safe and effective compounds are mainly chemocentric, merely relying on physicochemical properties of the molecules. While these methods provide crucial insight into relevant chemical properties, they might fail to clarify the biological mechanisms behind the effects induced by these compounds.

Toxicogenomics (TGx), on the other hand, makes use of omics technologies, such as transcriptomics and metabolomics, to study the effects of chemicals or pharmaceutical agents on human health [6–8]. Compared with classical toxicology, TGx provides a mechanistic understanding of the response that the exposure elicits in a biological system. TGx data can also be used to derive points of departure (POD) through benchmark dose modeling (BMD), namely the lowest dose at which no effect or low effects can be observed [9–11]. In this light, quantitative TGx approaches can identify early molecular events and biomarkers predictive of adverse outcomes for safety assessment, as well as help with the definition of safe dose ranges [12]. Finally, TGx can facilitate the use of in vitro data by identifying relationships between cellular models and in vivo experiments, guiding the development of new approach methodologies (NAMs) [5], which promotes the application of the 3R principles of animal experimentation.

TGx approaches alone are not able to link exposure mechanisms to a compound with specific chemical characteristics. Bridging this gap calls for integrated approaches that bring together cheminformatics and bioinformatics. Combining the strengths of both fields can lead to the development of methodologies that are able to shed light on relevant molecular structures for de novo compound design, i.e., compounds that are safe and effective by design.

The richness and variety of available data has opened new horizons for the development of data-driven methodologies for development of safe and effective compounds. For example, the availability of information regarding compounds biodistribution and metabolism has allowed the correct extrapolation of the posology in humans through pharmacokinetic models (ADME/PK) [13]. However, there are still major challenges to overcome before these methods can be fully exploited. These challenges include the development of an integrated knowledge space that collects relevant data, and the development of integrative models for their analysis. Data integration and multiview learning methodologies combine multiple datasets to increase the model accuracy and robustness and to find correlations between the different data layers. Various methods for data integration have been established, and the selection of appropriate approaches depends on the data and experiment at hand. The integration can be considered early, intermediate, or late, based on the stage of the analysis when integration is performed [14]. Early integration methods are efficient when the different data layers are natively comparable between each other. On the other hand, when the data layers are too different, late integration methods should be preferred. Late integration methods allow the identification of the optimal metrics and algorithms to be applied to each data layer, and only the final predictions from each model are merged.

In this chapter, we describe different compound representations and available databases useful for developing computational methodologies. Furthermore, we review existing and emerging computational strategies for compound efficacy and safety assessment (Fig. 5.1, Table 5.1). Each model makes use of distinct data types, ranging from physicochemical properties to molecular signatures, such as transcriptomics data. We show how these methods have been applied in recent research for developing effective compounds and to support their safety assessment. Finally, we describe how knowledge graphs, which are increasingly recognized in different fields, can be used for compound efficacy and safety assessment.

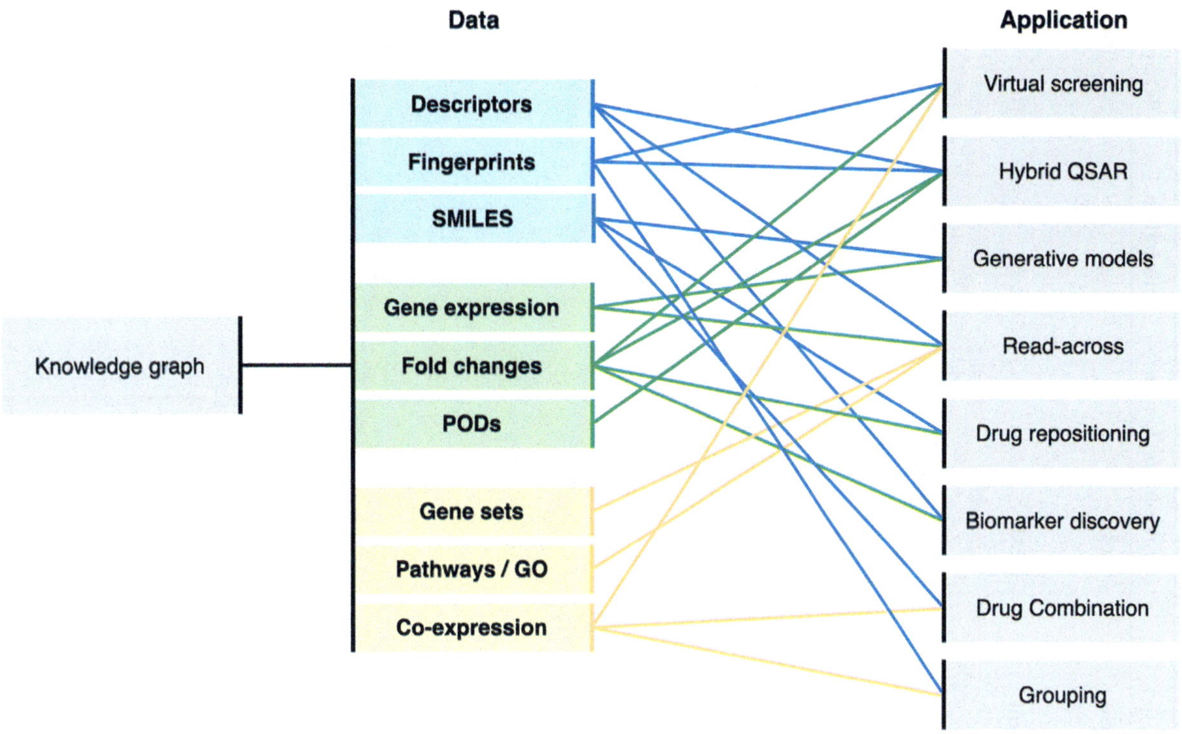

FIGURE 5.1 Primary data types and how the different approaches described in this chapter integrate them.

Compound representation

Multiple types of representation for compounds are available both in terms of their molecular structure and mechanism of action (MOA) (Fig. 5.2). One major challenge associated with the implementation of computational methods is the identification of the best molecular and mechanistic representation of compounds and their activities. Such representations need to be accurately described and manipulated by computers but are also easy to understand and interpret by chemists, biologists, and toxicologists. In the next paragraphs, we describe the different types of representations available and discuss the pros and cons of their usage.

Molecular representation

The chemical structures of compounds are usually expressed with data formats such as molecular graphs, SMILES (simplified molecular-input line-entry system), and InChI (international chemical identifier). A molecular graph is a 2D representation of the molecule as an undirected graph whose nodes represent atoms and the edges represent bonds between them. SMILES is a sequence-based notations system used to represent the chemical structures as ASCII strings. Similarly, the InChI is a textual representation for chemical substances that was designed to provide a standard encoding of the molecular information and to facilitate their search on available databases [27].

Although these representations are useful for storing, searching, and retrieving chemical compounds, they are not always suited for modeling directly. Therefore, a multitude of numerical formats are available for this goal. Molecular descriptors numerically summarize various chemical properties such as constitutional properties (molecular weight, logP, number of rotatable bonds) and topological descriptors, based on algebraic manipulations of the molecule interpreted as a graph. Several thousands of descriptors are available, each highlighting distinct characteristics of the molecules [28], as well as many software implementations for their calculations [29,30]. Alternatively, chemical compounds can also be represented as substructure fingerprints, binary vectors that represent the presence (1) or absence (0) of substructures within a molecule. Different sets of relevant substructures have been defined, such as the MACCS (Molecular ACCess System) keys [31] and PubChem fingerprints [32], as well as data-driven substructures such as ECFP (extended-connectivity fingerprint) [33]. The ease of transforming the molecular graphs into linear representation has made SMILES and InChI well-established representation for computational models.

TABLE 5.1 Integrative approaches for compound efficacy and safety assessment.

Approach	Data	Algorithm	Aim	Applicability domain	Integration	References
QSAR	Genes log fold-change and molecular descriptors	Linear regression with lasso regularization	Safety assessment	Drugs	Early	[15]
Nano-QSAR	Molecular descriptors to predict AOP point of departure (BMDL)	Linear regression	Safety assessment	ENMs	Early	[16]
Read across	Gene expression and physicochemical properties	Nearest neighbors' analysis	Safety assessment	Chemicals ENMs	Early	[17]
Biomarker discovery and toxicity prediction	Multiomics, physicochemical properties, protein corona, BAL counts	Genetic algorithm with RF classifier	Safety assessment	ENMs	Early	[18]
Virtual screening	Log fold-changes, drug targets, molecular descriptors	Read-across Gradient boosting regression Consensus modeling	Efficacy	Drugs	Late	[19]
Generative models	Gene expression, SMILES	Deep neural networks	Efficacy	Drugs	Intermediate	[20]
Drug combinations	Gene expression drug targets, SMILES	Network analysis	Efficacy	Drugs		[21]
Grouping	Scaffolds, physicochemical parameters, and mode-of-action	Network analysis	Efficacy and safety	Drugs		[22]
Drug repositioning	SMILES and gene expression profiles	SVM (support vector machine)	Efficacy	Drugs	Intermediate	[23]
Drug side effect/target prediction	Drug—side effects, drug—target and drug—indication and possible problem-specific datasets	KG based Node embedding + classifier	Efficacy and safety assessment	Drugs chemicals	Intermediate	[24]
Drug repositioning	Gene—gene, gene—disease, pathways, GO, drug targets	Connections in the KG	Efficacy	Drugs	Early	[25]
Clinical trials	Clinical trial and drug data	Integrated linked data in a KG	Efficacy and safety assessment	Clinical trials (drugs)	Intermediate	[26]

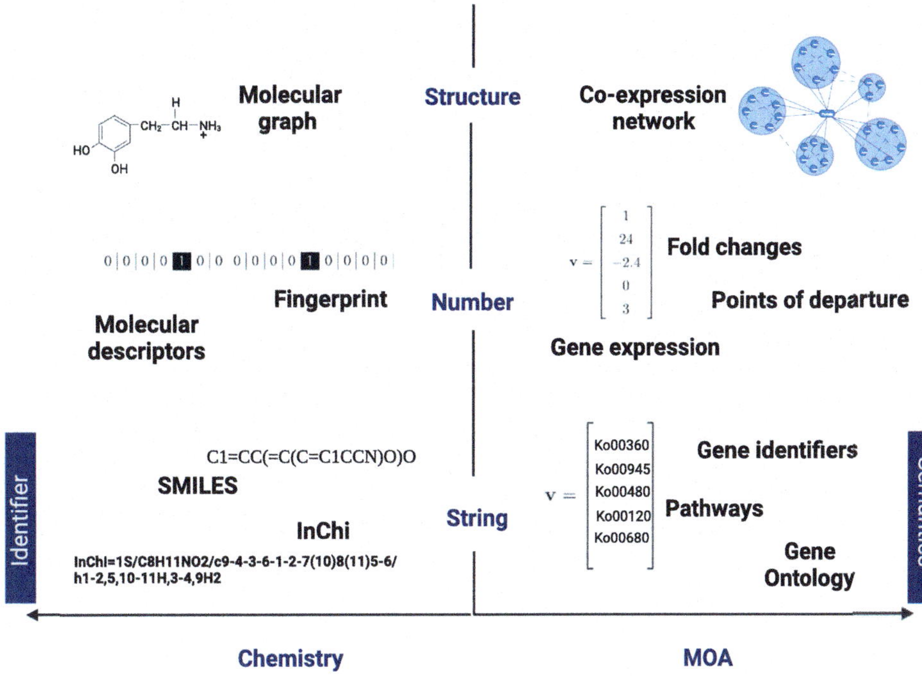

FIGURE 5.2 Different strategies to represent the chemical structure of compounds (left/chemistry) and their mechanism of action (right/MOA). In both cases, the compounds can be represented with strings or characters, with numerical values, or with network/graph structures.

However, the advent of deep learning methodologies, and of graph neural networks (GNN), has made raw molecular graphs a popular representation, especially in the context of deep generative models [34].

Recently, the InChI representation has also been suggested as a good standard for ENMs [35]. Besides that, ENMs are represented by various sets of predictors, known as nanomaterials descriptors, which cover the particle characteristics, the physicochemical properties of its components and impurities, the presence of a coating (if any), as well as extrinsic descriptors, which depend on external conditions [36]. Only on January 1, 2020, the Reach regulation has included a specific representation for nanoparticles, including the broader definition of "nanoform."[1]

MOA representation

The MOA of a compound can be defined as the set of molecular alterations that are induced by its exposure. The MOA can be studied by means of omics experiments, such as transcriptomics, metabolomics, and proteomics experiments. A standard way to analyze these data is to perform differential analysis. Shortly, this consists in comparing the treated samples with the untreated ones to identify the differences induced by the exposure [8]. This results in a fold-change and *P*-values for each molecule under investigation where the former expresses the amplitude of change, while the latter shows how statistically relevant their differences are.

In the case of gene expression analysis, the MOA of a compound can be represented by a vector whose length is the number of profiled genes, and each position of the vector contains its associated fold-change value. Another option is to represent the MOA as a list of characters representing the identifiers of the differentially expressed genes (classical thresholds used in this analysis are absolute fold-change >1.5 and *P*-value <.05). When multiple doses or concentrations are used, the MOA can be represented by the list of dose-dependent molecules and their associated points of departure (e.g., BMD, BMDL, BMDU values) [7]. Furthermore, the underlying mechanisms can be studied by mapping the list of relevant genes to functional pathways or gene ontology (GO) terms by means of enrichment analysis [8]. In this case, the MOA can be represented as a list of functional terms and their associated *P*-values.

1 Commission Regulation (EU) 2018/1881 of 3 December 2018 amending Regulation (EC) No 1907/2006 of the European Parliament and of the Council on the Registration, Evaluation, Authorisation and Restriction of Chemicals (REACH) as regards Annexes I, III,VI, V (European Commission, 2018).

Moreover, the MOA of an exposure can be represented and studied by gene coexpression networks generated from transcriptomics data [7]. From such networks, it is possible to identify biomarkers of exposure based on the relevance of their topological measurements in the network (e.g., most central genes). Network-based approaches offer the advantage of being able to represent the MOA in terms of interactions between molecules and allow the retrieval of relevant modules of associated genes [37].

Datasets for compound discovery

The development of computational models for drug efficacy and safety assessment is dependent on data availability. Table 5.2 summarizes most of the relevant datasets available for the research community.

Some of these datasets, including PubChem [38], DrugBank [42], and ChEMBL [41], contain information on both chemical structures and bioactivities of compounds. The PubChem database is maintained by the National Center for Biotechnology Information (NCBI) that is part of the US National Institutes of Health (NIH) [38]. It includes structural information for millions of compounds and integrates data from hundreds of different sources related to bioactivities and bioassays. The DrugBank database was developed by the University of Alberta and The Metabolomics Innovation Centre located in Alberta, Canada [42]. It contains structural, pharmacological, and pharmaceutical information about drugs and their targets. The ChEMBL database comes from the European Bioinformatics Institute (EBI) of the European Molecular Biology Laboratory (EMBL) based at the Wellcome Trust Genome Campus, United Kingdom [41]. It contains manually curated information related to bioactive molecules with drug-like properties, including their chemical structure, mechanism of action, and their targets.

Other datasets contain molecular alterations that can be specific of toxicogenomic studies on drugs and chemicals. For example, the Connectivity Map (CMap) dataset was developed by the Broad Institute to study the functional connections among diseases, genetic perturbations, and drug actions [47]. The dataset includes hundreds of gene expression profiles for multiple drugs tested on multiple cancer cell lines. The Open TG-GATEs is a toxicogenomic dataset compiled by the Japanese Toxicogenomics Project consortium (TGP) with the aim to improve the drug assessment process [48]. It includes gene expression profiles for 170 compounds tested in vivo (rat) and in vitro (primary rat and human hepatocytes). The LINCS 1000 was funded by the NIH with the aim of gaining a detailed understanding of cell pathways and aid efforts to develop innovative therapies. It includes millions of gene expression signatures profiled on a variety of cell lines exposed to thousands of small molecules as well as genetic perturbations [46]. Other sources for toxicogenomics data can be individual datasets deposited to general purpose repositories, such as the NCBI GEO (Gene Expression Omnibus), where the research community can upload their experimental data, if compliant with the MIAMI (Minimum Information About a Microarray Experiment) and MINSEQE (Minimum Information About a Next-generation Sequencing Experiment) standards [45]. GEO provides array- and sequence-based data as well as tools for the retrieval and download of specific experiments and curated expression profiles. Furthermore, the ToxCast (Toxicity Forecasting) project and Tox21 (Toxicology in the 21st Century) datasets are relevant resources to build predictive models for compound safety assessment [44]. The ToxCast project, developed by the United States Environmental Protection Agency (EPA), uses in vitro high-throughput screening (HTS) of nuclear receptors and stress response pathways to offer a cheaper and complementary method to traditional in vivo toxicity investigations. Tens of in vitro endpoints have been tested for about 10,000 different environmental toxins in the Tox21 10K collection.

The complementary information coming from multiple datasets could be integrated to develop computational models. However, it is important to note that this integrative process can be challenging since each dataset uses its own standard for unique identifiers. However, some work has been already done to address this, for example, the data fields of DrugBank link information to identifiers in other databases [52].

Furthermore, the gaps in the availability of data that represent both the chemistry and the MOA of the same compounds complicate the integration. When it comes to the availability of chemical representation of compounds, datasets such as PubChem, ZINC, or STICH include data for millions of compounds. There is also a reasonable amount of bioactivity data available in datasets such as ChEMBL or Tox21; however, when looking for expression profiles to represent the compounds MOA, the number decreases to thousands (e.g., L1000) or hundreds (e.g., Open TG-GATEs). This strongly influences the types of integrated models that can be developed. For example, deep learning models have been proven to be powerful predictive tools, which need large amounts of data for effective training. In addition, there is a heavy imbalance in the distribution of the activity data. Indeed, strong activities are not very common, and most of the datasets are skewed toward low or no activity. Moreover, a lot of missing data is present, and negative results are usually not recorded.

TABLE 5.2 List of available datasets.

Chemical structures/Assays/Effects		
PubChem	Contains chemical physical properties, but also biological activities and toxicological information [38].	https://pubchem.ncbi.nlm.nih.gov/
STITCH	Allows to perform chemical search based on their structure. Also contains chemicals protein connections [39].	http://stitch.embl.de/
ZINC	Provides 2D and 3D molecular structure of purchasable compounds [40].	https://zinc.docking.org/
ChEMBL	Brings together chemical, bioactivity, and genomic data for drug-like molecules [41].	https://www.ebi.ac.uk/chembl/.
DrugBank	Contains chemical information about drugs, their actions, and their targets [42].	https://go.drugbank.com/.
OpenTargets	Contains dataset and tools for drug target identification and prioritization [43].	https://www.opentargets.org/
ToxCast	Contain hundreds of high-throughput assays to describe potential hazard of chemicals.	https://www.epa.gov/chemical-research/exploring-toxcast-data-downloadable-data
Tox21	Allow access to chemical quality-control information. Furthermore, multiple datasets for training machine learning models are available [44].	https://tox21.gov/resources/
Molecular alterations		
NCBI GEOs (Gene Expression omnibus)	Functional genomic repository containing array and sequencing data complying with the MIAMI and MINSEQE standards [45].	https://www.ncbi.nlm.nih.gov/geo/
L1000	Library of drug signature profiled with high-throughput experiments on multiple cell lines and different concentrations. The dataset also includes knockdown profiles [46].	http://www.lincsproject.org
CMap (Connectivity Map)	Contains transcriptomic profiles of small molecules tested on multiple cell lines [47].	
Open TG-GATEs	Contains gene expression profiles of 170 compounds tested in in vivo (rats) and in vitro (primary rat hepatocytes, primary human hepatocytes) [48].	https://dbarchive.biosciencedbc.jp/en/open-tggates/desc.html (this is a mirror to the original database)
DrugMatrix	A comprehensive toxicogenomic database including drug exposure tested on rats in multiple tissues [49].	https://outage.niehs.nih.gov/drugmatrix/index.html

Continued

TABLE 5.2 List of available datasets.—cont'd

CTD (comparative Toxicogenomics Database)	An online dataset that links toxicological information for chemicals, with genes, diseases phenotypes, and different exposures [50].	http://ctdbase.org/
Toxicogenomic data collection for ENM exposure	A comprehensive collection of microarrays and RNASeq data related to ENMs exposures. It also includes physicochemical characterization of the ENMs [51].	https://zenodo.org/record/3949890
NANOSOLUTIONS	A collection of different types of data layers (including mRNA, miRNA, proteomics, protein corona, physicochemical properties, and BAL cell counts), for 31 ENMs [18].	https://zenodo.org/record/4247173

When it comes to the study of the ENMs mechanism of action by means of toxicogenomic data, a comprehensive repository is still lacking. However, a recent effort has been performed by Saarimäki et al. to collect, process, and manually curate a wide collection of gene expression studies related to ENMs exposures [51]. Unlike the small molecule case, which can be solely described by their chemical properties, ENMs require both structural descriptors and environmental descriptors [53]. Recently, Swirog et al. demonstrated that the external conditions affect both the MOA and the nanomaterial transformations, when predicting the toxicity of several metals in natural waters. However, a dataset collecting a comprehensive set of structural and environmental descriptors and their measurements is currently missing, limiting the development of models that are closer to real-world exposures scenario. Finally, one major challenge in building computational models for ENMs is the lack of datasets combining molecular data with endpoint assessment. Distinct types of data are often produced independently from each other with little crossover in the exposure setups. To this end, comprehensive datasets such as those produced in the context of the NANOSOLUTIONS project can serve as the starting point to build ENM-specific classifiers and computational models [18].

Virtual screening

Originally, the lead compound discovery phase was performed though random screening and empirical observations of the effects of the compounds. Technological advancement led to the improvement of random screening through the development of HTS methods. HTS can be used to test many thousands of compounds against a molecular target or cellular assay in a limited amount of time. However, such an approach is still expensive and requires a lot of experiments to be performed. Virtual screening is a fast and cost-effective computational technique that can be used to support HTS by prioritizing potential hits from large libraries of compounds. Classical virtual screening analysis can be performed by structure-based methods (docking) or ligand-based methods [54]. In both cases, the methods consider the structure of the compounds disregarding their biological effects. However, including the prediction of effects in virtual screening pipelines would significantly increase the amount of hits and reduce failure rates in the next steps of compound development.

Serra and Fratello et al. recently developed an integrative strategy bridging toxicogenomic and cheminformatics approaches to prioritize potentially effective drug treatments for the COVID-19 disease [19]. They integrated classical QSAR analysis with bioinformatics methods such as the differential expression analysis, dynamic dose-dependent modeling, gene coexpression analysis, and the connectivity map approach. The DrugBank database was virtually screened to prioritize potential candidate drugs for treating the COVID-19 disease. Further in vitro tests in human cells suggested 7-hydroxystaurosporine and bafetinib as potential COVID-19 treatments, and more interestingly, the combination of the two showed an increasing inhibitory effect on the SARS-CoV-2 infection.

Quantitative structure–activity relationship

QSAR models assume that similar compounds will induce similar biological responses. These relationships are identified as a function of their chemical structure. Despite the advancements obtained by QSAR models, they are not without limitations. First, given the extreme complexity of the chemical space, it is often the case that QSAR models are accompanied by an applicability domain, which depends on the specific training set used and defines the chemical subspace for which predictions are more confident. Second, QSAR modeling can be hindered by activity cliffs [55], i.e., pairs of structurally similar compounds with highly different response activities. Most likely, these discrepancies in the response activity are due to some unaccounted factors induced by the choice of specific numerical features used to represent the molecules. Third, chemocentric QSAR approaches often disregard what are the molecular alterations affecting the biological system after exposure, which are related to therapeutic efficacy and toxicity. Therefore, it is crucial to choose an adequate set of complementary descriptors. A promising approach to building more comprehensive numerical representations of molecules is to include bioactivity descriptors, derived from large-scale biological measurements such as transcriptomic profiles of chemical perturbations [46]. For example, Perualila-Tan et al. combined gene expression profiles and EFPC fingerprints to infer the association between drugs substructures and their effect on gene expression response as well as bioactivity data, represented as the inhibitory activity of the compounds in terms of their IC50 values [56]. Serra et al. developed a QSAR model for the prediction of the binding affinity of drugs to human serum albumin [15]. In this work, they integrated molecular descriptors representing the chemical structure of the drugs, with gene log fold-change patterns representing the MOA of the drugs using a sparse linear model. They showed that the integrative model achieved better performances compared with models fitted only on chemical or mechanistic descriptors, as well as providing deeper insights about the mechanistic behavior of compounds with specific chemical properties. Another example of integrative modeling has been applied to the prediction of gene deregulation induced by chemical perturbation [57]. In particular, the authors trained a deep neural network that takes as input the representations of the compound, described as an ECFP fingerprint, and the gene, represented as a collection of GO terms that describe the gene's function. Alternatively, point of departures estimated from dose-dependent transcriptional responses can be used to identify key structural properties that strongly correlate to adverse outcomes. Jagiello et al. [16] developed a Nano-QSAR modeling strategy to elucidate the contribution of physicochemical properties of carbon nanotubes in predicting the BMDL values of three pathways commonly associated with lung fibrosis. The knowledge obtained from this type of effort can be exploited to build more robust risk assessment tools as well as developing novel assay platforms.

Generative models

De novo molecular design aims at automatically generating novel chemical structures with desired biological properties [58]. Recent advancement in the artificial intelligence field resulted in an increased popularity of deep generative models. Such models can learn the relevant properties of the training compounds and automatically generate synthetic ones with similar characteristics.

These methodologies include recurrent neural networks (RNNs), variational autoencoders (VAEs), and generative adversarial networks (GANs) [59]. RNNs are neural networks with an internal memory that are suitable ML algorithms for sequential data. VAEs are probabilistic neural networks composed of two modules named encoder and decoder. The former encodes the molecules representation into a low-dimensional latent space as a continuous and smooth probability distribution. The latter converts these continuous vectors back to discrete molecular representations. GANs are composed of two competitive neural networks namely the generator and the discriminator. The generator is trained to produce new instances like those in the training set, while the discriminator is optimized to distinguish the synthetic instances from the real ones.

Classically, the generated molecules are evaluated based on quantitative criteria. For example, when generating therapeutic small molecules, drug likeliness and molecular characteristics are estimated [60,61]. However, generative models that only work by using the chemical structures of the compounds neglect the information related to the alterations happening into the biological systems after exposure/treatment to the compound. This lack of information can lead to the development of novel and easy to synthesize compounds that, however, might result in undesired effects. Thus, integrative models that consider both the chemical structures and the biological alterations are needed.

To this end, Méndez-Lucius et al. [20] proposed a new generative framework that bridges systems biology and drug design by using gene expression data to condition a GAN trained to generate new molecules from SMILES. With such an approach, they can automatically design active-like molecules to have a desired induced transcriptomic profile. They showed the effectiveness of their approach to generate inhibitor-like molecules integrating data for drugs present in the ChEMBL database and the gene expression signatures of their knocked-out target present in the L1000 database. This approach has the advantage that no prior knowledge of the molecular target of the compounds is required and can significantly speed up the screening phases of drug design since it can be used to generate chemical libraries with higher changes of hit in the HTS step.

Read-across

Read-across is used to infer knowledge on a substance, based on its similarity to other well-known substances. In the context of chemical safety, it can be used to infer adverse effects of a compound, using the information of a group of compounds with similar characteristics [62]. Compound read-across is classically based on structural characteristics; however, similar approaches can be applied to toxicogenomic data when assuming that a pair of exposure are related, if their mechanism of action is similar. Moreover, when multiple data layers are available (e.g., chemical structure, mechanistic information, biokinetic properties, etc.) integrative strategies can be applied to perform more robust and reliable analyses.

One successful example of integrative analysis for ENMs safety assessment is the work developed by Serra et al. [63] where they integrated data related to ENMs, drugs, human diseases, and chemicals to contextualize the ENM effects. In particular, the MOA of ENMs and drugs was represented as a ranked list of differentially expressed genes, ordered from the most upregulated to the most downregulated, and retrieved from the NanoMiner [64] and Connectivity Map (CMap) datasets [47], respectively. On the other hand, chemicals and human diseases were represented as sets of associated genes retrieved from the CTD database [50]. In this work, the authors identified a correlation between structural properties of the ENMs and their effects on human health. Indeed, their work highlighted significant connections between metal-based ENMs and neurodegenerative disorders. Further studies have confirmed these results in primary human dopaminergic neurons [65].

Furthermore, Varsou et al. [17] developed a web-based application, called ToxFlow, for the toxicity prediction of ENM by using a read-across methodology that combines the ENMs physicochemical properties with omics data, summarized at the level of functional pathways and gene ontology terms. ToxFlow is designed to give the user the possibility to filter omics data based on the scores obtained from a pathway enrichment analysis. Afterward, these findings are used to perform a correlation-based read-across analysis to predict compounds toxicity based on the toxicity scores of their analogs.

Biomarker discovery

The present development in molecular and cellular biology and its associated technical advancements has had a significant impact on chemical safety assessment strategies. Particularly, descriptive toxicology is evolving into mechanism-based predictive assessment through the quantitative examination of molecular and functional impacts of chemicals and ENMs that take place in biological systems. To this end, systems toxicology methods, which analyze different omics data, can aid in the discovery of novel mechanistic biomarkers. Integrating such approaches with the analysis of the chemical structure of compound can significantly improve their development toward "safe-and-sustainable-by-design" principles.

Recently, Fortino et al. [18] developed an integrative computational infrastructure for the safety assessment of ENMs. They tested their approach on a dataset including 31 industrially relevant ENMs. They grouped the ENMs in three hazard categories and used genetic algorithms and random forests classifiers to build predictive models from omics data, physicochemical properties, and protein corona data. They performed the analysis on each single layer individually and on their combinations. Higher predictive performances were obtained by ensemble models that combined different data layers. Finally, they performed RT-PCR validation of the identified biomarker on a panel of external ENMs.

Systems pharmacology

Efficient drug development that produces new effective medicines, accessible and affordable for all, is of enormous importance. Classical pharmacological approaches, also known as phenotypic drug discovery (PDD), aim at the definition of the dose at which a drug induces a therapeutic effect on the observed phenotype, with moderate or negligible undesirable side effects. More recently, a migration toward more comprehensive, semimechanistic pharmacological models has been observed. Such models take into consideration molecular details about drug–target binding, posing the roots of modern systems pharmacology. The evolution of these approaches gave rise to target-based drug discovery (TDD), where molecular perturbations induced by the drugs are investigated at a large-scale resolution. In addition, technological advances in the analysis and integration of multiple omics layers fostered an unprecedented gap filling in the knowledge of biological system responses to drug treatments. In this regard, systems pharmacology has found a niche in translational medicine, where drug discovery translates a novel drug from its research foundations, where in vitro efficacy testing and safety assessment are performed, onto the threshold of clinical trials, where human dose prediction and dosing regimen become paramount, that is, the "bench-to-bedside" bridge.

Systems pharmacology oftentimes makes use of pathway models, which are utilized to group and annotate set of genes into biological functions. The relevance of this annotation is that diseases are characterized by certain molecular perturbations, which exacerbate into pathological phenotypes. Such perturbation can be detected and interpreted thanks to biological pathways investigation. Therefore, it is crucial to define a disease map of molecular (and consequently pathway) perturbations to design tailored pharmacologic treatments. Given that human diseases arise from complex combination of events, this aspect poses the basis of personalized medicine, greatly advocated in cancer research. Such a complexity led to the shift from a *"one disease–one drug"* to a *"one disease–multiple drugs"* paradigm. Defining a combination therapeutic plan by employing multiple drugs allows to widen the activity spectrum toward the disease while minimizing the toxicity due to the lower doses used for the single drugs in the combination. As mentioned before, chemocentric approaches have dominated drug design. These approaches use the underlying assumption that an effective drug is capable of physically interacting, and possibly changing the function of specific targets, but they oftentimes neglect the MOA of the drugs. On the other hand, mechanistic-based approaches, also referred to as "connectivity mapping," rely on pattern matching and are based on the assumption that an effective drug can revert the molecular alterations underlying a specific disease [47]. So far, the success of drug discovery has been hampered by the lack of integration between these two approaches.

In the context of mechanistic drug discovery, network pharmacology approaches have shown to be a precious instrument that accelerates the discovery of novel associations between drugs and diseases. A long-known problem of multidrug interactions is the assessment of the combinatorial effect of a simultaneous treatment of multiple drugs on a biological system. Network-based representations of gene–gene coexpression interactions greatly enhanced the prediction of such synergistic effects on a large scale, in contrast with the experimental evaluation of single drug–drug interactions. Network pharmacology approaches simultaneously evaluate the molecular impairment of the disease under study and the mechanism of action of drugs, underlying network targets for a certain disease. Federico et al. developed an integrated method aimed at the identification of drug combinations, which considers both molecular characteristics of the disease under consideration and the intrinsic characteristics of the drugs, such as their mechanism of action and chemical structure [21]. Their method relies on the fact that (1) the repurposing is based on networks rather than on simple pattern matching between drug sensitivity and disease-related gene expression signatures, allowing a thorough characterization of the relationships among the deregulated genes; (2) their approach takes into consideration multiple aspects of drug–drug and drug–target interactions, including topological properties of target genes within the network, drugs' mechanism of action, and their chemical structure; (3) disease network models allow to evaluate the effect of drugs' combination. Another seminal study is the one performed by Sirci et al. [22]. The authors developed a network-based integrated approach to group drugs with similar scaffolds, physicochemical parameters, and mechanism of actions. They computed drug–drug networks where two pairs of drugs are connected if they have similar chemical characteristics. Furthermore, they computed drug–drug networks where two drugs are connected if they have similar transcriptional profile. The integration between chemocentric and mechanistic approaches has greatly benefited another field related to drug discovery, named drug repositioning. Drug repositioning is a method used to find new therapeutic outcomes for existing drugs. This process has the main advantage of speeding up the drug development process, since it finds candidate drugs that are already tested and approved. Classical drug repositioning methods involve the analysis and the comparison of the gene expression response after drug treatment and the transcriptomic alteration profile of the disease. However, including various levels of information, such as the chemical structure of the drugs or how the drug targets interact in a protein–protein interaction network, showed to increase the performance of the predictive models. Napolitano et al. [23] developed a drug-centered integrative method for the prediction of the therapeutic classes of FDA-approved drugs. They developed an SVM-based intermediate integration method that considers the drug chemical structures, the transcriptomic alteration profiles after drug treatments, and drug–target information. Their approach first builds a drug similarity matrix in each data layer and then combines them into a unified kernel that is finally used to train an SVM for the prediction of the therapeutic class. Their model reached a 78% of accuracy allowing to interpret the top misclassified drugs as candidate drugs for repositioning.

Knowledge graph–based approaches for chemical safety and drug design

The continuous production of independent datasets for compound design and safety assessment gives rise to many challenges on how to integrate, consolidate, and utilize them. Data-centered applications require the usage of sophisticated data management solutions, which are flexible enough for the highly heterogenous data available. Knowledge graphs (KGs) have recently emerged as a promising solution. They are knowledge bases to which a reasoning engine can be applied to generate new knowledge about the system under investigation [66]. They store the data in a graph-based format (Fig. 5.3), which allows, in addition to traditional queries, the usage of network topological metrics as well as the application of path-based queries [67].

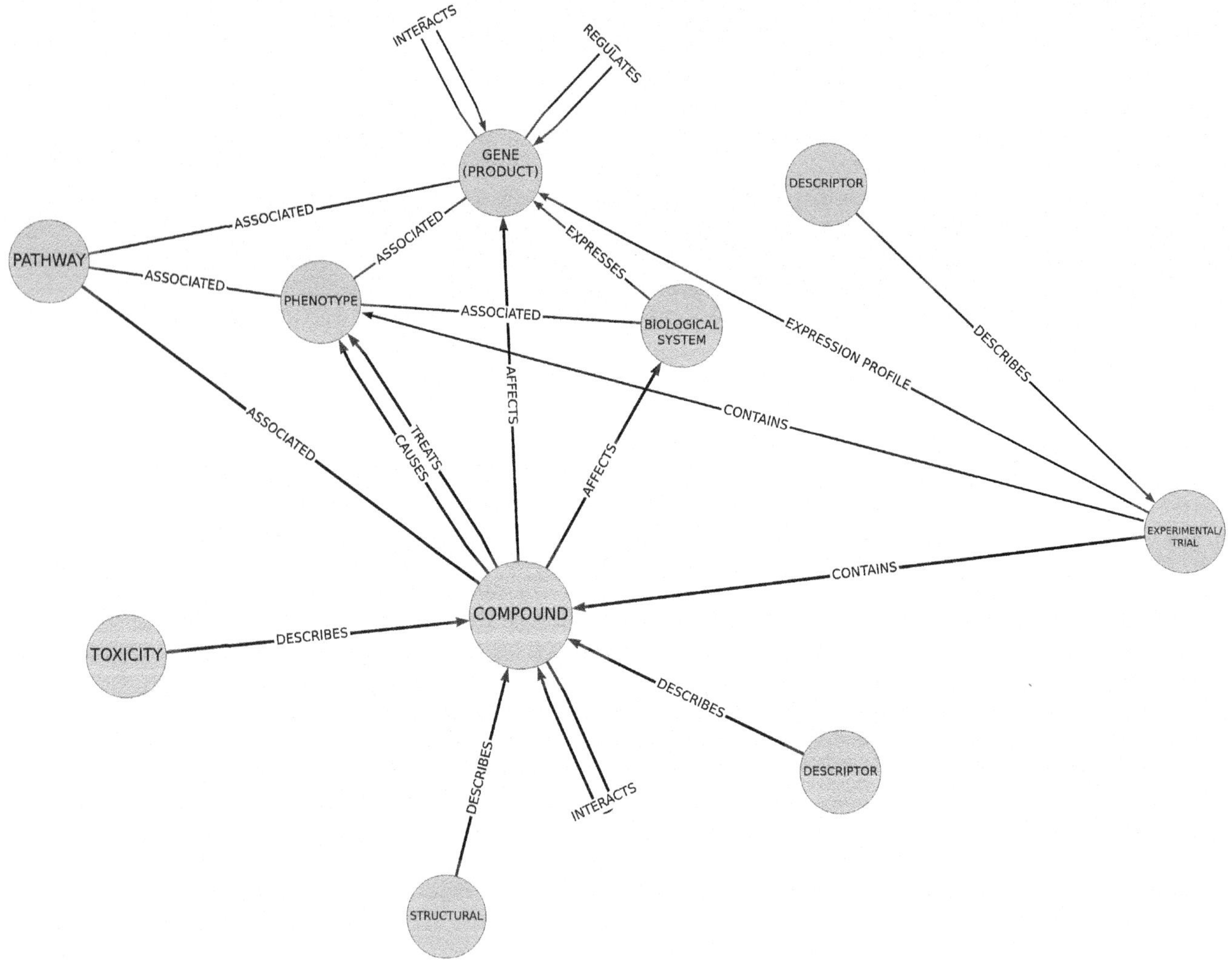

FIGURE 5.3 A possible higher-level schema of a KG suitable for task-related to compound efficacy and safety assessment. The KG contains entities (nodes) and relationships between them (edges), for a variety of data ranging from compound-specific information to genetic and clinical information.

KGs have been successfully applied for many tasks related to drug development and chemical safety assessment processes as well as their risk management [67]. Identifying the possible adverse reactions of a compound, as early as possible in the development step, preferably before extensive animal- or human-based trials are performed, can improve the speed as well as costs of compound development. KGs excel at this task, since they contain large-scale data from different views of the problem [67]. For example, compound structural information can be combined with systems biology data as well as experimental data and clinical trials. These integrated datasets in combination with different analysis methods can be used to flag possible successful drug candidates as well as instances that may have severe adverse reactions and therefore are unlikely to succeed during further testing. These prediction results can be further validated experimentally, where only likely successful candidates will be tested [19].

Integrated approaches and KGs have especially emerged as a fast method for the rapid response and understanding of COVID-19 [19,25,68]. Through the integration of previous knowledge about similar viruses, biological processes, and existing drugs, it was possible for researchers to provide predictions on long-term effects of the disease, its mechanism, and drug repositioning candidates.

For example, Pavel and del Giudice et al. [25] integrated different gene expression datasets of COVID-19, knowledge of viral–protein interactions, human protein—protein interactions, and drug—protein targets in a KG. Their analysis highlighted the underlying processes that a COVID-19 infection may cause and suggested candidate drugs. Furthermore, Zhang et al. [68] built a KG containing drug, phenotypes, target, and adverse outcome nodes. Through the combination of node embedding and classifiers, they were able to link drugs with side effects. In their case study, the authors showed the ability of their model to predict compounds that may or may not induce liver injury. However, the model was not able to predict the severity of this adverse outcome relationship.

KGs have been also successfully applied in clinical trial studies. For example, Chen et al. [26] developed a KG that combines information about clinical trials for multiple compounds and conditions. They applied node embedding methodologies to the KG to build a low representation environment suitable for drug repositioning–related predictive tasks.

Furthermore, the power of an integrated knowledge base for the management of hazardous compounds in a KG structure was shown by Zheng et al. [69], where a deep learning–based entity recognition system was used to extract knowledge from unstructured documents and to create an effective, extensive information system for the risk management of chemical compounds.

In all application areas and steps of the chemical safety assessment and development process, the goal is to make data-driven decisions, which are validated and confirmed with experimental setups, and their outcomes are fedback into the knowledge base for the next, improved, set of predictions. Therefore, integrated approaches are not a onetime setup, but a continuous loop of information collection, analysis, and confirmation. By providing informed decisions about the experiments to be performed, time, cost, and materials needed can be significantly reduced.

Conclusions

The need for a faster and more successful compound development process has led to the emergence of multiple integrative approaches for the assessment of compound safety and efficacy. These approaches are built on computational models that merge cheminformatics and bioinformatics for a holistic view that brings together multiple disciplines. While the increasing amounts of data expedite the development of integrative models, they also introduce several challenges. First, given the interdisciplinary nature of the field, there is a need to efficiently integrate, manage, and visualize the data [70]. This often requires combining data from various formats and resources, and with multiple levels of confidentiality. Even though KGs are effective tools for this task, some challenges remain. For example, missing data and imbalance in the number of data points for different conditions can hamper the group comparison analysis. Furthermore, the identification of the best data layers and robust learning algorithms to answer specific biological questions is a challenge when extrapolating relevant information from the integrated knowledge space.

It is important to note that to fully benefit from integrated in silico approaches in real-world applications, more efforts are needed to translate them from the research environment to clinical and regulatory settings. This can be facilitated by the establishment of standardized pipelines for data generation, analysis, and reporting. Although the growing amount of data supports the development of robust models, the guidelines and standards for regulatory use should be clearly defined and adopted by the community to build confidence toward in silico approaches.

Similarly, there is still a gap in how knowledge extracted from in vitro experiments can be translated to in vivo. For example, in the drug development process, the extrapolation of the best posology for human treatment from in vitro studies is a major challenge. So far, this has been addressed by the development of pharmacodynamic (PK) models based on ADME (absorption, distribution, metabolism, excretion) properties of compounds. ADME/PK is now considered an integral part of drug development, and it helps to predict pharmacological properties as well as off-targets and toxic effects in the preliminary stages of the development process [13].

In this chapter, we discuss the best representation compounds for computational models, we identify relevant datasets to be integrated, and we review several integrative methodologies that have been successfully applied in the biomedical field. Furthermore, we describe and discuss how the new emerging knowledge graph methodologies can be applied for compound efficacy and safety assessment. Despite the remaining challenges, the results reached so far are promising and show that integrative models can boost the compound development process.

References

[1] Hingorani AD, Kuan V, Finan C, et al. Improving the odds of drug development success through human genomics: modelling study. Sci Rep 2019;9:18911.
[2] Kimmelman J, Federico C. Consider drug efficacy before first-in-human trials. Nature 2017;542:25–7.
[3] Desai N. Challenges in development of nanoparticle-based therapeutics. AAPS J 2012;14:282–95.
[4] Zhou S-F, Zhong W-Z. Drug design and discovery: principles and applications. Molecules 2017;22(2):279. https://doi.org/10.3390/molecules22020279.
[5] Afantitis A, Melagraki G, Isigonis P, et al. NanoSolveIT Project: driving nanoinformatics research to develop innovative and integrated tools for in silico nanosafety assessment. Comput Struct Biotechnol J 2020;18:583–602.
[6] Kinaret PAS, Serra A, Federico A, et al. Transcriptomics in toxicogenomics, part I: experimental design, technologies, publicly available data, and regulatory aspects. Nanomaterials 2020;10(4):750. https://doi.org/10.3390/nano10040750.
[7] Serra A, Fratello M, Cattelani L, et al. Transcriptomics in toxicogenomics, part III: data modelling for risk assessment. Nanomaterials 2020; 10(4):708. https://doi.org/10.3390/nano10040708.
[8] Federico A, Serra A, Ha MK, et al. Transcriptomics in toxicogenomics, part II: preprocessing and differential expression analysis for high quality data. Nanomaterials 2020;10(5):903. https://doi.org/10.3390/nano10050903.
[9] Serra A, Saarimäki LA, Fratello M, Marwah VS, Greco D. BMDx: a graphical Shiny application to perform Benchmark Dose analysis for transcriptomics data. Bioinformatics 2020;36:2932–3.
[10] Phillips JR, Svoboda DL, Tandon A, et al. BMDExpress 2: enhanced transcriptomic dose-response analysis workflow. Bioinformatics 2019;35: 1780–2.
[11] Serra A, Fratello M, Del Giudice G, Saarimäki LA, Paci M, Federico A, et al. TinderMIX: time-dose integrated modelling of toxicogenomics data. GigaScience 2020;9(5). https://doi.org/10.1093/gigascience/giaa055.
[12] Halappanavar S, van den Brule S, Nymark P, et al. Adverse outcome pathways as a tool for the design of testing strategies to support the safety assessment of emerging advanced materials at the nanoscale. Part Fibre Toxicol 2020;17:16.
[13] Krüger A, Gonçalves Maltarollo V, Wrenger C, Kronenberger T. ADME profiling in drug discovery and a new path paved on silica. In: Drug discovery and development - new advances; 2019. https://doi.org/10.5772/intechopen.86174. Working Title.
[14] Serra A, Fratello M, Fortino V, Raiconi G, Tagliaferri R, Greco D. MVDA: a multi-view genomic data integration methodology. BMC Bioinf 2015;16:261.
[15] Serra A, Önlü S, Coretto P, Greco D. An integrated quantitative structure and mechanism of action-activity relationship model of human serum albumin binding. J Cheminf 2019;11:38.
[16] Jagiello K, Halappanavar S, Rybińska-Fryca A, Willliams A, Vogel U, Puzyn T. Transcriptomics-based and AOP-informed structure-activity relationships to predict pulmonary pathology induced by multiwalled carbon nanotubes. Small 2021;17:e2003465.
[17] Varsou D-D, Tsiliki G, Nymark P, Kohonen P, Grafström R, Sarimveis H. toxFlow: a web-based application for read-across toxicity prediction using omics and physicochemical data. J Chem Inf Model 2018;58:543–9.
[18] Fortino V, Kinaret PAS, Fratello M, et al. Biomarkers of nanomaterials hazard from multi-layer data. Nat Commun 2022;13:3798.
[19] Serra A, Fratello M, Federico A, et al. Computationally prioritized drugs inhibit SARS-CoV-2 infection and syncytia formation. Briefings Bioinf 2022;23(1). https://doi.org/10.1093/bib/bbab507.
[20] Méndez-Lucio O, Baillif B, Clevert D-A, Rouquié D, Wichard J. De novo generation of hit-like molecules from gene expression signatures using artificial intelligence. Nat Commun 2020;11:10.
[21] Federico A, Fratello M, Scala G, et al. Integrated network pharmacology approach for drug combination discovery: a multi-cancer case study. Cancers 2022;14(8):2043. https://doi.org/10.3390/cancers14082043.
[22] Sirci F, Napolitano F, Pisonero-Vaquero S, Carrella D, Medina DL, di Bernardo D. Comparing structural and transcriptional drug networks reveals signatures of drug activity and toxicity in transcriptional responses. NPJ Syst Biol Appl 2017;3:23.
[23] Napolitano F, Zhao Y, Moreira VM, Tagliaferri R, Kere J, D'Amato M, et al. Drug repositioning: a machine-learning approach through data integration. J Cheminf 2013;5:30.
[24] Zhang F, Sun B, Diao X, Zhao W, Shu T. Prediction of adverse drug reactions based on knowledge graph embedding. BMC Med Inf Decis Making 2021;21:38.
[25] Pavel A, del Giudice G, Federico A, Di Lieto A, Kinaret PAS, Serra A, et al. Integrated network analysis reveals new genes suggesting COVID-19 chronic effects and treatment. Briefings Bioinf 2021;22(2):1430–41. https://doi.org/10.1093/bib/bbaa417.
[26] Chen Z, Peng B, Ioannidis VN, Li M, Karypis G, Ning X. A knowledge graph of clinical trials ([Formula: see text]). Sci Rep 2022;12:4724.
[27] Heller S, McNaught A, Stein S, Tchekhovskoi D, Pletnev I. InChI - the worldwide chemical structure identifier standard. J Cheminf 2013;5:7.
[28] Todeschini R, Consonni V. In: Handbook of molecular descriptors. John Wiley & Sons; 2000.

[29] Mauri A, Consonni V, Pavan M, Todeschini R. Dragon software: an easy approach to molecular descriptor calculations. Match 2006;56: 237–48.
[30] Moriwaki H, Tian Y-S, Kawashita N, Takagi T. Mordred: a molecular descriptor calculator. J Cheminf 2018;10:4.
[31] Durant JL, Leland BA, Henry DR, Nourse JG. Reoptimization of MDL keys for use in drug discovery. J Chem Inf Comput Sci 2002;42:1273–80.
[32] Bolton EE, Wang Y, Thiessen PA, Bryant SH. PubChem: integrated platform of small molecules and biological activities. Elsevier; 2008. p. 217–41.
[33] Rogers D, Hahn M. Extended-connectivity fingerprints. J Chem Inf Model 2010;50:742–54.
[34] Cheng Y, Gong Y, Liu Y, Song B, Zou Q. Molecular design in drug discovery: a comprehensive review of deep generative models. Briefings Bioinf 2021;22(6). https://doi.org/10.1093/bib/bbab344.
[35] Lynch I, Afantitis A, Exner T, et al. Can an InChI for Nano address the need for a simplified representation of complex nanomaterials across experimental and nanoinformatics studies? Nanomaterials 2020;10(12):2493. https://doi.org/10.3390/nano10122493.
[36] Wyrzykowska E, Mikolajczyk A, Lynch I, et al. Representing and describing nanomaterials in predictive nanoinformatics. Nat Nanotechnol 2022;17:924–32.
[37] Marwah VS, Kinaret PAS, Serra A, Scala G, Lauerma A, Fortino V, et al. Inform: inference of network response modules. Bioinformatics 2018; 34:2136–8.
[38] Kim S, Chen J, Cheng T, et al. PubChem in 2021: new data content and improved web interfaces. Nucleic Acids Res 2021;49:D1388–95.
[39] Szklarczyk D, Santos A, von Mering C, Jensen LJ, Bork P, Kuhn M. Stitch 5: augmenting protein-chemical interaction networks with tissue and affinity data. Nucleic Acids Res 2016;44:D380–4.
[40] Irwin JJ, Tang KG, Young J, Dandarchuluun C, Wong BR, Khurelbaatar M, et al. ZINC20-A free ultralarge-scale chemical database for ligand discovery. J Chem Inf Model 2020;60:6065–73.
[41] Gaulton A, Hersey A, Nowotka M, et al. The ChEMBL database in 2017. Nucleic Acids Res 2017;45:D945–54.
[42] Wishart DS, Feunang YD, Guo AC, et al. DrugBank 5.0: a major update to the DrugBank database for 2018. Nucleic Acids Res 2018;46: D1074–82.
[43] Ochoa D, Hercules A, Carmona M, et al. Open Targets Platform: supporting systematic drug-target identification and prioritisation. Nucleic Acids Res 2021;49:D1302–10.
[44] Richard AM, Huang R, Waidyanatha S, et al. The tox21 10K compound library: collaborative chemistry advancing toxicology. Chem Res Toxicol 2021;34:189–216.
[45] Barrett T, Wilhite SE, Ledoux P, et al. NCBI GEO: archive for functional genomics data sets–update. Nucleic Acids Res 2013;41:D991–5.
[46] Subramanian A, Narayan R, Corsello SM, et al. A next generation connectivity map: L1000 platform and the first 1,000,000 profiles. Cell 2017; 171:1437–1452.e17.
[47] Lamb J, Crawford ED, Peck D, et al. The Connectivity Map: using gene-expression signatures to connect small molecules, genes, and disease. Science 2006;313:1929–35.
[48] Igarashi Y, Nakatsu N, Yamashita T, Ono A, Ohno Y, Urushidani T, et al. Open TG-GATEs: a large-scale toxicogenomics database. Nucleic Acids Res 2015;43:D921–7.
[49] Svoboda DL, Saddler T, Auerbach SS. An overview of national toxicology program's toxicogenomic applications: drugmatrix and toxfx. In: Hong H, editor. Advances in computational toxicology: methodologies and applications in regulatory science. Cham: Springer International Publishing; 2019. p. 141–57.
[50] Davis AP, Grondin CJ, Johnson RJ, Sciaky D, Wiegers J, Wiegers TC, et al. Comparative toxicogenomics database (CTD): update 2021. Nucleic Acids Res 2021;49:D1138–43.
[51] Saarimäki LA, Federico A, Lynch I, Papadiamantis AG, Tsoumanis A, Melagraki G, et al. Manually curated transcriptomics data collection for toxicogenomic assessment of engineered nanomaterials. Sci Data 2021;8:49.
[52] Sun M, Zhao S, Gilvary C, Elemento O, Zhou J, Wang F. Graph convolutional networks for computational drug development and discovery. Briefings Bioinf 2020;21:919–35.
[53] Swirog M, Mikolajczyk A, Jagiello K, Jänes J, Tämm K, Puzyn T. Predicting electrophoretic mobility of TiO2, ZnO, and CeO2 nanoparticles in natural waters: the importance of environment descriptors in nanoinformatics models. Sci Total Environ 2022;840:156572.
[54] Maia EHB, Assis LC, de Oliveira TA, da Silva AM, Taranto AG. Structure-based virtual screening: from classical to artificial intelligence. Front Chem 2020;8:343.
[55] Stumpfe D, Bajorath J. Exploring activity cliffs in medicinal chemistry. J Med Chem 2012;55:2932–42.
[56] Perualila-Tan N, Kasim A, Talloen W, Verbist B, Göhlmann HWH, QSTAR Consortium, et al. A joint modeling approach for uncovering associations between gene expression, bioactivity and chemical structure in early drug discovery to guide lead selection and genomic biomarker development. Stat Appl Genet Mol Biol 2016;15:291–304.
[57] Woo G, Fernandez M, Hsing M, Lack NA, Cavga AD, Cherkasov A. DeepCOP: deep learning-based approach to predict gene regulating effects of small molecules. Bioinformatics 2020;36:813–8.
[58] Meyers J, Fabian B, Brown N. De novo molecular design and generative models. Drug Discov Today 2021;26:2707–15.
[59] Mouchlis VD, Afantitis A, Serra A, Fratello M, Papadiamantis AG, Aidinis V, et al. Advances in de novo drug design: from conventional to machine learning methods. Int J Mol Sci 2021;22(4):1676. https://doi.org/10.3390/ijms22041676.
[60] Bickerton GR, Paolini GV, Besnard J, Muresan S, Hopkins AL. Quantifying the chemical beauty of drugs. Nat Chem 2012;4:90–8.
[61] Gómez-Bombarelli R, Wei JN, Duvenaud D, Hernández-Lobato JM, Sánchez-Lengeling B, Sheberla D, et al. Automatic chemical design using a data-driven continuous representation of molecules. ACS Cent Sci 2018;4:268–76.
[62] Escher SE, Kamp H, Bennekou SH, et al. Towards grouping concepts based on new approach methodologies in chemical hazard assessment: the read-across approach of the EU-ToxRisk project. Arch Toxicol 2019;93:3643–67.
[63] Serra A, Letunic I, Fortino V, Handy RD, Fadeel B, Tagliaferri R, et al. INSIdE NANO: a systems biology framework to contextualize the mechanism-of-action of engineered nanomaterials. Sci Rep 2019;9:179.
[64] Kong L, Tuomela S, Hahne L, Ahlfors H, Yli-Harja O, Fadeel B, et al. NanoMiner - integrative human transcriptomics data resource for nanoparticle research. PLoS One 2013;8:e68414.

[65] Gupta G, Gliga A, Hedberg J, Serra A, Greco D, Odnevall Wallinder I, et al. Cobalt nanoparticles trigger ferroptosis-like cell death (oxytosis) in neuronal cells: potential implications for neurodegenerative disease. Faseb J 2020;34:5262–81.
[66] Pavel A, Saarimäki LA, Möbus L, Federico A, Serra A, Greco D. The potential of a data centred approach and knowledge graph data representation in chemical safety and drug design. Comput Struct Biotechnol J 2022;20:4837–49.
[67] Pavel A, Serra A, Cattelani L, Federico A, Greco D. Network analysis of microarray data. Methods Mol Biol 2022;2401:161–86.
[68] Zhang R, Hristovski D, Schutte D, Kastrin A, Fiszman M, Kilicoglu H. Drug repurposing for COVID-19 via knowledge graph completion. J Biomed Inf 2021;115:103696.
[69] Zheng X, Wang B, Zhao Y, Mao S, Tang Y. A knowledge graph method for hazardous chemical management: ontology design and entity identification. Neurocomputing 2021;430:104–11.
[70] Martínez-García M, Hernández-Lemus E. Data integration challenges for machine learning in precision medicine. Front Med 2021;8:784455.

CHAPTER

6

Deep learning quantitative structure–activity relationship methods for chemical toxicity prediction and risk assessment

Shuheng Huang[1,2] *and Hu Mei*[1]

[1]Key Laboratory of Biorheological Science and Technology (Ministry of Education), College of Bioengineering, Chongqing University, Chongqing, China [2]Key Laboratory of Tropical Biological Resources of Ministry of Education and One Health Institute, School of Pharmaceutical Sciences, Hainan University, Haikou, China

Abbreviations

AE Autoencoder
AUC Area under the curve
CNN Convolutional neural network
DDIs Drug–drug interactions
DenseNet Dense convolutional network
DL Deep learning
DNN Deep neural network
GAN Generative adversarial network
GCN Graph convolutional network
GNN Graph neural network
GRU Gated recurrent unit
LSTM Long short-term memory
ML Machine learning
MLP Multilayer perceptron
QSAR Quantitative structure–activity relationship
RF Random forest
RNN Recurrent neural network
SMILES Simplified molecular-input line-entry system
SVM Support vector machine

Introduction

With the exponential increase in the number of new chemical entities, efficient and accurate prediction of molecular toxicities becomes urgent requirements for drug development and environmental risk assessment [1]. In the past 6 decades, quantitative structure–activity relationship (QSAR) has been one of the most important and widely used in silico methods for predicting biological, chemical, and environmental toxicities of small chemical molecules, which aims to explore the intrinsic relationships between molecular features/properties and toxicities by establishing qualitative or quantitative machine learning (ML) models.

Up to date, there are many ML methods intensively applied to QSAR researches, including naïve Bayes, logistic regression, decision trees (DTs), random forest (RF), partial least squares (PLS), support vector machines (SVMs), and multilayer perceptron (MLP). In a sense, ML techniques dominate the development of QSAR researches in the past decades. However, traditional ML methods have apparent limitations, especially in handling high-dimensional unstructured data or complex nonlinear problems. Recently, deep learning (DL) techniques have achieved great successes in speech processing, computer vision, natural language processing, etc. [2,3] and showed great prospect in the QSAR researches of chemical toxicity prediction [4,5]. In comparison with traditional ML methods, DL can process large-scale structured or unstructured data and can automatically extract meaningful high-level features from raw data, which make it more suitable for resolving complex real-world problems [6,7].

QSAR in Safety Evaluation and Risk Assessment
https://doi.org/10.1016/B978-0-443-15339-6.00019-9

Herein, after a brief introduction of the basic concepts of the most popular DL methods, we systematically review their recent applications in the QSAR researches of chemical toxicity prediction and risk assessment. Furthermore, the challenges and trends are also discussed.

Deep learning methods

The mainstream DL architectures consist of deep neural network (DNN), convolutional neural network (CNN), recurrent neural network (RNN), graph neural network (GNN), autoencoder (AE), generative adversarial network (GAN), etc. Herein, we just introduce the basic concepts of DNN, CNN, RNN, and GNN, which are most widely used in the QSAR domain.

Deep neural network

DNN can be considered as a stacked neural network with more than three hidden layers (Fig. 6.1A). Similar to traditional shallow neural networks, the core concepts of DNN architecture are weights, biases, nonlinear activation, and backward propagation of errors. DNN has been successfully applied to a variety of learning tasks such as image classification, object detection, and video classification. Due to its excellent applicability for different data types, DNN is also the most popular DL architecture used for the QSAR researches. However, the superior performances of DNN generally come at the cost of computational complexity, which greatly hinder its applications in many cases. Meantime, DNN is sensitive to data perturbations and tends to make incorrect predictions for adversarial examples [8].

Convolutional neural network

As one of the most successful deep learning architecture, CNN has become dominant in the tasks of computer vision, image recognition, and natural language processing. Taking the advantages of convolution and pooling operations, CNN can learn directly from raw data and is capable of automatically extracting meaningful high-level features while filtering redundant information (Fig. 6.1B). By a cascade of convolution and pooling operations, the obtained high-level features are then fed into one or more fully connected layers followed by an output layer. Thus, CNN can avoid manual feature extraction as traditional ML methods do [9,10]. Another distinctive feature of CNN is weight sharing among the neurons of the same layer, which can significantly reduce the number of training parameters. It should be mentioned that the convolution layers can be connected to each other in a feed-forward fashion, e.g., dense convolutional network (DenseNet) [11] or by shortcut connections, e.g., residual network [12]. Both DenseNet and residual network can alleviate gradient vanishing or degradation problems without adding extra parameters and computational complexity.

Recurrent neural network

RNN is another well-known DL architecture especially used for the learning tasks of unstructured data such as text, voice, and sequence. RNN can process sequential data in spatial or temporal spaces and can use current input information as well as the "memory" obtained from the previous steps for solving context-dependent prediction tasks. One of the most widely used RNN architecture is long short-term memory (LSTM) network. The LSTM [13] consists of a set of recurrently connected memory cells (Fig. 6.1C). The function of the memory cells is to update the memory and the hidden state of current cell over time through input gate, forget gate, and output gate. The input gate determines which information of the current input should be used for updating cell memory, the forget gate determines which information should be removed from the previous cell memory, and output gate can generate current output according to the previous hidden state, current input, and the updated cell memory. Several LSTM variants have been proposed, e.g., stacked LSTM Network [14], bidirectional LSTM (BLSTM) [15], and gated recurrent unit (GRU) [16].

Graph neural network

GNN is a widely used DL architecture for non-Euclidean data such as social media network, knowledge graph, and recommender system. A graph consists of a set of nodes and a set of edges, and each of nodes or edges has its

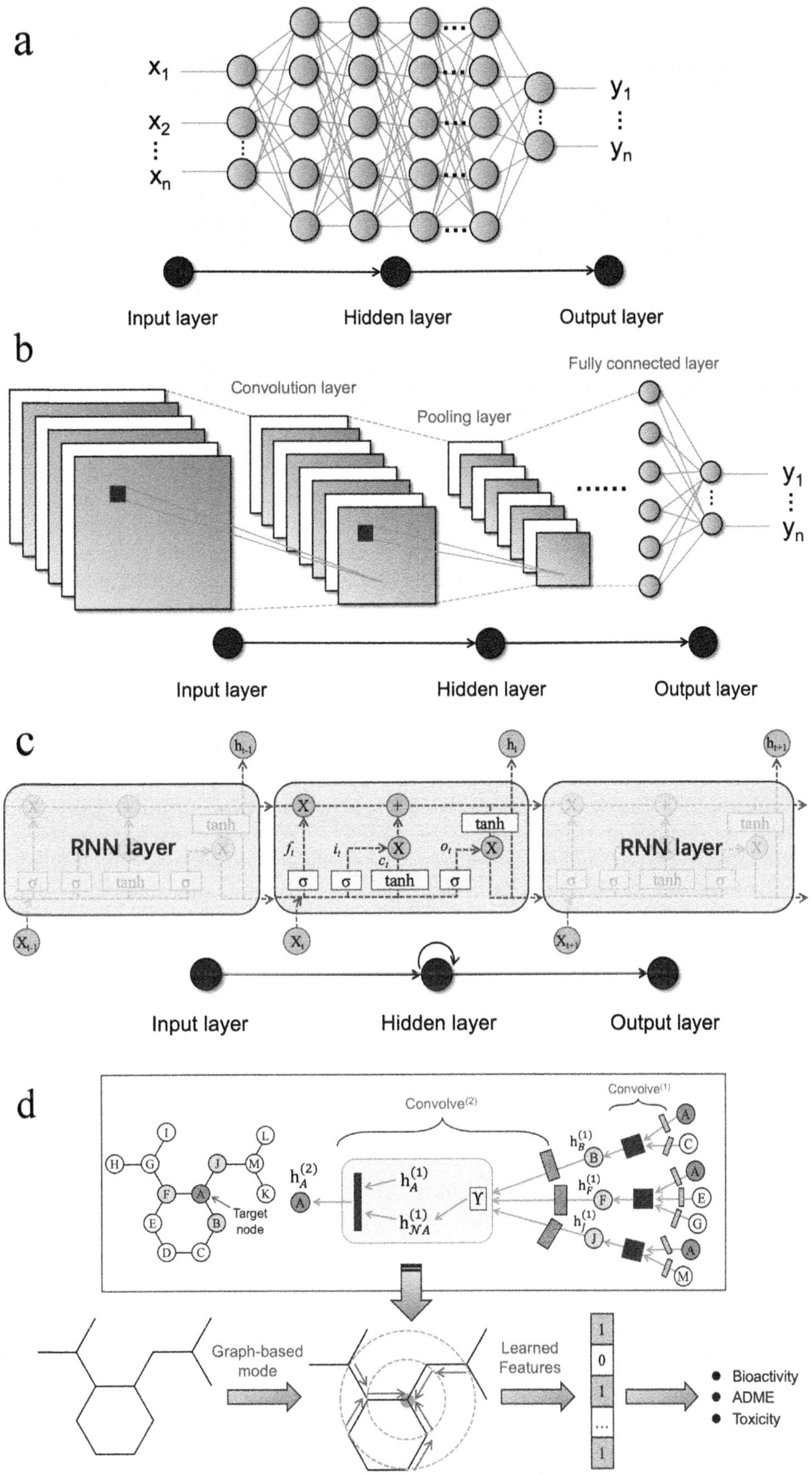

FIGURE 6.1 The illustrations of the principles of (A) DNN, (B) CNN, (C) LSTM, and (D) GCN networks.

respective feature representation. The core idea of GNN is to iteratively update the representations of nodes and edges via a message passing mechanism [17]. According to the research of Wu et al. [18], GNN can be categorized into four groups: recurrent graph neural network (RecGNN), graph convolutional network (GCN), graph autoencoder (GAE), and spatial–temporal graph neural network (STGNN). For example, GCN [19] adopts convolution operations to implement message passing between the adjacent nodes (Fig. 6.1D). By stacking multiple convolution layers, the final hidden representation of each node can receive the messages from remote nodes. According to the types of learning tasks, GNN can be designed to predict at node level, edge level, or graph level.

Key DL techniques for QSAR researches

Molecular representation

In the QSAR researches, molecular descriptors are generally derived from laborious computation of compositional, physicochemical, topological, quantum chemical, fingerprint, and grid-based molecular properties. With the development of DL, more types of molecular information can be directly used for QSAR modeling, which is of great benefit to the common QSAR researches [20]. For example, SMILES (simplified molecular-input line-entry system) notation of chemical compounds can be used in RNN network, 2D and 3D molecular images in CNN network, molecular graph and 3D coordinates in GNN network, etc. [21]. By one-hot encoding and autoembedding techniques, categorical and symbol variables can also be transformed easily into tensors for the use of DL networks [22]. One-hot encoding is a simple and fast way to create binary tensors. However, high-dimensional binary representation may lead to "curse of dimensionality" [23]. Autoembedding can map categorical or symbol variables into a continuous high-dimensional representation space, and the representations can be iteratively learned or updated by the following supervised learning process. Besides, the derived high-dimensional representations can also be used for related DL or ML tasks.

Hyperparameter optimization and overfitting controlling

Hyperparameters refer to the parameters that cannot be updated during a DL training process. There are two types of hyperparameters [24]. One is structural related hyperparameters determining the model structures and complexity, including the number of hidden layers, the number of neurons in each layer, activation function, etc. Another is training related hyperparameters determining the efficiency and accuracy of DL models, including optimizer, learning rate, batch size, etc. The methods of hyperparameter optimization can be divided into three classes: manual search, grid search, and Bayesian optimization [25]. As a powerful strategy for global optimization, Bayesian optimization targets to optimize expensive-to-evaluate functions and usually outperforms random search and grid search [24,26]. However, there are some drawbacks for Bayesian optimization, such as computationally costly and parallelism problem.

Same to traditional ML methods, one of the main challenges of DL is overfitting. There are a broad range of regularization strategies developed to alleviate the overfitting problem of DL, such as data augmentation, L1/L2 regularizations, dropout, and early stopping. Data augmentation [27] is a common method for preventing overfitting in image recognition domain, which creates fake samples by the operations of image translation, rotation, zoom, etc. L1/L2 regularizations can prevent overfitting by adding "absolute value of magnitude" (L1 norm) or "squared magnitude" (L2 norm) of coefficients as a penalty term of loss function [28]. Dropout can decrease the number of training parameters and avoid overfitting by randomly dropping out some neurons in the training process [29]. Early stopping is another form of regularization that aims at stopping training at a point when the performances on the validation set tend to decrease [30].

Model visualization and interpretation

Although most of DL models remain black boxes, some techniques have been developed for translating the network behaviors into outputs that can be interpreted [31], such as activations, class activation mapping (CAM), gradient-weighted CAM (Grad-CAM), gradient attribution, occlusion sensitivity, deep dream, etc., CAM [32] is a heat-map based visualization technique for CNN, which employs pooling layers to generate a map that highlights which parts of an image the network is using with respect to a particular class label. Grad-CAM [33] is a variant of the CAM approach, which determines the most important image area for classification by using the gradient of the classification score with respect to the convolution features. Gradient attribution [34] is derived from saliency map approach, which can provide pixel-resolution maps to explore which pixels are most important to the DL models.

Recent advances in DL-based QSAR researches in toxicity prediction and risk assessment

Recently, DL has been successfully applied to a variety of QSAR researches of chemical toxicity prediction and risk assessment and has been showing its tremendous potentials. In this section, we will give a comprehensive review of the latest researches (Table 6.1).

Deep neural network

Cytochrome P450 (CYP450) enzymes are responsible for eliminating exogenous chemical compounds from human body. Thus, inhibition of CYP450 enzymes may lead to serious adverse reactions. Wu et al. [35] established DNN and CNN classification models of the inhibitory activities of five dominant CYP450 isoforms (1A2, 2C9, 2C19, 2D6, and 3A4) by using PaDEL-1D and 2D descriptors and molecular fingerprints. For both of the DNN and CNN models, the average prediction accuracies for the five CYP450 isoforms were up to 85%. By using transcriptomic data, Wang et al. [36] adopted a DNN architecture to predict chemically induced biliary hyperplasia, liver fibrosis, and liver necrosis. The results showed high prediction performances of the DNN models with the cross-validated Matthews correlation coefficients ranging from 0.56 to 0.89 for the three endpoints. Interestingly, the DNN models also showed satisfying performances in predicting the injury phenotype for nonchemically induced injuries.

Based on the datasets for solubility, probe-likeness, hERG, KCNQ1, bubonic plague, Chagas, tuberculosis, and malaria, Korotcov et al. [37] compared DNN with traditional ML methods by using FCFP6 fingerprints as model inputs. For the external test datasets, the ranked normalized scores for the metrics of the obtained DNN models ranked higher than that of SVM, logistic regression, naïve Bayes, DT, and RF models. However, recent researches also suggested that DL methods may not always outperform the classic ML approaches [38,60]. For instance, Russo et al. [38] compared DNN with naïve Bayes, DT, RF, and SVM methods in the prediction tasks of estrogen receptor–binding activities of 24,305 compounds. The results showed that the cross-validated and test performances of the DNN model were equivalent to those of the classic ML models, although the training performances of the DNN model were relatively better.

Ryu et al. [39] proposed a computational framework, DeepDDI, to predict drug–drug and drug–food interactions. DeepDDI consists of a structural similarity profile generation pipeline and a multilabel DNN network. The structural similarity profile generation pipeline was used to capture the structural vectors of drug–drug pairs or drug–food pairs. Among 60,935 drug–drug pairs potentially showed adverse drug events, 60,509 (99.3%) were correctly predicted. Besides, DeepDDI also showed satisfying prediction performance on the 3,288,157 drug–food pairs. By using 1D, 2D, 3D, and fingerprint descriptors, Idakwo et al. [40] employed multilabel DNN and RF models to predict androgen receptor activities (agonist, antagonist, inactive, and inconclusive) of the 7665 compounds collected from Tox21 dataset. In comparison with the multilabel RF model, the multilabel DNN model achieved better prediction performances.

Besides, multitask DNN architectures have been also successfully applied to chemical toxicity prediction. Based on more than 50,000 compounds curated from public and corporate databases, Wenzel et al. [41] established a multitask DNN model to predict the ADMET properties including microsomal lability in different species, passive permeability in Caco-2/TC7 cells, and logD. The results showed that the multitask DNN model outperformed single-task DNN models on most of the datasets. A reasonable explanation is that the multitask DNN model can incorporate "signals" from the similar training samples of related or unrelated tasks [61]. Mayr et al. [42] developed a multiple-task DNN model to predict 12 toxicity endpoints of 12,707 chemical compounds. By standardization and nonlinearity treatments, a large number of static or dynamic features with different data types were used as the inputs of the DNN model. The results demonstrated the superiority of the multiple-task DNN model over the classic SVM, RF, and elastic net (ElNet) models. Based on the ensemble prediction results of the aforementioned models, the proposed DeepTox pipeline dominated by the DNN model won 9 out of the 15 Tox21 challenges.

TABLE 6.1 Recent DL-based QSAR researches of chemical toxicity prediction and risk assessment.

DL method	Endpoint	Task	Dataset	No. of samples	Input information	References
DNN/CNN	CYP450 inhibitory activities	Classification	Pubchem	17,143	PaDEL-1D and 2D descriptors, molecular fingerprints	[35]
DNN	Liver toxicities (biliary hyperplasia, fibrosis, and necrosis)	Single/multitask classification	TG-GATEs and DrugMatrix	8983	Gene and pathway/coexpression module features	[36]
DNN	Drug cytotoxicities of *Plasmodium falciparum*	Classification	CDD Public database	19,779	FCFP6 fingerprints	[37]
DNN	Estrogen receptor−binding activities	Classification	ChEMBL, Pubchem, EPA	24,305	ECFP6, FCFP6, ToxPrint or MACCS fingerprints, and RDKit descriptors	[38]
DNN	Drug−drug and drug−food interactions	Regression	DrugBank	191,878	SMILES-based structural similarity profile	[39]
DNN	Androgen receptor−binding activities	Classification	Tox21	7665	PaDEL-1D, 2D, 3D descriptors, and fingerprints	[40]
DNN	ADMET	Multitask regression	ChEMBL and Corporate database	>50,000	Physicochemical properties	[41]
DNN	12 toxicity endpoints	Multitask classification	Tox21	12,734	A large number of static or dynamic features	[42]
CNN	12 toxicity endpoints	Classification	Tox21	8014	2D molecular images	[43]
CNN	12 toxicity endpoints	Classification	Tox21	11,764	2D molecular images	[44]
CNN	CYP1A2 inhibitory activities, P-gp inhibitory activities, and Ames mutagenicities	Classification	PubChem	2380−14,993	2D molecular images	[45]
CNN	Reactivity with glutathione, cyanide, protein, and DNA	Multitask classification	Accelrys Metabolite Database	2803	Molecule-level and atom-level topological descriptors	[46]
CNN	12 toxicity endpoints	Classification	Tox21	>12,000	Grid-based descriptor	[47]
RNN	12 toxicity endpoints, HIV inhibitory activities, physical properties	Single/multitask classification	Tox21, HIV, FreeSolv, ESOL	8014, 41,193, 643, 1128	SMILES strings	[48]

RNN	Drug-induced liver injury	Classification	Dataset from NCTR and literature	2788	Directed acyclic graphs	[49]
GCN	Drug toxicities, side effects, toxicities in clinical trials	Single/multitask classification or regression	Tox21, ToxCast, SIDER, ClinTox	8014, 8615, 1427, 1491	Molecular graph	[50]
GCN	Drug−target interactions	Regression	ToxCast	530,605	Molecular graph	[51]
Semisupervised learning-based GCN	12 toxicity endpoints	Classification	Tox21	7831 and 50,527 (unlabeled)	Molecular graph	[52]
Attention-based spectral GCN	12 toxicity endpoints, hydration-free energies, lipophilicity, solubility, cytotoxicity	Classification or regression	Tox21, Freesolv, Lipophilicity, eSOL, CPAL	7831, 642, 4200, NA, NA	Molecular graph	[53]
Consensus DNN	39 toxicity endpoints	Multitask classification	Tox21	7857	Morgan, Avalon, and AtomPair fingerprints	[54]
Multimodal/consensus DNN	Half maximal growth-inhibitory concentration	Regression	IGC50 dataset	1792	SMILES strings, images, and physicochemical descriptors	[55]
Consensus DL models	Acute systemic toxicities	Multitask regression	Curated from public datasets	>80,000	Avalon fingerprints	[56]
AE-based DNN	CYP450 inhibitory activities	Multitask classification	Pubchem	53,179	PaDEL-1D and 2D descriptors, PubChem fingerprints	[57]
GCN combined with CNN	CYP450 inhibitory activities	Classification	PubChem	72,614	Molecular graph and AA sequences of binding pocket	[58]
Conditional GAN and DNN	Toxicities to zebrafish embryo	Classification	ToxCast	1059	Views and weights matrices derived from 3D structures	[59]

CDD, collaborative drug discovery; *CPAL*, cytotoxic profiling of annotated libraries using quantitative high-throughput screening; *EPA*, Environmental Protection Agency; eSOL, a database on the solubility of entire ensemble *E. coli* proteins; *Freesolv*, a database of experimental and calculated hydration free energies of small neutral molecules; *IGC50*, 50% inhibition growth concentration; *NCTR*, National Center for Toxicological Research.

Convolutional neural network

Goh et al. [43] established an image-based CNN architecture, Chemception, for predicting 12 Tox21 endpoints as well as biological and physical properties. In comparison with the MLP models by using ECFP fingerprints, Chemception slightly outperforms the MLP models in the prediction of biological and physical properties and slightly underperforms in toxicity prediction. Fernandez et al. [44] also used images of 2D molecular sketches to predict the same 12 Tox21 toxicity endpoints. The results showed that the performances of the image-based CNN models are comparable to that of the MLP models using classic QSAR descriptors. In addition, Shi et al. [45] applied image-based CNN models to predict a series of ADMET properties and achieved equivalent performances to the fingerprint-based ML models.

By using 15 molecule-level and 194 atom-level topological descriptors, Hughes et al. [46] proposed a multitask deep CNN model, XenoSite, to predict both sites and probabilities of molecular reactivity with glutathione, cyanide, protein, and DNA. For both prediction endpoints, the XenoSite model significantly outperformed the model of reactivity indices derived from quantum simulations. Yuan et al. [47] developed multichannel CNN models for predicting the 12 Tox21 endpoints. A total of six types of grid descriptors were used as models' inputs, which characterized hydrogen bond, positive/negative ionization, excluded volume, metallicity, and hydrophobicity properties. The results showed that the average area under the curve (AUC) of the multichannel CNN models for the 12 endpoints was 0.85, which was higher than that of Chemception [43], autoencoder [62], the associative neural network (ASNN) [63], and DNN [64] models.

Recurrent neural network

SMILES is a line notation method for describing chemical structures and is thus very suitable for RNN modeling. Goh et al. [48] proposed an RNN-based method, SMILES2vec, which can automatically learn molecular features from SMILES sequences. After hyperparameter optimization by Bayesian method, the SMILES2vec model outperformed Chemception [43] in two classification tasks (Tox21 and HIV datasets), but slightly underperformed in the regression task of solvation free energy. In addition, the performances of the SMILES2vec model were equivalent to that of the state-of-the-art GCN models [50].

Xu et al. [49] combined UGRNN molecular encoding scheme with a bisection algorithm to predict drug-induced liver injury. The UGRNN aimed to transform the undirected graphs (UGs) of molecular structures into directed acyclic graphs (DAGs) for the use of RNN network. First, a molecular structure was converted into a canonical SMILES string. Then, each of the atoms was sequentially defined as a root node. Lastly, the information of all other atoms was transferred to the root node along the shortest paths. Herein, each node was encoded by a vector with the information of atom and bond types. The results showed that the established models outperformed the previous drug-induced liver injury prediction models.

Graph neural network

Wu et al. [50] curated a large-scale benchmark called MoleculeNet for ML-based QSAR researches, which was built from multiple public databases such as Tox21, ToxCast, and ClinTox, and included over 700,000 compounds with a variety of experimental endpoints, e.g., physiochemical properties, biological activities, and ADMET properties. In this research, a systematic comparison among the classification and regression performances of different ML methods was performed. The results showed that the graph-based models including GCN, message passing neural network, and deep tensor neural network (TDNN) outperformed the conventional models in the test performances for 11 out of 17 datasets.

Feng et al. [51] developed a DL-based framework, PADME (Protein and Drug Molecule interaction prEdiction), to quantitatively predict drug–target interactions (DTIs). According to the representation method of small molecules, two variants of PADME with either molecular graph convolution (PADME-GraphConv) or ECFP fingerprint (PADME-ECFP) were proposed. Herein, target proteins were represented by using protein sequence composition (PSC) method. Then, the combined representation vectors of the small molecules and their targets were used as the inputs of the PADME network. The results showed that the PADME network achieved better performances on multiples datasets than SimBoost [65] and KronRLS [66] methods. In general, PADME-ECFP slightly outperforms PADME-GraphConv in many cases.

Chen et al. [52] developed a GCN model trained by a semisupervised learning algorithm, Mean Teacher, to predict the 12 Tox21 endpoints. The results indicated that the GCN models achieved an average AUC score of 0.757 for the test sets, which was better than that of the GCN models trained by supervised learning and conventional ML methods. The results revealed that semisupervised learning can promote prediction power by learning from the unannotated data.

In contrast to spatial GCN, Shang et al. [53] proposed an edge-aware multiview spectral GCN (EAGCN) model for the prediction tasks of five benchmark datasets (Tox21, Freesolv, Lipophilicity, eSOL, and CPAL). In consideration of the similarities in enthalpy and interatomic distances among the edges (chemical bonds) in different molecular graphs, EAGCN decomposed a molecular graph into multiple graphs according to edge attributes. Then, a constraint was enforced for each edge attribute by using consistent edge mapping (CEM) mechanism so that similar edges in the different molecular graphs can receive similar attention weights during message passing process. The results showed that the EAGCN models outperformed RF, message passing neural network, GCN, and GAT models for the three regression tasks (lipophilicity, eSOL, and CPAL) and two classification tasks (Tox21 and Freesolv).

Consensus, heterogenous, and other DL methods

Zakharov et al. [54] proposed a deep learning consensus architecture (DLCA) that combined consensus and multitask deep learning approaches together for the classification and regression tasks of the ChEMBL and Tox21 datasets. For the classification tasks of the Tox21 dataset, a total of five consensus multitask models were established, among which DLCA model based on three types of fingerprints achieved best performances with the average balanced accuracy of 0.67 and average AUC of 0.84 on the test datasets. Karim et al. [55] employed multiple neural network models, i.e., CNN, RNN, and MLP, to make consensus prediction of IGC50 toxicities by using ensemble averaging or meta neural network methods. To predict 59 acute systemic toxicity endpoints of more than 80,000 compounds, Jain et al. [56] also developed a consensus model composed of multitask DNN, GCN, and DLCA models with the prediction performances better than single prediction models, especially for the small datasets (less than 300 compounds).

Besides, there are also heterogenous DL architectures developed to predict chemical toxicities. Li et al. [57] developed a multitask AE-based DNN model for predicting the inhibitors of five CYP450 isoforms (1A2, 2C9, 2C19, 2D6, and 3A4). After layer-wise unsupervised pretraining of three-layer encoding network and fine-tuning using a supervised criterion, the multitask AE-based DNN model achieved better prediction performances than single-task AE-based DNN, multitask DNN, logistic regression, SVM, and DT models. By combining GCN with CNN architectures, Qiu et al. [58] also established a heterogenous GCNN model for predicting the inhibitors of the five dominant CYP isoforms. Herein, GCN with multihead attention mechanism and 1-D CNN were used for extracting the features of CPY ligands and pseudo-CYP sequences, respectively. The established GCNN model was superior to the state-of-the-art molecular fingerprint-based iCYP-MFE model [67].

By utilizing a 3D molecular vectorization algorithm [68], Green et al. [59] proposed a DNN model (Go-ZT) and a conditional GAN (GAN-ZT) model to predict molecular toxicities to zebrafish embryo. The results showed that the Go-ZT model significantly outperformed the GAN, SVM, RF, and MLP models in cross validation and independent tests. By leveraging the strengths of both the supervised GAN and DNN, the results indicated that the intersection between the two models was able to accurately predict the toxicities of the untested chemicals.

Free available DL-based tools for chemical toxicity prediction

Currently, there are a few of DL-based tools developed for predicting chemical toxicities (Table 6.2). Most of the tools provide precalibrated models and therefore do not require the prerequisite knowledge of computational chemistry for common users. DeepTox [42] is a prediction tool firstly proposed for the Tox21 Data Challenge. Besides the implemented DNN models, DeepTox also provides SVM, RF, and elastic net methods for the cross-checking of the DNN models. DeepDDI [39] is another DNN-based tool developed to predict drug–drug interactions (DDIs) and drug–food interactions (DFIs). DeepDDI can predict a total of 86 DDI types, which can be generated in the form of human-readable sentences.

Both NCATS Predictor [54] and XenoSite [46] are funded by the National Institutes of Health (NIH). NCATS Predictor employs DLCA method to make multitask prediction of biological activities and physicochemical properties and is especially suitable for the virtual screening researches, while XenoSite aims to predict reactivity sites and scores to assess the potential reactions of small molecules with macromolecules. OpenChem [69] toolkit is designed for drug design as well as toxicity prediction of the Tox21 endpoints, which provides several model types and can provide easy and fast modular design.

TABLE 6.2 The free available DL-based tools for predicting chemical toxicities.

Method	Name	Type	Description	Website	References
DNN	DeepTox	Software	Trained on Tox21 to predict chemicals toxicity by applying DNN combined with multiple machine learning methods	github.com/greenelab/deep-review/issues/72	[42]
DNN	DeepDDI	Software	Trained on DrugBank to predict DDIs types for given drug–drug pairs and drug–food pairs	bitbucket.org/kaistsystemsbiology/deepddi	[39]
DNN	NCATS Predictor	Webserver	Trained on Tox21 to predict 39 endpoints including different nuclear receptors signaling pathway toxicities and stress responses	predictor.ncats.io	[54]
CNN	XenoSite	Webserver	Trained on AMD dataset to predict both sites and probabilities of the molecular reactivity with glutathione, cyanide, protein, and DNA	swami.wustl.edu/xenosite/p/reactivity	[46]
Multiple DL methods	OpenChem	Software	PyTorch-based drug design toolkit for easy and fast model development and modular design	github.com/Mariewelt/OpenChem	[69]

Conclusions and future perspectives

The advances in the field of DL have made DL possible for large-scale and automated in silico toxicity prediction and risk assessment of chemical compounds, which is an urgent requirement with the exponential increased number of new chemicals exposure to environment. Although recent DL-based QSAR researches have achieved exciting performances in the prediction of a series of biological and environmental toxicities, there is still a long way ahead.

A free accessible large-scale benchmark dataset is a key prerequisite for successful development and evaluation of DL-based QSAR models. Although there are a lot of public databases such as Pubchem and ChEMBL for general purpose, data curation and standardization are difficult due to inconsistent data, different sources, and various experimental protocols. The toxicity benchmark datasets, e.g., Tox21 and ToxCast, are still limited in sample sizes, which impedes the technological advantages of the DL methods in a large degree. Also, there are challenges in the design of DL networks and hyperparameter optimization. The classic Bayesian optimization and its variants require a considerable amount of computation time and are difficult for parallelism. The population-based optimization methods can resolve parallelism problem; however, the effectiveness of which has not been fully explored.

For general QSAR researches of chemical toxicities, a relatively simple and explainable DL model would be more valuable than complicated and black box DL models. Due to the intrinsic network property, GNN has recently attracted more and more attentions in the QSAR domain. In principle, GNN encodes chemical compounds by natural molecular graphs and thus facilitates visualization analysis of molecular structures and determination of common toxicophores. Although there are a few of visualization and interpretation methods developed for CNN architecture, the visualization tools for other DL architectures are still limited. In addition, the development of evaluation methods for model complexity and robustness or stability should be focused. On the other hand, for the complicated multitask prediction, the integration of multimode chemical and biological information or design of combinational or hybrid DL architectures should be taken into consideration. It should be mentioned that some DL methods with great potential, such as GAN and deep reinforcement learning, are still underexplored due to high computational cost.

Overall, with the developments of large-scale benchmark datasets, computational power, visualization/interpretation methods, and multimode heterogenous DL architectures, DL-based QSAR methods will demonstrate their powerful technical advantages in the future.

Acknowledgments

The authors thank Dr. Huixiao Hong for critically reviewing the manuscript and providing helpful comments on deep learning methods for QSAR researches.

References

[1] Raies AB, Bajic VB. In silico toxicology: computational methods for the prediction of chemical toxicity. Wiley Interdiscip Rev Comput Mol Sci 2016;6(2):147–72.
[2] Russakovsky O, Deng J, Su H, Krause J, Satheesh S, Ma S, et al. Imagenet large scale visual recognition challenge. Int J Comput Vis 2015;115(3): 211–52.
[3] Deng L, Hinton G, Kingsbury B. New types of deep neural network learning for speech recognition and related applications: an overview. In: 2013 IEEE international conference on acoustics, speech and signal processing. IEEE; 2013. p. 8599–603.
[4] Lavecchia A. Deep learning in drug discovery: opportunities, challenges and future prospects. Drug Discov Today 2019;24(10):2017–32.
[5] Hemmerich J, Ecker GF. In silico toxicology: from structure–activity relationships towards deep learning and adverse outcome pathways. Wiley Interdiscip Rev Comput Mol Sci 2020;10(4):e1475.
[6] LeCun Y, Bengio Y, Hinton G. Deep learning. Nature 2015;521(7553):436–44.
[7] Zhang L, Tan J, Han D, Zhu H. From machine learning to deep learning: progress in machine intelligence for rational drug discovery. Drug Discov Today 2017;22(11):1680–5.
[8] Szegedy C, Zaremba W, Sutskever I, Bruna J, Erhan D, Goodfellow I, Fergus R. Intriguing properties of neural networks. 2013. arXiv preprint arXiv:1312.6199.
[9] Gu J, Wang Z, Kuen J, Ma L, Shahroudy A, Shuai B, et al. Recent advances in convolutional neural networks. Pattern Recogn 2018;77:354–77.
[10] Albawi S, Mohammed TA, Al-Zawi S. Understanding of a convolutional neural network. In: 2017 international conference on engineering and technology (ICET). IEEE; 2017. p. 1–6.
[11] Huang G, Liu Z, Van Der Maaten L, Weinberger KQ. Densely connected convolutional networks. In: Proceedings of the IEEE conference on computer vision and pattern recognition; 2017. p. 4700–8.
[12] He K, Zhang X, Ren S, Sun J. Deep residual learning for image recognition. In: Proceedings of the IEEE conference on computer vision and pattern recognition; 2016. p. 770–8.
[13] Gers FA, Schmidhuber J. Recurrent nets that time and count. In: Proceedings of the IEEE-INNS-ENNS international joint conference on neural networks. IJCNN 2000. Neural computing: new challenges and perspectives for the New Millennium. IEEE; 2000. p. 189–94.
[14] Fernández S, Graves A, Schmidhuber J. Sequence labelling in structured domains with hierarchical recurrent neural networks. In: Proceedings of the 20th international joint conference on artificial intelligence, IJCAI 2007; 2007.
[15] Schuster M, Paliwal KK. Bidirectional recurrent neural networks. IEEE Trans Signal Process 1997;45(11):2673–81.
[16] Cho K, Van Merriënboer B, Gulcehre C, Bahdanau D, Bougares F, Schwenk H, Bengio Y. Learning phrase representations using RNN encoder-decoder for statistical machine translation. 2014. arXiv preprint arXiv:1406.1078.
[17] Scarselli F, Gori M, Tsoi AC, Hagenbuchner M, Monfardini G. The graph neural network model. IEEE Trans Neural Netw 2008;20(1):61–80.
[18] Wu Z, Pan S, Chen F, Long G, Zhang C, Yu PS. A comprehensive survey on graph neural networks. IEEE Trans Neural Netw Learn Syst 2021; 32(1):4–24.
[19] Kipf TN, Welling M. Semi-supervised classification with graph convolutional networks. 2016. arXiv preprint arXiv:1609.02907.
[20] Khan AU. Descriptors and their selection methods in QSAR analysis: paradigm for drug design. Drug Discov Today 2016;21(8):1291–302.
[21] Xu Y, Yao H, Lin K. An overview of neural networks for drug discovery and the inputs used. Expert Opin Drug Discov 2018;13(12):1091–102.
[22] Akata Z, Perronnin F, Harchaoui Z, Schmid C. Label-embedding for image classification. IEEE Trans Pattern Anal Mach Intell 2015;38(7): 1425–38.
[23] Bengio Y, Paiement J-F, Vincent P, Delalleau O, Roux N, Ouimet M. Out-of-sample extensions for lle, isomap, mds, eigenmaps, and spectral clustering. Adv Neural Inf Process Syst 2003;16.
[24] Sun R-Y. Optimization for deep learning: an overview. J Oper Res Soc China 2020;8(2):249–94.
[25] Snoek J, Larochelle H, Adams RP. Practical Bayesian optimization of machine learning algorithms. Adv Neural Inf Process Syst 2012;25.
[26] Yu T, Zhu H. Hyper-parameter optimization: a review of algorithms and applications. 2020. arXiv preprint arXiv:2003.05689.
[27] Perez L, Wang J. The effectiveness of data augmentation in image classification using deep learning. 2017. arXiv preprint arXiv:1712.04621.
[28] Burden F, Winkler D. Bayesian regularization of neural networks. In: Artificial neural networks; 2008. p. 23–42.
[29] Srivastava N, Hinton G, Krizhevsky A, Sutskever I, Salakhutdinov R. Dropout: a simple way to prevent neural networks from overfitting. J Mach Learn Res 2014;15(1):1929–58.
[30] Prechelt L. Early stopping-but when? Neural networks: Tricks of the trade. Springer; 1998. p. 55–69.
[31] Samek W, Wiegand T, Müller K-R. Explainable artificial intelligence: understanding, visualizing and interpreting deep learning models. 2017. arXiv preprint arXiv:1708.08296.
[32] Zhou B, Khosla A, Lapedriza A, Oliva A, Torralba A. Learning deep features for discriminative localization. In: Proceedings of the IEEE conference on computer vision and pattern recognition; 2016. p. 2921–9.
[33] Selvaraju RR, Cogswell M, Das A, Vedantam R, Parikh D, Batra D. Grad-cam: visual explanations from deep networks via gradient-based localization. In: Proceedings of the IEEE international conference on computer vision; 2017. p. 618–26.
[34] Tomsett R, Harborne D, Chakraborty S, Gurram P, Preece A. Sanity checks for saliency metrics. Proc AAAI Conf Artif Intell 2020:6021–9.
[35] Wu Z, Lei T, Shen C, Wang Z, Cao D, Hou T. ADMET evaluation in drug discovery. 19. Reliable prediction of human cytochrome P450 inhibition using artificial intelligence approaches. J Chem Inf Model 2019;59(11):4587–601.
[36] Wang H, Liu R, Schyman P, Wallqvist A. Deep neural network models for predicting chemically induced liver toxicity endpoints from transcriptomic responses. Front Pharmacol 2019;10:42.
[37] Korotcov A, Tkachenko V, Russo DP, Ekins S. Comparison of deep learning with multiple machine learning methods and metrics using diverse drug discovery data sets. Mol Pharm 2017;14(12):4462–75.

[38] Russo DP, Zorn KM, Clark AM, Zhu H, Ekins S. Comparing multiple machine learning algorithms and metrics for estrogen receptor binding prediction. Mol Pharm 2018;15(10):4361–70.
[39] Ryu JY, Kim HU, Lee SY. Deep learning improves prediction of drug-drug and drug-food interactions. Proc Natl Acad Sci USA 2018;115(18): E4304–11.
[40] Idakwo G, Thangapandian S, Luttrell J, Zhou Z, Zhang C, Gong P. Deep learning-based structure-activity relationship modeling for multi-category toxicity classification: a case study of 10K Tox21 chemicals with high-throughput cell-based androgen receptor bioassay data. Front Physiol 2019;10:1044.
[41] Wenzel J, Matter H, Schmidt F. Predictive multitask deep neural network models for ADME-Tox properties: learning from large data sets. J Chem Inf Model 2019;59(3):1253–68.
[42] Mayr A, Klambauer G, Unterthiner T, Hochreiter S. DeepTox: toxicity prediction using deep learning. Front Environ Sci 2016;3.
[43] Goh GB, Siegel C, Vishnu A, Hodas NO, Baker N. Chemception: a deep neural network with minimal chemistry knowledge matches the performance of expert-developed QSAR/QSPR models. 2017. arXiv preprint arXiv:1706.06689.
[44] Fernandez M, Ban FQ, Woo G, Hsing M, Yamazaki T, LeBlanc E, et al. Toxic colors: the use of deep learning for predicting toxicity of compounds merely from their graphic images. J Chem Inf Model 2018;58(8):1533–43.
[45] Shi T, Yang Y, Huang S, Chen L, Kuang Z, Heng Y, Mei H. Molecular image-based convolutional neural network for the prediction of ADMET properties. Chemometr Intell Lab Syst 2019;194:103853.
[46] Hughes TB, Dang NL, Miller GP, Swamidass SJ. Modeling reactivity to biological macromolecules with a deep multitask network. ACS Central Sci 2016;2(8):529–37.
[47] Yuan Q, Wei Z, Guan X, Jiang M, Wang S, Zhang S, Li Z. Toxicity prediction method based on multi-channel convolutional neural network. Molecules 2019;24(18):3383.
[48] Goh GB, Hodas NO, Siegel C, Vishnu A. Smiles2vec: an interpretable general-purpose deep neural network for predicting chemical properties. 2017. arXiv preprint arXiv:1712.02034.
[49] Xu Y, Dai Z, Chen F, Gao S, Pei J, Lai L. Deep learning for drug-induced liver injury. J Chem Inf Model 2015;55(10):2085–93.
[50] Wu Z, Ramsundar B, Feinberg EN, Gomes J, Geniesse C, Pappu AS, et al. MoleculeNet: a benchmark for molecular machine learning. Chem Sci 2018;9(2):513–30.
[51] Feng Q, Dueva E, Cherkasov A, Ester M. Padme: a deep learning-based framework for drug-target interaction prediction. 2018. arXiv preprint arXiv:1807.09741.
[52] Chen J, Si Y-W, Un C-W, Siu SWI. Chemical toxicity prediction based on semi-supervised learning and graph convolutional neural network. J Cheminform 2021;13(1):93.
[53] Shang C, Liu Q, Tong Q, Sun J, Song M, Bi J. Multi-view spectral graph convolution with consistent edge attention for molecular modeling. Neurocomputing 2021;445:12–25.
[54] Zakharov AV, Zhao T, Nguyen DT, Peryea T, Sheils T, Yasgar A, et al. Novel consensus architecture to improve performance of large-scale multitask deep learning QSAR models. J Chem Inf Model 2019;59(11):4613–24.
[55] Karim A, Singh J, Mishra A, Dehzangi A, Newton M, Sattar A. Toxicity prediction by multimodal deep learning. In: Pacific Rim knowledge acquisition workshop. Springer; 2019. p. 142–52.
[56] Jain S, Siramshetty VB, Alves VM, Muratov EN, Kleinstreuer N, Tropsha A, et al. Large-scale modeling of multispecies acute toxicity end points using consensus of multitask deep learning methods. J Chem Inf Model 2021;61(2):653–63.
[57] Li X, Xu YJ, Lai LH, Pei JF. Prediction of human cytochrome P450 inhibition using a multitask deep autoencoder neural network. Mol Pharm 2018;15(10):4336–45.
[58] Qiu M, Liang X, Deng S, Li Y, Ke Y, Wang P, Mei H. A unified GCNN model for predicting CYP450 inhibitors by using graph convolutional neural networks with attention mechanism. Comput Biol Med 2022;150:106177.
[59] Green AJ, Mohlenkamp MJ, Das J, Chaudhari M, Truong L, Tanguay RL, Reif DM. Leveraging high-throughput screening data, deep neural networks, and conditional generative adversarial networks to advance predictive toxicology. PLoS Comput Biol 2021;17(7):e1009135.
[60] Liu R, Madore M, Glover KP, Feasel MG, Wallqvist A. Assessing deep and shallow learning methods for quantitative prediction of acute chemical toxicity. Toxicol Sci 2018;164(2):512–26.
[61] Xu Y, Ma J, Liaw A, Sheridan RP, Svetnik V. Demystifying multitask deep neural networks for quantitative structure–activity relationships. J Chem Inf Model 2017;57(10):2490–504.
[62] Galushka M, Browne F, Mulvenna M, Bond R, Lightbody G. Toxicity prediction using pre-trained autoencoder. In: 2018 IEEE international conference on bioinformatics and biomedicine (BIBM). IEEE; 2018. p. 299–304.
[63] Abdelaziz A, Spahn-Langguth H, Schramm K-W, Tetko IV. Consensus modeling for HTS assays using in silico descriptors calculates the best balanced accuracy in Tox21 challenge. Front Environ Sci 2016;4:2.
[64] Capuzzi SJ, Politi R, Isayev O, Farag S, Tropsha A. QSAR modeling of Tox21 challenge stress response and nuclear receptor signaling toxicity assays. Front Environ Sci 2016;4:3.
[65] He T, Heidemeyer M, Ban F, Cherkasov A, Ester M. SimBoost: a read-across approach for predicting drug–target binding affinities using gradient boosting machines. J Cheminform 2017;9(1):24.
[66] Pahikkala T, Airola A, Pietilä S, Shakyawar S, Szwajda A, Tang J, Aittokallio T. Toward more realistic drug-target interaction predictions. Brief Bioinform 2015;16(2):325–37.
[67] Nguyen-Vo T-H, Trinh QH, Nguyen L, Nguyen-Hoang P-U, Nguyen T-N, Nguyen DT, et al. iCYP-MFE: identifying human cytochrome P450 inhibitors using multitask learning and molecular fingerprint-embedded encoding. J Chem Inf Model 2021.
[68] d'Avezac M, Botts R, Mohlenkamp MJ, Zunger A. Learning to predict physical properties using sums of separable functions. SIAM J Sci Comput 2011;33(6):3381–401.
[69] Korshunova M, Ginsburg B, Tropsha A, Isayev O. OpenChem: a deep learning toolkit for computational chemistry and drug design. J Chem Inf Model 2021;61(1):7–13.

C H A P T E R

7

Predictive modeling approaches for the risk assessment of persistent organic pollutants (POPs): from QSAR to machine learning–based models

Prakrity Singh[1,2], *Shraddha Pandit*[1] *and Ramakrishnan Parthasarathi*[1,2]

[1]CSIR- Indian Institute of Toxicology Research, Lucknow, Uttar Pradesh, India [2]Academy of Scientific and Innovative Research (AcSIR), Ghaziabad, Uttar Pradesh, India

Abbreviations

3R replacement, reduction, and refinement
KNN K-Nearest Neighbor
ML machine learning
MLR multiple linear regression
OECD Organization for Economic Co-operation and Development
PBT persistent, bioaccumulative, and toxic
PCA principal component analysis
PLS partial least square
POPs persistent organic pollutants
QSAR quantitative structure–activity relationship
REACH Registration, Evaluation, Authorization, and Restriction of Chemical
SVMs support vector machines

Introduction

Urbanization and industrialization have increased the demand for consumer products, leading to increased production, utilization, and disposal of the diversity of chemical substances that burden the environment. These chemicals are actively utilized for industrial and agriculture purposes not limited to agro-chemicals, lifestyle products, etc. [1]. A class of chemicals referred to as persistent organic pollutants (POPs) [2] has gained wide attention due to their endurance and slow biodegradability. The distant transportation ability, long half-life, and growing concentration of such emerging contaminants in the environment have led to their biomagnification and bioaccumulation in organisms [3]. Due to the longevity associated with these chemicals, the magnitude of adverse impact on the human health and wildlife species remains high (Fig. 7.1).

In the early 1960s, after the publication of the book "Silent Spring," concern started growing about the effect of persistent chemicals [4,5]. From that moment, the involvement of different scientific communities was started toward the persistent, bioaccumulative, and toxic (PBT) behavior of POPs [2]. Therefore, many countries have endorsed national regulations and global agreements to regulate the synthesis and utilization of POPs [6,7]. Initially, the Basel Convention (on wastes) and the Rotterdam Convention (on previously informed agreement) started handling some aspects of POPs, such as their distribution and final disposition [8,9]. Further, in Stockholm, Sweden (in May 2001), the European Community and 90 other nations, along with the United States, signed a historic United Nations pact to address this worldwide issue, where a scientific review procedure was outlined. It contained specifications for the countries to confine or eliminate the manufacturing, usage, and release of the essential POPs. Moreover, in 2004, Stockholm Convention became the primary international and legally binding treaty for providing the

QSAR in Safety Evaluation and Risk Assessment
https://doi.org/10.1016/B978-0-443-15339-6.00026-6

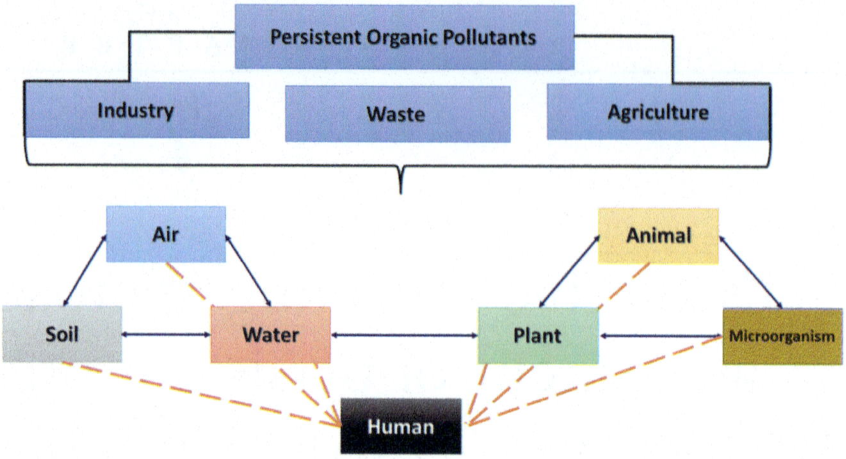

FIGURE 7.1 Different sources of persistent organic pollutants (POPs) and their distribution in the ecological system.

agreements and regulatory frameworks to restrict POPs [10]. Further, this Convention listed more than 20 chemicals under the category of POPs that are intentionally and unintentionally produced (Table 7.1).

Numerous challenges are associated with the toxicity assessment, traits, and properties to identify the associated risk of each existing chemical. Due to the broader categories of these chemicals, the assessment of physicochemical and toxicokinetic properties for each chemical remains compromised due to the great cost and time-consuming experiments. Hence, it becomes crucial to develop a competent risk assessment framework for the regulatory purposes of emerging contaminants. Computational toxicology provides vital support as it is an emerging field amalgam of mathematical, chemical, and biological approaches with high-throughput computational power to explore comprehensive relationships between chemical exposure and resulting toxicities [12,13].

In this scenario, the constant evolution in quantitative structure–activity relationship (QSAR) modeling has become one of the essential computational predictive modeling techniques. The basic notion is to develop an easy and reliable methodology that can help in the estimation of physicochemical, environmental, and toxicokinetic parameters for many existing and emerging POPs [2,14]. QSAR models can be categorized with the help of well-defined guidelines, statistical methods, and a diverse range of molecular connectivity indices that help in the enhancement of models' predictability for a different set of chemicals [15]. The selection of molecular connectivity indices for QSAR has been performed based on their simplicity in calculations and the possibilities to quantify them with the help of their specific structural properties. Finally, after connecting the structural parameters with physiological properties, a mathematical equation is obtained to perform the toxicity predictions of POPs. Here in this chapter, several predictive models used for the PBT analysis of POPs and different QSAR modeling techniques are thoroughly discussed.

Significant breakthroughs in QSAR modeling of POPs

Initially, QSAR was considered the logical extension of physical organic chemistry. It has evolved from a linear regression equation (for a small series of chemicals) to a wide variety of statistical and machine learning (ML) techniques for border range of chemicals [16]. Since Corwin Hansch's work of the QSAR more than 50 years ago [14], it has become one of the most widely used techniques for determining chemicals' physical and biological properties [17]. During the initial phase of QSAR research, to express a compound's lipophilicity, the log K_{ow} was assigned to predict the multimedia partitioning between two systems like sediment/water, biota/water, and water/air. Lipophilic POPs using QSAR were executed in 1988 by Hawker and Connell [18]. Further, for predicting the partitioning and chromatographic properties of POPs, a successful model was developed by Sabljic et al. in 2001 [12]. In the same year, another model came for predicting K_{oc} (soil sorption coefficients) of POPs using the soil/water partition coefficients of chemicals [19]. Further, for predicting the atmospheric persistence of POPs, Gramatica and their group have developed a genetic algorithm-based QSAR regression model [20].

TABLE 7.1 Class of chemicals listed under the category of persistent organic pollutants (POPs) by Stockholm Convention.

Classes of POPs	Name of chemicals	Type	References
Intentionally produced	Aldrin	Pesticide	[11]
	Chlordane	Pesticide	
	DDT	Pesticide	
	Dieldrin	Pesticide	
	Endrin	Pesticide	
	HCB	Pesticide	
	Heptachlor	Pesticide	
	Mirex	Pesticide	
	Toxaphene	Pesticide	
	Chlorodecone	Pesticide	
	Lindane	Pesticide	
	Hexabromobiphenyl	Industrial	
	Tetra-BDE	Industrial	
	Penta-BDE	Industrial	
	Hexa-BDE	Industrial	
	Hepta-BDE	Industrial	
	Endosulfan	Pesticide	
Unintentionally produced	PCBs	By-products	
	α-HCH	By-products	
	β-HCH	By-products	
	PCDDs and PCDFs	By-products	
	Dioxins	By-products	
	Furans	By-products	

Current advancements and guidelines for QSAR model development of POPs

The main issue associated with the risk assessment of POPs is the lack of proper categorization of physicochemical parameters, reactivity properties, biological and toxicological properties, and time-consuming testing methods [12]. The problem of POP's categorization was addressed by the Stockholm Convention, which has suggested the utilization of different computational method-based screening techniques for PBT analysis [10]. Organization for Economic Co-operation and Development (OECD), another regulatory agency, has given the five principles for the development of QSAR models [21], which are (1) a defined endpoint, (2) an unambiguous algorithm, (3) a defined domain of applicability, (4) appropriate measures of goodness of fit, robustness, and predictivity, and (5) a mechanistic interpretation, if possible. Similarly, under the Registration, Evaluation, Authorization, and Restriction of Chemicals (REACH) regulation, specific guidelines have been laid down for the evaluation of PBT and very persistent very toxic chemicals following the specific standards mentioned in Annexure 13 (ECA, 2008) [22,23]. REACH explicitly encourages the use of diverse approaches combining multivariate investigation with QSAR to bridge the data gap. Primary norms for the development and validation of a QSAR model to classify PBT-like compounds are based on (i) chemical structure, (ii) structural molecular descriptors, and (iii) available experimental data or commonly used persistence indicators [24]. Correlation is established between various descriptors of the chemicals and experimentally obtained endpoints to develop a model equation for the property prediction of new chemicals. Further, a number of validations are used to define the reliability of a QSAR model. With the help of goodness of fit, models must be evaluated to enhance predictivity [25].

Various methods that assist in risk assessment decision-making utilizing QSAR and expert models are discussed in further sections. These models can predict chemicals half-lives, biodegradation, bioaccumulation, and toxicity by using the existing experimental data along with read-across and grouping approaches that further performed chemical categorization on the basis of the shared structural features [26]. Complete workflow and procedures frequently used in toxicity identification and hazard assessment are explained in Fig. 7.2. Recent advancements made in the usage of computational toxicology are discussed further in this section.

Different molecular endpoints for the classification of POPs

To understand the PBT properties of "existing" or "new" commercially available chemicals, REACH suggested different pragmatic criteria with specified cut-off values (Fig. 7.3). POPs can be characterized by the vapor pressures, aqueous solubilities, lipophilicity, and half-lives in soil/water/air. Various toxicity end points like carcinogenicity, mutagenicity, and teratogenicity are also used for their classification. Such experimental values are used for the filtration of chemicals based on PBT and for defining the problem for which the QSAR model has been developed [27].

Persistence: The popular criteria for determining any chemical's environmental persistence is the degradation half-life in diverse matrices like water, sediment, air, and soil. Using the global half-life index (GHLI) is also suggested for POPs screening and model development [28]. The simplest method of analyzing the chemicals POP-like behavior is the United Nations Environment Program–based persistence approach (2), which defines an element is persistent if its half-lives exceed in any predetermined criteria in four environmental media (air >2 days, surface water >60 days, and soil or sediment >180 days) [29]. This method evaluates the single-media half-life and can be considered an experimental end point for model development. Besides these end points,

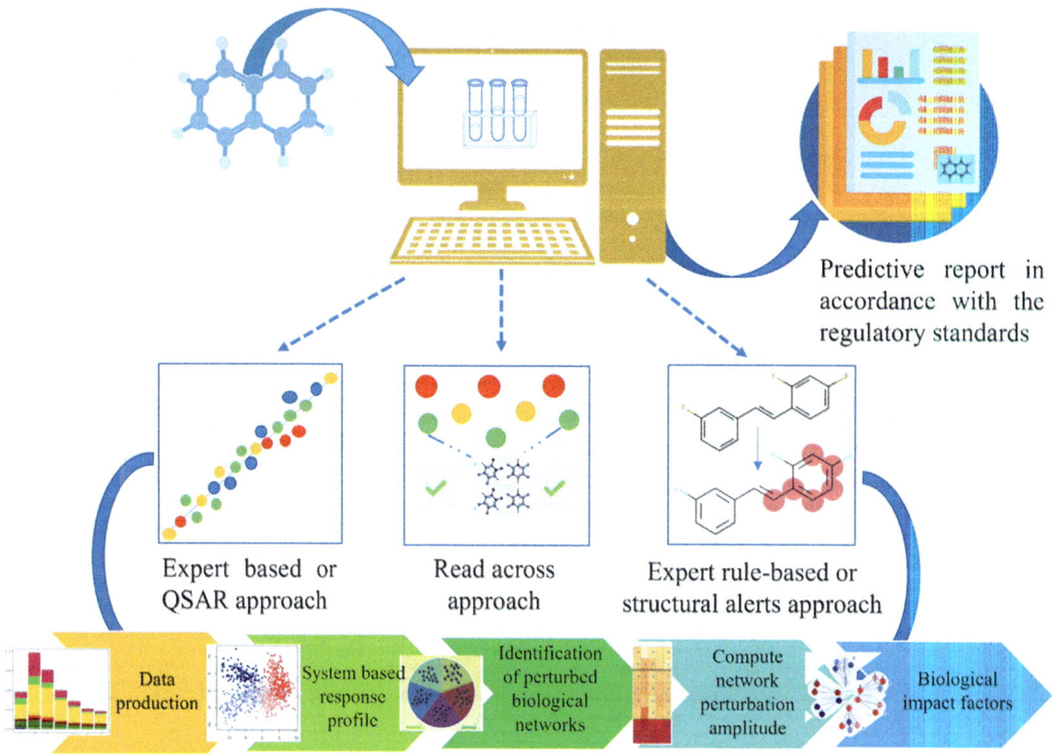

FIGURE 7.2 Nontesting in silico approaches utilized in the risk assessment of persistent organic pollutants (POPs).

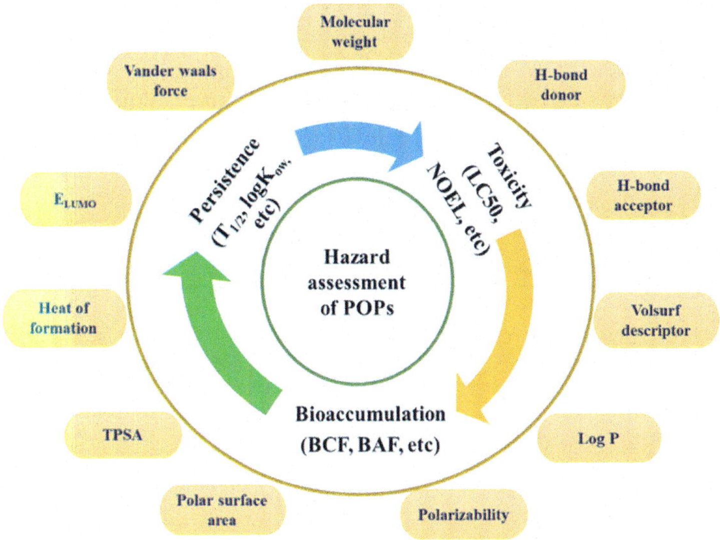

FIGURE 7.3 Different toxicity end points and molecular descriptors for the persistence, bioaccumulation, and toxicity analysis of persistent organic pollutants (POPs).

long-range transport or mobility around the globe can also be regarded as an experimental end point for QSAR development [30].

Bioconcentration/bioaccumulation: The process of increasing concentration and accumulation of chemicals through nondietary routes in the organism is known as bioconcentration or bioaccumulation, respectively. The tendency of a chemical to concentrate in aquatic species is typically stated as its bioconcentration factor (BCF). When the concentration of the chemical increases in the water bodies in comparison to water, it is known as bioaccumulation factor (BAF) [31]. For assessing these BCF/BAF factors, mostly fishes are considered the best model organism because of their high importance in human food sources and their easy accessibility to standardized testing procedures according to OECD-based guidelines [25].

In support of the 3 Rs principles (replacement, refinement, and reduction of animal testing), the log value of BCFs/BAFs (at standard experimental concentration) is mainly used for developing the bio-concentration/accumulation-based QSAR models. Other than these factors, some experimentally derived parameters, such as surface tension and water solubility, were also used for model development [32].

Toxicity: The presence of POPs residues in the adipose tissue of living beings makes it possible to determine the hazard imposed by these chemicals. The residues of these substances have the potential to create an adverse impact on human health and other species. The toxicity of these chemicals are also depending on their acute (short-term) or chronic (long-term) exposure level, that leads to skin irritation, eye irritation, skin sensitization, genotoxicity, reproductive and developmental toxicity, carcinogenicity, hepatotoxicity, and cardiotoxicity for human beings [33]. Further, QSAR models can be developed by using different end points such as median toxic dose (TD_{50}), lethal/effective concentration (LC_{50} or EC_{50}), no observable effect level—acute and chronic tests, and lowest observable effect level—acute and chronic tests [34].

Molecular descriptors utilized in the QSAR modeling of POPs

Different experimental and theoretical descriptors are used to capture the structural aspects of POPs. These molecular descriptors can be calculated from concise two-dimensional (2D) chemical structures in relatively simple forms to complex conditions representing three-dimensional (3D) chemical features or fingerprints containing several bit positions to characterize precise chemical details [35].

Numerous theoretical molecular descriptors are reported for the PBT analysis in the literature. For example, to predict the soot water partition coefficient of POPs molecular polarizability (α), standard heat of formation (ΔH_f), and energy of the lowest unoccupied molecular orbital show a dominant effect [36]. Simultaneously for modeling absorption, distribution, metabolism, and excretion properties of POPs, VolSurf descriptors are found useful [37]. Other than this, mean atomic van der Waals volumes, electro-topological positive variation, number of H-bond donors, information index related to the complementary information content, and fragment associated with heteroatom are essential. Variables that help in defining the environmental persistence are chemical sizes and various electronic features, such as electrostatic and dipole–dipole interactions in the surrounding media, that can directly impact the bioavailability and partitioning of substances into environmental media [20].

Various tools are also available which generate the diversity of descriptors with the help of chemical coordinates, SMILES, or chemical structures. Some standard tools and software are PaDEL [38], DRAGON [39], ChemAxon [40], etc.

Statistical and ML-based approaches for model development of POPs

The QSAR model development uses a 2D multivariate data table that comprehends the multidimensional data by linking the chemistry (i.e., the structure) and the biological consequences (i.e., the activity) of each chemical with the help of a QSAR equation [41].

$$\text{Experimental activity of POPs} = f\ (\text{Structural properties of POPs}) + \text{error} \tag{1}$$

Here, the experimental activity is followed by the experimental PBT endpoints, and the structural properties could be a comprehensive set of molecular features that must be able to define all the structural attributes of the selected POPs. Further model development procedures for the chemicals' PBT screening were performed using multivariate modeling techniques.

These can be divided into two broader categories: (i) statistical approaches, such as principal component analysis (PCA), partial least squares (PLS), cluster analysis, and multiple linear regression (MLR), and (ii) ML-based methods, such as naive Bayes, k-nearest neighbors (k-NN), and random forest, etc. These methods include different algorithms for classification and regression purpose (Fig. 7.4). Currently, the existing ML model includes binary classification of chemicals for the prediction and evaluation of biodegradability and inherent biodegradation in different conditions [42]. These modeling methods further help to understand the atmospheric degradation reactions, persistence, and other toxicity end points of POPs.

Classical approaches for QSAR model development

Principal component analysis (PCA): Due to the large and widespread dataset of POPs and their associated features, such as chemical reactivity and biological activity, it is essential to reduce their dimensionality without losing their information. For instance, when considering persistence, it is necessary to consider a chemical's duration (or half-life) in various environmental matrices (i.e., air, water, soil, and sediment). Chemical persistence in each media contributes to the total persistence in the environment [43]. To solve the purpose related to reducing the

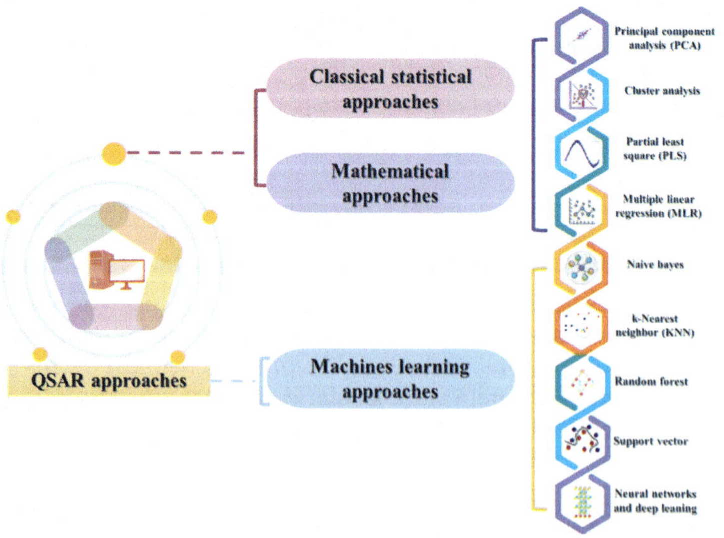

FIGURE 7.4 Various classical and machine learning (ML)-based predictive modeling approaches used for quantitative structure–activity relationship (QSAR) for the risk assessment of persistent organic pollutants (POPs).

dimensionality PCA plays a vital role, and it is also helpful in preserving the variability of data. PCA is a multivariate statistical analysis technique that contributes to understand the obtained correlations and covariance present in the data. It can combine the available data to produce a ranking or grouping of the tested substances according to their numerous properties simultaneously [44]. The original variables representing the examined system are combined linearly to create PCA. The contained data, as well as the main variance in the original data, are best understood by their combinations.

The PCA-based models predict the chemical's behaviors that depend on the changes in the cumulative index (for example, to understand the total persistent of the chemical in the environment), rather than on changes in each parameters (e.g., persistence in different environmental media). The possibility of using the QSAR-based prediction models for the cumulative indexes of the major end points from the chemical structure or screening the characteristics of chemicals before their synthesis is another advantage in the absence of data available for the individual variables. To date, more than 40 models have been created with single and cumulative end points integrated with the QSARINS program and QSAR model reporting format for PBT analysis of chemicals [45].

Cluster analysis (CA): CA is a method of allaying a set of chemicals so that similar chemicals form a cluster in a multidimensional space. It serves as an alternative to PCA for outlining the layout of data and classifying data. QSAR modeling by CA helps in the examination of data homogeneity along with their potential relationships. A similarity measure, such as distance, correlation, or a combination of the two properties, can be used by CA to group the items in a consistent manner [46]. This approach worked on the idea that similarity and sample separation have an inverse relationship. Therefore, in CA, a particular metric, such as the Manhattan distance, Euclidean distance, etc., is used to calculate the distances (or correlation) between all samples. Based on criteria used to determine the distance between two groups (linkage criterion), other clustering algorithms can be used to group the samples, including the centroid technique, Ward's approach, or average linkages. The tree's dendrogram-style representation of a hierarchical clustering of objects and table with several clustering are the two ways to cluster the data for representation purposes [47].

A study was conducted in 2008 for the classification of POPs based on an agglomerative clustering method [48]. In that study, three well-defined groups of persistence chemicals were generated by using hierarchical CA, for the grouping of available half-life data of POPs in air, water, soil, and sediment. The hierarchy of clusters was represented by a dendrogram, and the chemicals were classified into the highly persistent, medium persistent, and low persistent category. Further, these persistence categories were used as end points in the other QSAR modeling [48].

Partial least square: To get detailed information about the behavior of chemicals in the presence of different biological activity and additionally for designing new chemicals, the structure–property relationships play a vital role. In such a scenario, PLS-based regression analysis helps in identification of specific structural trends related to the differences in observed parameters [49]. PLS also combines the features obtained from PCA and turns very beneficial when there is a need to predict a set of dependent variables from a large group of independent variables.

In the case of POPs, multiple PLS-based models were developed for predicting the air half-life, bioconcentration, and different toxicological end points. A model developed by Khan et al. used simple and explainable 2D descriptors for the prediction of the air half-life of POPs. This model strengthens the data gap present in testing half-lives of POPs, when the emerging chemicals fall within the model's applicability domain [50]. Another PLS-based model was developed by Bordás et al. for predicting the bioconcentration factor of POPs. The prediction obtained remained statistically significant for the BCF [37].

Multiple linear regression: This method is known as the Hansch-Fujita strategy and has been the only statistical approach utilized in QSAR investigations. MLR technique helps develop an explanatory and predictive model equation between a chemical dataset and several descriptors (at least three). For MLR-based equation, data should be homogeneous, and the descriptors should be nonredundant; this equation requires a vital validation procedure. The developed model equation can be represented as:

$$y = b_0 + b_1x_1 + b_2x_2 + \cdots\cdots + b_Nx_N + e_N \quad (2)$$

In this equation, y depicts the dependent variable that represents the chemical activity, x defines the independent variable describing the molecular descriptor, and b is the coefficient of regression. For POPs, the MLR model was used by Venier et al. in 2010 for the concentration analysis in atmospheric vapor and particle phase. MLR is also used to determine the rate of change for the pollutant in people and the environment [51].

ML-based QSAR approaches

Different QSAR models are available for PBT evaluation, but they have limitations because of the narrow applicability domain and relatively small training set. So, robust models with high accuracy and interpretability are required. Here, ML-based models are viewed as a promising method [52]. Though the broad range of ML-based models has been used in medicinal chemistry for years, the use of these algorithms is limited in toxicity. We have explored the advancement in ML-based models for PBT analysis of POPs. We found that some robust ML-based models help in the toxicity evaluation, such as a mutagenicity model that was tested for the 20,000 molecules with the Ames test [52]. Another classification model is reported for analyzing the biodegradation of chemicals in water, with an accuracy of 85.1%, for predicting the persistence of chemicals [42]. Simultaneously, a model is developed using different ML algorithms such as support vector regression, random forest, and naïve Bayesian with 1579 chemicals and their reported half-life values [53]. Different ML-based algorithms that can be used to forecast POPs toxicity or other biological activities are explained below.

Naïve Bayes: This is an ML-based classifier method created using the Bayes theorem. Bayes theorem is one of the oldest approaches for classification and defines the prior probability distribution of data. In a naïve Bayes classifier, every feature's value is independent of others and calculates the correlation to particular labeled data. This algorithm uses a hybrid modeling approach by rule-based classifications and automatic systems to attain higher calculation accuracy [54].

K-Nearest Neighbor (kNN): kNN is a classification approach for the data distribution for the chemical with no previous knowledge. It is used for discriminant analysis when consistent parametric evaluations of probability densities are unidentified or hard to determine. The chemicals from the closest nodes are transferred to the query using a majority-voting algorithm in kNN, representing the data containing labeled and unlabeled nodes in a high-dimensional feature space. The value k indicates the number of nearest neighbors participating in the voting process in this case. As an extension of chemical similarity search to supervised learning, kNN is also used in ligand-based virtual screening to predict bioactivities from the top search results using chemical similarity metrics [55].

Support vector machines (SVMs): SVMs use nonlinear kernel functions to map data into high-dimensional space and identify an optimum separated hyperplane to solve the classification issue. The hyperplane is a linear combination of data points and is fitted to optimize the margin between support vectors and points closest to the decision boundary [56].

Random forest: Random forest is an ensemble learning technique where many decision trees are constructed based on training data, and predictions for classification or regression are made for new inputs using a majority-voting system similar to kNN. They attained superior interpretability accuracy to SVMs and neural networks [57].

Artificial neural networks (ANN): ANN is a set of ML methods inspired by the function of neurons working in the brain. In medicinal chemistry, the use of ANN started in the early twenties, and now it helps in solving the problems related to toxic chemicals by pattern classification, prediction, and optimization. We can categorize the ANN-based models into three groups (i) static ANN, (ii) dynamic ANN, and (iii) statistical ANN [58]. The multilayered ANN model has an input node and an output node; in between these nodes, many hidden layers are present. The inputs that proceed through each layer are progressively modified and do not resemble the original state. Further, this final representation became more predictive for quantified outcomes [59].

Contemporary QSAR tools for PBT analysis of POPs

Apart from the above-discussed models, some state-of-art tools have currently been utilized that stand relevant for PBT analysis. Certain contemporary QSAR tools have also made recent advancements in the field of computational predictive models.

Statistical approach: Statistical methods make predictions using important experimental data based on a fundamental ML framework [60]. Once the automated hierarchical model is built, it has to undergo validation procedures before performing the quantitative prediction [61]. The descriptors incorporated in the model are then used to generate a prediction that is considered significant only when the test chemical comes under the applicability domain [62]. The analysis of the statistical-based model as per the International Council for Harmonisation guideline M7 recommendation includes a critical evaluation of the model descriptors like chemical properties, reactive properties, and physicochemical features related to toxicity [63]. With this, the chemical space covered by the model, including experimental data covering the analog's chemical structure and overall mode performance, are critical elements of the QSAR model performance review to ensure model transparency.

Expert rule-based approach: This method utilizes literature and experimentally derived structural alerts of chemicals for making predictions about specific mechanisms of toxicity. The molecular substructures which can cause an adverse reaction leading to off-target interaction of the chemical substances are known as structural alerts. Models based upon expert rules describe the detailed toxicity mechanism to justify the presence of structural alert with appropriate references cited from the literature. A positive toxicity prediction is made when a specific alert is found to be associated with the test chemical [64]. Although special consideration is needed for the chemicals with negative toxicity results, most available models do not clarify the specific reason. In this scenario, the chemical may either be equivocal, negative, or out of the applicability domain of the predictive tool. With this, the expert-based tool's predictability can also be increased by providing an alert score, an explanation of negative prediction, a defined mode of mechanism, and the inclusion of experimental data matching the structural alert. The present knowledge has led to identifying the substructures that should be considered while designing any new chemical to avoid inevitable adversities associated with such endpoints.

Read across approach: Read across assessment is executed to report the data gaps involved in a category approach [65]. Toxicological, structural, and biological activity shared between the target and source compound is identified, and categories of chemicals are defined. This is a preferred predictive method, as an experimental result included in the read-across is considered more robust than the QSAR models. Generally, the prediction made by the read-across process is deemed more reliable as a solid scientific argument is required to prove the prediction false (rare case), even if otherwise predicted by the QSAR models.

A complete PBT analysis of every chemical is essential to protect human health and the environment. Therefore, industries are looking for low-cost and robust screening methods to prioritize the chemicals based on their hazard assessment. According to the REACH guideline, different nontesting methods have been used efficiently to bridge the gaps in the risk assessment of POPs. Such QSAR models have evolved considerably in the recent time for PBT analysis of POPs due to robust contribution by the scientific communities involved in the screening and prioritization of safer chemicals. Many of these models are incorporated into tools showing great predictive ability and high sensitivity and specificity. For POPs, PBT Profiler, and QSARINS [45] are some user-friendly tools in agreement with OECD principles. EPA has developed the PBT Profiler (available at http://www.pbtprofiler.net/) through a collaborative effort with the American Chemistry Council and the Society of Chemical Manufacturers and Affiliates. It follows the same principle and methods incorporated into EPISuite [66] and ECOSAR to make its predictions.

On the other hand, QSARINS is built on MLR-based QSAR models that use genetic algorithms for feature selection. It helps in the analysis of cumulative PBT behavior of emerging chemicals, and it also allows the validation of single models. Such models predict the toxicity of these chemicals by incorporating different properties, such as log $K_{ow/oa,}$ persistent time, long rang transportation, bioaccumulation, bioconcentration, and associated toxicity endpoints. Here, we briefly covered a comprehensive overview of different tools and modeling techniques utilized in QSAR development and risk assessment of POPs.

Conclusions

Chemicals with persistence, long-range transportation, and the ability to bio-magnify in ecosystems adversely affect human health and wildlife species. To control the overriding use of POPs, the regulatory agencies highly recommend their complete PBT screening analysis before coming into general use. QSAR models are fundamental computational and low-cost modeling approaches for these substances' rigorous chemical screening and risk assessment process. The development of the QSAR model for POPs can be divided into two broader groups to correspond with: classification methods and regression techniques. Several methods, such as PLS, PCA, ANN, and kNN, are effective at modeling both classification and regression-based analysis. The developed QSAR models, with the help of these statistical and ML-based techniques, have high predictive power in terms of testing the potential persistence and toxicity end points of POPs.

Acknowledgments

All the authors are thankful to the Council of Scientific and Industrial Research (CSIR), New Delhi, India, for the support under Mission Mode Project HCP-31(WP 5.1) and to CSIR, Indian Institute of Toxicology Research, Lucknow, for providing the computational resources. PS is thankful to CSIR, New Delhi, for the CSIR-SRF fellowship. The institutional manuscript number is IITR/SEC/MS/2022/93.

References

[1] Özkara A, Akyıl D, Konuk M. Pesticides, environmental pollution, and health. In: Environmental health risk-hazardous factors to living species. IntechOpen; 2016.
[2] Lipnick RL, Muir DC. History of persistent, bioaccumulative, and toxic chemicals. ACS Publications; 2000.
[3] Boethling R, Fenner K, Howard P, Klečka G, Madsen T, Snape JR, et al. Environmental persistence of organic pollutants: guidance for development and review of POP risk profiles. Integrated Environ Assess Manag: Int J 2009;5(4):539–56.
[4] Russell E. War and nature: fighting humans and insects with chemicals from World War I to Silent Spring. Cambridge University Press; 2001.
[5] Carson R (1962). Silent Spring Houghton Mifflin. Boston, MA, USA.
[6] Casper MJ. Synthetic planet: chemical politics and the hazards of modern life. Psychology Press; 2003.
[7] Stone DP. The changing Arctic environment. Cambridge University Press; 2015.
[8] Barrios P. The Rotterdam convention of hazardous chemicals: a meaningful step toward environmental protection. Geo. Int'l Envtl. L. Rev. 2003;16:679.
[9] VanDorn HM. The Rotterdam convention. Colo. J. Int'l Envtl. L. & Pol'y 1999;10:281.
[10] Stokke OS, Thommessen OB. Stockholm convention persistent organic pollutants (Stockholm convention: POPs). In: Yearbook of international cooperation on environment and development 2001-02. Routledge; 2013. p. 122–3.
[11] Pariatamby A, Kee YL. Persistent organic pollutants management and remediation. Procedia Environ Sci 2016;31:842–8.
[12] Sabljic A. QSAR models for estimating properties of persistent organic pollutants required in evaluation of their environmental fate and risk. Chemosphere 2001;43(3):363–75.
[13] Pandit S, Singh P, Parthasarathi R. Computational risk assessment framework for the hazard analysis of bisphenols and quinone metabolites. J Hazard Mater 2022;426:128031.
[14] Hansch C. Quantitative structure-activity relationships and the unnamed science. Acc Chem Res 1993;26(4):147–53.
[15] Tropsha A. Best practices for QSAR model development, validation, and exploitation. Mol inform 2010;29(6-7):476–88.
[16] Majumdar S, Basak SC. Beware of external validation!-a comparative study of several validation techniques used in QSAR modelling. Curr Comput Aided Drug Des 2018;14(4):284–91.
[17] Roy K, Narayan Das R. A review on principles, theory and practices of 2D-QSAR. Curr Drug Metabol 2014;15(4):346–79.
[18] Hawker DW, Connell DW. Octanol-water partition coefficients of polychlorinated biphenyl congeners. Environ Sci Technol 1988;22(4):382–7.
[19] Baker JR, Mihelcic JR, Sabljic A. Reliable QSAR for estimating Koc for persistent organic pollutants: correlation with molecular connectivity indices. Chemosphere 2001;45(2):213–21.
[20] Gramatica P, Papa E, Sangion A. QSAR modeling of cumulative environmental end-points for the prioritization of hazardous chemicals. Environ Sci J Integr Environ Res: Process Impacts 2018;20(1):38–47.
[21] Dearden JC, Rowe PH. Use of artificial neural networks in the QSAR prediction of physicochemical properties and toxicities for REACH legislation. In: Artificial neural networks. Springer; 2015. p. 65–88.
[22] Gabbert S, Hilber I. Time matters: a stock-pollution approach to authorisation decision-making for PBT/vPvB chemicals under REACH. J Environ Manag 2016;183:236–44.
[23] Merenyi S. Reach: regulation (EC) No 1907/2006. Munich: GRIN Verlag; 2012. Consolidated, version (June 2012) with an Introduction.
[24] Gramatica P, Cassani S, Sangion A. PBT assessment and prioritization by PBT index and consensus modeling: comparison of screening results from structural models. Environ Int 2015;77:25–34.
[25] Király P, Kiss R, Kovács D, Ballaj A, Tóth G. The relevance of goodness-of-fit, robustness and prediction validation categories of OECD-QSAR principles with respect to sample size and model type. Mol Inform 2022:2200072.
[26] Raunio H. In silico toxicology—non-testing methods. Front Pharmacol 2011;2:33.
[27] Chi Y, Zhang H, Huang Q, Lin Y, Ye G, Zhu H, et al. Environmental risk assessment of selected organic chemicals based on TOC test and QSAR estimation models. J Environ Sci 2018;64:23–31.
[28] Gramatica P, Papa E. Screening and ranking of POPs for global half-life: QSAR approaches for prioritization based on molecular structure. Environ Sci Technol 2007;41(8):2833–9.
[29] Junker T, Coors A, Schüürmann G. Compartment-specific screening tools for persistence: potential role and application in the regulatory context. Integrated Environ Assess Manag 2019;15(3):470–81.
[30] Lombardo A, Manganaro A, Arning J, Benfenati E. Development of new QSAR models for water, sediment, and soil half-life. Science of The Total Environment; 2022. p. 156004.
[31] Petoumenou MI, Pizzo F, Cester J, Fernández A, Benfenati E. Comparison between bioconcentration factor (BCF) data provided by industry to the European Chemicals Agency (ECHA) and data derived from QSAR models. Environ Res 2015;142:529–34.
[32] Nendza M, Kühne R, Lombardo A, Strempel S, Schüürmann G. PBT assessment under REACH: screening for low aquatic bioaccumulation with QSAR classifications based on physicochemical properties to replace BCF in vivo testing on fish. Sci Total Environ 2018;616:97–106.
[33] Nabarretti BH, Rigon RB, Burga-Sánchez J, Leonardi GR. A review of alternative methods to the use of animals in safety evaluation of cosmetics. Einstein (São Paulo) 2022;20.
[34] Gruiz K, Meggyes T, Fenyvesi É. Engineering tools for environmental risk Management: 2. Environmental toxicology. CRC Press; 2015.
[35] Hong H, Xie Q, Ge W, Qian F, Fang H, Shi L, et al. Mold2, molecular descriptors from 2D structures for chemoinformatics and toxicoinformatics. J Chem Inf Model 2008;48(7):1337–44.
[36] Zhang Q, Huang J, Yu G. Prediction of soot–water partition coefficients for selected persistent organic pollutants from theoretical molecular descriptors. Prog Nat Sci 2008;18(7):867–72.
[37] Bordás B, Bélai I, Kőmíves T. Theoretical molecular descriptors relevant to the uptake of persistent organic pollutants from soil by zucchini. A QSAR study. J Agric Food Chem 2011;59(7):2863–9.
[38] Yap CW. PaDEL-descriptor: an open source software to calculate molecular descriptors and fingerprints. J Comput Chem 2011;32(7):1466–74.
[39] Mauri A, Consonni V, Pavan M, Todeschini R. Dragon software: an easy approach to molecular descriptor calculations. Match 2006;56(2):237–48.
[40] ChemAxon L. Marvinsketch. Chem Axon Cambridge; 2013.

[41] Shayanfar S, Shayanfar A. Comparison of various methods for validity evaluation of QSAR models. BMC chemistry 2022;16(1):1–9.
[42] Huang K, Zhang H. Classification and regression machine learning models for predicting aerobic ready and inherent biodegradation of organic chemicals in water. Environ Sci Technol 2022;56(17):12755–64.
[43] Mackay D, Hughes DM, Romano ML, Bonnell M. The role of persistence in chemical evaluations. Integrated Environ Assess Manag 2014; 10(4):588–94.
[44] Wold S, Esbensen K, Geladi P. Principal component analysis. Chemometr Intell Lab Syst 1987;2(1–3):37–52.
[45] Gramatica P, Chirico N, Papa E, Cassani S, Kovarich S. QSARINS: a new software for the development, analysis, and validation of QSAR MLR models. Wiley Online Library; 2013.
[46] Pirhadi S, Shiri F, Ghasemi JB. Multivariate statistical analysis methods in QSAR. RSC Adv 2015;5(127):104635–65.
[47] Bu J, Liu W, Pan Z, Ling K. Comparative study of hydrochemical classification based on different hierarchical cluster analysis methods. Int J Environ Res Publ Health 2020;17(24):9515.
[48] Papa E, Gramatica P. Screening of persistent organic pollutants by QSPR classification models: a comparative study. J Mol Graph Model 2008; 27(1):59–65.
[49] Stanton DT. QSAR and QSPR model interpretation using partial least squares (PLS) analysis. Curr Comput Aided Drug Des 2012;8(2):107–27.
[50] Khan PM, Baderna D, Lombardo A, Roy K, Benfenati E. Chemometric modeling to predict air half-life of persistent organic pollutants (POPs). J Hazard Mater 2020;382:121035.
[51] Hites RA. Break point analyses of human or environmental temporal trends of POPs. Sci Total Environ 2019;664:518–21.
[52] Heo S, Safder U, Yoo C. Deep learning driven QSAR model for environmental toxicology: effects of endocrine disrupting chemicals on human health. Environ Pollut 2019;253:29–38.
[53] Lunghini F, Marcou G, Azam P, Enrici MH, Van Miert E, Varnek A. Publicly available QSPR models for environmental media persistence. SAR QSAR Environ Res 2020;31(7):493–510.
[54] Gautam J, Atrey M, Malsa N, Balyan A, Shaw RN, Ghosh A. Twitter data sentiment analysis using naive bayes classifier and generation of heat map for analyzing intensity geographically. In: Advances in applications of data-driven computing. Springer; 2021. p. 129–39.
[55] Manganaro A, Pizzo F, Lombardo A, Pogliaghi A, Benfenati E. Predicting persistence in the sediment compartment with a new automatic software based on the k-Nearest Neighbor (k-NN) algorithm. Chemosphere 2016;144:1624–30.
[56] Kafai M, Eshghi K. CROification: accurate kernel classification with the efficiency of sparse linear SVM. IEEE Trans Pattern Anal Mach Intell 2017;41(1):34–48.
[57] Wang MW, Goodman JM, Allen TE. Machine learning in predictive toxicology: recent applications and future directions for classification models. Chem Res Toxicol 2020;34(2):217–39.
[58] Malekian A, Chitsaz N. Concepts, procedures, and applications of artificial neural network models in streamflow forecasting. In: Advances in streamflow forecasting. Elsevier; 2021. p. 115–47.
[59] Choi RY, Coyner AS, Kalpathy-Cramer J, Chiang MF, Campbell JP. Introduction to machine learning, neural networks, and deep learning. Transl Vis Sci Technol 2020;9(2):14.
[60] Pandit S, Dhawan A, Parthasarathi R. Emerging computational methods for predicting chemically induced mutagenicity. In: Mutagenicity: assays and applications. Elsevier; 2018. p. 161–76.
[61] Myatt G, Beilke L, Cross K. In silico tools and their application. 2017.
[62] Nikolova-Jeliazkova N, Jaworska J. An approach to determining applicability domains for QSAR group contribution models: an analysis of SRC KOWWIN. Alternatives Lab Animals 2005;33(5):461–70.
[63] Amberg A, Beilke L, Bercu J, Bower D, Brigo A, Cross KP, et al. Principles and procedures for implementation of ICH M7 recommended (Q) SAR analyses. Regul Toxicol Pharmacol 2016;77:13–24.
[64] Yang C, Cheeseman M, Rathman J, Mostrag A, Skoulis N, Vitcheva V, et al. A new paradigm in threshold of toxicological concern based on chemoinformatics analysis of a highly curated database enriched with antimicrobials. Food Chem Toxicol 2020;143:111561.
[65] Wu S, Blackburn K, Amburgey J, Jaworska J, Federle T. A framework for using structural, reactivity, metabolic and physicochemical similarity to evaluate the suitability of analogs for SAR-based toxicological assessments. Regul Toxicol Pharmacol 2010;56(1):67 81.
[66] Boethling R, Costanza J. Domain of EPI suite biotransformation models. SAR QSAR Environ Res 2010;21(5–6):415–43.

CHAPTER

8

Machine learning–based QSAR for safety evaluation of environmental chemicals

Yuchen Gao[1,2], *Jiachen Zhang*[1], *Shixuan Cui*[1], *Yiqu Wu*[1], *Meiling Huang*[1] *and Shulin Zhuang*[1,2]

[1]Key Laboratory of Environment Remediation and Ecological Health, Ministry of Education, College of Environmental and Resource Sciences, Zhejiang University, Hangzhou, China [2]Women's Reproductive Health Key Laboratory of Zhejiang Province, Women's Hospital, School of Medicine, Zhejiang University, Hangzhou, China

Abbreviations

3R Replacement, reduction, and refinement
AD Applicability domain
AGBT Algebraic graph–assisted bidirectional transformer
ANN Artificial neural networks
AO Adverse outcome
AOP Adverse outcome pathway
APFP AtomPairs2DFingerprint
CAS Chemical abstract service
CNNs Convolutional neural networks
DL Deep learning
DNNs Deep neural networks
DT Decision trees
ECFP Extended connectivity fingerprint
EPA Environmental Protection Agency
FPs Fingerprints
GAN Generative adversarial networks
GraphFP GraphOnlyFingerprint
k-NN k-nearest neighbors
KEs Key events
LDA Linear discriminant analysis
LSTMs Long short-term memory networks
MACCS Molecular ACCess System
MIE Molecular initiating event
ML Machine learning
MLR Multilinear regression
PCA Principal component analysis
PubChem PubChem Substructure Fingerprint Description
QSAR Quantitative structure-activity relationships
RASAR Read-across structure–activity relationship
ReLeaSE Reinforcement Learning for Structural Evolution
RF Random forests
SSL Self-supervised learning
SVM Support vector machines
VAEs Variational autoencoders

Introduction

Anthropogenic chemical pollution poses a great threat to ecological sustainability and human health [1,2]. The safety management and risk assessment of environmental chemicals throughout their lifecycle are thus of paramount importance [3,4]. Conventional in vivo animal toxicity tests of chemicals are time-consuming, highly expensive, laborious, and significantly restricted to the principles of the Replacement, reduction, and refinement (3R) [5–8]. As advocated by the landmark report Toxicity Testing in the 21st Century: A Vision and a Strategy released by the U.S. National Academy of Sciences in 2007, the way of toxicity testing should be transformed from in vivo whole-animal testing with qualitative end points to a new toxicity testing paradigm consisting of in vitro toxicity pathway assays and in silico approaches [6,9]. As one efficient in silico approach on chemical safety assessment, quantitative structure–activity relationship (QSAR) was established by Corwin Hansch in 1962 and has been extensively employed in initial computer-aided drug screening [10]. Early QSAR models mostly relied on presumed functional relationships with statistical methods such as multiple linear regression, partial least squares, and principal component analysis to fit the relationship between molecular structure descriptors and adverse effects. Such models

QSAR in Safety Evaluation and Risk Assessment
https://doi.org/10.1016/B978-0-443-15339-6.00038-2

have relatively limited model generalization and interpretability. Currently, QSARs have significant roles in the prediction of exposure effect as well as screening and classification of risk environmental chemicals [11].

With the rapid increase of big data of various contaminants, machine learning (ML) has been widely employed in QSAR modeling for screening, property prediction, and category prediction [12]. ML methods have significantly advanced in recent years and are proven to be important alternatives to experimental testing for chemicals and able to estimate chemical toxic effects by establishing relationships between molecular structure and biological activity. ML-based QSAR is envisaged with more accurate predictions and opens new horizons to the chemical safety evaluation with discrimination performance and predictive ability, improving the initial linear regression analysis methods used in deriving QSAR equations in combination with sophisticated ML techniques. The ML-based QSAR models currently play a notable role in chemistry, material sciences, pharmaceutical, biomedicine, and toxicology [13–17]. The research papers on ML-based QSAR have been increasing as illustrated with the publications resulted from the search using keywords "QSAR" and "machine learning" in Web of Science (Fig. 8.1). Currently, the rapidly developed modern ML algorithms improve the performance of QSAR models, which could be used for screening of risk chemicals and facilitating chemical risk assessment. Herein, we review the evolution of ML-based QSAR, integrate different ML-based QSAR modeling technologies, and summarize recent research advances and applications of ML-based QSAR in research fields including toxicity screening and safety evaluation. With the description of current challenges on the application of ML-based QSAR models for chemical risk assessment, we provide future perspectives on refinement of ML-based QSAR models and the extension of their applications.

ML-driven QSAR modeling

The first QSAR model was developed to predict the biodegradability of organic environmental pollutants in an aqueous aerobic environment [18]. Since the 1990s, ML techniques have been widely used to address environmental risks. After decades of the development of molecular descriptors, fingerprints, and algorithms, QSAR modeling has gradually improved from the initial reaction mechanism framework to the classification or regression prediction using ML algorithms based on the structural and physicochemical indexes of chemicals. The process of ML-based QSAR modeling generally consists of the following steps: chemical dataset curation, descriptors generation, training

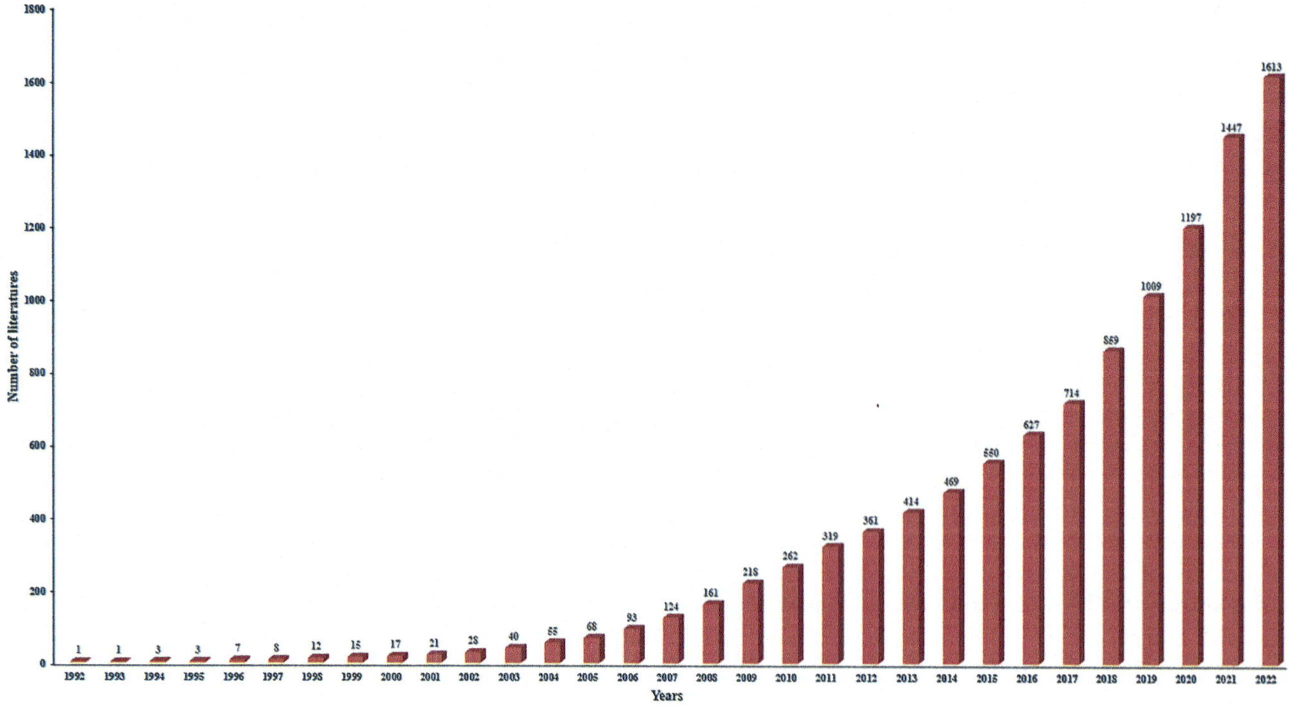

FIGURE 8.1 The statistics of publications with the boolean search of papers in web of Science using keyword "QSAR" and 'machine learning" from 1992 to 2022.

sets and test sets splitting, model building, prediction, and validation (Fig. 8.2). The development history of molecular structure descriptors, molecular fingerprints, and algorithms in QSAR modeling was summarized in this section.

Molecular structure descriptors

The preliminary step for ML model training is the transformation of molecular structures into molecular descriptors. Molecular descriptors are the digital representations of molecular properties, for the conversion of molecular structure information [19]. At present, there are more than 5000 kinds of molecular descriptors that can be generated by various software. Molecular descriptors can be classified in terms of origin (experimental descriptors and empirical descriptors), dimension (0D-4D), physical meaning (molecular property descriptors, composition descriptors, topological descriptors, and geometric descriptors), and numerical types (continuous descriptors and discrete descriptors). Fig. 8.3 illustrates molecular descriptors in different dimensions (0D, 1D, 2D, 3D, and 4D).

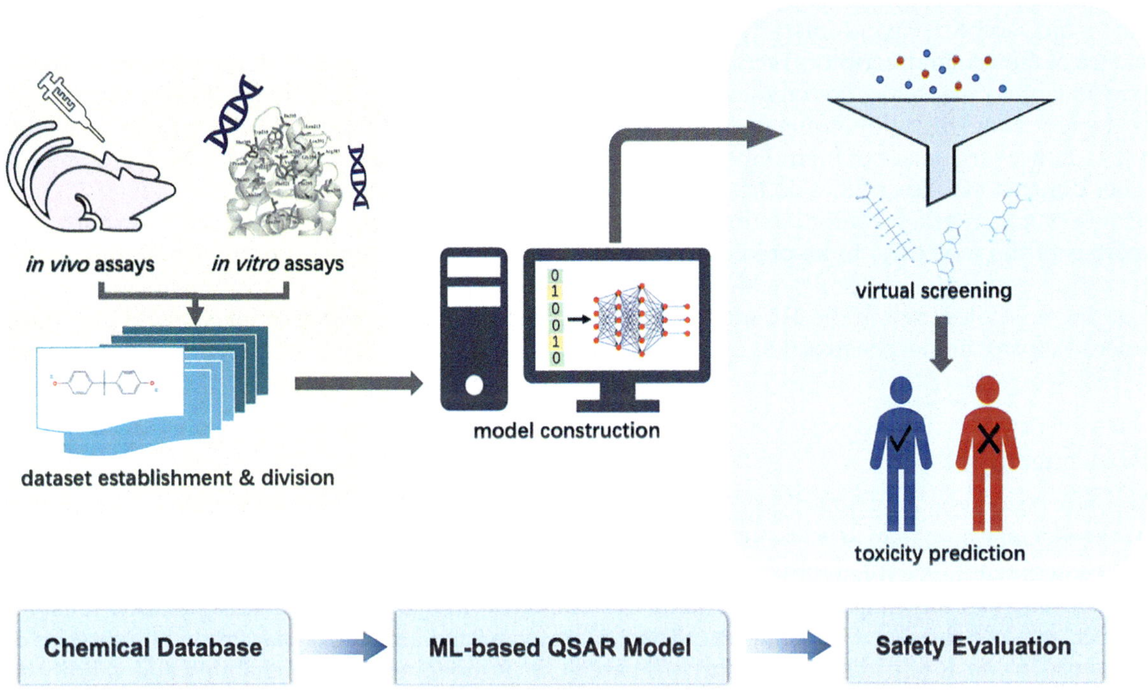

FIGURE 8.2 Workflow of MLs-based QSAR modeling for safety assessment of environmental chemicals.

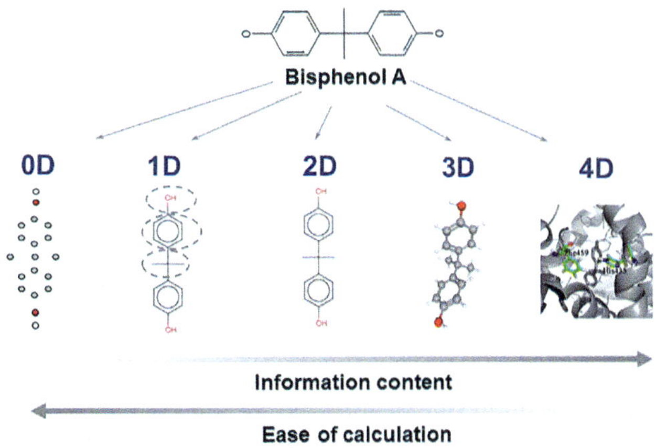

FIGURE 8.3 The representation of five kinds of 0D, 1D, 2D, 3D, and 4D molecular descriptors.

0D molecular descriptors are also called constitutional descriptors and count descriptors. Being constitutional descriptors, 0D descriptors include molecular weight (MW), molecular weight quantum-mechanical parameters, number of certain atoms, number of rotatable bonds, and type of bonds. 1D descriptors are mathematical representations of aggregated information such as numbers of partial functional groups, specific fragment structure, hydrogen bond donor, and acceptor atoms. Conventional 1D descriptors have equal values for isomers and usually demonstrate high degeneration. Both 0D and 1D descriptors are generally used with high-dimensional descriptors or combined with other descriptors to form a vector [20].

2D, 3D, and 4D molecular descriptors are mostly used for ML-based QSAR model construction. The emergence of 2D molecular descriptors promotes the expression of compound topology that defines the connectivity of atoms in a molecule according to the properties of chemical bonds, including topological parameters, molecular profiles, and 2D autocorrelation. The 3D chemical descriptors include 3D autocorrelation descriptors, surface and volume descriptors, substituent constants, quantum chemical descriptors, and specific molecular scaffolds, which have been employed in model construction and become one of the primary approaches to mechanism-based rational safety assessment. The combined descriptors with 2D or 3D descriptors are now used most widely during ML modeling. Higher dimensional (4D) models are developed to solve the conformation/orientation problem by simultaneously considering multiple structural conformations [21,22]. The receptor flexibility, inducible fit, and the effect of solvation on the main receptor–ligand interaction are added to higher dimensional descriptors on original basis [23].

The choice of molecular descriptors is critical since it has a great impact on the model performance. The structural complexity of a molecule cannot be completely explained by the molecular descriptors. This problem can be solved to some degree using high-dimensional descriptors or intergradation of multiple descriptors in a hierarchical manner [24]. However, the use of high-dimensional descriptors may increase the complexity of the problem; therefore, higher dimensional descriptors do not always have superior applicability.

There are several criteria for selection of descriptors to encode molecules [24,25]. (1) The descriptor is related to the properties of the molecules to be predicted; (2) even if the structural differences are small, the descriptor can generate distinct results for molecules with different structures; (3) the selected descriptors should be suitable for the size of the molecules used in the ML modeling; and (4) data quantity and the dimension of descriptors should be considered during model construction.

ML-driven fingerprints

Fingerprints (FPs) are a particular form of molecular descriptors in ML-based QSAR models and can facilitate a fast and effective representation of molecular structures. FPs indicate the existence and absence of substructures and functional groups with a fixed length. Molecules coded in this form are of great efficiency for storing, processing, and data analysis. FPs are considered as a dominant biometric trait due to their acceptability, reliability, high-security level, and low cost [26]. FPs create an efficient representation of the molecular graph with precise definition and rational use in QSAR research. Whether an ML-based QSAR model with high reliability and validity can be obtained largely depends on the quality of the input data, thus, it is of significance to rationally select the descriptors most closely related to the research objective.

The 2D fingerprints can be divided into substructure key-based fingerprints, topological or path-based fingerprints, circular fingerprints, and pharmacophore fingerprints [27]. They are one of the most popular similarity search tools and have been effectively used in a wide range of applications. PubChem Substructure Fingerprint Description (PubChem), extended connectivity fingerprints (ECFP), Molecular ACCess Systems (MACCS), and AtomPairs2D-fingerprint (APFP) are the most frequently used FPs [28–30]. PubChem possesses 881-bit long structural keys and can calculate a list of similar chemical structures for each compound, which are employed to carry out similarity searches and neighboring structures. ECFP and MACCS are a class of topological fingerprints for molecular characterization. APFP is defined in terms of the atomic environment and the shortest path separations between all pairs of atoms in the topological representation of a composite structure. It encodes 780 pairs of atoms at various topological distances. As for the calculation of molecular fingerprints used in ML models, RDKit, BlueDesc, CDK Descriptor, DRAGON, and MODEL are widely used for the representation of chemical structures. These open-source software packages are implemented with a wide variety of algorithms ranging from the canonicalization of chemical structure to molecular descriptor calculations [31].

Molecular structural complexity and high dimensionality are the key factors in designing efficient fingerprints [32,33]. 3D-structure-based fingerprints have been developed to capture 3D patterns of molecules with better performance than conventional fingerprints. An algebraic graph–assisted bidirectional transformer was used to construct molecular representations by combining the advantages of 3D element-specific-weighted colored algebraic graphs and deep bidirectional transformers [34]. The algebraic graph—based fingerprints significantly reduced the molecular structural complexity while retaining indispensable physical/chemical information and physical insight. GraphOnlyFingerprint, which encodes the 1024 path of a fragment in the composite structure, is a specialized version of the molecular fingerprint and is also not affected by the bond orders [35]. Despite much efforts in the past decade, accurate and reliable prediction of numerous molecular properties remains a challenge [36]. The algebraic graph characterized with advanced ML algorithms is effective in simplifying molecular structural complexity for studying various properties of small molecules in environmental science and safety assessment [37].

Algorithms

Classical QSAR models were traditionally built via multilinear regression (MLR) algorithm. Since 1994, ML-based QSAR approaches and their application to the explanation of chemical environmental behavior have been expanding beyond their original frames [38]. ML has been evolved into supervised learning, unsupervised learning, and semi-supervised learning. Supervised ML addresses two typical problems: classification and regression. Algorithms of supervised learning such as k-nearest neighbors (k-NN), linear discriminant analysis (LDA), decision trees (DT), random forests (RF), artificial neural networks (ANN), support vector machines (SVM), naïve Bayes models, and ensemble models are used most frequently for ML-based QSAR modeling [39,40]. Supervised learning can map chemical features to physiochemical properties and toxic effects, via the calculation of the relationship in iterative training. Common unsupervised learning algorithms, including k-means, principal component analysis (PCA), and other typical cluster methods, can play a role in dealing with large numbers of data with unknown labels. Unsupervised learning can identify specific chemical structure information on the basis of chemical structure similarity. Semi-supervised learning is a learning technique between supervised learning and unsupervised learning, using both labeled and unlabeled samples to learn. Semi-supervised learning algorithms are commonly used for QSAR model building with a small number of labeled samples and a large number of unlabeled samples.

The continuous expansion of modeling data and the needs to address image information spawned the advent of deep-learning (DL) networks. As a highly sophisticated subset of ML, DL is based on neural networks with more than two hidden layers and enables multiple independent tasks. The popular DL algorithms include capsule networks, generative adversarial networks (GAN), Variational AutoEncoders (VAEs), long short-term memory networks (LSTMs), transfer learning, deep neural networks (DNNs), and convolutional neural networks (CNNs). DL eliminates some of data preprocessing that typically involved in ML algorithms. DL does not need to extract features but automatically performs high-dimensional abstract learning on data through neural networks, which reduces the composition of feature engineering. The development of DL, together with numerous accessible chemical databases, has improved the QSAR performance [41]. DL algorithms are constantly evolving with the increasing demand for safety evaluation and the complexity of research chemicals, which have been used to construct QSAR models to assess the environmental risks of chemicals since the early 2000s [42–44]. The demand for DL models is constantly increasing, resulting in the creation of numerous frameworks and platforms for the development of custom applications [45]. Due to the introduction of deeper and more complex network model structure, the parameter adjustment becomes more strenuous. It is necessary to analyze the data scale and feature complexity to select the most appropriate algorithm.

ML-driven QSAR applications

High-throughput screening toxicity data-driven QSAR model

High-throughput screening (HTS) assays produced a tremendous amount of assay data for the construction of extensive datasets, which are beneficial to training ML-based QSAR [46]. Many freely accessible databases such as PubChem, ACToR, ChEMBL, and ChemIDplus contain a large amount of HTS bioassay data. ML/DL approaches can be used to build predictive (classification or regression) models based on the data in such comprehensive databases for chemical risk ranking [47,48]. ML-based QSAR models constructed with large datasets generated by HTS assays can improve model generalization ability, which is an issue for traditional QSAR models due to the small training datasets [49].

Read-across coupled with ML-based QSAR

As one of the alternative nontesting approaches, read-across is a widely employed data gap-filling approach to predict the same endpoint for the target substance [50]. As an alternative method to in vivo assessment, read-across is based on the hypothesis that chemicals having similar structures should have similar biological activity such as toxicokinetic and toxicodynamic properties as well as similar chemical properties such as reactivity and metabolic activity. Similarity between two chemicals can be measured by their Euclidean distance in the chemical space that is defined using a set of structural descriptors or the Tanimoto index that is usually calculated using their fingerprints. ML-based chemical read-across tool has been used to estimate the relevance of the selected structural features [51]. A large database makes the automation of the read-across approach feasible and assesses the quality of test data by allowing ML-based QSAR modeling of chemicals and chemical analogs [52]. Read-across software "Read-across v4.1" selects the most similar source compounds for a target chemical and predicts toxicity of the target chemical by taking the weighted average of toxicity of source compounds [51]. The combination of similarity-based read-across and ML-based QSAR (k-NN and RF) resulted in an emerging cheminformatics method, read-across structure–activity relationship (RASAR), which has been used for the prediction of various chemical toxic endpoints including reproductive toxicity, carcinogenicity, acute/chronic aquatic toxicity, etc [53,54]. The data fusion extended the simple RASAR by building similarity-based features for every chemical and created large feature vectors [55]. With the gradual improvement of efficiency, reliability, and interpretability, integrating chemical read-across algorithm into ML-based QSAR model can provide the critical information to strengthen the similarity rationale and has been widely applied for chemical hazard assessment.

Adverse outcome pathway-based modeling

The adverse outcome pathway (AOP) framework is a conceptual knowledge framework for capturing, organizing, evaluating, and visualizing scientific knowledge of relevant toxicological effects, which describe the mechanism connection among a molecular initiating event (MIE), a series of intermediate key events (KEs), and the adverse outcomes (AO). It assumes that the generation of AO in an organism requires an MIE, that is, the interaction of an exogenous compound and a biological macromolecule, and is subsequently triggered with a series of Kes [56]. The AOP framework provides formalized, transparent, and quality-controlled information to improve the development of alternative testing strategies. The application of AOP makes a collection of existing knowledge including the biologically plausible and empirically supported foundations and prediction of endpoints toxicity from mechanistic data [57]. Modeling methodologies include the identification, extraction, and high-throughput data and information to support the inclusion of quantitative considerations in the AOP framework (Fig. 8.4). The quantitative chemical activity can be predicted for MIEs and the understanding of the chemistry behind the MIE can be gained for AOP-based chemical risk assessments.

A toxicologically relevant biological perturbation can be triggered through the determination of the key chemical–biological interaction. The complex biology process can be simulated using an AOP theoretical framework and further achieve high-throughput prediction of contaminant toxicity. Unlike molecular simulations that

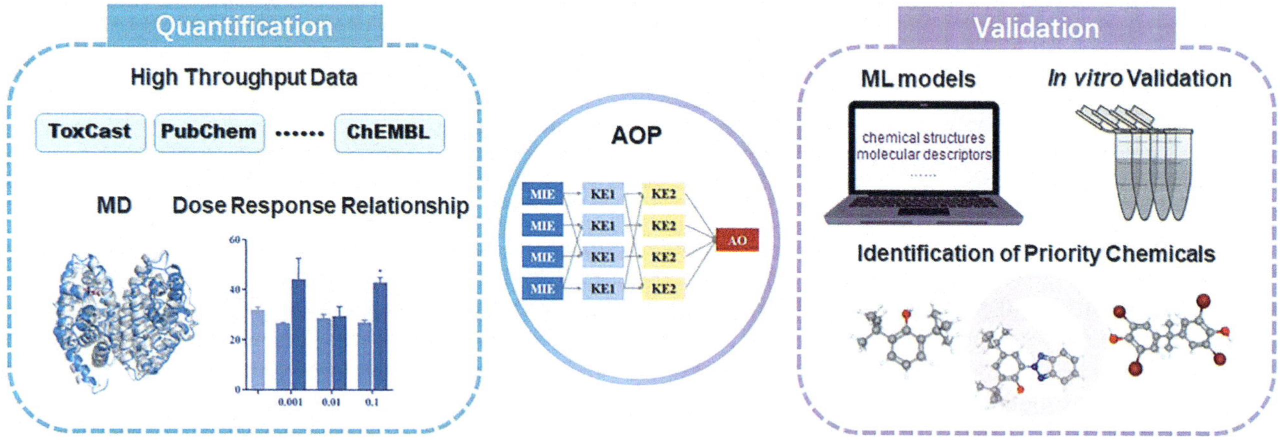

FIGURE 8.4 Validation or quantification of adverse outcome pathway (AOP) framework with the aid of computational models.

target the ligand–receptor competitive binding process, ML modeling can build HTS models based on the AOP framework of the entire biological response process (from molecule to cell to individual), which is useful for the definition of the related chemical space and development of QSAR models [58–60]. Current toxicity databases such as ToxCast and Tox21 involve contaminant–biomolecule interaction information via molecular docking and key toxicity pathway. Therefore, the combination of ML/DL approaches with chemical and biological public data into an explainable AOP context can achieve high accuracy for HTS toxicity predictions [61]. These models tend to provide molecular activity estimation that can be visualized to better understand the MIE and corresponding KEs, and necessary information for hazard classification, labeling, and risk management of chemicals [62].

Multiomics-based modeling

The rapid advances of high-content omics technologies lead to the production of different omics data such as gene expression, microRNA expression, and protein–protein interactions [63]. The development of computational methods and tools for omics data analysis has been proposed, and several algorithms have been successfully applied to the analysis of omics data, providing additional evidence of human health risk assessment. Models constructed by ML techniques can incorporate pathway analysis and random search variable selection into algorithms and simultaneously identify direct and indirect genomic effects on phenotypes. The alterations of DNA modification, aberrant mRNA, and microRNA expression can be utilized as feature characterization and functional map in ML/DL modeling [64].

Calculation tools such as "ATHENA" can identify features associated with adverse outcomes by integrative analysis of gene expression data from RNA sequence analysis, RT-qPCR, immunofluorescence staining, and the signaling pathways [65]. Various unsupervised multiview clustering and supervised classification and regression methodologies have been applied in analysis of this biochemical information, in order to provide clarified information for mechanism explanation of potential toxic effect of compounds [66].

Challenges in QSAR model construction

Data Curation. Data inadequacy includes heterogeneity, use of inappropriate units, and lack of information. It can also include the use of incorrect, misspelled, or insufficiently defined labels like chemical names, CAS numbers, and chemical structures. Imbalance dataset poses a challenge for predictive QSAR modeling because most ML algorithms for classification are designed with the assumption that each class has a similar number of samples. A large number of compounds were screened in *in vitro*/in vivo assays; however, only a small fraction of tested compounds turns out to be active. HTS data are typically imbalanced in general with a small ratio of active compounds to inactive ones [67]. ML-based QSAR models are more sensitive to the misclassification of minority chemicals than majority chemicals.

Model interpretability. For traditional QSAR models, the interpretation of prediction results for an apical endpoint appeared and even still exists currently. The mechanisms are complex, and thus it is more valuable to develop a model with a reliable scientific basis than a model with good performance. The interpretation of modeling results is likely to be restricted to the so-called "black-box" phenomenon of the input data and calculation process [68]. Integration of an ML model with a traditional mechanistic model may retain their respective merits and can be used to solve actual environment exposure scenarios.

Overfitting. Data redundancy and overfitting can be a problem when dealing with small data sets, and a large number of compounds with thousands of descriptors, despite the statistical test and parameter tuning. The existence and complex correlation of thousands of descriptors could lead to redundancy problems [11]. Therefore, feature selection methods have been recommended to select the optimum chemical representation in subsequent modeling rather than using all input descriptors [69]. Various neural networks algorithms, based on feature selection methods and learning algorithms, were devised to avoid these predicaments in chemical risk assessment [41].

Applicability domain. The applicability domain (AD) of a model could be defined as "the response and chemical structure space in which the model makes predictions with a given reliability" [70]. Chemical must be reasonably similar to some training-set chemicals for valid prediction. However, some QSAR research does not provide sufficient information (e.g., training set structures and descriptor values) for the assessment of model AD. The appearance of ML calculation, parameter optimization, and model validation impact model performance. QSAR modeling should follow the well-established statistical practices and "best practice" rules for standardization of data processing, model interpretation to construct rigorously validated and externally predictive ML-based QSAR models. There is no specific algorithm well-matched for all toxicity datasets. Given one specific dataset, model accuracy is affected by the feature selection method and model development technique. Models built with too many or too few descriptors are inappropriate, and both finding the optimal feature subset and selecting appropriate algorithms are of great importance for the building of the optimization model [69].

Limited co-exposure prediction models. QSAR can be a good alternative to toxicity assessment of mixtures, which can be predicted based on various structural features of mixture components. Recently, multiple groups have applied QSAR to mixture toxicity assessment and successfully proposed numerous mathematical models for different environmental end points. ML-based QSAR modeling of co-exposure to chemicals is gradually becoming a research hotspot. Animal experiments usually focus on the description of the phenotypic response or apparent phenomena, which is restricted to reveal underlying mechanisms and predict the quantitative impact of a chemical that is not among the original components or factors. It seems that mechanism-based in silico models might be the only promising tools to estimate the joint toxic effects of mixtures with arbitrary composition. Current explanatory or prediction models/schemes for joint toxic effects are coarse and predicted toxic thresholds can only be adopted in a conservative manner over a large safety factor. If reliance on so-called safety factors is to be diminished and models based on transparent mechanisms are to be emphasized in future toxicology, novel models with associated in silico objects have to be developed, which shall allow reasonable characterization of the dynamic interaction network of multiple queried chemicals [11,66].

Perspectives

Machine learning–based QSAR is developed rapidly in the aspects of algorithm development and model application. An enhancement in the access to data and ML techniques has contributed to the use of in silico QSAR model establishment methods for risk evaluation of environmental chemicals. The molecular structure descriptors covering the chemical–biological space are helpful to improve the predictive and mechanistic properties of QSAR models. Future molecular descriptors will focus on characterizing the structure of compounds and ligands more accurately and expanding the scope of application. ML-based QSAR is used for qualitative screening and quantitative prediction of molecular events, significantly improving its mechanism transparency and prediction effect. The perspective development of ML-based QSAR will ultimately achieve higher efficiency and lower cost in the fields of safety assessment of chemicals.

From single theory to multi-disciplinary implementation. Nowadays, ML QSAR model has become an advanced computational tool in the scientific research field. Single-disciplinary and interdisciplinary and community-driven developments have made QSAR one of the commonly used methods for modeling the physical and biological properties and predicting the potential environmental risk of chemicals. Unprecedented convergence and communication have been witnessed in the field of QSAR modeling based on ML [69]. The highest level of improvement can be achieved through combination of computational and experimental achievements, which each has an important contribution of their respective fields, meanwhile striving to understand and develop working knowledge of complementary fields.

From private to public. The development of ML-based QSAR models is moving toward data and source code sharing. Many databases and model construction techniques are available for download and use. Some software tools facilitate grouping and category building and to assist in the development of predictive toxicology, supporting the use of ML-based QSAR models in different regulatory frameworks [11]. Currently more and more online platforms and software have been developed supporting by ML/DL algorithms that provide easy access to the comprehensive, accurate, and efficient prediction of environmental risk for chemicals. The open-source development of ML-based QSAR software is conducive to accelerating the popularization of computing technology applications in the research and industry fields, so to improve the efficiency of development.

From minority to majority. Although ML approaches have shown promising applications in chemical risk evaluation and other research fields, many of the cited applications are still relatively new. Now some prediction models have been used in industry or governmental agencies to support public health decision-making [69]. Government regulatory agencies of many countries have supported the development and public release of various QSAR and decision-support tools for ecological and health hazard assessments. There is a need for further education to spread relevant knowledge across multiple fields, so that it can be applied more widely and sustainably. Developing a common agenda and approach in QSAR regulation is one of the most important goals of the scientific community.

The modeling framework of QSAR with advanced ML algorithms has envisioned an attractive paradigm for future toxicity testing and risk assessment. With previously described in silico models, general rules that are transferable among similar cases at the same spatial level can be modeled by well-understood mathematics or logic rather than obscurely descriptive paragraphs. New methodologies and applications are to be standardized, validated, and then eventually adopted by the industry and regulatory agencies. Although challenges remain for computational toxicology, the endeavor to overcome these challenges will influence continuous innovation and prosperous development for the field of chemical risk assessment. ML-based QSAR models provide a new paradigm for toxicological research to replace time-consuming and laborious experimental approaches and decipher intrinsic risk mechanisms from in vitro or in vivo experiments, which will be increasingly applied in environmental chemicals safety evaluation.

Acknowledgments

The authors appreciated the financial support from National Key Research and Development Program of China (SQ2022YFC3700187), the National Natural Science Foundation of China (No. 22276166, 22136001), and the Zhejiang Provincial Natural Science Foundation of China (No. LZ22B070001).

References

[1] Johnson AC, Jin XW, Nakada N, Sumpter JP. Learning from the past and considering the future of chemicals in the environment. Science 2020; 367(6476):384–7.
[2] Rappaport SM, Smith MT. Environment and disease risks. Science 2010;330(6003):460–1.
[3] Schwarzman MR, Wilson MP. New science for chemicals policy. Science 2009;326(5956):1065–6.
[4] Wang ZY, Walker GW, Muir DCG, Nagatani-Yoshida K. Toward a global understanding of chemical pollution: a first comprehensive analysis of national and regional chemical inventories. Environ Sci Technol 2020;54(5):2575–84.
[5] Flecknell PA. Replacement, reduction and refinement. ALTEX 2002;19:47–53.
[6] Hartung T. Toxicology for the twenty-first century. Nature 2009;460(7252):208–12.
[7] Shin HM, Ernstoff A, Arnot JA, Wetmore BA, Csiszar SA, Fantke P, et al. Risk-based high-throughput chemical screening and prioritization using exposure models and in vitro bioactivity assays. Environ Sci Technol 2015;49(11):6760–71.
[8] Judson R, Richard A, Dix DJ, Houck K, Martin M, Kavlock R, et al. The toxicity data landscape for environmental chemicals. Environ Health Perspect 2009;117(5):685–95.
[9] Collins FS, Gray GM, Bucher JR. Toxicology - transforming environmental health protection. Science 2008;319(5865):906–7.
[10] Hansch C, Maloney PP, Fujita T. Correlation of biological activity of phenoxyacetic acids with hammett substituent constants and partition coefficients. Nature 1962;194(4824):178–80.
[11] Muratov EN, Bajorath J, Sheridan RP, Tetko IV, Filimonov D, Poroikov V, et al. Qsar without borders. Chem Soc Rev 2020;49(11):3525–64.
[12] Miller TH, Gallidabino MD, MacRae JI, Hogstrand C, Bury NR, Barron LP, et al. Machine learning for environmental toxicology: a call for integration and innovation. Environ Sci Technol 2018;52(22):12953–5.
[13] Kitano H. Computational systems biology. Nature 2002;420(6912):206–10.
[14] Hirschberg J, Manning CD. Advances in natural language processing. Science 2015;349(6245):261–6.
[15] Esteva A, Kuprel B, Novoa RA, Ko J, Swetter SM, Blau HM, et al. Dermatologist-level classification of skin cancer with deep neural networks. Nature 2017;542(7639):686.
[16] Mustapha IB, Saeed F. Bioactive molecule prediction using extreme gradient boosting. Molecules 2016;21(8):983.
[17] Kermany DS, Goldbaum M, Cai WJ, Valentim CCS, Liang HY, Baxter SL, et al. Identifying medical diagnoses and treatable diseases by image-based deep learning. Cell 2018;172(5):1122–31.
[18] Kompare B. Estimating environmental pollution by xenobiotic chemicals using QSAR (QSBR) models based on artificial intelligence. Water Sci Technol 1998;37(8):9–18.
[19] Klekota J, Roth FP. Chemical substructures that enrich for biological activity. Bioinformatics 2008;24(21):2518–25.
[20] Sheffield TY, Judson RS. Ensemble qsar modeling to predict multispecies fish toxicity lethal concentrations and points of departure. Environ Sci Technol 2019;53(21):12793–802.
[21] Hopfinger AJ, Wang S, Tokarski JS, Jin BQ, Albuquerque M, Madhav PJ, et al. Construction of 3D-QSAR models using the 4D-QSAR analysis formalism. J Am Chem Soc 1997;119(43):10509–24.
[22] Andrade CH, Pasqualoto KFM, Ferreira EI, Hopfinger AJ. 4D-QSAR: perspectives in drug design. Molecules 2010;15(5):3281–94.
[23] Mao JS, Akhtar J, Zhang X, Sun L, Guan SH, Li XY, et al. Comprehensive strategies of machine-learning-based quantitative structure-activity relationship models. iScience 2021;24(9):103052.
[24] Winter R, Montanari F, Noe F, Clevert DA. Learning continuous and data-driven molecular descriptors by translating equivalent chemical representations. Chem Sci 2019;10(6):1692–701.
[25] Karelson M, Lobanov VS, Katritzky AR. Quantum-chemical descriptors in QSAR/QSPR studies. Chem Rev 1996;96(3):1027–43.
[26] Sanches-Neto FO, Dias-Silva JR, Queiroz LHK, Carvalho-Silva VH. "pySiRC": machine learning combined with molecular fingerprints to predict the reaction rate constant of the radical-based oxidation processes of aqueous organic contaminants. Environ Sci Technol 2021; 55(18):12437–48.
[27] Gao KF, Nguyen DD, Sresht V, Mathiowetz AM, Tu MH, Wei GW. Are 2D fingerprints still valuable for drug discovery? Phys Chem Chem Phys 2020;22(16):8373–90.

[28] Mahmoud RS, Yousef AH. Using molecular fingerprints as descriptors in toxicity prediction: a survey. Ieee Int C Bioinform 2019:2649–54.
[29] Liu M, Zhang L, Li SM, Yang TZ, Liu LL, Zhao J, et al. Prediction of herg potassium channel blockage using ensemble learning methods and molecular fingerprints. Toxicol Lett 2020;332:88–96.
[30] Hemmerich J, Troger F, Fuezi B, Ecker GF. Using machine learning methods and structural alerts for prediction of mitochondrial toxicity. Mol Inf 2020;39(5):2000005.
[31] Ayed M, Lim H, Xie L. Biological representation of chemicals using latent target interaction profile. BMC Bioinf 2019;20(674):1–10.
[32] Verma J, Khedkar VM, Coutinho EC. 3D-QSAR in drug design - a review. Curr Top Med Chem 2010;10(1):95–115.
[33] Nguyen DD, Gao KF, Wang ML, Wei GW. Mathdl: mathematical deep learning for d3r grand challenge 4. J Comput Aided Mol Des 2020;34(2): 131–47.
[34] Garcia-Hernandez C, Fernandez A, Serratosa F. Ligand-based virtual screening using graph edit distance as molecular similarity measure. J Chem Inf Model 2019;59(4):1410–21.
[35] Smieja M, Warszycki D. Average information content maximization-A new approach for fingerprint hybridization and reduction. PLoS One 2016;11(1):1–16.
[36] Williams C. Reverse fingerprinting, similarity searching by group fusion and fingerprint bit importance. Mol Divers 2006;10(3):311–32.
[37] Chen D, Gao KF, Nguyen DD, Chen X, Jiang Y, Wei GW, et al. Algebraic graph-assisted bidirectional transformers for molecular property prediction. Nat Commun 2021;12(1):1–9.
[38] Sternberg MJE, King RD, Lewis RA, Muggleton S. Application of machine learning to structural molecular-biology. Philos Trans R Soc Lond Ser B Biol Sci 1994;344(1310):365–71.
[39] Khlebnikov AI, Schepetkin IA, Quinn MT. Computational structure-activity relationship analysis of small-molecule agonists for human formyl peptide receptors. Eur J Med Chem 2010;45(11):5406–19.
[40] Tropsha A. Best practices for qsar model development, validation, and exploitation. Mol Inf 2010;29(6–7):476–88.
[41] Ghasemi F, Mehridehnavi A, Perez-Garrido A, Perez-Sanchez H. Neural network and deep-learning algorithms used in QSAR studies: merits and drawbacks. Drug Discov Today 2018;23(10):1784–90.
[42] Zhao J, Hou XY, Pan MQ, Zhang H. Attention-based generative adversarial network in medical imaging: a narrative review. Comput Biol Med 2022;149(105948):1–12.
[43] Chen YZ, Yang XH, Wei ZH, Heidari AA, Zheng NG, Li ZC, et al. Generative adversarial networks in medical image augmentation: a review. Comput Biol Med 2022;58(101552):1–20.
[44] Gohlke H, Hendlich M, Klebe G. Knowledge-based scoring function to predict protein-ligand interactions. J Mol Biol 2000;295(2):337–56.
[45] Hu SS, Chen P, Gu PY, Wang B. A deep learning-based chemical system for QSAR prediction. IEEE J Biomed Health Inform 2020;24(10): 3020–8.
[46] Liu X, Lu DW, Zhang AQ, Liu Q, Jiang GB. Data-driven machine learning in environmental pollution: gains and problems. Environ Sci Technol 2022;56(4):2124–33.
[47] Zhao FR, Li L, Chen Y, Huang YC, Keerthisinghe TP, Chow A, et al. Risk-based chemical ranking and generating a prioritized human exposome database. Environ Health Perspect 2021;129(4):1–12.
[48] Cheng WX, Ng CA. Using machine learning to classify bioactivity for 3486 per- and polyfluoroalkyl substances (PFASs) from the oecd list. Environ Sci Technol 2019;53(23):13970–80.
[49] Zhu H, Zhang J, Kim MT, Boison A, Sedykh A, Moran K. Big data in chemical toxicity research: the use of high-throughput screening assays to identify potential toxicants. Chem Res Toxicol 2014;27(10):1643–51.
[50] van Leeuwen K, Schultz TW, Henry T, Diderich B, Veith GD. Using chemical categories to fill data gaps in hazard assessment (vol 20, pg 207, 2009). SAR QSAR Environ Res 2009;20(5–6):591.
[51] Chatterjee M, Roy K. Chemical similarity and machine learning-based approaches for the prediction of aquatic toxicity of binary and multicomponent pharmaceutical and pesticide mixtures against aliivibrio fischeri. Chemosphere 2022;308(136463):1–10.
[52] Hartung T. Making big sense from big data in toxicology by read-across. ALTEX 2016;33(2):83–93.
[53] Luechtefeld T, Marsh D, Rowlands C, Hartung T. Machine learning of toxicological big data enables read-across structure activity relationships (RASAR) outperforming animal test reproducibility. Toxicol Sci 2018;165(1):198–212.
[54] Helma C, Vorgrimmler D, Gebele D, Gutlein M, Engeli B, Zarn J, et al. Modeling chronic toxicity: a comparison of experimental variability with (Q)SAR/read-across predictions. Front Pharmacol 2018;9:413.
[55] Banerjee A, Roy K. First report of Q-RASAR modeling toward an approach of easy interpretability and efficient transferability. Mol Divers 2022;26(5):2847–62.
[56] Ankley GT, Bennett RS, Erickson RJ, Hoff DJ, Hornung MW, Johnson RD, et al. Adverse outcome pathways: a conceptual framework to support ecotoxicology research and risk assessment. Environ Toxicol Chem 2010;29(3):730–41.
[57] Tan H, Wu J, Zhang R, Zhang C, Li W, Chen Q, et al. Development, validation, and application of a human reproductive toxicity prediction model based on adverse outcome pathway. Environ Sci Technol 2022;56(17):12391–403.
[58] Allen TEH, Goodman JM, Gutsell S, Russell PJ. Defining molecular initiating events in the adverse outcome pathway framework for risk assessment. Chem Res Toxicol 2014;27(12):2100–12.
[59] Allen TEH, Liggi S, Goodman JM, Gutsell S, Russell PJ. Using molecular initiating events to generate 2D structure-activity relationships for toxicity screening. Chem Res Toxicol 2016;29(10):1611–27.
[60] Wittwehr C, Aladjov H, Ankley G, Byrne HJ, de Knecht J, Heinzle E, et al. How adverse outcome pathways can aid the development and use of computational prediction models for regulatory toxicology. Toxicol Sci 2017;155(2):326–36.
[61] Jeong J, Garcia-Reyero N, Burgoon L, Perkins E, Park T, Kim C, et al. Development of adverse outcome pathway for PPAR gamma antagonism leading to pulmonary fibrosis and chemical selection for its validation: toxcast database and a deep learning artificial neural network model-based approach. Chem Res Toxicol 2019;32(6):1212–22.
[62] Ciallella HL, Russo DP, Aleksunes LM, Grimm FA, Zhu H. Revealing adverse outcome pathways from public high-throughput screening data to evaluate new toxicants by a knowledge-based deep neural network approach. Environ Sci Technol 2021;55(15):10875–87.
[63] Sato K, Akiyama M, Sakakibara Y. RNA secondary structure prediction using deep learning with thermodynamic integration. Nat Commun 2021;12(1):1–9.

[64] Barel G, Herwig R. Network and pathway analysis of toxicogenomics data. Front Genet 2018;9:35–54.
[65] Kim D, Li R, Dudek SM, Ritchie MD. Athena: identifying interactions between different levels of genomic data associated with cancer clinical outcomes using grammatical evolution neural network. BioData Min 2013;6(23):1–14.
[66] Jeong J, Choi J. Artificial intelligence-based toxicity prediction of environmental chemicals: future directions for chemical management applications. Environ Sci Technol 2022;56(12):7532–43.
[67] Li QL, Wang YL, Bryant SH. A novel method for mining highly imbalanced high-throughput screening data in pubchem. Bioinformatics 2009; 25(24):3310–6.
[68] Zhong SF, Zhang K, Bagheri M, Burken JG, Gu A, Li BK, et al. Machine learning: new ideas and tools in environmental science and engineering. Environ Sci Technol 2021;55(19):12741–54.
[69] Cherkasov A, Muratov EN, Fourches D, Varnek A, Baskin II, Cronin M, et al. QSAR modeling: where have you been? Where are you going to? J Med Chem 2014;57(12):4977–5010.
[70] Netzeva TI, Worth AP, Aldenberg T, Benigni R, Cronin MTD, Gramatica P, et al. Current status of methods for defining the applicability domain of (quantitative) structure-activity relationships - the report and recommendations of ecvam workshop 52. ATLA 2005;33(2):155–73.

CHAPTER

9

Advances in QSAR through artificial intelligence and machine learning methods

Chandrabose Selvaraj[1], *Elango Elakkiya*[2], *Paulraj Prabhu*[2], *Devadasan Velmurugan*[3] *and Sanjeev Kumar Singh*[4]

[1]Center for Transdisciplinary research, Department of Pharmacology, Saveetha Dental College and Hospital, Saveetha Institute of Medical and Technical Sciences (SIMATS), Saveetha University, Chennai, Tamil Nadu, India [2]Department of Computer Applications, Alagappa University, Karaikudi, Tamil Nadu, India [3]Department of Biotechnology, School of Bioengineering, SRM Institute of Science and Technology, Kattankulathur, Tamil Nadu, India [4]Department of Data Sciences, Centre of Biomedical Research, SGPGIMS Campus, Lucknow, Uttar Pradesh, India

Abbreviations

AD Applicability domain
AI Artificial intelligence
ANN Artificial neural network
CV Cross-validation
DNNs Deep neural networks
EPA Environmental Protection Agency
GA Genetic algorithm
GAN Generative adversarial network
GB Gradient boosting machines
GDB Gradient-boosted decision trees
ID3 Iterative Dichotomiser 3
KNN k-nearest neighbors
LN Linear notation
MD Molecular dynamics
ML Machine learning
MMP Matching molecular pair
NN Neural network
PCR Polymerization chain reaction
QSAR/QSPR Quantitative structure–activity/property relationship
RBFNNs Radial basis function neural networks
RBM Restricted Boltzmann Machine
RF Random forest
SOMs Self-organizing maps
SVM Support vector machine

Introduction

Quantitative structure–activity/property relationship (QSAR/QSPR) models make use of chemometric and cheminformatic methods in order to link the molecular structure information of chemicals, pharmaceuticals, toxicants, and environmental pollutants with their biological activity (either therapeutic or detrimental) or other properties of the substances. In particular, QSAR has been helpful for medicinal chemists and physicists in understanding how changes to the molecular structure at specific substituent sites can affect the reported biological activity and physicochemical features of a substance [1]. Since QSAR is capable of accurately and rapidly analyzing massive datasets, it is being considered as a viable tool in the process of developing new drugs. QSAR modeling aims to study the mathematical correlations between a target activity (or property) and a collection of structural and physicochemical factors. These variables are typically referred to as descriptors. Because of its impressive predictive ability, QSAR has garnered a lot of attention, and much effort have been put into developing models that have excellent prediction performance. In the absence of experimental evidence, the QSAR methodology has become an increasingly popular method for the risk assessment of chemical components [2]. The Environmental Protection Agency (EPA) in the United States takes this strategy, and the European Union's REACH legislation has strongly

https://doi.org/10.1016/B978-0-443-15339-6.00033-3

recommended it. The design of drugs, the modeling of ADME and T, the design of food chemicals and agrochemicals, the design of nanomaterials, and risk assessment are only few of the core essential sectors that can benefit from these models. Predictions that are obtained from models have the potential to be utilized in the pursuit of regulatory goals in the chemical business as well as in the creation of new analogs in the pharmaceutical, food, and agricultural industries. However, in order for QSAR models to be useful, they need to be informative and easy to interpret. This means that the model should be able to show the underlying molecular features that lead to an increase or decrease in biological activity [3].

In traditional QSAR, there is a set of predictor variables (X) that are linked to the response variable (Y). In addition, QSAR makes use of molecular descriptors including theoretical (graph) and physicochemical descriptors to provide a more complete picture. Finding better software to compute both mathematical and chemical descriptors is important to elucidate the relationships between these descriptors and their associated activities and structural characteristics (chemical groups). However, finding and creating a QSAR with the right descriptors is more difficult when dealing with a big dataset of bioassays [4]. Recent improvements in fragment-based descriptors are being used to address these challenges; in this way, the fragments directly represent the chemical structure to be created easily. Therefore, it is more important than ever to implement a new practice in the QSAR method by computing a set of descriptors for training and test set compounds with suitable algorithms and statistical methods for descriptors, as this makes it much easier to select the right fragments from large datasets, which in turn makes the situation worse. QSAR models numerically correlate the chemical structures of molecules with their biological activity, allowing, via the mathematical system, to find both biological and physicochemical properties of the novel molecule from the information of its structure [5].

QSAR models are heavily incorporated into computer and statistical methods to find the theoretical prediction of biological activity of molecules, paving the way for the theoretical design of potential future new drugs. QSAR is the most efficient method for discovering novel compounds when experimental data and facilities are unavailable because it focuses on chemical and structural features and biological activity to forecast new lead molecules at a considerably low cost and quickly. Since the QSAR methodology draws from various fields, it incorporates data from disciplines like organic chemistry and pharmacology. By generating equations that have demonstrated a high chance of pharmacological success, QSAR is able to reward this circumstance by guiding the directed design of ligands that do not yet exist [6]. Based on the equations and available data, it permits a prediction of the biological activity. Multiple linear regressions are used to analyze data gathered from the literature or the lab, with the values of biological activity of ligands serving as the dependent variable and the derived descriptors serving as the independent variables. Statistically demonstrating the connection between biological activity and chemical structures is the goal of the empirical method known as QSAR, which makes use of regression and classification on a large set of structure–property data. This model can be trained and applied for high efficiency, and it can be used to predict novel chemical features and their biological qualities based on the chemical features without having to resort to molecular time [7]. As a result, QSAR has gained popularity and is now used in a wide range of fields such as drug design and toxicity prediction.

Roadmap of QSAR method

Multiple laboratories collaborated over the past decades to develop QSAR modeling as a ligand-based drug design strategy. In 1863, Cros made the first step toward developing QSAR when he noticed that water solubility inversely correlated with toxicity. The lower water solubility of alcohols was directly correlated with an increase in the toxicity of alcohols toward mammals [8]. A little time later, in 1868, Crum-Brown and Fraser observed that different chemical substituents had different physiological effects [9]. Hans Horst Meyer showed in the later 1890s that the toxicity of organic substances was inversely proportional to their lipophilicity [10]. After then, research was conducted on the linear relationship between lipophilicity (which may be determined, for example, by oil–water partition coefficients) and biological features. Louis Hammett did research on the electrical properties, equilibrium constants, and reactivity of organic acids and bases [11]. These preliminary investigations laid the groundwork for later QSAR research by showing that molecular structures had a direct impact on the outcome (i.e., biological activity and chemical property) [12]. While studying the dependence of plant growth regulators and pesticides on Hammett constants and hydrophobicity, Hansch formally created the term QSAR in 1962 and laid the fundamental foundations for the field [13]. When it comes to quantifying SAR, the Free-Wilson model provides a straightforward and effective approach. As a molecular descriptor, the presence or absence of substituents or functional groups is used to account for the differences among a set of structurally similar compounds. In contrast to Hansch analysis which correlates physicochemical attributes with biological activity values, this numerical method establishes direct relationships between structural features and biological properties. Still, from a theoretical and practical point of view, these two ways of thinking are very similar [14]. Both models were often used together to make a mixed approach. This mixed approach uses parameters of the Free-Wilson type to describe the activity contributions of certain structural changes and parameters of the physicochemical type to describe the effects, substituents have on the biological activity. This integrated model gives exceptional performance for traditional QSAR, as evidenced by numerous successful applications, especially the work of Hansch and his group on the SAR of enzyme inhibitors. Different iterations of the Free-Wilson method have lately been put to good utilization in fragment-based drug design [15].

The applications of QSAR methods are enormous, but main few are listed below.

I. QSAR efficiently reduces the time and cost.
II. QSAR predicts the biological, physical, chemical, and pharmaceutical activities/properties rationally.
III. QSAR provides collective information from big data.
IV. QSAR provides molecular insights of mechanism of action for biological activity.

QSAR methods based on molecular descriptors

QSAR relies heavily on molecular descriptors, which are used to extract quantitative features from chemical structures that are rich in information. Quantitative molecular descriptors like field, molecular shape, and molecular descriptors and qualitative molecular descriptors like fingerprints MACCS keys, MDL, and daylight descriptors have emerged over the past few years [16]. Molecular descriptors can be broken down into a number of different categories depending on the type of data being described, such as Boolean, vector, tensor, real, scalar field, and vector field, or on the underlying physical features, such as composition, topology, geometry, and molecular properties. Data mining, compound activity prediction, and compound diversity analysis are just some of the many applications that make use of molecular descriptors, which can have any number of dimensions from one to six (1D-6D) depending on the structure [17].

Artificial intelligence in drug screening

Algorithms based on artificial intelligence (AI) have attracted a lot of attention recently due to their efficiency in both time and money, making them a highly sought-after technology. When the two stages of a deep neural nets (DNNs) data (generative and protective) are trained simultaneously, bias can be given to the output, where it is either rewarded or penalized for a specific property, allowing for reinforcement learning. The matching molecular pair (MMP) algorithm, which is connected to a single charge in the unique lead molecule and affects the compound's bioactivities, has also been widely used in QSAR [18]. Along with MMP, other machine learning (ML) methods like DNN, random forest (RF), and gradient boosting machines (GBM) show some changes. It has been seen that DNN can make better predictions than RF and GBM. Large databases like PubChem and ChEMBL contain information about millions of compounds, such as their structures and targets [19]. With MMP and ML, the bioactivity of a drug can be predicted, including its intrinsic clearance, ADMET, oral exposure, and method of action. In the process of finding new drugs, the toxicity optimization of a compound takes a lot of time and money. However, it is a crucial step that adds more value to the development process. The most important part of the feature is getting more information out of an input database through mapping, which is also the most useful part of the selection [20]. It has also been involved in preparing network input variables based on the input database. Later, when the input has unrelated and redundant information, it is important to choose the right technique to reduce the risk of overfitting.

Three feature selection categories are employed in QSAR studying: (i) filter, (ii) wrapper, and (iii) hybrid of both.

In the filter method, data with fewer dimensions are only used for processing the data itself. Features are chosen based on their scores and features with low scores are ignored. The filter approach is further classified into two categories: (i) feature extraction, which is used to predict descriptors from huge datasets, and (ii) feature extraction, which is used to manage input data from high-level dimensions. Filter techniques used in the drug discovery process include fisher scoring, F-test, and Kolmogorov–Smirnov statistics [21]. The main advantages of this filter approach are the enhanced speed and reduced processing time, whereas the main downside is the absence of interaction in categorization. The wrapper strategy, which is more expensive and time-consuming, is another important way for estimating the ideal subset of features based on issues in classifier algorithm reduction. Depending on the algorithm's strategy, the wrapper approach is divided into deterministic and randomized strategies. Deterministic techniques create the same output dependent on this unmodified input, while randomized approaches produce a different conclusion based on their input [22]. For the deterministic strategy, the wrapper method uses algorithms like k-nearest neighbors (KNN), support vector machine (SVM), and step-wise regression. For the random strategy, genetic algorithm (GA), self-organizing maps (SOM), and RF method are used. Another method to forecast the optimum descriptors and the power of subsequent models for the model interpretations is the maximization algorithm using a sparse. In the hybrid approach, it was hard to understand the complex data, so a merge filter and wrapper methods were used to keep the benefits. McCulloch and Pitts were the first to use neural algorithms in Chemoinformatics in 1940. At first, it had two major problems: it was slow to converge and the solutions it came up with during training were hard to predict [23].

Neural network (NN) uses both supervised and unsupervised learning algorithms, which makes it the most effective and versatile method for finding new drugs. All learning neural algorithms are based on the feed-forward NN and a single hidden layer with a number of neurons. They also have a minimum number of layers: input layer, hidden layer, and output layer. When molecular descriptors are used in the input neurons to predict biological activity, they can also be used to predict the hidden values or output values. This is because the hidden values and output values are dependent on the biological activity [24]. In the field of computational chemistry, NN-based methods have been utilized. Backpropagation, often known as back-propagation (BP), was the first approach to artificial neural network (ANN) learning that was presented by Aoyama and colleagues. It was designed to forecast biases, weights, and mistakes based on derivative computation. The output error is predicted in the same way as feed-forward NN. One of the most conventional approaches of optimizing and minimizing error in a function in order to obtain the global error is to use the gradient method. The BP algorithm suffers from not one but two significant flaws. One issue is that whenever there are a lot of data variables and network variables, it tends to become trapped in the local minimum. The other issue is that many inaccuracies appear on the surface due to the complicated and nonlinear relationship between the descriptors [25]. The output of NNs is directly influenced by how close local minima are to their corresponding global minima. Initial predictions of the network parameters are made randomly, and the selection of codons leads to the arrival at local minima. Second, it has problems in terms of vanishing gradients in initial layers in cases when the parameters were modified in the BP stage, which is a significant limitation. Counterpropagation is a hybrid NN approach used in QSAR that was successfully utilized to forecast the Kovats indices for substituted phenols. The Kohenen network and the out-star structure of the Grossberg layer are utilized in counterpropagation (CP) networks to forecast the winning units for input variables and to map the winners in their class. In other words, the learning producers are working on two important steps: first, finding the winner by measuring the distance between input neurons and weights, and second, updating the winner's weight [26]. The linear notation (LN) is a shortcoming of critical concentration (CC) in that it only provides a limited number of different solutions. Radial basis function networks and NN were employed to predict boiling points from structural factors [27].

Iterative integration of different models in QSAR

The standard QSAR approach is relatively straightforward and applies to chemicals that are close analogs. In comparative molecular field analysis, its aesthetic attractiveness increased molecular activity; hence, it was exceedingly successful. Due to its increased sophistication, obscurity of descriptors, and precision, the deployment of the atom or fragment model on example molecules has now been widespread. It also increases the scope of prediction at the expense of interpretability in order to comprehend the chemical foundation for the activity of improved-design molecules [28]. The context-dependent selection of the most relevant subsets of descriptors is a crucial step in QSAR modeling for a better pool. This enhances the generalizability of models and makes interpretation simpler due to fewer descriptors. Typically, models are evaluated in two fundamental ways. Initially, it identifies the descriptors that are the most crucial step for improving the properties of molecules, also known as feature importance or descriptors importance for ML [29]. Another method is appropriate for training the model on substructure-type descriptors to project the significant features of the model and to highlight the characteristics connected with the more favorable activity. Molecules whose atoms are colored according to their contribution constitute the molecular "heat map". The prediction of descriptors and model-independent approaches for feature interpretation is applied to minor perturbations to the input descriptors one at a time, observing the effect on the modeled property. Even though this method has its limitations, it has found widespread use in the field of interpretation. When the variables selected have the same sign and size of their coefficient across the multiple QSARs, this may help mechanistic to understand the causation, so it is important to remember the statistical method that can distinguish correlation from causation and interpretation, which are not always related to a mechanism and are used for mechanistic interpretability. There are various common ML methods in QSAR, however, due to their relative newness and popularity across many disciplines, DNN are now the most widely utilized ML methods when compared to other prominent ML approaches covered in this chapter. DNN approaches are widely used in a variety of applications. In the 1990s, single hidden layer neural nets were used as an ML method to construct the QSAR model, and they have seen a revival in the last decade. DNNs are now viable and computationally tractable thanks to recent developments in hardware and GPUs. In AI applications like as image classification and speech recognition, DNN outperforms any techniques that were first applied to QSAR following the Merck molecular activity challenges in 2012. The DNN's output into QSAR, as well as numerous other ML methods utilized in QSAR modeling such as kNN, partial least squares (PLS), SVM, RF, and relevance vector machines. In the pharmaceutical sector, DNN and ML are employed in accuracy prediction, sensitive tunable hyper-parameter, descriptor selection, enhanced training, and model interpretability [30]. For many years, RF techniques have been one of the most prominent approaches in QSAR because they make meaningful predictions with customizable parameters and may be parallelized. Furthermore, the level of agreement in prediction between distinct RF trees can be utilized to determine the applicability domain (AD). Boosting is also frequently employed as an accurate and fast method, particularly with the most recent implementation and light gradient boosting machine [31].

Neural network learning

The study of NNs has gotten a lot of attention in recent years. Today lot of modern technology is built on artificial NNs, which are computational models. Numerous NNs operate using the same principles as the nervous system in the human body. A subset of DL (deep learning) known as ANN fits within the broad domain of AI [32]. Recent advances in AI have launched a new revolution that has opened a seemingly endless number of possibilities, including the solution of genetic riddles and the exploration of the human brain. These emerging technologies attempt to build AI that, in some situations, outperforms human intelligence. They also aspire to produce faster, more effective, and smarter ways to conduct numerous daily tasks [33]. A class of ML algorithms with numerous applications is called NNs. Computer vision, speech recognition, natural language processing, and pattern identification are some of the most often used NN applications. Today, NNs are used in various applications and are receiving a great deal of attention in the research community. Basically, all we need do is to train the machine to use some NNs. One area of AI known as ML deals with a system's capacity to learn from data using supervised or unsupervised statistics and ML techniques such as decision trees, NNs, SVM, and support vector regression. Using its own capabilities, DL is a subclass of ML, learns sequential data that maximize the success of the learning process [34].

Artificial neural network—quantitative structure–activity relationship

In order to create QSAR models, ANN have become a popular ML algorithm. ANN is a type of ML that was initially developed to mimic the structure and function of biological NNs to make predictions and classify data. A standard ANN is built from three primary building blocks: the transfer function, the learning rule, and the connection formula [35]. Feed-forward ANNs (FF-ANNs) have thus far proven to be the most widely deployed ANNs in practical applications. BP approach, which finds the ideal parameters by minimizing the error of the network via the derivatives of the error function, is the most widely used among the numerous learning algorithms for estimating the parameter of FF-ANN. When the output or anticipated value is compared to the actual value, the BP algorithm attempts to reduce the gap between the two as much as possible [36].

Machine learning models—Quantitative structure–activity relationship

Recent years have seen a significant increase in the use of ML in QSAR models, with the goal of bettering the procedures involved in molecular design and screening, chemical structure prediction, and category forecasting. Methods from the ML family, such as SVM and RF, are frequently employed in the search for a new and effective treatment [37]. For example, the scoring function is calculated in molecular docking. Chemical fingerprints are utilized in a wide variety of ML models and prediction tasks and are generated through graph evaluation of a set of 2D or 3D chemical descriptors. The integration of large datasets and ML represents a major advance in the study of gene sequencing, single-cell sequencing, multi-omics interaction data, protein structure, and gene–protein interactions. ML models are used to forecast the discrimination, and these ML models are split into discriminating prediction and generative prediction variants. This is because traditional QSAR primarily focused on the predictive ability of candidate compounds, which is a popular usage in ML models. Pattern recognition is related to the employment of discriminative models, which map the association between the properties of the sample and the target of the prediction based on the data [38]. The discriminative model is used to evaluate the distribution of the sample data, while the generative model is used to forecast new compounds based on the learned distribution. Both models are used in conjunction with each other. Supervised learning, semi-supervised learning, unsupervised learning, and self-supervised learning are some of the most used classifications for these strategies. Due to its inherent characteristics, this model is only useful for supervised learning, if discriminant models are employed. These models are commonly used for feature selection, as well as algorithm and training set quality. Quantum chemistry, semiempirical approaches, and molecular dynamic simulation were used to evaluate the research with conformation-related indicators in QSAR and conformational calculation for isomers. The ML-based QSAR method is utilized to evaluate the energy changes that occur during conformational changes in order to foresee how the solvent environment impacts the conformational changes and the molecular energy [39]. Because ML can learn from data, it simply needs to exhaust as many of the features relating to the nature of the QSAR as is feasible before applying standard QSAR algorithms to learn from the training set. The essential features change depending on the structure–activity or prediction goal. ML-based methods are an efficient way to study physical and chemical properties that are difficult to compute using theoretical methods such as density functional theory (DFT) or molecular dynamics (MD) but can typically be computed by cheminformatics models. Examples of these properties include water solubility and logP (logarithm of octanol: water partition coefficient), both of which are directly related to drug activity. Calculating chemoinformatic parameters of compounds that are only tangentially related to therapeutic efficacy such as melting point, sublimation energy, and solubility requires the application of QSAR algorithms. On the other hand, QSAR techniques should be avoided in situations where the underlying mechanisms are already thoroughly understood. Quantum chemistry is a method that can be utilized to ascertain characteristics of an atom or molecule such as its dipole moment, vibration frequency, and polarizability [40].

Deep learning—Quantitative structure–activity relationship

As a result of its successful application in imaging, machine translation, and speech, DL has spread throughout the biological sciences, where it is being used for applications as diverse as the study of population-wide genetic variation, the development of synthetic biology drugs, and the diagnosis of disease. Multitask learning, capsule networks, self-encoding decoders, generative adversarial networks (GANs), long short team, variational AutoEncoders, DNNs, and convolutional neural networks (CNNs) are all examples of DL approaches [41]. Multitask learning acquires a key structural representation of information-rich biological data. A GAN, for example, forms and generates the metabolic network and loop network of microorganisms. Variational AutoEncoder can be used to encode compound structures and can be combined with GAN for learning purposes. DL is able to predict the protein structure based on the gene sequence. While the atoms, strategy networks, and Monte Carlo trees are used to predict the chemical reaction, a CNN is used to do learning based on the force field and the protein–ligand complex. In one-shot learning, a matching network is utilized to make predictions regarding novel compound matches and the receptors to which they are associated. It is impossible for DL to replace the traditional methods of ML because ML methods have higher interpretability and lower data requirements. DL requires a tremendous amount of data and processing, and it is effectively a black box. DL approaches are frequently utilized in the field of computational chemistry. These methods consist of one or more hidden layers and a limited number of units in each layer for the purpose of feature change. However, this tactic works far better when applied to resolving issues that are less significant than those that occur in the real world. In QSAR, the DL methods including CNN, DNN, and NN with more than two layers are the initial parameter. DBN is a type of generative unsupervised learning method in which higher level features are built from lower-level ones. The energy-based Restricted Boltzmann Machine (RBM) method is used as a discriminate and generative model for both labeled and unlabeled data. Similarly, DBN has an unsupervised learning algorithm applied layer by layer. The structure of autoencoder has two main parts, one to encode the input data and the other to decode the hidden units to reconstruct the input data. The autoencoder training algorithms are utilized very efficiently as a building component in order to pretrain the NN. As a consequence of this, it significantly cuts down on redundancy difficulties, such as becoming trapped in a local minimum, and also boosts the overall speed of the model [42]. CNN explains the NN approach as neurons that have weights and biases associated with them. For instance, in-house databases with a limited number of descriptors are able to make use of the entire network. However, in the case of a large dataset with a significant number of descriptors, such as the ChEMBL database, and network parameters that are quickly assayed and rapidly lead to overfitting, which is the goal of the connected redundancy problem. In contrast to a feed-forward, an NN makes use of all of the connected layers. The conventional layer, the pooling layer, and the fully connected layer make up the architecture of CNN. The fully linked layer is the topmost layer. Every map is then sub- and down-sampled with average sum or max pooling over pxp neighbor pixels, where p denotes the changes between the small and large input data. The conventional and connected layers are used to extract features from neurons, and it has K-filters and convNet with small size data [43].

Decision tree algorithms

When working with tabular data, decision trees are a type of ML algorithm that compete with NNs in terms of quality and performance and frequently outperform NNs. For the purpose of decision analysis, a decision tree is a useful tool for graphically and explicitly depicting decisions and the decision-making process. A decision tree is a supervised learning algorithm. Similar as other types of supervised learning algorithms, decision tree can be utilized to tackle issues involving both regression and classification [44]. The objective of utilizing a decision tree is to develop a training model that can be used to forecast the class or the value of the target chemical by learning fundamental choice rules that are derived from the training data. The decision tree begins its analysis at the node that represents the tree's root to make an educated guess on the record's class label. Check whether the value of the root attribute is the same as the value of the attribute in the record. Follow the path of the branch that corresponds to that value determined by the comparison and go to the following node. It is capable of solving problems in either the discrete or continuous setting, depending on the nature of the data being worked with [45]. The WEKA tool is the only one that can replicate the ID3 (Iterative Dichotomiser 3) approach, and the data set only has categorical information in it. In terms of decision tree characteristics, this is important to note [46]. ID3 does not allow continuous datasets in the simulation environment. CART and C4.5 both have characteristics that are analogous to those of ID3's [47,48]. The primary difference between C4.5 and CART is that the former has the capability of accepting continuous data sets as simulation input, while the latter does not. Fig. 9.1 explains the working principle of decision tree and Fig. 9.2 explains the physiognomies of data type available in decision tree [49].

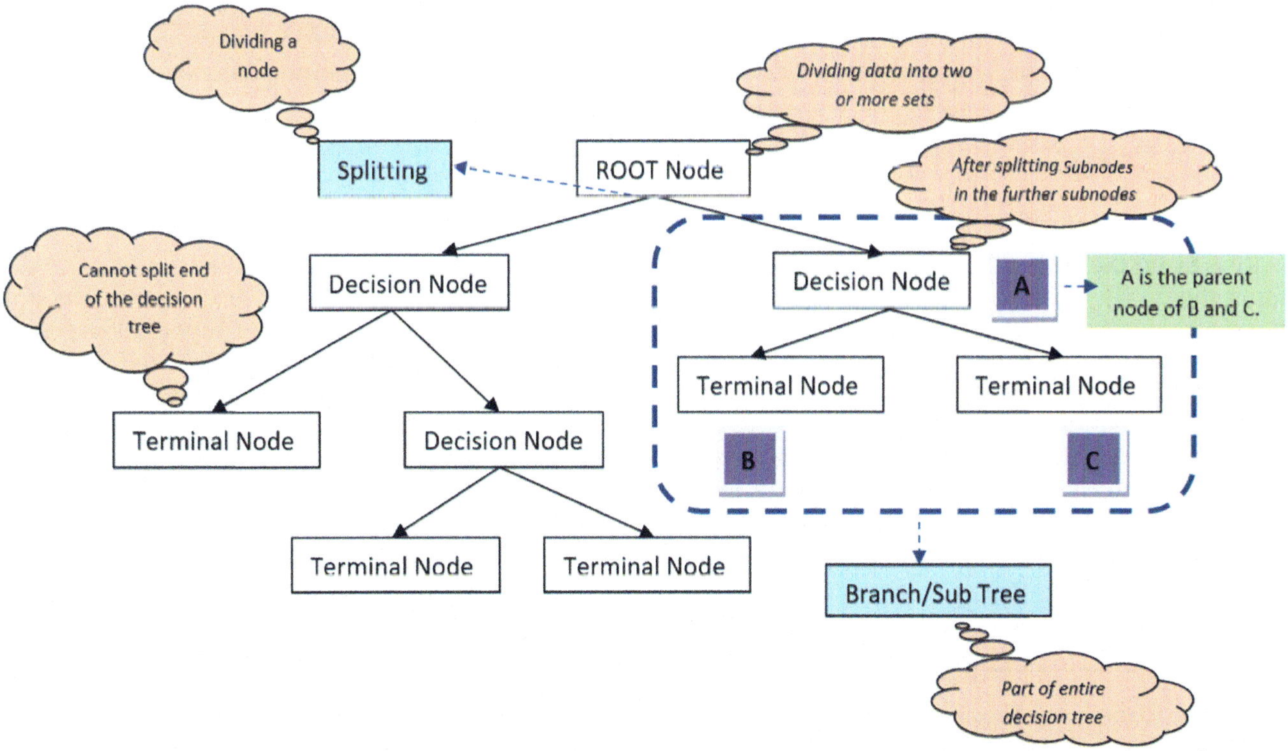

FIGURE 9.1 Working principle of decision tree.

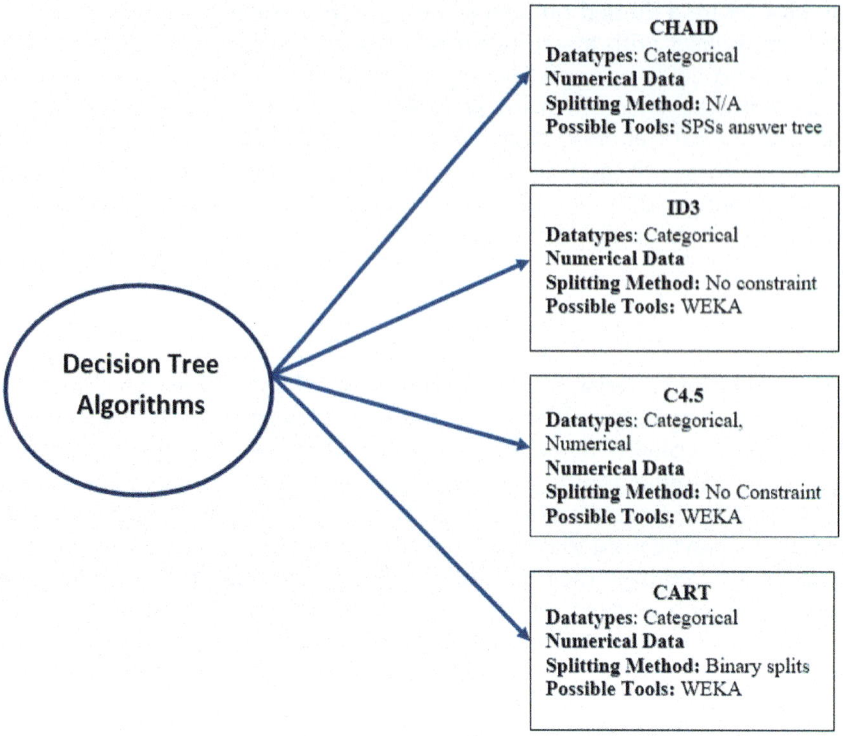

FIGURE 9.2 Physiognomies of data types in decision tree algorithm.

The decision tree explicitly states each conceivable alternative and follows each one through to completion in a single view to simplify comparisons between the multiple choices. Transparency is one of its advantages. Two more key advantages are the ability to select more biased physiognomies and the nature of their intuitiveness. Furthermore, it is straightforward to categorize and interpret. Variable screening and feature sections perform admirably in decision trees. In terms of performance, nonlinearity has no effect on any of the decision tree's properties. In order to determine whether or not a split is the "best," decision tree techniques are used to categorize the characteristics into distinct groups. The resulting partition is as pure as feasible because the splitting criterion for each branch must be the same. Table 9.1 elucidates the comparison of various decision tree algorithms [50].

Random forest

RF is by far the most popular and important ML algorithm due to its effectiveness and success in a wide range of applications [51]. The first RF algorithm was proposed by Brieman, and its derivatives have been used successfully in water resource applications. Because of limitations in the categorization and computation of important metrics, RF is employed for quantile prediction, survival analysis, and causal inference. Furthermore, it is utilized in different scientific fields such as agriculture, land cover classification, remote sensing, ecology, wetland classification, bioinformatics, genomics, and QSAR. The theoretical and practical elements of RF-based algorithms have been thoroughly investigated. In summary, RF is a necessary ensemble learning method that uses decision trees as base learners. A number of studies have been conducted to demonstrate the RF algorithms, and the basic concept of algorithms is derived from earlier works such as Biau and Scornet. In essence, RF is an ML algorithm that has both classification and regression tree concepts, as well as bagging with another degree of randomization [52].

Supervised learning

In order to forecast a variable, supervised learning techniques are utilized to locate the function and combine the set of variables. A dependent variable is a variable for which predictions are being made, while a predictor variable refers to the arguments of the function. For the quantitative study of dependent variables, researchers used

TABLE 9.1 Decision tree algorithms.

Algorithm name	Classification	Description
CART—regression and classification	It measures using Gini index.	Numerical splitting is used to build the CART-based tree.
ID3 (Iterative Dichotomize 3)	It utilizes information gain and entropy functions metrics.	Those are the only difficulties with discrete values. The continuous dataset must therefore be classed as a discrete data collection.
C4.5	The updated version is ID3.	It involves both continuous and discrete datasets. Additionally, it can handle Incomplete datasets. Overfiltering is solved by the method known as "PRUNNING."
C5.0	The updated version is C4.5	With C5.0, we may choose between statistically allocating the case among the outcomes.
CHAID (chi-square automatic interaction detector).	Precedes the initial implementation of ID3.	This kind of decision tree is utilized for a nominal scaled variable. This method identifies the dependent variable from a datasets categorized variables.
MARS (multiadaptive regression splines).	To determine the best split	To obtain the optimal split, we can use a regression tree based on MARS.

regression algorithms, while for the qualitative analysis, they turned to classification algorithms, both of which fall under the umbrella of supervised learning [53]. In the latter, the dependent variables are ranked without the measure required to statistically evaluate the differences between them. RF has been employed as a powerful method in QSAR recently, but its adoption has been slow. The consensus nonlinear model formed from several individual RF models is what makes them so useful. However, RF has a number of benefits, including that RF models are highly resistant to overfitting and do not require a relatively complex and time-consuming process for the selection of variables. The advanced QSAR model, that is, the simple representation of molecular structure QSAR approach, which is incorporated in HiT QSAR software [54], has been used and results in previous QSAR studies, and this is in comparison to the number of the mechanism if the action could be stored with training sets. Due to the high interpretability in simple representation of molecular structure model extraction, it was utilized to probe the creation of simplex-descriptor-dependent models that are both eternally predictive and easily interpretable. The impact of tuning parameters and noise insertion on the RF model's behavior and optimization is studied in relation to other common modeling techniques, such as least-squares, latent-structure, and kNN models, using a straightforward description and a collaborative consensus model [55]. Prediction of a continuous response in the QSAR model is analogous to regression, whereas prediction of a categorical response is analogous to classification. Commonly used identifiers for this purpose include p and n, where p typically surpasses the number of samples and n stands for the statistical techniques used to make the prediction. Multiple linear regressions (MLR), kNN, and linear discriminant analysis (LDA) are examples of conventional statistical approaches that have been shown to be ineffective due to their lack of accuracy and sophistication in their variable selection filters. For instance, in evolutionary algorithms, alternative methods such as decision trees, ANN, SVM, and PLS are often needed for the pre-detection because of the enormous number of irrelevant descriptors. A decision tree has no limits compared to other methods; yet, it has low accuracy. As a result, tree ensembles are utilized as natural selection for QSAR modeling, where desired decision tree features are paired with good prediction performance [56]. The most promising ensemble learning approaches are boosting and RF. The numerous ML and AI algorithms utilized in drug development methods are presented in Table 9.2.

In methods like as bagging, RF is also an ensemble of unpruned trees, with each tree trained as a bootstrap sample of the training set and used in the forecast, which is predicted as the majority vote of trees or by averaging their output. RF differs from bagging in that each node is distinct; hence, this approach is regarded as dividing candidates from a random sample of variables (Fig. 9.3).

Unlike traditional tree-based QSAR models, which have limitations due to factors like a hyperrectangular partition of the descriptor space and the greediness of tree-growing algorithms, RF models have an extra feature that greatly improves the QSAR modeling's applications in the drug design process. Other methods such as out-of-the-bag prediction are frequently used for error estimation and threshold setting; furthermore, variable measurement is critical for model interpretation [57]. Because this model is useful for determining the intrinsic proximity of two compounds, the neighbor molecules are estimated using it. Because each tree in the ensembles is a bootstrap, the out-of-bag prediction is used to assess the tree's test set. Out-of-bag data will account for roughly one-third of the entire training set for each tree. As a result, one-third of the training set molecules will be absent from the ensemble

TABLE 9.2 Various machine learning and artifical intelligence algorithms are used in the drug discovery process.

AI-category	Algorithms	Task
ML/supervised	Decision tree	Classification
	SVM	Classification/regression
	RF kNN	classification/regression
	Naïve	classification
	GB	Bayesian-classification
	ANN	Classification/regression
	Support vector	classification/clustering
		Regression
ML/unsupervised	Principal component analysis	Clustering
	k-mean clustering	Clustering
	SOM	Clustering
	Hierarchical clustering	Clustering
	t-SNE	Clustering
	Singular value decomposition	Dimensionality reduction
	Linear discriminant analysis	Dimensionality reduction
DL	CNNs	Representation-learning techniques
	RNNs	Representation-learning techniques
	Deep belief networks (DBN)	Representation-learning techniques
	Stacked autoencoders	Representation-learning techniques
	LSTMs	Representation-learning techniques
	Deep Boltzmann machine (DBM)	Representation-learning techniques

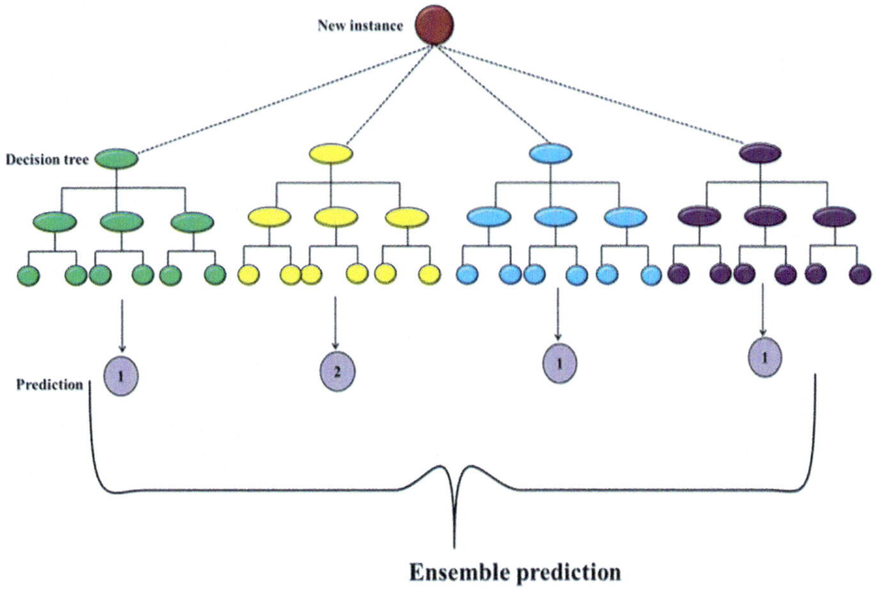

FIGURE 9.3 Depicted the random forest (RF) descriptors with ensemble prediction.

tree used to compose the prediction for the computation of an error rate. The ensemble of B trees in RF is denoted by [T1(X), ..., TB(X)], where x is a dimensional vector of molecular descriptors or properties associated with the molecule. The output of the ensembles is [Y1) T1(X), ..., YB) TB(X)] where B is the prediction of the molecule, and the results of all the trees are highly aggregated to produce the final prediction Y. In the case of classification, Y represents the class prediction for the majority of trees, whereas in regression, it is used to predict the average of an individual tree. In comparison to developing a single decision tree, RF techniques are more effective at examining a large number of descriptors in a dataset when using a combination of two separate procedures [58]. To get the right model complexity for the optimal prediction strength, which is typically required for a single decision tree, we first put the standard tree growing algorithms and their descriptors through their paces in terms of splitting performance at each node, while RF tests the subset of the tree (in other words, the RF algorithms are effectively used in p descriptors). Bagging and boosting are two more popular ensembles and learning methods that have been demonstrated to significantly outperform a single tree. Bagging is formed from RF with the letter "p," whereas boosting is a sequence of trees, each trained on all of the data, but the data points are derived from each tree based on their misclassification by the preceding test. The ensemble output is predicted by weighted vote and weighted average of categorization and regression, respectively. Decision forest and random forest are the two ensembles of the tree, and the tree formed through training data. Decision forest is another ensemble constructed and used to investigate each tree and does not fall below the threshold. The earliest documented comparison of decision forest (DF) and random forest (RF) performance showed similar results.

Intrinsic proximity measure

In QSAR modeling, the clustering of the molecule in a descriptor space and a number of assessments of the nearby molecule are difficult. To address this issue, an unsupervised model with some measurements of the similarity between the molecules in a description space is used. In numerous applications, researchers focus on proximity measurements determined by a subset of descriptors that are efficiently relevant to the activity of interest. Intrinsic proximity measures are made available by RF and can be computed between any two molecules in the same or different sets, for example, a training set and a test set. In the two-molecule RF model, prediction closeness was determined by counting the number of times each molecule appeared at the same terminal node across all ensemble trees. One key benefit of proximity measurement is that it is supervised since the structure of trees is determined by the activity of interest. Furthermore, they contribute little to the closeness measure because of the impact of irrelevant descriptors of ensembles [59]. However, if both molecules are included in the training set, this proximity metric may be flawed unless only out-of-bag data are utilized for the calculation. QSAR modeling makes extensive use of tree-based ensembles that are implemented in ML methods. These include the XGBoost classifier, the bagging classifier, the RF classifier, and the extra trees classifier. The decision tree classifier's ensemble of classifiers, also known as its base classifiers, goes by both names. One of the sought-after valuable forecasts is the decision tree, which organizes rules in the form of a tree. Class labels and tests on attributes are represented as nonleaf nodes. Using the test's output and path, it determines how to categorize the new observation and then follows that path along the tree. Decision theory is intuitive and simple to teach because it mimics the way humans make decisions. Groups of decision nodes are built into the RF classifier to boost the decision tree's stability and efficiency [60]. This approach utilizes Breiman's bagging sampling alongside random feature selection approaches to construct a set of decision trees with a predetermined degree of variation. Bagging is used in each tree constructed by a sample with replacement of the training data; these trees can then function as basis estimates to build the class label via majority vote. With base decision tree mode, the instance's class is predicted using the tree's internal structure. Uses of the RF in data analysis include label ranking, stock selection, clinical risk prediction for survival, and multivariate data analysis [61].

AdaBoost classifier

AdaBoost is another ML algorithm that uses a weighted linear combination of numerous weak learners and a unique classifier. Learning algorithms frequently use AdaBoost to reweight subsets of the original training data. Further, it is an iterative method, with the misclassified instance from the previous iteration being given additional weight with each iteration. In this method, each instance is given a fixed weight, and when the algorithm iterates, the mistakenly categorized instances weights are increased, while the appropriate instance's weight is decreased. The iterative technique is widely used as a baseline classifier on the training set with updated parameters. Iteratively combining many linear models yields the final categorization model. However, the AdaBoost technique is used in a wide variety of literature reviews despite its sensitivity to outliers and noisy data. The quantile sketch algorithm handles weighted data has a column block structure for parallel layers, and employs a cache aware-algorithm to fetch and store the gradient statistics. XGBoost is also a popular ML technique that employs a regulated model to prevent problems with overfitting and sparse patterns in the data. More recently, the gradient-boosted decision trees (GBDT) and a lightweight version of the GBM algorithm has been utilized to predict the stability of hard rock pillars. The XGBoost technique takes into account the gene expression values and uses a vote system to determine the class label of a new instance based on the results of many ML classifiers.

Partial least squares regression

The linear regression method of PLS is commonly used in cheminformatics for calibrating the model variables, and it is also frequently employed in the analysis of valuable, even necessary, for typical data, such as QSAR and QSPR. This strategy is most effective when used to datasets with highly correlated independent variables such as the following. Overfitting is prevented and a large number of possible test results are generated, all thanks to the fact that the components utilized in this method allow for complete control over the model computing [62]. Principal component analysis and PLS regression are two examples of approaches that use an intermediate set of linear latent variables to accomplish regression. PLS gives the maximum covariance between the modeled properties and scores the empirical covariance between two mean-centered variables. A score with the maximum variation is generated using polymerization chain reaction (PCR), whose concept and components differ slightly from loading vectors [63]. In addition, both orthogonal loading and correlated score vectors are techniques utilized for PLS. PLS and principal components are determined by repeating strategy. In most cases, QSAR data only have a few of objects that serve as both the calibration and test sets, making it a good candidate for using a cross-validation (CV) or bootstrap approach. This conventional CV method is employed to determine the optimal number of PLS components by means of the training data. For instance, from the calibration set nCALIB objects have their size arbitrarily divided up into a set number of segments that are all around the same size [64].

Software's available for QSAR

A limited collection of well-known databases, websites, and software tools, both free and commercial, that can be utilized in QSAR research is provided in Table 9.3.

TABLE 9.3 QSAR programs and link for the open source and commercial packages.

S. No	Program	Link
I.	ECOSAR	https://www.epa.gov/tsca-screening-tools/ecological-structure-activity-relationships-ecosar-predictive-model
II.	ChemACE	https://www.epa.gov/tsca-screening-tools/chemical-assessment-clustering-engine-chemace
III.	ChemSTEER	https://www.epa.gov/tsca-screening-tools/chemsteer-chemical-screening-tool-exposures-and-environmental-releases
IV.	EPI Suite	https://www.epa.gov/tsca-screening-tools/epi-suitetm-estimation-program-interface
V.	QSPR-Thesaurus	https://cadaster.eu/node/118.html
VI.	ReachScan	https://www.epa.gov/tsca-screening-tools/reachscan-exposure-assessment-model
VII.	OSIRIS property Explorer	https://www.organic-chemistry.org/prog/peo/
VIII.	OECD QSAR Toolbox	https://www.oecd.org/chemicalsafety/risk-assessment/oecd-qsar-toolbox.htm
IX.	ToxPredict	http://old.opentox.org/dev/testing/testcasedevelopment/toxpredict
X.	VEGA	https://www.vegahub.eu/portfolio-item/vega-qsar/
XI.	ToxTree	https://toxtree.sourceforge.net/
XII.	QSAR TOOLBOX	https://qsartoolbox.org/
XIII.	Dragon	http://www.talete.mi.it/products/dragon_description.htm
XIV.	CORAL	http://www.insilico.eu/coral/
XV.	ChemDes	http://www.scbdd.com/chemdes/
XVI.	CoFFer	http://coffer.informatik.uni-mainz.de/
XVII.	VIDEAN	http://lidecc.cs.uns.edu.ar/VIDEAN/
XVIII.	ChemProp	https://www.ufz.de/index.php?en=34598
XIX.	CODESSA	http://www.codessa-pro.com/
XX.	AZOrange	https://github.com/AZcompTox/AZOrange
XXI.	McQSAR	http://users.abo.fi/mivainio/mcqsar/
XXII.	Open3dqsar	https://open3dqsar.sourceforge.net/
XXIII.	Phase	https://www.schrodinger.com/products/phase
XXIV.	VolSurf+	https://www.moldiscovery.com/software/vsplus/
XXV.	Cyprotex	https://www.cyprotex.com/insilico

Concluding remarks

In the process of discovering new drugs, the application of AI and ML algorithms is showing significant promise for reducing both the amount of time required and the associated costs, despite the fact that AI algorithms are not the answer to every problem in the drug research industry. DL approaches address the limitations of AI and ML by integrating data across several levels using nonlinear models. Despite the fact, this layered data integration delivers great accuracy and precision. In comparison, AI and ML algorithms are more adaptable to construct NNs for specific problems. Because of this, applying AL and NNs is now quite simple. The power of the AL and ML algorithms is widely employed to supplement human intellect, enhance our capabilities, and improve the methodology used in drug discovery. In the early stages, drug repurposing is the only method for clinical observation. However, recently developed AI and ML models learn task-related features from raw data for molecular representation, and these algorithm-based methods can improve the screening process from large data and reduce the false positive rates due to imbalance in positive and negative data. For instance, lead optimization is a difficult phase in the drug development process because it improves ADMET characteristics and target activities; yet, these parameters are independent of one another and are mutually incompatible, and these problems are resolved by a better model. The use of vectors and their relationships in constituent element analysis can aid in understanding the significance of local structures, and research into AI and ML algorithms will help the field of QSAR attain high efficiency and low cost. In conclusion, this chapter has presented developments in QSAR alongside supporting AI and ML backends for the forecasting of new drug candidates.

Acknowledgments

CS thankfully acknowledge the Saveetha University for providing the infrastructure facility to conduct this research work. S.K.S. thankfully acknowledge the Tamil Nadu State Council for Higher Education (TANSCHE) for the research grant (Au/S.o. (P&D): TANSCHE Projects: 117/2021).

References

[1] Perkins R, Fang H, Tong W, Welsh WJ. Quantitative structure-activity relationship methods: perspectives on drug discovery and toxicology. Environ Toxicol Chem 2003;22(8):1666–79.
[2] Cherkasov A, Muratov EN, Fourches D, Varnek A, Baskin II, Cronin M, et al. QSAR modeling: where have you been? Where are you going to? J Med Chem 2014;57(12):4977–5010.
[3] Rovida C, Barton-Maclaren T, Benfenati E, Caloni F, Chandrasekera PC, Chesne C, et al. Internationalization of read-across as a validated new approach method (NAM) for regulatory toxicology. ALTEX 2020;37(4):579–606.
[4] Cramer RD. The inevitable QSAR renaissance. J Comput Aided Mol Des 2012;26(1):35–8.
[5] Wang T, Wu MB, Lin JP, Yang LR. Quantitative structure-activity relationship: promising advances in drug discovery platforms. Expert Opin Drug Discov 2015;10(12):1283–300.
[6] Vilar S, Costanzi S. Predicting the biological activities through QSAR analysis and docking-based scoring. Methods Mol Biol 2012;914:271–84.
[7] Schneider A, Hommel G, Blettner M. Linear regression analysis: part 14 of a series on evaluation of scientific publications. Dtsch Arztebl Int 2010;107(44):776–82.
[8] Acharya C, Coop A, Polli JE, Mackerell Jr AD. Recent advances in ligand-based drug design: relevance and utility of the conformationally sampled pharmacophore approach. Curr Comput Aided Drug Des 2011;7(1):10–22.
[9] Livingstone DJ. Computational techniques for the prediction of toxicity. Toxicol In Vitro 1994;8(4):873–7.
[10] Kar S, Roy K, Leszczynski J. Impact of pharmaceuticals on the environment: risk assessment using QSAR modeling approach. Methods Mol Biol 2018;1800:395–443.
[11] Kavlock RJ. Structure-activity relationships in the developmental toxicity of substituted phenols: in vivo effects. Teratology 1990;41(1):43–59.
[12] Lipnick RL. Hans Horst Meyer and the lipoid theory of narcosis. Trends Pharmacol Sci 1989;10(7):265–9.
[13] Sharma BK, Singh P, Kumar R, Sharma S. Quantitative structure-activity relationship study of novel rhinacanthins and related naphthoquinone esters as anticancer agents. J Enzyme Inhib Med Chem 2008;23(1):50–5.
[14] Verma J, Khedkar VM, Coutinho EC. 3D-QSAR in drug design—a review. Curr Top Med Chem 2010;10(1):95–115.
[15] Kubinyi H. Quantitative structure-activity relationships. 2. A mixed approach, based on Hansch and Free-Wilson analysis. J Med Chem 1976; 19(5):587–600.
[16] Mao J, Akhtar J, Zhang X, Sun L, Guan S, Li X, et al. Comprehensive strategies of machine-learning-based quantitative structure-activity relationship models. iScience 2021;24(9):103052.
[17] Keyvanpour MR, Shirzad MB. An analysis of QSAR research based on machine learning concepts. Curr Drug Discov Technol 2021;18(1): 17–30.
[18] Nag S, Baidya ATK, Mandal A, Mathew AT, Das B, Devi B, et al. Deep learning tools for advancing drug discovery and development. 3 Biotech 2022;12(5):110.
[19] Gupta R, Srivastava D, Sahu M, Tiwari S, Ambasta RK, Kumar P. Artificial intelligence to deep learning: machine intelligence approach for drug discovery. Mol Divers 2021;25(3):1315–60.
[20] Yang X, Wang Y, Byrne R, Schneider G, Yang S. Concepts of artificial intelligence for computer-assisted drug discovery. Chem Rev 2019; 119(18):10520–94.
[21] Salvi M, Acharya UR, Molinari F, Meiburger KM. The impact of pre- and post-image processing techniques on deep learning frameworks: a comprehensive review for digital pathology image analysis. Comput Biol Med 2021;128:104129.
[22] Askr H, Elgeldawi E, Aboul Ella H, Elshaier Y, Gomaa MM, Hassanien AE. Deep learning in drug discovery: an integrative review and future challenges. Artif Intell Rev 2022:1–63.
[23] Kavakiotis I, Tsave O, Salifoglou A, Maglaveras N, Vlahavas I, Chouvarda I. Machine learning and data mining methods in diabetes research. Comput Struct Biotechnol J 2017;15:104–16.
[24] Uddin S, Khan A, Hossain ME, Moni MA. Comparing different supervised machine learning algorithms for disease prediction. BMC Med Inform Decis Mak 2019;19(1):281.

[25] Agany DDM, Pietri JE, Gnimpieba EZ. Assessment of vector-host-pathogen relationships using data mining and machine learning. Comput Struct Biotechnol J 2020;18:1704–21.
[26] Kourou K, Exarchos TP, Exarchos KP, Karamouzis MV, Fotiadis DI. Machine learning applications in cancer prognosis and prediction. Comput Struct Biotechnol J 2015;13:8–17.
[27] Sarkar K, Bonnerjee D, Srivastava R, Bagh S. A single layer artificial neural network type architecture with molecular engineered bacteria for reversible and irreversible computing. Chem Sci 2021;12(48):15821–32.
[28] Kwon S, Bae H, Jo J, Yoon S. Comprehensive ensemble in QSAR prediction for drug discovery. BMC Bioinform 2019;20(1):521.
[29] Liu R, Glover KP, Feasel MG, Wallqvist A. General approach to estimate error bars for quantitative structure-activity relationship predictions of molecular activity. J Chem Inf Model 2018;58(8):1561–75.
[30] Hu S, Chen P, Gu P, Wang B. A deep learning-based chemical system for QSAR prediction. IEEE J Biomed Health Inform 2020;24(10):3020–8.
[31] Yin Z, Ai H, Zhang L, Ren G, Wang Y, Zhao Q, et al. Predicting the cytotoxicity of chemicals using ensemble learning methods and molecular fingerprints. J Appl Toxicol 2019;39(10):1366–77.
[32] Alzubaidi L, Zhang J, Humaidi AJ, Al-Dujaili A, Duan Y, Al-Shamma O, et al. Review of deep learning: concepts, CNN architectures, challenges, applications, future directions. J Big Data 2021;8(1):53.
[33] Abdelhafiz D, Yang C, Ammar R, Nabavi S. Deep convolutional neural networks for mammography: advances, challenges and applications. BMC Bioinf 2019;20(Suppl. 11):281.
[34] Shin HC, Roth HR, Gao M, Lu L, Xu Z, Nogues I, et al. Deep convolutional neural networks for computer-aided detection: CNN architectures, dataset characteristics and transfer learning. IEEE Trans Med Imaging 2016;35(5):1285–98.
[35] Pantic I, Paunovic J, Cumic J, Valjarevic S, Petroianu GA, Corridon PR. Artificial neural networks in contemporary toxicology research. Chem Biol Interact 2023;369:110269.
[36] Raza A, Chohan TA, Buabeid M, Arafa EA, Chohan TA, Fatima B, et al. Deep learning in drug discovery: a futuristic modality to materialize the large datasets for cheminformatics. J Biomol Struct Dyn 2022:1–16.
[37] Sousa GHM, Gomes RA, de Oliveira EO, Trossini GHG. Machine learning methods applied for the prediction of biological activities of triple reuptake inhibitors. J Biomol Struct Dyn 2022:1–10.
[38] Zhou H, Shan M, Qin LP, Cheng G. Reliable prediction of cannabinoid receptor 2 ligand by machine learning based on combined fingerprints. Comput Biol Med 2022;152:106379.
[39] Lin J, Wen L, Zhou Y, Wang S, Ye H, Su J, et al. PepQSAR: a comprehensive data source and information platform for peptide quantitative structure-activity relationships. Amino Acids 2022.
[40] Soares TA, Nunes-Alves A, Mazzolari A, Ruggiu F, Wei GW, Merz K. The (re)-evolution of quantitative structure-activity relationship (QSAR) studies propelled by the surge of machine learning methods. J Chem Inf Model 2022;62(22):5317–20.
[41] Ota R, Yamashita F. Application of machine learning techniques to the analysis and prediction of drug pharmacokinetics. J Control Release 2022;352:961–9.
[42] Gautam V, Gupta R, Gupta D, Ruhela A, Mittal A, Mohanty SK, et al. deepGraphh: AI-driven web service for graph-based quantitative structure-activity relationship analysis. Brief Bioinform 2022;23(5).
[43] Pasrija P, Jha P, Upadhyaya P, Khan MS, Chopra M. Machine learning and artificial intelligence: a paradigm shift in big data-driven drug design and discovery. Curr Top Med Chem 2022;22(20):1692–727.
[44] Fernandes PO, Martins DM, de Souza Bozzi A, Martins JPA, de Moraes AH, Maltarollo VG. Molecular insights on ABL kinase activation using tree-based machine learning models and molecular docking. Mol Divers 2021;25(3):1301–14.
[45] Bugeac CA, Ancuceanu R, Dinu M. QSAR models for active substances against *Pseudomonas aeruginosa* using disk-diffusion test data. Molecules 2021;26(6).
[46] Hajjej F, Alohali MA, Badr M, Rahman MA. A comparison of decision tree algorithms in the assessment of biomedical data. Biomed Res Int 2022;2022:9449497.
[47] Teixeira A. Discriminant analysis with binary decision tree (CART: classification and regression tree). Rev Mal Respir 2004;21(6 Pt 1):1174–6.
[48] Velu SR, Ravi V, Tabianan K. Data mining in predicting liver patients using classification model. Health Technol (Berlin) 2022;12(6):1211–35.
[49] Nayarisseri A, Khandelwal R, Tanwar P, Madhavi M, Sharma D, Thakur G, et al. Artificial intelligence, big data and machine learning approaches in precision medicine & drug discovery. Curr Drug Targets 2021;22(6):631–55.
[50] Wang MWH, Goodman JM, Allen TEH. Machine learning in predictive toxicology: recent applications and future directions for classification models. Chem Res Toxicol 2021;34(2):217–39.
[51] Aqeel I, Bilal M, Majid A, Majid T. Hybrid approach to identifying druglikeness leading compounds against COVID-19 3CL protease. Pharmaceuticals (Basel) 2022;15(11).
[52] Collins SP, Barton-Maclaren TS. Novel machine learning models to predict endocrine disruption activity for high-throughput chemical screening. Front Toxicol 2022;4:981928.
[53] Moran KR, Dunson D, Wheeler MW, Herring AH. Bayesian joint modeling of chemical structure and dose response curves. Ann Appl Stat 2021;15(3):1405–30.
[54] Kuz'min VE, Artemenko AG, Muratov EN. Hierarchical QSAR technology based on the Simplex representation of molecular structure. J Comput Aided Mol Des 2008;22(6–7):403–21.
[55] Jaganathan K, Tayara H, Chong KT. An explainable supervised machine learning model for predicting respiratory toxicity of chemicals using optimal molecular descriptors. Pharmaceutics 2022;14(4).
[56] Melge AR, Parate S, Pavithran K, Koyakutty M, Mohan CG. Discovery of anticancer hybrid molecules by supervised machine learning models and in vitro validation in drug resistant chronic myeloid leukemia cells. J Chem Inf Model 2022;62(4):1126–46.
[57] Suay-Garcia B, Falco A, Bueso-Bordils JI, Anton-Fos GM, Perez-Gracia MT, Aleman-Lopez PA. Tree-based QSAR model for drug repurposing in the discovery of new antibacterial compounds against *Escherichia coli*. Pharmaceuticals (Basel) 2020;13(12).
[58] Sadgrove NJ, Oliveira TB, Khumalo GP, Vuuren SFV, van Wyk BE. Antimicrobial Isoflavones and derivatives from Erythrina (Fabaceae): structure activity perspective (Sar & Qsar) on experimental and mined values against *Staphylococcus aureus*. Antibiotics (Basel) 2020;9(5).
[59] Akamatsu M, Katayama T, Kishimoto D, Kurokawa Y, Shibata H, Ueno T, et al. Quantitative analyses of the structure-hydrophobicity relationship for N-acetyl di- and tripeptide amides. J Pharm Sci 1994;83(7):1026–33.

[60] Osakai T, Hirai T, Wakamiya T, Sawada S. Quantitative analysis of the structure-hydrophobicity relationship for di- and tripeptides based on voltammetric measurements with an oil/water interface. Phys Chem Chem Phys 2006;8(8):985–93.
[61] Tsantili-Kakoulidou A, Panderi I, Piperaki S, Csizmadia F, Darvas F. Prediction of distribution coefficients from structure. Comparison of calculated and experimental data for various drugs. Eur J Drug Metab Pharmacokinet 1999;24(3):205–12.
[62] Stanton DT. QSAR and QSPR model interpretation using partial least squares (PLS) analysis. Curr Comput Aided Drug Des 2012;8(2):107–27.
[63] Zhou YP, Cai CB, Huan S, Jiang JH, Wu HL, Shen GL, et al. QSAR study of angiotensin II antagonists using robust boosting partial least squares regression. Anal Chim Acta 2007;593(1):68–74.
[64] Edlund C, Jackson TR, Khalid N, Bevan N, Dale T, Dengel A, et al. LIVECell-A large-scale dataset for label-free live cell segmentation. Nat Methods 2021;18(9):1038–45.

CHAPTER

10

Advances of the QSAR approach as an alternative strategy in the environmental risk assessment

Laura M. Saavedra[1], *Juan C. Garro Martinez*[2] *and Pablo R. Duchowicz*[1]

[1]Instituto de Investigaciones Fisicoquímicas Teóricas y Aplicadas (INIFTA), La Plata, Argentina [2]Centro Científico Tecnológico San Luis (CCT-CONICET), Facultad de Química, Bioquímica y Farmacia, Universidad Nacional de San Luis, San Luis, Argentina

Abbreviations

AC_{50} Half-maximal active concentration
AOPs Adverse outcome pathways
CoMFA Comparative molecular fields analysis
ERA Environmental risk assessment
MLR Multivariable linear regression
MoAs Mode of actions
MoEs Margin of exposure
NAMs New approach methodologies
OECD Organisation for Economic Co-operation and Development
QSAR Quantitative structure–activity relationships
TSCA Toxic Substances Control Act

Introduction

In the modern world, a proper environmental risk assessment (ERA) of chemicals represents an important step in premarketing inspections of regulated substances (e.g., drugs, pesticides, feed additives, and cosmetics) as well as the postmarket surveillance system (e.g., commercial chemicals, emerging contaminants or natural compounds) [1–3]. ERA comprises a wide range of disciplines and research areas including toxicology, chemistry, cellular biology, biochemistry, ecology, physiology, statistics, and mathematics.

The main application of ERA consists in quantifying levels of chemical exposure through proper exposure assessments to given specie, population, or ecosystem, considering relevant sources (air, water, or soil) and routes of exposure (oral, inhalation, and dermal), to define safe levels for such chemicals by means of hazard identification processes, and to establish the risk associated with such exposures through rigorous risk characterization procedures [4–6]. Thus, ERA has a relevant role during research activities, defining biodiversity and ecosystem protection goals by regulatory and advisory entities.

The diversity and complexity of chemicals available in the environment can be large and includes pesticides, biocides, feed additives, pharmaceuticals, and emerging contaminants from anthropogenic activities such as persistent organic pollutants, heavy metals, dioxins, and antibiotics, among others [7]. Thereupon, different methodologies are part of the standard higher-tier options in the ERA field, covering both exposure components, through emission and data models, as well as ecological components on the hazard identification and risk characterization by means of population models [8], which requires extensive confidential data and entails multidisciplinary efforts.

In this context, continuous advances in biology, chemistry, and computational fields have contributed to the accelerated development of computer simulation studies, also known as computer-assisted modeling methods, which were initially proposed to guide the search for new drugs or identify the polluting agents present in the

https://doi.org/10.1016/B978-0-443-15339-6.00032-1

environment, by adjusting a model [9]. Though, currently, computer-assisted modeling methods are not yet fully implemented during the decision-making process, new approach methodologies (NAMs) have played an important role to support a mechanistic understanding of toxicity endpoints, providing relevant information to move toward such mechanistic understanding and bridging existing data gaps [10,11]. These NAMs include different in silico tools to estimate biological and toxicological effects of chemicals in data poor and data-rich situations at different levels of biology organization. Thus, NAMs have contributed of holistic background risk assessment models, generating a complete risk characterization outcome informative for environmental impact assessment, promoting enhancement of the regulatory and authorization processes [12,13].

Nowadays, there are multiple benefits from theoretical exposure estimations at an environment or health level to support a wide range of risk management decisions if it properly implemented; through well-recognized in silico strategies, it can determine the risk posed by chemicals on the ecosystem by combining available data on the chemical exposure and hazard identification [11,14].

Within different computational methods employed in ERA approach, the quantitative structure–activity relationships (QSAR) theory proposed by Corwin Hansch and Toshio Fujita in 1964 have stood out for offering useful elements to address different chemical issues involved in the complex processes and mechanisms of biotic and abiotic systems [15–17].

The principal aim of QSAR models consists in explaining the chemical response of a substance in terms of its structural features. A QSAR theoretical model proposes rational parallelism between the molecular structure of a chemical, which is represented by means of hundreds or even thousands of molecular descriptors, and the targeted end point that such chemical produces in a certain organism or environment. This is how the QSAR formalism allows estimating several final effects such as the activity, toxicity, or property of a new or hypothetical chemical, based on the knowledge of the molecular structure involved.

For more than 58 years, the exponential growth of the number of scientific publications based upon QSAR models has exposed its constant evolution, mainly attributed to the continuous increase in freely available chemical databases, as well as the inclusion of both simple and complex statistical techniques with high capability to establish correlations between several structural characteristics of molecules and a defined physicochemical, organoleptic, biological, or toxicological response [18]. As a result, QSAR models have been widely applied in academia, industry, and government agencies as successful computational tools for predicting adverse or favorable effects that chemicals, materials, or nanomaterials produce on human health and ecosystems [19].

The QSAR approach is one of the major computational strategies employed in environmental sciences, which has been included by several governmental entities worldwide, such as the Organization for Economic Co-operation and Development (OECD) [20], the European Chemicals Agency [21,22], the German Federal Institute: Bundesinstitut für Risikobewertung (BfR) [23], U.S. Food and Drug Administration (US FDA) [24], U.S. Environmental Protection Agency (US EPA) [25], and U.S. Department of Health and Human Services (US DHHS-NTS) [26], as part of their regulatory and authorization procedures, providing a complete theoretical tool with the ability to identify and predict multiple chemical exposures and hazardous end points related to the risk characterization (safety, efficacy, quality, and performance) of chemicals, especially in data-poor situations. Until now, there are numerous QSAR models focused on classifying and identifying the main structural features associated with hazard or toxic responses generated by millions of commercial chemical products.

However, it should be noted that the principal hypothesis of the QSAR methodology does not consist in explaining the mode of action (MoA) or the toxicokinetic, pharmacokinetic, and toxicodynamic effects of the molecules; in contrast, it focuses on measuring the targeted end point that a chemical produces on an organism or defined environment. Although, if the QSAR model achieves reliable activity predictions, it is able to provide a brief knowledge of the structural features or mechanism basis that may affect a specific end point, even can guide the rational design of new chemicals, or proposes a useful computational tool based on overall weight-of-evidence [27,28].

Several applications of the QSAR approach have demonstrated its ability for analyzing large molecular assemblages and predicting numerous end points defined by difficult-to-measure test protocols. Those conditions have allowed researchers to discern trends in the existing data with relevant information about public health or the environment, thus extrapolating trends to not-synthesized compounds or lacking chemical, physical, toxicological, and biological measures [29]. Currently, the QSAR approach has been contributed to innovative research aimed at enhancing the productivity of pesticides, feed additives and supplements, pharmaceuticals, and cosmetic products, which play an important role in the global economy and international marketing [30].

The QSAR theory paradigm does not only provide predictive models with the ability to quantify the toxicological and biological responses of different molecules. Besides, this paradigm has been considered to guide the rational synthesis of new agents based on the 12 principles of green chemistry, reducing time, and experimental costs involved in vivo and in vitro tests. In addition, the QSAR approach is a computer-assisted modeling method commonly used by animal welfare legislation in both Europe and the United States statements with the goal of limiting animal experiments to the absolutely necessary minimum level [31,32].

As part of the main scene of this chapter, the principal aspects of ERA and the role of NAMs in the regulation and authorization processes are described. Thereupon, the general features of the development and validation of QSAR models are detailed. The QSAR approach and its fundaments as well as general methodologies of the QSAR models are introduced. Moreover, the features, contributions, and advances of QSAR modeling in ERA processes are included. Last, but not least, the future perspectives of QSAR modeling as an alternative computational strategy in the ERA approach are given.

The principal aspects of ERA

In the modern world, environmental protection and sustainability are the main aims of developed and underdeveloped countries within their governing strategies. In this context, the ERA helps to inform decision-making in such regard, through tiered approaches that are applied depending on the assessment question(s), resources, and data available to produce a proper fit-for-purpose risk assessment, in order to answer the assessment question initially established.

For instance, tiering is a process that allows for the planning of a risk assessment [1–3]. The principal aim of ERA consists on evaluate the likelihood and extend of harmful effects resulting by mean of exposure of a chemical on several organism, such as mammalians, birds, plants or even microbes, and different levels of biological organization like single or multiple cells, individual population, and ecosystems [33].

The application of tiers depends on whether the assessment in based on data-poor cases allowing for simple and conservative approaches; in contrast, when data become available, more accurate and realistic approaches can be applied; particularly in a high tier for the re-evaluation of a regulated product. In similar manner to human health risk assessment, the process of ERA in typically composed of the three principal steps: (1) problem formulation, (2) hazard and exposure assessment, and (3) risk characterization. Each of these steps is described below:

- *Problem formulation*: it consists in a systematic planning step to identify the major features to be considered in a particular assessment in relation to preliminary hypotheses with regard to hazard assessment (likelihood and severity of adverse effect which might occur or have occurred) and exposure assessment (likelihood and relevance of exposure) [1]. Given the multiplicity of possible assessment end points and exposure paths, it is fundamental that the problem formulation is performed at the beginning of the process to clearly define the ecological entity to be evaluated and plan the direction of the assessment.
- *Hazard and exposure assessment*: the hazard assessment involves proper hazard identification and characterization by means of the real determination of the potential stressors that are associated with climate change and biological entities (e.g., viruses, bacteria, fungi, etc.), which cause adverse effects, due to their environmental fate properties or their intrinsic toxicity, and under what circumstances occur. Likewise, the exposure assessment evaluates both the potential exposure paths and scenarios as well as the extent and frequency of exposure.
- *Risk characterization*: it provides evidence that the protection goals defined in the problem formulation are reached as well as details sufficient protection for the organisms, and such assessment can be concluded. Alternatively, it indicates when sufficient protection cannot be achieved; in those cases, risk management or higher tiers should be considered (e.g., risk mitigation measures).

In a proper hazard identification and characterization for a regulated substance (e.g., pesticides, feed additives, cosmetics, or drugs) or environmental contaminants (e.g., heavy metals, persistent organic pollutants, or emerging contaminants), the first tier may consider ecotoxicity studies from standardized laboratory experiments with relevant animal experiments to extrapolate reference points/points of departure such as the no-observed-effect-concentration and the half-maximal lethal dose or lethal concentration for specific exposure time(s). Hence, exposure data are compared to such reference points to metric risk level.

New approach methodologies (NAMs)

Nowadays, several regulatory entities have gradually demanded acute and chronic toxicity data from proper and well-established toxicity assays as part of their procedures to register new chemicals, pharmaceuticals, pesticides, cosmetics, feed additives, supplements, and other industrial products. In consequence, such toxicological information is commonly obtained from regular in vivo assays, which are material-expending, time-consuming, and technically demanding as well as ethically unacceptable [1].

Particularly, the ERA research efforts have been focused on the development and application of the NAMs, commonly known for their efficiency to estimate targeted biological or toxicological information, considering any conceptual framework and modern toxicological strategies that serve as (replacement, reduction, or refinement) alternatives to animal testing [34,35].

The novel use of NAMs for regulatory decision-making refers to any technologies, methodologies, approaches, or combinations thereof that can be applied to provide information on chemical hazard and potential human exposure [36]. Such methodologies can avoid or significantly reduce the use of animals testing to the absolutely necessary minimum level, accomplishing with the goals of animal welfare legislation in Europe and the United States, which demand replace or reduce the use of vertebrate animals in the testing of chemical substances or mixtures and promote the development and timely incorporation of alternative test methods or strategies that do not require new vertebrate animal testing, such as in silico, in chemico, in vitro, and ex vivo approaches, which can also be split into different areas, for instance:

- Computational toxicology and bioinformatics
- High-throughput screening methods
- Testing of substances in chemical categories
- Tiered testing methods
- In vitro studies
- Systems biology
- New or revised methods identified by validation bodies such as the OECD and the Interagency Coordinating Committee on the Validation of Alternative Methods (ICCVAM)
- Industry consortia that develop information submitted under this title

For years, corporations have pursued a number of approaches to ensure the products they sell are safe without testing on animals. Hitherto, there are more than 40 in vitro methods approved by international regulatory bodies. However, there are other strategies that are also considered as NAMs, which are briefly detailed below:

- *Use of available information*: Existing data on target or similar chemicals can be used instead of conducting new assays. Some physicochemical properties can lead to avoid of assays for certain end points. It is noteworthy that a good hazard and exposure assessment starts with gathering of data, and the eChemPortal (https://www.echemportal.org/echemportal/) or the adverse outcome pathway (AOP) Knowledge Base (https://aopkb.oecd.org/) are suitable sources to begin [37].
- *Threshold of toxicological concern*: Based on information about other chemicals with similar structures and some data about a target chemical, it may be possible to establish a level below which a chemical is unlikely to be harmful.
- *Computer-based methods*: The unknown toxicity value of a chemical or a chemical set can be calculated by modeling the molecular structure of a well-defined molecular database, and its reliable toxicological data, using unequivocal mathematical algorithms and accurate validation techniques. There are several commercial or freely sources of useful computer-based methods that will be expressed in next section.
- In vitro *methods*: From simple cell cultures to 3D models of human skin, liver, and other organs, exist many methods available to assess chemical safety [38,39].
- *Education programs*: Within these programs, the New Approach Methodology Use for Regulatory Application (https://www.pcrm.org/ethical-science/animal-testing-and-alternatives/nura) can be used for training and resources to chemical companies and industry, government, academia, and other interested stakeholders on how to incorporate NAMs into testing strategies, promoting the use of nonanimal approaches within various regulatory frameworks.

Currently, there are different NAMs available to support the regulatory processes of companies, which could be included within the testing strategy programs as part of the alternative techniques to reduce animal experiments. Many of these methods have been approved by international validation bodies, which are appropriate when used alone and others should be applied as an integrated approach to testing and assessment in order to make a regulatory decision [40].

Successful implementation of NAMs to meet regulatory requirements for both intra- and interagencies collaborations involves more emphatic prioritization schemes to comprise criteria that allow decision-making and avoid simply having lower-priority compounds set aside awaiting exhaustive testing, once the higher-priority compounds pass through more emphatic testing of toxicity pathways [41]. For that reason, there are four basic multilevel strategies to employ NAMs and high-throughput exposure tools for context-dependent safety assessments.

The levels are characterized, in part, by the investment in information generated by each level and by confidence in the results. Obtaining confidence in their use will be essential to make NAMs acceptable by regulatory agencies [42]. For instance, the US EPA strategic plan based on the Toxic Substances Control Act (TSCA) has three important components: (1) identifying, developing, and integrating NAMs for TSCA decisions, (2) building confidence that the NAMs are scientifically reliable and relevant for TSCA decisions, and (3) implementing the reliable and relevant NAMs for TSCA decisions. However, more approaches to establish confidence (validation) will need to be developed within the four testing levels. Output of testing and exposure assessment at each level is suitable for informing different decisions [43]. In order to define the role of the testing levels of NAMs, each level is described below:

Level 1—Computational screening: high-throughput predictions of exposure, theoretical toxicity and expected metabolites are obtained using computational methods and would support chemical categorization and decision-making in limited contexts. The optimal suite of computational tools in this level should estimate physicochemical properties, infer possible toxicity by means of proper and reliable techniques, such as QSAR modeling, MoA, AOP, threshold of toxicological concern, and read-across methods, which can predict or may take into account likely metabolites, and assess exposures that would be expected to arise during anticipated use conditions.

The goal in Level 1 is computational tools that are developed with as large a range of compounds as possible in order to have confidence when calculating similar properties for a new compound or new class of compounds. Level 1 is particularly relevant in ERA key processes, due to the fact that the toxicological information is commonly obtained from conventional animal tests, which are material-expending, time-consuming, technically demanding, and ethically unacceptable. Such key processes are critical for chemical toxicity made it possible to identify where NAMs development is particularly required at different levels of the toxicological and ecotoxicological assays.

In some cases where Level 1 analyses fail to give a sufficiently large margin of safety, the next step should be to use Level 2 to test for biological responses in high-throughput cellular or subcellular assay platforms, first considering NAMs that target possible toxicity identified in Level 1.

Level 2—High-throughput in vitro *screening*: This level would be completed by rapid, high-throughput dose—response screening of agent bioactivity and high-throughput in vitro-in vivo assays extrapolation. Level 2 comprises several assays that can be easily run on a large number of chemicals in high-throughput mode, such as Toxcast (https://www.epa.gov/chemical-research/exploring-toxcast-data-downloadable-data) or Tox21 (https://ncats.nih.gov/tox21/) assays from US EPA and US National Institutes of Health, respectively [44]. Ideally, such experiments should be tailored toward endpoints of regulatory concern, with a proper understanding of how the assay fits known MoA or AOPs.

Level 2 was proposed to provide directly analogous data for a large set of chemicals in order to facilitate the grouping of chemicals by particular MoA and generate a platform where unknowns can be subjected to the same battery of assays, and MoA assigned based on the activity features seen in the results, even though many of these assays still show promise for grouping chemicals as part of a process known as biological read across.

In vitro approach is similar to hazard identification in the traditional risk assessment process, where activity assays and/or similarity to chemicals with known hazard profiles can be used to justify a finding that further study is or is not justified. Moreover, within this level, absorption, distribution, metabolism, and elimination data would also be combined with bioactivity assay data to convert AC_{50} values to human equivalent doses using in vivo or in vitro extrapolation techniques and compared with exposure estimates to generate a MoE.

A relevant challenge in respect of response interpretation in Level 2 assays, which search for distinguishing biologically relevant pathway responses from exposition effect that can arise from compounds with a lack of specific affinity for cellular assays, and at relatively high concentrations, elicit undetectable affinity.

Now, there are different gene expression platforms such as BioSpyder (https://www.biospyder.com/products) that reduce the cost of whole genome differential gene expression analysis. Furthermore, pathway visualization methods for MoA analysis such as MoAviz (https://scitovation.com/lri-tools-and-methods-mode-of-action-visualization-moaviz/) assess both potency and biological functions/pathways affected by treatment. In Level 2, confidence in subsequent MoE calculations will increase, and some Level 2 results may be accepted for decision-making, partially due to narrowing uncertainty regarding the MoE and knowledge of likely MoA.

Level 3—Fit-for-purpose assays and safety assessment: Fit-for-purpose assays would be considered with read-outs that involve key signaling processes for cellular pathways associated with a chemical's MoA and include markers for correlating with or directly measure adversity. In this sense, Level 3 would be designed so that the output from the NAMs in Level 2 and from more refined exposure assessments increase confidence in the calculated MoE and support proper risk assessment without leading up to more complex assays.

Ideally, dose—response value from the fit-for-purpose assays as well as computational pathway models could provide a mechanistic understanding of the shape of the dose—response curve and support more informed extrapolation to relevant human exposure [45]. Thereby, the quantitative in vitro-in vivo tool considers human-relevant metabolism coupled with dose—response relationships from the fit-for-purpose assays; thus quantitative in vitro-in vivo would provide more reliable predicted MoEs and define regions of safety, that is, exposure concentrations that do not represent a risk to human health.

Based on the MoE defined with these more comprehensive fit-for-purpose assays and better knowledge of use-specific exposures, a decision-maker might still consider this information to be insufficient, leading to consideration of more complex assays and detailed chemical-specific exposure data, such as ones included in Level 4.

Level 4—More Intact System: During the last years, the goal of delimiting multiple levels for context-appropriate testing is that over time the tools employed for targeted in vivo assays would be regarded as studies of last resource. Then, the possibility to accumulate relevant and context-dependent data at each level of NAMs should substantially reduce the number of chemicals analyzed through regular-used complex experiments.

Alternative multidimensional and multicellular experiments, providing a variety of biologically inspired test systems for conducting more integrated toxicity testing, could provide tissue-based dose responses rather than moving to studies in animal models. Moreover, MoA information obtained by means of high-throughput assay and fit-for-purpose analysis would support more limited testing in animal models targeted to the specific MoA. It is noteworthy that in a decision context, such as ranking liabilities of possible substitutes, lead candidate selection, prioritization for remediation, and formal regulatory decision-making, chemicals would not have to be tested sequentially at all four levels in a tiered approach.

In most cases, the NAMs assays are focused on human health risk-based decision-making; however, the principles and approaches included here also apply to environmental risk evaluation, highlighting that the experiments, in silico tools, and hazard and exposure assessments may need to be tailored to provide targeted ecological risk assessment decision context. Further, the context-dependent testing approach should have a wide chemical applicability domain, for which computational tools (QSAR modeling, high-throughput or read-across methods) could estimate different physicochemical, toxicology, and biology end points with reasonable accuracy and proper understanding of when more testing and exposure assessment are required, following the four levels detailed above [46].

The QSAR approach and its fundaments

In the last years, computer-assisted modeling methods, also known in silico methods, have been a key for accessing, deducing, modeling, exploring, and projecting wide chemical data sets, without requiring the synthesis, characterization, and testing of each compound. Due to the fact that computer simulations provide many advantages, they have been postulated as essential tools to strategically identify and design new chemical entities with specific effects [47].

The QSAR theory constitutes one of the main computational tools designed for developing rational correlation models based on the end point data of chemical compounds and their structural features. Thus, QSAR applications can find the relationship between one or more molecular structure aspects and the toxicity or activity produced by molecules, in order to identify the biological or toxicological capability of new or hypothetic compounds.

In that context, it is possible to infer that QSAR models are the final result of the computational process that begins with the extensive characterization of the molecular structures, using molecular descriptors, which encode the information of the physicochemical properties as well as the simple (constitutional and topological) and complex (electronic and geometric) structural features of each evaluated molecule; to finally obtain a model that raises some inferences, hypotheses and predictions of the final effect that a chemical compound could generate on a targeted organism or environmental resource [48].

Overall, the obtained correlations from a QSAR modeling are expressed for a mathematical equation defined through different linear or nonlinear statistical techniques, such as univariate regression, multivariable linear regression (MLR), partial least square, linear discriminant analysis, or cluster analysis, which propose rational parallelisms of quantifying or classifying a final end point.

QSAR theory has the ability to guide the rational, selective, and rapid design of new chemical precursors with unedited properties, considering only the information contains in the molecular structure. Thus, the development and use of the QSAR models could guide the search for alternative chemicals that may replace the indiscriminate use of commercial pesticides or improve the therapeutic effect of current drugs. Recently, the regular increase of QSAR research has shown their relevant ability to determinate, by a logic mathematical equation, the effect of the structural characteristics on the activity or toxicity of new molecules. This approach is particularly useful when the experimental assays are extensive, laborious, and expensive as well as involve the animal testing.

The QSAR theory was initially designed for estimating the end point that a molecular set generates against a defined organism or ecosystem. The most recent advances in QSAR modeling include the development of new molecular descriptors based upon available experimental information, such as receptor binding potential, the activation and inhibition enzymatic capability, or the rate of formation of a protein complex, in order to elucidate key processes associated with the final toxicological response from a molecule.

Thereby, this chapter includes the principal concepts of QSAR modeling as well as the most common QSAR applications. Likewise, several paradigms related to the usefulness, efficacy, scope, and limitations of QSAR modeling are described.

QSAR theory

The QSAR theory is based on the basic postulate of presenting the final effect generated by a chemical in terms of its structural features and physicochemical properties, which are defined by theoretical, semi-empirical, and experimental information. Certainly, the physicochemical parameters such as the boiling point, melting point, or surface tension are specific characteristics of a chemical that provide relevant information to explain a targeted end point. Therefore, from a mathematical view, the activity or toxicity of a compound can be represented as a function of its molecular structure and physicochemical properties, which are properly encoded by means of the so-called molecular descriptors, Eq. (10.1) [49].

$$\textbf{Activity / Property} = f(\textbf{molecular descriptors}) \tag{10.1}$$

This concept was raised in 1964 through the proposal of Corwin Hansch and Toshio Fujita to design mathematical regression models, which correlate the structural features from small molecular sets with some simple and known physicochemical properties, with the aim of estimating their therapeutic capability to design novel drugs [15–17].

In this sense, a QSAR model expresses a multivariable mathematical parallelism between several physicochemical and structural attributes encoded by molecular descriptors (X_{ij}), and an experimental function of biological, toxicological, or physicochemical type (Y_i). Thus, a model has the capability to estimate the end point value of each molecule. For instance, to one molecule i, the linear equation establishes a rational parallelism among molecular descriptors ($X_1, X_2, ..., X_n$) and required toxic or biologic response (Y), as is shown below:

$$\boldsymbol{Y_i = X_{i1}b_1 + X_{i2}b_2 + ... + X_{in}b_n + e_i} \tag{10.2}$$

Where b_n corresponds to the linear slopes that express the correlation between a given molecular feature (X_{in}) and the activity of compound $i(Y_i)$; e_i parameter represents a constant value. The slopes and constants are calculated by means of the regression analysis. There are different QSAR models based on a dependent variable (Y) of activity or toxicity, which has been useful during the regulatory and authorization processes of chemicals and will be detailed in the next section [10,27].

The QSAR models are exhaustive and rigorous computational strategies to anticipate active or inactive physicochemical, biological, organoleptic, or toxicological effects by analyzing a wide number of independent variables (molecular descriptors and/or fingerprints), processed through an unequivocal algorithm for selecting variables that can predict end point values from known or still unexplored chemicals. For this reason, the computational tools based upon QSAR modeling have gradually increased their role in academic research and industry as well as within regulatory processes to decision-making on the safety of chemical products.

Similarity influence in the QSAR models

The central paradigm underlying QSAR modeling is the principle of structure–activity similarity. This paradigm establishes that homologous chemicals contain similar activities or properties. Therefore, it can be deduced that the analysis and estimation of end points from similar chemical products lead to the final response of many others without such endpoint values. Within QSAR theory, the similarity term is associated with all compounds that have similar molecular structures, MoA, or functions. Fig. 10.1 exhibits two molecular structures belonging to deltamethrin (A) and cypermethrin (B) pesticides. These chemicals are structural and functionally analogous; both compounds share the common cyclopropanecarboxylic acid substructure in pyrethroids and have similar insecticidal power.

The possibility to define structural analogous involves a common mathematical approach that uses similarity indices to measure the degree of accordance among the patterns detailed in Fig. 10.2, which are responsible for encoding the structural features of molecules. It should be noted that a high degree of similarity based on the mathematical similarity indices does not indicate that there is similarity in the MoA or that they are functional analogous.

Similarity indices employ two-dimensional or three-dimensional information from chemical structures to define different types of correlation indices, such as the Tanimoto index, the Ricards Hodgkin index, the cosine similarity index linked to the Hamman's similarity and difference measure, and the Euclidean distance index based on the ordinary distance between the considered structural features [50].

Particularly, functional analogous have been identified due to the fact that they show similar physicochemical or biological behavior, without the need to be structural analogous. A relevant example of functional analogous compounds is pesticides, which belong to different chemical classes with functional similarity, as shown in Table 10.1.

As a result of the constant increase of QSAR studies, molecular similarity has been considered a valid strategy to generate molecular descriptors. The first application of molecular similarity coupled to QSAR modeling was proposed by Martin et al. [51]. Since then, molecular similarity has been used in several QSAR kinds of research, either within the classic formulation of Carbó indices or the topological similarity indices, among other QSAR applications [52].

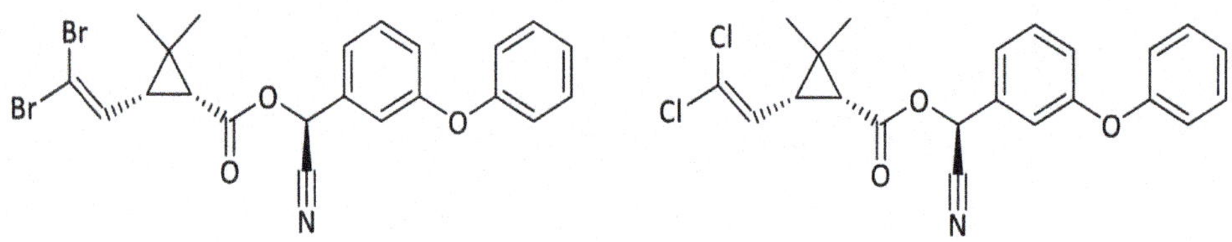

A B

FIGURE 10.1 Structural and functional analog compounds. Deltamethrin (A) and cypermethrin (B) pesticides are structural and functionally analogous.

FIGURE 10.2 Considered patterns in molecular similarity indices [10]. The A, B, C, and D patterns are commonly considered in molecular similarity indices.

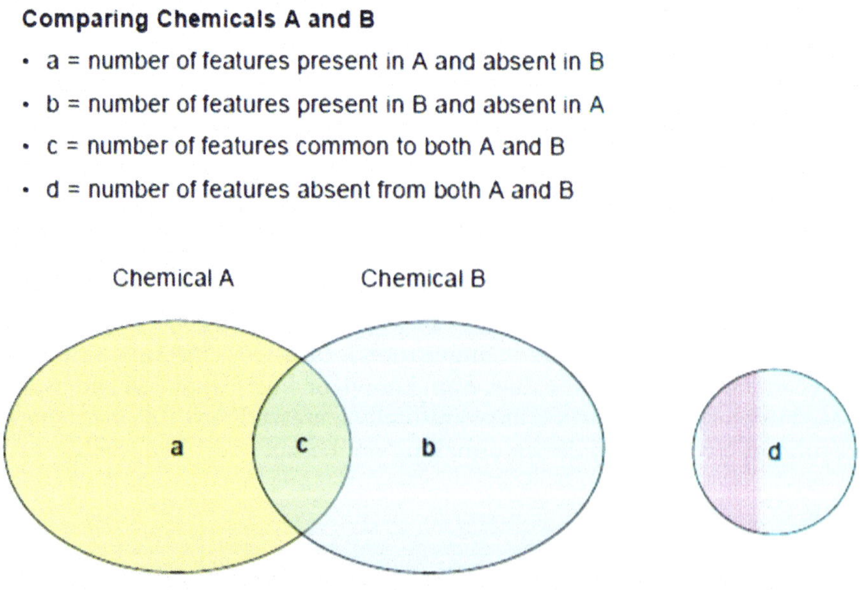

TABLE 10.1 Mode of action (MoA) related to chemical classes of commercial pesticides [20].

Mode of action	Chemical class
Acetylcholinesterase inhibitors (AChEIs)	Carbamates
	Organophosphates
GABA-activated chloride channel antagonist	Cyclodiene chloride derivatives
	Phenylpyrazole
Sodium channel modulators	Organochlorides
	Pyrethroids
Nicotinic acetylcholine receptor agonist	Neonicotinoids
Juvenile hormone mimic	Juvenile hormone analogous
	Carbamates
	Pyridines of insect growth regulator

Aims of QSAR modeling

The main goal of a QSAR model is predicting a targeted end point from new or untested chemicals, avoiding the time and costs required during the synthesis, instrumental characterization, and biological assays of large inactive or dangerous molecular sets. However, QSAR modeling has other goals that can help during ERA processes and are detailed below:

1. Guide the selective and rapid design of new chemical entities with the desirable final effects, reducing the experimental practices necessary to evaluate any biological, physicochemical, or toxicological effect caused by a molecule against a target organism or delimited ecosystem.
2. Reveal the theoretical, semi-empirical, or experimental variables associated with the active behavior of different molecular sets.
3. Reduce or replace the complex and expensive in vivo assays, especially such that involve animal slaughter or mistreatment.
4. Provide the range of concentration or dose necessary for a chemical substance to generate a specific final effect.
5. Elucidate an approximation of the mechanism of action that a chemical compound can produce in the environment or living beings.
6. Establish ligand-based or receptor-based selective search of large molecular sets with prominent candidates for drugs.
7. Carry out classification analysis based on the predictions, considering the activity categories established experimentally for the evaluated end point.
8. Optimize molecular structures with prominent properties or activities toward a particular and effective response profile.

General methodologies of the QSAR models

As part of the common methodologies employed during QSAR modeling, the structure–activity relationships, MLR analysis of Hansch, and comparative molecular fields analysis (CoMFA) have been highlighted among the most used strategies to estimate qualitative or quantitative property or activity data from different molecules. These mathematical methods are briefly detailed below.

Structure–activity relationship approach

The structure–activity relationship approach consists of the basic knowledge of estimating biological or physicochemical responses produced by the chemical products, in general, organic molecules, in the function of the structural features variation of them. Thus, it is possible to deduce that analogous molecules have similar final responses; if these are measured under the equal experimental conditions following the same test protocol [53].

Particularly, the structure–activity relationship study compares qualitative and quantitative experimental aspects through a mathematical analysis of variance, such as regression analysis, discriminant analysis, factor analysis, or pattern recognition method. According to the experimental values of the evaluated molecules, the correlation with the best fit is used to develop a mathematical expression that has the ability to estimate the activity or property of the unexplored or untested analogous molecules.

However, although the molecular structure contains all the chemical information, only part of that information can be quantified by experimental techniques. Therefore, it is necessary to use numerical variables so-called molecular descriptors, which have the ability to encode numerous structural features of simple or complex molecular systems.

Now, it can be inferred that a QSAR model seeks to quantify the structure–activity relationships by means of rational mathematical parallelism, which can be generated by different statistical techniques of mathematical variance analysis coupled to the computational tools recognized by exploring several molecular representations such as Euclidean geometry, graph theory, and vector representation.

Multivariable linear regression

The analysis proposed by Hansch and Fujita [17] is known as the extra-thermodynamic method, which is a linear free energy approximation for designing drugs inspired by a series of contemplated compounds within the formulation of mathematical prediction equations based on MLR combinations of simple physicochemical parameters.

This methodology employs MLR analysis to determine the best least squares that engage the dependent variable (activity or property) within a linear combination of independent variables (molecular descriptors). The best-achieved correlation, expressed in Eq. (10.3), is employed for quantifying the activity of chemical compounds without experimental data [55].

$$\log\frac{1}{C} = -a\pi^2 + b\pi + \rho\sigma + cE_s + dS + e \tag{10.3}$$

Where C corresponds to the evaluated activity or property; a, b, ρ, cyd are the regression coefficients derived from the linear curve that fits them; π, σ y E_s are hydrophobicity, electronic, and steric constants of the substituent, conceptualized by Hansch, Hammet, and Taft, respectively [17,54,55]; S indicates a topological factor that links the shapes variations that occur in molecules due to the interaction with receptors; e is the constant of the equation.

In this approach, the final response is expressed as $\log 1/C$. Therefore, the lower the $\log 1/C$ values, the greater the biological, pharmacological, or toxicological effect of molecules. For Hansch and Fujita model, the use of π, σ y E_s, or molar refractivity descriptors, allowed obtaining pharmacological information about simple organic compounds, and making predictions in more complex molecular systems reliably, quickly, and at low cost. Likewise, quantitative information on the activity of new molecules can be obtained within the reliability limit of the MLR technique.

The main advantage of the linear regression methods is that they present fewer over fitting problems in the selection of the most relevant molecular descriptors for the activity, due to the fact that they do not require the optimization of several parameters during the design of the model (just one regression coefficient per descriptor); thus, simpler-to-interpret mathematical equation can be obtained.

Comparative molecular fields analysis

Different research based on the QSAR approach suggests that biological selectivity results from each targeted compound performing highly specific interactions, such as hydrogen bonding with a ligand. However, other observations have indicated that ligand-binding preferences are mainly due to noncovalent field effects exerted in the spatial area of the ligands [56].

In the 1980s, this approach motivated the proposal of QSAR-3D studies, focusing on the molecular interaction field's space, where the type of representation of molecular structures, the number of molecular descriptors required to characterize a ligand field, and the availability of final effect data were decisive when establishing models with the capability to estimate different biological end points.

In particular, CoMFA analysis helps to define when drug−receptor interactions are noncovalent and if variations in activity depend on changes in the steric or electrostatic fields of drugs. This method employs the optimized molecular structure of a chemical substance to define the three-dimensional lattice around it, in order to measure the interactions of the atoms that act as a probe at each point of the lattice, and thus define the contour that joins the points with similar interaction values. In CoMFA, steric and electronic interaction variables associated with thousands of points of the lattice are used to describe the phenomenon that leads to the final response. The results obtained generated mathematical prediction expressions and contour maps that are used to interpret the most important regions for the activity.

Nowadays, there are a series of available programs to develop QSAR-3D models based on CoMFA, CoMFA with hydrophobicity, comparative molecular similarity index analysis, and comparative molecular active site analysis methods [57,58].

QSAR modeling

QSAR modeling explores different statistical techniques to design correlation models, which can correlate the quantitative information of the molecular structures with a delimited behavior of the same. The information requires for developing a QSAR model raises from two main sources:

- Hundreds of molecular descriptors encode several structural and physicochemical features of the chemical products.
- The available and well-defined experimental values of the required property or activity.

In general, the development, validation, and application of QSAR models consist of seven general steps, illustrated in Fig. 10.3. Such general steps should be considered during QSAR modeling, splitting into stages such as preparation, processing, validation and predictive performance, and interpretation of the model (only if it is possible). A brief description of each stage is included below.

The stages in QSAR modelling

Data preparation: collect experimental data and digitalize molecular structures.

Data control: data filtering and normalization.

Molecular descriptors calculating: by means of freely available or commercial programs the most relevant structural features of the molecules can be encoded.

Split of the molecular set: define balanced molecular subsets, in general, so-called training, validation and test sets.

Development of QSAR model: Involves several statistical techniques with well-defined variable selection strategies.

Internal and external validation model: uses different validation probes to define the predictive power, accuraccy, robustness and other validation criteria.

Interpretability: defines the Applicability Domain (ADom) of the model and its level of interpretation.

FIGURE 10.3 General stages in quantitative structure−activity relationship (QSAR) modeling [46].

Training

Proper data preparation begins with the selection of a molecular set with reliable experimental data, processed under strict laboratory measurements and assay protocols that guarantee to obtain of well-founded information on a delimited final response.

In the modeling of a delimited activity, property, or toxicity, the concentration required to show a punctual response is mainly evaluated. In this case, the most commonly used measurement parameters are the medium inhibitory concentration (IC_{50}) measure, which indicates the concentration that produces 50% inhibition or antagonism to the receptor; the medium effective concentration (EC_{50}), which details the concentration responsible for 50% of effective action; and medium lethal concentration, exhibiting concentration that causes the death of 50% evaluated population.

Each above mentioned measure is obtained by means of several test protocols that differ both in the recommended experimental conditions and in the evaluated animal model or cell cultures [59]. A fundamental concept about in vivo or in vitro biological measurements is that they tend to present greater uncertainty in relation to physicochemical measurements, due to the fact that they involve more adjustable variables, which include experimental uncertainty when evaluating the response produced by one or more living systems [59].

Modeling processes

In QSAR studies, there are three types of molecular descriptors classified by their calculating form: the experimental descriptors that quantify the obtained laboratory test values; semi-empirical descriptors from physicochemical measures data such as boiling point, melting point, and octanol/water partition coefficient (log *P*); and theoretical descriptors obtained by calculating molecular descriptors programs, which can generate tens of thousands of them. Table 10.2 includes different sources for calculating molecular descriptors that are commonly used during QSAR modeling, and such can be obtained through commercial or freely available software.

The QSAR theory is commonly supervised by a statistical approach that provides reliable and consistent predictions of the modeled property or activity. For that reason, it is important to verify the contribution and suitability of the experimental data incorporated into the modeling processes through the data pretreatment criterion, where indeterminate, constant, or almost constant (with a single different value) molecular descriptors are excluded as well as molecule duplicates are also eliminated of the modeling processes.

For modeling purposes, if it is possible the total molecular set should be partitioned into three molecular subsets commonly known as the training set, validation set, and test set, which are used to design, manage, and probe the model, respectively [27]. Furthermore, when there is additional information about required activity or property from unexplored or nontested molecules is available, it is proper to establish a new molecular set called application set, which is used to measure the application capability of the obtained model.

The initial molecular set can be split up evenly, considering structural features or property values. However, various chemometric techniques, such as *k*-means cluster analysis or sphere exclusion, are commonly employed for obtaining balanced molecular subsets, in accordance with the structural–activity relationships.

Validation and interpretation of the QSAR models

To consider a QSAR model as valid and reliable, it is essential to accomplish multiple internal and external validation criteria such as cross-validation, leave-one-out, leave-more-out, and other new and complex validation proof. Besides, during the QSAR model validation, proper external validation should be done, which guarantees the model's predictive power on unknown or nontested molecular sets.

Specifically, a QSAR model is not enabling the entire universe of existing or still unexplored molecules. Then, it is necessary that each QSAR model defines its applicability domain, which reveals the chemical scope of its predictions. Until now, QSAR modeling has numerous proposed validation techniques. Moreover, QSAR validation criteria are continuously increased, including more rigorous and difficult proofs for accomplishing [60].

Nowadays, the newest QSAR models are focused on predicting the modeled toxic activity, providing a possible interpretation of the phenomenon responsible for the final effect, and deducing some aspects of the mechanism of action involved based on the experimental or predicted data of other final effects associated with the evaluated activity [61].

Thereby, it should be mentioned that the aim of a QSAR analysis is to generate proper computational tools for predicting a targeted final effect generated by a set of new, hypothetical, or unexplored molecules, using information contained in molecular structure attributes. Although, if the QSAR model achieves a solid statistical base,

TABLE 10.2 List of the principal available programs and web platforms for calculating molecular descriptors and fingerprints.

Program	Description	Website
ACD/LogP freeware	It calculates Log *P* from structural fragments	https://www.acdlabs.com/products/percepta/predictors/logp/index.php
ACD/Tox Suite	Predicting toxicity end points	https://www.acdlabs.com/products/percepta/
ADAPT	Calculating both topological descriptors as well as geometric and electronic variables	http://research.chem.psu.edu/pcjgroup/adapt.html
ADMET predictor	It offers information about ADMET properties	https://www.simulations-plus.com/
AFGen	It calculates descriptors based on topological fragments	http://glaros.dtc.umn.edu/gkhome/afgen/overview
BioPPsy	Calculating pharmacokinetic properties to estimate targeted drugs	https://sourceforge.net/projects/bioppsy/
BioTriangle	Expecting molecular descriptors from 0D to 3D dimensionality	http://biotriangle.scbdd.com/
BlueDesc	It reports several descriptors from the main molecular structure features	http://www.ra.cs.uni-tuebingen.de/software/bluedesc/welcome_e.html
CACTVS	It calculates 0D-2D molecular descriptors and 3D geometric variables	https://www2.chemie.uni-erlangen.de/software/cactvs/tools.html
CDK	Calculating topological, geometric, constitutional, and charge descriptors	https://cdk.github.io/
ChemDes	It reports 3679 molecular descriptors and 59 fingerprints	http://www.scbdd.com/chemdes/
CODESSA PRO	It estimates quantum-chemical, topological and electronic descriptors	http://www.codessa–pro.com/
CoFFer	It estimates molecules and makes alerts for the fragments associated with the evaluated property	http://coffer.informatik.uni-mainz.de/
CORAL	It can build QSAR models based on flexible descriptors	http://www.insilico.eu/coral/CORALSEA.html
Danish	It provide semi-empirical descriptors based on toxicological information from 166.000 molecules	http://qsar.food.dtu.dk/
DRAGON	DRAGON 7 calculates 5270 0D-3D descriptors as well as fingerprints	http://www.talete.mi.it/products/dragon_description.htm
ECOSAR	It defines relevant properties that are associated to aqueous toxicity	https://www.epa.gov/tsca-screening-tools/ecological-structure-activity-relationships-ecosar-predictive-model

Continued

TABLE 10.2 List of the principal available programs and web platforms for calculating molecular descriptors and fingerprints.—cont'd

Program	Description	Website
E-DRAGON	It calculates 1600 freely available molecular descriptors from DRAGON program	https://michem.unimib.it/news/virtual-computational-chemistry-laboratory-vcc-lab/
EPI Suite	EPA program that estimates physic-chemical properties focused on environmental sources	https://www.epa.gov/tsca-screening-tools/epi-suitetm-estimation-program-interface
eTOXlab	For modeling purposes based on relevant information to control new chemical products	https://github.com/phi-grib/eTOXlab
GRID	It calculates the GRID empirical force field at the grid point	http://www.moldiscovery.com/soft_grid.php
GUSAR	It provides acute and chronic toxicity property information from in vivo assays on rats	http://pharmaexpert.ru/GUSAR/antitargets.html
HYBOT-PLUS	It can estimate hydrogen bonds and free energy factors	https://www.timtec.net/software/hybot-plus/
ISIDA-Fragmentor	It generates substructural fragments and marked fragments with the ISIDA property	http://infochim.u-strasbg.fr/spip.php?rubrique41
MD-LOVIs	It provides 352.000 molecular descriptors from the vector of atomic weights and local vertex invariants	http://tomocomd.com/md-lovis
MolCode Toolbox	Calculating toxicological and biological properties for environmental and sanitary controlling	https://www.compudrug.com/molcode_toolbox
MOLCONN-Z	Estimating common topological descriptors, such as electro-topological variables	http://www.edusoft-lc.com/molconn/
Mold^2	Calculating 777 descriptors based upon molecular topology	https://www.fda.gov/science-research/bioinformatics-tools/mold2
MOLE db	It offers 1124 molecular descriptors from 234.773 molecules	http://michem.disat.unimib.it/mole_db/
MOLGEN	Generating 708 arithmetic, topological and geometric descriptors	http://www.molgen.de/
MolInfo	It estimates structural and physicochemical properties from molecular structure	http://cdb.ics.uci.edu/cgibin/tools/MolInfoWeb.py
MolSig	Calculating descriptors according to the molecular stereochemistry	http://molsig.sourceforge.net/
OCHEM	It provides theoretical or experimental information of several physicochemical and biological properties	https://ochem.eu/home/show.do

TABLE 10.2 List of the principal available programs and web platforms for calculating molecular descriptors and fingerprints.—cont'd

Program	Description	Website
Open3dqsar	It makes chemometric analysis considering high performance of molecular interaction fields	http://open3dqsar.sourceforge.net/
PaDEL	Calculating 1.875 1D, 2D y 3D descriptors as well as 12 types of fingerprints (16.092 bits)	http://www.yapcwsoft.com/dd/padeldescriptor/
Pattern Match Counter	Counting the functional groups or structural fragments of the molecule	http://chemdb.ics.uci.edu/cgibin/PatternMatchCounterWeb.py
PowerMV	It makes statistical analysis, draw and visualization of the molecule, calculation of molecular descriptors, and search for structural similarity	https://www.niss.org/research/software/powermv
QSARpro	Calculating more than 1000 topological and geometric molecular descriptors	https://www.vlifesciences.com/products/QSARPro/Product_QSARpro.php
QuBILS-MAS	It estimates algebraic descriptors of bilinear (2D), quadratic, and linear forms	http://tomocomd.com/qubils-mas
SiRMS	It generates 2D descriptors to isolates molecules, mixtures, semi-mixtures and reactions	http://qsar4u.com/pages/sirms.php
SYBYL-X	Estimating EVA geometric descriptors associated to CoMFA and CoMSIA field analysis	https://www.certara.com/pressreleases/certara-enhances-sybyl-x-drug-design-and-discovery-software-suite/
T.E.S.T.	It reports different toxicity endpoints from QSAR models and their predictions	https://www.epa.gov/chemical-research/toxicity-estimation-software-tool-test
TMACC	It generates 2D descriptors considering topological maximum cross-correlation	http://comp.chem.nottingham.ac.uk/download/tmacc/index.html
VEGA HUB	In silico tool to predict relevant properties for regulatory or academic purposes	https://www.vegahub.eu/download/
VIDEAN	Calculating visual and interactive descriptors	http://lidecc.cs.uns.edu.ar/VIDEAN/

accomplishing several internal validation criteria and providing correct prediction of the activity or toxicity values from the external validation set, it could assist the rational and selective design of new chemical entities with specific and effective properties, and also guide ERA processes in order to avoid in vivo assays, which are extensive, laborious, and expensive as well as involve the use of animal testing.

Features, contributions, and advances of QSAR modeling in ERA processes

Within the evaluation strategies and determination of the active response of the molecules, different in vivo and in vitro assays are used to obtain information about the harmful or benign effects that these chemicals generate on an animal model or a cell culture. These types of assays are considered highly complex due to the fact that when evaluating large molecular libraries, they require more time and high-cost input. In addition, most in vivo test protocols are ethically unacceptable to the international scientific community because they involve animal sacrifice or mistreatment [29].

In particular, the need to design applicable, reproducible, simple, and accurate computational tools to anticipate various final toxicological or biological effects generated by one or several chemical substances in a defined organism or environment, has promoted the application of QSAR models, as an alternative to the complex tests, for predicting several toxicological endpoints of large molecular sets, highly heterogeneous from natural, semi-synthetic, or synthetic chemicals, in order to streamline the authorization processes by regulatory entities [1,31].

Furthermore, the implementation of new computational methods has promoted the development of specific strategies in ERA processes, supporting chemical categorization and decision-making in limited contexts, and defining the risk posed by chemicals in the environment by combining available data on chemical exposure and hazard identification.

In the last case, the experimental data for the toxicological modeling could be collected from different scientific studies or web databases, depending on the targeted toxicity endpoint. Nowadays, there are more than 100 of web pages that provide reliable information about different endpoints with human health or environmental care purposes. Some these are detailed below [62]:

- **Aquatic toxicity:**
  ECOTOX: http://cfpub.epa.gov/ecotox/
  TOXNET: https://toxnet.nlm.nih.gov/
  ESIS: https://old.datahub.io/dataset/esis
- **Carcinogenesis and genotoxicity:**
  LAZAR: http://www.in-silico.de/
  RITA: https://reni.item.fraunhofer.de/reni/public/rita/
  ISSCAN: http://alttox.org/resource-center/databases/
- **Developmental toxicity:**
  BDSM:https://kundoc.com/pdf-data-inputmodule-for-birth-defects-systemsmanager-.html
  Devtox: https://www.devtox.org/index_en.php
- **Drug biological products, food and agrochemical induced toxicity:**
  AERS: https://healthdata.gov/dataset/adverseevent-reporting-system-aers
  CEDI/ADI Database: https://www.fda.gov/food/ingredientspackaginglabeling/packagingfcs/cedi/default.htm
  Toxtree: http://toxtree.sourceforge.net/
  HERA: https://www.heraproject.com/RiskAssessment.cfm
  ACToR: https://actor.epa.gov/actor/home.xhtml

For QSAR modeling, rational mathematical parallelisms are considered in order to contemplate the more relevant aspects of each molecular structure, codifying thousands of topological, constitutional, electronic, and geometric molecular descriptors, which can be generated by freely available and commercial programs.

In general terms, a program is one of the major components of the study required for enhancing different tasks involved in QSAR modeling. Such tasks where the software is required are: (1) a sufficient amount of high-quality data, (2) proper selection of molecular descriptors, (3) QSAR matrix preprocessing that can be performed in one or several steps, depending on the nature, quantity, and quality of the dataset, (4) appropriate filtering strategy to remove indeterminate, constant, or near-constant descriptors (with a single different value); as well as exclude duplicate or wrongly imported molecules, (5) balanced molecular set division, (6) model development, employing statistical strategies to select the optimal variables simply, rapidly, and systematically; finally, (7) validation process, defining applicability domain (AD), and accomplishing different validation techniques. Currently, exist in a wide variety of computational tools based on QSAR theory for modeling and predicting different targeted toxicity endpoints, some of them are briefly detailed in Table 10.3.

Most of the in silico tools included in Table 10.3 are freely available, enabling researchers can run, copy, and distribute such tools anywhere in the scientific community. For instance, OpenFoodTox tools from the European Food Safety Authority chemical hazard database have provided low-cost and rapid means for quantitative and qualitative assessment of chemical toxicity [6], including 1650 risk assessment for over 4500 chemical products, which are considered within ERA processes, the premarketing inspections of regulated substances (e.g., drugs, pesticides, feed additives, and cosmetics) and the postmarket surveillance system (e.g., commercial chemicals, emerging contaminants, or natural compounds).

TABLE 10.3 Freely available or commercial software tools for QSAR modeling and predictions.

Program	Developer and description	Website
ACD/ChemSketch	It allows you to digitize molecular structures in various molecular representation formats and calculates some atomic properties, e.g., polarizability, molar volume, molecular weight, molar refractivity, dielectric constant, refractive index, surface tension.	https://www.acdlabs.com/resources/freeware/chemsketch/
CORAL-QSAR/QSPR	A freely available tool that allows the design of QSAR models that involve only flexible descriptors.	http://www.insilico.eu/coral/
EPA's EPI Suite and ECOSAR	It based on physicochemical properties and toxicological estimations that are obtained by QSAR models and have environmental applications, e.g., Henry's law (log H_k *EPI*) and the sorption coefficient of soil and sediments (log K_{oc} *EPI*).	https://www.epa.gov/tsca-screening-tools/download-epi-suitetm-estimation-program-interface-v411
QSARINS	A software for the development and validation of multivariable linear regression (MLR) models by ordinary least squares (OLS) and genetic algorithm (GA) for variable selection.	http://www.qsar.it/
BioTransformer3	It is an open access software tool, and freely accessible web service for accurate, and comprehensive in silico metabolism prediction and metabolite identification.	https://biotransformer.ca/
ADMET Predictor	It is the flagship machine learning platform for ADMET modeling with extended capabilities for data analysis, metabolism prediction, and AI-driven drug design.	https://www.simulations-plus.com/software/admetpredictor/toxicity/
CERAPP	It allows large-scale modeling using 32,464 structures showed the possibility of screening large libraries of chemicals using a consensus of different in silico approaches.	http://ehp.niehs.nih.gov/15-10267/
OpenFoodTox	EFSA's chemical hazard database It contains hazard data used in over 1650 risk assessment for over 4500 chemicals.	https://www.efsa.europa.eu/en/data-report/chemical-hazards-database-openfoodtox
OECD QSAR Toolbox	It allows the use of QSAR models using a category approach for the prediction of toxic properties of chemicals, supports testing through intelligent testing strategies when data gaps are identified and sustainable development and green chemistry through toxicity predictions before chemical production.	https://www.oecd.org/chemicalsafety/risk-assessment/oecd-qsar-toolbox.htm

Continued

TABLE 10.3 Freely available or commercial software tools for QSAR modeling and predictions.—cont'd

Program	Developer and description	Website
VEGA Hub	It provides a platform with a large number of QSAR models and read across tools predicting a wide range of biological, environmental, and physicochemical properties of chemicals in relation to their structure.	www.vegahub.eu
OPERA	It obtains predictions for physicochemical properties, environmental fate parameters and toxicity endpoints as well as ecotoxicity parameters including fish BCF, soil adsorption coefficient, and biodegradability.	https://github.com/kmansouri/OPERA

In addition, the Estimation Program Interface (EPI suite) developed by US EPA [63] is a freeware that estimates physical/chemical and environmental fate properties, using a single input to run estimation programs, such as KOWWIN (estimates the log octanol-water partition coefficient, log K_{OW}, of chemicals using an atom/fragment contribution method), MPBPWIN (melting point, boiling point, and vapor pressure of organic chemicals are estimated using a combination of techniques), KOCWIN (estimates the organic carbon-normalized sorption coefficient for soil and sediment), BCFBAF (estimates fish bioconcentration factor and its logarithm using two different methods, the first is the traditional regression based on log K_{OW}, and the second involves the Arnot-Gobas method, which calculates BCF from mechanistic first principles), among other physicochemical properties. Moreover, the EPI suite includes the ECOSAR module that estimates ecotoxicity, for example, aquatic toxicity [64].

Finally, the most recognized open sources of QSAR modeling platforms (included in Table 10.3) with a relevant role in the ERA approach are the following:

- *The OECD QSAR Toolbox* is commonly used by governments, the chemical industry, and other stakeholders in filling gaps in ecotoxicity data needed for assessing the hazards of chemicals [65]. QSAR Toolbox is focused on the prediction of toxic properties of chemicals, supporting testing by means of intelligent testing strategies when data gaps are identified and sustainable development and green chemistry through toxicity predictions before chemical production.
- *The VEGA Hub* includes QSAR models about bees and daphnia (acute toxicity predictions) and fish (e.g., predictions of bioaccumulation factors and acute toxicity) toxicological assays, which include appreciated information for ERA processes [66].
- The OPEn Structure–Activity/Property Relationship App is a suite of open source developed by the US-EPA and now supported by the US National Institute of Environmental Health Sciences, which have been built on curated data and QSAR-ready chemical structures standardized using open-source workflow. Overall, the datasets used to build OPERA models range from 150 chemicals for biodegradability half-life to 14,050 chemicals for log K_{OW}, with an average of 3222 chemicals across all endpoints [67].

Future perspective of QSAR modeling within ERA approach

New horizons for ERA process to move toward holistic approaches which aim at protecting humans, animals, and the environment. A typical example of that includes the consideration of critical aspects of species sensitivities in relation to chemical toxicity in risk assessment, especially when considering the balance among farming, safety, and sustainability [8–10].

In order to facilitate this paradigm, constant development in ERA studies is ongoing through the use of NAMs approaches, which have played an important role to support mechanistic understanding of toxicity, providing a means to move toward such mechanistic understanding as well as bridging data gaps [68]. These NAMs include different in silico tools to predict toxicity endpoints from chemicals in data-poor and data-rich situations of different levels of biology organization.

In a regulatory context, the QSAR approach has provided ways to cover different stressors that will not substitute but will complement the assessment of single chemicals, offering the possibility to connect prospective and retrospective models, and use monitoring data for the calibration and verification of the predictions. This approach proposes to update the premarketing risk assessment of each regulated chemical, considering the actual decision, and then integrate the risk in well-recognized QSAR model platforms at regional, national, and international levels as appropriate.

Overall, the evaluation resulting from the application of QSAR models should be integrated based on the evidence available, if it is possible [69]. In fact, the integration of different models has offered reliability in the ERA processes, considering multiple lines of evidence. However, this also requires the user to be familiar with QSAR modeling approaches, which have been detailed throughout this chapter, and to take into consideration the detailed information provided by each tool to interpret the output and make conclusions with confidence.

Finally, it is remarkable that the continuous development of new substances with novel properties or activities will promote the permanent modification of the environmental impact. In consequence, new environmental risk management strategies will open for regulators, developers, and practitioners, which would encourage accurate computational tools, in order to assess the real environmental impact of new chemicals and establish strategic practices to define it. Moreover, as part of relevant future perspectives, the development of open-source databases in reporting structured hazard and chemical exposure data will provides a means of investigating species sensitivities at the individual and collective levels.

Acknowledgments

We acknowledge the electronic library facilities from Ministerio de Ciencia, Tecnología e Innovación Productiva (MINCyT). LMS, JCG, and PRD are members of the scientific researcher career of the National Scientific and Technical Research Council of Argentina (CONICET).

References

[1] Astuto MC, Di Nicola MR, Tarazona JV, Rortais A, Devos Y, Djien Liem AK, et al. *In silico* methods for environmental risk assessment: principles, tiered approaches, applications, and future perspectives (chapter 23). In: Benfenati E, editor. *In silico* methods for predicting drug toxicity. Methods in molecular biology. New York, USA: Humana; 2022. p. 589–636.

[2] Bopp SK, Kienzler A, Richarz AN, van der Linden SC, Paini A, Parissis N, Worth AP. Regulatory assessment and risk management of chemical mixtures: challenges and ways forward. Crit Rev Toxicol 2019;49:174–89.

[3] Ingenbleek L, Lautz LS, Dervilly G, Darney K, Astuto MC, Tarazona J, et al. Risk assessment of chemicals in food and feed: principles, applications and future perspectives. In: Harrison RM, editor. Environmental pollutant exposures and public health. London, United Kingdom: The Royal Society of Chemistry; 2021. p. 1–38.

[4] U.S. EPA. Concepts, methods, and data sources for cumulative health risk assessment of multiple chemicals, exposures and effects: a resource document. Washington, DC: U.S. Environmental Protection Agency; 2008. EPA/600/R-06/013F.

[5] Organisation for Economic Co-operation and Development (OECD). Guidance on grouping of chemical. Environment health and safety publications, series on testing and assessment no. 80, report no. ENV/JM/MONO(2007) 28, JT03232745. Paris: OECD; 2007.

[6] EFSA (European Food Safety Authority). Modern methodologies and tools for human hazard assessment of chemicals. EFSA J 2014;12(4): 3638–87. https://doi.org/10.2903/j.efsa.2014.3638.

[7] European Medicines Agency. Guideline on the environmental risk assessment of medicinal products for human use. Pre-authorisation evaluation of medicines for human use. EMEA/CHMP/SWP/4447/00 corr 2. 2006. https://www.ema.europa.eu/.

[8] Tarazona JV. Site-specific environmental risk assessment. In: Wexler P, editor. Encyclopedia of toxicology. 3rd ed. New York, USA: Academic Press; 2014. p. 278–82.

[9] Myatt GJ, Ahlberg E, Akahori Y, Allen D, Amberg A, Anger LT, et al. *In silico* toxicology protocols. Regul Toxicol Pharmacol 2018;96:1–17.

[10] Kruhlak NL, Benz RD, Zhou H, Colatsky TJ. (Q)SAR modeling and safety assessment in regulatory review. Clin Pharmacol Ther 2012;91(3): 529–34.

[11] Scotti L, Ishiki HM, Cavalcante Duarte M, Branquinho Oliveira T, Scotti MT. Computational approaches in multitarget drug discovery. In: Nicolotti H, editor. Computational toxicology: methods and protocols. Bari, Italy: Humana Press; 2018. p. 327–45.

[12] Punt A, Bouwmeester H, Blaauboer BJ, Coecke S, Hakkert B, Hendriks DF, et al. New approach methodologies (NAMs) for human-relevant biokinetics predictions. Meeting the paradigm shift in toxicology towards an animal-free chemical risk assessment. ALTEX 2020;37(4): 607–22.

[13] European Chemical Agency (ECHA). New approach methodologies in regulatory science: proceedings of a scientific workshop. Helsinki, Finland: European Chemicals Agency; 2016.

[14] Venkatapathy R, Wang NCY. Developmental toxicity prediction (chapter 14). In: Reisfeld B, Mayeno AN, editors. Computational toxicology: methods in molecular biology. vol II. Heidelberg, Germany: Springer; 2013. p. 305–40.
[15] Hansch C, Maloney PP, Fujita T, Muir RM. Correlation of biological activity of phenoxyacetic acids with Hammett substituent constants and partition coefficients. Nature 1962;194:178–80.
[16] Hansch C. Substituent constants for correlation analysis in chemistry and biology. New York: John Wiley and Sons; 1979.
[17] Hansch C, Leo A. Exploring QSAR, fundamentals and applications in chemistry and biologyvol. 1. Washington DC, USA: American Chemical Society (ACS), Professional Reference Book; 1995. p. 139–205.
[18] Cronin M. Quantitative structure-activity relationships (QSARs)—applications and methodology (chapter 10). In: Puzyn T, Leszczynski J, Cronin MTD, editors. Recent advances in QSAR studies: methods and applications. Heidelberg, Germany: Springer Press; 2010. p. 3–11.
[19] Muhammad U, Uzairu A, Arthur DE. Review on: quantitative structure activity relationship (QSAR) modelling. J Anal Pharm Res 2018;7(2): 240–2.
[20] OECD environment health and safety publications series on testing and assessment no. 69 guidance document on the validation of (quantitative) structure–activity relationship [(Q)SAR] models. 2007. Available on: http://www.oecd.org/officialdocuments/publicdisplaydocumentpdf/?cote5env/jm/mono(2007)2&doclanguage5en. (Accessed 21 September 2022).
[21] European Chemical Agency (ECHA). Guidance on information requirements and chemical safety assessment chapter R.6: QSARs and grouping of chemicals. 2008. Available on: https://echa.europa.eu/documents/10162/13632/information_requirements_r6_en.pdf/77f49f81-b76d-40ab-8513-4f3a533b6ac9.
[22] European Chemical Agency (ECHA). Practical guide: how to use and report (Q)SARs. 2016. Available on: https://echa.europa.eu/documents/10162/13655/pg_report_qsars_en.pdf.
[23] The German Federal Institute: Bundesinstitut für Risikobewertung (BfR). Research on risk identification, early risk detection and risk reduction. Research program, research areas: uses of QSAR in chemical assessment. Berlin, Germany: Bundesministerium für Ernährung und Landwirtschaft (BMEL); 2007. Available on: https://www.bfr.bund.de/en/use_of_qsar_in_chemical_assessment-10429.html. [Accessed 26 September 2022].
[24] U.S. Food and Drug Administration (U.S. FDA). Applied regulatory science: regulatory science at CDER. Department of Health and Human Services; 2017. Available on, https://www.fda.gov/drugs/scienceresearch/ucm294603.htm#Computational. [Accessed 23 September 2022].
[25] United State Environmental Protection Agency (US EPA). Q)uantitative structure activity relationship [(Q)SAR] guidance document. In: Manibusan M, Peterson J, Kent R, Chen J, et al., editors. North American Free Trade Agreement (NAFTA) Technical Working Group on Pesticides (TWG). Atlanta, USA: U.S. EPA; 2012. Available on: https://www.epa.gov/pesticide-registration/quantitative-structure-activity-relationshipqsar-guidance-document. [Accessed 23 September 2022].
[26] U.S. Department of Health and Human Services–National Toxicological Program (NTS). Computational toxicology projects. 2020. Available on: https://ntp.niehs.nih.gov/go/niceatm-comptox.
[27] Benfenati E, Clook M, Fryday S, Hart A. QSARs for regulatory purposes: the case for pesticide authorization. In: Benfenati E, editor. Quantitative structure-activity relationships (QSAR) for pesticide regulatory purposes. Amsterdam, Netherlands: Elsevier Science Ltd; 2007. p. 1–57.
[28] Chen B, Zhang T, Bond T, Gan Y. Development of quantitative structure activity relationship (QSAR) model for disinfection by product research: a review of methods and resources. J Hazard Mater 2015;299:260–79. https://doi.org/10.1016/j.jhazmat.2015.06.054.
[29] Cherkasov A, Muratov EN, Fourches D, Varnek A, Baskin II, Cronin M, et al. QSAR modeling: where have you been? Where are you going to? J Med Chem 2014;57:4977–5010.
[30] Höfer T, Gerner I, Gundert-Remy U, Liebsch M, Schulte A, Spielmann H, et al. Animal testing and alternative approaches for the human health risk assessment under the proposed new European chemicals regulation. Arch Toxicol 2004;78:549–64.
[31] Burello E. Review of (Q)SAR models for regulatory assessment of nanomaterials risks. NanoImpact 2017;8:48–58.
[32] Saavedra LM, Duchowicz PR. Predicting zebrafish (*Danio rerio*) embryo developmental toxicity through a non-conformational QSAR approach. Sci Total Environ 2021;796:148820.
[33] ECHA (European Chemicals Agency). The use of alternatives to testing on animals for the REACH regulation. ECHA article 117(3) of the REACH regulation report no.: ECHA-17-R-02-EN. Helsinki, Finland: ECHA; 2017.
[34] Dusinska M, Rundén-Pran E, Schnekenburger J, Kanno J. Toxicity tests: in vitro and in vivo. In: Fadeel B, Pietroiusti A, Shvedova AA, editors. Adverse effects of engineered nanomaterials: exposure, toxicology, and impact on human health. 2nd ed. London, United Kingdom: Academic Press; 2017. p. 51–82.
[35] Stucki AO, Barton-Maclaren TS, Bhuller Y, Henriquez JE, Henry TR, Hirn C. Use of new approach methodologies (NAMs) to meet regulatory requirements for the assessment of industrial chemicals and pesticides for effects on human health. Front Toxicol 2022;1(4):964553. https://doi.org/10.3389/ftox.2022.964553.
[36] Miller-Holt J, Behrsing HP, Clippinger AJ, Hirn C, Stedeford TJ, Stucki AO. Use of new approach methodologies (NAMs) to meet regulatory requirements for the assessment of tobacco and other nicotine-containing products. Front Toxicol 2022;9(4):943358. https://doi.org/10.3389/ftox.2022.943358.
[37] EChemPortal. Organisation for economic co-operation and development (OECD). 2022. Available on: https://www.echemportal.org/echemportal/content/participants. [Accessed 15 October 2020].
[38] Melvin EA, McMullen PD, Phillips MB, Yoon M, Pendse SN, Clewell HJ, et al. Developing context appropriate toxicity testing approaches using new alternative methods (NAMs). ALTEX 2019;36(4):523–34. https://doi.org/10.14573/altex.1906261.
[39] Crofton KM, Mundy WR. External scientific report on the interpretation of data from the developmental neurotoxicity in vitro testing assays for use in integrated approaches for testing and assessmentvol. 18. EFSA Support; 2021. p. 1–42.
[40] Thomas RS, Philbert MA, Auerbach SS, Wetmore BA, Devito MJ, Cote I, et al. Incorporating new technologies into toxicity testing and risk assessment: moving from 21st century vision to a data-driven framework. Toxicol Sci 2013;136:4–18. https://doi.org/10.1093/toxsci/kft178.
[41] Avila AM, Bebenek I, Bonzo JA, Bourcier T, Davis Bruno KL, Carlson DB, et al. An FDA/CDER perspective on nonclinical testing strategies: classical toxicology approaches and new approach methodologies (NAMs). Regul Toxicol Pharmacol 2020;114:104662. https://doi.org/10.1016/j.yrtph.2020.104662.

[42] FDA. Advancing regulatory science at FDA: focus areas of regulatory science (FARS). U.S. Food & Drug Administration; 2021. Available on: https://www.fda.gov/media/145001/download. [Accessed 23 October 2022].
[43] US EPA. Strategic plan to promote the development and implementation of alternative test methods within the TSCA program. 2018. Available on: http://tiny.cc/tufdez. [Accessed 23 October 2022].
[44] Richard AR, Judson K, Houck C, Grulke P, Volarath I, Thillainadarajah C, et al. The ToxCast chemical landscape—paving the road to 21st century toxicology. Chem Res Toxicol 2016;29(8):1225–51.
[45] Yoon M, Blaauboer BJ, Clewell HJ. Quantitative *in vitro* to *in vivo* extrapolation (QIVIVE): an essential element for *in vitro*-based risk assessment. Toxicology 2015;332:1–3. https://doi.org/10.1016/j.tox.2015.02.002.
[46] Roy K, Kar S, Nas RN. Understanding the basics of QSAR for applications in pharmaceutical sciences and risk assessment. London, United Kingdom: Elsevier; 2015.
[47] Dearden JC. The history and development of quantitative structure-activity relationships (QSARs). Int J Quantum Struct Prop Relat 2016;1(1): 1–44.
[48] Illing P. General overview of the safety evaluation of chemicals. In: Hester RE, Harrison MR, editors. Alternatives to animal testing. Environmental science and technology. vol. 23. Cambridge, United Kingdom: The Royal Society of Chemistry; 2006. p. 3–24.
[49] Walker JD, Jaworska J, Comber MHI, Schultz W, Dearden JC. Guidelines for developing and using quantitative structure–activity relationships. Environ Toxicol Chem 2003;22(8):1653–65.
[50] Rum G, Herndon WC. Molecular similarity concepts. Analysis of steroid-protein binding constants. J Am Chem Soc 1991;113:9055–60.
[51] Martín M, Sanz F, Campillo M, Pardo L, Péres J, Turmo J, et al. Quantum chemical structure-activity relationship on *b*-carbolines as natural monoamine oxidase inhibitors. Int J Quantum Chem 1983;83:1643–52.
[52] Carbó-Dorca R, Robert D, Amat LI, Jirones X, Besalú E. Application of Quantum Similarity to QSAR. In: Molecular Quantum Similarity in QSAR and Drug Design. Lecture Notes in Chemistry. 73. Berlin, Germany: Springer-Verlag; 2000. p. 26–38. https://doi.org/10.1007/978-3-642-57273-9_3.
[53] Lessigiarska I. Development of structure-activity relationship for pharmacotoxicological endpoints relevant to European union legislation. 2006. Thesis of the Liverpool John Moores University for the PhD Degree. School of Pharmacy and Chemistry & Institute of Health and Consumer Protection: Liverpool & Ispra.
[54] Hammett LP. Some relations between reaction rates and equilibrium constants. Chem Rev 1935;17(1):125–36.
[55] Taft Jr RW. Linear free-energy relationships from rates of esterification and hydrolysis of aliphatic and ortho-substituted benzoate esters. J Am Chem Soc 1952;74:2729–32.
[56] Cramer III RD, Patterson DE, Bunce JD. Comparative molecular field analysis (CoMFA). Effect of shape on binding of steroids to carrier proteins. J Am Chem Soc 1988;110:5959–67.
[57] Klebe G. Comparative molecular similarity indices analysis: CoMSIA. In: Kubinyi H, Folkers G, Martin YC, editors. 3D QSAR in drug design. Three-dimensional quantitative structure activity relationships. vol. 3. Dordrecht, Germany: Springer Press; 1998. p. 87–104.
[58] Kotani T, Higashiura K. Comparative molecular active site analysis (CoMASA). An approach to rapid evaluation of 3D QSAR. J Med Chem 2004;47(11):2732–42.
[59] Peter SC, Dhanja JK, Malik V, Radhakrishnan N, Jayakanthan M, Sundar D. Quantitative structure-activity relationship (QSAR): modeling approaches to biological applications. In: Ranganathan S, Gribskov M, Nakai K, Schönbach C, editors. Encyclopedia of bioinformatics and computational biology. vol. 2. Amsterdam, Netherlands: Elsevier; 2019. p. 661–76.
[60] Gramatica P. Principles of QSAR models validation: internal and external. Mol Inform 2007;26:694–701.
[61] Chirico N, Gramatica P. Real external predictivity of QSAR models: how to evaluate it? Comparison of different validation criteria and proposal of using the concordance correlation coefficient. J Chem Inform Model 2011;51:2320–35.
[62] Aher RB, Khan K, Roy K. A brief introduction to quantitative structure-activity relationships as useful tools in predictive ecotoxicology. In: Roy K, editor. Ecotoxicological QSARs. Methods in pharmacology and toxicology. New York, NY: Humana; 2020.
[63] US EPA. Estimation programs interface suite™ for Microsoft® windows, v 4.11. Washington, DC: United States Environmental Protection Agency; 2019.
[64] Saavedra LM, Romanelli GP, Duchowicz PR. A non-conformational QSAR study for plant-derived larvicides against Zika *Aedes aegypti* L. vector. Environ Sci Pollut Res 2020;27:6205–14.
[65] Dimitrov SD, Diderich R, Sobanski T, Pavlov TS, Chankov GV, Chapkanov AS. QSAR Toolbox—workflow and major functionalities. SAR QSAR Environ Res 2016;27:203–19. https://doi.org/10.1080/1062936X.2015.1136680.
[66] Roncaglioni A, Lombardo A, Benfenati E. The VEGAHUB platform: the philosophy and the tools. Altern Lab Anim 2022;50(2):121–35. https://doi.org/10.1177/02611929221090530.
[67] Mansouri K, Grulke CM, Judson RS, Williams AJ. OPERA models for predicting physicochemical properties and environmental fate endpoints. J Cheminform 2018;10:10. https://doi.org/10.1186/s13321-018-0263-1.
[68] European Food Safety Authority (EFSA), Devos Y, Auteri D, de Seze G, Fabrega J, Heppner C, et al. Building a European partnership for next generation, systems-based environmental risk assessment (PERA)vol. 9. EFSA Supporting Publication; 2022. p. e200503.
[69] Benfenati E, Chaudhry Q, Gini G, Dorne JL. Integrating *in silico* models and read-across methods for predicting toxicity of chemicals: a step-wise strategy. Environ Int 2019;131:105060.

CHAPTER

11

QSAR modeling based on graph neural networks

Haobo Wang, Wenjia Liu and Jingwen Chen

Key Laboratory of Industrial Ecology and Environmental Engineering (Ministry of Education), Dalian Key Laboratory on Chemicals Risk Control and Pollution Prevention Technology, School of Environmental Science and Technology, Dalian University of Technology, Dalian, China

Abbreviations

ACs activity cliffs
AD applicability domains
A_{PR} area under precision-recall curve
A_{ROC} area under receiver operating characteristic curve
BCF bioconcentration factor
DCNN deep convolutional neural network
DL deep learning
EC_{50} **or** LC_{50} median effective (lethal) concentrations
ECHA European Chemicals Agency
EPA United States Environmental Protection Agency
GAT graph attention network
GNN graph neural network
k_B base-catalyzed hydrolysis
K_{OA} octanol–air partition coefficient
K_{OW} octanol–water partition coefficient
MACCS molecular access system
ML machine learning
MLR multiple linear regression
P_{AW} attention weight
PBT persistent, bio-accumulative, and toxic
PFASs per- and polyfluorinated alkyl substances
POPs persistent organic pollutants
QSAR quantitative structure-activity relationship
R_A accuracy
R_{FP} false-positive rate
R_P precision
R_{TP} true-positive rate
SALs structure–activity landscapes
S_W water solubility

Introduction

While improving nearly all aspects of people's livelihoods either directly or indirectly, chemicals may be inevitably released into the environment, potentially adversely affecting human and ecosystem health [1,2]. Prioritization of chemicals based on their proclivity to harm humans or ecosystems is a guiding principle in environmental management [3,4]. Persistent, bio-accumulative, and toxic (PBT) chemicals are of great concern due to their environmental risks, and identification of PBT chemicals that are utilized in the market is a prerequisite for the management of chemicals [4,5]. However, over 350,000 chemicals and their mixtures have production and usage registrations [6]. Their transformation products in the environment are even more. Due to the large amount of chemicals and limited resources, it is difficult to evaluate their environmental risks by experimentally testing their PBT attributes [7,8]. Therefore, it is necessary to develop computational methods for efficiently complementing data and identifying PBT chemicals [9–12].

Quantitative structure–activity relationship (QSAR) models for predicting chemical attributes of environmental interest have been extensively constructed [13–21]. For example, the Center for Computational Toxicology and Exposure of the United States Environmental Protection Agency (EPA) constructed a series of prediction models on end points like bioconcentration factor (*BCF*), octanol–air partition coefficient (K_{OA}), and octanol–water partition coefficient (K_{OW}) [17]. EPA developed the widely used Estimation Programs Interface (EPI) Suite [21], which is a physical/chemical property and environmental fate estimation program including individual models like KOWWIN, BIOWIN, etc. Some studies focused on integrating these attributes or estimation programs to identify potential PBT chemicals. For example, Strempel et al. developed a PBT screening framework based on a comparison of

https://doi.org/10.1016/B978-0-443-15339-6.00012-6

estimated PBT attributes of each chemical with the PBT thresholds defined under the REACH regulation of the European Union [22].

Recent studies constructed QSAR models for screening of PBT chemicals directly based on existing PBT/non-PBT chemical lists [23–25]. Graph neural network (GNN), a novel neural network architecture, was introduced to construct QSAR models for screening of PBT chemicals [24]. Unlike most previous QSAR models based on manually selected molecular descriptors, QSAR models based on the GNN algorithm can directly learn from the original structures of molecules, giving full play to the feature extraction advantages of deep learning [26]. GNN is appropriate for data types with clear topology, such as molecular structures and can fully considers and utilizes their structural information, and thus its prospects in QSAR modeling are highly expected [27–29]. Comparison of descriptor-based QSAR models and GNN-based QSAR models is depicted in Fig. 11.1.

The development of QSAR modeling tends to allow advanced models to fit the complex relationships between chemical structures and their attributes that lie behind data. GNN-based QSAR models may revolutionize the descriptor-based QSAR paradigm and are expected to achieve high performances in general tasks. This chapter summarizes the existing QSAR modeling methods for the management of chemicals and introduces the basic principles and applications of the GNN algorithm.

QSAR models for the management of chemicals

Along with growing lists of commercial chemicals, QSAR models are expected to perform data complement and screening-level prioritization for synthetic chemicals, providing references for the management of chemicals.

QSAR models for predicting chemical properties

PBT attributes of chemicals relate to their migration, distribution, degradation, and transformation in multimedia environments, as well as impacts on the public health. Since experimental data are typically not accessible for the bulk of commercial chemicals, the parameters describing these environmental behavior and toxicities need to be calculated or predicted.

The evaluation standard of environmental persistence is based on the degradation half-life ($t_{1/2}$) or reaction rate constant (k) of chemicals. Xu et al. selected density functional theory methods and semi-empirical quantum chemical methods by comparing experimental second-order rate constant for base-catalyzed hydrolysis (k_B) values and those calculated by quantum chemistry [13,15]. Accordingly, QSAR models for k_B prediction were developed by combining experimental and calculated values.

The *BCF*, an essential parameter for evaluating chemical bioaccumulation, is defined as the ratio of chemical concentration in an organism to the concentration in surrounding environment at equilibrium. Previous *BCF* prediction models were constructed based on physical and chemical properties [30], molecular fragments [31] or

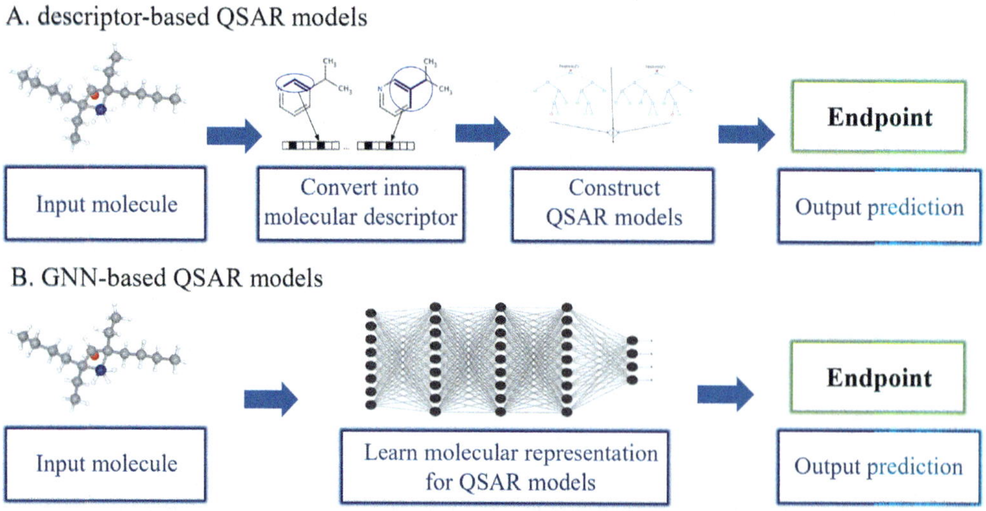

FIGURE 11.1 Comparison of descriptor-based and graph neural network (GNN)-based quantitative structure–activity relationship (QSAR) models.

multiparameter linear free energy relationships [32]. In recent years, QSAR models based on machine learning (ML) and molecular descriptors have gradually achieved better prediction results on large datasets [33]. Other properties, including the K_{OW}, water solubility (S_W), and biodegradability, etc., have also been extensively modeled using ML algorithms [17,34].

Median effective (lethal) concentrations (EC_{50} or LC_{50}) of aquatic organisms were generally employed to describe acute toxicities of chemicals by the REACH regulation [35]. Previous models for predicting acute toxicities on aquatic organisms were based on structurally similar chemicals. For example, Cassani et al. employed the multiple linear regression (MLR) algorithm to construct acute toxicity prediction models of triazole chemicals on *Daphnia Magna* and rainbow trout (*Oncorhynchus mykiss*) [36]. ML-based models can be constructed on datasets with diverse chemical structures and have a broad domain of applications. For example, Mansouri et al. compiled an acute toxicity data inventory for 11,992 chemicals and constructed consensus prediction models, demonstrating higher accuracy and stronger robustness than previous ones [37].

QSAR models for screening PBT chemicals

Traditional methods for determining whether a chemical has PBT attributes are to compare its PBT attribute values to criteria thresholds [38]. The EPA waste minimization prioritization tool [39] combines scoring algorithms with expert judgments for screening PBT chemicals among thousands of chemicals. Recently, deep learning (DL), an advancing field of ML, has been used in QSAR modeling with large-scale datasets [40–42]. The rapidly developed DL enabled the discovery of complicated patterns underlying big data. With more PBT/non-PBT chemicals available, it is expected to screen PBT chemicals in high throughput by constructing QSAR models based on the basic principle that molecular structures determine their attributes.

Sun et al. developed a deep convolutional neural network (DCNN) model for screening of PBT chemicals [25]. Take the advantage of DCNN model's ability to handle high-dimensional feature vectors, their model encodes the molecular structures using molecular descriptor matrixes which are a two-dimensional molecular descriptor representation matrix consisting of 2424 molecular descriptors. This DCNN model had an accuracy of 90.4% on an external validation set and achieved better PBT screening performance than the previous studies.

However, the DCNN model can only accept molecular structures in the form of two-dimensional matrixes where molecular descriptors are not placed by their physicochemical relevance. This operation inevitably introduces noise into the training process, which could affect robustness and prediction performance of the model. Previous studies also suggested that the manually extracted molecular descriptors can significantly influence selection of modeling algorithms and prediction performance [43]. *In silico* screening of PBT chemicals could be perceived to involve various chemical properties with broad structural diversities in existing organic chemicals. Therefore, DCNN models developed based on molecular descriptors might not fully characterize the structural features of PBT chemicals and lead to inaccurate screening.

In recent years, GNN, an emerging DL algorithm, has been successfully applied to predict physicochemical properties (S_W, solvation-free energy, etc.) and toxicity (androgen receptor, estrogen receptor, and aryl hydrocarbon receptor activity, etc.) of chemicals [44]. In the framework of GNN algorithm, chemicals are represented in the form of molecular graphs. In a molecular graph, the feature vectors of edges encode the features corresponding to chemical bonds, and the feature vectors of nodes encode the features corresponding to nonhydrogen atoms. Hydrogen atoms are usually represented as the number of hydrogen atoms attached to the atom in the nonhydrogen atom feature vectors. The GNN models take the molecular graphs as input and learn the relationships between the input graphs and the predicted endpoints, thus enabling so-called "end-to-end" learning. In other words, GNN can automatically learn concise and representative molecular features from original molecular structures, avoiding human intervention in molecular feature selections. The automatically optimized molecular features allow GNN to simplify QSAR modeling processes and achieve better prediction performance in a wide range of prediction tasks [44].

By introducing a parameter of attention weight (P_{AW}) for each node (atom) and automatically optimizing the P_{AW} during training, graph attention network (GAT), as an improved GNN, can promote model prediction by focusing on local molecular structures that contribute significantly to tasks of interest [45,46]. Wang et al. [24] applied the GAT algorithm for the first time to PBT chemical screening. In the PBT chemical screening model based on the GAT algorithm, the P_{AW} essentially reflects the contribution of each node to the PBT properties of the molecule to be predicted. Therefore, environmental chemists can infer the basis for the predictions made by models through the P_{AW} and thus judge whether the predictions are reliable. The P_{AW} can also be employed to identify the alert substructures associated with the molecular PBT attributes and avoid the appearance of these structures in future chemical designs. This data-driven model predicts complex end points directly from molecular structures and is efficient, mechanistically interpretable, and easy to implement programmatically. These advantages allow GAT models to have important application prospects in the field of chemical environmental risk assessments.

GNN algorithm

The GAT algorithm, a subset of the GNN algorithm, is a DL algorithm. The introduced P_{AW} can improve the accuracy and interpretability of GAT models, making it more suitable for tasks related to chemical attribute prediction. The principles of GNN and GAT as well as their advantages and related studies in QSAR modeling are presented here.

Basic principles of GNNs

GNNs can be divided into spectral GNNs and spatial GNNs. Spectral GNNs have a solid mathematical foundation but are computationally complex and have strict restrictions on structure of graphs. They are still employed in fields such as action recognition [47] but not in the field of QSAR modeling. Therefore, this chapter focuses on spatial GNNs commonly utilized in molecular attribute predictions.

The core idea of spatial GNNs is to update the feature representation of a central node by aggregating the features of its neighboring nodes and performing convolutional operations on neighboring graphs [48]. The process of information flow between the neighbors and the central node is also called message passing. Depending on how messages are passed to central nodes, different propagation rules are developed, and the different rules integrate the structural features of the network in different ways.

Message passing of GNNs can be divided into two phases: message passing phase and readout phase. The message passing phase is described in Eqs. (11.1) and (11.2):

$$n_v^{t+1} = \sum_{w \in N_v} \mathrm{M}_t\left(h_v^t, h_w^t, e_{v,w}^t\right) \tag{11.1}$$

$$h_v^{t+1} = \mathrm{U}_t\left(h_v^t, n_v^{t+1}\right) \tag{11.2}$$

Here, h_v^t and h_w^t denote the feature vectors of nodes v and w after t iterations (e.g., types of atoms, number of connected hydrogen atoms, etc.), respectively, and e_{vw}^t represents feature vectors of the edges of the connected node pairs v and w after t iterations (e.g., types of chemical bonds, etc.); N_v indicates a set of neighbor nodes of the central node v; $\mathrm{M}_t(\cdot)$ expresses the message passing function that passes the feature vectors of the neighbor nodes to the central node; n_v^{t+1} means the features of neighboring nodes and adjacent edges after integration by the $\mathrm{M}_t(\cdot)$ function. $\mathrm{M}_t(\cdot)$ function is generally a function that takes the maximum, minimum or average value of the features, etc. The update function $\mathrm{U}_t(\cdot)$ is used to update the feature vectors of the central node. The $\mathrm{U}_t(\cdot)$ function is generally implemented by a fully connected layer, where the node features in the neighborhood essentially share the same set of learnable weight value matrices $\boldsymbol{W}$.

The readout phase uses the readout function $\mathrm{R}(\cdot)$ to compute the predicted end point $\hat{y}$ for the entire graph G based on all node feature vectors h_v^T updated after the last message passing phase.

$$\hat{y} = \mathrm{R}\left(\left\{h_v^T \middle| v \in G\right\}\right) \tag{11.3}$$

Depending on choice of the message passing function $\mathrm{M}_t(\cdot)$, the update function $\mathrm{U}_t(\cdot)$, and the readout function $\mathrm{R}(\cdot)$, variants of the GNNs can be adapted to different task requirements.

GNNs generally have a so-called local feature-aware mechanism, which means that during the iterations of GNNs, a node in one layer only carries out message passing with some neighbor nodes in the previous layer and integrates local information in the next layer to obtain the global information of the network. The local feature-aware mechanism enables the GNN-based QSAR models to focus on the important local structural features of chemicals while providing complete learning of the overall structures of chemicals, thus achieving better prediction results.

Principles of graph attention networks

During the training process of GNNs, with an increase in the number of network layers and iterations, the features of each node tend to converge to the same value. This phenomenon, known as over-smooth, leads to a decrease in the fitting ability. To address this problem, Veličković et al. introduced attention mechanism to GNNs and developed the GAT algorithm [49]. During the training, P_{AW} allows the GAT model to focus on the important information of the training samples and learn them fully. The attention mechanism adds P_{AW} to the message passing process of the GNN model so that the model can identify the important nodes in the molecular graphs and gradually increase the P_{AW} of the important nodes by back propagation and gradient descent algorithm. At the same time, the attention mechanism also makes the P_{AW} of the nodes that contribute less to the prediction gradually decrease to avoid their negative interference.

Unlike the message passing function $M_t(\cdot)$ in other GNN models, aggregation of node features in GAT models is implemented by the attentive layer. The input of the attentive layer is a set of node features, and the output is a set of node features multiplied by P_{AW}. The implementation process can be divided into three steps as described in Eqs. (11.4)–(11.6):

$$P_{A,v,w} = \text{relu}\left(\mathbf{W} \cdot h_v, \mathbf{W} \cdot h_w\right) \tag{11.4}$$

Here, h_v and h_w denote the feature vectors of nodes v and w, respectively. The feature vector of each node is multiplied by a vector of learnable weight parameters **W**. Subsequently, the $P_{A,v,w}$ is calculated according to the node feature vectors by nonlinear activation function relu(·).

$$P_{AW,vw} = \text{softmax}\left(P_{A,v,w}\right) = \frac{\exp\left(P_{A,v,w}\right)}{\sum_{k\in N_v} \exp\left(P_{A,v,k}\right)} \tag{11.5}$$

Here, the attention parameter $P_{A,v,w}$ represents the importance of the features of node w to node v. The GAT model calculates the importance of all nodes in the neighborhood N_v and normalizes $P_{A,v,w}$ using the softmax(·) function to obtain the attention weight parameter $P_{AW,v,w}$ of node w to node v.

$$h_v{}' = \sum_{w\in N_v} P_{AW,v,w} \cdot \mathbf{W} \cdot h_w \tag{11.6}$$

Finally, the linear combination of $P_{AW,v,w}$ and the corresponding feature vector are calculated as a final output feature vector $h_v{}'$ of each node.

Fig. 11.2 is an example illustrating the computation process of updating node feature vectors in the attentive layer. The feature vector of each node is represented by a vector. For node #6, there are three adjacent nodes #1, #5, and #7. These three nodes plus node #6 form the neighborhood of node #6. The attention parameters $P_{A,v,w}$ are calculated between node #6 and each node in the neighborhood using the attention mechanism. $P_{A,6,1}$ between node 6 and node 1 is calculated as:

$$P_{A,6,1} = \text{relu}\left(\mathbf{W} \cdot h_6, \mathbf{W} \cdot h_1\right) \tag{11.7}$$

Here, h_6 and h_1 denote the feature vectors of node #6 and node #1, respectively. **W** represents parameter vectors that can be gradually optimized in training. Finally, the attention parameters of node #6 and node #1 are calculated using the activation function relu(·). The attention parameters of node #6 with the rest of the nodes in the neighborhood, that is, $P_{A,6,5}$, $P_{A,6,7}$, and $P_{A,6,6}$, are calculated in the same way.

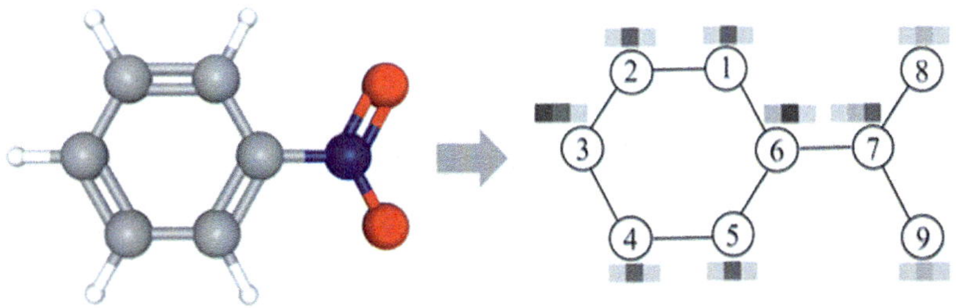

FIGURE 11.2 Example of a molecular graph. Squares of different colors represent different atomic characteristics.

Subsequently, the function softmax(·) is used to normalize $P_{A,v,w}$ of node 6 and the four nodes in the neighborhood to obtain four $P_{AW,v,w}$ for subsequent node feature updates. The calculation of $P_{AW,6,1}$ between node 6 and node 1 in Eq. (11.8) is taken as an example.

$$P_{AW,6,1} = \text{softmax}\left(P_{A,6,1}\right) = \frac{\exp\left(P_{A,6,1}\right)}{\exp\left(P_{A,6,1}\right) + \exp\left(P_{A,6,5}\right) + \exp\left(P_{A,6,7}\right) + \exp\left(P_{A,6,6}\right)} \tag{11.8}$$

Finally, using $P_{AW,v,w}$ of node #6 with its four neighbor nodes and the optimizable parameter vector W, a differentiated aggregation of the neighborhood is performed to complete this node feature vector update by Eq. (11.9).

$$h_6' = P_{AW,6,1} \cdot W \cdot h_1 + P_{AW,6,5} \cdot W \cdot h_5 + P_{AW,6,7} \cdot W \cdot h_7 + P_{AW,6,6} \cdot W \cdot h_6 \tag{11.9}$$

It is worth noting that W of the four nodes is shared here.

Compared with other GNN models, GAT introduces the attention mechanism that can dynamically describe the relationships of each neighboring node and achieve differentiated aggregation of neighboring nodes. If $P_{AW,v,w}$ and W are combined as a coefficient, GAT essentially assigns different learning parameters to each node in neighborhood implicitly. Therefore, the GAT model can alleviate the oversmoothing problem to some extent.

In addition, during the training of GAT model, the attention mechanism is continuously used to compute $P_{AW,v,w}$ for each node to infer the relationship between nodes based on model performance. This continuous learning of the relationships between nodes allows the model to dynamically focus on local graph structures that are more relevant to particular tasks. In QSAR models constructed with GAT, local graph structures are equivalent to the local molecular structures. This property of the attention mechanism allows the GAT model to pay more attention to important local molecular structures and to improve model performance without prior knowledge. The $P_{AW,v,w}$ parameters after model training facilitate interpretability of models.

QSAR model based on graph attention network

In constructing QSAR models based on GAT, molecular structures are firstly read from SMILES codes and encoded into molecular graphs. Each atom in the molecular graphs, as well as the chemical bonds, is encoded with a feature vector of the same length. All molecular graphs contain an atomic feature matrix and a bond feature matrix for subsequent training. In addition, the molecular graphs need the connection relationships between the nodes for determining the neighborhood range for feature aggregation and update in each iteration. The connection relationships are represented by adjacency matrixes. After converting an SMILES string into a molecular graph, the feature vectors of all atomic nodes are updated using the attention layers to encode the molecular structural features. The node feature aggregation based on attention mechanism is illustrated in Fig. 11.3.

In a GAT model for chemical attribute prediction, the endpoint to be predicted is usually a label of the entire molecule, that is, the label of the whole graph. Therefore, predictions need to be made based on the features of the whole graph, rather than the features of each atomic node in the attention layer. A general practice is to introduce a virtual node connected to all atomic nodes to represent the molecule after the attentive layer for atomic node features.

$P_{AW,v,w}$ of the virtual nodes is calculated concerning all atomic nodes using the same attentive layer for virtual nodes as for atomic nodes (later called the molecular attentive layer for distinction), and the feature vectors of the virtual nodes are updated by differentially aggregating the feature vectors of all atomic nodes based on $P_{AW,v,w}$.

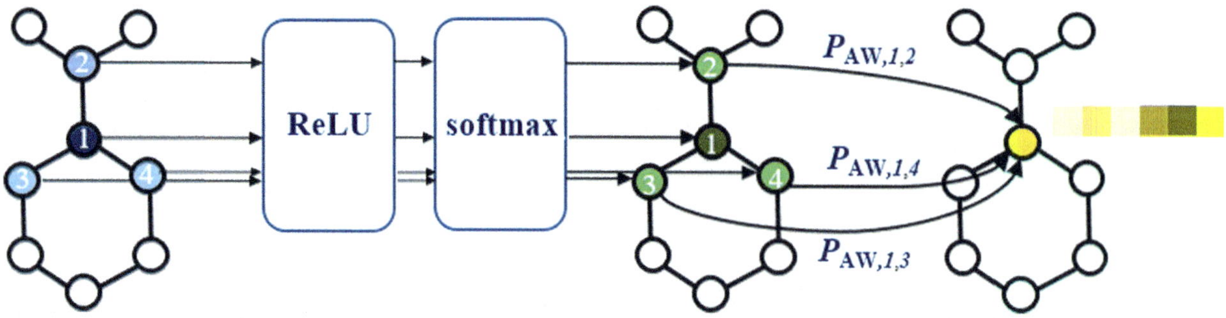

FIGURE 11.3 Node feature aggregation based on attention mechanisms.

As a result, the feature vectors of the virtual nodes contain highly representative molecular features related to the predicted attributes. Using Eq. (11.3), the end point $\hat{y}$ of a molecule is calculated based on the molecular feature vectors from the molecular attentive layer.

Subsequently, a loss function is used to measure the difference between the predicted end point $\hat{y}$ and the true label. The larger the difference, the larger the value of the loss function. Training of the model is the optimization of all learnable parameters, including $P_{AW,v,w}$ and W, to minimize the value of the loss function. Based on the value of the loss function and the gradient descent algorithm, a new iteration of optimization can be performed by the optimizer to gradually reduce the value of the loss function and achieve "end-to-end" training. The optimized $P_{AW,v,w}$ directly reflects the weights assigned to specific atoms by the GAT model for tasks in question and therefore can be used to explain the training process and the model prediction. In summary, the core idea of the GAT, which is differentiated aggregation of nodes within a neighborhood, increases the weight of important substructures within chemicals. The overall framework of GAT-based QSAR models is shown in Fig. 11.4.

GNN for QSAR modeling

The QSAR models constructed based on GNN for chemical property predictions and PBT chemical screening are introduced here.

GNN models for chemical property predictions

In recent years, applications of GNNs in constructing QSAR models have been gradually increased. Stokes et al. trained a GNN model for predicting antimicrobial activity of chemicals [50]. This GNN model was trained on a dataset containing 2335 molecules. Some of the molecules have the capability to inhibit *E. coli* growth. The model was applied to several databases containing 107 million molecules to identify potential lead compounds with *E. coli* inhibitory activity. Compounds were ranked according to the predictions. This approach identified eight antimicrobial compounds that significantly differed in structures from known antibiotics. Two of the molecules were found to have potent antibiotic broad-spectrum activity.

Different from traditional predefined molecular descriptors, the approach of GNNs to characterize molecular structures incorporates a task-specific relevant training process to optimize feature extraction schemes. This is the reason that GNN-based models can identify antimicrobial compounds that are structurally different from known antibiotics. The GNN models have strong generalization capability, meaning that the models can recognize a broader range of general patterns based on samples outside the training set, which is important for QSAR model construction.

Jiang et al. [44] compared predictive accuracy and computational efficiency of traditional descriptor-based ML models and GNN-based models. The tests were launched on 11 public datasets and covered various chemical attributes. Their results showed that the GNN models performed well in complex tasks. Compared with other models, the GNN models have the advantage of "end-to-end" learning for molecular attribute prediction and can automatically extract structural features that are more relevant to the tasks. Therefore, the GNN models not only simplify the construction process of QSAR models but also can obtain high prediction accuracy when completely new prediction tasks are encountered and molecular features cannot be selected empirically.

GAT models were constructed by Xiong et al. for various molecular attribute prediction tasks and achieved better prediction performance than previous models [46]. In addition, the P_{AW} was proved to reflect the degree of contribution of each node in molecular graphs to model prediction. Based on P_{AW}, Ryu et al. identified several substructures by labeling atoms or substructure groups of great importance as the basic structural features for molecules of high photovoltaic efficiency [50,51]. By comparing with the results of quantum chemical calculations, the structural features extracted by P_{AW} coincided with the regions of donor and acceptor orbitals for charge-transfer excitations.

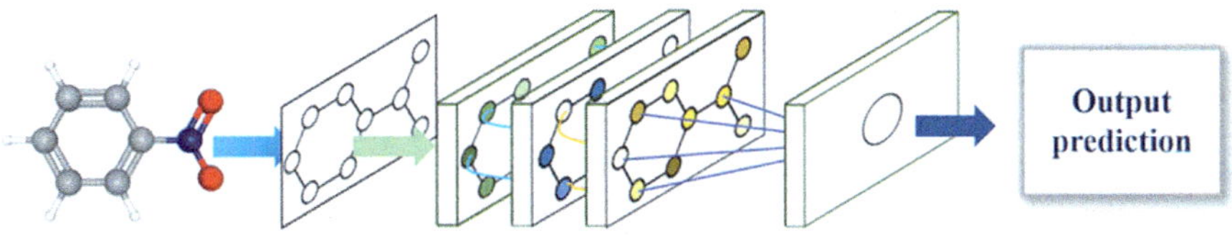

FIGURE 11.4 Overall framework of quantitative structure–activity relationship (QSAR) models based on graph attention network.

The constructed model framework not only exhibits excellent prediction performance but also provides a certain degree of interpretation of the prediction results.

The Tox21 Data Challenge launched in 2014 by the National Center for Advancing Translational Sciences at the US National Institutes of Health encouraged development and comparison of different QSAR models for chemical toxicity prediction. The bioassay data included 12 toxicological end points mainly related to nuclear receptor effects such as activation of the estrogen receptor and stress response effects such as the heat shock response effect. Li et al. constructed QSAR models based on GNNs for the 12 end points and applied them to toxicity prediction of per- and polyfluorinated alkyl substances (PFASs) detected in human serum samples [52]. Their prediction results were consistent with those from previous experimental studies [52]. Since PFASs are widely used in industrial and consumer products, some studies specifically constructed QSAR models based on GNN algorithms and successfully identified bioactivities of 3486 PFASs [53]. The screening results focused on different target receptors or enzymes such as the neuropeptide S receptor, CYP2C9, and the aldehyde dehydrogenase 1.

GNN training generally requires a large amount of task-relevant labeled data, and there is often a shortage of labeled data for many environmentally relevant endpoints. To address this issue, some researchers explored new training paradigms on GNNs, including self-supervised learning and pretraining [54]. Wang et al. proposed a self-supervised GNN framework, MolCLR, to optimize the representation of molecular features using a large amount of unlabeled data for pretraining [54]. After the pretraining, the models were fine-tuned in downstream molecular property predictions. MolCLR showed very competitive performance on datasets with small data volume compared to models trained by supervised learning only.

GAT models for screening PBT chemicals

A previous study constructed a database of PBT chemicals containing 1588 PBT/persistent organic pollutants (POPs) chemicals and 13,406 nonPBT/POPs chemicals [24]. According to the definition of POPs, all POPs are PBT chemicals. In addition, POPs also have long-range transport characteristics, and some POPs are byproducts of human activities rather than purposefully synthesized chemicals [55]. Therefore, POPs can be adopted as supplementary data to this database. Based on this database, the GAT algorithm was employed to construct QSAR models for PBT screening [24]. Accuracy (R_A), true-positive rate (R_{TP}), false-positive rate (R_{FP}), precision (R_P), area under receiver the operating characteristic (ROC) curve (A_{ROC}), and area under the precision-recall (PR) curve (A_{PR}) were used to evaluate the GAT model performance and were calculated using Eqs. (11.10)–(11.13).

$$R_A = \frac{(N_{TP} + N_{TN})}{(N_{TP} + N_{TN} + N_{FP} + N_{FN})} \tag{11.10}$$

$$R_{TP} = \frac{N_{TP}}{(N_{TP} + N_{FN})} \tag{11.11}$$

$$R_{FP} = \frac{N_{FP}}{(N_{TN} + N_{FP})} \tag{11.12}$$

$$R_P = \frac{N_{TP}}{(N_{TP} + N_{FP})} \tag{11.13}$$

Here, N_{TP}, N_{TN}, N_{FP}, and N_{FN} denote the number of true-positive, true-negative, false-positive, and false-negative cases in model predictions, respectively. The horizontal and vertical coordinates of the ROC curves are R_{FP} and R_{TP} under different classification thresholds. The horizontal and vertical coordinates of the PR curves are R_{TP} and R_P under different classification thresholds. The area under the ROC curve and the area under the PR curve are calculated to obtain A_{ROC} and A_{PR}.

Fig. 11.5 shows the learning curve of this GAT model. It can be seen that after only five iterations, the R_A of the GAT model on the training set increased from about 30% to more than 90%. Subsequently, the R_A increases at a relatively steady rate, finally up to more than 95%. The validation set is not directly involved in the training of the model so that it can provide a more objective reference for the model performance. The performance of the model on the validation set and the training set is generally consistent, indicating that the GAT model is not overfitted or underfitted.

The P_{AW} quantifies the contribution of atoms to the PBT attributes of particular molecules. As an example, the variations in P_{AW} during the training for two expert-verified PBT chemicals are shown in Fig. 11.5. It can be seen that the value of each P_{AW} considerably changes after certain training. With the rapid increase of R_A in the first

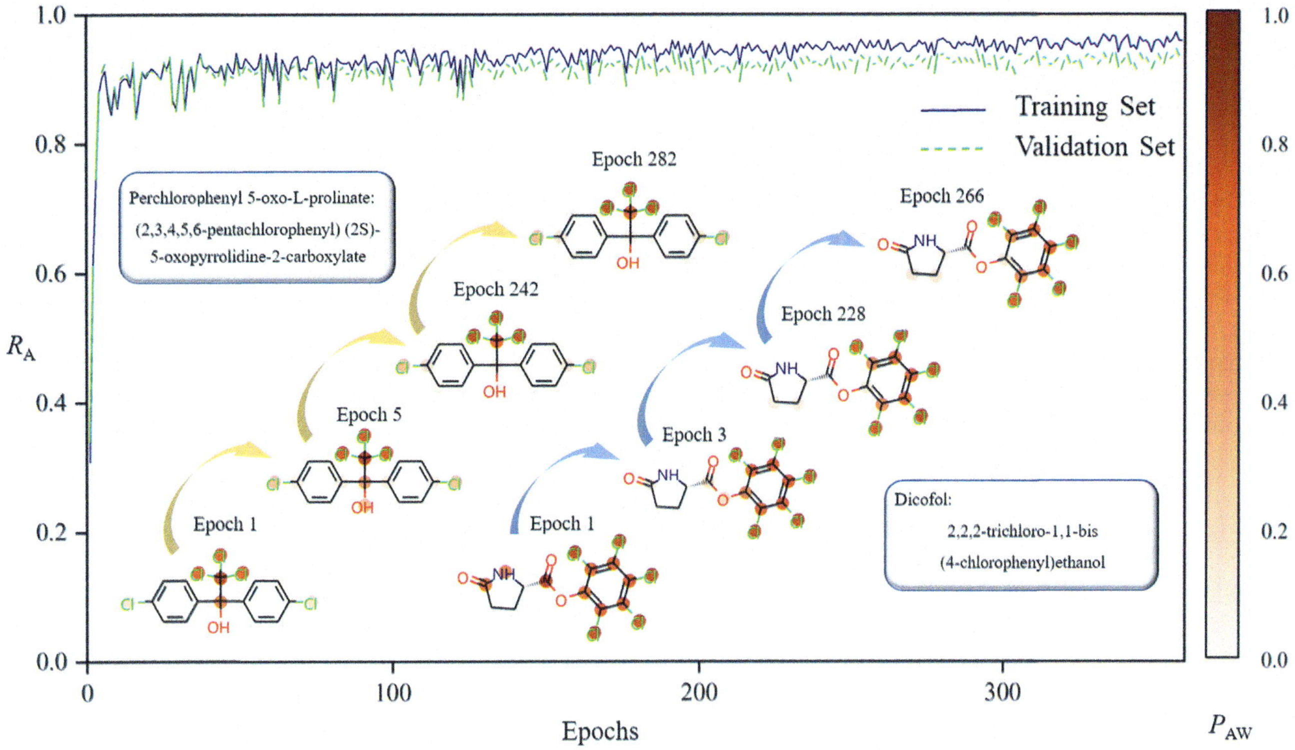

FIGURE 11.5 Learning curve (variation of prediction accuracy R_A with learning epochs) of the graph attention network (GAT) model and visualization of P_{AW} for two expert-verified persistent, bio-accumulative, and toxic (PBT) molecules.

five iterations, the P_{AW} of atoms gradually concentrates on halogen and aromatic atoms, while oxygen and nitrogen atoms are progressively neglected. After 200 iterations, the halogen and aromatic atoms were steadily assigned high P_{AW} values. Khan et al. also found longer $t_{1/2}$ for molecules containing halogenated aromatic structures in constructing a QSAR model to predict chemical degradation half-lives ($t_{1/2}$) [56]. This visualization enhances the transparency of the model mechanism and meets the requirement of openness and transparency of QSAR models, which facilitates the practical application of the GAT model in PBT chemicals screening.

The training and validation sets are directly and indirectly involved in the GAT model training, respectively, so the test set was used to evaluate the model prediction performance. The training set, validation set, and test set were yielded by randomly dividing the same database as adopted in the previous study [25] for five times to evaluate performance and robustness of the GAT model. The mean and standard deviation of the model performance metrics obtained from the five training sessions were calculated. The performance indicators of the GAT model on the test set are shown in Fig. 11.6.

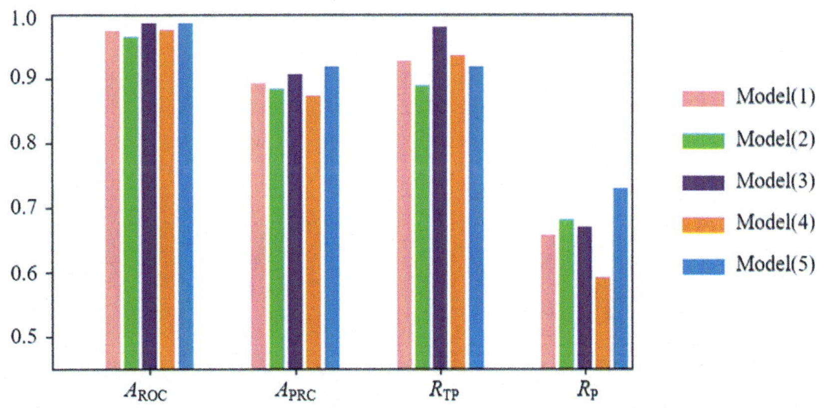

FIGURE 11.6 Performance of graph attention network (GAT) models on the test groups.

R_{TP} reflects the ability of the model to correctly predict actual positive samples (PBT chemicals), and R_{FP} demonstrates the probability of misclassifying negative cases as positive. The significance of A_{ROC} is to pursue the maximum R_{TP} value and the minimum R_{FP} value. Thus, A_{ROC} reflects the model's ability to classify positive and negative samples. As shown in Fig. 11.6, A_{ROC} of the GAT models was consistently high and stable, and the A_{ROC} value on the test set (97.9%) was also better than that of the previous best model (96.8%) [25], indicating the GAT model's satisfactory performance in screening of PBT chemicals.

R_P indicates the proportion of correctly predicted positive samples, and R_{TP} also focuses on the identification of positive samples. The significance of the PR curve and A_{PR} lies in the simultaneous pursuit of maximum R_P and R_{TP}. Therefore, A_{PR} is more suitable for assessing the ability of classification models to identify positive samples, especially for evaluating GAT models trained with an imbalanced database. The PR curve of the GAT model is slightly higher than that of the previous model [24].

Since the GAT model aims to screen out all PBT chemicals, R_{TP} is an important indicator in practical application. The mean R_{TP} of the GAT model on the test set (93.2%) was significantly higher than the corresponding value in the previous model (80.0%) [25], indicating that the screening results of the GAT model are more reliable.

Applicability domains for GNN-based QSAR models

Artificial intelligence, ML, and big data analysis techniques have rapidly advanced in recent years, providing powerful QSAR modeling algorithms and tools. However, QSARs are essentially data-driven models and principles presented in the training set and may not be applicable to compounds of arbitrary structures [46]. As reliability of QSARs is always limited to a specific range, the concept of applicability domain (AD) was proposed. The requirements for AD characterization of QSAR models were proposed in the guidance document on the validation of (quantitative) structure activity relationships models by the Organization for Economic Co-operation and Development [47].

The functionality of AD is to define compounds that can be reliably predicted by QSAR models, which can also be understood as the set of compounds to which the model applies. The structural characterization of compounds is generally represented by descriptors. Accordingly, the descriptor space covered by the training compounds can be used as the AD, also known as the descriptor domain. As shown in Fig. 11.7, the term implies that there is a region in descriptor space where the "normal" compounds are located, and borders of this space should be defined. Many methods for defining descriptor domains have been developed over the years, including hyper-rectangularity, convex hull, kernel density estimation, distance-based methods, and so on [48]. However, these methods are either not applicable to high-dimensional spaces or can be challenging for AD characterization with classical descriptor domains because the descriptor selection phases are implicitly internal to the ML training process, limiting development of AD characterization for QSAR models built with ML.

An AD characterization method based on molecular fingerprints (AD_{FP}) has been proposed and verified for QSAR models constructed with various ML algorithms [20]. The AD_{FP} characterization method divides chemical space of the database into different clusters based on pairwise structural similarity between all molecules in the dataset and sets a threshold S_{cutoff} to exclude those molecules with lower similarity values than other molecules from the

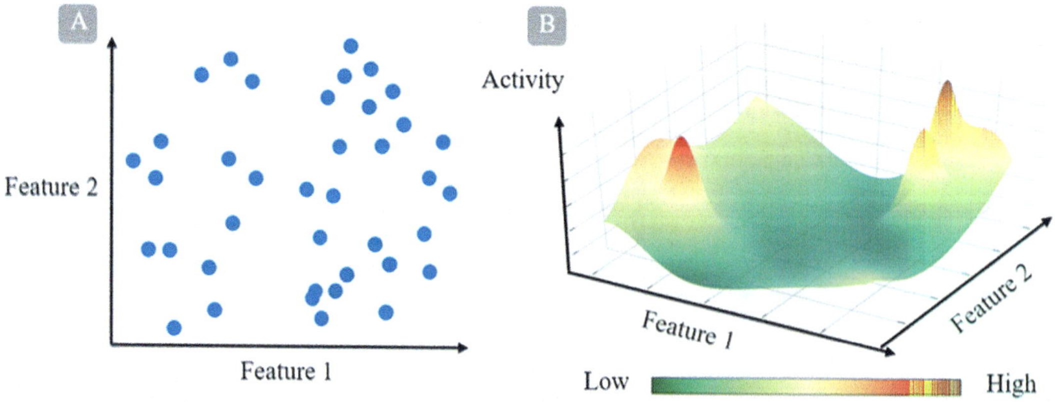

FIGURE 11.7 Chemical space and structure activity landscape.

same clusters. The similarity values between molecules can be calculated based on molecular access system (MACCS) and Tanimoto similarity coefficients. In a previous study for screening of PBT chemicals, with increase in the S_{cutoff} value, more chemicals with lower similarity values are excluded from the ADs [20]. When $S_{cutoff} = 0.90$ for the AD_{FP} characterization, the values of A_{ROC}, A_{PR}, R_{TP}, and R_A increased significantly, indicating a prominent ability of the AD characterization.

Molecules with similar structures but conflicting activities exhibit "local discontinuity" in their structure–activity landscapes (SALs) (Fig. 11.7), also known as activity cliffs (ACs). Previous studies proved that the local discontinuity of chemicals positively correlates with absolute prediction errors in cross-validation [20]. Although QSAR models can be relatively unreliable in predicting chemicals on the ACs, traditional AD characterization methods never considered ACs.

An improved AD characterization method, termed $AD_{FP\text{-}AC}$, was developed to solve this problem, which detects chemicals that may be located on ACs with local discontinuity scores (S_{LD}) [24]. The S_{LD} value of a molecule m is calculated by Eq. (11.14).

$$S_{LD}(m) = \frac{1}{K}\sum\nolimits_{\{n|S(m,n)\geq S_{cutoff},\ m\neq n\}} S(m,n)\cdot D(m,n) \tag{11.14}$$

Here, n denotes a molecule in a community set whose similarity to the molecule m $S(m,n)$ is larger than S_{cutoff}. $D(m,n)$ indicates the absolute difference between the activities of molecule m and molecule n, and K is the number of elements in the set. In principle, molecules with larger S_{LD} values are likely to stay on ACs.

$AD_{FP\text{-}AC}$ sets a threshold C_{cutoff} to select molecules with higher S_{LD} values and to locate the ACs. In the framework of $AD_{FP\text{-}AC}$, SALs can be considered as a contour of S_{LD} values with S-values. Based on the S-values for the existing molecular clusters, the molecules that need to be predicted can be inserted into the SAL to determine whether the unlabeled molecules are on the AC.

The previous study indicated that the prediction performance of the ML models could be further improved when the C_{cutoff} value decreases, and the number of chemicals covered in the AD decreases [24]. It can be observed that if a similar number of molecules are excluded from the AD, there is usually a more significant improvement in the model performance with the assistance of the $AD_{FP\text{-}AC}$.

Conclusions

Rapid and efficient complement of data and screening of PBT chemicals are necessary for management of chemicals. It is difficult to experimentally test PBT attributes for a large number of chemicals due to the limited resources. Development of QSAR models is an important way to solve this problem.

Organic chemicals are structurally diverse, and the performance of descriptor-based QSAR models can be limited. Comprehensive screening of PBT chemicals involves various attributes of organic chemicals simultaneously. Therefore, it is challenging to select a suitable set of descriptors for screening models of PBT chemicals. Although some DL-based models show good screening ability, weak interpretability may hinder their practical applications.

The GNN is an advanced DL modeling technique, can take the original structure of chemicals as inputs, and automatically optimize extraction schemes for molecular structure features. The GAT algorithm enhances the accuracy of chemical property predictions and PBT screening by introducing P_{AW}. Visualizations of P_{AW} variations enable interpretability of the screening results. Structural alerts of PBT chemicals are also expected to be identified based on P_{AW}.

Because QSAR models are data-driven, rules learned from training data may not be necessarily applicable to molecules with arbitrary structures. Therefore, characterizing ADs of models can enhance the reliability of model predictions. The $AD_{FP\text{-}AC}$, which considered both FP similarities and compounds at ACs, improved the reliability of the prediction and screening results.

The GAT model, in conjunction with the $AD_{FP\text{-}AC}$, may become efficient tools for the management of chemicals. Given the concise molecular structure representations and the "end-to-end" architecture that can be easily programmed and computerized, GAT models can also be developed to predict values on physicochemical properties, environmental behavioral, and toxicological parameters for the vast and ever-increasing number of synthetic chemicals.

Acknowledgments

This study was supported by the National Natural Science Foundation of China (22136001) and the National Key R&D Program of China (2022YFC3902100).

References

[1] Wu S, Powers S, Zhu W, Hannun YA. Substantial contribution of extrinsic risk factors to cancer development. Nature 2016;529(7584):43–7.
[2] Walsh M. Antibiotics for emerging pathogens. Science 2009;325(5944):1089–93.
[3] United Nation. Chemicals and waste; sustainable development goals knowledge platform. 2016. Available from: https://sustainabledevelopment.un.org/topics/chemicalsandwaste.
[4] Muir DCG. Are there other persistent organic pollutants? A Challenge for environmental chemists. Environ Sci Technol 2006;40:7157–66.
[5] Weisbrod AV, Burkhard LP, Arnot J, Mekenyan O, Howard PH, Russom C, et al. Workgroup report: review of fish bioaccumulation databases used to identify persistent, bioaccumulative, toxic substances. Environ Health Perspect 2007;115(2):255–61.
[6] Wang ZY, Walker GW, Muir DCG, Nagatani-Yoshida K. Toward a global understanding of chemical pollution: a first comprehensive analysis of national and regional chemical inventories. Environ Sci Technol 2020;54(5):2575–84.
[7] Scheringer M, Strempel S, Hukari S, Ng CA, Blepp M, Hungerbuhler K. How many persistent organic pollutants should we expect? Atmos Pollut Res 2012;3(4):383–91.
[8] Reppas-Chrysovitsinos E, Sobek A, MacLeod M. In silico screening-level prioritization of 8468 chemicals produced in OECD countries to identify potential planetary boundary threats. Bull Environ Contam Toxicol 2018;100(1):134–46.
[9] Brown TN, Wania F. Screening chemicals for the potential to be persistent organic pollutants: a case study of arctic contaminants. Environ Sci Technol 2008;42:5202–9.
[10] Oberg T, Iqbal MS. The chemical and environmental property space of REACH chemicals. Chemosphere 2012;87(8):975–81.
[11] Ng CA, Scheringer M, Fenner K, Hungerbuhler K. A framework for evaluating the contribution of transformation products to chemical persistence in the environment. Environ Sci Technol 2011;45:111–7.
[12] Howard PH, Muir DC. Identifying new persistent and bioaccumulative organics among chemicals in commerce. III: byproducts, impurities, and transformation products. Environ Sci Technol 2013;47(10):5259–66.
[13] Xu T, Chen JW, Wang ZY, Tang WH, Xia DM, Fu ZQ, et al. Development of prediction models on base-catalyzed hydrolysis kinetics of phthalate esters with density functional theory calculation. Environ Sci Technol 2019;53(10):5828–37.
[14] Mansouri K, Grulke CM, Judson RS, Williams AJ. OPERA models for predicting physicochemical properties and environmental fate endpoints. J Cheminf 2018;10(1):10.
[15] Xu T, Chen JW, Chen X, Xie HJ, Wang ZY, Xia DM, et al. Prediction models on pK_a and base-catalyzed hydrolysis kinetics of parabens: experimental and quantum chemical studies. Environ Sci Technol 2021;55(9):6022–31.
[16] Tang WH, Li Y, Yu Y, Wang ZY, Xu T, Chen JW, et al. Development of models predicting biodegradation rate rating with multiple linear regression and support vector machine algorithms. Chemosphere 2020;253:126666.
[17] Mansouri K, Cariello NF, Korotcov A, Tkachenko V, Grulke CM, Sprankle CS, et al. Open-source QSAR models for pK_a prediction using multiple machine learning approaches. J Cheminf 2019;11(1):60.
[18] Chen J, Wu N, Qu R, Xu X, Shad A, Pan X, et al. Photodegradation of polychlorinated diphenyl sul fi des (PCDPSs) under simulated solar light irradiation: kinetics, mechanism, and density functional theory calculations. J Hazard Mater 2020;398:122876.
[19] Acharya K, Werner D, Dolfing J, Barycki M, Meynet P, Mrozik W, et al. A quantitative structure-biodegradation relationship (QSBR) approach to predict biodegradation rates of aromatic chemicals. Water Res 2019;157:181–90.
[20] Wang ZY, Chen JW, Hong HX. Developing QSAR models with defined applicability domains on PPARγ binding affinity using large data sets and machine learning algorithms. Environ Sci Technol 2021;55(10):6857–66.
[21] EPI USEP. Suite™-estimation program interface. U.S. Environmental Protection Agency; 2012. Available from: https://www.epa.gov/tsca-screening-tools/epi-suitetm-estimation-program-interface.
[22] Strempel S, Scheringer M, Ng CA, Hungerbuhler K. Screening for PBT chemicals among the "existing" and "new" chemicals of the EU. Environ Sci Technol 2012;46(11):5680–7.
[23] Wassenaar PNH, Rorije E, Janssen NMH, Peijnenburg WJGM, Vijver MG. Chemical similarity to identify potential substances of very high concern—an effective screening method. Computational Toxicology 2019;12:100110.
[24] Wang HB, Wang ZY, Chen JW, Liu WJ. Graph attention network model with defined applicability domains for screening PBT chemicals. Environ Sci Technol 2022;56(10):6774–85.
[25] Sun X, Zhang X, Muir DCG, Zeng EY. Identification of potential PBT/POP-like chemicals by a deep learning approach based on 2D structural features. Environ Sci Technol 2020;54(13):8221–31.
[26] Chuang KV, Gunsalus LM, Keiser MJ. Learning molecular representations for medicinal chemistry. J Med Chem 2020;63(16):8705–22.
[27] Nguyen TH, Nguyen LH, Truong TN. Application of machine learning in developing quantitative structure-property relationship for electronic properties of polyaromatic compounds. ACS Omega 2022;7(26):22879–88.
[28] Wu J, Kang Y, Pan P, Hou T. Machine learning methods for pK_a prediction of small molecules: advances and challenges. Drug Discov Today 2022;27(12):103372.
[29] Fassio AV, Shub L, Ponzoni L, McKinley J, O'Meara MJ, Ferreira RS, et al. Prioritizing virtual screening with interpretable interaction fingerprints. J Chem Inf Model 2022;62(18):4300–18.
[30] Commission E. Technical guidance document (TGD) on risk assessment Brussels, Belgium. The European Community; 2003.
[31] Tao S, Hu HY, Lu XX, Dawson RW, Xu FL. Fragment constant method for prediction of fish bioconcentration factors of non-polar chemicals. Chemosphere 2000;41(10):1563–8.
[32] Fatemi MH, Abraham MH, Haghdadi M. Prediction of biomagnification factors for some organochlorine compounds using linear free energy relationship parameters and artificial neural networks. SAR QSAR Environ Res 2009;20(5–6):453–65.
[33] Lunghini F, Marcou G, Azam P, Patoux R, Enrici MH, Bonachera F, et al. QSPR models for bioconcentration factor (*BCF*): are they able to predict data of industrial interest? SAR QSAR Environ Res 2019;30(7):507–24.
[34] Zang Q, Mansouri K, Williams AJ, Judson RS, Allen DG, Casey WM, et al. *In silico* prediction of physicochemical properties of environmental chemicals using molecular fingerprints and machine learning. J Chem Inf Model 2017;57(1):36–49.
[35] Organization for Economic Co-operation and Development. Test No. 203: fish, acute toxicity test. 2019.

[36] Cassani S, Kovarich S, Papa E, Roy PP, van der Wal L, Gramatica P. Daphnia and fish toxicity of (benzo)triazoles: validated QSAR models, and interspecies quantitative activity-activity modelling. J Hazard Mater 2013;258:50–60.

[37] Mansouri K, Karmaus AL, Fitzpatrick J, Patlewicz G, Pradeep P, Alberga D, et al. CATMoS: collaborative acute toxicity modeling suite. Environ Health Perspect 2021;129(4):47013.

[38] Moermond CT, Janssen MP, de Knecht JA, Montforts MH, Peijnenburg WJ, Zweers PG, et al. PBT assessment using the revised annex XIII of REACH: a comparison with other regulatory frameworks. Integrated Environ Assess Manag 2012;8(2):359–71.

[39] Ralston MDFDL, Jon JH, et al. In the U.S. Environmental Protection Agency waste minimization prioritization tool: computerized system for prioritizing chemicals based on PBT characteristics. American Chemical Society; 2000.

[40] Goh GB, Hodas NO, Vishnu A. Deep learning for computational chemistry. J Comput Chem 2017;38(16):1291–307.

[41] Gawehn E, Hiss JA, Schneider G. Deep learning in drug discovery. Mol Inform 2016;35(1):3–14.

[42] LeCun Y, Bengio Y, Hinton G. Deep learning. Nature 2015;521(7553):436–44.

[43] Fujita T, Winkler DA. Understanding the roles of the "two QSARs". J Chem Inf Model 2016;56(2):269–74.

[44] Jiang D, Wu Z, Hsieh C-Y, Chen G, Liao B, Wang Z, et al. Could graph neural networks learn better molecular representation for drug discovery? A comparison study of descriptor-based and graph-based models. J Cheminf 2021;13(1):12.

[45] Yu J, Wang J, Zhao H, Gao J, Kang Y, Cao D, et al. Organic compound synthetic accessibility prediction based on the graph attention mechanism. J Chem Inf Model 2022;62(12):2973–86.

[46] Xiong Z, Wang D, Liu X, Zhong F, Wan X, Li X, et al. Pushing the boundaries of molecular representation for drug discovery with the graph attention mechanism. J Med Chem 2020;63(16):8749–60.

[47] Tang Y, Tian Y, Lu J, Li P, Zhou J. Deep progressive reinforcement learning for skeleton-based action recognition. In: 2018 IEEE/CVF conference on computer vision and pattern recognition; 2018. p. 5323–32.

[48] Bruna J, Szlam A, LeCun Y. Spectral networks and deep locally connected networks on graphs. arXiv; 2013. 1312.6203.

[49] Veličković P, Cucurull G, Casanova A, Romero A, Liò P, Bengio Y. Graph attention networks. arXiv; 2017. 1710.10903.

[50] Stokes JM, Yang K, Swanson K, Jin W, Cubillos-Ruiz A, Donghia NM, et al. A deep learning approach to antibiotic discovery. Cell 2020;180(4): 688–702.

[51] Ryu S, Lim J, Hong HH, Kim WY. Deeply learning molecular structure-property relationships using attention and gate-Augmented graph convolutional network. arXiv; 2018. 1805.10988vol. 3.

[52] Li LH, Yu NY, Wang XB, Shi W, Liu HL, Zhang XW, et al. Comprehensive exposure studies of per- and polyfluoroalkyl substances in the general population: target, nontarget screening, and toxicity prediction. Environ Sci Technol 2022;56(20):14617–26.

[53] Cheng WX, Ng CA. Using machine learning to classify bioactivity for 3486 per- and polyfluoroalkyl substances (PFASs) from the OECD list. Environ Sci Technol 2019;53(23):13970–80.

[54] Wang YY, Wang JR, Cao ZL, Farimani AB. Molecular contrastive learning of representations via graph neural networks. Nat Mach Intell 2022; 4(3):279–87.

[55] United Nation Environment Program. What are POPs?. 2019. Available from: http://www.pops.int/TheConvention/ThePOPs/tabid/673/Default.aspx.

[56] Khan PM, Baderna D, Lombardo A, Roy K, Benfenati E. Chemometric modeling to predict air half-life of persistent organic pollutants (POPs). J Hazard Mater 2020;382:121035.

PART II

Tools and data sources for QSAR

CHAPTER

12

Modeling safety and risk assessment with VEGAHUB

Emilio Benfenati, Alessandra Roncaglioni and Erika Colombo

Istituto di Ricerche Farmacologiche Mario Negri IRCCS, Department of Environmental Health Sciences, Milano, Italy

Abbreviations

3R Replacement, reduction, and refinement
ADI Applicability Domain Index
AOP Adverse outcome pathway
CAS Chemical abstract service
CMR Carcinogenic, mutagenic, and reprotoxicant
DILI Drug-induced liver injury
DQD Danish QSAR Database
EDs Endocrine disruptors
EFSA European Food Safety Authority
FCM Food contact materials
FG Functional groups
GUI Graphical user interface
MCS Maximum common substructures
MG Molecular groups
MoS Margin of safety
NOAEL No observed adverse effect level
NTM Nontesting methods
PBT Persistent, bioaccumulative, and toxic
QMRF QSAR Model Reporting Format
QSAR Quantitative structure–activity relationship
SAs Structural alerts
SCCS Scientific Committee for Consumer Safety
SSbD Safe and sustainable by design
VEGA Virtual models for property evaluation of chemicals within a global architecture

The global needs of modern society about risk assessment and safety

Modern society is more and more aware of the need to address in a holistic way the challenges related to the health impact of the multiple substances we are exposed to. This necessity deals with the need to cope with multiple toxicological endpoints to proper characterize hazard. Unfortunately, we do not have the capacity to perform all the assays suitable to cover all the aspects related to toxicity. Huge efforts are made to address the so-called adverse outcome pathways (AOPs), with the aim to elucidate the mechanism provoking the final adverse effect [1]. This can enable the toxicologist to develop individual assays and tools to identify if a substance can trigger one or more steps along the adverse process. In addition to the need to cope with our knowledge gaps related to a single substance, there is the need to address coexposure and the effects of mixtures. In this larger vision, AOP, as well as toxicodynamics, and mixtures have to be addressed jointly with the information regarding the toxicokinetics, and the precise internal and external exposure pathways. Only at this point, it will be possible to make a correct risk assessment.

Furthermore, addressing the risk is no more sufficient. The urgent need is to identify solutions, minimize the adverse impact, and substitute critical substances with alternative ones that are safer. These efforts go into the direction of supporting the idea of a cleaner environment since the very beginning of the development of a new substance through a safe and sustainable by design (SSbD) approach [2].

This task is a complex one. Multiple factors are involved, and some of them are linked. To fix a single aspect may be misleading because the solution has to be acceptable from multiple points of view. The solution has to be provided at the system level, and not on partial, individual components. This complexity requires specific methodologies, with a key role in information technology approaches. Indeed, the particularly difficult task is better addressed

QSAR in Safety Evaluation and Risk Assessment
https://doi.org/10.1016/B978-0-443-15339-6.00005-9

by taking advantage of computerized tools. This is because there are many factors, there are interdependencies between them, the assessment has to be quantitative, and we should also take into account the uncertainty associated with the data. For all these reasons, the use of in silico tools offers a unique perspective. Indeed, it would be very difficult to solve the problem without computers. It is important to underline that this innovative approach is not simply a methodological one. The use of the in silico tools offers, for the first time, the possibility to introduce at the conceptual level the perspective to organize the task systematically, and not step by step.

The novelty of the methodology is introducing an alternative view compared with the traditional scheme, which is sequential, proceeding in a tiered approach. Indeed, if the single steps have to be done manually, what is commonly done is to identify the faster, easier steps, and the assays that require less resources. If the assessment can be done in a simplified, cheap way, this solution is adopted, and successive experiments are planned progressively, in case refinement is needed. However, for the computer to predict one property or 100 is not an issue. Thus, the perspective is completely different, the overview much larger, and a number of assumptions in the workflow can be avoided. Similarly, the frequent debate if it is more convenient to start from the exposure or the hazard to make a risk assessment is overruled by the possibility to cope at the same time with all components of the risk assessment process.

The VEGAHUB components

The development of the components of the tools within VEGAHUB [3] started from the classical in silico models, based on collections of quantitative structure–activity relationship (QSAR) models. However, we added more tools, which overall are moving the in silico tools within the more ambitious perspective as we have discussed before. The same tool can be used per se or integrated within a more complex software system. Here we characterize these components and how they interact.

The VEGA in silico models

A key pillar within VEGAHUB is the group of in silico models called VEGA models. There are currently about 100 models for many properties: toxicological, ecotoxicological, environmental, physicochemical, and toxicokinetics areas (see Table 12.1).

In some cases, there is more than one model for the same endpoint. This is useful, because the user has more results, and thus, it is easier to evaluate the value based on a weight-of-evidence approach. Indeed, in the case of multiple models, each model is different, and thus the fact that different models provide the same result increases the

TABLE 12.1 List of all endpoint and models in the current version of the VEGA platform (version 1.2.3).

Area	Endpoint	Nr. of models
Human toxicity	Mutagenicity (Ames test)	6
	Micronucleus assay	2
	Chromosomal aberration	1
	Carcinogenicity	10
	Skin sensitization	6
	Skin irritation	3
	Eye irritation	3
	Acute toxicity (LD50)	1
	Cramer classification	1
	Estrogen receptor effect	2
	Androgen receptor effect	1
	Thyroid receptor effect	2

TABLE 12.1 List of all endpoint and models in the current version of the VEGA platform (version 1.2.3).—cont'd

Area	Endpoint	Nr. of models
	Glucocorticoid Receptor	1
	Thyroperoxidase Inhibitiory Activity	1
	Endocrine disruptor activity	1
	Developmental toxicity	2
	NOAEL	3
	LOAEL	2
	Hepatotoxicity	1
Ecotoxicity	BCF	4
	Aquatic acute toxicity	15
	Aquatic chronic toxicity	3
	Mode of action	3
	Terrestrial acute toxicity	1
	Sludge toxicity	2
	Zebrafish embryo activity	1
Fate and distribution	Ready biodegradability	1
	Persistence (sediment)	2
	Persistence (soil)	2
	Persistence (water)	2
	Persistence (air)	1
Physicochemical properties	Octanol/water partition coefficient (logP)	3
	Vapour Pressure	1
	Water Solubility	1
	Melting Point	2
	Hydrolysis	1
	Henry's law constant	1
	Octanol/air partition coefficient (KOA)	1
	Soil adsorption coefficient of organic compounds (KOC)	1
Human PBPK	Plasma protein binding	2
	Aromatase activity	2
	P-glycoprotein activity	1
	Hepatic steatosis MIE	4
	Skin permeation (LogKp)	2
	Adipose tissue–blood partition	1
	Body elimination half-life	1
Ecological PBPK	kM/half-life	1

reliability of this result. Conversely, if there are conflicting results, this fact is an indication of uncertainty in the final result. As we will see, beyond the result as prediction, VEGA provides elements to evaluate the quality of the result and multiple elements for the evaluation of the prediction.

The multiple models may provide different perspectives also regarding the endpoint. Indeed, they may be based on different approaches to address the endpoint. For instance, VEGA has models for carcinogenicity, which are based on animal data, or data processed and evaluated by human experts and also based on evidence different from in vivo animal studies. Thus, the CAESAR model in VEGA for carcinogenicity is based on data on rat (male and female) species, while the ISSCAN-CGX model is based on data where the positive or negative call has been assessed by human experts based on multiple, heterogenous data.

Each model is provided with a guide, a QSAR Model Reporting Format (QMRF), and its dataset, which includes the substances used in the training and test set. This information is available at the page where the user selects the model, or from the Internet [4]. Please notice that each model is usually based on a specific dataset, and thus the applicability domain of the specific model is based on its own dataset. This is at the basis of the Applicability Domain Index (ADI). The ADI is an original approach to evaluate the reliability of the prediction. Of course, in the case of an unknown substance, we cannot know a priori if the prediction is correct or wrong. Thus, we base this evaluation on the results obtained on the most similar substances. In practice, VEGA shows for each substance to be predicted the six most similar substances. The similarity is based on the similarity index optimized within VEGA [5]. It ranges from 0 to 1, where 1 is the identity. VEGA shows the structure of the six most similar values, their similarity value (typically a good value is above 0.85, while below 0.75 the similarity is modest), and their experimental and predicted values. Based on these data, VEGA runs, for each prediction, a separate software program, providing the ADI. The ADI is based on multiple factors, which are also measured and provided to the user. The presence of similar substances in the training set is one of these factors. Another factor evaluates the correctness of the prediction of similar substances. A third factor compares the predicted value of the target substance with the experimental (not the predicted value, as in the point above) values of similar substances. There are other factors, depending on the model, assessing how large the error is in prediction, the presence of rare fragments, and other components. Basically, the ADI considers the three main elements at the basis of the QSAR model: the chemical information, the property, and the algorithm. In other terms, the ADI is not simply based on the chemical structure, and the similarity between the target and similar substances, but the assessment is specific for the endpoint and the chemical substance.

We recommend the user to examine with attention the warning provided by VEGA within the ADI. As a help to the user, VEGA provides a simplified evaluation of the ADI in three categories. The ADI is labeled as good, moderate, or low. This information is represented as stars in the PDF output: three, two, or one. In our experience, the statistics of the predictions for the substances predicted with good ADI values are better than those with lower ADI. It means that this program is able to identify weak areas of prediction. At the same time, we underline that the user should carefully evaluate the results. VEGA provides assistance and identifies warnings. It is up to the user to double-check these elements. It may be that the user takes a different decision regarding the reliability of the model. The automatic process is useful in the case of the evaluation of many substances, but for single substances, it is preferable to evaluate them carefully.

Indeed, as we said, VEGA shows the six most similar substances. These are shown with their experimental value. These data can be used to make a read-across evaluation, regardless of the prediction. In other terms, VEGA should be used not only for its prediction but as a tool for read-across. The presence of very similar substances with conflicting values compared with that provided by VEGA may represent a strong basis to overrule the prediction.

Beyond the prediction from the QSAR model and the read-across system, VEGA offers a third line of evidence, very useful for the overall final assessment: the presence of rules associated with the effect, such as structural alerts.

The read-across tools

As we have seen before, the VEGA models already include a way to make read-across, since the report provides the six most similar compounds. However, VEGAHUB offers a more versatile tool for read-across: ToxRead. ToxRead makes available separate modules for different endpoints. Indeed, read-across should be done specifically for each endpoint, and not simply based on the chemical structure. ToxRead shows the N most similar compounds and the presence of rules associated with the effect. The user has to choose the number of similar substances, and the software identifies the N most similar compounds. Typically, six is a reasonable number, but the user can try different numbers to dig deeper in the set of analog(s).

ToxRead is an interactive tool, useful to explore similar substances, and this is the main difference compared with VEGA, which provides a fixed outcome. ToxRead shows the target substance at the center of a graph, and its structure is also represented in the upper right corner of the screen (see Fig. 12.1). Circles indicate chemical substances. Thus, linked to the target substance, ToxRead shows the N most similar substances, using exactly the same program for similarity used for VEGA. Thus, this is the first level of screening of the most similar substances, to be used in the case that there are no rules associated with the effect. However, in most of the cases, ToxRead indicates a number of rules related to the specific property of interest, such as mutagenicity.

These rules, in the case of human toxicity, are typically in the form of structural alerts. The rules are represented by triangles in the graph. Linked to each rule, the software shows the N most similar substances, which contain that rule. Thus, the graph is represented by clusters of similar substances, which belong to the same family since they contain a certain structural alert. The color of the similar substance is red or green, whether the substance is toxic or not. Similarly, the triangle is red or green according to the rule, if it represents toxicity or lack of toxicity. Indeed, ToxRead contains both rules indicating adverse effect or lack of effect. The intensity of the color of the triangle depends on the prevalence of toxic or nontoxic substances in the overall training set containing that rule. The size of the circles representing similar substances is related to the similarity compared with the target substance under investigation. In this way, the user has immediately an overall view of the presence of rules for toxicity or lack of toxicity, and the presence of similar substances for each rule, which are toxic or not. More details about ToxRead have been described by Gini et al. [6].

VEGAHUB also offers some new other tools for the read-across: RAXpy and VERA.

RAXpy is a read-across tool based on structural, biological, and metabolic similarities [7]. Given a target compound, the best analogs are identified with an algorithm that integrates three orthogonal similarity metrics, considering toxicologically relevant aspects: (1) structural similarity based on structural fingerprints; (2) biological similarity based on biological fingerprints codifying the outcomes of high-throughput screening (HTS) assays from PubChem; (3) metabolic similarity based on the presence of common metabolic pathways between the target and the analog as simulated by SyGMa library [8].

Structural filters based on the presence of maximum common substructures (MCS) and common functional groups (FGs) can be also applied to narrow the chemical space for the analog(s) search. Read-across prediction is made for the target based on the activity of analog(s) identified with multiple similarity methods.

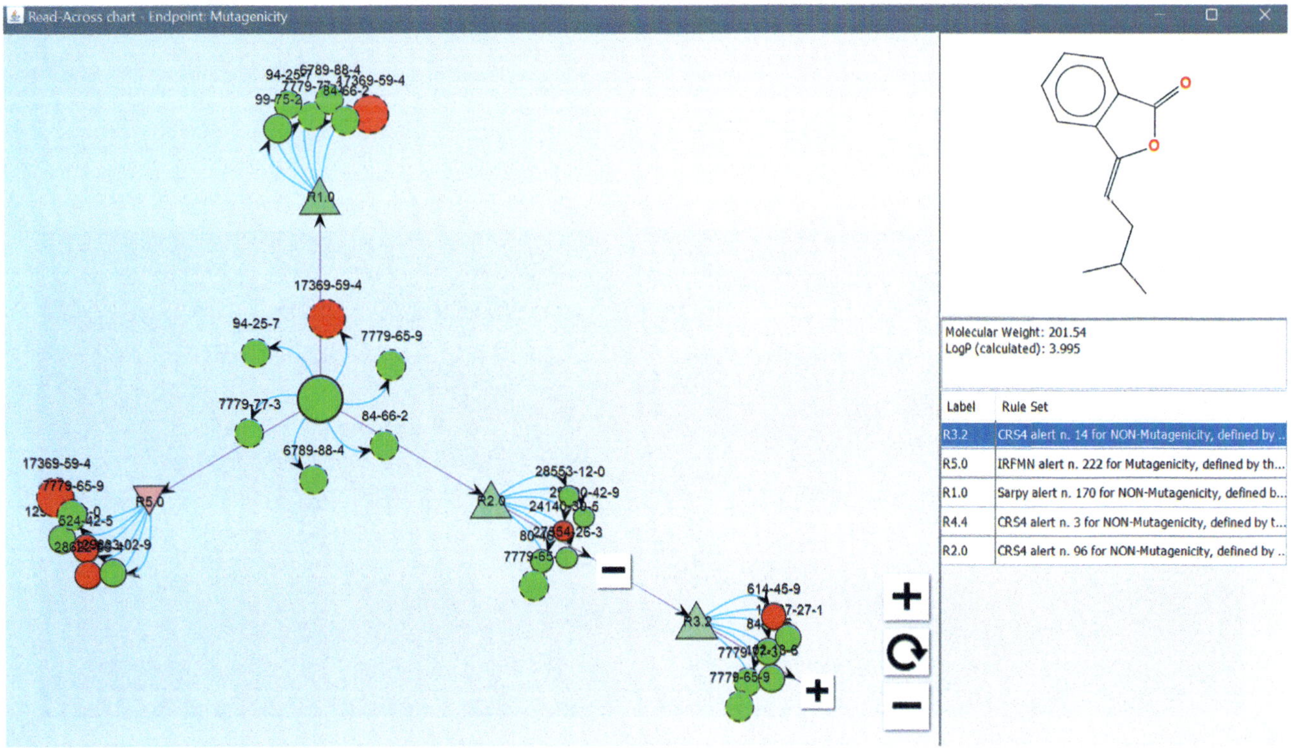

FIGURE 12.1 The ToxRead graphical output.

Currently the RAXpy software includes database related to two different endpoints: human hepatotoxicity (as from ToxRef database) and drug-induced liver injury (DILI).

VERA (Virtual Extensive Read-Across) provides means to assess the similarity between chemicals using structural alerts specific to the property, predefined molecular groups, and structural similarity (see Fig. 12.2). VERA aims to process many possible features (SAs, structural alerts and MGs, molecular groups) present in target substances and uses similar substances with these features to assess whether any of them may be responsible for a given adverse effect, even if this feature may not be present in the most similar substances. In this way, the software can go beyond the classical read-across process and help the expert to find also exception rules specific for each endpoint.

More details for these tools are described in the respective papers [7,9].

The VERMEER tools

The tools described before, for prediction with QSAR models and read-across, are addressing typically hazard properties. However, hazard and exposure must be addressed jointly to perform risk assessment.

The VERMEER tools are specifically designed for this task. Within the same software platform, the user can get all the elements necessary for risk assessment. Previously, separate programs had to be run, to model the exposure or

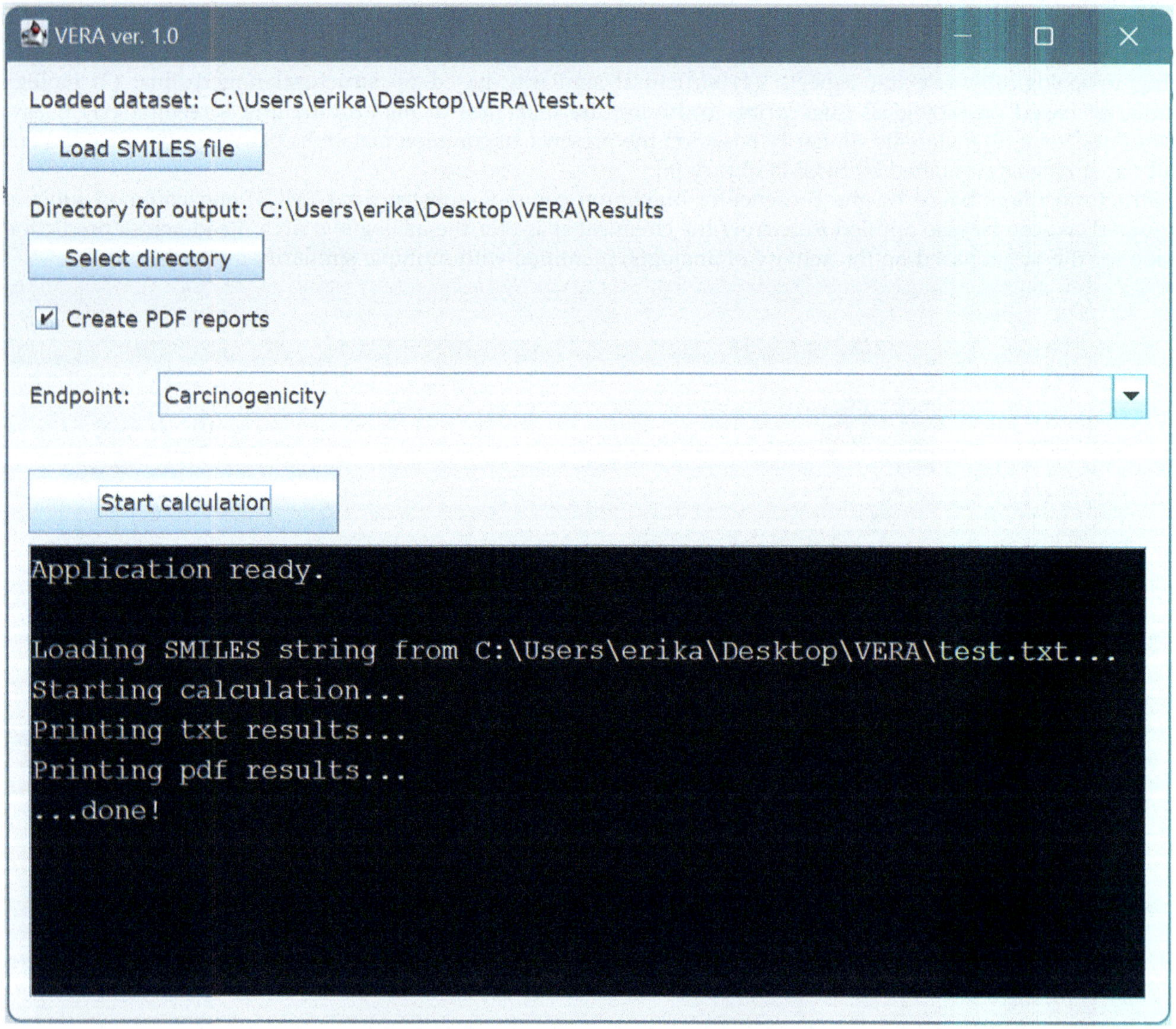

FIGURE 12.2 Vera graphical user interface (GUI).

the hazard. The VERMEER tools represent something unique for this reason. Of course, the risk is context-specific, and thus there are different tools for specific situations, such as the cosmetic scenario, or the food contact materials (FCMs). The cosmetic scenario is for cosmetic products. The user is requested to indicate the ingredients of the product, their concentration, and the kind of product (Fig. 12.3). We notice here some main differences compared with what we have previously seen. Here, the overall product is evaluated, with all its ingredients. Furthermore, the concentration has to be indicated. It is possible that a certain ingredient is safe or not depending on the concentration. Furthermore, the result will depend on the product because the exposure scenarios are different for different products, such as shampoo or body cream. The tool is designed to provide a number of data specific for cosmetics and to strictly follow the conceptual scheme described by the European authorities for cosmetics, i.e., the Scientific Committee for Consumer Safety (SCCS), as in their Notes [10].

The typical exposure route for cosmetics is the skin. Thus, the software considers the concentration of the ingredient (external exposure) and then predicts the internal exposure using two predictive models. The default value as indicated by SCCS is also provided (50% of the external dose). For the hazard, the software contains its database with experimental values, and the substance is also searched, if present in the lists of substances, which are not allowed for cosmetics, or if a threshold is indicated for the maximum allowed concentration. If the ingredient is not present in the database, the VEGA models are used. The reliability of the prediction is also reported. The toxicological endpoints that are considered are those relevant for cosmetics, such as skin sensitization, mutagenicity, no observed

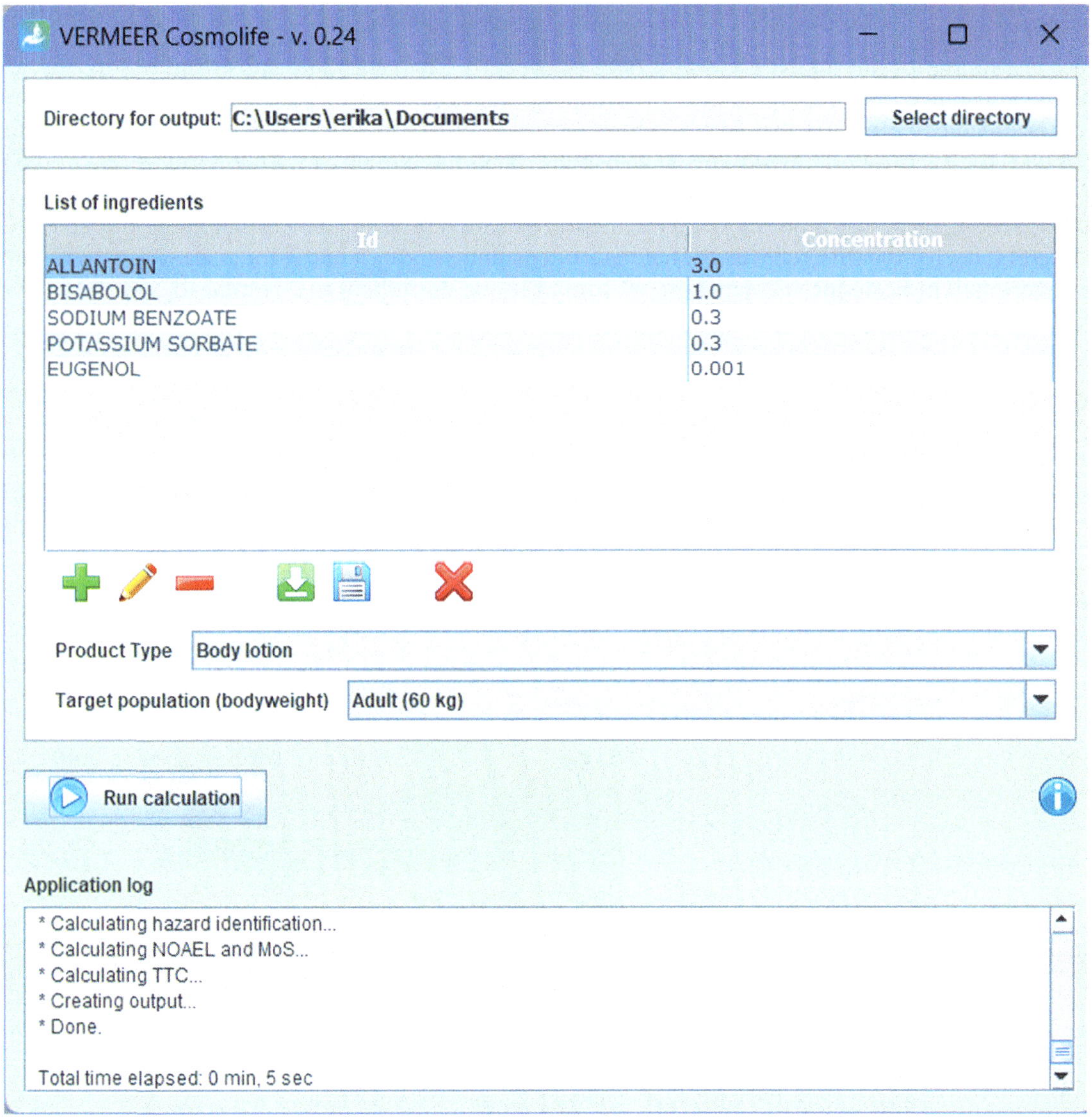

FIGURE 12.3 VERMEER Cosmolife input GUI.

adverse effect level (NOAEL), etc. The endpoints are used to identify toxicological issues, and to calculate the margin of safety (MoS). In this way, the user has a complete matrix of values for all the ingredients, using experimental values, if available, or predicted ones (see Fig. 12.4). The details of the tool have been described [11].

There are other VERMEER tools, such as that related to the FCMs. This tool includes a specific component related to the migration from the plastic container into the food. This is implemented in the platform MERLIN-Expo and requires inputs such as the duration and the temperature of the exposure, the kind of plastic material, and food. Then, the tool provides hazard property values, which are those specific for the evaluation of FCM, according to the European regulation. These endpoint (mutagenicity, genotoxicity, etc.) values are taken from the VEGA library, or predicted by the VEGA models if the experimental values are not available. Also in this case, the reliability of the prediction is reported. The details of the tool have been described in the literature [12]. We notice that this VERMEER tool, and the others, can be easily applied to related cases. For instance, this tool for FCM can be applied to medical devices where there may be the release of contaminants from the plastic. The food is mimicked considering a mixture of water and ethanol. Thus, the material should be characterized using this parameter.

The ToxEraser tool

ToxEraser represents the step forward after the risk assessment process. We have seen that the VERMEER tools can identify a risky substance, for instance, in a cosmetic product. The risk can be avoided by reducing the concentration, for instance, but this is not always possible. Thus, how to move toward safer substances? ToxEraser wants to proceed in this direction.

ToxEraser is looking for alternatives to substances of concern. The candidate substitute has to provide a similar function as the substance to be replaced; thus, we need another kind of information: the functional use of the substances. This is specific to the particular sector in the market. The ToxEraser tool for cosmetics, for instance, includes a number of uses for the different ingredients in a cosmetic product. Indeed, if we have found that a dye is toxic, we cannot replace it with a safe surfactant. The software contains libraries of substances, which are used as ingredients in the cosmetic sector, grouped according to their function and then ordered according to their adverse impact. Thus, there are several lists, for the different functional uses of cosmetic ingredients. If the user wants to replace a surfactant, the software will indicate the substances most similar to the ingredient to be replaced, which are safer than the critical substance belonging to the same functional group.

The JANUS tool

The German environmental agency funded the JANUS project, with the purpose to prioritize substances of concern for a series of purposes. The prioritization scheme covers the substances, which are carcinogenic, mutagenic and reprotoxicant (CMR), persistent, bioaccumulative and toxic (PBT), and endocrine disruptors (EDs).

Processed product

Product type: **Body lotion**

Ingredients:

		Ingredient Id	CAS	INCI	Conc. %	Annex	Mutagenicity	Skin Sensitization	Dermal abs.	MoS	TTC
Details	PDF	ALLANTOIN	97-59-6	ALLANTOIN	3.0	-	NON-Mutagen (moderate reliability)	Sensitizer (low reliability)	40%	268.64	0.0023 mg/kg bw/day
Details	PDF	BISABOLOL	23089-26-1	BISABOLOL	1.0	-	NON-Mutagen (good reliability)	Sensitizer (moderate reliability)	40%	133.5	0.0023 mg/kg bw/day
Details	PDF	SODIUM BENZOATE	532-32-1	SODIUM BENZOATE	0.3	V	NON-Mutagen (EXPERIMENTAL value)	NON-Sensitizer (EXPERIMENTAL value)	80%	93.55	0.046 mg/kg bw/day
Details	PDF	POTASSIUM SORBATE	24634-61-5	POTASSIUM SORBATE	0.3	V	NON-Mutagen (EXPERIMENTAL value)	Sensitizer (good reliability)	40%	3382.03	0.046 mg/kg bw/day
Details	PDF	EUGENOL	97-53-0	EUGENOL	0.001	III	NON-Mutagen (EXPERIMENTAL value)	Sensitizer (EXPERIMENTAL value)	80%	233360.39	0.046 mg/kg bw/day

Processed with VERMEER Cosmolife version 0.24

FIGURE 12.4 VERMEER Cosmolife output table.

Furthermore, the software evaluates the degradation products generated into the environment, addressing the CMR, PBT, and ED properties of the parental substances and their degradation products. This is a complex system, running 48 separate models together. Also in this case, the VEGA models are used.

The JANUS tool has the purpose to prioritize substances, thus identifying, for instance, the top 10 out of the list of the most toxic ones. This purpose is different from the simple screening process, because it requires a quantitative assessment, to get an ordered list. JANUS evaluated experimental data, if available, and if not available makes the predictions. The uncertainty of the final assessment is given. The results can be explored, and filtered, providing specific views and details on the reason why a certain substance is in a certain position in the ranking, and which are the most influential properties for the final outcome.

The architecture and the conceptual basis within VEGAHUB

The overall philosophy behind the VEGAHUB platform is to facilitate the reduction of the adverse impact of chemical substances on the environment and human health. This process is a complex one. There is a fundamental scientific component related to the environmental and health targets. This is the key pillar. The effort is to handle the different endpoints affecting the human health and the environment within the same system, in agreement with the One Health approach. This has been done within the JANUS system, for instance.

Another fundamental aspect is the link to the regulation. The conceptual scheme has to comply with the regulation, and this means to refer to the specific situation. This has been described in particular within the VERMEER tools. These tools are focused on the specific regulations (cosmetic or FCM), which define the list of endpoints to be used for a particular situation; thus, these tools implement the information regarding legal threshold values applicable to the specific context. The idea behind these tools is to help users to comply with regulation, saving time and resources. Within the same platform, all the components necessary for the assessment are present. Thus, the user is facilitated to verify compliance. The possibility to apply the ToxEraser tool promotes the reduction of the adverse impact, whenever necessary.

However, ToxEraser introduces a further level of complexity and an additional player: industry. The interests of authorities and industry are not identical. Both have to address safety and correct use of the substance regarding the potential adverse effects. Thus, VEGAHUB promotes the use of the tools for risk assessment (including tools for hazard and exposure), which must be the same, used by authorities and industry. It is not logical to have two systems, one for industry and one for authorities. It is not convenient to have conflicting results because different tools are used. It is possible to use multiple tools, but they should be used by both stakeholders in the same way. VEGAHUB promotes accessibility and transparency, as fundamental components of the confidence in the results. For this reason, the VEGAHUB tools are freely available, to optimize their use at all levels, without economic barriers. The tools run on local computer, to avoid confidentiality issues when the user is studying substances not yet on the market. The documentation on the training sets used to build up the models, and the algorithms, is also available, again for transparency. This is an advantage, compared with the use of commercial tools, because their documentation may be less accessible, and this may represent an issue in case of conflicting results because the cause of disagreement cannot be verified.

The industry has multiple activities, not only those specific to satisfy the regulatory requirements such as those related to the registration of products. For research activities, it is appropriate to use internally specific software programs. ToxEraser is a tool, which most likely can be used by the industry, screening better ingredients for safer products. Thus, conceptually, we have some tools, regarding adversities, which are in common between industry and regulators, and other tools, which are for the industry. However, from the information technology point of view, the two systems, for adversity and functional use, have to talk to each other, to optimize the results avoiding safety issues. Thus, the system has to be unified, and the optimization of the structure of novel substances has to cope with adverse effects to be avoided.

At the research level, for the exploration of properties, VEGAHUB offers several useful software programs, available in the download section. These tools can help industry, for instance, if they want to investigate particular substances and develop models based on confidential data. One example is the SARpy software, which is very easy to use and generates models. The SARpy software has been described [13].

The use of VEGAHUB for safety and risk assessment

We have seen that for research and development, the industry has purposes that are different from those related to the authorization of the substance, and the industry has a higher level of flexibility for the different needs. Conversely, in the process of regulatory requirements, authorities require a strict procedure. The documentation must be detailed and understandable, to be convincing. The simple predicted value is not considered sufficient. The user should carefully assess if the prediction is reliable. Thus, beyond the quality of the result, there is the need to provide elements suitable for the judgment regarding the acceptability of the result.

We will address here the case of the results of the VEGA models, because they are the basis of the other tools, and because they are usually applied for the registration of the substances. It is important to say that the VEGA models provide three lines of evidence, and all of them should be used.

The first line of evidence is the prediction itself. The model provides the predicted value, with its reliability. This should be discussed. VEGA identifies the presence of critical aspects, and if present they should be discussed. The user may argue that these warnings can be disregarded explaining why. This is possible because the user may have further lines of evidence.

It is also possible that there are multiple models for the same endpoint. This is an advantage because it offers support to the discussion. In some cases, for instance, for the mutagenic impurities in pharmaceutical substances, it is recommended to use more than one model to predict mutagenicity [14]. This is because one model should be expert-based, while the second one should be statistical-based. In this way, there are two orthogonal approaches, which increase the reliability of the overall assessment. Of course, if the results from multiple models are conflicting, this is an indication of uncertainty.

However, it is possible to deal with conflicting results, and for this, it is important to analyze the robustness of the individual results. The individual results can be simply counted assigning to each the same reliability, or it is possible to integrate them in multiple ways, as we discussed [15].

Another line of evidence provided by the VEGA models is the information about the experimental values of similar substances. This can be used for read-across, and the fact that there are similar substances in agreement with the prediction is a good indication of the reliability of the result. It is possible that similar substances have experimental values in disagreement with the predicted value of the target compound. In this case, the similarity of the substances is important. If there is a very similar substance with a conflicting value, this may overrule the prediction. Conversely, if the similarity is moderate, the in silico prediction is probably more reliable.

However, it is important to take into consideration if the target substance and the similar one share the same toxic mechanism. For instance, if there is a similar substance, which contains a structural alert associated to toxicity, and this structural alert is not present in the target substance, the similar substance is not relevant for our purposes. This allows us to introduce the third line of evidence provided by VEGA: the reasoning, in the form of rules associated to the effect.

The ideal case is when all the three lines of evidence are in agreement. If this is not the case, the reliability and relevance of each individual line of evidence should be evaluated. We discussed the matter addressing the weight of evidence of the in silico and read-across tools [15].

The role of VEGAHUB within a larger network

VEGA is committed to providing freely available a number of tools to help and speed up the assessment of substances, moving toward safer products and reduced pollution. The effort is huge, several other platforms are providing free tools, and it is necessary to establish a network between the different initiatives. We do not think that having a monolithic tool is the right solution. Many projects are ongoing and will start in the future, and all will contribute to the overall goal. Thus, it is important to facilitate the synergy and exchange of tools and data.

VEGA is active in this networking process. VEGA has been linked to the OECD QSAR Toolbox [16], and it has been donated to the AMBIT platform [17] and to the CLICC platform for life cycle assessment [18]. Work is in progress to link VEGA to the Danish QSAR Database (DQD) of the Danish Environmental Protection Agency [19] and to insert the VEGA predictions within the OpenFoodTox [20] database of the European Food Safety Authority (EFSA). Similar networking initiatives are ongoing with US platforms.

The EC-funded project LIFE CONCERT [21] is establishing a network of in silico tools in Europe. It involves these four platforms: VEGA, OCHEM [22], DQD, and AMBIT. It will offer a large collection of tools for read-across and in silico models, for a large number of the endpoints to be addressed for chemical safety, including toxicological, ecotoxicological, environmental, physicochemical, and toxicokinetic properties. The philosophy behind each platform is different, and this increases the utility to have multiple tools. We already discussed VEGA. AMBIT is specific for read-across and deals with substance in its complete meaning, thus also including impurities. It covers many approaches to explore similar substances considering multiple properties. The DQD is a large repository of predicted values obtained using many models, including commercial tools. In this way, the user has available predictions with tools, which may be expensive. OCHEM offers the possibility to develop new models using several sophisticated methods, beyond a list of models already developed.

Conclusions

Modern society has the need to evaluate an increasing number of substances, in a fast way, for many properties. There are even more ambitious targets, to eliminate from the market substances of concern, reducing the adverse impact of substances. The complexity of this challenge requires an intelligent approach to integrate into the same conceptual architecture many parameters. The availability of advanced in silico systems offers unprecedented possibilities to address this complex goal, reshaping the research in the field. VEGA contributed to this scientific progress by introducing several tools, aiming to reduce the adverse impact of substances. The statistical quality of the tools is a key factor, but we also address other aspects. We explain as much as possible the results and provide several levels of information, beyond the prediction, such as the link to the experimental evidence associated with the data at the basis of the model, and the information about the mechanism of effect, codified into explicit rules. These factors are important to increase the acceptability of the in silico approach. The toxicologists are familiar with a language and methods, which are different from those offered by the in silico tools. What we did is an attempt to make understandable the results. However, this process is also important from a scientific point of view, to clarify the different lines of evidence produced by the software tools and reduce the uncertainty associated with the predicted value.

The transparency of the result is also an important aspect, related to the credibility of the approach. The full documentation of the model, with its guide, dataset, QMRF, is requested by the regulation and by the stakeholders in general.

Finally, VEGA is part of a broader research activity where several initiatives are ongoing, and it is important to establish a network between different systems.

Acknowledgments

We acknowledge the financial support of the LIFE CONCERT REACH project (LIFE17 GIE/IT/000461).

References

[1] Ankley GT, Edwards SW. The adverse outcome pathway: a multifaceted framework supporting 21st century toxicology. Curr Opin Toxicol 2018;9:1–7.

[2] Joint Research Centre (European Commission), Caldeira C, Farcal R, et al. Safe and sustainable by design chemicals and materials: review of safety and sustainability dimensions, aspects, methods, indicators, and tools. Luxembourg: Publications Office of the European Union; 2022.

[3] VEGA HUB—virtual models for property evaluation of chemicals within a global architecture. https://www.vegahub.eu/. (Accessed 14 October 2022).

[4] QSAR Model Reporting Format (QMRF) of VEGA models—VEGA HUB. https://www.vegahub.eu/portfolio-item/vega-qsar-models-qrmf/. (Accessed 14 October 2022).

[5] Floris M, Manganaro A, Nicolotti O, et al. A generalizable definition of chemical similarity for read-across. J Cheminform 2014;6.

[6] Gini G, Franchi AM, Manganaro A, et al. ToxRead: a tool to assist in read across and its use to assess mutagenicity of chemicals. SAR QSAR Environ Res 2014;25(12):999–1011.

[7] Gadaleta D, Bakhtyari AG, Lavado GJ, et al. Automated integration of structural, biological and metabolic similarities to improve read-across. ALTEX-Altern Anim Exp 2020;37(3):469–81.

[8] Ridder L, Wagener M. SyGMa: combining expert knowledge and empirical scoring in the prediction of metabolites. ChemMedChem 2008; 3(5):821–32.

[9] Viganò EL, Colombo E, Raitano G, et al. Virtual extensive read-across: a new open-access software for chemical read-across and its application to the carcinogenicity assessment of botanicals. Molecules 2022;27(19):6605.

[10] SCCS notes of guidance for the testing of cosmetic ingredients and their safety evaluation—11th revision. https://health.ec.europa.eu/publications/sccs-notes-guidance-testing-cosmetic-ingredients-and-their-safety-evaluation-11th-revision_en. (Accessed 14 October 2022).

[11] Selvestrel G, Robino F, Baderna D, et al. SpheraCosmolife: a new tool for the risk assessment of cosmetic products. ALTEX 2021;38(4):565–79.

[12] Ciffroy P, Mertens B, Van Hoeck E, et al. Modeling the migration of chemicals from food contact materials to food: the MERLIN-expo/VERMEER toolbox. Food Chem Toxicol 2022;166:113118.
[13] Ferrari T, Cattaneo D, Gini G, et al. Automatic knowledge extraction from chemical structures: the case of mutagenicity prediction. SAR QSAR Environ Res 2013;24:365–83.
[14] EMA. ICH M7 assessment and control of DNA reactive (mutagenic) impurities in pharmaceuticals to limit potential carcinogenic risk. European Medicines Agency; 2018. https://www.ema.europa.eu/en/ich-m7-assessment-control-dna-reactive-mutagenic-impurities-pharmaceuticals-limit-potential. [Accessed 14 October 2022].
[15] Benfenati E, Chaudhry Q, Gini G, Dorne JL. Integrating in silico models and read-across methods for predicting toxicity of chemicals: a stepwise strategy. Environ Int 2019;131:105060.
[16] The OECD QSAR Toolbox—OECD. https://www.oecd.org/chemicalsafety/risk-assessment/oecd-qsar-toolbox.htm. (Accessed 14 October 2022).
[17] AMBIT—cheminformatics data management system—Cefic-Lri. http://cefic-lri.org/toolbox/ambit/. (Accessed 14 October 2022).
[18] Chemical Life Cycle Collaborative | UC Santa Barbara. https://clicc.ucsb.edu/. (Accessed 14 October 2022).
[19] The Danish Environmental Protection Agency (EPA). https://eng.mst.dk/. (Accessed 14 October 2022).
[20] OpenFoodTox: chemical hazards database | EFSA. https://www.efsa.europa.eu/de/discover/infographics/openfoodtox-chemical-hazards-database. (Accessed 14 October 2022).
[21] Life concert reach—concerting experimental data and in silico models for REACH. https://www.life-concertreach.eu/. (Accessed 14 October 2022).
[22] Online chemical modeling environment. https://ochem.eu/home/show.do. (Accessed 14 October 2022).

CHAPTER

13

Recent advancements in QSAR and machine learning approaches for risk assessment of organic chemicals

Pathan Mohsin Khan[1], *Gopala Krishna Jillella*[1] *and Kunal Roy*[2]

[1]National Institute of Pharmaceutical Education and Research-Kolkata, Kolkata, India [2]Drug Theoretics and Cheminformatics Laboratory, Department of Pharmaceutical Technology, Jadavpur University, Kolkata, India

Abbreviations

AD applicability domain
ALOGP Ghose–Crippen Octanol-water Partition Coefficient
AMW average molecular weight
ANN artificial neural network
AOP adverse outcome pathway
AUC area under the curve
CSCL Japanese Chemical Substances Control Law
ECOTOX ECOTOXicology Knowledgebase
EDS endocrine-disrupting substances
ESIS European Chemical Substances Information System
EU European Union
GA genetic algorithm
GFA genetic function approximations
GPR Gaussian process regression
HC hierarchical clustering
I–QSTR interspecies–quantitative structure–toxicity relationship
LDA linear discriminant analysis
LOO leave one out
LV latent variables
MAE mean absolute error
MLOGP Moriguchi octanol-water partition coefficients
MLOGP2 Squared Moriguchi Octanol-water partition coefficients
MLR multiple linear regression
Mw molecular weight
NB Naive Bayes
OCs organic chemicals
OECD Organization for Economic Co-operation and Development
OLS ordinary least squares
PAN pesticide action network
PCRA principal component regression analysis
PCs principal components
PFAS polyfluoroalkyl compounds
PLS partial least squares
PPCPs pharmaceuticals and personal care products
PPPs plant protection products
QSPR quantitative structure–property relationship
QSTR quantitative structure–toxicity relationship
R^2_{pred} predictive R^2
RASAR read-across structure–activity relationship
REACH Registration, Evaluation, Authorization and Restriction of Chemicals
RMSE root mean square error
$RMSE_P$ root means square error of prediction
ROC receiver operating characteristic
RR ridge regression
SEE standard error of estimation
SVM support vector machine
TCPS tissue culture polystyrene
TSCA toxic substances control act
Uc unsaturation count
USEPA U.S. Environmental Protection Agency
VEM valence electron mobile
XGBR extreme gradient boosting-regression

Introduction

The rise of sustainable and innovative chemistry propels the consumer and industrial product landscape, with viable applications in diverse sectors (e.g., health, technology, energy, etc.). There is a clear relationship between advanced global chemistry research and the quantity of commercially available chemical compounds with a substantial increase in the number of synthetic organic chemicals designed and produced, commercialized, and later

QSAR in Safety Evaluation and Risk Assessment
https://doi.org/10.1016/B978-0-443-15339-6.00035-7

identified in the environment [1]. Since the early 1800s, the CAS REGISTRY^{SM} (www.cas.org, accessed April 2, 2022) has recorded over 198 million organic and inorganic substances, which include alloys, coordination compounds, minerals, mixtures, polymers, and salts. If the current trends continue, the CAS REGISTRY^{SM} will contain approximately 650 million new chemical substances by 2065 [2]. However, not all of the organic chemicals that have been identified are used in daily necessities and/or pose a threat to humans, animals, or the climate. The currently marketed organic chemicals are estimated to have anywhere from 75,000 to 140,000 [3]. The insufficient information on the potential health and ecological risks posed by the large percentage of these chemicals is a cause for concern. For example, available experimental aquatic toxicity data of synthetic organic chemicals against various species are estimated to be 11%, similar bioconcentration data of organic chemicals are approximately 1%, and persistence data of approximately 0.2% of chemical substances commercialized in the European Union (EU) and the United States (US) [3]. This emphasizes the importance of comprehensive and efficient risk assessments and the various regulatory bodies of all new chemical substances to avoid or minimize the potentially hazardous effects of these chemical compounds on both human health and the environment. Therefore, to promote decision-making about the commercialization of new organic chemicals (e.g., pharmaceuticals and personal care products (PPCPs), household items, polymers, feed and food additives, plant protection products (PPPs), and other innovative chemical-based products), several regulators, such as the EU REACH [4], the US Toxic Substances Control Act (TSCA) [5], and the Japanese Chemical Substances Control Law (CSCL), [6] require an environmental risk assessment (ERA) data on all newly registered compounds or imported substances with an annual volume of one tone (10 tones for CSCL) [6]. Although several paradigms for performing ERAs have been proposed, they all share the same goals: the identification and characterization of chemicals of global concern, the prioritization of substances for further evaluation testing, and the management of potentially dangerous health and safety risks associated with the organic chemicals. Identification of acute toxicity effects is the initial stage in carrying out an ERA [7]. Traditional acute toxicity testing is supported by computer-aided methods, particularly quantitative structure—activity relationships (QSARs) and read-across procedures, which aim to fill up data gaps [8]. These techniques are founded on the similarity principle, which states that chemical compounds with comparable properties will exhibit similar behavior and that a small change in a compound's chemical structure will affect its endpoint value. In order to comprehend the complex interactions between chemical structures, physical and chemical properties, mechanisms, and modes of action, these methodologies make use of data on in vitro/in vivo toxicity and chemoinformatic tools. Several traditional QSAR and numerous machine-learning algorithms have been created and effectively used to forecast the acute toxicity of new (or untested) organic compounds to aquatic plants, invertebrates, fish, reptiles, amphibians, birds, and humans. One of the primary goals of traditional QSAR and machine learning (ML) approaches is to predict the toxicity of specific organic compounds using existing organic chemical toxicity data. This can be done by creating a mathematical model that links the structural characteristics of organic substances (also known as numerical descriptors) to the desired toxicity end point. It may be difficult to address the complicated relationships between input chemical structural features and output response end points using conventional QSAR approaches. Therefore, the most effective way to comprehend and portray these complicated relationships is through the application of cutting-edge ML techniques. Several studies predicted the acute toxicity of organic chemicals using a range of classical and ML algorithms such as multiple linear regression [MLR], principal component regression [PCR], partial least squares regression [PLS], and polynomial and spline regression models) [1,9], similarity-based approaches [10] (e.g., *k*-nearest neighbor [*k*-NN] models), and more complex and sophisticated ML techniques (e.g., random forests [RF] [11], neural networks [12,13]), and support vector machines [SVM]) [14].

This chapter primarily focuses on traditional QSAR and ML methodologies used to evaluate possible risks connected with organic compounds and their more recent usage in predicting the toxicity of organic. A brief summary of the approaches used to create predictive quantitative structure—property relationship (QSPR) and ML models in computational predictive toxicology is provided in "Brief overview of the methodologies used for QSAR modeling in predictive toxicology" section.

Brief overview of the methodologies used for QSAR modeling in predictive toxicology

The similarity principle serves as the foundation for QSAR modeling. According to this theory, even slight modifications to the molecular composition of organic chemicals or any other compounds will alter the observed micro- and macroscopic response endpoints [15]. In general, a QSAR model is constructed using one or more classical or ML algorithms that extract the frequently complex relationships between the molecular structure of organic chemicals and their toxicity response end point against various species such as aquatic plants, invertebrates, fish, reptiles, amphibians, birds, and humans. Data preparation and collection, computation of molecular descriptors, feature selection, model development (using either a linear or nonlinear learning method), applicability domain analysis, and thorough model validation are all essential aspects of the QSAR modeling process [15,16] (as shown in Fig. 13.1). The importance of the in silico techniques in the toxicity prediction is shown in Fig. 13.2, while the types of different ML tools are shown in Fig. 13.3. Following is a detailed explanation of each of the aforementioned steps.

Data collection and data preparation

The most essential prerequisite for developing both QSAR and ML models is a reliable data set of organic compounds with clearly defined chemical structures and experimentally measured toxicity response endpoints. The primary source for a sizable amount of reliable organic toxicity data can be compiled from different academic papers, printed handbooks, and online web-based repositories, depending on the intended toxicity end points. Currently, several toxicity databases have been digitalized, allowing easy access to toxicity data of organic substances and giving accurate data for computational modeling of various toxicity endpoints. Table 13.1 shows representative examples of some of the publicly available databases that contain toxicity data. The majority of the databases were created by gathering a large amount of data from the literature.

However, experimental data from many publications suffer an undetected problem: data leaking. Nonideal experimental results are frequently discarded without publication, even though such negative data are also vital for model training. Data collected from various sources or publications can suffer substantially due to poor data quality and poor model performance caused by experimental errors, uncertainties, or inadequate criteria.

The generation of a robust, reliable QSAR, and ML model with significant accuracy is accomplished by removing errors and uncertainties in the data during data extraction and preparation. Care should be taken to ensure that the chemical structures of polymers and experimentally measured properties of interest are accurate. Because all measured data are associated with errors, and these errors will affect the reliability of models; these can be avoided by using data obtained from the same experimental protocols and conditions in the same laboratory. The collected data must be carefully checked for duplicates, salts, and so on. QSAR frequently assumes a normal distribution of experimentally measured property values of interest, which is not always the case. Because the property of interest being modeled ranges over several orders of magnitude, sometimes it will deviate from a normal distribution.

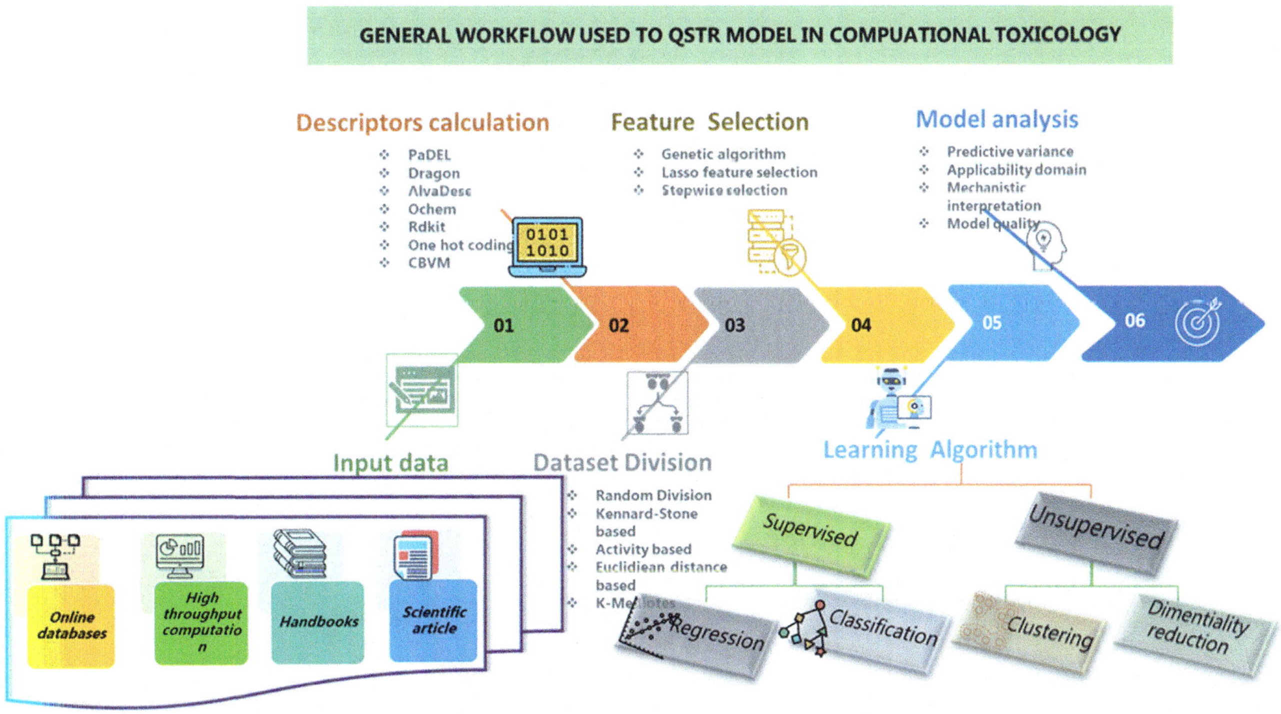

FIGURE 13.1 A schematic overview of the general methodology used for quantitative structure–property relationship (QSAR) modeling in computational toxicology.

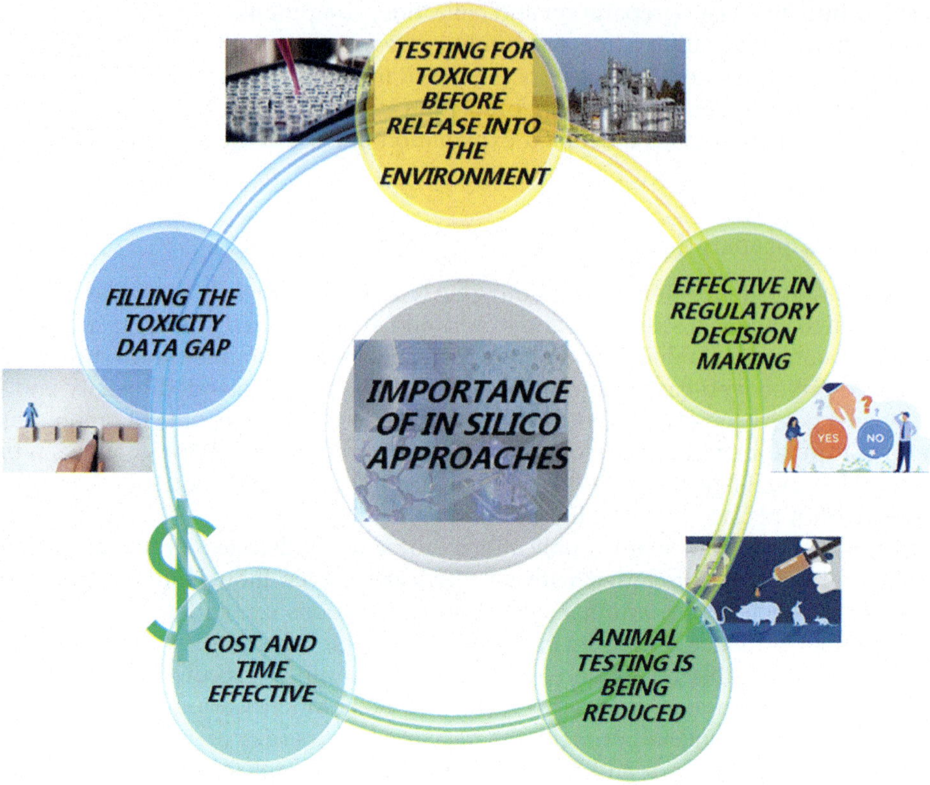

FIGURE 13.2 The importance of in silico approaches in toxicity prediction.

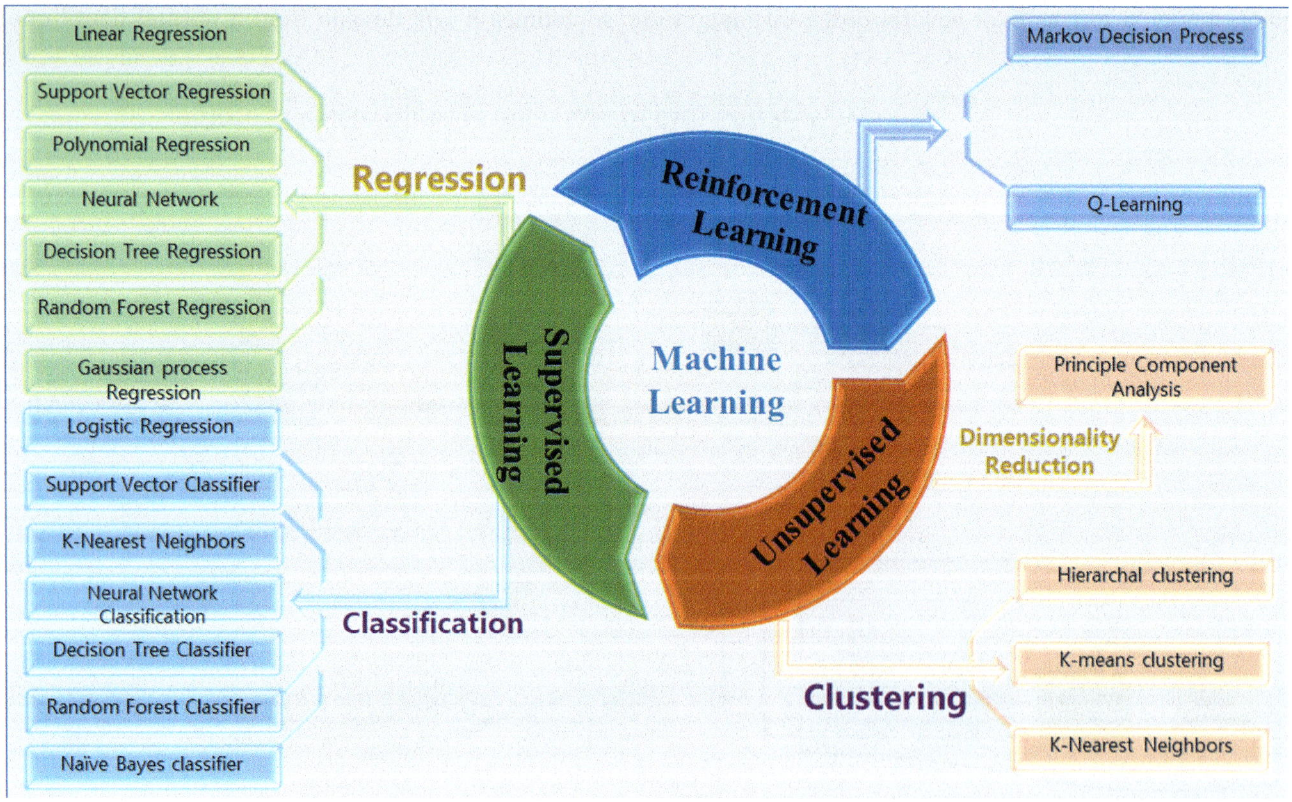

FIGURE 13.3 Types of machine learning algorithms.

TABLE 13.1 List of the central toxicological databases.

Sr. no	Name of databases	Descriptions	Websites
1	ECOTOX	The ECOTOXicology Knowledgebase (ECOTOX) is a database that contains https://www.cdc.gov/niosh/rtecs/rtecsaccess.html information on particular chemical toxicity for aquatic life, terrestrial vegetation, and wildlife.	https://cfpub.epa.gov/ecotox/
2	ESIS	An IT system called the European Chemical Substances Information System gives one access to data on chemicals, including inventories of commercially available chemicals, priority lists for persistent, bio-accumulative, and toxic chemicals, and the process for assessing the risks associated with chemicals.	https://old.datahub.io/dataset/esis
3	Pubchem	The greatest repository of freely available chemical data. Chemicals can be found by name, structure, molecular formula, and other identifiers. Discover information about biological activities, safety and toxicity, chemical and physical qualities, patents, literature citations, and more.	https://pubchem.ncbi.nlm.nih.gov/
4	RTECS	RTECS is a collection of information taken from publicly available scientific literature. The information is presented in the format created by the RTECS staff and is ordered alphabetically by the name of the primary chemical. The file contains data on six different categories of toxicity: There are six types of multiple dosage toxicity: (1) primary irritation, (2) mutagenesis effects, (3)	https://www.cdc.gov/niosh/rtecs/rtecsaccess.html

Continued

TABLE 13.1 List of the central toxicological databases.—cont'd

Sr. no	Name of databases	Descriptions	Websites
		reproductive effects, (4) tumorigenic effects, and (5) acute toxicity. There are also notes on the species investigated and the method of administration utilized, as well as specific numerical toxicity values like LD_{50}, LC_{50}, TDL_o, and TCL_o.	
5	EXTOXNET	It contains information about the various types of pesticide toxicity.	http://extoxnet.orst.edu/etn.txt.html
6	PAN	Pesticide action network (PAN) is a database of information on pesticides that includes data on a wide range of topics, including human toxicity (chronic and acute), ecotoxicity, regulatory information, and more for roughly 15,300 pesticide active ingredients and the products that are created from them, including solvents and other additives used in pesticide products.	https://www.pesticideinfo.org/about/overview

In such a situation, logarithmic transformation of the experimental toxicity of organic chemicals is used for QSAR modeling (i.e., Log(toxicity of organic chemicals) or Log(1/(toxic concentration of organic chemicals)) [17]. Similarly, while processing the data, the hierarchical clustering (HC) method should be employed, which divides information into clusters based on chemical similarity, and produces high prediction accuracy of the generated model [18]. Finally, when dealing with an uneven chemical dataset in terms of active and inactive substances, extreme caution is required, and excessively unbalanced datasets create issues for both regression and classification modeling. QSAR and ML models trained on unbalanced datasets tend to assign more cases to more categories, resulting in poor category resolution and low model sensitivity [19].

Molecular descriptor computation

Molecular descriptors are numerical values acquired by characterizing the molecular structures of molecules for QSAR and ML model construction. The molecular descriptors (features) of organic molecules serve as an independent variable in the risk assessment studies of organic chemicals. Because of advances in computational power, it is now possible to estimate thousands of different types of numerical descriptors in a matter of minutes using a variety of both free and paid software packages such as PaDEL-Descriptor, Dragon, SiRMS, AlvaDesc, CODESSA, OChem, RDkit, and others. Furthermore, many forms of novel molecular descriptors and methods for estimating novel descriptors are being designed and developed. They are fast-evolving, and explaining them is beyond the scope of this chapter. More information on molecular descriptors can be found in recent books and review papers [20–22].

Feature selection

Feature selection is an important step in the QSAR modeling process. The features chosen will significantly impact the accuracy of the model's predictions. It also impacts the model's ability to demonstrate how chemical structures connect to the desired attribute. Because of the high computational cost, prolonged processing times, and detrimental impacts on model quality, the complete pool of numerical descriptors directly cannot be employed for QSAR modeling. Therefore, to develop robust and reliable QSAR models, the subsets of features are chosen from the entire pool of descriptors by removing redundant, noisy, or pointless descriptors—all without losing important data. Factor analysis, all-possible subset selection, genetic algorithms, step-wise forward selection, and backward elimination are the few most widely used techniques for selecting descriptors from the entire pool of descriptors. The review paper by Khan and Roy provides further information on various feature selection techniques [23].

Modeling algorithms

Based on the response end points' type (continuous or categorical), computational modeling methodologies may be roughly divided into two categories: regression-based and classification-based QSAR modeling. Regression-based QSPR modeling is used when both the dependent and independent variables have continuous numerical values; classification-based QSAR modeling is used when the response variable is discrete, such as a class, category, or group.

Additionally, the regression QSAR modeling is divided into two groups based on the type of modeling functions: linear regression QSAR modeling and nonlinear regression-based QSAR modeling (mostly ML approaches). Traditional ordinary least squares, MLR, PLS, principal component regression analysis (PCRA), ridge regression, and genetic function approximations (GFA) are examples of linear regression approaches [15,24]. On the other hand, SVM, decision tree, artificial neural network (ANN), deep neural network, Gaussian regression process, Naive Bayes algorithm, RF, and k-NN algorithm are examples of nonlinear regression approaches.

MLR is the most often used linear algorithm for QSAR modeling or model generation due to its simple methodology, reliability, and ability to quickly comprehend the characteristics utilized in model development [25]. It generates property models based on a set of numerical molecular descriptors' linear functions. The generic MLR equation is shown in Eq. (13.1).

$$Y = a_0 + a_1X_1 + a_2X_2 + a_3X_3 + \ldots + a_nX_n \tag{13.1}$$

Here, Y is the response end point (i.e., toxicity in case of risk assessment of organic chemicals), $X_1, X_2, X_3, \ldots X_n$ are the numerical descriptors calculated from molecular graphs of organic chemicals, and $a_1, a_2, a_3, \ldots a_n$ are regression coefficients of the model, while a_0 is a regression constant.

The main drawback of adopting an MLR model is that it introduces multicollinearity into the final model and produces several unique, comparable models using various descriptor combinations [26]. It will lessen the accuracy of the independent variables' interpretation and impact the variance related to the prediction. Another drawback of MLR models is that they are frequently overfitted, which has a detrimental influence on the accuracy of their predictions and their capacity to generalize.

To address the constraints of the basic MLR methodology, QSAR models can be developed using alternative linear methods such as PLS [27] and PCRA [28]. Instead of directly calculating the potential danger of organic compounds from molecular descriptors, these approaches use extracted latent variables or principal components containing relevant information from a large number of molecular variables as predictors. Simply said, the linear regression model is formed by projecting the dependent and independent variables to a new space (in the case of PLS) defined by latent variables. In PCRA, the predictor variable data matrix is treated to obtain several principal components. The best component is chosen based on the variance in the X data matrix, implying that PCRA captures the most variation solely in the X data matrix. On the other hand, PLS attempts to work with both by optimizing the covariance between X and Y.

All linear regression models are designed to estimate the potential risks of organic compounds as a function of molecular descriptors. However, it will completely ignore the nonlinear connections between molecular descriptors derived from organic chemicals and their toxicity end points. The previously mentioned ML nonlinear regression approaches can be employed to address this issue. The present chapter discusses ML basics and different applications of ML approaches in predicting potential hazards of organic chemicals.

Basics of ML

It is a branch of artificial intelligence that enables computer systems to automatically learn and enhance their ability to address a particular issue based on past data and current knowledge and predict results for brand-new data without having to be explicitly programmed [29]. Three categories of ML may be made in general: reinforcement learning, unsupervised learning, and supervised learning (as shown in Fig. 13.1).

In order to apply supervised learning, the sample training data must be labeled, which implies that each input variable's (i.e., independent variable (X)) information and its user-defined target output's (i.e., dependent variable (Y)) information must be included. Y might be a continuous quantitative value or a discrete class, group, or category. To put it simply, the model is trained using labeled data to comprehend the relationship between the input and target variables; after training and processing, the model's quality is tested by feeding sample data to see whether it accurately predicted the target output or not [30]. Regression and classification are two subcategories of algorithms that may be used in supervised learning based on the types of target variables (Y). On the other hand, an unsupervised learning strategy is when a computer learns from the data without any supervision when it is given unlabeled training data or missing target output values for each input variable in the data [31]. The main objective of unsupervised learning is to explore the raw data and find new features, groups/clusters of objects with similar characteristics, or patterns. It is further subdivided into two types of algorithms: association and clustering. Reinforcement learning is an iterative learning process where learning agents adjust their states and engage with their environments in an effort to maximize the reward value they are aiming for. The learning agent rewards the right action and penalizes the wrong action in a certain context. Instead of looking for specific patterns in the input data, it often concentrates on the interactions between algorithms and their environments. [31], which is an inadequate method for the studies of possible risks of organic compounds. Therefore, we would not cover it in this chapter.

ML algorithms

SVM

SVM was introduced at the fifth annual workshop on computational learning theory (COLT-92) by Boser, Guyon and Vapnik based on the statistical learning frameworks later published by Vapnik and Chervonenkis [32]. SVM is a prominent supervised learning technique that is used for both classification and regression issues [33] by identifying the optimum decision boundary (hyperplane) in an N-dimensional space that distinctly classifies the data points while reducing error and increasing geometric margin [34]. There are several different hyperplanes for separating the two groups of data points. The objective of SVM is to discover a plane with the greatest margin, that is, the greatest distance between data points from both classes. Maximizing the margin distance gives some reinforcement, allowing subsequent data points to be categorized with more certainty. SVMs are divided into two types based on the dataset domain used to train the models [35]. Linear SVM is used when the domain of the training dataset can be linearly separated into two different groups. Nonlinear SVM, on the other hand, is utilized when the domain of the training dataset is linearly inseparable. In SVM, the low-dimensional feature space data are projected into the high-dimensional feature space data using a kernel trick technique [36], which makes the data linearly separable. Finding the best decision boundary in an N-dimensional space to categorize each data point in the dataset is the main objective of the SVM method. The ideal boundary of the plane is a hyperplane (as shown in Fig. 13.4). In the case of regression issues, SVM finds the analytical expression of the hyperplane within a reasonable error [37].

Decision tree

It is one of the supervised ML approaches. It is used to address both classification and regression issues. However, it is mostly used to solve categorization difficulties. The primary purpose of using a decision tree approach is to create a training model that can be used to predict the class or value of the target variable by learning basic decision rules from previous data. A dendrogram or tree-like structure of distinct nodes is used to depict decision tree models. The root node holds the feature sample dataset used for modeling; the root node is separated into numerous branches called decision nodes, which contain the modeling decision rules; and the last node is the leaf node, which displays the modeling conclusion [38]. Readers are directed to other publications written on the decision tree method for further information [23].

k-NN

The k-NN algorithm is a basic supervised learning technique that is used to solve classification and regression issues using similarity metrics. In simple words, it classifies the query chemical based on its closeness to the number

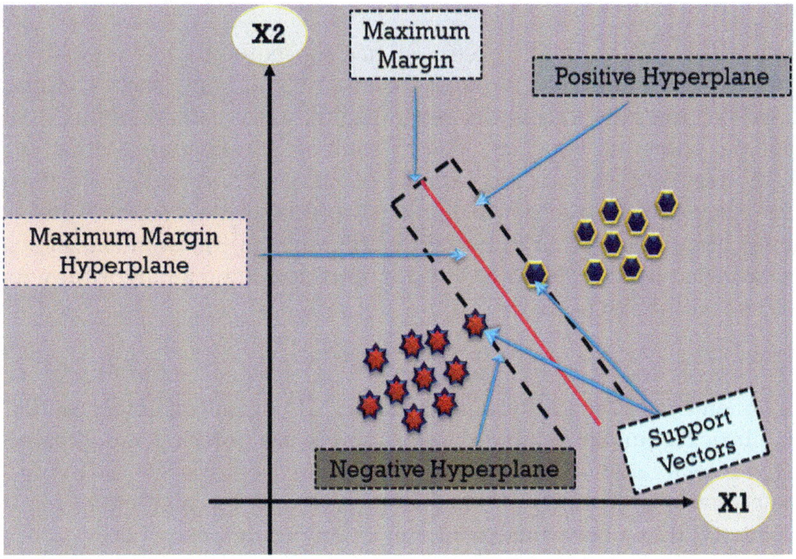

FIGURE 13.4 The basic structure of classification of data by SVM.

of training compounds (because comparable compounds occur close to training compounds) (as shown in Fig. 13.5). Assume that we have certain data points representing two separate classes and that we are required to classify a new object into one of the two groups. The k-NN algorithm in this situation first determines the *k* number of neighbor objects in the dataset, then uses distance functions to calculate the k-closest distance to the new object (the most popular being Euclidian distance), and finally uses the majority of those neighbors to determine the class of the new object [23,39].

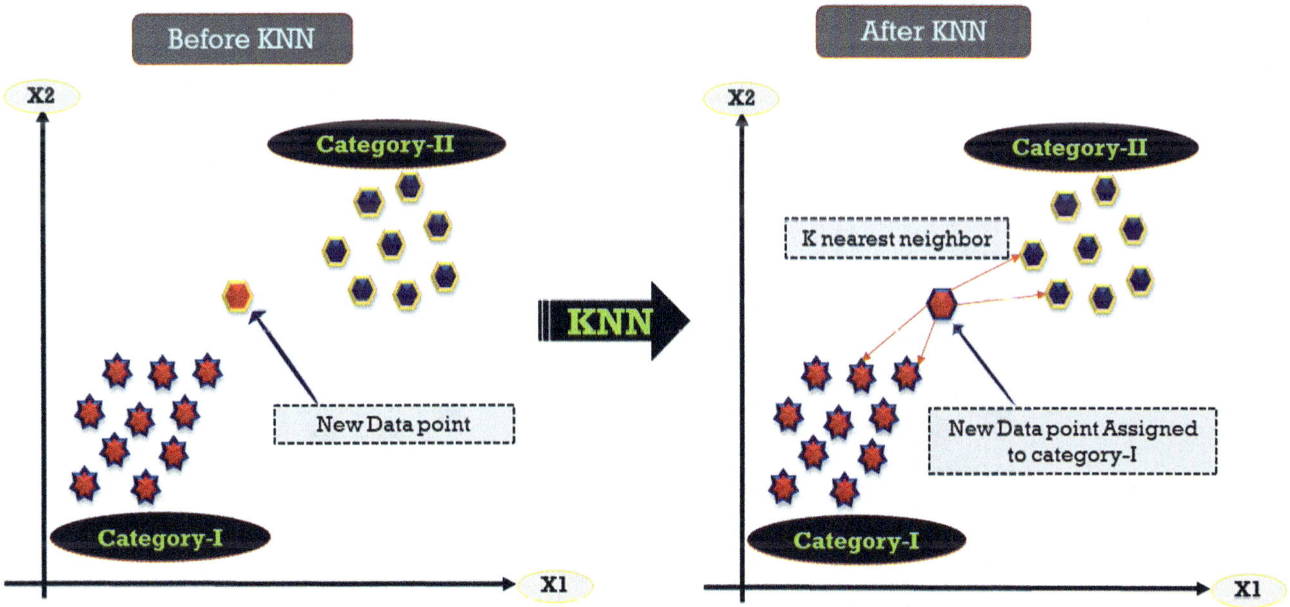

FIGURE 13.5 Basic overview of k-nearest neighbors algorithm. The *right side* of figure shows that the entire dataset is divided into two categories, with category I consisting of the *dark red* dataset points and with category II including the *dark blue points* and the *light red point* serving as the query point that has to be categorized. On the *left side* of the figure shows that the query point is assigned to category I (*dark red points*) based on the straightforward nearest-neighbor approach (k = 1) since it is situated closest to the category I data points.

RF

L. Breiman first developed the RF method in 2001. RF has subsequently been widely applied to solve general-purpose classification and regression issues. It is one of the ensemble ML approaches; it incorporates the predictions of several randomized decision trees to increase the generalizability/robustness of the RF model in comparison to a single decision tree model (as shown in Fig. 13.6). The RF estimator, in layman's terms, is a model that fits several decision tree classifiers/regressors on various sub-samples of a dataset and makes use of averaging to improve projected accuracy and control over-fitting. In order to eliminate correlations across trees and boost the model's predictive accuracy, it makes use of randomization in two different ways: (a) random subset selection from the original dataset and (b) random variable selection. In order to learn more about the RF algorithm, readers are recommended to different articles and/or book chapters [23].

Naive Bayes

The NB algorithm is a member of the group of supervised ML algorithms built on the well-accepted Bayesian theorem. It uses a probabilistic technique to solve classification issues, offering a mathematical way to comprehend how data observation affects each target class. This approach involves properly classifying sample data points from the dataset based on the posterior probability of that class, which may be calculated from the prior probability of a certain class and the likelihood that a new data point will belong to that class [23].

Gaussian process regression

The Gaussian process (GP) is a supervised, nonparametric ML approach based on Bayesian probability theory. It has several applications, including regression, classification, and parameter optimization. However, Gaussian process regression is the most often used method for dealing with regression issues. Unlike other methods, the Gaussian process regression (GPR) model can predict unknown data points based on current information (kernels) and offer uncertainty measurements for projected unknown data points. To understand GPR, one must be familiar with ideas about a variety of fundamental concepts such as multivariate normal distribution, kernels, nonparametric models, joint and conditional probability, and so on. The previous reports are suggested for people interested in learning more about the concept of GPR [40].

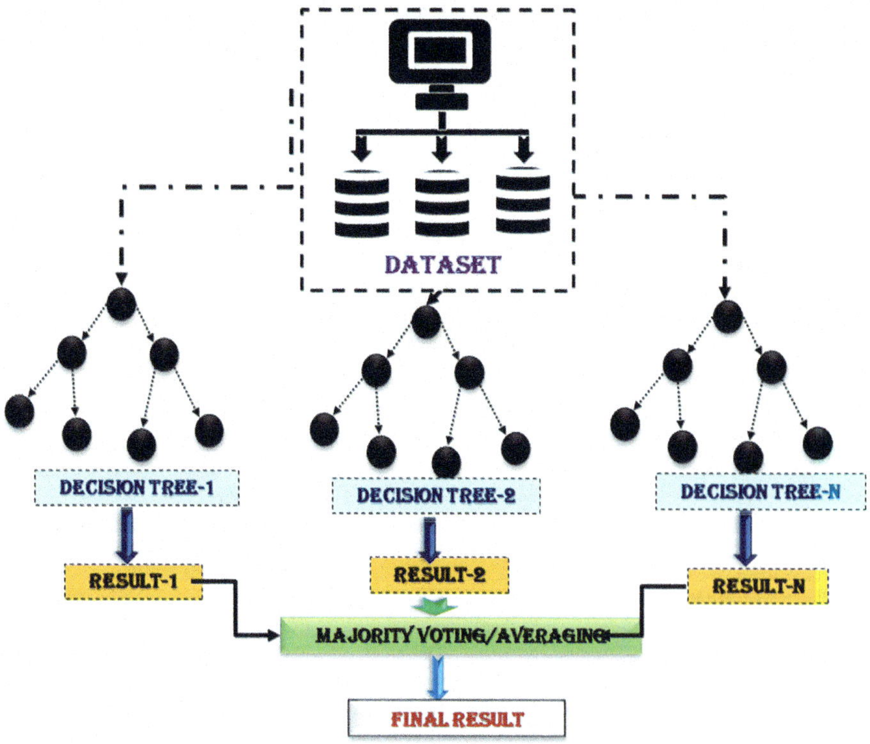

FIGURE 13.6 The RF algorithm's basic structure. At each branching point, the input dataset is divided depending on the yes/no rule, resulting in several sub-tree structures. Individual tree forecasts are summed and integrated to make a single consensus model prediction based on voting (for classification) or averaging (for regression).

Neural networks

A neural network is a statistical or computational technique that replicates the architecture and learning capabilities of a neuronal network in the human brain [41]. It is the most widely used method for modeling difficult nonlinear modeling problems (i.e., the nonlinear relationship between independent and dependent variables) and predictions. ANN was originally proposed by McCulloch and Pitts in 1943 and is the most fundamental and earliest form of the neural network method [42]. Since then, several neural network architectures have been described, and research is currently ongoing. Every neural network model's design is built on a large number of interconnected input, hidden, and output neurons that communicate with one another. A basic ANN model has three layers in its architecture: the input layer, which holds the input molecular descriptors, the hidden layer, and the output layer (which contains the response variable). The independent predictors (X1 ......Xn) form the input layer; the hidden layer provides a nonlinear modification to the network's input, assigning weights to the input and directing it to the output via the activation function. The internal structure of the hidden layer of a neural network varies depending on its purpose. The dependent variable "y" makes up the output layer (response variables). However, multilayer perceptrons, also known as deep neural networks, are neural networks with more than one hidden layer and are frequently used to accurately replicate the complex nonlinear interaction between dependent and independent variables (as shown in Fig. 13.7). The most popular deep neural networks are convolutional neural networks, recurrent neural networks, long/short-term networks, and many others [43].

QSAR model validation

To strategically validate the final QSAR model, the complete dataset is first partitioned into training and test sets using a variety of dataset division approaches or random partition methods [44]. Following that, the training set is utilized for model training and hyperparameter tweaking, while the test set is used to determine the final trained model's external set prediction capabilities. The quality of any regression QSPR model is evaluated using a variety of internal and external validation metrics. The internal quality/validation metrics of the trained model include a coefficient of determination (R^2), leave-one-out cross-validation (Q^2_{LOO}) or leave-many-out cross-validation (Q^2_{LMO}), the standard error of estimation, mean absolute error, and root mean square error (RMSE) [45]. On the other hand, the external predictive power of any QSPR model is calculated by taking into account external predictive

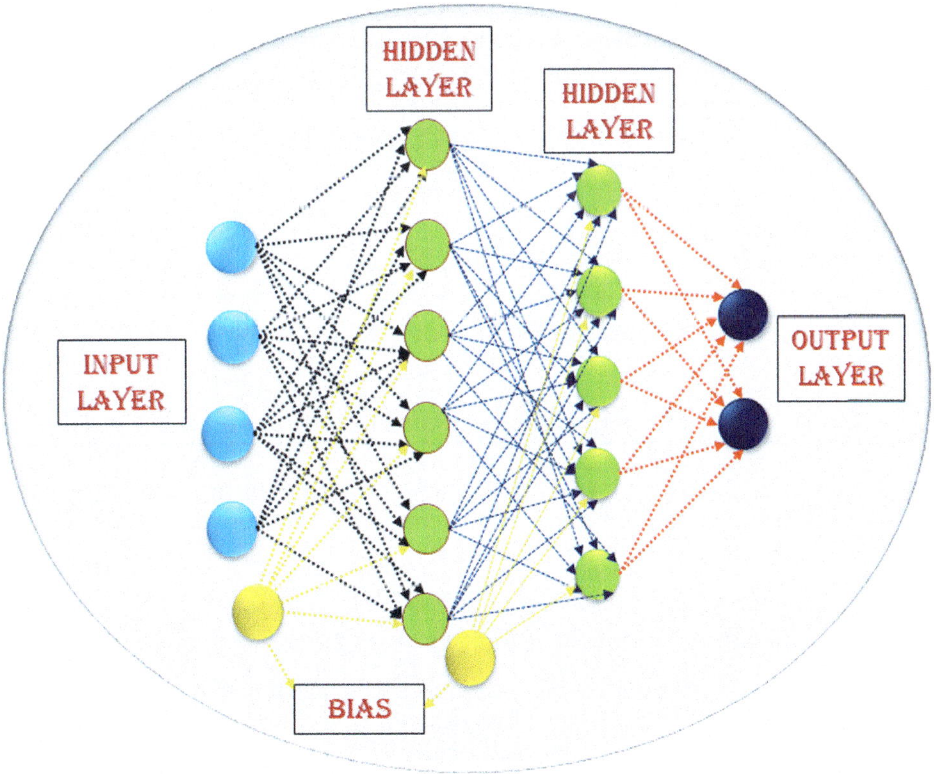

FIGURE 13.7 Simple overview of multilayer perceptron neural network with one input layer (comprising of four descriptors) connected to the two hidden layers and the final output layer.

variance (R^2_{pred}) and other parameters such as standard error of prediction, mean absolute error of test set and root mean square error of prediction [46].

Other methods

Read-across

QSAR modeling is prominent in silico tool for data gap filling [47]. QSAR is a statistical model creation approach that requires a large number of data points for meaningful model construction. Furthermore, in most circumstances, the relevant data points must be separated into training and test sets for validation purposes in order to comply with the OECD's recommendations. As a result, some of the available experimental data cannot be utilized for model construction and must be reserved for model validation. Such waste, in the case of limited datasets, may result in statistically less accurate model construction. Such a scenario can be better handled using read-across, a grouping method based on chemical similarity rather than statistical model creation [47]. It is an alternative, nonanimal approach for filling data gaps that reveals information on the biological activity and toxicological dangers of the target substances developed from data on parent compound(s) with known action or toxicity and comparable characteristics or chemical profiles. It is one of the most significant nonanimal techniques used today, and it is mostly used to generate ecotoxicological data to fill data gaps and make regulatory decisions. Although quantitative read-across approaches have also been adopted recently, regulatory bodies prefer and frequently employ the qualitative read-across approach. The phrase "source chemicals" refers to chemical analogs with known toxicity data rather than the "target compounds," which are typically referred to as the query chemicals. Read-across predictions are often produced using analogous and categorical methods. The analog technique uses a single or limited source chemical for prediction, but the category approach uses several source chemicals; consequently, it is more resilient and trustworthy. The read-across technique's main goal is to offer prediction data that is deemed to be (more or less) similar to the omitted standard experimental test, and it has therefore been used mostly for toxicity/ecotoxicity data gap filling of chemicals in a regulatory environment.

Read-across structure-activity relationship

Read-across structure—activity relationship (RASAR) is a recently developed discipline that leverages the read-across chemical similarity ideas (an unsupervised phase) and eventually produces a supervised learning model [48]. In simple words, it is an integration of read-across and QSAR approach for the risk assessment of organic chemicals. The issues associated with animal experimentation such as gathering information on the toxicity of thousands of compounds against hundreds of distinct endpoints is nearly unfeasible; mechanistic explanation of adverse outcome pathway (AOP) consisting of key events, molecular initiating events, and adverse outcomes are the need of the hour. In this regard, computational prediction approaches that include mechanistic explanations, such as AOP consideration, appear to be extremely promising. RASAR may be used for a large number of end points at the same time and can be linked to various toxicological targets targeting molecular initiating events (MIEs) in the adverse outcome pathway of many toxicity endpoints [49]. Given these considerations, RASAR appears to be a potential method for constructing expert systems for predicting the toxicity and ecotoxicity of pharmaceuticals and organic compounds in general. The RASAR approach has been utilized for both classification and quantitative prediction problems; however, it has mostly been employed in graded predictions or classification models.

QSPR applications in toxicity prediction of organic chemicals

Here, we discuss some representative examples of QSAR and ML modeling applications for (mainly eco-) toxicity of organic chemicals. Considering the emerging concern for endocrine disruptor chemicals, we have discussed them separately.

Prediction of endocrine disruptive chemical toxicity using QSTR

The production, secretion, transport, metabolism, binding, or elimination of hormones are all disrupted by endocrine-disrupting substances (EDS). EDCs have been shown to have negative impacts on the immunological, reproductive, neurological, cardiovascular, and developmental systems in humans [50]. Countless different chemical compounds, both natural and synthetic, have the potential to disrupt endocrine systems and pose risks to both human health and the environment. In the majority of developed nations, expensive testing of chemicals for the possibility of endocrine disruption is necessary. A variety of in vitro and in vivo assays are performed to discover EDCs since the mechanisms of endocrine disruption are broad and complicated (e.g., interactions with hormone and nonsteroid receptors, activation of enzyme and signaling pathways, etc.). Therefore, researchers have made multiple attempts to forecast the endocrine disruption potential of organic compounds using various classic and machine-learning techniques. For instance, Zhang et al. [50] created a k-NN QSAR model to estimate ligand binding affinity to the estrogen receptor because modulation of estrogen-like activity is the most prevalent adverse impact observed by EDCs. In this study, the authors used two separate datasets: (a) 232 compounds tested against Erα and (b) 137 compounds tested against Erβ. The Dragon program was used to compute the molecular descriptors, and both datasets were then subjected to fivefold external cross-validation and model building using the k-NN method. The finally derived QSAR for ER-mediated endocrine disruption (Er model) was based on 87 molecular descriptors with statistical significance model quality ($R^2 = 0.73$). On the other hand, the QSAR model for Erβ was below the acceptance criteria. The authors have suggested a few essential features which are responsible for positive or negative impact on ER-mediated endocrine disruption such as aromatic ratio and the presence of nitrogen and oxygen atom pair at distance one will reduce the binding affinity; however, B09[C−O] and nArOH increase the binding affinity of organic chemicals.

In another study, Liu et al. [51] have developed classification as well as regression QSAR models for predicting the binding affinity of endocrine-disrupting chemicals to human sex hormone-binding globulin. The classification and regression-based QSAR models have been developed using datasets of 125 and 87 compounds, respectively, employing Dragon and quantum chemical descriptors. The final dataset was randomly divided into training and test sets. The classification model is developed using the k-NN algorithm, however, the regression model was generated using the stepwise multiple linear regression techniques. The classification model is based on the three descriptors (i.e., nR09, nR10 (number of the 9-membered ring and 10-membered ring, respectively), and RDF155v (spheres of a radius of 15.5 Å, and it provides information on the interatomic distances in an entire molecule)) with statistically acceptable values of different quality metrics such as sensitivity, specificity, the accuracy of 86.4%, 80.0%, 84.4% and 85.7%, 87.5%, 86.2% for the training set and the validation set, respectively. On the other hand, the regression-based QSAR model is based on four descriptors (i.e., R4p, SssssC, RDF155u and L3s) with an adjusted determination coefficient (R^2_{adj}) of 0.810, a root mean square error (RMSE) of 0.616, a leave-one-out cross-validation squared correlation coefficient (Q^2_{LOO}) of 0.777, a bootstrap method (Q^2_{BOOT}) of 0.756, an external validation coefficient (Q^2_{ext}) of 0.544, and an $RMSE_{ext}$ of 0.859.

Khan et al. [52] developed QSTR and i-QSTTR models using the toxicity data of endocrine disruptor chemicals (EDCs) including perfluoroalkyl and polyfluoroalkyl compounds (PFAs), dioxins, bisphenol, polychlorinated biphenyls (PCB), and phthalates. Using a genetic algorithm followed by the PLS regression, a model was developed following the demanding OECD standards. The models were developed utilizing 2D descriptors with discrete physicochemical interpretations and validated using a range of generally accepted validation techniques. By estimating the models' application domain, the scope of predictions was determined. As evidenced by the constant repetition of 2D atom pair descriptors, the presence of halogens, sulfur, and phosphorus in EDC compounds had a substantial effect on their toxicity. The lipophilic contributions as indicated by logP terms (mainly ALOGP2 and XlogP) were the second most important factor governing EDC hazards. The toxicity of EDCs was reduced by hydrophilic functional groups, such as esters, aliphatic ethers, branching, and a greater oxygen concentration. Interspecies models were used to fill in data gaps by utilizing the hierarchical link between species. Using the "prediction reliability indicator" tool, prediction accuracy was determined.

Similarly, Kar et al. [53] proposed QSAR models to predict the endocrine-disrupting activity of per- and polyfluoroalkyl substances using the dataset of 24 chemicals. In that study, the final classification, as well as regression QSAR models, were generated using $N_{Training}$: 16 and 10, N_{test}: 8 and 5 using the LDA technique and multiple linear regression techniques, respectively. The classification model was based on three descriptors such as Me (mean atomic Sanderson electronegativity for scaled on carbon atom), nCsp2 (number of sp2 hybridized carbon atom), and H-050 (hydrophobicity of H atoms attached to heteroatom). On the other hand, the regression model was based on the two descriptors, that is, IC3 or information content index (neighborhood symmetry of 3-order) and $\sum \beta_s$ which signifies the sigma average valence electron mobile (VEM) count.

Several other studies reported linear and nonlinear QSPR models to predict the endocrine-disrupting potential of organic chemicals which one might find interesting, please refer to the original articles. He et al. [54] have proposed a novel control strategy for the endocrine-disrupting effect of PAEs on pregnant women in the traffic system. Heo et al. [55] have proposed an ML-based QSAR model to predict the effect of endocrine-disrupting chemicals on human health. Gomatam et al. [56] have proposed an ML model to predict the toxicity of endocrine disruptors and blood−

brain barrier permeability using chirality-sensitive descriptors. Similarly, Kovarich et al. [57] have also purposed a QSAR model to predict the endocrine-disrupting activity of perfluorinated compounds.

Prediction of miscellaneous chemical toxicity using QSTR

Khan et al. [58] developed a QSAR model to model the toxicity of organic chemicals (OCs) in water bodies, which are mostly present in sewage treatment facilities through industrial discharge. In that study, the authors generated robust local QSAR models for several chemical classes and moderately robust global QSAR models for the fish mortality end point employing the PLS regression technique using a large dataset of 1121 organic chemicals. For the selection of the features, genetic algorithm and stepwise regression were utilized, while model validation was conducted utilizing multiple demanding validation criteria following the stringent OECD QSAR validation guidelines. The descriptors employed in the study were obtained from the software packages such as SiRMS (simplex representation of molecular structures) (Version 4.1.2.270), Dragon (Version 7.0), and PaDEL-descriptor (Version 2.20). The final global model is based on the 10 molecular descriptors obtained from the PaDEL and Dragon software tool with a statistical significance of $R^2 = 0.600$ and leave-one-out cross-validation ($Q^2 = 0.590$) and external predictive variance $R^2_{pred} = 0.600$.

Khan and Ojha [59] (2020) studied the relationship between the toxicity of organic chemicals (OCs) and their structural and physicochemical characteristics. The authors developed a linear discriminant analysis (LDA) model for the classification of organic chemicals according to their acute observed toxicity in algae. This study gives robust consensus QSAR models generated from 334 organic compounds spanning a broad chemical domain for the prediction of effective chemical doses for 50% and 10% inhibition of algal growth. For the QSAR model construction, only 2D descriptors with a clear physicochemical meaning were used, and development, validation, and interpretation were conducted following the OECD's rigorous criteria. Using a genetic algorithm and a step-by-step method, the final QSAR models were created using the partial least squares regression technique. Also examined was the application domain of the created models. The generated consensus models were then used to forecast 64 organic substances with no definite observed responses, and the "prediction reliability indicator" was used to assess the accuracy of predictions.

In 2020, Jillella et al. [60] developed QSAR models using two distinct *Salmonella typhimurium* mutagenicity end points with or without the presence of liver metabolic microsomal enzymes (S9), namely TA98-S9 and TA98 + S9, in an attempt to describe the underlying causes of mutagenicity primarily due to organic chemicals. The models were created using 2D variables with clear physicochemical significance that were estimated using the Dragon, SiRMS, and PaDEL-descriptor software programs. Model construction was conducted using stepwise regression followed by PLS regression under the stringent OECD criteria for QSAR model development and validation. They have proposed that the primary contributing characteristics to mutagenicity are lipophilicity, electronegativity, branching, and unsaturation, among others.

In 2022, Jillella et al. [61] developed QSAR models for predicting the toxicity of 22 organic dyes against water flea *Daphnia magna* [$\log(1/EC_{50})$]. In this modeling study, the authors utilized only 2D descriptors and the multilayered stepwise regression and linear discriminant analysis (LDA) method for the generation of regression and classification-based models, respectively; however, the final regression-based QSAR models were obtained via PLS regression.

In 2019, Khan et al. [62] modeled *Daphnia magna* toxicity produced by a variety of agrochemicals, including herbicides, fungicides, insecticides, and microbicides. The models for both the individual data sets and the combined data set were created using just simple and interpretable two-dimensional descriptors and were then rigorously validated using test set compounds. Individual validated models were utilized to construct consensus models with the aim of improving prediction quality and reducing prediction errors. All individual models of various classes of agrochemicals and the global set of agrochemicals exhibited good statistical quality and prediction ability. In general, the generated models indicate that toxicity increases with lipophilicity and decreases with polarity.

To accurately predict the toxicity of organic aromatic compounds, Keshavarz et al. [63] introduced a new QSAR model. This model is based on the logarithm of the 50% growth inhibitory concentration of *Tetrahymena pyriformis* (log(*IGC50−1*) which is widely used in ecotoxicology and environmental safety applications. A novel model was developed and tested using the largest experimental data set of log(*IGC50−1*) for 892 organic aromatic chemicals. The number of nitro groups, carbon, and halogen atoms as well as some particular polar groups and molecular weight develop a fundamental correlation based on additive variables. The dependability of the core correlation is increased by developing a better correlation based on two nonadditive correction functions. Two of the best techniques now in use, which call for intricate descriptors, are tested against the enhanced correlation to determine their dependability. Internal and external validation tests have verified the projected outcomes for the training and test sets, i.e., 661 and 231, respectively. In comparison to more complex models with few data, the values of the correlation coefficient (R^2), mean error (ME), root mean square error (RMSE), and maximum of errors (Max Error) for 661/231 of training/test aromatic compounds are 0.8442/0.7771, 0.0000/0.0149, 0.3166/0.3603, and 0.9732/0.9825, respectively.

Lavado et al. [64] have proposed two QSAR models for the prediction of reproductive toxicity in *Folsomia candida* using the dataset of 54 organic compounds. In the present study, the final models were generated using the CORAL software and the PLS regression technique using 2D descriptors generated by the Dragon software tool. The final PLS model for the prediction of reproductive toxicity of organic chemicals is based on the eight molecular descriptors, which are B06[N−O (Presence/absence of N−O at topological distance 6), SaasC (Sum of aasC E-states), B06 [N−F](Presence/absence of N−F at topological distance 6), GATS1m (Geary autocorrelation of lag 1 weighted by mass), VE1sign_B(e) (coefficient sum of the last eigenvector from Burden matrix weighted by Sanderson electronegativity), B03[N−Cl] (presence/absence of N−Cl at topological distance 3), B08[N−Cl] (presence/absence of N−Cl at topological distance 8) and S-107 (Counts of R2S/RS-SR groups).

Gui et al. [65] conducted a study on the adsorption characteristics of 13 organic chemicals by polyethylene (PE) and chlorinated polyethylene (CPE) microplastics in freshwater. The bulk of organic compounds demonstrates no discernible variations in their adsorption on two microplastics (MPs), according to the results. Nevertheless, the adsorption of polycyclic aromatic hydrocarbons and chlorobenzene on CPE is evidently stronger than that on PE, resulting in a countermeasure for two pesticides. For the prediction of adsorption capacity, a quantitative structure−property relationship (QSPR) analysis was conducted. A QSPR model with satisfactory performance ($R^2 = 0.8586$) was developed to predict the adsorptive affinity (expressed as $\log K_d$) of organic compounds on PE MPs using MLR on 49 determined and collected data points. The model is dominated by the octanol/water partition coefficient ($\log K_{ow}$) and the excess molar refractive index (E). Using 13 determined adsorption data, a QSAR model with satisfactory performance ($R^2 = 0.9310$) was developed for $\log K_d$ values of CPE MPs in freshwater. The $\log K_{ow}$ and the highest negative charge on the Cl atom (*Q-max,cl*) are decisive for adsorption. The findings can provide a scientific basis for the risk evaluation of MP- and organic-polluted waterways.

Gao et al. [66] developed predictive models for the abiotic reduction of 60 organic compounds with a variety of reducible functional groups, including nitroaromatic compounds (NACs), aliphatic nitro-compounds (ANCs), aromatic N-oxides (ANOs), isoxazoles (ISXs), polyhalogenated alkanes (PHAs), sulfoxides and sulfones (SOs), and others. Using Fe(II)-tiron as a model reductant system, rate constants for their reduction were measured. The rates qualitatively followed the pattern NACs > ANOs > ISXs > PHAs > ANCs > SOs. For the development of QSARs, both conventional chemical descriptor-based and ML (ML)-based methodologies were examined. Within classes, conventional univariate QSARs based on the molecular descriptor E_{LUMO} (energy of the lowest unoccupied molecular orbital) demonstrated strong associations. A combination of E_{LUMO} and Abraham descriptors for physicochemical features in multivariate QSARs resulted in somewhat enhanced correlations within classes for NCs and NACs but not within other classes or between classes. The resultant ML model predicts reduction rates for all classes of compounds and all conditions tested with an accuracy comparable to that of conventional QSARs for individual classes ($r^2 = 0.41-0.98$ for univariate QSARs, 0.71−0.99 for multivariate QSARs, and 0.83 for the ML model). Both approaches necessitated an a priori classification strategy for the chemicals in order to train the models. This study provides two alternative modeling approaches for full abiotic reactivity prediction for evaluating the survival of organic molecules in anoxic water conditions.

Ai et al. [71] developed a regression model by combining the recursive feature elimination (RFE) and SVM methods for the bioconcentration factor and toxicity of organic compounds. In the case of the regression model, the authors computed 2D molecular descriptors from a dataset encompassing 450 different compounds, whereas, in the classification models, 12 molecular fingerprints were derived from a dataset encompassing 400 distinct chemicals using three ensemble models constructed using 3 ML algorithms. In the regression model, the statistical parameters R^2 and R^2_{pred} were 0.860 and 0.757, respectively. Other factors suggested that the regression model made accurate predictions and was capable of accurately predicting a new set of compounds according to the standards established by Golbraikh, Tropsha, and Roy [44,45,67]. The ensemble-SVM classification model has an overall accuracy, sensitivity, specificity, and area under the receiver operating characteristic curve of 92.2, 95.1, 86.0, and 0.965, respectively, in fivefold cross-validation, and 87.3, 92.6, 74.0, respectively, in external validation. These metrics revealed that our ensemble-SVM model was more stable and provided more precise forecasts than earlier models. Therefore, the model might be used to accurately forecast aquatic toxicity and assess threats to aquatic ecosystems. Through predictions generated by the two types of models, the authors found a number of structures most significant to acute aquatic toxicity; this information may be useful for aquatic toxicology research and aquatic system management.

In 2021, Gajewicz-Skretna et al. [68] validated the usefulness of similarity-based ML approaches for predicting the acute aquatic toxicity of several organic compounds to *Daphnia magna* and *Oryzias latipes*. Two similarity-based approaches (i.e., kernel-weighted local polynomial method and distance-weighted k-NN technique) that use a small training dataset, most similar to a specified fitting point, as opposed to the whole dataset that includes a wide variety of compounds, were evaluated. The kernel-weighted local polynomial method was superior to the distance-weighted k-NN technique in several ways. The findings demonstrate the significance of lipophilicity, electrophilic reactivity, molecule polarizability, and size in influencing acute toxicity. This method appears to be a valuable tool for evaluating the toxicity of new or untested substances due to the model's rigorous validation.

Xu et al. [69] obtained 5952 molecules from an examination of cyclic adenosine monophosphate (cAMP), a major signaling route in thyrocytes, which were utilized to develop a binary classification model by comparing RF, extreme gradient boosting (XGB), and logistic regression (LR). The RF-based prediction model had the highest identification accuracy for revealing substances that may block thyroid-stimulating hormone receptor (TSHR). The RF model's recall was calculated to be 0.89, its balance accuracy was 0.85, and its receiver operating characteristic (ROC) curve area under (AUC) was 0.92, showing that it had a very high predictive capacity. The lowest CDocker energy (CE) and CDocker interaction energy (CIE) for drugs and TSHR were identified and then used as descriptors in the predictive models. A regression model, extreme gradient boosting-regression (XGBR), with an R^2 of 0.65 was successfully developed to predict inhibitory action for active substances. Important components in the predicted model included dissociation properties, molecule structure, and binding energy. In this study, the authors have shown that the QSAR models are beneficial not only for discovering substances that inhibit TSHR but also for forecasting the inhibitory activity of active molecules.

In another study, Li. et al. [70] investigated the acute contact toxicity of organic compounds to honey bees. For this, widely used 12 types of physical-chemical properties (i.e., molecular weight (MW), (AlogP) Ghose-Crippen LogKow, logS (solubility), number of hydrogen bond acceptors (nHBA), and number of hydrogen bond donors (nHBD), molecular number of hydrogen bond donors, logD, molecular surface area (MSA), molecular polar surface area (MPSA), the number of molecular fractional polar surface area (MFPSA), number of rotatable bonds (nRTB), number of rings (nR), and number of aromatic rings (nAR)) were calculated and evaluated. Besides this, seven types of molecular fingerprints were also included in the study. The included fingerprints are as follows Estate fingerprint (79 bits), CDK fingerprint (1024 bits), CDK extended fingerprint (1024 bits), KlekotaRoth fingerprint (4860 bits), MACCS keys (166 bits), PubChem fingerprint (881 bits), and Substructure fingerprint (307 bits). On the basis of molecular descriptors and fingerprints, a variety of binary classification models have been constructed. In this study, the authors employed 5 ML algorithms, including SVM, k-NN, NB, decision tree, and RF, from a variety of available modeling algorithms. They generated a total of 40 classification models by combining five machine-learning algorithms, seven fingerprint types, and a set of molecular descriptors. Following fivefold cross-validation and external validation, several models demonstrated strong predictive ability. Also explored was the significance of 12 major physical-chemical features and chemical honey bee toxicity (HBT). Five properties, including AlogP, logD, MW, MSA, and nRTB showed positive correlation coefficients with HBT, whereas molecular solubility (logS) and the number of hydrogen bond donors (nHBD) showed negative correlation coefficients. From the KRFP and SubFP fingerprints, seven privileged substructures responsible for chemical HBT were discovered. The results of this work should provide vital knowledge and valuable tools for chemical HBT estimates in risk assessment.

In 2020, Chen et al. [14] performed a study using ML prediction of toxicity of organic compounds toward fat head minnow. This study consists of 963 organic compounds that are acutely poisonous to fathead minnows which were divided into a training set (482 compounds) and a test set (481 compounds) in a ratio of around 1: 1. SVM and genetic algorithm were utilized to create the quantitative structure–activity/toxicity relationship (QSAR/QSTR) models for 96-hour pLC_{50} using only six chemical descriptors. Internal (leave-one-out cross-validation) and external validation tests were employed to confirm the best SVM model (R2 = 0.756). The validation results ($q^2_{int} = 0.699$ and $q^2_{ext} = 0.744$) were satisfactory for predicting acute toxicity in fathead minnows in comparison to other models reported in the literature, even though the SVM model has only six molecular descriptors (CLOGP, SM6_B(P), NDB, nHM, SPMAD_EA, and MOR10E) and a large dataset for the test set containing 481 compounds. Among the modeled descriptors, CLOGP indicates lipophilicity, SM6_B(P) has an effect on molecular polarizability, NDB and nHM indicate molecular size, and MOR10E defining three-dimensional shapes links to fathead minnow molecular permeability.

Conclusion

Diverse statistical and ML techniques have been employed to successfully and precisely model various toxicity of organic chemicals against different species. In order to predict the toxicity of organic chemicals in a less invasive manner, replace animal toxicity experiments, prioritize and rank the organic chemicals based on their toxicity potential, and assist in the data gap fillings and regulatory decision-making, these data-driven approaches have become an essential component of computational toxicology. In this chapter, we reviewed representative published studies on modeling different types of toxicity using linear or nonlinear QSPR and ML algorithms. The majority of the reported models have substantial statistical values and appropriate prediction performances; nevertheless, few pieces of research have been undertaken with smaller or homologous, or congeneric series of datasets, which influences the application domain of the trained models in the chemical space.

Acknowledgment

PMK and GK thank NIPER, Kolkata, for fellowships.

References

[1] Wang Z, Walker G, Muir D. Toward a global understanding of chemical pollution: a first comprehensive analysis of national and regional chemical inventories. Environ Sci Technol 2020;54(5):2575–84.
[2] CAS. CAS assigns the 100 millionth CAS registry number® to a substance designed to treat acute myeloid leukemia. 2015. 2015.
[3] Johnson A, Jin X, Nakada N, Sumpter J. Learning from the past and considering the future of chemicals in the environment. Science 2020; 367(6476):384–7.
[4] Nuernberg RG, Schöpf TJ, Renner N. Regulation (EC) No. 1907/2006 of the European Parliament and of the Council of 18 December 2006 concerning the registration, evaluation, authorisation. 2019. 2019.
[5] USEPA. EPA: TSCA chemical substance inventory—google Scholar. 2003. 2003.
[6] CSCL Act on the regulation of manufacture and evaluation of chemical substances—English—Japanese Law translation.
[7] Habitat Branch. Environmental risk assessment (ERA): an approach for assessing and reporting environmental conditions. 2000. p. 1–84.
[8] MTD Cronin. Q) SARs to predict environmental toxicities: current status and future needs. Environ Sci Process Impacts 2017;19(3):213–20.
[9] Netzeva TI, Dearden JC, Edwards R, Worgan ADP, Cronin MTD. QSAR analysis of the toxicity of aromatic compounds to Chlorella vulgaris in a novel short-term assay. J Chem Inf Comput Sci 2004;44(1):258–65.
[10] Cassotti M, Ballabio D, Consonni V, Mauri A, Tetko IV, Todeschini R. Prediction of acute aquatic toxicity toward Daphnia magna by using the GA-kNN method. Altern Lab Anim 2014;42(1):31–41.
[11] Polishchuk PG, Muratov EN, Artemenko AG, Kolumbin OG, Muratov NN, Kuz'min VE. Application of random forest approach to QSAR prediction of aquatic toxicity. J Chem Inf Model 2009;49(11):2481–8.
[12] Louis B, Agrawal V. QSAR modeling of aquatic toxicity of aromatic aldehydes using artificial neural network (ANN) and multiple linear regression (MLR). J Indian Chem Soc 2011;88(1):99.
[13] Wang Y, Chen X. A joint optimization QSAR model of fathead minnow acute toxicity based on a radial basis function neural network and its consensus modeling. RSC Adv 2020;10(36):21292–308.
[14] Chen X, Dang L, Yang H, Huang X, Yu X. Machine learning-based prediction of toxicity of organic compounds towards fathead minnow. RSC Adv 2020;10(59):36174–80.
[15] Roy K, Kar S, Das RN. Understanding the basics of QSAR for applications in pharmaceutical sciences and risk assessment. Understanding the basics of QSAR for applications in pharmaceutical sciences and risk assessment. New York: Academic press; 2015. p. 1–479. 2015.
[16] Toropov AA, Toropova AP, Benfenati E, Nicolotti O, Carotti A, Nesmerak K, et al. QSPR/QSAR analyses by means of the CORAL software: results, challenges, perspectives. In: Quantitative structure-activity relationships in drug design, predictive toxicology, and risk assessment. IGI Global; 2015. p. 560–85. 2015.
[17] Katritzky AR, Kuanar M, Slavov S, Hall CD, Karelson M, Kahn I, et al. Quantitative correlation of physical and chemical properties with chemical structure: utility for prediction. Chem Rev 2010;110(10):5714–89.
[18] Ban Z, Zhou Q, Sun A, Mu L, Hu X. Screening priority factors determining and predicting the reproductive toxicity of various nanoparticles. Environ Sci Technol 2018;52(17):9666–76.
[19] Ciallella HL, Zhu H. Advancing computational toxicology in the big data era by artificial intelligence: data-driven and mechanism-driven modeling for chemical toxicity. Chem Res Toxicol 2019;32(4):536–47.
[20] Karelson M, Lobanov VS, Katritzky AR. Quantum-chemical descriptors in QSAR/QSPR studies. Chem Rev 1996;96(3):1027–43.
[21] Helguera A, Combes R, Gonzalez M, Cordeiro MN. Applications of 2D descriptors in drug design: a dragon tale. Curr Top Med Chem 2008; 8(18):1628–55.
[22] Todeschini R, Consonni V. Handbook of molecular descriptors, vol 11. Wiley-VCH. John Wiley and Sons; 2000. 2000. 688 pp.
[23] Khan PM, Roy K. Current approaches for choosing feature selection and learning algorithms in quantitative structure–activity relationships (QSAR). Expet Opin Drug Discov 2018;13(12):1075–89.
[24] Ma R, Luo T. PI1M: a benchmark database for polymer informatics. J Chem Inf Model 2020;60(10):4684–90.
[25] Khan PM, Roy K. QSPR modelling for investigation of different properties of aminoglycoside-derived polymers using 2D descriptors. SAR QSAR Environ Res 2021;32(7):595–614.

[26] Weisberg S. Applied linear regression: third edition. Hoboken, New Jersey: John Wiley and Sons; 2005. Applied Linear Regression: Third Edition. 2005. 1–310 pp.
[27] Wold S, Sjöström M, Eriksson L. PLS-regression: a basic tool of chemometrics. Chemometr Intell Lab Syst 2001;58(2):109–30.
[28] Jolliffe I. Principal component analysis. In: Lovric M, editor. International Encyclopedia of statistical science. Berlin, Heidelberg: Springer Berlin Heidelberg; 2011. p. 1094–6. 2011.
[29] Sha W, Guo Y, Yuan Q, Tang S, Zhang X, Lu S, et al. Artificial intelligence to power the future of materials science and engineering. Adv Intell Syst 2020;2(4):1900143.
[30] Gomes CP, Selman B, Gregoire JM. Artificial intelligence for materials discovery. MRS Bull 2019;44(7):538–44.
[31] Celebi ME, Aydin K. Unsupervised learning algorithms. Berlin: Springer International Publishing; 2016. 2016. Unsupervised Learning Algorithms. 1–558 pp.
[32] Cortes C, Vapnik V. Support-vector networks. Mach Learn 1995;20(3):273–97.
[33] Noble WS. What is a support vector machine? Nat Biotechnol 2006;24:1565–7. 2006.
[34] Andreola R. Support Vector Machines na classificação de imagens hiperespectrais 2009;16(1):130.
[35] Ruppert D. The elements of statistical learning: data mining, inference, and prediction. J Am Stat Assoc 2004;99. New York: Springer; 2004. 567–567 pp.
[36] Schölkopf B, Mika S, Burges CJC, Knirsch P, Müller KR, Rätsch G, et al. Input space versus feature space in kernel-based methods. IEEE Trans Neural Network 1999;10(5):1000–17.
[37] Sha W, Li Y, Tang S, Tian J, Zhao Y, Guo Y, et al. Machine learning in polymer informatics. InfoMat 2021;3(4):353–61.
[38] Wang XZ, Perston B, Yang Y, Lin T, Darr JA. Robust QSAR model development in high-throughput catalyst discovery based on genetic parameter optimisation. Chem Eng Res Des 2009;87(10):1420–9.
[39] Den?ux T. Pattern classification. In: Handbook of neural computation. 2nd ed. New York: John Wiley; 2004. 2004.
[40] Deringer VL, Bartók AP, Bernstein N, Wilkins DM, Ceriotti M, Csányi G. Gaussian process regression for materials and molecules. Chem Rev 2021;121(16):10073–141.
[41] Kim B, Lee S, Kim J. Inverse design of porous materials using artificial neural networks. Sci Adv 2020;6(1):eaax9324.
[42] McCulloch WS, Pitts W. A logical calculus of the ideas immanent in nervous activity. Bull Math Biophys 1943;5(4):115–33.
[43] Sainath TN, Vinyals O, Senior A, Sak H. Convolutional, long short-term memory, fully connected deep neural networks. In: ICASSP, IEEE international conference on acoustics, speech and signal processing - proceedings; 2015. p. 4580–4. 2015.
[44] Golbraikh A, Tropsha A. Beware of q2! J Mol Graph Model 2002;20(4):269–76.
[45] Roy K, Mitra I. On various metrics used for validation of predictive QSAR models with applications in virtual screening and focused library design. Comb Chem High Throughput Screen 2011;14(6):450–74.
[46] Roy K, Das RN, Ambure P, Aher RB. Be aware of error measures. Further studies on validation of predictive QSAR models. Chemometr Intell Lab Syst 2016;152(March):18–33.
[47] Berggren E, Amcoff P, Benigni R, Blackburn K, Carney E, Cronin M, et al. Chemical safety assessment using read-across: assessing the use of novel testing methods to strengthen the evidence base for decision making. Environ Health Perspect 2015;123(12):1232–40.
[48] Luechtefeld T, Marsh D, Rowlands C, Hartung T. Machine learning of toxicological big data enables read-across structure activity relationships (RASAR) outperforming animal test reproducibility. Toxicol Sci 2018;165(1):198–212.
[49] Schug T, Janesick A, Blumberg B. Endocrine disrupting chemicals and disease susceptibility. J Steroid Biochem Mol Biol 2011;127(3–5): 204–15.
[50] Zhang L, Sedykh A, Tripathi A, Zhu H, Afantitis A. Identification of putative estrogen receptor-mediated endocrine disrupting chemicals using QSAR-and structure-based virtual screening approaches. Toxicol Appl Pharmacol 2013;272(1):67–76.
[51] Liu H, Yang X, Lu R. Development of classification model and QSAR model for predicting binding affinity of endocrine disrupting chemicals to human sex hormone-binding globulin. Chemosphere 2016;156:1–7.
[52] Khan K, Roy K, Benfenati E. Ecotoxicological QSAR modeling of endocrine disruptor chemicals. J Hazard Mater 2019;369:707–18.
[53] Kar S, Sepúlveda M, Roy K, Leszczynski J. Endocrine-disrupting activity of per-and polyfluoroalkyl substances: exploring combined approaches of ligand and structure based modeling. Chemosphere 2017;184:514–23.
[54] He W, Yang H, Pu Q, Li Y. Novel control strategies for the endocrine-disrupting effect of PAEs to pregnant women in traffic system. Sci Total Environ 2022;851:158269.
[55] Heo S, Safder U, Yoo C. Deep learning driven QSAR model for environmental toxicology: effects of endocrine disrupting chemicals on human health. Environ Pollut 2019;253:29–38.
[56] Gomatam A, Joseph B, Gawde U, Raikuvar K. Predicting toxicity of endocrine disruptors and blood–brain barrier permeability using chirality-sensitive descriptors and machine learning. Comput Toxicol 2022;24:100240.
[57] Kovarich S, Papa E, Li J, Gramatica P. QSAR classification models for the screening of the endocrine-disrupting activity of perfluorinated compounds. SAR QSAR Environ Res 2012;23(3–4):207–20.
[58] Khan K, Baderna D, Cappelli C, Toma C, Lombardo A. Ecotoxicological QSAR modeling of organic compounds against fish: application of fragment based descriptors in feature analysis. Aquat Toxicol 2019;212:162–74.
[59] Khan K, Ojha P. Ecotoxicological modeling of organic chemicals for their acute toxicity in algae using classification and toxicophore-based approaches. Int J Quant Struct Prop Relatsh 2020;5(2):17–72.
[60] Jillella G, Khan K, Roy K. Application of QSARs in identification of mutagenicity mechanisms of nitro and amino aromatic compounds against *Salmonella typhimurium* species. Toxicol Vitro 2020;65:104768.
[61] Jillella GK, Roy K. QSAR modelling of organic dyes for their acute toxicity in Daphnia magna using 2D-descriptors. SAR QSAR Environ Res 2022;33(2):111–39.
[62] Khan P, Roy K, Benfenati E. Chemometric modeling of Daphnia magna toxicity of agrochemicals. Chemosphere 2019;224:470–9.
[63] Keshavarz M, Shirazi Z, Sheikhabadi P. Risk assessment of organic aromatic compounds to Tetrahymena pyriformis in environmental protection by a simple QSAR model. Process Saf Environ Protect 2021;150:137–47.
[64] Lavado G, Baderna D, Carnesecchi E, Toropova A, Toropov A, Dorne J, et al. QSAR models for soil ecotoxicity: development and validation of models to predict reproductive toxicity of organic chemicals in the collembola Folsomia. J Hazard Mater 2022;423:127236.

[65] Gui B, Xu X, Zhang S, Wang Y, Li C, Zhang D, et al. Prediction of organic compounds adsorbed by polyethylene and chlorinated polyethylene microplastics in freshwater using QSAR. Environ Res 2021;197:111001.
[66] Gao Y, Zhong S, Torralba-Sanchez T, Tratnyek P, Weber E, Chen Y, et al. Quantitative structure activity relationships (QSARs) and machine learning models for abiotic reduction of organic compounds by an aqueous Fe (II) complex. Water Res 2021;192:116843.
[67] Roy K, Mitra I, Ojha PK, Kar S, Das RN, Kabir H. Introduction of rm2(rank) metric incorporating rank-order predictions as an additional tool for validation of QSAR/QSPR models. Chemometr Intell Lab Syst 2012;118(August):200–10.
[68] Gajewicz-Skretna A, Furuhama A, Yamamoto H, Suzuki N. Generating accurate in silico predictions of acute aquatic toxicity for a range of organic chemicals: towards similarity-based machine learning methods. Chemosphere 2021;280:130681.
[69] Xu X, Wang C, Gui B, Yuan X, Li C, Zhao Y, et al. Application of machine learning to predict the inhibitory activity of organic chemicals on thyroid stimulating hormone receptor. Environ Res 2022;212:113175.
[70] Li X, Zhang Y, Chen H, Li H, Zhao Y. Insights into the molecular basis of the acute contact toxicity of diverse organic chemicals in the honey bee. J Chem Inf Model 2017;57(12):2948–57.
[71] Ai H, Wu X, Zhang L, Qi M, Zhao Y, Zhao Q, et al. QSAR modelling study of the bioconcentration factor and toxicity of organic compounds to aquatic organisms using machine learning and ensemble methods. Ecotox Environ Saf 2019;179:71–8.

CHAPTER

14

admetSAR—A valuable tool for assisting safety evaluation

Yaxin Gu, Chaofeng Lou and Yun Tang

Laboratory of Molecular Modeling and Design, School of Pharmacy, East China University of Science and Technology, Shanghai, China

Abbreviations

ACC Accuracy
ADMET Absorption, distribution, metabolism, excretion, toxicity
AUC Area under the curve
FN False negative
FP False positive
GCNN Graph convolutional neural network
GNN Graph neural network
HOB Human oral bioavailability
PPB Plasma protein binding
QSAR Quantitative structure–activity relationship
SE Sensitivity
SP Specificity
TN True negative
TP True positive

Introduction

There are more than 150,000 kinds of chemicals closely related to people's daily life, such as medicines, pesticides, food ingredients and additives, cosmetics, industrial chemicals, etc. These chemicals greatly improve the quality of people's life. However, they also bring potential safety problems to human health and the environment [1,2].

Before entering the market, most commercial chemicals have not undergone the same rigorous and systematic toxicity tests as medicines, so it is necessary and urgent to conduct safety evaluation on these chemicals. Traditional chemical safety evaluation methods are based on animal tests, but these tests are expensive, time-consuming, and laborious. Therefore, a variety of in silico tools have been developed as alternatives to assess chemical safety, because they are cheap, fast, and accurate and could be done before a compound is synthesized.

Absorption, distribution, metabolism, excretion, and toxicity (ADMET) are tightly related to pharmacokinetic and toxicokinetic properties and safety of chemicals. Hence, ADMET plays a key role in safety evaluation [3,4]. Facilitated by rapid development of computer power and enormous ADMET experimental data, a number of in silico methods and tools are developed for ADMET evaluation [5]. Moreover, the application of artificial intelligence technology, such as machine learning and deep learning methods, provides efficient and convenient solutions for construction of quantitative structure–activity relationship (QSAR) models (Fig. 14.1B).

A variety of tools and webservers have been applied in chemical safety assessment, such as admetSAR 2.0 [6], ProTox II [7], ADMETlab 2.0 [8], SwissADME [9], OECD QSAR Toolbox, etc. However, the models in these webservers differ, as do the endpoints focused. These differences include three aspects. (1) Datasets: the datasets for these studies have different sources and different data preprocessing approaches. (2) Methods: the QSAR models are widely used, while read-cross and expert system-based methods are also used in OECD QSAR Toolbox. (3) Parameters: well-tuned model parameters can improve model performance. In this chapter, we will describe our webserver admetSAR.

Our admetSAR was first released in 2012, armed with more than 210,000 high-quality experimental data for 96,000 compounds [10]. admetSAR has a user-friendly interface that allows users to retrieve chemical information

QSAR in Safety Evaluation and Risk Assessment
https://doi.org/10.1016/B978-0-443-15339-6.00004-7

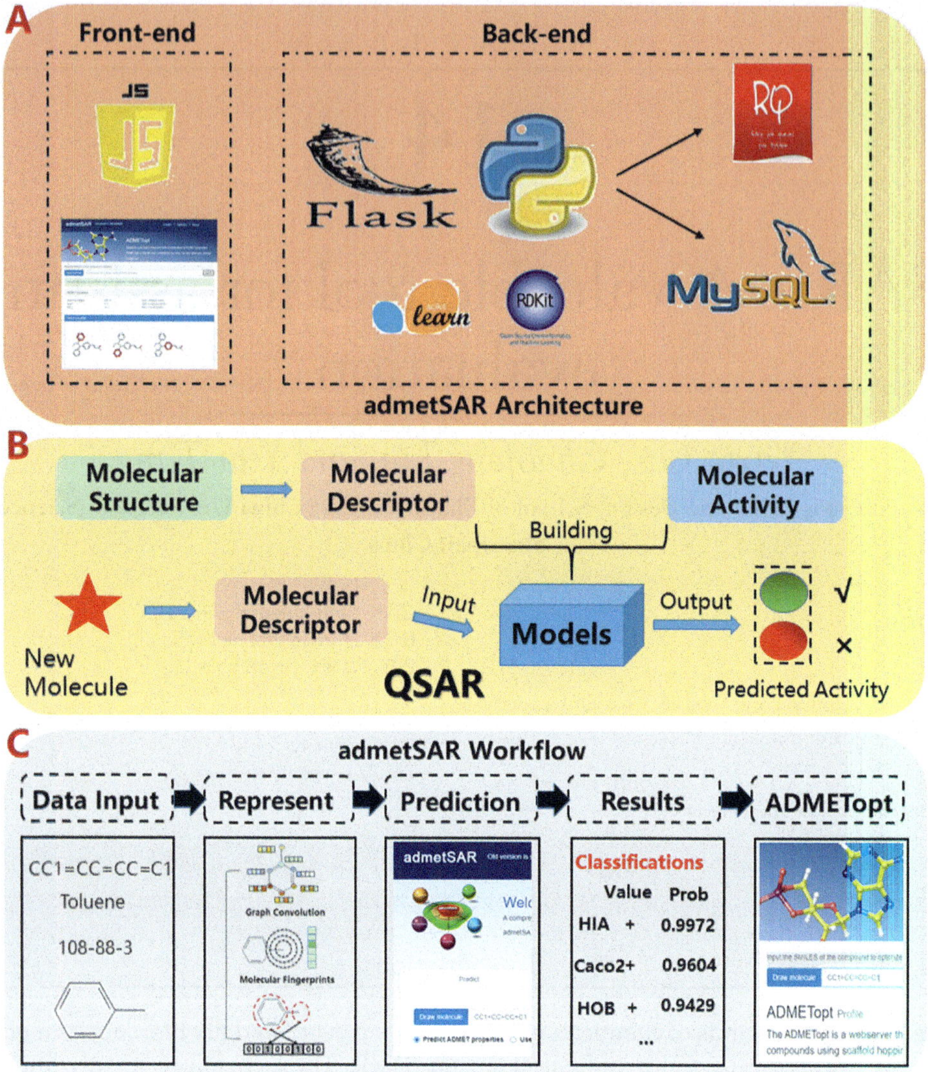

FIGURE 14.1 (A) The architecture of admetSAR 2.0, including front-end and back-end. (B) The construction of QSAR models and the prediction workflow of a new molecule. (C) The workflow of admetSAR 2.0.

using its CASRN, common names or structures easily. It also provides 27 QSAR models for chemical ADMET prediction. It has been widely used in drug discovery and risk assessment and served thousands of users around the world. Since version 4.0 in 2014, DrugBank has presented our predictions for each of the drug in the database [11]. In 2019, we updated the server to version 2.0. Currently, admetSAR 2.0 supports single-molecule prediction and batch evaluation, allowing for customized endpoint prediction service. It provides evaluation for 51 ADMET-related endpoints using 38 binary classification models, six multilabel models, three multiclassification models, and four regression models. Notably, each final model is backed up by the optimal one trained by various machine learning algorithms and molecular representation. In addition, a module named ADMETopt to optimize the ADMET properties of a query compound by scaffold hopping is available in admetSAR 2.0 [12]. We believe that admetSAR 2.0 would be a valuable and powerful tool for safety evaluation and risk assessment.

Basic architecture of admetSAR

As shown in Fig. 14.1A, the basic architecture of the webserver includes front-end and back-end. Users interact with the back-end server through a browser, and the front-end uses JavaScript to achieve an asynchronous dynamic interactive presentation. admetSAR 2.0 was built using the Python web framework of Flask and task scheduling control with Python-rq to prevent server overload. The pretrained models and model predictions were stored in MySQL. The RDKit and Pybel package were employed to provide cheminformatics support such as data preprocessing and representation. Additionally, Scikit-learn and DeepChem packages were used in model implementation. Users can easily access admetSAR through Mozilla Firefox, Google Chrome, and other browsers.

The workflow of admetSAR 2.0 consists of five steps, as shown in Fig. 14.1C. The webserver provides a variety of data input methods and enables users to search for chemical information by SMILES, common name, CASRN, or structure easily. Property prediction is performed after data representation, and the result presentation page is simple and clear. In addition, a CSV-formatted result file is returned for download. Furthermore, ADMETopt module provides support for the optimization of target molecules.

Details of admetSAR

Data collection and preparation

The comprehensive data were obtained from various data sources via searching ADMET-related keywords, including open-access databases such as DrugBank [11], ChEMBL [13], and CPDB [14], peer-reviewed literature, and high-throughput screening projects such as Tox21 [15] and CYP450 [16]. In data preparation, all compounds were preprocessed by Pipeline Pilot (Version 2017 R2). Organometallic compounds, inorganic compounds, mixtures, and salts were removed. Duplicates were also removed. The remaining compounds were represented as canonical SMILES. A high-quality dataset of 0.96 million chemical entries covering 51 ADMET-related endpoints was constructed. More details of the data were shown in Table 14.1.

Data representation and model building

The SMILES string of a compound can characterize the structural features of the compound, but it cannot be recognized by a computer and thus cannot be directly used as model input. Therefore, after receiving the compound information, admetSAR 2.0 webserver will characterize the input as molecular fingerprints, a variety of digital features that can be fed into machine learning models. The molecular fingerprint of a compound is defined based on a substructure dictionary or a graph-based hashing algorithm and consists of a binary string, each bit of which corresponds to one or more specific substructures. A bit is set to "1" if a molecule contains the substructure, and "0" otherwise. Totally six types of molecular fingerprints were used in admetSAR 2.0, namely RDK Fingerprint, Morgan Fingerprint, MACCS Fingerprint, Substructure Fingerprint, AtomPairs Fingerprint, and TopoTorsion Fingerprint. These fingerprints were implemented with RDKit and Pybel packages. Machine learning algorithms support vector machine, random forest, k nearest neighbors, and Naive Bayesian were used to build models, and these algorithms were implemented by Scikit-learn package with Python scripts [17–20].

In addition to molecular fingerprints, for individual endpoints with large amount of data, the admetSAR 2.0 development team constructed graph convolutional neural network (GCN) models, in which molecular graphs were used as input. Compared with molecular fingerprint, automated feature extraction by GCN can effectively avoid the influence of human factors on model feature extraction. In addition, the graph neural networks (GNNs) are deep learning methods that outperform machine learning methods in terms of learning ability on endpoints with large amount of data [21,22]. Here, an open-source toolchain, DeepChem, provides support for deep learning models.

For each specific ADMET endpoint, the best performing model was preserved in admetSAR 2.0 by exploring different combinations of molecular representations and algorithms. After decades of exploration, the admetSAR 2.0 development team has obtained a series of high-quality and high-performance prediction models whose performance has been proven to be better than the other reported models (Fig. 14.2).

Model evaluation and prediction

Prior to model evaluation, admetSAR has an applicability domain judgment process by calculating six physicochemical properties of the input molecule. The six physicochemical descriptors are molecular weight, AlogP, the number of atoms, the number of rings, the number of hydrogen bond acceptors, and the number of hydrogen bond donors. For example, if the molecular weight of a molecule is greater than or less than 99% of the compounds

TABLE 14.1 admetSAR 2.0 endpoints.

Endpoint (abb.)	Total number	Positive/negative	Model type
CYP1a2 inhibitor (CYP1a2i)	14,903	7415/7488	Binary
CYP2d6 inhibitor (CYP2d6i)	14,741	3060/11,681	Binary
CYP2c9 inhibitor (CYP2c9i)	14,709	4978/9731	Binary
CYP2c19 inhibitor (CYP2c19i)	14,576	6041/8535	Binary
CYP3a4 inhibitor (CYP3a4i)	18,561	6707/11,854	Binary
CYP2d6 substrate (CYP2d6s)	671	191/480	Binary
CYP2c9 substrate (CYP2c9s)	673	142/531	Binary
CYP3a4 substrate (CYP3a4s)	674	357/317	Binary
P-glycoprotein substrate (Pgps)	1565	718/847	Binary
P-glycoprotein inhibitor (Pgpi)	1943	1172/771	Binary
Breast cancer resistance protein inhibitor (BCRPi)	970	432/538	Binary
Bile salt export pump inhibitor (BSEPi)	607	317/290	Binary
Organic cation transport protein 1 inhibitor (OCT1i)	184	60/124	Binary
Organic cation transport protein 2 inhibitor (OCT2i)	904	244/633	Binary
Multidrug and toxin extrusion transporter 1 inhibitor (MATE1)	818	80/738	Binary
OATP1b1 inhibitor1 (oatp1b1i)	1855	1657/198	Binary
OATP1b3 inhibitor1 (oatp1b3i)	1873	1743/130	Binary
OATP2b1 inhibitor1 (oatp2b1i)	219	44/175	Binary
Blood brain barrier (BBB)	2358	1811/1547	Binary
Plasma protein binding (PPB)	1209	–	Regression
Caco-2 (caco2)	674	303/371	Binary
Human intestinal absorption	578	500/78	Binary
Water solubility (WS)	1708	–	Regression

TABLE 14.1 admetSAR 2.0 endpoints.—cont'd

Endpoint (abb.)	Total number	Positive/negative	Model type
Acute oral toxicity (AO)	10,207	–	Quaternary and regression
Hepatotoxicity (Hepa)	3708	3115/593	Binary
Micronucleus (MN)	576	237/339	Binary
Human either-a-go-go (hERG)	978	717/261	Binary
Ames mutagenesis (Ames)	8348	4866/3482	Binary
Eye irritation (EI)	5220	3874/1346	Binary
Eye corrosion (EC)	2299	887/1412	Binary
Carcinogenesis (Carc)	916	476/440	Ternary and binary
Biodegradation (Biod)	1605	591/1014	Binary
Avian toxicity (Avian)	574	94/484	Binary
Tetrahymena pyriformis toxicity (TPT)	1571	–	Regression
Honey bee toxicity (HBT)	676	174/502	Binary
Androgen receptor (AR)	5919	1059/4860	Binary
Peroxisome proliferator –activated receptor gamma (PPARg)	5456	576/4880	Binary
Aromatase inhibitor (Aro)	6154	697/5457	Binary
Glucocorticoid receptor modulator (GR)	6055	601/5454	Binary
Thyroid receptor modulator (TR)	5372	373/4999	Binary
Estrogen receptor modulator (ER)	6055	1338/4252	Binary
Fish aquatic toxicity (FAQ)	554	366/188	Binary
Respiratory toxicity (Resp)	2336	1304/1032	Binary
Reproductive toxicity (Repro)	1823	892/961	Binary
Mitochondrial toxicity (Mito)	3407	1538/1869	Binary
Chemical neurotoxicity (Neural)	422	–	Regression
Human oral bioavailability (HOB)	995	509/486	Binary
UDP-glucuronosyl transferases (UGT) inhibitor	1248	643/605	Binary
Subcellular localization (Subc)	614	–	Regression

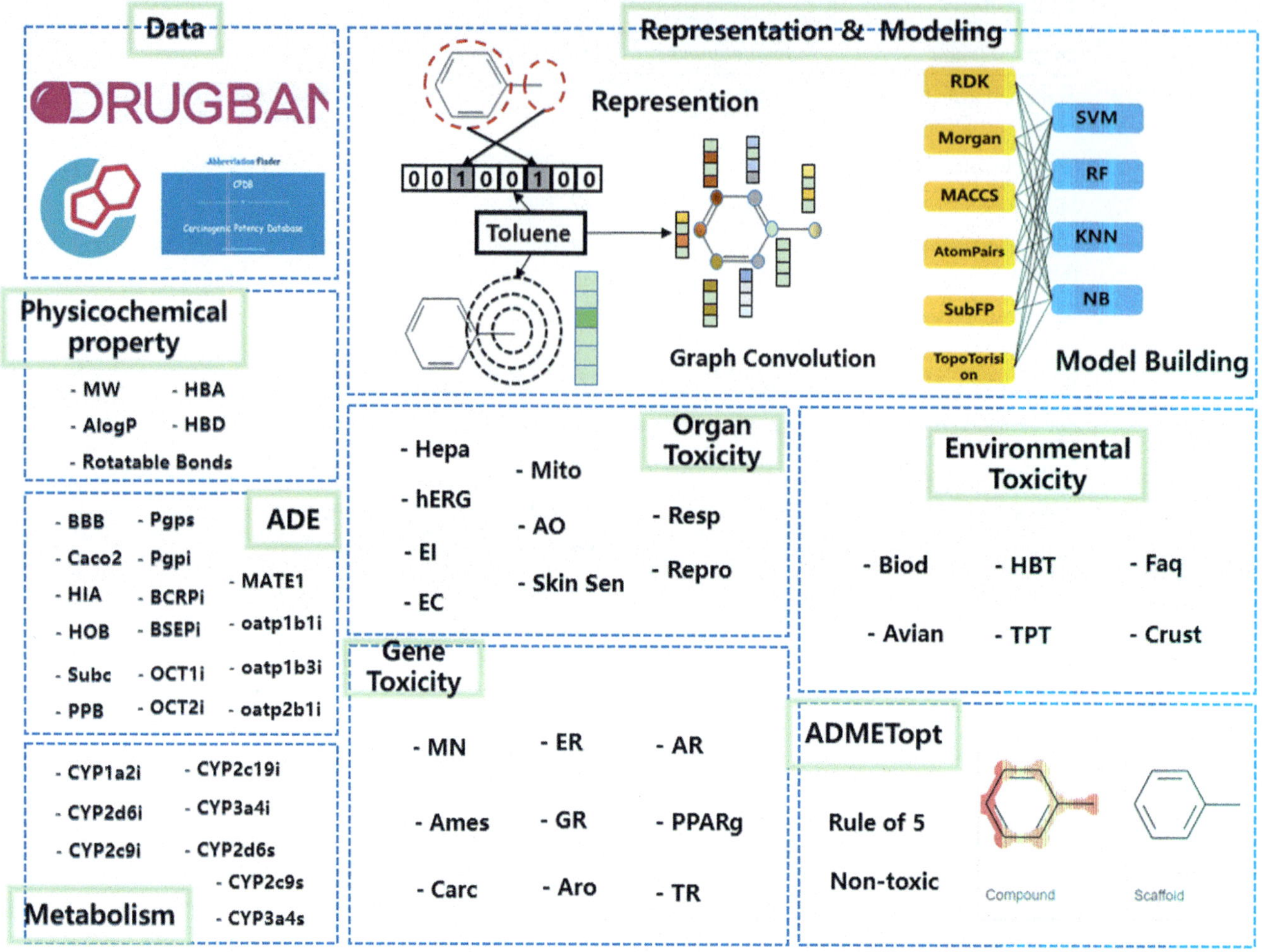

FIGURE 14.2 The modeling details of admetSAR 2.0, and the endpoints available in admetSAR 2.0.

in the training set, it will be marked as "Warning," which means that the compound is at the edge of the applicability domain and the prediction result should be considered with caution. If the molecular weight is greater than the maximum value of the molecules in the training set, it will be marked as "Out of domain," which means that the compound is not in the applicability domain and the prediction result is not credible. If the molecular weight is not consistent with the aforementioned two conditions, the compound will be considered to be in the applicability domain of the model and will be marked as "In domain." The applicability domain of plasma protein binding (PPB) model was plotted in Fig. 14.3A. As shown in Fig. 14.3B, the blue color ranges in the distributions of the six physicochemical properties of the training compounds characterize the chemical space of the applicability domain of the PPB model.

Five-fold cross-validation was employed to select the best model. Furthermore, there are different performance metrics for classification and regression models. Classification models were evaluated by sensitivity (SE), specificity (SP), accuracy (ACC), and the area under the recursive operating curve (AUC). SE, SP, and ACC are calculated through the confusion matrix of true positive (TP), false positive (FP), true negative (TN), and false negative (FN). AUC is equal to the probability that a classifier will rank a randomly chosen positive instance higher than a randomly chosen negative one. Unlike classification models, regression models were evaluated by determination coefficient (R^2) and root mean square error (RMSE) defined as following. In this study, multilabel models for endocrine disruption chemicals are evaluated via Eqs. (14.7)–(14.10).

$$\mathrm{ACC} = \frac{\mathrm{TP} + \mathrm{TN}}{\mathrm{TP} + \mathrm{TN} + \mathrm{FP} + \mathrm{FN}} \tag{14.1}$$

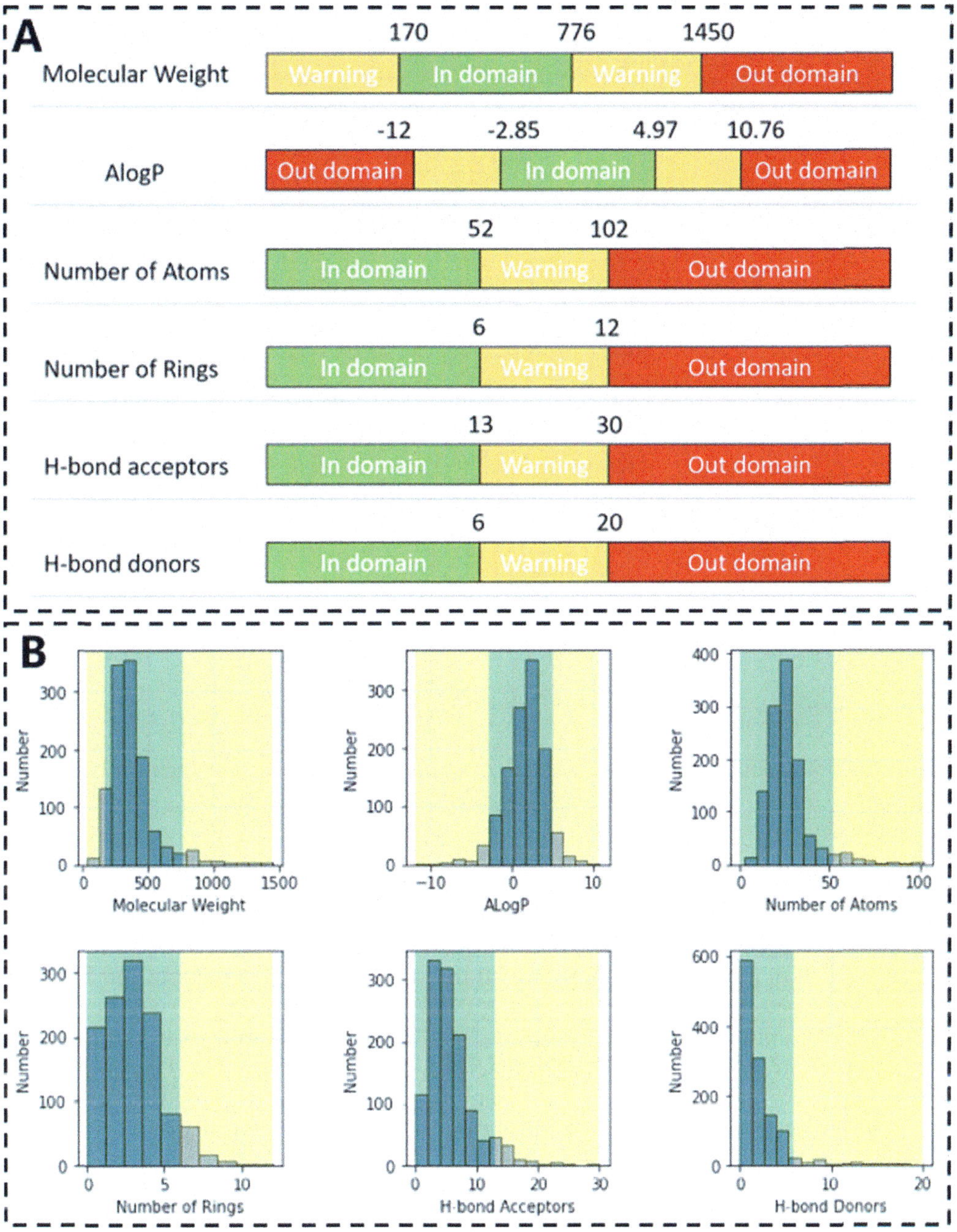

FIGURE 14.3 (A) The applicability domain of plasma protein binding (PPB) model. (B) Distribution of physicochemical parameters of compounds in the training set of PPB data set.

$$SE = \frac{TP}{TP + FN} \tag{14.2}$$

$$SP = \frac{TN}{TN + FP} \tag{14.3}$$

$$AUC = \int_{\infty}^{-\infty} TPR(T)(-FPR'(T))dT \tag{14.4}$$

$$R^2 = 1 - \frac{\sum_i (\hat{y}_i - y_i)^2}{\sum_i (\overline{y}_i - y_i)^2} \tag{14.5}$$

$$\text{RMSE} = \sqrt{\frac{1}{m}\sum_{i=1}^{m} (y_i - \hat{y}_i)^2} \tag{14.6}$$

where y_i presents true values, $\hat{y}_i$ presents predicted values, and m is the total number of datasets.

$$\text{SubsetAccuracy} = \frac{1}{n}\sum_{i=1}^{n} [\![Y_i = Z_i]\!] \tag{14.7}$$

$$\text{Hamming Loss} = \frac{1}{n}\frac{1}{k}\sum_{i=1}^{n} |Y_i \Delta Z_i| \tag{14.8}$$

$$\text{Recall}_{\text{micro}} = \frac{1}{n}\sum_{i=1}^{n} \frac{|Y_i \cap Z_i|}{|Y_i|} \tag{14.9}$$

$$\text{Precision}_{\text{micro}} = \frac{1}{n}\sum_{i=1}^{n} \frac{|Y_i \cap Z_i|}{|Z_i|} \tag{14.10}$$

where Y_i represents the real label set of the ith instance, Z_i is the predicted one, n is the number of instances, and k is the number of labels.

Property optimization module—ADMETopt

In addition to property prediction service, admetSAR also provides property optimization service through ADMETopt module, which applies scaffold hopping strategy and ADMET screening for lead optimization. The chemical scaffolds were collected from two databases, ChEMBL and Enamine (http://www.enamine.net/), containing 570,000 and 1.75 million unique molecules, respectively. All scaffolds matching the definition were extracted using RDKit and represented in canonical SMILES strings. Duplicates were removed. The similarity of two scaffolds was evaluated based on the Tanimoto similarity coefficient, which is defined in Eq. (14.11). The core idea of scaffold hopping strategy is to replace a scaffold of a query compound with another similar one.

$$S_{i,j} = \frac{\sum_{k=1}^{m} [\text{FP}_k(i) = \text{FP}_k(j) \neq 0]}{\sum_{k=1}^{m} [\text{FP}_k(i) \neq 0 \text{ or } \text{FP}_k(j) \neq 0]} \tag{14.11}$$

where i and j are the scaffold numbers, m is the length of the scaffold fingerprint (1021 bits), and $\text{FP}_k(i)$ is the kth bit of the fingerprint of scaffold i. The expression in square brackets yields 1 if true and 0 otherwise.

ADMETopt provides screening and optimization of seven physicochemical properties and eight biological properties. Users can determine the chemical space of the optimized molecules by setting the thresholds of physicochemical properties or defining the favorable ADMET properties. Overall, ADMETopt can help search the surrounding chemical space of a given lead compound to obtain a new chemical entity with better physicochemical properties and ADMET properties.

Usage of admetSAR

admetSAR has a user-friendly interface to search or predict ADMET profiling of a compound, and it is freely accessible at http://lmmd.ecust.edu.cn/admetsar2/. The webserver includes a submission web page, an easy-to-understand result page, and a convenient property optimization module for nonexperts in computer-aided drug design. Fig. 14.4 displays the user homepage of the webserver, where users can access detailed information about admetSAR through various toolbars.

Submission web page

The query molecule should be represented as SMILES string and submitted to the webserver for prediction. In addition to direct SMILES submission, InDraw plugin as a molecular editor is embedded to convert the molecular structure to SMILES. After the structure is drawn, click "save" button, and the SMILES of the molecule will be automatically displayed in the inputbox.

Predict and *Advanced Predict* are available to support single molecule and batch prediction. In *Predict* service, once a job is submitted, the webserver will automatically predict all the ADMET endpoints and return a successful message with task id and URL address accessing to the result. *Advanced Predict* allows batch evaluation and endpoint customization. The query molecules should be prepared by representing them as SMILES and keeping one SMILES per line in the textbox. Notably, at most 10 molecules can be submitted in a batch.

We have predefined two sets of endpoints, *ADMET properties for drug discovery* and *Predict for eco-toxicity*. For *ADMET properties for drug discovery*, it contains the models of admetSAR 2.0 for compound ADMET properties prediction. *Predict for eco-toxicity* invokes the endpoints related to eco-toxicity in admetSAR 2.0 to make predictions, including honey bee toxicity, biodegradation, crustacean aquatic toxicity, fish aquatic toxicity, and *Tetrahymena pyriformis* toxicity. In *Custom endpoint* service, users can select the endpoints of interest from the 51 endpoints supported. The submission page of admetSAR 2.0 is shown in Fig. 14.5.

FIGURE 14.4 The homepage of admetSAR 2.0 webserver. The *blue* toolbar at the *top* of the web page allows users to navigate within different tools. A second bar gives access to different property prediction service and provides compound search service based on admetSAR 1.0.

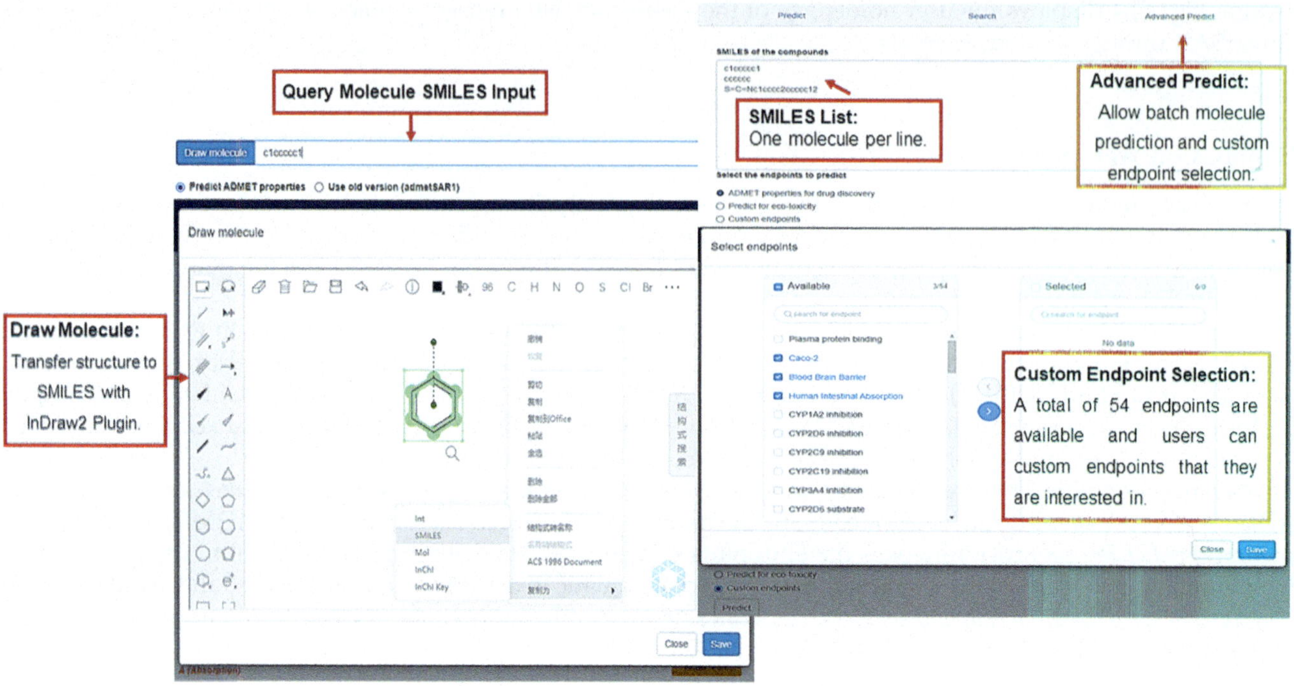

FIGURE 14.5 admetSAR 2.0 property prediction submission page. Users can use the SMILES format as well as molecular structures as input. The webserver supports single molecule prediction and batch prediction, and in *Advanced Predict* mode, custom prediction endpoints can be implemented.

Output web page

When the task of model prediction is completed, it will automatically jump to the prediction results page as shown in Fig. 14.6. The output result page provides unambiguous interpretation of results to facilitate the users' follow-up study. The single-molecule prediction is headed according to the compound number. When multiple molecules are submitted in *Advanced Predict* mode, their results will be listed on the left as "compound 1", "compound 2", ... as their input order. The first part of the output page shows the two-dimensional structure of the query molecule, SMILES, and physicochemical property parameters such as molecular weight, AlogP, hydrogen bond donor, hydrogen bond acceptor, etc. Moreover, the webserver displays the applicability domain of the given compound, and users are recommended to be cautious about the results of compounds with "Out of domain" label. The ADMET-predicted profile section includes classification and regression results. For the classification models, it presents the endpoint names, values, and predicted probabilities. For example, for human oral bioavailability (HOB) endpoint, the value "+" represents Positive/Good HOB, while "−" represents Negative/Unsatisfactory HOB. The probability is related to the value, which has a threshold of 0.5; if the probability is less than 50%, it would be predicted as an unsatisfactory HOB. For toxicity endpoints, the token "+" usually represents the query molecule is toxic, while "−" represents nontoxic. A higher or lower prediction probability indicates that the model is more confident on the prediction. To facilitate visual analysis, certain classification endpoint results such as toxicity are visualized with different color bands: green for satisfactory properties and red for unsatisfactory properties.

In addition to the web presentation, admetSAR 2.0 also provides a downloadable CSV file for users by clicking the "Download the results" button, which has the same content as the output prediction interface. If users wish to modify the structure of the compound, they can connect to the ADMETopt online service by clicking on the "Optimize" button to perform a backbone leap to optimize the ADMET properties of the compound. If users are unsatisfied with the compound properties and want to modify the compound structure, they can click the "Optimize" button to the ADMETopt online service.

ADMETopt web page

ADMETopt is available at http://lmmd.ecust.edu.cn/admetsar2/admetopt/. As shown in Fig. 14.7, the process of ADMETopt optimizing molecules is presented. Similar to admetSAR, ADMETopt supports SMILES format and

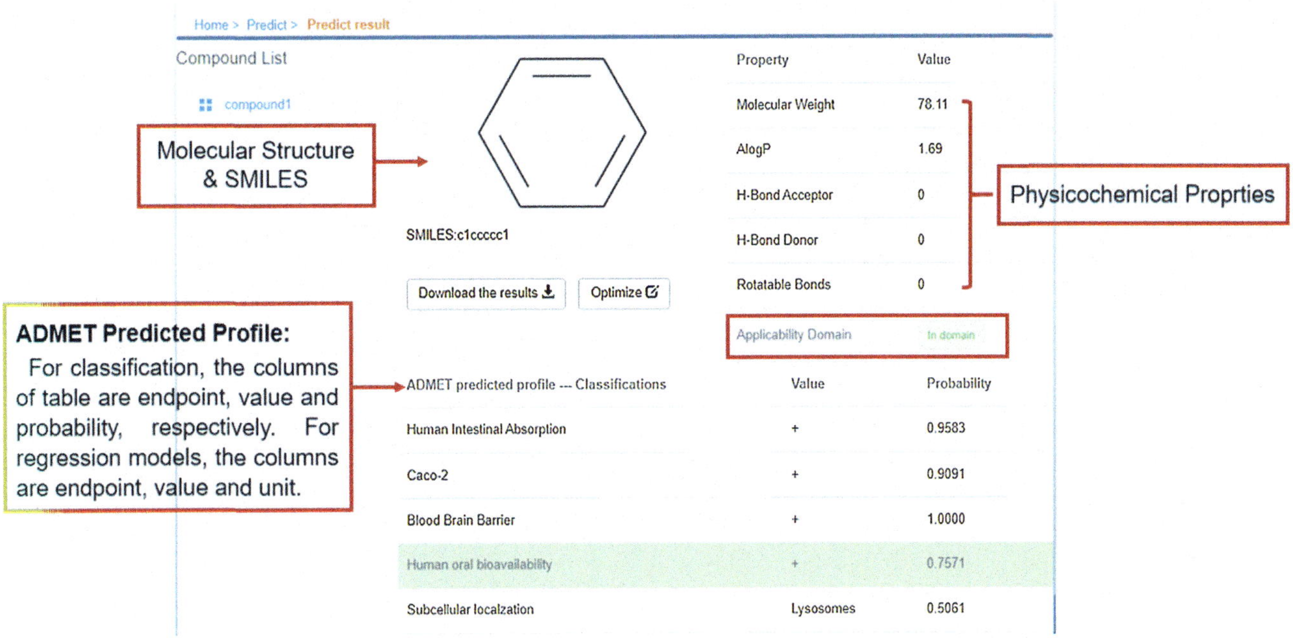

FIGURE 14.6 Calculated parameters are grouped in different sections of the output page. The *left* section includes two-dimensional chemical structure and canonical SMILES. The physicochemical and ADMET properties results are also compiled in other sections.

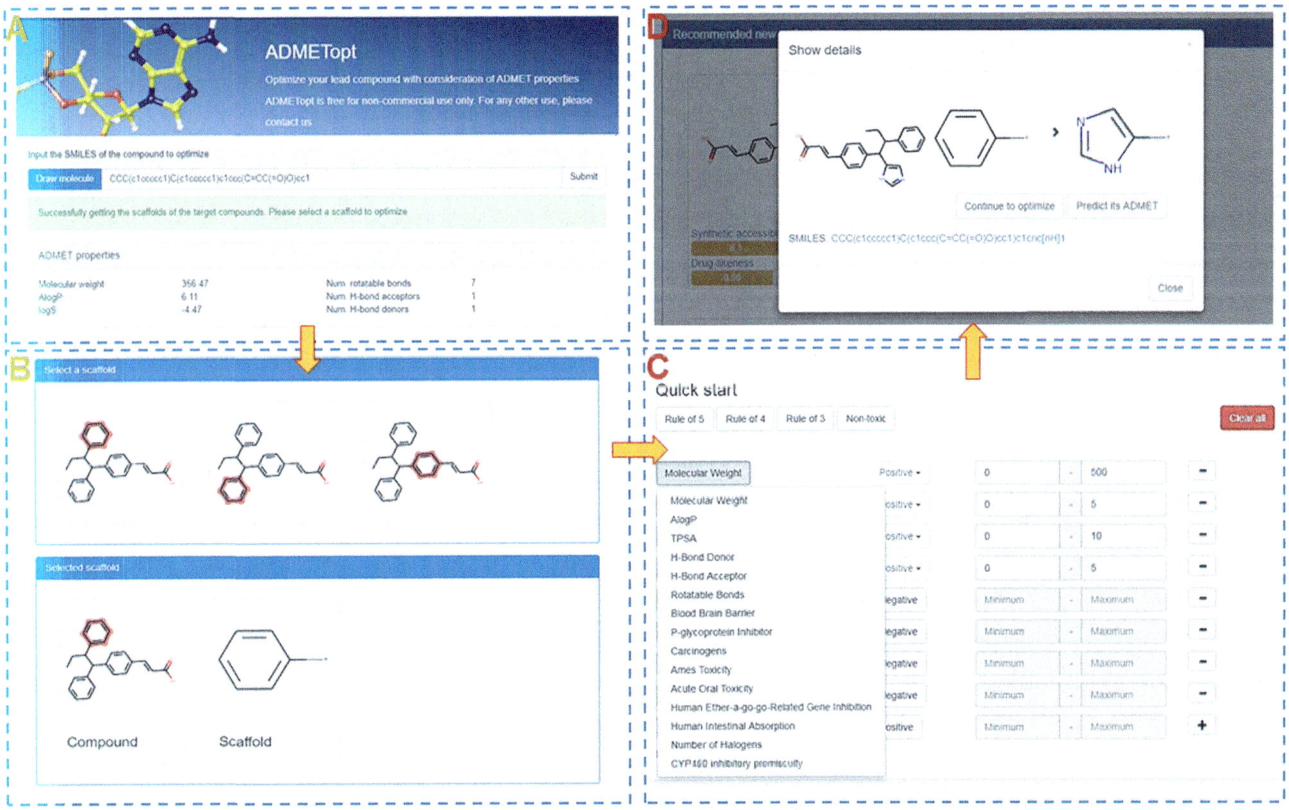

FIGURE 14.7 The workflow of ADMETopt. (A) Input SMILES and basic chemical properties of molecule will be shown. (B) Display all scaffolds and select one molecular scaffold of interest for further optimization. (C) Restrictions for optimization. (D) The query compound information and result of the optimized compound.

molecule structure as input. After successfully inputting a query molecule, users will get all the scaffolds and then need to select one scaffold for further optimization (Fig. 14.7B). Subsequently, users can set up to 15 restrictions including seven physicochemical properties and eight biological properties to narrow the chemical space of the optimized compound. For some binary properties such as acute oral toxicity, carcinogenicity, and Ames toxicity, users need to select "Positive/Negative" to set the optimization target. However, for continuous properties such as molecular weight, AlogP, and rotatable bonds, users need to set the maximum and minimum range of values to achieve optimization. The submitted job will be displayed in the page like Fig. 14.7D when the optimized process is completed.

Applications of admetSAR

admetSAR provides reliable and remarkable models, which have been reported in 24 papers and have been highly recognized and acclaimed by the scientific community [23–46]. To date, it has been widely employed in drug discovery and environmental risk assessment. The original publications of both admetSAR 1.0 and 2.0 are highly cited with 1022 and 393 citations, respectively (Web of Science, assessed on November 8, 2022). admetSAR not only has been recommended by DrugBank but also has contributed to many scientific studies.

Evaluation of newly synthesized molecules

After the synthesis of a new molecule with a specific function, spectral characterization, pharmacokinetic properties and safety assessment, and bioactivity testing are required. In the ADMET evaluation of newly synthesized molecules, computational methods are widely used for their rapid and low-cost characteristics, and quite a few of these studies were benefited from admetSAR. The admetSAR webserver was used to evaluate the drug-like properties and toxicity of small molecules with different structures such as spiro quinoxaline-β-lactam-based heterocyclic compounds, synthetic curcumin analogs, and compounds with azomethine structures, which are expected to play a therapeutic role in the treatment of cancer, Alzheimer's disease, etc. [47–49].

ADMET evaluation in virtual screening

It is well known that virtual screening is a computer-aided technique for rapid access to lead compounds for a target of interest, which is usually divided into structure-based virtual screening and ligand-based virtual screening. An ideal drug candidate should have not only specific target activity but also promising ADMET properties. admetSAR is widely used by researchers for ADMET evaluation in lead screening because of its excellent and robust performance. The newly emerged coronavirus disease COVID-19 has become a big threat to human health. It was declared as a global health emergency by the World Health Organization. In search for drugs to treat this disease, admetSAR tool facilitated rapid ADMET evaluation of potentially active compounds to improve drug discovery results [50,51]. Wu et al. successfully identified seven retinoid-related orphan receptor γt inverse agonists, using a novel network-based virtual screening approach, and admetSAR also played an active role in this study [52].

ADMET evaluation in mechanism exploration

In addition to providing compound screening services in drug discovery, the prediction of ADMET properties also contributes to mechanistic studies, especially in network pharmacology-based mechanistic explorations. Di et al. predicted skin sensitization of compounds first and then explored the allergic mechanism for haptens [53,54]. In herbal mechanism studies, researchers filtered the ADMET properties of herbal components before conducting subsequent studies. Wang et al. probed the molecular mechanisms of Polygonum multiflorum Thunb–induced liver injury; a human intestinal absorption model was used to filter the components and subsequently determine the compounds that interact with CYP450 enzymes by admetSAR [55]. In addition, to explore arctiin and arctigenin as potential therapeutic compounds for silicosis, admetSAR and molecular docking were used to analyze the better efficacy of arctiin. It shows high oral availability, low toxicity, and better multimolecular active metabolic stability.

TABLE 14.2 Comparison of functions of admetSAR 2.0 with other web-based tools.

	admetSAR 2.0	ADMETlab2.0	SwissADME	pkCSM
Endpoints				
Physicochemical property	5	17	12	6
ADME	35	23	9	20
Organ toxicity	9	9	0	7
Eco toxicity	6	3	0	2
Gene toxicity	10	12	0	1
Molecular search	Yes	No	No	No
Structural alerts	No	Yes	Yes	No
Optimization tools	Yes	No	Yes	No

Comparison with other tools

Nowadays, there are numerous excellent online services available for ADMET evaluation, such as ADMETlab 2.0, SwissADME, and pkCSM. We compared the functions available in these web-based tools, with details shown in Table 14.2.

More specifically, these tools help researchers evaluate ADMET properties in a time- and cost-efficient manner, filter out undesirable compounds, and provide guidance on lead optimization. Thereinto, ADMETlab 2.0 provides most predictable endpoints, including toxicophore rules, toxicity pathways, and some medicinal chemistry measures that researchers would be interested in. ADMETlab 2.0 applied a multitask graph attention framework to handle regression and classification tasks simultaneously with fast prediction speed. SwissADME excels in physicochemical properties and has the most extensive in-house methods such as BOILED-Egg and iLOGP. In contrast, admetSAR 2.0 is a comprehensive service platform and provides integrating molecular search, ADMET evaluation, analysis, and optimization systems. A wrapper has been added in the new version of admetSAR to redirected at admetSAR 1.0 for molecular search. In terms of endpoints, admetSAR provides the most diverse metabolism properties and special eco-toxicity prediction services that contribute to environmental risk assessment. For query molecules with undesirable ADMET properties, users can easily use ADMETopt to optimize for favorable properties.

Conclusions and outlook

Here, we presented a valuable web-based tool, admetSAR, for chemical safety assessment, which incorporates in silico prediction models of ADMET properties and physicochemical properties. The webserver has a user-friendly graphical interface and provides users with a comprehensive and efficient prediction service for 51 ADMET-related endpoints, as well as accessible information retrieval capabilities. Equipped with the robust QSAR models that are the optimal models constructed using the state-of-the-art machine learning methods, the models in admetSAR have been studied for many years and got peer-recognized. High-quality model clusters provide stable and reliable support for prediction results. In addition, batch evaluation and customizable predictive endpoints services support flexible and convenient use. Barriers for nonexpert users are eliminated by intuitive output results and effective interpretation. More than prediction, the optimization service of chemical ADMET properties is available in admetSAR. These practical tools are designed to better aid drug discovery and risk assessment.

Although ADMET predictions are playing an increasingly significant role in drug development, they still face a series of challenges. With the rapid growth of digital information, ADMET prediction has entered the era of big data. On one hand, how to use big data including clinical data and omics data (e.g., genomics, metabolomics) to gain insight into biological phenomena and disease occurrence is a problem. On the other hand, how to ensure the quality of data is also a key challenge for ADMET prediction. Dealing with the effects of different noisy data is highly complicated. In this field, reasonable recourse to artificial intelligence methods is deemed to be promising and valuable. In addition, improving the interpretability of models is one of the future directions. For example, ProTox-II gives a pioneering attempt and provides possible mechanisms of molecular toxicity based on adverse outcome pathways. Last, but not least, there is a requirement and an inevitable trend to develop computational assessment methods that take into consideration of the real exposure environment and exposure dose of chemicals.

References

[1] Khan EA, Syeda SR. Chemical substitution in processes for inherently safer design: pros and cons. Pure Appl Chem 2022;94(7):889–99.
[2] Landrigan PJ, Fuller R, Acosta NJR, Adeyi O, Arnold R, Basu NN, et al. The Lancet Commission on pollution and health. Lancet 2018;391(10119):462–512.
[3] Vamathevan J, Clark D, Czodrowski P, Dunham I, Ferran E, Lee G, et al. Applications of machine learning in drug discovery and development. Nat Rev Drug Discov 2019;18(6):463–77.
[4] Hodgson JJN. ADMET—turning chemicals into drugs. Nat Biotechnol 2001;19(8):722–6.
[5] Yamashita F, Hashida M. In silico approaches for predicting ADME properties of drugs. Drug Metab Pharmacokinet 2004;19(5):327–38.
[6] Yang H, Lou C, Sun L, Li J, Cai Y, Wang Z, et al. admetSAR 2.0: web-service for prediction and optimization of chemical ADMET properties. Bioinformatics 2019;35(6):1067–9.
[7] Banerjee P, Eckert AO, Schrey AK, Preissner R. ProTox-II: a webserver for the prediction of toxicity of chemicals. Nucleic Acids Res 2018;46(W1):W257–63.
[8] Xiong G, Wu Z, Yi J, Fu L, Yang Z, Hsieh C, et al. ADMETlab 2.0: an integrated online platform for accurate and comprehensive predictions of ADMET properties. Nucleic Acids Res 2021;49(W1):W5–14.
[9] Daina A, Michielin O, Zoete V. SwissADME: a free web tool to evaluate pharmacokinetics, drug-likeness and medicinal chemistry friendliness of small molecules. Sci Rep 2017;7(1):1–13.
[10] Cheng F, Li W, Zhou Y, Shen J, Wu Z, Liu G, et al. admetSAR: a comprehensive source and free tool for assessment of chemical ADMET properties. J Chem Inf Model 2012;52(11):3099–105.
[11] Law V, Knox C, Djoumbou Y, Jewison T, Guo AC, Liu Y, et al. DrugBank 4.0: shedding new light on drug metabolism. Nucleic Acids Res 2014;42(D1):D1091–7.
[12] Yang H, Sun L, Wang Z, Li W, Liu G, Tang Y. ADMETopt: a web server for ADMET optimization in drug design via scaffold hopping. J Chem Inf Model 2018;58(10):2051–6.
[13] Gaulton A, Hersey A, Nowotka M, Bento AP, Chambers J, Mendez D, et al. The ChEMBL database in 2017. Nucleic Acids Res 2017;45(D1):D945–54.
[14] Gold LS, Manley NB, Slone TH, Rohrbach L, Garfinkel GB. Supplement to the carcinogenic potency database (CPDB): results of animal bioassays published in the general literature through 1997 and by the National Toxicology Program in 1997–1998. Toxicol Sci 2005;85(2):747–808.
[15] Attene-Ramos MS, Miller N, Huang R, Michael S, Itkin M, Kavlock RJ, et al. The Tox21 robotic platform for the assessment of environmental chemicals—from vision to reality. Drug Discov Today 2013;18(15–16):716–23.
[16] Veith H, Southall N, Huang R, James T, Fayne D, Artemenko N, et al. Comprehensive characterization of cytochrome P450 isozyme selectivity across chemical libraries. Nat Biotechnol 2009;27(11):1050–5.
[17] Noble WS. What is a support vector machine? Nat Biotechnol 2006;24(12):1565–7.
[18] Domingos P, Pazzani M. On the optimality of the simple Bayesian classifier under zero-one loss. Mach Learn 1997;29(2):103–30.
[19] Belgiu M, Dragut L. Random forest in remote sensing: a review of applications and future directions. ISPRS J Photogramm 2016;114:24–31.
[20] Zhang ML, ZZ H. ML-KNN: a lazy learning approach to multi-label learning. Pattern Recognit 2007;40(7):2038–48.
[21] Liu R, Madore M, Glover KP, Feasel MG, Wallqvist A. Assessing deep and shallow learning methods for quantitative prediction of acute chemical toxicity. Toxicol Sci 2018;164(2):512–26.
[22] Playe B, Stoven V. Evaluation of deep and shallow learning methods in chemogenomics for the prediction of drugs specificity. J Cheminform 2020;12(1):11.
[23] Cheng F, Shen J, Yu Y, Li W, Liu G, Lee PW, et al. In silico prediction of Tetrahymena pyriformis toxicity for diverse industrial chemicals with substructure pattern recognition and machine learning methods. Chemosphere 2011;82(11):1636–43.
[24] Sun L, Yang H, Li J, Wang T, Li W, Liu G, et al. In silico prediction of compounds binding to human plasma proteins by QSAR models. ChemMedChem 2018;13(6):572–81.
[25] Li F, Fan D, Wang H, Yang H, Li W, Tang Y, et al. In silico prediction of pesticide aquatic toxicity with chemical category approaches. Toxicol Res 2017;6(6):831–42.
[26] Zhao P, Peng Y, Xu X, Wang Z, Wu Z, Li W, et al. In silico prediction of mitochondrial toxicity of chemicals using machine learning methods. J Appl Toxicol 2021;41(10):1518–26.
[27] Chen J, Yang H, Zhu L, Wu Z, Li W, Tang Y, et al. In silico prediction of human renal clearance of compounds using quantitative structure-pharmacokinetic relationship models. Chem Res Toxicol 2020;33(2):640–50.
[28] Xu X, Zhao P, Wang Z, Zhang X, Wu Z, Li W, et al. In silico prediction of chemical acute contact toxicity on honey bees via machine learning methods. Toxicol Vitro 2021;72:105089.
[29] Zhang C, Zhou Y, Gu S, Wu Z, Wu W, Liu C, et al. In silico prediction of hERG potassium channel blockage by chemical category approaches. Toxicol Res 2016;5(2):570–82.
[30] Wang Q, Li X, Yang H, Cai Y, Wang Y, Wang Z, et al. In silico prediction of serious eye irritation or corrosion potential of chemicals. RSC Adv 2017;7(11):6697–703.
[31] Sun L, Yang H, Cai Y, Li W, Liu G, Tang Y. In silico prediction of endocrine disrupting chemicals using single-label and multilabel models. J Chem Inf Model 2019;59(3):973–82.
[32] Zhang C, Cheng F, Li W, Liu G, Lee PW, Tang Y. In silico prediction of drug induced liver toxicity using substructure pattern recognition method. Mol Inform 2016;35(3–4):136–44.

[33] Cheng F, Yu Y, Zhou Y, Shen Z, Xiao W, Liu G, et al. Insights into molecular basis of cytochrome p450 inhibitory promiscuity of compounds. J Chem Inf Model 2011;51(10):2482–95.
[34] Cheng F, Yu Y, Shen J, Yang L, Li W, Liu G, et al. Classification of cytochrome P450 inhibitors and noninhibitors using combined classifiers. J Chem Inf Model 2011;51(5):996–1011.
[35] Yang H, Li X, Cai Y, Wang Q, Li W, Liu G, et al. In silico prediction of chemical subcellular localization via multi-classification methods. MedChemComm 2017;8(6):1225–34.
[36] Wang Z, Zhao P, Zhang X, Xu X, Li W, Liu G, et al. In silico prediction of chemical respiratory toxicity via machine learning. Comput Toxicol 2021;18:100155.
[37] Jiang C, Yang H, Di P, Li W, Tang Y, Liu G. In silico prediction of chemical reproductive toxicity using machine learning. J Appl Toxicol 2019; 39(6):844–54.
[38] Fan D, Yang H, Li F, Sun L, Di P, Li W, et al. In silico prediction of chemical genotoxicity using machine learning methods and structural alerts. Toxicol Res 2018;7(2):211–20.
[39] Li X, Du Z, Wang J, Wu Z, Li W, Liu G, et al. In silico estimation of chemical carcinogenicity with binary and ternary classification methods. Mol Inform 2015;34(4):228–35.
[40] Cheng F, Ikenaga Y, Zhou Y, Yu Y, Li W, Shen J, et al. In silico assessment of chemical biodegradability. J Chem Inf Model 2012;52(3):655–69.
[41] Jiang C, Zhao P, Li W, Tang Y, Liu G. In silico prediction of chemical neurotoxicity using machine learning. Toxicol Res 2020;9(3):164–72.
[42] Cao Q, Liu L, Yang H, Cai Y, Li W, Liu G, et al. In silico estimation of chemical aquatic toxicity on crustaceans using chemical category methods. Environ Sci Process Impacts 2018;20(9):1234–43.
[43] Wang Z, Yang H, Wu Z, Wang T, Li W, Tang Y, et al. Silico prediction of blood-brain barrier permeability of compounds by machine learning and resampling methods. ChemMedChem 2018;13(20):2189–201.
[44] Zhang C, Cheng F, Sun L, Zhuang S, Li W, Liu G, et al. In silico prediction of chemical toxicity on avian species using chemical category approaches. Chemosphere 2015;122:280–7.
[45] Xu C, Cheng F, Chen L, Du Z, Li W, Liu G, et al. In silico prediction of chemical Ames mutagenicity. J Chem Inf Model 2012;52(11):2840–7.
[46] Li X, Chen L, Cheng F, Wu Z, Bian H, Xu C, et al. In silico prediction of chemical acute oral toxicity using multi-classification methods. J Chem Inf Model 2014;54(4):1061–9.
[47] Dabhi RA, Dhaduk MP, Bhatt VD, Bhatt BS. Synthetic approach toward spiro quinoxaline-β-lactam based heterocyclic compounds: spectral characterization, SAR, pharmacokinetic and biomolecular interaction studies. J Biomol Struct Dyn 2022:1–17.
[48] Chowrasia D, Jafri A, Azad I, Rais J, Sharma N, Khan F, et al. In vitro and in silico growth inhibitory, anti-ovarian & anti-lung carcinoma effects of 1, 5 diarylpenta-1, 4-dien-3-one as synthetically modified curcumin analogue. J Biomol Struct Dyn 2021:1–18.
[49] Chowrasia D, Jafri A, Azad I, Rais J. A single-molecule with multiple investigations: synthesis, characterization, computational methods, inhibitory activity against Alzheimer's disease, toxicity, and ADME studies. Comput Biol Med 2022;146:105514.
[50] Rolta R, Yadav R, Salaria D, Trivedi S, Imran M, Sourirajan A, et al. In silico screening of hundred phytocompounds of ten medicinal plants as potential inhibitors of nucleocapsid phosphoprotein of COVID-19: an approach to prevent virus assembly. J Biomol Struct Dyn 2021;39(18): 7017–34.
[51] Malik A, Kohli M, Jacob NA, Kayal A, Raj TK, Kulkarni N, et al. In silico screening of phytochemical compounds and FDA drugs as potential inhibitors for NSP16/10 5′methyl transferase activity. J Biomol Struct Dyn 2021:1–13.
[52] Wu Z, Ma H, Liu Z, Zheng L, Yu Z, Cao S, et al. wSDTNBI: a novel network-based inference method for virtual screening. Chem Sci 2022; 13(4):1060–79.
[53] Di P, Yin Y, Jiang C, Cai Y, Li W, Tang Y, et al. Prediction of the skin sensitising potential and potency of compounds via mechanism-based binary and ternary classification models. Toxicol Vitro 2019;59:204–14.
[54] Di P, Wu Z, Yang H, Li W, Tang Y, Liu G. Prediction of the allergic mechanism of haptens via a reaction-substructure-compound-target-pathway network system. Toxicol Lett 2019;317:68–81.
[55] Wang YY, Li J, Wu ZR, Zhang B, Yang HB, Wang Q, et al. Insights into the molecular mechanisms of Polygonum multiflorum Thunb-induced liver injury: a computational systems toxicology approach. Acta Pharmacol Sin 2017;38(5):719–32.

C H A P T E R

15

QSAR tools for toxicity prediction in risk assessment—Comparative analysis

Wei Shi, Jing Guo and Tong Bao

State Key Laboratory of Pollution Control and Resource Reuse, School of the Environment, Nanjing University, Nanjing, China

Abbreviations

ANN Artificial neural network
ASNN Associative neural network
CAS Chemical abstract service
CPDB Carcinogenic potency database
EC_{50} Half maximal effective concentration
EPA Environmental Protection Agency
FDA Food and Drug Administration
GALAS Global, Adjusted Locally According to Similarity
GHS Globally harmonized system
HC Hierarchical clustering
HSDB Hazardous substances data bank
IUCLID International Uniform Chemical Information Database
kNN k-nearest neighbor
Lazar Lazy structure: activity relationships
LC_{50} Median lethal concentration
LD_{50} Median lethal dose
LDA Linear discriminative analysis
LOAEL The lowest observed adverse effect level
LR Logistic regression
MLR Multiple linear regression
MRDD Maximum recommended daily dose
NBC Naive Bayesian classifier
NOAEL No observed adverse effect
NTP National toxicology program
OCHEM Online chemical modeling environment
OECD Organisation for Economic Co-operation and Development
OPERA Open (quantitative) structure–activity/property relationship app
PLS Partial least squares
QSAR Quantitative structure–activity relationship
RAIS Risk assessment information system
RAS Russian Academy of Sciences
REACH Registration, Evaluation, Authorisation and Restriction of Chemicals
RF Random forest
RTECS Registry of toxic effects of chemical substances
SMILES Simplified molecular-input line-entry specification
SVM Support vector machine
T.E.S.T. Toxicity estimation software package tool
TD_{50} Median toxic dose

Introduction

With the rapid growth of industry and the increasing demands of life, the number of chemicals continues to grow. By early 2022, there were 193 million organic and inorganic substances listed in the Chemical Abstracts Service (CAS). Chemicals can threaten human health and ecological safety through various exposure pathways such as environmental media and food [1,2]. Therefore, there is a great need for chemical management to protect public health and ecological safety. Risk assessment is a prerequisite for the management of chemicals [3], the key to which lies in the acquisition of exposure data and toxicity data [4]. Traditional chemical risk assessment methods heavily rely on experimental tests. However, it is difficult to acquire exposure and toxicity data in an efficient and low-cost way by experimental tests, which fails to meet the demand for risk assessment of a large number of chemicals [5].

To reduce animal tests, save costs and time, and fill the data gaps in risk assessment, countries and organizations have started to focus on the development of nontesting methods [6]. The Registration, Evaluation, Authorisation and Restriction of Chemicals (REACH) regulation issued by the EU in 2007 and the report of the US National Committee

QSAR in Safety Evaluation and Risk Assessment
https://doi.org/10.1016/B978-0-443-15339-6.00016-3

published in the same year "Toxicity Testing in the 21st Century: A Vision and a Strategy" consider alternatives to animal experimentation as an indispensable tool for chemical risk assessment and toxicity studies [7,8]. The quantitative structure–activity relationship (QSAR) is the core tool for toxicity prediction. QSAR is a method to establish quantitative relationships between various structural parameters of molecules and their biological effects [9], to further understand the relationship between the microstructure and macroscopic activity of molecules at the molecular level, to infer key influencing properties, and to provide a basis for compound structure design, mechanism studies, and toxicity prediction of compounds.

Currently, numerous QSAR software tools are available for predicting the health effects and ecotoxicity of compounds, and they are widely used in various fields such as cheminformatics, chemical biology, medicinal chemistry, and environmental management [10]. In Canada, under the new substance regulations of the Canadian Environmental Act, regulatory agencies are required to perform a prescreening and ranking of the Canadian domestic substances list by QSAR [11]. QSAR has also been widely used by the US Food and Drug Administration (FDA) to perform toxicological assessments of drug impurities [12]. In this chapter, eight QSAR tools for toxicity prediction that are commonly used in chemical risk assessment were selected. The basic information of these QSAR tools, including the development background, the prediction endpoints, and the user experience, was first introduced. Then, for different endpoints, the QSAR models were compared and analyzed in five dimensions, including the dataset, the endpoint types, the characterization methods of molecular structures, the algorithm for model construction, and the application domain, thus supporting the selection of appropriate QSAR tools.

The basic information of toxicity prediction software package

The development of toxicity prediction software packages is closely related to the needs of each country or organization for chemical toxicity information at different times. Since the 21st century, the number of chemicals has been increased extremely rapidly, but traditional toxicity testing experiments are time-consuming, labor-intensive, and hardly applicable to the toxicity assessment of a large number of chemicals. For this reason, various countries and organizations are paying increasing attention to the development of alternative methods to animal experiments and efficient, fast, and accurate QSAR toxicity prediction methods [13]. The toxicity endpoints and the user experience varied among different toxicity prediction software packages. Thus, in this section, the basic information of each toxicity prediction software package is presented in terms of the development background, prediction endpoints, and user experience (Table 15.1).

Development background

The eight commonly used toxicity prediction software packages presented in this study were developed in different countries or regions, of which five (VEGA, QSAR TOOLBOX, OCHEM, Lazar, and Danish (Q)SAR Database) were developed by EU Member States or organizations, and three (T.E.S.T., OPERA, and BIOVIA) were developed in the United States. The development of toxicity prediction software packages is closely related to the needs of each country or organization for chemical toxicity information at different times.

The REACH regulation issued by the EU is the most stringent chemical management regulation in the world, which introduces the management concept based on "chemical risk" and the regulation core of "no data, no market" for the first time [14]. The process of chemical management requires a large amount of toxicity data. However, traditional toxicity tests are generally time-consuming and labor-intensive, and fail to meet the massive need for the toxicity evaluation of growing amounts of chemicals. For this reason, the EU developed a comprehensive technical guidance document in 1996 to support the "Directive on New Substances and the Regulation on Existing Substances" which for the first time guided the use of QSAR in the chemical risk assessment process [15], indicating the beginning of the use of computational toxicology as a complementary and alternative approach. In addition, the Guidance on Information Requirements and Chemical Safety Assessment issued by European Chemicals Agency claims that nontesting methods (e.g., QSAR) may be used when conducting new toxicity tests for REACH registration [16]. Driven by these regulations, a series of toxicity prediction software were developed. In 2004, the Technical University of Denmark extensively used the QSAR model for the collection of reliable quality and safety analysis guidelines, leading to the development of the Danish (Q)SAR Database [17]. In 2006, the Organisation for Economic Co-operation and Development (OECD) developed the QSAR TOOLBOX by integrating knowledge and data for grouping of chemicals and filling data gap filling. VEGA was developed by Istituto di Ricerche

TABLE 15.1 Basic information on toxicity prediction software packages.

Software package	Country/organization	Release time	Model number	License	Running platform	Input forms	Batch predictions	URL
VEGA	Italy	N/A	>90	Free	Windows/Linux/Mac	SMILES, SDF file	√	https://www.vegahub.eu/
QSAR TOOLBOX	OECD	2006	>900	Free	Windows	CAS, SMILES, name molecule drawing	×	https://qsartoolbox.org/
T.E.S.T.	USA	2010	N/A	Free	Windows/Linux/Mac	CAS, SMILES, name, molecule drawing	√	https://www.epa.gov/chemical-research/toxicity-estimation-software package-tool-test
OCHEM	Sweden	N/A	>190	Free	Web	CAS, SMILES, name, EXCEL/MOL/SDF file	√	https://www.ochem.eu/home/show.do
OPERA	USA	N/A	>50	Free	Windows/Linux	SMILES, CAS, molecule drawing, name	×	https://ntp.niehs.nih.gov/whatwestudy/niceatm/comptox/ct-opera/opera.html
Lazar	German	2008	N/A	Free	Web	SMILES, molecule drawing	×	https://lazar.in-silico.ch/predict
Danish (Q)SAR Database	Denmark	2004	>200	Free	Web	SMILES, molecule drawing, name, MOL/SDF file	×	http://qsar.food.dtu.dk/
BIOVIA	USA	N/A	N/A	Commercial	Windows/Linux/Mac	N/A	N/A	https://www.3ds.com/products-services/biovia/

Lazar, lazy structure–activity relationships; *N/A*, not available; *OCHEM*, online chemical modeling environment; *OECD*, Organisation for Economic Co-operation and Development; *OPERA*, open (quantitative) structure–activity/property relationship app; *T.E.S.T.*, toxicity estimation software package tool.

Farmacologiche Mario Negri by integrating the QASR models provided by T.E.S.T., CAESAR, and other contributors. With the rapid development of the QSAR software package, more attentions have been given to the standardization of QSAR modeling methods and the quality of models. In 2007, the OECD issued a guidance document on validating QSAR models, which specifies five basic principles for using QSAR and standardizes the QSAR modeling methods [18]. For this reason, the OCHEM platform with the explicit data collection process and standardized modeling methods has been developed [19]. In addition, to reduce the operational difficulty of the toxicity prediction software packages to facilitate users, the German Federal Institute for Risk Assessment developed Lazar in 2008 [20], which simplified the development and validation processes of models and created prediction models that can be used directly without prior processing of input data.

In 1976, the Toxic Substances Control Act issued by the United States authorized the Environmental Protection Agency (EPA) to conduct risk assessments of all new chemicals [21]. However, physicochemical and toxicological data on new chemicals are scarce. Traditional animal experiments are costly and time-consuming, far from meeting the needs of risk assessment for a large number of chemicals. For this reason, the EPA developed the framework for the computational toxicology research program in 2003, which planned to use QSAR models to enhance dose effect assessment and quantitative risk assessment [22]. Subsequently, in 2007, the National Committee released a report, *Toxicity Testing in the 21st Century: a Vision and a Strategy,* which proposed the use of alternatives to animal experimentation [23]. With the release of this landmark report, the development of toxicity prediction methods was given a major boost. EPA utilized its extensive knowledge and experience to develop T.E.S.T., which can predict multiple toxicity endpoints and fill the data gaps efficiently. In addition, the stringent regulations in the United States have also promoted the development of commercial companies that offer chemical risk assessment services. As a result, commercial toxicity prediction software packages, such as BIOVIA, have been developed. Although a large number of QSAR models have been developed for various toxicity endpoints, not all QSAR models are suitable for regulatory purposes due to the variability of the data and algorithms among different models and the lack of documentation and transparency. To develop QSAR models that can be used in regulatory decision-making, the National Institute of Environmental Health Science developed a high-quality and transparent OPERA prediction software package.

Toxicity prediction endpoints

According to the risk assessment technical guidance issued by the EU and EPA, there are eight health endpoints (acute toxicity, carcinogenicity, mutagenicity, reproductive toxicity, repeated dose toxicity, sensitivity, eye irritation/corrosion, and skin irritation/corrosion) and two ecotoxicity endpoints (aquatic toxicity and terrestrial toxicity) that need to be focused in the risk assessment of chemicals. The coverage of the aforementioned health endpoints and ecotoxicity endpoints by each of the toxicity prediction software packages was compared (Table 15.2).

Health endpoints evaluate the harmful effects of chemicals on humans and are used to assess the health risks of chemicals. The results showed that all the eight health endpoints could be covered by the eight toxicity prediction software packages. Among them, mutagenicity had the greatest coverage, which could be predicted by seven toxicity prediction software packages. Acute toxicity and carcinogenicity had the second-greatest coverage, which could be predicted by five toxicity prediction software packages. The consistency and availability of experimental data drove the development of QSAR prediction models for the aforementioned three health endpoints. The coverage of repetitive dose toxicity, sensitization, and skin irritation/corrosion by these eight toxicity prediction software packages was less than 50%. Among them, repetitive dose toxicity and skin irritation/corrosion could be predicted by three toxicity prediction software packages, and sensitization was predicted by four toxicity prediction software packages. Moreover, the coverage of reproductive toxicity and eye irritation/corrosion was the lowest in this study, which could only be predicted by two toxicity prediction software packages. The complexity of the toxicological mechanism of reproductive toxicity limits the construction of relevant QSAR models [24]. In addition, the lack of experimental data on eye irritation/corrosiveness also limits the construction of relevant models. Further comparison of the health endpoints covered by each toxicity prediction software package showed that QSAR TOOLBOX and BIOVIA covered the most health endpoints. QSAR TOOLBOX, as a comprehensive prediction software package [25] that integrates comprehensive theoretical knowledge, experimental data, and computational tools from different sources, could predict all the eight health endpoints. BIOVIA can predict all the health endpoints except reproductive toxicity. However, OCHEM and OPERA can only predict one of the eight health endpoints. Both OCHEM and OPERA focus more on the prediction of physicochemical properties [26].

TABLE 15.2 Prediction endpoints covered by toxicity prediction software packages.

	Health endpoints								Ecotoxicological endpoints	
	Acute toxicity	Carcinogenicity	Mutagenicity	Reproductive toxicity	Repeated dose toxicity	Sensitization	Skin irritation/corrosion	Eye irritation/corrosion	Aquatic toxicity	Terrestrial toxicity
VEGA	O	O	O	O	O	O			O	O
QSAR TOOLBOX	O	O	O	O	O	O	O	O	O	O
T.E.S.T.	O		O						O	
OCHEM			O						O	
OPERA	O									
Lazar		O	O						O	
Danish (Q)SAR Database		O	O			O	O		O	
BIOVIA	O	O	O		O	O	O	O	O	

Note: 1≤Number of models<10; 10≤Number of models<100; Number of models>100; Number of models is unknown

Ecotoxicity endpoints evaluate the ability of chemicals to cause harm to nonhuman species and are used to assess the ecological risks of chemicals. The coverage of ecotoxicity endpoints was analyzed, and the statistical results showed that all two ecotoxicity endpoints were covered by the eight toxicity prediction software packages. Among them, the coverage of aquatic toxicity is greater, which could be predicted by seven toxicity prediction software packages. However, the terrestrial toxicity could only be predicted by two toxicity prediction software packages. On the one hand, this is because the amount of experimental data on terrestrial toxicity is much less than that on aquatic toxicity. On the other hand, policies supporting conservation measures in terrestrial ecosystems have far lagged behind those in aquatic ecosystems [27]. Further comparison of the ecotoxicity endpoints covered by each toxicity prediction software package shows that VEGA and QSAR TOOLBOX have the greatest coverage of ecotoxicity endpoints, which could predict both terrestrial and aquatic toxicity. T.E.S.T., OCHEM, Lazar, Danish (Q)SAR Database, and BIOVIA could only predict aquatic toxicity endpoints. OPERA, which focused more on the prediction of physicochemical endpoints failed to predict any of the ecotoxicity endpoints [26].

The above 10 health and ecotoxicity endpoints could be covered by the eight toxicity prediction software packages. For health effect prediction, QSAR TOOLBOX and BIOVIA cover the most health endpoints and are recommended. For ecotoxicity prediction, VEGA and QSAR TOOLBOX are recommended because they cover the most ecotoxicity endpoints. In the future, the coverage of the prediction endpoints of the current toxicity prediction software package should be expanded, especially for reproductive toxicity, eye irritation/corrosion, and terrestrial toxicity, thus expanding the application of toxicity prediction software packages.

User experience

The user experience, including the input format, whether it is free or not, and whether it can be run on multiple platforms, all have a significant impact on the use of toxicity prediction software packages. In this section, a comparative analysis on the aforementioned three aspects of user experience was conducted for each of the toxicity prediction software packages to provide software package selection suggestions to users with different needs and knowledge bases.

To meet the different needs of users, toxicity prediction software packages often support multiple input forms. The input forms of these toxicity prediction software packages can be divided into two categories: (1) chemical names and identifiers, and (2) chemical structures. The multiple input forms not only enhance the user-friendliness of the software packages but also compensate for incomplete data (e.g., the user only has information about the chemical CAS number). Searching by chemical names and identifiers is the most basic search method for toxicity prediction software packages. Identifiers mainly include CAS number, SMILES, etc. All the software packages support searching by SMILES. Chemical structures can be input by uploading structure files or drawing structures of target chemicals through the software package-integrated chemical structure editor. All the eight software packages support the prediction of chemical toxicity by chemical structure, while only QSAR TOOLBOX, T.E.S.T., OPERA, Lazar, and Danish (Q)SAR Database support drawing structures of chemicals through their structure editors. In addition to the basic single-substance input, VEGA, T.E.S.T., and OCHEM also provide the function of batch prediction, which greatly improves the prediction efficiency of the software packages and better meets the users' needs.

Another important factor affecting the user experience is whether the toxicity prediction software package is freely accessible. The toxicity prediction software packages collected in this study were developed by nonprofit organizations or teams, including government agencies or organizations, research institutes, and universities. Thus, all these toxicity prediction software packages are free to access by the public, except for BIOVIA, which can only be accessed through a fee-based subscription.

In addition, there are differences in the platforms on which the software packages run. OCHEM, Lazar, and Danish (Q)SAR Database are web-based and thus are more convenient for predicting toxicity without downloading. The remaining five software packages all need to be downloaded and installed before use. VEGA, T.E.S.T., and BIOVIA support all the operating systems (Windows/Linux/Mac), which enabled a wider application. However, OPERA only supports Windows/Linux systems, and QSAR TOOLBOX only supports Windows systems. Users may choose the appropriate toxicity prediction software package according to their computer configuration.

Modeling methods of the toxicity prediction software packages

In 2007, the OECD issued a guidance document on validating QSAR models, which specifies five basic principles for developing QSAR: (1) a defined endpoint; (2) an unambiguous algorithm; (3) a defined domain of applicability; (4) appropriate measures of goodness-of-fit, robustness, and predictivity; and (5) a mechanistic interpretation (if possible) [18]. Determination of prediction endpoints, collection of raw datasets, characterization of chemical structures, selection of modeling algorithms, and characterization of application domains are key steps to developing QSAR models, which greatly affect the transparency of its prediction mechanism and reliability of results. Therefore, in this section, a comparative analysis on prediction endpoint types (Table 15.3), raw datasets (Table 15.4), structure characterization methods, modeling algorithms, and application domain characterization methods was conducted for these eight software packages.

Type of endpoints

Acute toxicity describes the adverse effects of a substance that result either from a single exposure or from multiple exposures in a short period (usually less than 24 h) [32]. Acute toxicity is usually expressed as the "median lethal dose (LD_{50})," which is the dose that causes death in 50% of animals. The acute toxicity could be predicted by

TABLE 15.3 Types of prediction endpoints of QSAR models in toxicity prediction software packages.

Software package	Endpoints								
	Acute toxicity	Carcinogenicity	Mutagenicity	Reproductive toxicity	Repeated dose toxicity	Sensitization	Skin irritation/ corrosion	Eye irritation/ corrosion	Aquatic toxicity	Terrestrial toxicity
VEGA	LD_{50}	Classification/ OSF/IUR	Classification	Classification	NOAEL	Classification-skin	Classification	—	Classification/ LC_{50}/EC_{50}	LD_{50}
QSAR TOOLBOX	LD_{50}	Classification	Classification	Classification	LOAEL	Classification-skin/ respiratory	Classification	Classification	LC_{50}/EC_{50}	LC_{50}
T.E.S.T.	LD_{50}	—	Classification	—	—	—	—	—	Classification	—
OCHEM	—	—	Classification	—	—	—	—	—	LC_{50}	—
OPERA	LD_{50}/GHS categories/ EPA categories	—	—	—	—	—	—	—	—	—
Lazar	—	TD_{50}/ classification	Classification	—	—	—	—	—	LC_{50}	—
Danish (Q)SAR Models	—	Classification	Classification	—	—	Classification-respiratory	Classification	—	LC_{50}/EC_{50}	—
BIOVIA	LD_{50}	TD_{50}/ classification	Classification	—	LOAEL	Classification-skin	Classification	N/A	LC_{50}	

— the software package cannot predict this endpoint; *EC_{50}*, half maximal effective concentration; *GHS*, globally harmonized system; *LC_{50}*, median lethal concentration; *LD_{50}*, median lethal dose; *LOAEL*, the lowest observed adverse effect level; *N/A*, not available; *NOAEL*, no observed adverse effect; *TD_{50}*, median toxic dose.

TABLE 15.4 Raw datasets of QSAR models in toxicity prediction software packages.

Software package	Endpoints									
	Acute toxicity	Carcinogenicity	Mutagenicity	Reproductive toxicity	Repeated dose toxicity	Sensitization	Skin irritation/ corrosion	Eye irritation/ corrosion	Aquatic toxicity	Terrestrial toxicity
VEGA	N/A	CPDB/RAIS	Kazius et al. [28]	TERIS/FDA/ Wu et al. [29]	N/A	N/A	RTECS/HSDB	—	ECOTOX	OpenFoodTox
QSAR TOOLBOX	RTECS/IUCLID	CPDB	Kazius et al. [28]	Literature	FDA-MRDD	Graham et al. [30]	RTECS/HSDB	RTECS/HSDB	ECOTOX	N/A
T.E.S.T.	ChemIDplus	—	Hansen et al. [31]	—	—	—	—	—	Literature	—
OCHEM	—	—	Hansen et al. [31]	—	—	—	—	—	RAS	—
OPERA	eChemPortal/ HSDB/ChemIDplus	—	—	—	—	—	—	—	—	—
Lazar	—	CPDB	Kazius et al. [28]/ Hansen et al. [31]/ EFSA	—	—	—	—	—	N/A	—
Danish (Q) SAR Models	—	CPDB	Kazius et al. [28]	—	—	Graham et al. [30]	N/A	—	N/A	—
BIOVIA	RTECS	CPDB/NTP	Kazius et al. [28]/ ISSCAN	—	EPA database/ literature	Literature	RTECS/HSDB	N/A	N/A	

—, the software package cannot predict this endpoint; *CPDB*, Carcinogenic potency database; *FDA*, Food and Drug Administration; *HSDB*, Hazardous substances data bank; *IUCLID*, International Uniform Chemical Information Database; *MRDD*, maximum recommended daily dose; *N/A*, not available; *NTP*, National Toxicology Program; *RAIS*, risk assessment information system; *RAS*, Russian Academy of Sciences; *RTECS*, registry of toxic effects of chemical substances.

VEGA, QSAR TOOLBOX, T.E.S.T., OPERA, and BIOVIA. The acute toxicity endpoints of all the five toxicity prediction software packages are expressed as LD_{50} under the oral or subcutaneous route in mammals. Among them, OPERA also enables hazard classification based on the predicted LD50 values.

Carcinogenicity means the properties of substances that induce cancer or increase its incidence [33]. Traditionally, the standard experimental method for assessing the carcinogenicity of chemicals according to the OECD guidelines is the 2-year rodent carcinogenicity bioassay [34]. Long-term animal testing assays are too costly, and thus, multiple models have been developed to predict carcinogenicity [35]. VEGA, QSAR TOOLBOX, Lazar, Danish (Q)SAR Database, and BIOVIA provide QSAR models for predicting carcinogenicity. The predicted endpoints can be classified into two categories: (1) classification and (2) continuous values (TD_{50}, oral slope factor, inhalation unit risk). The models for predicting carcinogenicity in the QSAR TOOLBOX and Danish (Q)SAR Database are qualitative models, which predict chemicals as "carcinogenic" or "noncarcinogenic." VEGA, Lazar, and BIOVIA contain both qualitative and quantitative models for carcinogenicity prediction.

Mutagenicity means a permanent change in the amount or structure of the genetic material in a cell [36]. The term "mutation" applies both to heritable genetic changes that may be manifested at the phenotypic level and to the underlying DNA modifications when known (including specific base pair changes and chromosomal translocations). The reverse mutation assay (*Ames* test), or *Salmonella typhimurium*, is a commonly used and validated in vitro test and is used as an initial screen for mutagenicity [37]. The mutagenicity could be predicted by seven software packages VEGA, QSAR TOOLBOX, T.E.S.T., OCHEM, Lazar, Danish (Q)SAR Database, and BIOVIA. The mutagenicity QSAR models in all the seven software packages are classification models, which classify the chemicals as "mutagen" or "nonmutagen."

Repeated dose toxicity reflects the adverse effects that may be induced by repeated exposure to a compound over a limited period [38]. Repeated dose toxicity is generally evaluated by the level of no observed adverse effect (NOAEL), which is the dose at which no adverse effect is observed, or the lowest observed adverse effect level (LOAEL), which is the lowest dose for repeated dose toxicity measures [39]. Among the four software packages that could predict repeat dose toxicity, VEGA, QSAR TOOLBOX, and BIOVIA use LOAEL as the endpoint, and VEGA predicts the NOAEL in rodents. Notably, QSAR TOOLBOX could directly predict repeated dose toxicity in humans.

Sensitization is usually defined as an autoimmune reaction caused by inhalation or leakage of an external substance through the skin or respiratory tract [40]. A compound is classified as a sensitizer if there is evidence that it can cause sensitization in a large population through dermal or respiratory exposure, or if there are positive results in appropriate animal tests. The sensitization could be predicted by VEGA, QSAR TOOLBOX, Danish (Q)SAR Database, and BIOVIA. Among them, VEGA and BIOVIA could predict dermal sensitization in rodents, the Danish (Q) SAR Database could predict respiratory sensitization in humans, and QSAR TOOLBOX predicts both skin sensitization and respiratory sensitization in humans. The QSAR models for sensitization prediction in the aforementioned four software packages are all classification models, which classify chemicals as "sensitizers" or "nonsensitizers."

Adverse skin/eye reactions caused by chemical exposure mainly include skin/eye irritation and skin/eye corrosion [41]. Skin irritation is defined as local reversible damage produced by skin exposure to chemicals for more than 4 h, while skin corrosion is defined as visible irreversible necrosis from the epidermis to the dermis produced by skin exposure to chemicals for more than 4 h [42]. Eye irritation is defined as reversible inflammatory changes in the eye and its surrounding mucosa produced by contact with the test chemical on the surface of the eye, and eye corrosion is defined as irreversible tissue damage caused by contact with the test chemical on the surface of the eye. Only QSAR TOOLBOX, Danish (Q)SAR Database, and BIOVIA could predict skin/eye irritation/corrosion. Among them, the QSAR TOOLBOX and Danish (Q)SAR Database could predict the skin irritation/corrosion of rabbits, while BIOVIA can predict both the skin irritation/corrosion and the eye irritation/corrosion of rabbits. The skin/eye irritation/corrosion prediction QSAR models in the aforementioned three software packages are all classification models, which classify chemicals as "sensitizers" or "nonsensitizers."

Ecotoxicity includes aquatic toxicity and terrestrial toxicity. Aquatic toxicity is mainly divided into acute aquatic toxicity and chronic aquatic toxicity. Acute aquatic toxicity is defined as the inherent property of a chemical that causes harm to aquatic organisms for a short period. Chronic aquatic toxicity is defined as the inherent property of a chemical that causes harmful effects on aquatic organisms during the exposure period determined by the life cycle of the organism. The aquatic toxicity could be predicted by VEGA, QSAR TOOLBOX, T.E.S.T., OCHEM, Lazar, Danish (Q)SAR Database, and BIOVIA. Among the QSAR models for aquatic toxicity prediction, the *fathead minnow* and *daphnia magna* are the most widely used aquatic species. Moreover, QSAR TOOLBOX can predict the aquatic toxicity of most species, which could also predict the aquatic toxicity of *algae*, and its prediction endpoints are mainly 96 h LC_{50} or 48 h/72 h/96 h EC_{50}. T.E.S.T., which differs from the quantitative models of other software packages, predicts the mode of action of chemicals in *fathead minnow* within 96 h (e.g., acetylcholinesterase inhibition, anticoagulation, acetylcholine receptor antagonism, narcosis, neurotoxicity). Terrestrial toxicity is also classified into acute toxicity and chronic toxicity. VEGA and QSAR TOOLBOX could predict the terrestrial toxicity, for example, VEGA could predict the 48 h LD_{50} of *honeybee*, and QSAR TOOLBOX could predict the 14 days LC_{50} of *earthworm*.

Raw datasets

Collecting and obtaining experimental data are the basis for developing QSAR toxicity prediction models, and the quality of raw data can significantly affect the performance of QSAR models [43]. The sources of raw datasets for the QSAR models include three categories: (1) public databases; (2) literature; and (3) government documents and reports. For the QSAR models predicting acute toxicity, except that VEGA did not provide the raw dataset, the raw datasets were mainly obtained from public databases developed or regulated by the government, such as the Registry of Toxic Effects of Chemical Substances (RTECS), Hazardous Substances Data Bank (HSDB), and ChemIDplus databases. These public databases provide a large amount of toxicity data with high confidence. For the QSAR models predicting carcinogenicity, the raw data of VEGA, QSAR TOOLBOX, Lazar, Danish (Q)SAR Database, and BIOVIA were obtained from the Carcinogenic Potency Database (CPDB). The CPDB is a unique and widely used international resource that contains the results of animal cancer tests performed on chemicals, convenient access to the relevant bioassay kinds of literature, and qualitative and quantitative analysis of positive and negative experiments published in the general literature over the past 50 years [44]. In addition to the CPDB, the raw carcinogenicity data of VEGA are obtained from the Risk Assessment Information System (RAIS), and the raw carcinogenicity data of BIOVIA are obtained from the carcinogenicity experimental reports of the National Toxicology Program (NTP). For the QSAR models predicting mutagenicity, the raw mutagenicity data are mainly from the literature, including the Kazius mutagenicity dataset [45] and the Benchmark mutagenicity dataset [28], which are high-quality datasets. In addition, raw mutagenicity data are obtained from Chemical Carcinogens Structures and Experimental Data and government reports. For the QSAR models that predict reproductive toxicity, the raw datasets for the reproductive toxicity of VEGA were obtained from reproductive toxicity data that were carefully reviewed by Wu et al. [31]. For the QSAR models predicting sensitization, the raw datasets are mainly from the dataset compiled by Graham et al. [29]. They extracted relevant data from the medical literature over the past 30 years. However, as different criteria are used in identifying positive (sensitizers) and negative (nonsensitizers), there may be heterogeneity in the dataset. The raw datasets for sensitization prediction of BIOVIA mainly include the data published in *Contact Dermatitis* and the literature. For the QSAR models predicting skin/eye irritation/corrosion, the raw dataset for both the QSAR TOOLBOX and Danish (Q)SAR Database were obtained from the RTECS and HSDB databases, while the dataset for the BIOVIA model was obtained by a critical review of the literature. For the QSAR models predicting aquatic and terrestrial toxicity, the raw dataset is mainly obtained from the ECOTOX database, which is currently the world's largest compilation of ecotoxicity data and supports chemical management through a systematic and transparent literature review process [30]. In addition to the ECOTOX, the raw datasets are also obtained from government documents and the literature.

Structural characterization method

Before QSAR modeling, it is often necessary to convert the structure of a compound into a computer-readable form such as molecular descriptors and molecular fingerprints. Molecular descriptors are measures of the properties of a molecule and can be the physicochemical properties of the molecule or the numerical indexes derived by various algorithms based on the molecular structure. Molecular descriptors are the most commonly used molecular structure characterization method and are widely used in QSAR modeling. Molecular descriptors mainly include molecular properties, constitutional descriptors, topologic descriptors, geometric descriptors, and quantum chemical descriptors. All the eight toxicity prediction software packages use molecular descriptors to characterize chemical structures (Table 15.5), which indicates that molecular descriptors are the most widely used chemical representation. However, developing QSAR models using molecular descriptors is facing challenges, including the difficulty in mechanism interpretation and the possibility of the "activity cliff" effect, which refers to the significantly different activity of molecules with similar structures [46]. Molecular fingerprints are a special qualitative molecular description method originally designed for structural similarity search and clustering of molecules, and they are widely used in drug discovery and virtual screening [47]. A fingerprint is a representation of the structural properties of a molecule that are encoded by an established algorithm to generate an array of specified lengths consisting of

TABLE 15.5 Structural characterization methods of each software package.

Software package	Molecular descriptors			Molecular fingerprints	SMILES
	Molecular properties descriptors	Constitutional descriptors	Topologic descriptors		
VEGA	O	O	O		O
QSAR TOOLBOX	O	O			
T.E.S.T.	O	O	O		
OCHEM	O	O			
OPERA		O	O		
Lazar	O	O		O	
Danish (Q)SAR Database	O	O			
BIOVIA	O	O		O	

0 and 1. Compared with traditional molecular descriptors, molecular fingerprints provide information on substructures and enable the structural similarity search on this basis [48]. Molecular fingerprints highlight their unique superiority in the field of large-scale compound toxicity prediction due to their explicit algorithm design, definition, and encoding patterns, which are well suited to machine learning and deep learning. Lazar and BIOVIA use molecular fingerprints to characterize the molecular structure. In addition to molecular descriptors and molecular fingerprints, SMILES can be applied as a new form of compound characterization in QSAR studies. VEGA uses SMILES to develop QSAR models.

Modeling algorithms

In the process of modeling QSAR, statistical methods and machine learning algorithms are widely used (Table 15.6). The widely used statistical methods include multiple linear regression (MLR), logistic regression (LR), and partial least squares (PLS). These three statistical modeling methods were used in VEGA, QSAR TOOLBOX, T.E.S.T., Danish (Q)SAR Database, and BIOVIA. All the three statistical modeling methods aim to

TABLE 15.6 Modeling algorithms used in the QSAR model in the software packages.

Software package	Modeling techniques												
	MLR	LR	PLS	kNN	RF	LDA	SVM	ANN	ASNN	HC	NBC	GALAS	Consensus
VEGA	O			O			O	O					O
QSAR TOOLBOX		O	O			O						O	
T.E.S.T.	O			O	O					O			O
OCHEM					O				O				
OPERA				O									O
Lazar				O	O		O						
Danish (Q)SAR Database		O	O										
BIOVIA			O								O		

ANN, artificial neural network; *ASNN*, associative neural network; *GALAS*, Global, Adjusted Locally According to Similarity; *HC*, hierarchical clustering; *kNN*, k-nearest neighbor; *LDA*, linear discriminative analysis; *LR*, logistic regression; *MLR*, multiple linear regression; *NBC*, naive Bayesian classifier; *PLS*, partial least squares; *RF*, random forest; *SVM*, support vector machine.

find a linear relationship between the molecular descriptors of training chemicals and their toxicity. PLS is the generalization of the MLR, which is a multivariate linear method for assessing the relationship between independent and dependent variables. PLS could assess the linear relationship between the molecular structure and activity of compounds in small sample sizes with multiple dimensions. LR is a generalized linear regression method, whose derivation process and calculation are similar to that of regression, but it is mainly used to solve dichotomous or multiclass classification problems. Machine learning algorithms used in these packages include k-nearest neighbor (kNN), random forest (RF), support vector machine (SVM), artificial neural network (ANN), and naive Bayesian classifier (NBC). The kNN method was used by four of these toxicity prediction software packages, VEGA, T.E.S.T., OPERA, and Lazar. The RF method was used in T.E.S.T., OCHEM, and Lazar. The SVM, ANN, and NBC algorithms are used by one toxicity prediction software package. The limited application of SVM may be due to the long time required to compute large amounts of data. ANN tends to overfit, which limits their application in QSAR modeling. In addition, for NBC, the assumption of conditional independence among attributes can hardly be satisfied in practical applications. Different machine algorithms have different basic principles, features, applications, etc. The kNN is an algorithm that classifies samples by measuring the distance between different feature values, which means that the class with the k closest values to the sample to be measured represents the class of that sample. Its main advantages are simplicity and efficiency, and it is usually suitable for small datasets. The RF is a classification algorithm based on integrated learning, in which several decision trees are assembled to learn and predict the data individually, and it uses the average method to merge the predicted values of all decision trees to obtain a more accurate prediction value. The RF algorithm has an excellent ability to process large data with a high accuracy. SVM is a structured machine learning algorithm based on the principle of structural risk minimization, which is general, robust, and computationally simple. It is commonly used to solve classification and regression problems. The ANN is an algorithm that simulates the function of the human brain to process data, which has the characteristics of nonlinearity, self-organization, and self-learning. It is powerful in solving nonlinear problems, so it is often used to build nonlinear QSAR models. The NBC is a classification algorithm based on Bayes' theorem with the assumption of conditional independence of features. It is viewed as an optimal classifier when there is no dependency between a particular feature and other features.

In practical modeling applications, the combination of multiple methods is often required to obtain better prediction performance. Therefore, hybrid optimization modeling methods have been used in the QSAR modeling process by these toxicity prediction software packages, including Global, Adjusted Locally According to Similarity (GALAS), hierarchical clustering (HC), consensus, associative neural network (ASNN), and linear discriminant analysis (LDA). QSAR TOOLBOX uses the GALAS modeling approach to develop predictive models, which is a combination of the global baseline QSAR model and local similarity-based corrections [49]. This approach significantly improves the model performance compared with the traditional PLS method [50]. T.E.S.T. predicts acute toxicity by multiple algorithms, including consensus, HC, and kNN. The HC method uses a weighted average of predictions from different models to estimate toxicity. The kNN method estimates the toxicity by taking the average of the three chemicals in the training set that are most similar to the test chemical. The consensus method uses the mean of predicted toxicity from HC and kNN methods to predict LD_{50} [51]. Similar to T.E.S.T., OPERA combined 139 acute toxicity prediction models into a consensus model to predict the acute toxicity of chemicals, which is called the Collaborative Acute Toxicity Modeling Suite (CATMoS) [52]. Lazar combined kNN learning with SVM to generate local models by automatically screening similar compounds as a training set during prediction. OCHEM uses ASNN to develop models for mutagenicity prediction. ASNN is a combination of ANN and kNN algorithms, which greatly improves the predictive performance compared with traditional ANN and kNN methods. The ASNN approach could explain the result of the neural networks by analyzing the correlation between data cases in the model space.

Traditional mathematical statistics and machine learning methods such as kNN, MLR, PLS, and SVM have good transparency, can further illustrate the mechanism of toxicity through the interpretation of descriptors, and have good predictive performance on specific types or small chemical datasets. For large-scale datasets, due to the variability of compound structures, it is often necessary to combine or optimize traditional statistical methods and machine learning algorithms to obtain a better prediction performance.

Application domain

Every QSAR model has its application domain. The application domain of a QSAR model is generally considered the chemical space determined by the molecular properties used in the model, molecular descriptors, and response values of the training set, which is used to help users to assess the reliability of prediction results. The application

TABLE 15.7 Application domain characterization methods for each software package.

Software package	Applicability domain characterization methods				
	Descriptor ranges	**Descriptor space distance matrix**	**Structure similarity matrix**	**Structure feature screening**	**Analog prediction performance**
VEGA	O		O	O	O
QSAR TOOLBOX			O	O	
T.E.S.T.					O
OCHEM			O		
OPERA		O	O		
Lazar			O		
Danish (Q)SAR Database			O		
BIOVIA	O			O	

domain of the QSAR models in these toxicity prediction software packages could be characterized in five forms, including the descriptor ranges, the descriptor space distance matrix, the structure similarity matrix, the structure feature screening, and the analog prediction performance (Table 15.7). The descriptor range of the training set is used to characterize the application domain of part of the QSAR models in VEGA and BIOVIA. When the descriptor of the chemical to be predicted is out of the range of the training set, an alert is issued to the input compound to indicate that the input chemical is out of the application domain of the QSAR model. The descriptor space distance matrix is used to characterize the application domain of QSAR models in OPERA, which uses the leverage value to calculate the distance between the target and the training set in the descriptor space. The descriptor space distance matrix based on the leverage method is a commonly used application domain definition method, which defines the chemical space of application by assessing the similarity between the chemicals to be predicted and the chemicals in the training set through molecular descriptors and response values. The structural similarity matrix is another widely applied method to characterize the application domains, which was used by VEGA, QSAR TOOLBOX, OCHEM, OPERA, Lazar, and Danish (Q)SAR Database. The structural similarity metric is measured by calculating the structural similarity between the target chemicals and the chemicals in the training set. The distance to model method is a commonly used method to calculate the structural similarity matrix, which essentially calculates the similarity between the training set and the test set. The structural feature screening method was adopted by VEGA, QSAR TOOLBOX, and BIOVIA to characterize the application domain, which compares the structural features of the chemicals in the training set and those of the chemicals in the test set. When the substructures that are not included in the training set are present in the chemicals to be predicted, they are considered outside the application domain of the model. However, subtle changes in molecular structure may also cause mutations in activity [53], making the first four application domain characterization methods unable to guarantee the reliability of prediction. VEGA and T.E.S.T. can identify subtle changes in molecular structure that cause mutations in activity by predicting analogs of the target chemical.

Perspectives

The weaknesses of the toxicity prediction software package include the following:

(1) Insufficient information completeness
The insufficient completeness of the information provided by the toxicity prediction software package will likely limit its practical application. On the one hand, according to the guidance document issued by the OECD on validating QSAR models [18], QSAR models should provide clearly defined prediction endpoints, explicit algorithms, model application domains, matrix for model performance, and, to the extent possible, mechanistic

explanations of the model. In addition, the quality and quantity of the raw datasets have a large impact on the model performance, so whether the raw datasets are provided or not is also considered. BIOVIA does not provide endpoint type information for eye irritation/corrosion. Some of the software packages including VEGA, QSAR TOOLBOX, Lazar, Danish (Q)SAR Models, and BIOVIA lacked the raw datasets of QSAR models for some of the endpoints, and the mechanistic explanations of the models were not provided by any of the software packages. On the other hand, OPERA, Lazar, and BIOVIA do not provide development background information, user manuals, interface operation tips, etc., which results in a lack of friendliness to nonexpert users.

(2) Uneven quality of the raw dataset

The availability of reliable and specific datasets has become a critical point limiting the development of QSAR toxicity prediction modeling. The raw datasets usually come from many different sources such as public databases, reports, and literature, resulting in the uneven quality of raw data. Specifically, although large databases provide data on a large number of chemicals, data quality often lack effective assessment and management. In addition, data collected in the literature usually are obtained using different experimental methods and judging criteria, thus making it difficult to compare with each other. QSAR toxicity prediction modeling requires a credible raw dataset. The low-quality raw data may lead failure in QSAR modeling. In the field of computational toxicology, the lack of high-quality public datasets has become a bottleneck problem.

(3) Poor mechanistic interpretability

QSAR models are often considered "black boxes," and it is extremely challenging to interpret the QSAR model results. With the continuous development of QSAR technology, QSAR has moved from simple and interpretable linear methods using a few descriptors to complex multiparameter and nonlinear methods. However, with the increasing complexity of QSAR modeling methods, the interpretability of their models has deteriorated. For example, ANN models have better predictive power but do not provide more information on the effect of descriptors on activity. The interpretation of QSAR models can help to understand the complexity of biological or physicochemical processes, guide structure optimization, or perform knowledge-based validation of QSAR models. However, the lack of transparency and interpretability of QSAR models has limited their use in regulatory science.

For the future development of QSAR models in toxicity prediction software packages, the following three aspects should be enhanced:

(1) Enhancing the information completeness of the model

The information completeness of the QSAR model will affect the understanding and application of the model by users to a great extent. Therefore, in the future, each software package should strengthen the collection and integration of basic information of QSAR models, which at least contains five aspects of the information according to the guidance document issued by the OECD on validating QSAR models, including clear prediction endpoints, clear algorithms, a clear scope of use, appropriate measures of fitness, robustness and predictability, and explanation of the mechanism. In addition, for the existing QSAR models in this toxicity prediction software package, the aforementioned information should be supplemented and provided in a standardized way. Additionally, these toxicity prediction software packages should try to provide their development background information, user manuals, interface operation tips, etc. to increase the user-friendliness of the software packages.

(2) Strengthening the evaluation and integration of raw datasets

High-quality toxicity data are essential to enhance the performance of QSAR models. In future studies, the evaluation and integration of toxicity databases and datasets should be strengthened. Methods such as the weight-of-evidence method can be used to score the data quality of the collected datasets in terms of test methods, endpoint characterization methods, exposure time, and experimental species, thus removing low-quality experimental data and obtaining large-scale, high-quality raw datasets.

(3) Increasing the mechanistic interpretability of the model

The mechanistic interpretability of QSAR is one of the desirable attributes to promote its acceptance at the regulatory level according to the five OECD principles mentioned before. The mechanistic interpretability of QSAR is greatly influenced by the choice of descriptors, so the selection of appropriate descriptors is a key step to increasing the mechanistic interpretability of QSAR. Currently, the field of QSAR toxicity prediction should enhance the development of descriptor selection tools so that descriptor sets with high interpretability, low redundancy, and high statistical performance can be selected. In addition, the novel QSAR modeling approach based on the mechanism of toxic action not only increases the mechanistic interpretability of the models but also improves the prediction accuracy [12]. The development of this approach should be enhanced in the future.

Acknowledgments

This work was supported by the National Natural Science Foundation of China (21922603, 22276094, and 42207329), the Jiangsu Province Key Research and Development Program (BE2022837), the Natural Science Foundation of Jiangsu Province for Young Scholars (BK20210190), and the Environmental Protection Research Project of Jiangsu Province (2021003 and 2018001).

References

[1] Drakvik E, Altenburger R, Aoki Y, Backhaus T, Bahadori T, Barouki R, et al. Statement on advancing the assessment of chemical mixtures and their risks for human health and the environment. Environ Int 2020;134:105267.
[2] Pelletier M, Bonvallot N, Glorennec P. Aggregating exposures & cumulating risk for semivolatile organic compounds: a review. Environ Res 2017;158:649–59.
[3] Verdonck FAM, Van Sprang PA, Vanrolleghem PA. Uncertainty and precaution in European environmental risk assessment of chemicals. Water Sci Technol 2015;52(6):227–34.
[4] Embry MR, Bachman AN, Bell DR, Boobis AR, Cohen SM, Dellarco M, et al. Risk assessment in the 21st century: roadmap and matrix. Crit Rev Toxicol 2014;44(3):6–16.
[5] Birnbaum LS, Burke TA, Jones JJ. Informing 21st-century risk assessments with 21st-century science. Environ Health Perspect 2016;124(4): A60–3.
[6] Toropov AA, Toropova AP, Raska I, Leszczynska D, Leszczynski J. Comprehension of drug toxicity: software and databases. Comput Biol Med 2014;45:20–5.
[7] Andersen ME, Krewski D. The vision of toxicity testing in the 21st century: moving from discussion to action. Toxicol Sci 2010;117(1):17–24.
[8] Rusyn I, Daston GP. Computational toxicology: realizing the promise of the toxicity testing in the 21st century. Environ Health Perspect 2010; 118(8):1047–50.
[9] Fjodorova N, Novich M, Vrachko M, Smirnov V, Kharchevnikova N, Zholdakova Z, et al. Directions in QSAR modeling for regulatory uses in OECD member countries, EU and in Russia. J Environ Sci Health C 2008;26(2):201–36.
[10] Piir G, Kahn I, Garc ía SAT, Sild S, Ahte P, Maran U. Best practices for QSAR model reporting: physical and chemical properties, ecotoxicity, environmental fate, human health, and toxicokinetics endpoints. Environ Health Perspect 2018;126(12):126001.
[11] Hughes K, Paterson J, Meek ME. Tools for the prioritization of substances on the domestic substances list in Canada on the basis of hazard. Regul Toxicol Pharmacol 2009;55(3):382–93.
[12] Rouse R, Kruhlak N, Weaver J, Burkhart K, Patel V, Strauss DG. Translating new science into the drug review process: the US FDA's division of applied regulatory science. Ther Innov Regul Sci 2018;52(2):244–55.
[13] Seidle T, Stephens ML. Bringing toxicology into the 21st century: a global call to action. Toxicol Vitro 2009;23(8):1576–9.
[14] Rim KT. In silico prediction of toxicity and its applications for chemicals at work. Toxicol Environ Health Sci 2020;12(3):191–202.
[15] Bodar C, de Bruijn J, Vermeire T, van der Zandt P. Trends in risk assessment of chemicals in the European union. Hum Ecol Risk Assess 2002; 8(7):1825–43.
[16] Kovarich S, Ceriani L, Fuart Gatnik M, Bassan A, Pavan M. Filling data gaps by read-across: a mini review on its application, developments and challenges. Mol Inform 2019;38(8–9):1800121.
[17] Cronin MTD, Jaworska JS, Walker JD, Comber MHI, Watts CD, Worth AP. Use of QSARs in international decision-making frameworks to predict health effects of chemical substances. Environ Health Perspect 2003;111(10):1391–401.
[18] OECD. Guidance document on the validation of (quantitative) structure-activity relationship [(Q)SAR] models. Paris: Organisation for Economic Co-operation and Development; 2014.
[19] Sushko I, Novotarskyi S, Körner R, Pandey AK, Rupp M, Teetz W, et al. Online chemical modeling environment (OCHEM): web platform for data storage, model development and publishing of chemical information. J Comput Aided Mol Des 2011;25(6):533–54.
[20] Maunz A, Gütlein M, Rautenberg M, Vorgrimmler D, Gebele D, Helma C. Lazar: a modular predictive toxicology framework. Front Pharmacol 2013;4:38.
[21] Locke PA, Myers DBJ. A replacement-first approach to toxicity testing is necessary to successfully reauthorize TSCA. ALTEX-Altern Anim Exp 2011;28(4):266–72.
[22] Kavlock R, Dix D. Computational toxicology as implemented by the U.S. EPA: providing high throughput decision support tools for screening and assessing chemical exposure, hazard and risk. J Toxicol Environ Health B 2010;13(2–4):197–217.
[23] Krewski D, Acosta D, Andersen M, Anderson H, Bailar JC, Boekelheide K, et al. Toxicity testing in the 21st century: a vision and A Strategy. J Toxicol Environ Health B 2010;13(2–4):51–138.
[24] Martin MT, Knudsen TB, Reif DM, Houck KA, Judson RS, Kavlock RJ, et al. Predictive model of rat reproductive toxicity from ToxCast high throughput screening1. Biol Reprod 2011;85(2):327–39.
[25] Dimitrov SD, Diderich R, Sobanski T, Pavlov TS, Chankov GV, Chapkanov AS, et al. QSAR toolbox—workflow and major functionalities. SAR QSAR Environ Res 2016;27(3):203–19.
[26] Mansouri K, Grulke CM, Judson RS, Williams AJ. OPERA models for predicting physicochemical properties and environmental fate endpoints. J Cheminform 2018;10(1):10.
[27] O'Halloran K. Toxicological considerations of contaminants in the terrestrial environment for ecological risk assessment. Hum Ecol Risk Assess 2016;12(1):74–83.
[28] Hansen K, Mika S, Schroeter T, Sutter A, ter Laak A, Steger-Hartmann T, et al. Benchmark data set for in silico prediction of Ames mutagenicity. J Chem Inf Model 2009;49(9):2077–81.
[29] Graham C, Rosenkranz HS, Karol MH. Structure–activity model of chemicals that cause human respiratory sensitization. Regul Toxicol Pharmacol 1997;26(3):296–306.
[30] Olker JH, Elonen CM, Pilli A, Anderson A, Kinziger B, Erickson S, et al. The ECOTOXicology knowledgebase: a curated database of ecologically relevant toxicity tests to support environmental research and risk assessment. Environ Toxicol Chem 2022;41(6):1520–39.

[31] Wu S, Fisher J, Naciff J, Laufersweiler M, Lester C, Daston G, et al. Framework for identifying chemicals with structural features associated with the potential to act as developmental or reproductive toxicants. Chem Res Toxicol 2013;26(12):1840–61.
[32] Holzhütter HG, Genschow E, Diener W, Schlede E. Dermal and inhalation acute toxic class methods: test procedures and biometric evaluations for the globally harmonized classification system. Arch Toxicol 2003;77(5):243–54.
[33] Terry C, Yan Z, Corvaro M, Gehen SC. A retrospective study on EU harmonised classifications for carcinogenicity to guide future research. Regul Toxicol Pharm 2021;119:104800.
[34] Marone PA, Hall WC, Hayes AW. Reassessing the two-year rodent carcinogenicity bioassay: a review of the applicability to human risk and current perspectives. Regul Toxicol Pharm 2014;68(1):108–18.
[35] Benfenati E, Benigni R, Dmarini DM, Helma C, Kirkland D, Martin TM, et al. Predictive models for carcinogenicity and mutagenicity: frameworks, state-of-the-art, and perspectives. J Environ Sci Health C 2009;27(2):57–90.
[36] Sundar R, Jain MR, Valani D. Mutagenicity testing: regulatory guidelines and current needs. In: Kumar A, Dobrovolsky VN, Dhawan A, Shanker R, editors. Mutagenicity: assays and applications. New York: Academic Press; 2018. p. 191–228.
[37] Ames BN. Identifying environmental chemicals causing mutations and cancer. Science 1979;204(4393):587–93.
[38] Desprez B, Birk B, Blaauboer B, Boobis A, Carmichael P, Cronin MTD, et al. A mode-of-action ontology model for safety evaluation of chemicals: outcome of a series of workshops on repeated dose toxicity. Toxicol Vitro 2019;59:44–50.
[39] Pizzo F, Benfenati E. In silico models for repeated-dose toxicity (RDT): prediction of the no observed adverse effect level (NOAEL) and lowest observed adverse effect level (LOAEL) for drugs. Methods Mol Biol 2016;1425:163–76.
[40] Rodford R, Patlewicz G, Walker JD, Payne MP. Quantitative structure-activity relationships for predicting skin and respiratory sensitization. Environ Toxicol Chem 2003;22(8):1855–61.
[41] Ishii S, Ishii K, Nakadate M, Yamasaki K. Correlation study in skin and eye irritation between rabbits and humans based on published literature. Food Chem Toxicol 2013;55:596–601.
[42] Mancebo A, Hernández O, González Y, Aldana L, Carballo O. Assessment of skin and eye irritation of 14 products under the stepwise approach of the OECD. Cutan Ocul Toxicol 2008;27(3):173–85.
[43] Fourches D, Muratov E, Tropsha A. Trust, but verify: on the importance of chemical structure curation in cheminformatics and QSAR modeling research. J Chem Inf Model 2010;50(7):1189–204.
[44] Fitzpatrick RB. CPDB: carcinogenic potency database. Med Ref Serv Q 2008;27(3):303–11.
[45] Kazius J, McGuire R, Bursi R. Derivation and validation of toxicophores for mutagenicity prediction. J Med Chem 2005;48(1):312–20.
[46] Maggiora GM. On outliers and activity cliffs why QSAR often disappoints. J Chem Inf Model 2006;46(4). 1535–1535.
[47] Rogers D, Hahn M. Extended-connectivity fingerprints. J Chem Inf Model 2010;50(5):742–54.
[48] Cereto-Massagué A, Ojeda MJ, Valls C, Mulero M, Garcia-Vallvé S, Pujadas G. Molecular fingerprint similarity search in virtual screening. Methods 2015;71:58–63.
[49] Didziapetris R, Dapkunas J, Sazonovas A, Japertas P. Trainable structure–activity relationship model for virtual screening of CYP3A4 inhibition. J Comput Aided Mol Des 2010;24(11):891–906.
[50] Sazonovas A, Japertas P, Didziapetris R. Estimation of reliability of predictions and model applicability domain evaluation in the analysis of acute toxicity (LD 50). SAR QSAR Environ Res 2010;21(1–2):127–48.
[51] Zhu H, Martin TM, Ye L, Sedykh A, Young DM, Tropsha A. Quantitative structure–activity relationship modeling of rat acute toxicity by oral exposure. Chem Res Toxicol 2009;22(12):1913–21.
[52] Mansouri K, Karmaus AL, Fitzpatrick J, Patlewicz G, Pradeep P, Alberga D, et al. CATMoS: collaborative acute toxicity modeling suite. Environ Health Perspect 2021;129(4):047013.
[53] Lo YC, Rensi SE, Torng W, Altman RB. Machine learning in chemoinformatics and drug discovery. Drug Discov Today 2018;23(8):1538–46.

C H A P T E R

16

Fast and efficient implementation of computational toxicology solutions using the *FlexFilters* platform

Suman K. Chakravarti and Sai Radha Mani Alla

MultiCASE Inc., Mayfield Heights, OH, United States

Abbreviations

ADME Absorption, distribution, metabolism, and excretion
AI Acceptable intake
BCF Bioconcentration factor
BP Boiling point
Caco Colorectal adenocarcinoma
CPD Conditional probability distribution
CPDB Carcinogenicity potency database
CV Cross-validation
DA Domain applicability
GPMT Guinea pig maximization test
HBA Hydrogen bond acceptor
HBD Hydrogen bond donor
HL Henry's law constant
HOMO Highest occupied molecular orbital
LC50 Lethal concentration to cause the death of 50% of test animals
LLNA Local lymph node assay
LMO Leave many out
LogP Octanol–water partition coefficient
LUMO Lowest unoccupied molecular orbital
MP Melting point
NNA N-nitrosamine
PAMPA Parallel artificial membrane permeability assay
QSAR Quantitative structure–activity relationship
QSMR Quantitative structure–metabolite relationship
QSPR Quantitative structure–property relationship
SMARTS SMILES arbitrary target specification
SMILES Simplified molecular-input line-entry system
TD50 Median toxic dose (dose at which toxicity occurs in 50% of cases)
VP Vapor pressure
WS Water solubility

Introduction

Computational toxicology or "in silico toxicology" is a rapidly developing discipline that is based on the integration of various resources to create methods for assessing adverse effects (health and environmental) of chemicals [1]. Novel computational approaches may potentially help us to move away from in vivo and in vitro studies, ultimately leading to decrease in animal usage, suffering, cost, time, and more robust assessments [2,3]. In silico approaches often include multiple computational resources working together, e.g., in silico DNA reactivity assessment of drug impurities usually requires a combination of a statistical QSAR model and a rule-based system to predict Ames mutagenicity and to identify structural alerts, a reference database of compounds for surrogate search and a database of assay conditions and scientific bibliography [4–6]. To remain effective, such systems should permit frequent updates to keep up with the evolving needs and advancements in the scientific understanding. For example, the in silico systems for mutagenicity assessments are continuously improved by the incorporation of new experimental Ames data and expert knowledge-based structural alerts [4]. Also, at the initial stages of development, rapid iteration and experimentations are vital for testing various hypothesis and for building robust systems. Therefore, it is important to rethink how these systems are implemented computationally. The speed of implementation, modifiability, modularity, flexibility, reusability, and scalability of the in silico systems are crucial.

https://doi.org/10.1016/B978-0-443-15339-6.00055-2

Although the overall computational architecture of a specific in silico system depends heavily on the specific toxicity endpoint, various cheminformatic functions are commonly needed. These modules can broadly be placed in two tiers; the first-tier modules are essentially the fundamental building blocks, and the second-tier modules are built from the first-tier modules:

1. First-tier cheminformatic modules
   - Primary molecular descriptor calculators, e.g., molecular weight, HOMO, LUMO, surface area, molecular volume, partial charge, E-state, topological descriptors, etc.
   - Molecular fragment generators.
   - Substructure search engines.
   - Databases of experimental assay data.
   - Molecular similarity calculators.
   - Structure curators.
   - Unit converters.
   - Miscellaneous.
2. Second-tier cheminformatic modules
   - Higher-level descriptor calculators, e.g., octanol–water partition coefficient, water solubility, vapor pressure, etc.
   - Molecular fingerprint generators.
   - Nearest neighbor search engines.
   - Multiparameter QSAR models.
   - Expert rule-based modules.
   - Metabolite generators.
   - Higher level predictive networks, e.g., Bayesian networks.
   - Domain applicability evaluators.
   - Miscellaneous.

The ease of use and effectiveness of these modules may depend on how they are implemented as software components. Traditionally, an in silico system consists of a parent software program and a collection of models and databases. The parent software primarily serves the purpose of storing and executing the logic for performing computations. Such systems are rigid in nature, and modifications in the parent software are usually slow and expensive. In addition, optimization of various associated hyperparameters is difficult. For example, molecular fragments, a commonly used descriptor, can be linear and atom-centered and have many possible atom type definitions. Similarly, molecular fingerprints and similarity measures can be of many different types. Therefore, it is next to impossible to decide all combinations of these variables beforehand to get the best models. Modifiability and modularity are severely compromised if such conditions are hard-coded in the parent program.

Reusability of the modules is another very desirable factor, however, often overlooked. It is the ability to use an existing resource in different problems without modification. For example, some toxicity structural alerts can be used in very different toxicity endpoints, e.g., alerts representing Schiff base formation, alkylation, and Michael addition are common between fish toxicity and skin sensitization. At a different level, one toxicity module can be used as a descriptor in another module, e.g., mutagenicity as a descriptor in a carcinogenicity assessment module. Even operations such as data curation and domain applicability check can be considered as reusable components. Independence of the modules and the ability to control their behavior without changing their core is essential for reusability.

The "filter" concept

In this chapter, we describe an in silico toxicology platform [7], in which important portions of computational logic are embedded in the modules instead of the parent software. The *FlexFilters* methodology and modules are the main components of the commercial software *QSAR Flex* [8]. The modules, also referred to as *filters*, are stored as scripts in editable text files and have qualities like a computer programming language (Fig. 16.1). The scripts are interpreted and executed by the parent software. The modules are like *functions* in computer programming, i.e., execute a sequence of instructions to perform a task and produce outputs. They are reusable and can be invoked using a one-line text script from any place. The behavior of the modules can be controlled via parameters, supplied in the call, much like function arguments in computer programming. The *fathead minnow* acute toxicity module,

```
FILTER_TYPE ORDINARY_REGR
VER 1
FILTER  666 CURATION_FILTER VER 1   BASIC   NEUTRALIZE_NEG_CHARGES  NEUTRALIZE_NITROGEN_POS_CHARGES KEEP_LARGEST_COMP
FILTER  1   CONSTANT     VER 1   REGR_PARAMS 3   2.921   0   1
FILTER  2   MODEL_FILE   VER 1   logp.filter REGR_PARAMS 3   0.057   0   1
FILTER  3   MODEL_FILE   VER 1   Water Solubility.filter REGR_PARAMS 3   -0.373  0   1
FILTER  4   IN_BUILT_DESCR  VER 1   MOL_WT  REGR_PARAMS 3   0.157   182.734 89.404
FILTER  5   ATOM_CENTERED_FRAG  VER 1   0   0 0 0 0 1 0 COUNT_1 [C2H]   REGR_PARAMS 3   0.166   0   1
FILTER  6   ATOM_CENTERED_FRAG  VER 1   1   0 0 0 0 1 0 COUNT_1 [c]-[Cl3]   REGR_PARAMS 3   0.031   0   1
FILTER  7   ATOM_CENTERED_FRAG  VER 1   1   0 0 0 0 1 0 COUNT_1 [C2H]=[O2]  REGR_PARAMS 3   0.110   0   1
FILTER  8   ATOM_CENTERED_FRAG  VER 1   1   0 0 0 0 1 0 COUNT_1 [c]-[C2H]=[O2]  REGR_PARAMS 3   0.008   0   1
FILTER  9   ATOM_CENTERED_FRAG  VER 1   0   0 0 0 0 1 0 COUNT_1 [C3]    REGR_PARAMS 3   -0.016  3.608   3.795
FILTER  10  ATOM_CENTERED_FRAG  VER 1   0   0 0 0 0 1 0 COUNT_1 [N2_2]  REGR_PARAMS 3   0.053   0   1
FILTER  11  ATOM_CENTERED_FRAG  VER 1   0   0 0 0 0 1 0 COUNT_1 [N3]    REGR_PARAMS 3   -0.097  0   1
FILTER  12  ATOM_CENTERED_FRAG  VER 1   1   0 0 0 0 1 0 COUNT_1 [C3]-[C3]   REGR_PARAMS 3   -0.019  0   1
FILTER  13  ATOM_CENTERED_FRAG  VER 1   2   0 0 0 0 1 0 COUNT_1 [c]:[c]:[c]:[c](:[c])-[C2]  REGR_PARAMS 3   -0.043  0   1
FILTER  14  ATOM_CENTERED_FRAG  VER 1   3   0 0 0 0 1 0 COUNT_1 [C2]-[O3]-[C3]-[C3]-[C3]-[C3]-[C3]  REGR_PARAMS 3   -0.180  0   1
FILTER  15  ATOM_CENTERED_FRAG  VER 1   1   0 0 0 0 1 0 COUNT_1 [C2]-[Cl3]  REGR_PARAMS 3   0.044   0   1
FILTER  16  ATOM_CENTERED_FRAG  VER 1   1   0 0 0 0 1 0 COUNT_1 [C2]=[C2](-[C3])-[Cl3]  REGR_PARAMS 3   0.535   0   1
.
.
.
FILTER  155 ATOM_CENTERED_FRAG  VER 1   5   0 0 0 0 1 0 COUNT_1 [C3]-[C3]-[C3]-[C3]-[C3]-[C3]-[C3]-[C3]-[C3]-[O3H]  REGR_PARAMS 3   0.096   0   1
FILTER  156 ATOM_CENTERED_FRAG  VER 1   1   0 0 0 0 1 0 COUNT_1 [C3]-[N3+]  REGR_PARAMS 3   -0.109  0   1
FILTER  157 ATOM_CENTERED_FRAG  VER 1   1   0 0 0 0 1 0 COUNT_1 [C3]-[N3+](-[C3])(-[C3])-[C3]  REGR_PARAMS 3   -0.389  0   1
```

FIGURE 16.1 Contents of the file of a filter module for computing acute toxicity in fathead minnow. Each line represents a subfilter operating on the query molecule, e.g., filter #666 is for structure curation, filter #1 is the regression constant, filters #2 and #3 compute LogP and water solubility, respectively, filter #4 is to compute molecular weight, filter #5—157 is to identify and count various molecular fragments in the input molecule.

shown in Fig. 16.1, takes an input molecule and gives fish toxicity (−Log of LC50 in mol/L) as the output. It has the following attributes:

- The module is stored as a text file named as *fathead_minnow.filter.*
- This module is based on ordinary regression modeling method, specified by the `FILTER_TYPE` keyword as the first line.
- First submodule (ID #666) is a curation filter, which performs three curation steps on the input molecule, i.e., neutralizes negative charges on anions, neutralizes positive charges on some cations, and only retains the largest component in a salt or a mixture.
- Filter #1 is the regression coefficient, returns a value of 1, which is multiplied by the regression coefficient.
- Filters #2 and #3 are calls to modules for calculating LogP and water solubility, stored separately as *logp.filter* and *water solubility.filter* files.
- Filter #4 computes molecular weight of the input molecule.
- Filters #5—157 identify and count various toxicity-related molecular fragments in the input molecule. These fragment filters are *in-place* functions, i.e., they do not have an associated file and directly perform the desired computations.
- The keyword `REGR_PARAMS` followed by a series of arguments are regression coefficients, centering and scaling parameters for individual filters, which operate on and modify the output value from the filter.
- This *fathead minnow* acute toxicity module itself can be invoked from other modules via a one-line call, e.g., `FILTER xx MODEL_FILE VER 1 fathead_minnow.filter.`

The syntax strings for the filters can be generated programmatically. Most importantly, the scripts can be moved around and combined with other filters for performing more complex and sophisticated functions. We borrowed the term "filter" from convolutional neural networks [9], in which the so-called "filters" are applied to obtain successively more sophisticated abstractions from the input data. The programming language like nature of the *FlexFilters* lends themselves for easy construction, modification, and recombination to create highly flexible and modular systems.

Syntax for the filter calls

The function of majority of the filters can be invoked using a one-line call from anywhere. The calls have specific syntax, usually starting with the keyword `FILTER`, followed by an integer ID number, a keyword for the type of the filter, a version number after the keyword `VER`, and a series of parameters that control the behavior of the filter. For example, the following is a call to identify and count the aliphatic amine molecular fragments in the query molecule:

```
FILTER 7 ATOM_CENTERED_FRAG VER 1 1 000010 COUNT_1 [C3]-[N3H2]
```

Optionally, call syntax may also contain parameters for regression coefficients, Bayesian conditional probabilities, and scaling and centering parameters, identified by the keywords `REGR_PARAMS` or `NAIVE_BAYES_PARAMS`. It is important to note that the call will still be functional even without the additional parameters. The task of the parameters is to only modify the output of the filter. For example, the following call computes the total polar surface area, and the extra parameters are highlighted in bold:

```
FILTER 9 IN_BUILT_DESCR VER 1 TPSA REGR_PARAMS 3 -0.580 28.854 30.461
```

Frequently used filters in *FlexFilters* platform

We have created a variety of filters to carry out commonly used functions, which are listed in Table 16.1. However, some of them need special mention here:

1. **Molecular fragment detectors (*FragFilter*):** Molecular fragments are very commonly used as descriptors in QSAR/QSPR models, as they are a good representation of molecular substructures, transparent, and interpretable. However, we found it very challenging to simultaneously achieve effective encoding of structural environment, human readability, and unique representation for fragments. Therefore, molecular fragments are encoded in *FlexFilters* with varying levels of atom level details and a variant of Weininger's canonical representation algorithm [10]. Majority of these representations are human readable. The representation is controlled using six binary flags, which are supplied as parameters to the fragment filters:
   1. *Count_H_Flag:* When true, count of hydrogens on all atoms is explicitly shown, else, hydrogen counts are shown only when the atom is not an aliphatic saturated carbon or hetero aromatic, e.g., [C2]([O2])([C3H2][C3H2])[O3H], where [C2] and [O2] represent sp^2 carbon and oxygen, respectively. [O3H] represents sp^3 oxygen with one hydrogen, and [C3H2] is an sp^3 carbon with two attached hydrogens. Branching is represented by parentheses.
   2. *StereoFlag:* When true, local stereo information is shown in the fragments, i.e., *R*, *S*-configuration and cis/trans double-bond configurations, e.g., [C3_r]([C3_s])([C3][O3][P3_1]([O2])([O3H])[O3H])[O3], where [C3_r] and [C3_s] are sp^3 carbons with *R* and *S* configurations, respectively, [P3_1] is a tetrahedral phosphorus with a double bond attached to it.
   3. *IsostericFlag:* When true, certain atoms are replaced with their isosteric representations, e.g., Cl is replaced with [F, Cl, Br, I] and −O− is replaced with [O, S, NH], e.g., [C2]([O,S,NH])([C3][C3][C3][C3][C3])[O,S,NH].
   4. *InternalReprsnFlag:* The fragment representation is not human-readable when this flag is set to true; however, much more information about the structural environment of the atoms is encoded. The atoms are represented by a series of binary flags, e.g., [01010000000000800800000][03010000000000600600000], which is a two-atom carbonyl group fragment (i.e., >C=O).
   5. *ShowBondsFlag:* Using this flag, the fragments show the bonds between the atoms, e.g., [C2](=[O2])(−[C3])−[O3].
   6. *WholeMolFlag:* When true, atoms are shown only using their symbols and the fragments essentially become SMILES notation. This type of representation preserves minimum atomic structural environment, e.g., OC(Nc(cc)cc)C(N)C.
2. **Predictive models (*ModelFilter*):** These filters represent various types of models for predicting physicochemical property or biological activity of the input molecule. Following model types are implemented so far:
   1. Read-across filters (*ReadAcrossFilter*): These filters find a specified number of similar chemicals and estimate the property of the input molecule using the properties of these closest neighbors. These filters are a combination of a molecular fingerprint database filter (*FP_DB_Filter*) and a molecular property database filter (*MolPropDBFilter*). A special type of read-across filter is the alert environment similarity filter (*AlertEnvSimFilter*), which finds nearest neighbors containing a specific structural alert, and the similarity is measured using the structural environment in the vicinity of the alert.
   2. Regression models: These filters include classification and regression-based models built using logistic and ordinary multivariable regression methods, respectively. They are composed of a list of other filters as independent variables for the model, e.g., *ConstantFilter, InBuiltDescrFilters, FragFilters, SMARTSFilters*, or even other *ModelFilters*. The subfilters are appended with coefficients/weights and scaling and centering factors.

TABLE 16.1 Details of some of the implemented filters in *FlexFilters* platform.

Filter type, name	Function and use	Call parameters	Output
Constant (*ConstantFilter*)	Serves as the constant term in regression and naïve Bayes models.	None	Always gives an output of 1.
Molecular fragment detectors *(FragFilter)*	Detect a specific molecular fragment in the input molecule. Used in QSAR/QSPR as descriptors.	– Type of the fragment, e.g., linear or atom centered. – 6 binary flags to specify atomic level information and fragment representation. – The hit counting method. – A unique representation of the fragment.	Count of the fragment, and atoms of the input molecule that are part of the fragment.
Primary molecular descriptor calculators *(InBuiltDescrFilter)*	Compute various first-tier molecular descriptors for the input molecule, e.g., molecular weight, surface area, volume, E-state, charge descriptors, etc.	Keyword for the name of the descriptor.	Computed value of the descriptor.
SMARTS search *(SMARTSFilter)*	Searches a SMARTS substructure pattern in the input molecule. Used in QSAR/QSPR models to encode functional groups or as structural alerts in expert rule-based systems	– The SMARTS pattern. – The hit counting method.	Count of the substructure and atoms of the input molecule that are part of the substructure.
Structure curators *(CurationFilter)*	Perform a series of specified curation steps on the input molecule, e.g., formal charge neutralizations, detection of incorrect valence on atoms, treatment of salts/mixtures, etc.	List of curation steps.	The curated molecule.
Structural similarity filter *(StructSimFilter)*	Computes similarity between the query chemical and a second chemical (target) stored in the filter.	– SMILES code for the target molecule. – Type of fingerprint and fingerprint parameters. – The fingerprint of the target molecule.	Similarity between the input and the target chemical.
Layered structural similarity filter *(LayeredStructSimFilter)* and alert environment similarity	Compute similarity between the query and the target molecule, however, using only the structural features around a specified	– SMILES code for the target molecule. – Type of fingerprint and fingerprint parameters.	"Local" similarity between the input and the target chemical.

Continued

TABLE 16.1 Details of some of the implemented filters in *FlexFilters* platform.—cont'd

Filter type, name	Function and use	Call parameters	Output
(AlertEnvSimFilter) filter	substructure on the molecules (e.g., around a structural alert).	– The fingerprint of the target molecule. – The atom indices of the substructure on the target atom. – Name of the file that contains different atom types needed to define the substructural environment.	
Molecular property finder *(MolPropDBFilter)*	Searches for bioassay outcomes or experimental properties of the input molecule in a database. Used extensively with many modules, including nearest neighbor searches.	Name of the property database file.	Returns the property of the query chemical, if found.
Molecular fingerprint finder *(FP_DB_Filter)*	Searches for similar chemicals using a database of molecular fingerprints (*StructSimFilter* filters). Mainly used in nearest neighbor searches.	Name of the fingerprint database file.	Returns a list of most similar chemicals.
Predictive models *(ModelFilter)*	Represent various types of predictive models, e.g., regression, naïve Bayes, random forest, etc. They predict physicochemical property or biological activity of the input molecule.	Name of the model filter file.	Return calculated property/activity.
Metabolite generator *(MetabolismFilter)*	Predicts and searches human metabolites for the input molecule.	– A threshold offset to screen metabolites. – Name of folder containing metabolite database and QSMR (quantitative structure −metabolism relationship) models.	Returns a list of Phase I and Phase II metabolites.
Applicability domain checker *(DomainCheckFilter)*	Defines a chemical space for a collection of chemicals. This in turn helps to define the domain of applicability (*DA*) of models.	– Method for checking *DA*. – Name of various supporting files.	Status of the input molecule if it belongs to the applicability domain or not.

TABLE 16.1 Details of some of the implemented filters in *FlexFilters* platform.—cont'd

Filter type, name	Function and use	Call parameters	Output
Threshold applier (*ThresholdApplier*)	Converts the numerical output value of a model to different class labels, e.g., probability values to active/inactive class.	List of threshold values and corresponding class labels.	Class label.
Concentration to response converters (*Concentration 2Response*)	Converting log concentration-based outcome values to a regular format, e.g., LC50 to mg/kg.	– Logarithmic concentration. – Molecular weight of the input molecule.	Transformed outcome.
Annotation supplier (*AnnotationFilter*)	Finds various information connected to a record of any kind, e.g., details of a biological assay for a compound, mechanistic details for an expert alert, bibliographic information, etc.	– IDs of the record to be searched. – The fields whose detailed need to be retrieved.	A list of values of indicated records from the database.

3. Naïve Bayes models: These filters are primarily used for classification (binary or multiclass) tasks. They are very similar to the regression models; however, instead of regression coefficients, various conditional probabilities are appended to the filters. Also, the constant term is composed of prior probabilities of various classes.
4. Random forest models: Both random forest classifiers and regressors can be created. The filter is composed of an ensemble of decision trees.
5. Bayesian network modules: These are higher-level modules, composed of other model and database filters connected in the configuration of a network. Bayesian network modules contain a **.bif* file that specifies the network structure, i.e., the nodes and the edges between them. The network is essentially a directed, acyclic graph. The nodes contain the conditional probability distribution (CPD) values for connected nodes.

Building *FlexFilters* modules

1. **Module construction:** The methodology for creating *FlexFilters* modules is different for different types of modules. The following is the outline for the regression-based modules:
   1. *Train and test set construction:* The first step in building a predictive model is to assemble a training and a test set of compounds along with their associated property/activity. The training set is for building the module and the test set is for validation.
   2. *Creation of the initial candidate filters:* The one-line call strings for the candidate filters are generated manually or automatically and saved in a text file. These candidate filters are usually selected based on our understanding of the endpoint, for example, molecular fragments, primary molecular descriptors (e.g., molecular weight, HBA, HBD, surface and E-state descriptors etc.), or list of expert knowledge-based alerts as SMARTS filters. Molecular fragments are generated using the training set compounds and could potentially be in thousands, therefore, always generated programmatically. A constant filter is always added. At this stage, none of the filter strings contain any regression coefficients, which will be generated later during the variable selection step. An example of a candidate filter list is shown in Fig. 16.2.
   3. *Generating the X-matrix:* The filters at this stage cannot be used for making predictions; however, they are fully capable of producing output values by operating on input molecules. Therefore, the training set molecules were processed by these filters to produce the "*X-matrix*," which is an $N \times M$ matrix. Each of the N rows of the *X-matrix* is for a training compound, and each of the M columns is for a filter. Each member of this matrix is a computed descriptor value for a molecule. The regression constant (*ConstantFilter*) column contains all "1s."

```
FILTER_TYPE ORDINARY_REGR
VER 1
FILTER  1   CONSTANT    VER 1
FILTER  2   MODEL_FILE  VER 1   logp.filter
FILTER  3   MODEL_FILE  VER 1   Water Solubility.filter
FILTER  4   MODEL_FILE  VER 1   VP.filter
FILTER  5   MODEL_FILE  VER 1   BP.filter
FILTER  6   MODEL_FILE  VER 1   MP.filter
FILTER  7   IN_BUILT_DESCR  VER 1   MOL_WT
FILTER  8   IN_BUILT_DESCR  VER 1   HBA
FILTER  9   IN_BUILT_DESCR  VER 1   HBD
FILTER  10  IN_BUILT_DESCR  VER 1   VDW_SURF
.
.
.
FILTER  28  IN_BUILT_DESCR  VER 1   ATOM_ESTATE_[O3H]
FILTER  29  IN_BUILT_DESCR  VER 1   ATOM_ESTATE_[N3H]
FILTER  30  IN_BUILT_DESCR  VER 1   ATOM_ESTATE_[Cl3]
.
.
.
FILTER  76  ATOM_CENTERED_FRAG  VER 1   0   0 0 0 0 1 0 COUNT_1 [Cl3]
FILTER  77  ATOM_CENTERED_FRAG  VER 1   0   0 0 0 0 1 0 COUNT_1 [c]
FILTER  78  ATOM_CENTERED_FRAG  VER 1   0   0 0 0 0 1 0 COUNT_1 [C3]
FILTER  79  ATOM_CENTERED_FRAG  VER 1   1   0 0 0 0 1 0 COUNT_1 [c]-[Cl3]
FILTER  80  ATOM_CENTERED_FRAG  VER 1   1   0 0 0 0 1 0 COUNT_1 [C3]-[Cl3]
.
.
.
FILTER  452 ATOM_CENTERED_FRAG  VER 1   3   0 0 0 0 1 0 COUNT_1 [Br3]-[c]1:[c]:[c]:[c]:[c]:[c]1-[O3]-[c](:[c]):[c]
FILTER  453 ATOM_CENTERED_FRAG  VER 1   4   0 0 0 0 1 0 COUNT_1 [Br3]-[c]1:[c]:[c]:[c](:[c](:[c]1)-[Br3])-[O3]-[c](:[c]):[c]
FILTER  454 ATOM_CENTERED_FRAG  VER 1   3   0 0 0 0 1 0 COUNT_1 [Br3]-[c]1:[c]:[c]:[c](:[c]:[c]1)-[O3]-[c]
```

FIGURE 16.2 A list of 454 initial candidate filters used in building the bioconcentration factor (*BCF*) module.

Optionally, scaling and centering parameters are computed for the columns and added to the corresponding filters, and the *X-matrix* columns are also scaled and centered.

4. *Variable selection:* The *X-matrix*, along with the values of the property being modeled (the *Y*-column), is subjected to a variable selection process for selecting a subset of a few relevant descriptors. We have used the L1 regularization/Lasso regression variable selection method [11]; however, any other method can also be used. The variable selection method usually also generates the regression coefficients.
5. *The finished module:* The list of selected filters, fully decorated with the regression coefficients, scaling and centering factors, constitutes the finished module, which is saved as a **.filter* text file (Fig. 16.3). This module is now fully capable of making predictions for a query chemical.

```
FILTER_TYPE ORDINARY_REGR
VER 1
FILTER  1   CONSTANT    VER 1   REGR_PARAMS 3   1.943   0   1
FILTER  2   MODEL_FILE  VER 1   logp.filter REGR_PARAMS 3   0.475   3.458   2.184
FILTER  3   MODEL_FILE  VER 1   Water Solubility.filter REGR_PARAMS 3   -0.530  -3.789  2.599
FILTER  12  IN_BUILT_DESCR  VER 1   TPSA    REGR_PARAMS 3   -0.037  34.284  35.068
FILTER  16  IN_BUILT_DESCR  VER 1   VSA_PNEG    REGR_PARAMS 3   -0.099  12.116  13.095
FILTER  18  IN_BUILT_DESCR  VER 1   VSA_POL REGR_PARAMS 3   -0.060  16.795  18.978
FILTER  23  IN_BUILT_DESCR  VER 1   ATOM_ESTATE_[O2]    REGR_PARAMS 3   -0.083  7.091   12.185
FILTER  30  IN_BUILT_DESCR  VER 1   ATOM_ESTATE_[Cl3]   REGR_PARAMS 3   0.053   6.681   11.378
.
.
.
FILTER  88  ATOM_CENTERED_FRAG  VER 1   3   0 0 0 0 1 0 COUNT_1 [c]1:[c]:[c]:[c](:[c]:[c]1)-[Cl3]   REGR_PARAMS 3   0.152   0   1
FILTER  94  ATOM_CENTERED_FRAG  VER 1   1   0 0 0 0 1 0 COUNT_1 [C2]=[O2]   REGR_PARAMS 3   -0.005  0   1
FILTER  96  ATOM_CENTERED_FRAG  VER 1   1   0 0 0 0 1 0 COUNT_1 [c]:[c](:[c])-[C2]  REGR_PARAMS 3   -0.071  0   1
FILTER  98  ATOM_CENTERED_FRAG  VER 1   2   0 0 0 0 1 0 COUNT_1 [c]:[c](:[c])-[C2](=[O2])-[O3H] REGR_PARAMS 3   -0.200  0   1
FILTER  101 ATOM_CENTERED_FRAG  VER 1   1   0 0 0 0 1 0 COUNT_1 [c]-[O3H]   REGR_PARAMS 3   -0.064  0   1
FILTER  109 ATOM_CENTERED_FRAG  VER 1   2   0 0 0 0 1 0 COUNT_1 [c]:[c](:[c]:[c](:[c])-[N2_2])-[N2_2]   REGR_PARAMS 3   -0.031  0   1
.
.
.
FILTER  447 ATOM_CENTERED_FRAG  VER 1   3   0 0 0 0 1 0 COUNT_1 [Br3]-[c]1:[c]:[c]:[c](:[c](:[c]1)-[Br3])-[O3]-[c]  REGR_PARAMS 3   0.052   0   1
FILTER  449 ATOM_CENTERED_FRAG  VER 1   3   0 0 0 0 1 0 COUNT_1 [Br3]-[c]1:[c]:[c]:[c](:[c](:[c]1)-[Br3])-[O3]  REGR_PARAMS 3   0.688   0   1
FILTER  454 ATOM_CENTERED_FRAG  VER 1   3   0 0 0 0 1 0 COUNT_1 [Br3]-[c]1:[c]:[c]:[c](:[c]:[c]1)-[O3]-[c]  REGR_PARAMS 3   0.500   0   1
```

FIGURE 16.3 List of filters in a finished module, in this case a module for estimating bioconcentration factor (*BCF*). Note the IDs of the filters are not continuous due to the removal of some filters by the variable selection step. A total 97 filters are in the finished module, reduced from the 454 candidates. Also note the regression, scaling, and centering parameters at the end of the filter strings.

6. *Building a property database for the module:* Optionally, the training chemicals along with their activity values can be used for constructing a property database filter (a *MolPropDBFilter*) and saved as a **.txtdb* text file. The two filter files, i.e., the main module and the property database, are given the same name, for example, files for the LogP assessment modules are named as *logp.filter* and *logp.txtdb.*

Modules that are based on modeling methods other than regression may have some unique building steps of their own. For example, for the random forest–based modules, the *X*- and *Y*-matrices are subjected to the *RandomForestRegressor* and *RandomForestClassifier* Python methods [12] for random forest construction. The random forest models are then converted to text files with custom format (as **.rff* files) and used in the *FlexFilters* platform. Thus, the finished filter is composed of three files: **.filter, *.txtdb,* and **.rff.*

For the read-across modules, a property database filter (*MolPropDBFilter* as a **.txtdb* text file) and a **.filter* file for fingerprint database (a *FP_DB_Filter*) are created from the training set compounds and stored as two separate filter files.

The final models are usually combined with additional filters, e.g., domain applicability, unit conversion filters, etc.

2. **Module validation:** *FlexFilters* modules are validated using internal cross-validation (*CV*) and/or external validation. Cross-validation is performed by creating a training and test sets by randomly withholding a certain percentage of the available compounds, rebuilding the module using the training compounds and then evaluating the test compounds. This process is repeated several times, each time with a different test/training set combination. The module is completely rebuilt from scratch in each *CV* cycle, i.e., regenerating the initial list of filters and *X*- and *Y*-matrices, full variable selection to get relevant descriptors, and reweighting the descriptors. This is essentially a rigorous leave-many-out (*LMO*) *CV* process. We have usually performed 10%-out-10 times or 5%-out-20 times *CV*s.

External validation is done when all the steps of the module building are completed, i.e., hyperparameter tuning, internal *CV*, and selection of the best model. The external validation is done by testing a set of compounds that were never part of the model building process and was kept external from the start.

3. **Discovering the best version of a module (hyperparameter tuning):** *FlexFilters* modules are generally associated with several hyperparameters, which can be tweaked to produce multiple versions of the module with potentially varying performance. The hyperparameter space can be searched to get the best performing module. For example, fragment representations may affect model performance. As mentioned before, every fragment filter contains six binary flags, i.e., *Count_H_Flag, StereoFlag, IsostericFlag, InternalReprsnFlag, ShowBondsFlag,* and *WholeMolFlag.* State of these flags can be treated as hyperparameters. The modeling methodology itself can be a hyperparameter. For the modules shown in this study, we have mainly tried two methodologies, i.e., multiparameter regression or random forest. However, if the performance is not significantly different, we have sided with the multiparameter regression over random forest due to its simplicity and ease of interpretation. On average, a dozen variation for every module were built and cross-validated to find the best performing one. For the LogP module, the result of the hyperparameter search is shown in Table 16.2.

Applying the modules for prediction

When a module is applied on a query compound, a check is first made to see if the query is already present in the property database of this module (*MolPropDBFilter, *.txtdb* file). If found, the experimental activity/property is returned, otherwise the query is subjected to a domain applicability (*DA*) check (in *DomainCheckFilter*). If the query passes the *DA* test, then it is subjected to the activity/property calculation. The specifics of the calculation depend on the type of the module. For the regression and naïve Bayes modules, the query is subjected to all the subfilters, one by one. The output from each filter is centered/scaled and subjected to the filter regression/naïve Bayes parameters. The final calculated value is a function of all the outputs from the individual filters. The individual outputs of the subfilters also remain available with the overall output. This is useful for interpretation of QSAR evaluations, e.g., identification of alerts and value of other descriptors in the query molecule.

For random forest modules, the query molecule is evaluated with all the trees in the forest (stored in the **.rff* file), and the predicted value is the consensus/majority/average from all the trees. When read-across modules are used, the query molecule is first searched in the *FP_DB_Filter* database to retrieve the specified numbers of most similar neighbors. Subsequently, these neighbors are searched in the *MolPropDBFilter* database to get their target properties.

TABLE 16.2 Effect of various combinations of fragment flags and two modeling methodologies for finding the best module for octanol-water partition coefficient (LOGP).

	Ordinary regression 10% out 5 times CV		Random forest 10% out 5 times CV *1000 trees, min samples leaf* = 1	
Fragment flags	r^2	*RMSE*	r^2	*RMSE*
Count_H_Flag, ShowBondsFlag	0.929	0.489	0.859	0.683
ShowBondsFlag	**0.931**	**0.482**	0.871	0.654
WholeMolFlag	0.925	0.503	0.876	0.641
IsostericFlag, ShowBondsFlag	0.909	0.553	0.857	0.688
IsostericFlag, Count_H_Flag	0.905	0.566	0.844	0.718
InternalReprsnFlag, Count_H_Flag	0.928	0.490	0.863	0.674

The best performance metrics is indicated by the bold letters, i.e., performance from the *ShowBondsFlag* fragment flag combined with ordinary regression.

If a *ThresholdApplier* or a *Concentration2Response* filter is associated with a module, the appropriate unit conversions are performed on the calculated values.

Examples of computational toxicology solutions built using the *FlexFilters* methodology

To demonstrate effectiveness of the *FlexFilters* methodology for creating useful in silico systems, we have built a set of statistical modules, expert rule-based systems and databases for assessing various physicochemical, ADME properties, and toxicological endpoints, commonly evaluated in computational toxicology studies. The details of these modules are shown in Tables 16.3 and 16.4. In this section, we will discuss their development process.

We started from the physicochemical properties, then to ecotoxicity and ADME properties, and finally, complex apical toxicity endpoints such as skin sensitization and ototoxicity. They were built in a way so that modules from earlier steps can be used as descriptors for subsequent modules. For example, the octanol–water partition coefficient (*LOGP*) module, built in the first step, is used as a descriptor in the water solubility (*WS*) module. The *WS* module itself was used in several subsequent modules. All the listed modules also contain a *MolPropDBFilter* for yielding experimental data, if available.

Physicochemical properties

1. *Octanol–water partition coefficient (LOGP)*: LogP is one of the oldest and extensively used QSAR/QSPR descriptors. It correlates with several toxicological endpoints and is an important factor in drug discovery [13,14]. Due to its central role, we have selected it as the first module for development. The data was assembled from the publication of Mansouri et al. [15] and consists of 14,050 compounds.
2. *Water solubility (WS):* WS is also a very important property of molecules and used in environmental chemistry and drug discovery for orally administered drugs to ensure adequate bioavailability [16,17]. The data for this module was assembled from the publication of Mansouri et al. [15] and consists of 4222 compounds.
3. *Vapor pressure (VP):* VP is a useful indicator of chemical volatility and distribution in various phases, i.e., water, air, and soil. Therefore, *VP* is needed in environmental toxicity assessments. The dataset consists of 2032 compounds and assembled from Mansouri et al. [15].
4. *Henry's law constant (HL):* HL, also referred to as the air–water partition coefficient, is an important descriptor for environmental toxicology as it describes the transfer of chemicals between air and water [18]. The data for this module was assembled from Mansouri et al. [15] and consists of 590 compounds.
5. *Boiling point (BP):* BP of a compound indicates its physical state and volatility [19] and is related to various other molecular properties [20]. The data for this module was assembled from Mansouri et al. [15] and consists of 5433 compounds.

TABLE 16.3 List of toxicity modules built using regression methods for endpoints with continuous activity values.

Module	Training set size	Test set size	Modeling method[a]	Cross-validation performance r^2	Cross-validation performance *RMSE*	Test set performance r^2	Test set performance *RMSE*	Descriptors used[b]
LogP	12,645	1405	*OLSR*	0.931	0.482	0.936	0.469	*FRAGS*
Water solubility	3800	422	*RF_REGR*	0.893	0.722	0.895	0.794	*FRAGS, LOGP, MW*
Vapor pressure	1829	203	*OLSR*	0.901	1.099	0.940	0.939	*FRAGS, MW, HBA, HBD, WS, VOL_DESCR, SURF_DESCR, ESTATE*
Henry's law constant	531	59	*OLSR*	0.896	0.700	0.849	0.890	*FRAGS, LOGP, WS, VP*
Boiling point	4890	543	*OLSR*	0.923	23.781	0.915	26.077	*FRAGS, MW, VP*
Melting point	7787	865	*OLSR*	0.744	50.060	0.711	52.827	*FRAGS, LOGP, WS, BP, VP*
Bioconcentration factor	563	62	*OLSR*	0.820	0.595	0.847	0.552	*FRAGS, LOGP, WS, VP*
Fathead minnow toxicity	931	103	*OLSR*	0.662	0.817	0.686	0.941	*FRAGS, LOGP, WS, MW*
Aquatic toxicity expert alerts (Daphnia)	2155	239	*OLSR*	0.521	1.032	0.496	0.964	*LOGP, EXPRT_ALERTS*
Skin permeation	244	27	*OLSR*	0.856	0.377	0.882	0.310	*FRAGS, LOGP, HBA, SURF_DESCR, ESTATE*
PAMPA effective permeability	171	19	*OLSR*	0.649	0.721	0.605	0.624	*FRAGS, LOGP, HBD, SURF_DESCR, ESTATE*
Caco-2 permeability	1167	129	*OLSR*	0.471	0.679	0.490	0.716	*FRAGS*
N-nitrosamine carcinogenicity potency read-across module	135	–	*READ_ACROSS*	–	–	–	–	–

[a]OLSR, *ordinary least square regression;* READ_ACROSS, *surrogate search and read-across;* RF_REGR, *random forest regressor.*
[b]BCF, *bioconcentration factor;* BP, *boiling point;* ESTATE, *E-state descriptors;* EXPRT_ALERTS, *knowledge-based expert alerts;* FRAGS, *atom-centered molecular fragments;* HBA, *count of hydrogen bond acceptors;* HBD, *count of hydrogen bond donors;* LOGP, *octanol–water partition coefficient;* MP, *melting point;* MW, *molecular weight;* SURF_DESCR, *surface descriptors;* VOL_DESCR, *volume descriptors,* VP, *vapor pressure;* WS, *water solubility.*

TABLE 16.4 *FlexFilters* toxicity modules built using classification methods for endpoints with categorical (binary or multiclass) activity labels.

				Cross-validation performance		Test set performance		
Module	**Training set size**	**Test set size**	**Modeling method**[a]	*AUC*	*Accuracy*	*AUC*	*Accuracy*	**Descriptors used**[b]
Microsomal stability	2276	252	*LOGIST_REGR*	0.802	0.756	0.832	0.786	*FRAGS, LOGP, HBD, ESTATE*
Skin sensitization, binary (GPMT)	1115	123	*RF_CLASSF*	–	0.750	–	0.846	*FRAGS*
Skin sensitization expert alerts, binary (GPMT)	740	498	*LOGIST_REGR*	0.753	0.719	0.728	0.783	*EXPRT_ALERTS*
Skin sensitization, 3-class (LLNA)	854	209	*RF_CLASSF*	–	0.762	–	0.713	*FRAGS, LOGP, MW, WS, BP, VP, MP*
Ototoxicity	2376	264	*RF_CLASSF*	–	0.765	–	0.716	*FRAGS, LOGP, MW, WS, BP, VP, MP*

[a]LOGIST_REGR, *logistic regression;* RF_CLASSF, *random forest classifier.*
[b]BP, *boiling point;* ESTATE, *E-state descriptors;* EXPRT_ALERTS, *knowledge-based expert alerts;* FRAGS, *atom-centered molecular fragments;* HBD, *count of hydrogen bond donors;* LOGP, *octanol–water partition coefficient;* MP, *melting point;* MW, *molecular weight;* SURF_DESCR, *surface descriptors;* VP, *vapor pressure;* WS, *water solubility.*

6. *Melting point (MP):* Like boiling point, *MP* is also used to identify a compound. It is used extensively for assessing the environmental fate and transport and key in assessing environmental toxicity due to its relationship with solubility [21,22]. The dataset for this module is from Mansouri et al. [15] and consists of 8652 compounds.

Ecotoxicity

1. *Bioconcentration factor (BCF): BCF* is the ratio of a chemical's concentration in an organism and in its surrounding environment, e.g., in water. *BCF* is an important factor in aquatic toxicity as it provides an estimate of levels of toxic chemicals in an organism based on concentration in the aquatic environment [23–25]. The training data was assembled from Mansouri et al. [15] and consists of 625 compounds.
2. *Acute toxicity in fathead minnow:* Various species of fish are used as models to assess potential risk of chemicals to the aquatic environment. Among them, fathead minnow (*Pimephales promelas*) is the most commonly used for assessing acute toxicity. The dataset for this module was assembled from various publications [26–30] and consists of 1034 compounds. The toxicity values are in the form of 96 h. LC_{50}.
3. *Aquatic toxicity expert alert-based system:* We have also developed a special module to assess aquatic toxicity, using expert knowledge-based structural alerts. A total of 71 alerts were assembled from various published work [31–34], covering toxicity mechanisms including acylation, alkylation, electrophile precursors, acetylcholinesterase inhibitors, nucleophilic substitution, Michael addition, Schiff base formation, etc. *Daphnia magna* acute toxicity (48 h. LC_{50}) dataset of 2399 compounds, collected from various publications [33,35–38], was used for validating these alerts. After variable selection, 45 alerts were found to be predictive and were used for building the model.

ADME modules

1. *Skin permeation:* Assessing penetration of chemicals through human skin is important for the development and safety assessment of pharmaceuticals and cosmetics [39]. The data for this module is from Zeng et al. [40] and consists of 271 compounds.
2. *PAMPA effective permeability:* In vitro parallel artificial membrane permeability assay (PAMPA) is used for assessing passive, transcellular permeation of compounds, mainly for evaluating crossing of barriers such as the GI tract and the blood-brain barrier, potentially useful for screening out poorly orally absorbed drugs at the earlier stages of drug discovery. The data for this module was assembled from Chi et al. [41] and consists of 190 compounds.
3. *Microsomal stability:* The microsomal stability assay is an indicator of hepatic metabolic stability of chemicals and is a key parameter in early identification of compounds that have poor oral bioavailability as drugs, inadequate therapeutic concentrations, short half-life, and other undesirable characteristics [42–44]. The data for this module was assembled from the publication of Siramshetty et al. [44] and PubChem bioassay AID #1508591 [45] and consists of 2528 compounds. The activity labels of this dataset are binary in nature and was treated as a classification problem.
4. *Caco-2 permeability (CACO_2):* The Caco-2 cell monolayer-based in vitro permeability assay is commonly used to model intestinal epithelial barrier for assessing oral absorption of drugs and highly valuable in drug discovery. Both active and passive transport can be studied using this assay [46]. The data for this module was assembled from the publication of Hai et al. [47] and consists of 1296 compounds.

Others

1. *Skin sensitization (binary):* Skin sensitizers have the potential to elicit an immunological response and cause allergic contact dermatitis (ACD) upon repeated exposure, which can significantly diminish quality of life [48–50]. Assessment of skin sensitization potential is an integral part of chemical safety, especially those used in cosmetics. The data for this module was assembled from Alves et al. [51] and consists of 1238 compounds. The activity labels of this dataset are from in vivo Guinea Pig Maximization Test (GPMT).
2. *Skin sensitization expert alert system (binary):* We have also developed a separate module to assess skin sensitization using expert knowledge-based structural alerts. A total of 85 alerts were assembled from various publications [52–56], covering various sensitization mechanisms, e.g., protein alkylation and arylation, pro-SN2, protein acylation/sulfonation, protein nitrosation, Michael addition, Schiff base formation, etc. The same dataset of 1238

compounds tested in GPMT assay was used to validate these alerts. Forty-five alerts were found to be predictive after being subjected to variable selection.

3. *Skin sensitization (3-class):* In addition to the aforementioned two modules, which classify compounds as sensitizers and nonsensitizers, we have developed another module that assigns one of three sensitization classes, i.e., strong to extreme sensitizers, weak to moderate sensitizers, and nonsensitizers. The data for this module was assembled from Wilm et al. [50] and consists of 1063 compounds. The activity labels of this dataset are from local lymph node assay (LLNA).
4. *N-Nitrosamine carcinogenicity potency module:* *N*-Nitrosamines (*NNAs*) are classified as "cohort-of-concern" compounds due to their potent carcinogenic nature. Recently, *NNA* impurities have been detected in several active pharmaceutical ingredients and have been recalled [57,58] for potentially serious health and economic effects. Performing the 2-year rodent carcinogenicity assay for the *NNAs* is not possible due to ethical and practical considerations. However, not all *NNA*s are equally potent, as their carcinogenicity TD_{50} values range over 4 log units [59]. Therefore, there is an urgent need for in silico tools to assess the carcinogenic potency of *NNA* impurities. In this respect, the *FlexFilter* module for assessing carcinogenicity TD_{50} for an untested *NNA* is based on search of suitable surrogate *NNA*s [60]. The surrogates are searched using a local similarity concept [61], which measures similarity between *NNAs* by checking the immediate structural environment around the nitrosamine moiety. The module is based on a combination of layered structural similarity filter (*LayeredStructSimFilter*) and a molecular property database (*MolPropDBFilter*) filter. The retrieved surrogates are usually reviewed by human experts for calculating acceptable intake (AI) limit for the query *NNA*. The surrogate database is mainly assembled from Carcinogenicity Potency Database (CPDB) [62] and consists of 135 *NNAs* with rodent carcinogenicity assay data. The module is also combined with a *MolPropDBFilter* for 463 *NNAs* with Ames mutagenicity test data, the *LogP* and the water solubility modules, and *AnnotationFilters* for supplying experimental data from carcinogenicity and mutagenicity assays. Therefore, each retrieved analog is presented with carcinogenicity TD_{50}, Ames mutagenicity data, *LogP*, *WS* values, and assay data.
5. *Ototoxicity:* Drug-induced ototoxicity is a common adverse drug effect and potentially causes hearing loss or balance disturbance or both [63–65], impacting the quality of life and sometimes increase the risk of falls. The data for this module was assembled from Huang et al. [63] and consists of 2640 compounds with binary activity labels.

Conclusions and future directions

In this chapter, we described a new software platform called *FlexFilters* for implementing in silico modules for computational toxicology problems. The modules are deployed as simple editable text files, resembling programming language scripts. Usually, every line in the script performs a specific computation, e.g., descriptor computation, database search, similarity calculation, data curation, etc. Modularity, flexibility, and scalability are some of the important characteristics of these modules, leading to development of in silico solutions that are flexible, easy to maintain, and adaptable to new needs. We have shown the implementation of several in silico toxicology modules to demonstrate the immediate usefulness of the approach.

Future directions for *FlexFilters* include implementation of more complex toxicological endpoints, e.g., cardiotoxicity, hepatotoxicity; and development of filters for deep neural networks and expansion of the list of the first-tier filters. Another direction is to increase the computation speed of *FlexFilters* assessments by eliminating repetitive computations, as calls to some filters may exist multiple times in various submodules; therefore, repetitive execution of the same module by the same query molecule can occur. This can potentially be prevented by maintaining a dictionary of the history of the calls and their outcomes. Currently, we are working on this problem. Despite some repetitiveness in the computing steps, *FlexFilters* is already very fast and highly usable, proving its usefulness in development and application of computational toxicology solutions.

References

[1] Reisfeld B, Mayeno AN. What is computational toxicology? In: Reisfeld B, Mayeno AN, editors. Computational toxicology. Methods in molecular biology, vol 929. Totowa, NJ: Humana Press; 2012. p. 3–7.

[2] Benigni R. In silico approaches to genetic toxicology: progress and future. Mutagenesis 2019;34(1):1–2. https://doi.org/10.1093/mutage/gey018.

[3] Krewski D, Acosta D, Andersen M, Anderson H, Bailar JC, Boekelheide K, et al. Staff of committee on toxicity test. Toxicity testing in the 21st century: a vision and a strategy. J Toxicol Environ Health Part B 2010;13(2–4):51–138. https://doi.org/10.1080/10937404.2010.483176.

[4] Honma M, Kitazawa A, Cayley A, Williams RV, Barber C, Hanser T, et al. Improvement of quantitative structure–activity relationship (QSAR) tools for predicting Ames mutagenicity: outcomes of the Ames/QSAR International Challenge Project. Mutagenesis 2019;34(1): 3–16. https://doi.org/10.1093/mutage/gey031.

[5] Jayasekara PS, Skanchy SK, Kim MT, Kumaran G, Mugabe BE, Woodard LE, et al. Assessing the impact of expert knowledge on ICH M7 (Q) SAR predictions. Is expert review still needed? Regul Toxicol Pharmacol 2021;125:105006. https://doi.org/10.1016/j.yrtph.2021.105006.

[6] Barber C, Amberg A, Custer L, Dobo KL, Glowienke S, Van Gompel J, et al. Establishing best practise in the application of expert review of mutagenicity under ICH M7. Regul Toxicol Pharmacol 2015;73(1):367–77. https://doi.org/10.1016/j.yrtph.2015.07.018.

[7] Chakravarti SK. FlexFilters: a scalable and flexible QSAR platform for addressing complex and diverse types of in silico safety assessment of chemicals, poster. In: Society of toxicology annual meeting, Anaheim; 2020.

[8] Chakravarti S. K. QSAR Flex. v1.0. https://www.multicase.com; 2021. [Accessed 16 April 2023].

[9] Duvenaud D, Maclaurin D, Aguilera-Iparraguirre J, Gómez-Bombarelli R, Hirzel T, Aspuru-Guzik A, Adams RP. Convolutional networks on graphs for learning molecular fingerprints. arXiv; November 3, 2015.

[10] Weininger D, Weininger A, Weininger JL. SMILES. 2. Algorithm for generation of unique SMILES notation. J Chem Inf Comput Sci 1989;29(2): 97–101. https://doi.org/10.1021/ci00062a008.

[11] Friedman J, Hastie T, Tibshirani R. Regularization paths for generalized linear models via coordinate descent. J Stat Softw 2010;33(1):1–22.

[12] Pedregosa F, Varoquaux G, Gramfort A, Michel V, Thirion B, Grisel O, et al. Scikit-learn: machine learning in Python. 2012. https://doi.org/10.48550/ARXIV.1201.0490.

[13] Walker MJ. Training ACD/LogP with experimental data. QSAR Comb Sci 2004;23(7):515–20. https://doi.org/10.1002/qsar.200410010.

[14] Zhu H, Sedykh A, Chakravarti S, Klopman G. A new group contribution approach to the calculation of LogP. Curr Comput Aided Drug Des 2005;1(1):3–9. https://doi.org/10.2174/1573409052952323.

[15] Mansouri K, Grulke CM, Judson RS, Williams AJ. OPERA models for predicting physicochemical properties and environmental fate endpoints. J Cheminform 2018;10(1):10. https://doi.org/10.1186/s13321-018-0263-1.

[16] Kim J, Jung DH, Rhee H, Choi S-H, Sung MJ, Choi WS. Aqueous solubility of poorly water-soluble drugs: prediction using similarity and quantitative structure-property relationship models. Korean J Chem Eng 2008;25(4):865–73. https://doi.org/10.1007/s11814-008-0143-x.

[17] Eros D, Keri G, Kovesdi I, Szantai-Kis C, Meszaros G, Orfi L. Comparison of predictive ability of water solubility QSPR models generated by MLR, PLS and ANN methods. Mini-Rev Med Chem 2004;4(2):167–77. https://doi.org/10.2174/1389557043487466.

[18] Wang Q, Yao A, Shokri M, Dineva AA. Predictive modeling of Henry's law constant in chemical structures using LSSVM and ANFIS algorithms. Math Comput Sci 2020. https://doi.org/10.20944/preprints202002.0248.v1. preprint.

[19] Yalkowsky SH, Krzyzaniak JF, Myrdal PB. Relationships between melting point and boiling point of organic compounds. Ind Eng Chem Res 1994;33(7):1872–7. https://doi.org/10.1021/ie00031a030.

[20] Zhang J, Liu Z, Liu W. QSPR study for prediction of boiling points of 2475 organic compounds using stochastic gradient boosting: predicting boiling points using SGB. J Chemom 2014;28(3):161–7. https://doi.org/10.1002/cem.2587.

[21] Watkins M, Sizochenko N, Rasulev B, Leszczynski J. Estimation of melting points of large set of persistent organic pollutants utilizing QSPR approach. J Mol Model 2016;22(3):55. https://doi.org/10.1007/s00894-016-2917-0.

[22] Deeb O, Goodarzi M, Alfalah S. Prediction of melting point for drug-like compounds via QSPR methods. Mol Phys 2011;109(4):507–16. https://doi.org/10.1080/00268976.2010.532164.

[23] Grisoni F, Consonni V, Villa S, Vighi M, Todeschini R. QSAR models for bioconcentration: is the increase in the complexity justified by more accurate predictions? Chemosphere 2015;127:171–9. https://doi.org/10.1016/j.chemosphere.2015.01.047.

[24] Grisoni F, Consonni V, Vighi M, Villa S, Todeschini R. Expert QSAR system for predicting the bioconcentration factor under the REACH regulation. Environ Res 2016;148:507–12. https://doi.org/10.1016/j.envres.2016.04.032.

[25] Grisoni F, Consonni V, Vighi M, Villa S, Todeschini R. Investigating the mechanisms of bioconcentration through QSAR classification trees. Environ Int 2016;88:198–205. https://doi.org/10.1016/j.envint.2015.12.024.

[26] Wu X, Zhang Q, Hu J. QSAR study of the acute toxicity to fathead minnow based on a large dataset. SAR QSAR Environ Res 2016;27(2): 147–64. https://doi.org/10.1080/1062936X.2015.1137353.

[27] Chen X, Dang L, Yang H, Huang X, Yu X. Machine learning-based prediction of toxicity of organic compounds towards fathead minnow. RSC Adv 2020;10(59):36174–80. https://doi.org/10.1039/D0RA05906D.

[28] Wang Y, Chen X. A joint optimization QSAR model of fathead minnow acute toxicity based on a radial basis function neural network and its consensus modeling. RSC Adv 2020;10(36):21292–308. https://doi.org/10.1039/D0RA02701D.

[29] Cassotti M, Ballabio D, Todeschini R, Consonni V. A similarity-based QSAR model for predicting acute toxicity towards the fathead minnow (*Pimephales promelas*). SAR QSAR Environ Res 2015;26(3):217–43. https://doi.org/10.1080/1062936X.2015.1018938.

[30] Jia Q, Zhao Y, Yan F, Wang Q. QSAR model for predicting the toxicity of organic compounds to fathead minnow. Environ Sci Pollut Res 2018; 25(35):35420–8. https://doi.org/10.1007/s11356-018-3434-8.

[31] Hermens JL. Electrophiles and acute toxicity to fish. Environ Health Perspect 1990;87:219–25. https://doi.org/10.1289/ehp.9087219.

[32] Russom CL, Bradbury SP, Broderius SJ, Hammermeister DE, Drummond RA. Predicting modes of toxic action from chemical structure: acute toxicity in the fathead minnow (*Pimephales promelas*). Environ Toxicol Chem 1997;16(5):948–67. https://doi.org/10.1002/etc.5620160514.

[33] von der Ohe PC, Kühne R, Ebert R-U, Altenburger R, Liess M, Schüürmann G. Structural AlertsA new classification model to discriminate excess toxicity from narcotic effect levels of organic compounds in the acute Daphnid assay. Chem Res Toxicol 2005;18(3):536–55. https://doi.org/10.1021/tx0497954.

[34] Barron MG, Lilavois CR, Martin TM. MOAtox: a comprehensive mode of action and acute aquatic toxicity database for predictive model development. Aquat Toxicol 2015;161:102–7. https://doi.org/10.1016/j.aquatox.2015.02.001.

[35] Moosus M, Maran U. Quantitative structure–activity relationship analysis of acute toxicity of diverse chemicals to *Daphnia magna* with whole molecule descriptors. SAR QSAR Environ Res 2011;22(7–8):757–74. https://doi.org/10.1080/1062936X.2011.623317.

[36] Furuhama A, Hayashi TI, Tatarazako N. Acute to chronic estimation of *Daphnia magna* toxicity within the QSAAR framework. SAR QSAR Environ Res 2016;27(10):833–50. https://doi.org/10.1080/1062936X.2016.1243151.

[37] Bouhedjar K, Benfenati E, Nacereddine AK. Modelling quantitative structure activity–activity relationships (QSAARs): auto-pass-pass, a new approach to fill data gaps in environmental risk assessment under the REACH regulation. SAR QSAR Environ Res 2020;31(10):785–801. https://doi.org/10.1080/1062936X.2020.1810770.
[38] Lunghini F, Marcou G, Azam P, Enrici MH, Van Miert E, Varnek A. Consensus QSAR models estimating acute toxicity to aquatic organisms from different trophic levels: algae, *Daphnia* and fish. SAR QSAR Environ Res 2020;31(9):655–75. https://doi.org/10.1080/1062936X.2020.1797872.
[39] Alves VM, Muratov E, Fourches D, Strickland J, Kleinstreuer N, Andrade CH, Tropsha A. Predicting chemically-induced skin reactions. Part II: QSAR models of skin permeability and the relationships between skin permeability and skin sensitization. Toxicol Appl Pharmacol 2015;284(2):273–80. https://doi.org/10.1016/j.taap.2014.12.013.
[40] Zeng R, Deng J, Dang L, Yu X. Correlation between the structure and skin permeability of compounds. Sci Rep 2021;11(1):10076. https://doi.org/10.1038/s41598-021-89587-5.
[41] Chi C-T, Lee M-H, Weng C-F, Leong MK. In silico prediction of PAMPA effective permeability using a two-QSAR approach. Int J Mol Sci 2019;20(13):3170. https://doi.org/10.3390/ijms20133170.
[42] Hu Y, Unwalla R, Denny RA, Bikker J, Di L, Humblet C. Development of QSAR models for microsomal stability: identification of good and bad structural features for rat, human and mouse microsomal stability. J Comput Aided Mol Des 2010;24(1):23–35. https://doi.org/10.1007/s10822-009-9309-9.
[43] Shah P, Siramshetty VB, Zakharov AV, Southall NT, Xu X, Nguyen D-T. Predicting liver cytosol stability of small molecules. J Cheminform 2020;12(1):21. https://doi.org/10.1186/s13321-020-00426-7.
[44] Siramshetty VB, Shah P, Kerns E, Nguyen K, Yu KR, Kabir M, et al. Retrospective assessment of rat liver microsomal stability at NCATS: data and QSAR models. Sci Rep 2020;10(1):20713. https://doi.org/10.1038/s41598-020-77327-0.
[45] National Center for Biotechnology Information. PubChem bioassay record for AID 1508591, NCATS rat liver microsome stability profiling. Source: National Center for Advancing Translational Sciences (NCATS); 2022. Retrieved October 12, 2022 from, https://Pubchem.Ncbi.Nlm.Nih.Gov/Bioassay/1508591.
[46] Hubatsch I, Ragnarsson EGE, Artursson P. Determination of drug permeability and prediction of drug absorption in caco-2 monolayers. Nat Protoc 2007;2(9):2111–9. https://doi.org/10.1038/nprot.2007.303.
[47] Pham-The H, González-Álvarez I, Bermejo M, Garrigues T, Le-Thi-Thu H, Cabrera-Pérez MÁ. The use of rule-based and QSPR approaches in ADME profiling: a case study on caco-2 permeability. Mol Inform 2013;32(5–6):459–79. https://doi.org/10.1002/minf.201200166.
[48] Borba JVB, Braga RC, Alves VM, Muratov EN, Kleinstreuer N, Tropsha A, et al. Pred-skin: a web portal for accurate prediction of human skin sensitizers. Chem Res Toxicol 2021;34(2):258–67. https://doi.org/10.1021/acs.chemrestox.0c00186.
[49] Na M, O'Brien D, Gerberick GF, Kern PS, Lavelle M, Lee I, et al. Benchmarking performance of SENS-IS assay against weight of evidence skin sensitization potency categories. Regul Toxicol Pharmacol 2022;130:105128. https://doi.org/10.1016/j.yrtph.2022.105128.
[50] Wilm A, Norinder U, Agea MI, de Bruyn Kops C, Stork C, Kühnl J, et al. Skin Doctor CP: conformal prediction of the skin sensitization potential of small organic molecules. Chem Res Toxicol 2021;34(2):330–44. https://doi.org/10.1021/acs.chemrestox.0c00253.
[51] Alves V, Borba JVVB, Braga RC, Korn D, Kleinstreuer N, Causey K, et al. Guinea pig maximization test dataset. 2022. https://doi.org/10.6084/M9.FIGSHARE.17708714.V2. 140341 Bytes.
[52] Payne MP, Walsh PT. Structure-activity relationships for skin sensitization potential: development of structural alerts for use in knowledge-based toxicity prediction systems. J Chem Inf Comput Sci 1994;34(1):154–61. https://doi.org/10.1021/ci00017a019.
[53] Gerner I, Barratt MD, Zinke S, Schlegel K, Schlede E. Development and prevalidation of a list of structure–activity relationship rules to be used in expert systems for prediction of the skin-sensitising properties of chemicals. Altern Lab Anim 2004;32(5):487–509. https://doi.org/10.1177/026119290403200505.
[54] Barratt MD, Basketter DA, Chamberlain M, Admans GD, Langowski JJ. An expert system rulebase for identifying contact allergens. Toxicol Vitro 1994;8(5):1053–60. https://doi.org/10.1016/0887-2333(94)90244-5.
[55] Enoch SJ, Ellison CM, Schultz TW, Cronin MTD. A review of the electrophilic reaction chemistry involved in covalent protein binding relevant to toxicity. Crit Rev Toxicol 2011;41(9):783–802. https://doi.org/10.3109/10408444.2011.598141.
[56] Enoch SJ, Madden JC, Cronin MTD. Identification of mechanisms of toxic action for skin sensitisation using a SMARTS pattern based approach. SAR QSAR Environ Res 2008;19(5–6):555–78. https://doi.org/10.1080/10629360802348985.
[57] Cross KP, Ponting DJ. Developing structure-activity relationships for N-nitrosamine activity. Comput Toxicol 2021;20:100186. https://doi.org/10.1016/j.comtox.2021.100186.
[58] Dobo KL, Kenyon MO, Dirat O, Engel M, Fleetwood A, Martin M, et al. Practical and science-based strategy for establishing acceptable intakes for drug product *N*-nitrosamine impurities. Chem Res Toxicol 2022;35(3):475–89. https://doi.org/10.1021/acs.chemrestox.1c00369.
[59] Thresher A, Foster R, Ponting DJ, Stalford SA, Tennant RE, Thomas R. Are all nitrosamines concerning? A review of mutagenicity and carcinogenicity data. Regul Toxicol Pharmacol 2020;116:104749. https://doi.org/10.1016/j.yrtph.2020.104749.
[60] Kruhlak NL, Chakravarti S, Kumaran G, Saiakhov R. A new structural similarity method to identify surrogate compounds for assessing the carcinogenicity of nitrosamine impurities. In: Poster, society of toxicology annual meeting, San Diego; 2022.
[61] Chakravarti SK, Saiakhov RD. Computing similarity between structural environments of mutagenicity alerts. Mutagenesis 2019;34(1):55–65. https://doi.org/10.1093/mutage/gey032.
[62] Gold LS, Slone TH, Manley NB, Garfinkel GB, Hudes ES, Rohrbach L, Ames BN. The carcinogenic potency database: analyses of 4000 chronic animal cancer experiments published in the general literature and by the U.S. National Cancer Institute/National Toxicology Program. Environ Health Perspect 1991;96:11–5. https://doi.org/10.1289/ehp.919611.
[63] Huang X, Tang F, Hua Y, Li X. In silico prediction of drug-induced ototoxicity using machine learning and deep learning methods. Chem Biol Drug Des 2021;98(2):248–57. https://doi.org/10.1111/cbdd.13894.
[64] Rizk HG, Lee JA, Liu YF, Endriukaitis L, Isaac JL, Bullington WM. Drug-induced ototoxicity: a comprehensive review and reference guide. Pharmacother J Hum Pharmacol Drug Ther 2020;40(12):1265–75. https://doi.org/10.1002/phar.2478.
[65] Zhou S, Li G-B, Huang L-Y, Xie H-Z, Zhao Y-L, Chen Y-Z, et al. A prediction model of drug-induced ototoxicity developed by an optimal support vector machine (SVM) method. Comput Biol Med 2014;51:122–7. https://doi.org/10.1016/j.compbiomed.2014.05.005.

C H A P T E R

17

DILIrank dataset for QSAR modeling of drug-induced liver injury

Minjun Chen, Tsung-Jen Liao, Dongying Li, Baitang Ning, Huixiao Hong and Weida Tong

Division of Bioinformatics and Biostatistics, National Center for Toxicological Research, United States Food and Drug Administration, Jefferson, AR, United States

Abbreviations

AR Adverse Reactions
ATC Anatomical Therapeutic Chemical
BW Boxed Warning
CFR Code of Federal Regulation
DILI Drug-induced liver injury
DILIN DILI network
FAERS FDA's adverse event reporting system
NSAIDs Nonsteroidal anti-inflammatory drugs
QSAR Quantitative structure–activity relationship
REACH Registration, Evaluation, Authorization and Restriction of Chemicals
WP Warnings and Precautions

Introduction

Drug-induced liver injury (DILI) is a major drug safety concern for clinicians, drug developers, and regulators [1,2]. It can cause serious clinical outcomes and is the leading cause of acute liver failure in the United States and Europe [3]. DILI is also one of the most important liabilities encountered during drug development and reviewing process and was recorded as a major cause for discontinuation of clinical trials [4,5]. So far, over 50 drugs have been withdrawn after marketing approval in the past decades due to the concern for hepatic adverse effects [6]. The unmet need to accurately predict DILI in humans from animal data has long been recognized. A large survey conducted by a collaborative consortium of the pharmaceutical industry reported that the standardized in vivo toxicological studies can identify only about half of the liver toxicity found in humans [7]. Therefore, new tools and approaches to better predict DILI risk in humans, especially at the early stage of drug development, are still urgently needed [8].

Tremendous efforts have been invested by the research community for developing new alternative methods including in silico methodologies to assess DILI beyond classical animal testing. Employing quantitative structure–activity relationship (QSAR) modeling for DILI assessment is widely used because it does not require physical substances, is high throughput, and quickly yields results [9]. Recently, rapid progress has been achieved in emerging technologies like high-throughput screening or high-content assays, which permit hundreds of compounds to be tested in a single run of an experiment for systematically assessing biological responses in cells. Moreover, novel approaches including in vitro assays [10], "omics" studies [11,12], 3D cell culture models [13–15], and organ-on-a-chip models [16,17], and pluripotent stem cells models [18–22] are continually adopted for assessing drug safety and efficacy.

Meanwhile, some large government-initiated efforts, such as the Registration, Evaluation, Authorization and Restriction of Chemicals (REACH) program [23] in Europe and Tox21 [24] and ToxCast [25] in the United States, have been deployed to evaluate these new approaches for their potential as alternative methodologies for toxicity assessment. Importantly, all these efforts, including QSAR modeling, require a reference list in place with a large set of

https://doi.org/10.1016/B978-0-443-15339-6.00014-X

drugs accurately annotated on their DILI risk in humans, which could help derive robust and reliable biomarkers or models. The DILI dataset presented herein represent a classification of the DILI risk in humans associated with specific drugs. In this era of data-driven science, model development depends on the accurate and consistent labeling of human phenotypes to improve the accuracy of model predictive performance [26].

Basic concepts for DILI annotations

Some drugs are more likely to cause liver injury or hepatotoxicity in humans than others, with those causing severe DILI being of the most concern. Drugs could present significantly different DILI risks even if they have similar chemical structures. For example, troglitazone and pioglitazone are both used for the treatment of diabetes by activating peroxisome proliferator-activated receptors. These two thiazolidinediones share similar chemical structures but not hepatotoxicity; troglitazone was withdrawn from the market due to severe hepatotoxicity, while pioglitazone is still used in clinical practice, with limited liver toxicity observed.

However, not all drugs present clear evidence for hepatotoxicity. Annotation of DILI risk for individual drugs poses significant challenges for clinicians [27]. Firstly, DILI does not frequently occur in clinical practice and has diverse and complicated clinical manifestations, which can mimic almost all other liver diseases. Secondly, there are no existing biomarkers for accurate diagnosis of DILI. Thirdly, causality adjudication of DILI is complicated and mostly relies on the exclusion of other diseases or causes. Lastly, DILI cases are severely under-reported, and most are potentially ignored in clinical practice.

When evaluating a drug's potential to cause DILI in humans, at least three attributes (i.e., causality, severity, and incidence) of DILI events need to be comprehensively investigated [27]. Whether the liver injury observed is caused by the investigated drug or other diseases and whether the symptoms are clinically relevant and could produce severe outcomes has to be evaluated. How many clinically significant cases were observed also needs to be considered. In reality, these questions are not easy to answer, and evaluation has to be conducted on a case-by-case basis to determine the potential of an individual drug to cause DILI in humans.

It is critical to obtain robust DILI data to provide consistent annotations; however, there is no existing database or information source currently available that takes the aforementioned factors into consideration. At present, the data sources available for annotating DILI risk are either monographs or case reports. Monographs are summarized by the in-field expert(s) by comprehensively collecting scientific evidence from a variety of sources. The advantages of using a monograph include the broad coverage of numerous drugs with relatively consistent criteria. The representative monographs are the FDA drug labeling database (FDALabel) [27], Physicians' Desk Reference [28], and US pharmacopeia [29].

Case reports are also widely used as the major data sources to support DILI annotations. There are several ongoing projects, such as the US DILI network (DILIN) [30] and the Spanish DILI registry [31], which collect DILI cases in the clinical setting and government-run postmarking surveillance systems, such as the FDA's adverse event reporting system (FAERS) [32,33] and the World Health Organization's (WHO's) Program for the International Drug Monitoring (VigiBase) [34]. Compared with monographs, the information from case reports is updated more frequently so they are therefore more current. Together with spontaneous literature reports, these systems provide a network to monitor DILI and other drug safety issues caused by drug products. Although case reports are more current, the data quality is highly dependent on the provider. However, because of serious under-reporting [35,36], the number of case reports likely do not reflect the true incidence of hepatic injury [37]. The consensus from experts is that another approach is warranted, but its development would be costly and possibly limited to only a small number of drugs.

Drug labeling for DILI annotation

FDA-approved drug labeling can be used for categorizing a drug's DILI risk in humans. Drug labeling, also referred to as prescription labeling, is required and approved by the FDA with a printed or graphic document product packaging [38]. It contains prescribing information, such as medication guides, patient package inserts, and instructions for use. Drug labeling is mandated by the Code of Federal Regulation (CFR) Title 21 (Food and Drugs) Part 201 (21CFR201.56) (http://www.accessdata.fda.gov/scripts/cdrh/cfdocs/cfCFR/CFRSearch.cfm?fr=201.57), which indicates information must be included pertaining to any adverse drug reactions observed in clinical trials or reported from postmarketing surveillance. The FDA has published guidance on how to draft and finalize drug labels [39–41]. The guidance for prescription drug labeling was first designed in 1979 [42] and later revised to include more information in 2006 [43].

The drug labeling is usually drafted by the sponsor and then approved by the FDA. A labeling document is composed of 17 sections, including a highlighted section summarizing the most important information. It provides comprehensive and balanced information, including the data from clinical trials, spontaneous reports via the Adverse Event Reporting System, and case reports published in literature [36]. Most importantly, it summarizes the essential scientific information from the preclinical studies, clinical trials, postmarketing monitoring, and case reports to provide detailed guidance for the safe and effective use of the drug. The summary information considers the severity, causality, and incidence of specific adverse events and integrates the serious thoughts and agreements from the experts at the time when drug labeling was developed. Moreover, the labeling documents are updated when new data are generated, and each drug approved by the agency is required to provide safety information. For these reasons, drug labeling can be a good data source for quality and consistent information for individual drugs [44].

Annotation schema for assessing DILI risk

Here, FDA-approved labels for prescription drugs were used to assess the hepatotoxic risk associated with marketed drugs. The overall classification schema to assess the potential of drugs to cause DILI is presented in Fig. 17.1 [6].

Grouping drugs based on the labeling sections

Drug safety data are mainly presented in three labeling sections: "Boxed Warning" (BW) or "Black Box Warning," "Warnings and Precautions" (WP), and "Adverse Reactions" (AR). In some labeling, "Warnings and Precautions" are further separated into separated "Warnings" and "Precaution" sections.

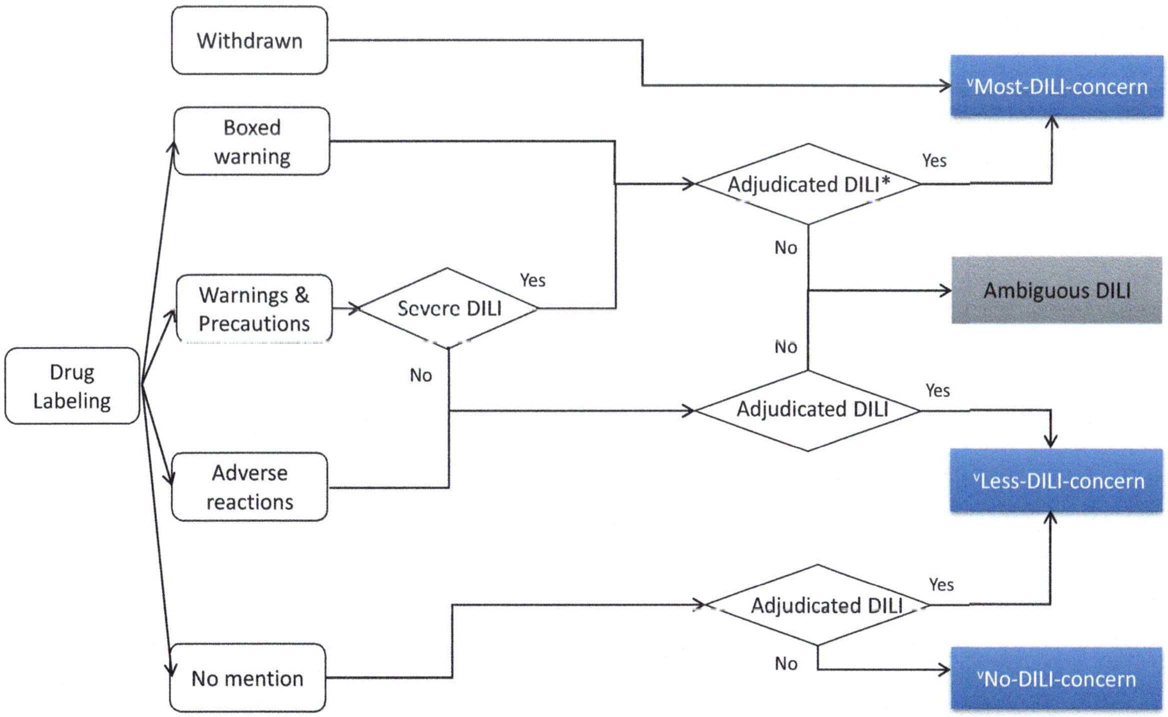

FIGURE 17.1 Flowchart for classifying drug-induced liver injury (DILI) annotations by weighing the causality evidence from adjudicated DILI cases. In this schema, three solid categories (verified most-, less-, no-DILI-concern) were obtained with one ambiguously annotated category, which did not correlate with causality.

A BW is a specially highlighted medication-related safety warning required for marketing an approved drug that the FDA categorizes as possessing "certain contraindications or serious warnings, particularly those that may lead to death or serious injury" [45]; the serious implications associated with the prescribed drug are issued by the FDA to stakeholders, including licensed practitioners, pharmacists, patients, pharmaceutical manufacturers, and distributors [46].

WP is a general warning that the FDA issues, which is considered less severe than BW. Per 21CFR201.57, the drug labeling must be revised to include a warning about a clinically significant hazard once reasonable evidence of a causal association with a drug is established [47]. In other words, the agency could impose a warning for a clinically significant event in certain emerging situations when the causality cannot be established in a timely fashion. Notably, the guidance does not provide a specific definition for clinically significant events attributed to DILI; this issue will be further discussed in the next section.

In the AR section, the manufacturers are required to indicate adverse events with some basis that there is a causal relationship between the drug and the occurrence of the adverse event. Additionally, adverse reactions occurring with drugs in the same pharmacologically active and chemical-related class are also listed. The adverse events in this section might lack causality or are considered guilty-by-association, and therefore are of minimal concern for DILI risk.

Based on the above descriptions, drugs were first grouped into the labeling sections they belonged to. A set of predefined DILI keywords were then used to screen the contents in the sections. If the keywords were found in any of the three sections, the drug was listed in that specific section (i.e., BW, WP, and AR). If a drug was found in multiple sections, it was assigned to the highest risk section following the order BW > WP > AR. If a drug was not associated with any of the predefined DILI keywords in these sections, it was assigned to the "No Match" group. Additionally, for those drugs without labeling but had been withdrawn from market or discontinued drug development due to hepatotoxicity, we assigned them into the category of Withdrawn or "WD." Using this approach, each drug was assigned to either WD, BW, WP, AR, or "No Match." The grouping results were manually reviewed by reading the full text of the labels to ensure that the right context of the DILI keywords was considered.

Determination of DILI severity

Next, DILI severity was determined according to the description language in the labeling sections of BW, WP, or AR. Considering the diverse language used in these descriptions, an 8-level DILI severity system as shown in Table 17.1 was used for categorizing the severity of DILI based on the labeling language. The highest level 8 was "fatal hepatotoxicity," which included death, fatal liver failure, or the need of a liver transplant. For example, a label mentioned that "when used orally, ketoconazole has been associated with hepatic toxicity, including some fatalities"; therefore, a severity level of 8 was assigned to ketoconazole. The next level was "acute liver failure," including liver/hepatic failure and fulminant hepatic necrosis. For example, a label described that "elevations of liver aminotransferases (ALT, AST) and liver failure have been reported with Tracleer," so a severity level of 7 was assigned to Tracleer (Bosentan). Similarly, the severity levels 1–6 were assigned for steatosis, cholestasis or steatohepatitis, liver aminotransferases increase, hyperbilirubinemia, jaundice, and necrosis, respectively. More details about the DILI severity levels can be found in Table 17.1.

Additionally, the 8-level DILI severity was further categorized into three main groups: severe risk included grades 6, 7, and 8; moderate risk included grades 4 and 5; and mild risk included grades 1, 2, and 3. Our analysis of the drugs grouped by sections (BW, WP, or AR) found that 80% of the BW drugs, 35% of WP drugs, and 15% of AR drugs were assigned to the severe risk group for DILI, which is consistent with the severity order of BW > WP > AR. In other words, the hierarchical nature of the labeling sections accurately correlated with the severity of DILI risk associated with the drugs.

Determination of DILI causality

As discussed previously, a drug safety warning can be issued when the agency identifies a clinically significant hazard with reasonable evidence of causality associated with a drug. Therefore, incorporating the causality information from clinical studies and public databases could further improve the DILI annotation process.

The DILI causality data were therefore collected from several quality sources including the large DILI registries [48,49] and the NIH LiverTox database [50]. The first source is a large international collaborative study by Suzuki et al. [49], which consists of 225 drugs with data collected from three major international DILI registries: the Spanish

TABLE 17.1 An 8-level system to categorize drug-induced liver injury (DILI) severity from the descriptions in drug labeling.

DILI category	Severity level	Specification and keywords	Examples of labeling language
Fatal hepatotoxicity	8	Death; fatal liver failure or needed liver transplant	When used orally, ketoconazole has been associated with hepatic toxicity, including some fatalities
Acute liver failure	7	Liver/hepatic failure; fulminant hepatic necrosis	Elevations of liver aminotransferases (ALT, AST) and liver failure have been reported with Tracleer
Liver necrosis	6	Histologically confirmed liver necrosis caused by drug	Rare instances of severe liver injury, including hepatic necrosis, have been reported in association with mexiletine treatment
Jaundice	5	Jaundice (clinically apparent), if caused by drug-induced hepatocellular injury	There is a low incidence of altered liver function or jaundice in patients treated with Marplan
Hyperbilirubinemia	4	Hyperbilirubinemia without visible jaundice, if not due to other causes like Gilbert syndrome or cholestasis	Ticlopidine therapy has been associated with elevations of alkaline phosphatase, bilirubin, and transaminases, which generally occurred within 1–4 months of therapy initiation
Liver aminotransferases increase	3	Liver aminotransferases increase (e.g., ALT, AST, transaminase, aminotransferase); abnormal liver/hepatic function test; liver/hepatic injury	Persistent increases (to more than 3 times the ULN (upper limit of normal)) in serum transaminases have occurred in approximately 1% of patients who received simvastatin in clinical studies
Cholestasis; steatohepatitis	2	Steatohepatitis, if probably caused by the drug; cholestasis, cholestatic hepatitis if caused by the drug; liver/hepatic damage/disorder/impairment/toxicity/reaction/hepatitis; hepatopathy	Jaundice of the cholestatic type of hepatitis or liver damage has been reported (in patients receiving trifluoperazine)
Steatosis	1	Steatosis; fatty liver; liver/hepatic steatosis	Lactic acidosis and severe hepatomegaly with steatosis have been reported with the use of nucleoside analogs including zidovudine

DILI Registry, the Swedish Adverse Drug Reactions Advisory Committee Database, and the United States Drug-Induced Liver Injury Network (DILIN). In the Suzuki et al. study, the causality of DILI cases was verified using the Roussel Uclaf Causality Assessment Method (RUCAM) scoring system [51] or expert opinions [52], with the cases classified as "possible," "probable," "highly probable," and "definite." The US DILIN study [48] reported 163 drugs as causing 800 DILI cases. All these cases were adjudicated by expert opinion with causality categories of "probable" or above. The NIH Livertox database included 344 drugs that were verified by three or more DILI cases in literature by causality assessments by DILI experts [45]. These sources together consisted of a total of 399 unique drugs with DILI causality adjudicated. Additionally, manual literature searches to identify other drugs that had been adjudicated as causing DILI were also performed.

Determination of DILI potential

DILI is a complicated disease and can represent a variety of different etiologies and attributes. For instance, mild elevations in liver enzymes might not be a concern in large-scale clinical trials but can signal the potential for significant liver injury at the early stage of drug development. This chapter focused on FDA-approved drugs with the potential to cause severe liver injury, resulting in a three-step approach to determine the DILI potential of a drug. In the first step, risk groups were based on where the DILI description was mentioned in the labeling (e.g., WD, BW, WP, AR, and No Match). In the second step, the severity of the DILI description was assessed based on an 8-level grading system as previously mentioned. In the third step, the number of drugs with DILI causality adjudicated was assessed. Based on the findings for a drug's DILI risk group, DILI severity, and DILI causality, the final DILI classification was determined. Notably, a severe DILI event mentioned in the drug labeling was considered as the justification for clinically significant events. The intent of this rationale was that, despite its scarcity, the findings of a small number of severe DILI cases would provide enough to enable identification of safety concerns after marketing approval [36].

This three-step approach, as illustrated in Fig. 17.1, was developed to determine the DILI potential of an individual drug by complementing the drug labeling with the available evidence of verified causality. Drugs were ultimately grouped into four categories of DILI potential: the verified most-, less-, no-DILI-concern groups, and the ambiguous DILI group. The specific methodology is outlined below:

1) Withdrawn or discontinued drugs with strong evidence of hepatotoxicity were consistently classified as high DILI risk among several published datasets. Therefore, these drugs were classified as the verified "most-DILI-concern" group.
2) A BW is a stronger warning than a WP and was also consistently classified as high DILI risk in published datasets [53]. If the causality was verified, these drugs were classified as the verified "most-DILI-concern" group; if causality could not be verified, drugs were classified as the "ambiguous DILI" group.
3) For drugs with a WP, the DILI descriptions were first designated as severe or moderate/mild DILI (i.e., diclofenac). If DILI was severe and causality was established, the drug would be placed in the verified "most-DILI-concern" group. Alternatively, if DILI was severe but causality was not verified, it would be placed in the "ambiguous DILI" group. However, if the DILI classification was moderate/mild and causality was established, the drug would be placed the in the verified "less-DILI-concern" group. Alternatively, if the DILI classification was moderate/mild, but causality was not verified, it would be placed in the "ambiguous DILI" group.
4) For AR drugs, if causality was established it would be placed in the verified "less-DILI-concern" group; otherwise the "ambiguous DILI" group was designated.
5) For drugs that did not fall into a labeling group, unless it was verified as a cause of DILI in literature, it would be placed in the verified "no-DILI-concern" group; otherwise, the verified "less-DILI-concern" group was designated.

Develop a DILIrank dataset to support the development of QSAR and other predictive models

DailyMed [54] was used to retrieve drug labeling information because it is updated regularly, is in the public domain, and can be freely accessed.

Drugs approved by the US FDA before January 1, 2010, were utilized. Only drugs intended for human use that contained a single active molecule and were administered systemically (i.e., oral, parental) were selected after food supplements, minerals, and cosmetics were excluded. Following these criteria, data were collected on a total of 1036 FDA-approved unique drugs [6]. Using the schema discussed in the previous section, these drugs were classified into the most- (192), less- (278), and no-DILI-concern (312) groups with verified causality or the ambiguous DILI (254) group. The entire list of these 1036 drugs, termed the "DILIrank dataset," and their related information can be found at the Liver Toxicity Knowledge Base website (https://www.fda.gov/science-research/liver-toxicity-knowledge-base-ltkb/drug-induced-liver-injury-rank-dilirank-dataset).

The WHO Anatomical Therapeutic Chemical (ATC) classification system was then used to analyze the landscape of the DILIrank dataset. The ATC's therapeutic categories (second level) and chemical subgroups (fourth level) were analyzed to identify the drug subgroups associated with a higher DILI risk using the Fisher exact test.

A total of seven therapeutic subgroups (second level), that is, antigout, antimycotics, antineoplastics, nonsteroidal anti-inflammatory drugs (NSAIDs), antivirals, psychoanaleptics, and immunostimulants, were found to be associated with a higher risk of DILI in humans. Most of these drug subgroups were widely reported to cause DILI cases in the clinical setting [55]. Interestingly, although several of the antibiotics were assigned to the "most-DILI-concern" group ($N = 9$), as a class they did not present a significantly higher hepatotoxic risk; as many as $N = 67$ antibiotic drugs were as assigned to the verified "less-DILI-concern" or "ambiguous DILI" groups.

Among the chemical subgroups (fourth level), some groups (i.e., acetic acid derivative NSAIDs, interferons, marcolide antibiotics, nucleoside/nucleotide reverse transcriptase inhibitors [NRTIs], and protein kinase inhibitors, triazole antifungals, etc.) were found to have a high hepatotoxicity risk. The NRTIs for example could be linked to underlying DILI mechanisms such as mitochondrial toxicity, which could lead to lactic acidosis and liver failure [56].

Concluding remarks

Over the years, tremendous efforts have been made to promote significant improvements in drug safety assessments, such as shifting from traditional animal testing to the use of high-throughput assays combined with computational bioinformatic tools. In this data-driven era, data quality is critical for accurately developing new predictive models and approaches using in silico methods such as QSAR. Although QSAR models have been widely used for predicting DILI in humans, a quality DILI annotation method has not yet been established but is imperative for improving the predictive performance of QSAR modeling.

Considering the importance of DILI potential in the drug development industry and for regulatory agencies, a schema for DILI annotation was established utilizing a drug labeling-based approach by weighing the causality assessment reported in literature, and the DILIrank dataset consisting of 1036 drugs was generated and classified into either three groups ("most-," "less-," and "no-DILI concern" groups) or the "ambiguous DILI" group based on causality verification [6]. The DILIrank dataset is one of the largest datasets with DILI annotations that is publicly available. Additionally, this annotation is more relevant in the regulatory context because it comes from agency-approved labeling documents. For instance, the verified "most-DILI-concern" group (e.g., troglitazone) could be considered as the drugs with a high potential to cause severe clinical outcomes, while those in the verified "less-DILI-concern" group (e.g., statins) can cause liver injury but rarely lead to a severe outcome, and the verified "no-DILI-concern" group has minimal risk and therefore is of little clinical concern.

The DILIrank dataset has been widely adopted as an essential component for developing other DILI datasets [57–62]. For example, Thakkar et al. incorporated DILIrank data with four other published DILI datasets to develop a binary classification for human hepatotoxicity of 1279 drugs. These annotated datasets supported the development of QSAR models for predicting DILI in humans [62–66] and were applied to emerging technologies such as high-throughput screening and high content assays for identifying DILI risk at the early stage of drug development. Moving forward, the DILIrank dataset will additionally be helpful in supporting clinical and epidemiological investigations determining the causality of suspected DILI cases in the future.

Acknowledgments

This research was supported in part by an appointment to the Research Participation Program at the National Center for Toxicological Research (Tsung-Jen Liao), administered by the Oak Ridge Institute for Science and Education through an interagency agreement between the US Department of Energy and the US Food and Drug Administration.

Disclaimer

This chapter reflects the views of the authors and does not necessarily reflect those of the US Food and Drug Administration. Any mention of commercial products is for clarification only and is not intended as approval, endorsement, or recommendation.

References

[1] Chen M, Borlak J, Tong W. Predicting idiosyncratic drug-induced liver injury-some recent advances. Expert Rev Gastroenterol Hepatol 2014; 8(7):721–3.

[2] Chen M, Zhang J, Wang Y, Liu Z, Kelly R, Zhou G, et al. Liver Toxicity Knowledge Base (LTKB)—a systems approach to a complex endpoint. Clin Pharmacol Ther 2013;95(5):409–12.

[3] Weaver RJ, Blomme EA, Chadwick AE, Copple IM, Gerets HH, Goldring CE, et al. Managing the challenge of drug-induced liver injury: a roadmap for the development and deployment of preclinical predictive models. Nat Rev Drug Discov 2020;19(2):131–48.

[4] Temple RJ, Himmel MH. Safety of newly approved drugs: implications for prescribing. JAMA 2002;287(17):2273–5.

[5] Kaplowitz N. Drug-induced liver disorders: implications for drug development and regulation. Drug Saf 2001;24(7):483–90.

[6] Chen M, Suzuki A, Thakkar S, Yu K, Hu C, Tong W. DILIrank: the largest reference drug list ranked by the risk for developing drug-induced liver injury in humans. Drug Discov Today 2016;21(4):648–53.

[7] Olson H, Betton G, Robinson D, Thomas K, Monro A, Kolaja G, et al. Concordance of the toxicity of pharmaceuticals in humans and in animals. Regul Toxicol Pharmacol 2000;32(1):56–67.

[8] Kaplowitz N. Idiosyncratic drug hepatotoxicity. Nat Rev Drug Discov 2005;4(6):489–99.

[9] Rodgers AD, Zhu H, Fourches D, Rusyn I, Tropsha A. Modeling liver-related adverse effects of drugs using knearest neighbor quantitative structure-activity relationship method. Chem Res Toxicol 2010;23(4):724–32.

[10] Obach RS, Kalgutkar AS, Soglia JR, Zhao SX. Can in vitro metabolism-dependent covalent binding data in liver microsomes distinguish hepatotoxic from nonhepatotoxic drugs? An analysis of 18 drugs with consideration of intrinsic clearance and daily dose. Chem Res Toxicol 2008; 21(9):1814–22.
[11] Zidek N, Hellmann J, Kramer PJ, Hewitt PG. Acute hepatotoxicity: a predictive model based on focused illumina microarrays. Toxicol Sci 2007;99(1):289–302.
[12] Jiang J, Pieterman CD, Ertaylan G, Peeters RL, de Kok TM. The application of omics-based human liver platforms for investigating the mechanism of drug-induced hepatotoxicity in vitro. Arch Toxicol 2019;93(11):3067–98.
[13] Proctor WR, Foster AJ, Vogt J, Summers C, Middleton B, Pilling MA, et al. Utility of spherical human liver microtissues for prediction of clinical drug-induced liver injury. Arch Toxicol 2017;91(8):2849–63.
[14] Cox CR, Lynch S, Goldring C, Sharma P. Current perspective: 3D spheroid models utilizing human-based cells for investigating metabolism-dependent drug-induced liver injury. Front Med Technol 2020;2:14.
[15] Hiemstra S, Ramaiahgari SC, Wink S, Callegaro G, Coonen M, Meerman J, et al. High-throughput confocal imaging of differentiated 3D liver-like spheroid cellular stress response reporters for identification of drug-induced liver injury liability. Arch Toxicol 2019;93(10):2895–911.
[16] Moradi E, Jalili-Firoozinezhad S, Solati-Hashjin M. Microfluidic organ-on-a-chip models of human liver tissue. Acta Biomater 2020;116: 67–83.
[17] Kanabekova P, Kadyrova A, Kulsharova G. Microfluidic organ-on-a-chip devices for liver disease modeling in vitro. Micromachines 2022; 13(3):428.
[18] Soldatow VY, LeCluyse EL, Griffith LG, Rusyn I. In vitro models for liver toxicity testing. Toxicol Res 2013;2(1):23–39.
[19] Kia R, Sison RL, Heslop J, Kitteringham NR, Hanley N, Mills JS, et al. Stem cell-derived hepatocytes as a predictive model for drug-induced liver injury: are we there yet? Br J Clin Pharmacol 2013;75(4):885–96.
[20] Shinozawa T, Kimura M, Cai Y, Saiki N, Yoneyama Y, Ouchi R, et al. High-fidelity drug-induced liver injury screen using human pluripotent stem cell–derived organoids. Gastroenterology 2021;160(3):831–846. e10.
[21] Goldring C, Antoine DJ, Bonner F, Crozier J, Denning C, Fontana RJ, et al. Stem cell–derived models to improve mechanistic understanding and prediction of human drug-induced liver injury. Hepatology 2017;65(2):710–21.
[22] Maepa SW, Ndlovu H. Advances in generating liver cells from pluripotent stem cells as a tool for modeling liver diseases. Stem Cell 2020; 38(5):606–12.
[23] Schoeters G. The REACH perspective: toward a new concept of toxicity testing. J Toxicol Environ Health B 2010;13(2–4):232–41.
[24] Shukla SJ, Huang R, Austin CP, Xia M. The future of toxicity testing: a focus on in vitro methods using a quantitative high-throughput screening platform. Drug Discov Today 2010;15(23–24):997–1007.
[25] Benigni R, Bossa C, Giuliani A, Tcheremenskaia O. Exploring in vitro/in vivo correlation: lessons learned from analyzing phase I results of the US EPA's ToxCast Project. J Environ Sci Health C 2010;28(4):272–86.
[26] Temple R. Hy's law: predicting serious hepatotoxicity. Pharmacoepidemiol Drug Saf 2006;15(4):241–3.
[27] Chen M, Vijay V, Shi Q, Liu Z, Fang H, Tong W. FDA-approved drug labeling for the study of drug-induced liver injury. Drug Discov Today 2011;16(15–16):697–703.
[28] Sakatis MZ, Reese MJ, Harrell AW, Taylor MA, Baines IA, Chen L, et al. Preclinical strategy to reduce clinical hepatotoxicity using in vitro bioactivation data for >200 compounds. Chem Res Toxicol 2012;25(10):2067–82.
[29] Whitsett M, Halegoua-De Marzio D, Rossi S. SlimQuick™-associated hepatotoxicity resulting in fulminant liver failure and orthotopic liver transplantation. ACG Case Rep J 2014;1(4):220.
[30] Fontana RJ, Watkins PB, Bonkovsky HL, Chalasani N, Davern T, Serrano J, et al. Drug-induced liver injury network (DILIN) prospective study. Drug Saf 2009;32(1):55–68.
[31] Stephens C, Robles-Diaz M, Medina-Caliz I, Garcia-Cortes M, Ortega-Alonso A, Sanabria-Cabrera J, et al. Comprehensive analysis and insights gained from long-term experience of the Spanish DILI registry. J Hepatol 2021;75(1):86–97.
[32] Ursem CJ, Kruhlak NL, Contrera JF, MacLaughlin PM, Benz RD, Matthews EJ. Identification of structure-activity relationships for adverse effects of pharmaceuticals in humans. Part A: use of FDA post-market reports to create a database of hepatobiliary and urinary tract toxicities. Regul Toxicol Pharmacol 2009;54(1):1–22.
[33] Zhu X, Kruhlak NL. Construction and analysis of a human hepatotoxicity database suitable for QSAR modeling using post-market safety data. Toxicology 2014;321:62–72.
[34] Ortland I, Mirjalili M, Kullak-Ublick GA, Peymani P. Drug-induced liver injury in Switzerland: an analysis of drug-related hepatic disorders in the WHO pharmacovigilance database VigiBase™ from 2010 to 2020. Swiss Med Wkly 2021;151:w20503.
[35] Sgro C, Clinard F, Ouazir K, Chanay H, Allard C, Guilleminet C, et al. Incidence of drug-induced hepatic injuries: a French population-based study. Hepatology 2002;36(2):451–5.
[36] Trontell AE. How the US Food and Drug Administration defines and detects adverse drug events. Curr Ther Res 2001;62(9):641–9.
[37] Zimmerman HJ. Drug-induced liver disease. Clin Liver Dis 2000;4(1):73–96. vi.
[38] Fang H, Harris SC, Liu Z, Zhou G, Zhang G, Xu J, et al. FDA drug labeling: rich resources to facilitate precision medicine, drug safety, and regulatory science. Drug Discov Today 2016;21(10):1566–70.
[39] CDER. Labeling for human prescription drug and biological products—implementing the new content and format requirements. 2006. http://www.fda.gov/downloads/Drugs/GuidanceComplianceRegulatoryInformation/Guidances/ucm075082.pdf.
[40] CDER. Adverse reactions section of labeling for human prescription drug and biological products—content and format. 2006. http://www.fda.gov/downloads/Drugs/GuidanceComplianceRegulatoryInformation/Guidances/ucm075057.pdf.
[41] CDER. Warnings and precautions, contraindications, and boxed warning sections of labeling for human prescription drug and biological products—content and format. 2006. http://www.fda.gov/downloads/Drugs/GuidanceComplianceRegulatoryInformation/Guidances/ucm075096.pdf.
[42] Liang BA. FDA use of the black box warning: time for reevaluation as a safety tool. J Clin Anesth 2002;14(8):561–3.
[43] Lal R, Kremzner M. Introduction to the new prescription drug labeling by the Food and Drug Administration. Am J Health Syst Pharm 2007; 64(23):2488–94.

[44] Murphy S, Roberts R. "Black box" 101: how the Food and Drug Administration evaluates, communicates, and manages drug benefit/risk. J Allergy Clin Immunol 2006;117(1):34–9.
[45] Cheng CM, Guglielmo BJ, Maselli J, Auerbach AD. Coverage of FDA medication boxed warnings in commonly used drug information resources. Arch Intern Med 2010;170(9):831–3.
[46] Beach JE, Faich GA, Bormel FG, Sasinowski FJ. Black box warnings in prescription drug labeling: results of a survey of 206 drugs. Food Drug Law J 1998;53(3):403–11.
[47] CDER. Drug-induced liver injury: premarketing clinical evaluation. 2009. http://www.fda.gov/downloads/Drugs/GuidanceComplianceRegulatoryInformation/Guidances/UCM174090.pdf.
[48] Chalasani N, Bonkovsky HL, Fontana R, Lee W, Stolz A, Talwalkar J, et al. Features and outcomes of 899 patients with drug-induced liver injury: the DILIN prospective study. Gastroenterology 2015;148(7):1340–1352. e7.
[49] Suzuki A, Andrade RJ, Bjornsson E, Lucena MI, Lee WM, Yuen NA, et al. Drugs associated with hepatotoxicity and their reporting frequency of liver adverse events in VigiBase: unified list based on international collaborative work. Drug Saf 2010;33(6):503–22.
[50] Hoofnagle JH, Serrano J, Knoben JE, Navarro VJ. LiverTox: a website on drug-induced liver injury. Hepatology 2013;57(3):873–4.
[51] Yang H, Guo D, Xu Y, Zhu M, Yao C, Chen C, et al. Comparison of different liver test thresholds for drug-induced liver injury: updated RUCAM versus other methods. Front Pharmacol 2019;10:816.
[52] Hayashi PH, Barnhart HX, Fontana RJ, Chalasani N, Davern TJ, Talwalkar JA, et al. Reliability of causality assessment for drug, herbal and dietary supplement hepatotoxicity in the Drug-Induced Liver Injury Network (DILIN). Liver Int 2015;35(5):1623–32.
[53] Chen M, Bisgin H, Tong L, Hong H, Fang H, Borlak J, et al. Toward predictive models for drug-induced liver injury in humans: are we there yet? Biomark Med 2014;8(2):201–13.
[54] de Leon J. Highlights of drug package inserts and the website DailyMed: the need for further improvement in package inserts to help busy prescribers. J Clin Psychopharmacol 2011;31(3):263–5.
[55] Tucker RA. Drugs and liver disease: a tabular compilation of drugs and the histopathological changes that can occur in the liver. Drug Intell Clin Pharm 1982;16(7–8):569–80.
[56] CDER. Human liver adverse effects database. 2009. http://www.fda.gov/AboutFDA/CentersOffices/CDER/ucm092203.htm.
[57] Nguyen-Vo T-H, Nguyen L, Do N, Le PH, Nguyen T-N, Nguyen BP, et al. Predicting drug-induced liver injury using convolutional neural network and molecular fingerprint-embedded features. ACS Omega 2020;5(39):25432–9.
[58] Chen Z, Jiang Y, Zhang X, Zheng R, Qiu R, Sun Y, et al. ResNet18DNN: prediction approach of drug-induced liver injury by deep neural network with ResNet18. Briefings Bioinf 2022;23(1):bbab503.
[59] Ivanov SM, Lagunin AA, Filimonov DA, Poroikov VV. Relationships between the structure and severe drug-induced liver injury for low, medium, and high doses of drugs. Chem Res Toxicol 2022;35(3):402–11.
[60] Jia X, Wen X, Russo DP, Aleksunes LM, Zhu H. Mechanism-driven modeling of chemical hepatotoxicity using structural alerts and an in vitro screening assay. J Hazard Mater 2022:129193.
[61] Thakkar S, Li T, Liu Z, Wu L, Roberts R, Tong W. Drug-induced liver injury severity and toxicity (DILIst): binary classification of 1279 drugs by human hepatotoxicity. Drug Discov Today 2020;25(1):201–8.
[62] He S, Ye T, Wang R, Zhang C, Zhang X, Sun G, et al. An in silico model for predicting drug-induced hepatotoxicity. Int J Mol Sci 2019;20(8):1897.
[63] Ancuceanu R, Hovanet MV, Anghel AI, Furtunescu F, Neagu M, Constantin C, et al. Computational models using multiple machine learning algorithms for predicting drug hepatotoxicity with the DILIrank dataset. Int J Mol Sci 2020;21(6):2114.
[64] Wu L, Liu Z, Auerbach S, Huang R, Chen M, McEuen K, et al. Integrating drug's mode of action into quantitative structure–activity relationships for improved prediction of drug-induced liver injury. J Chem Inf Model 2017;57(4):1000–6.
[65] Hong H, Thakkar S, Chen M, Tong W. Development of decision forest models for prediction of drug-induced liver injury in humans using a large set of FDA-approved drugs. Sci Rep 2017;7(1):1–15.
[66] Williams DP, Lazic SE, Foster AJ, Semenova E, Morgan P. Predicting drug-induced liver injury with Bayesian machine learning. Chem Res Toxicol 2019;33(1):239–48.

CHAPTER

18

Application of QSAR models based on machine learning methods in chemical risk assessment and drug discovery

Tuan Xu, Deborah K. Ngan and Ruili Huang

Division of Preclinical Innovation (DPI), National Center for Advancing Translational Sciences (NCATS), National Institutes of Health (NIH), Rockville, MD, United States

Abbreviations

3CL 3-Chymotrypsin-like
ACh Acetylcholine
AChE Acetylcholinesterase
ACS American Chemical Society
AD Applicability domain
AhR Aryl hydrocarbon receptor
AR-LBD Androgen receptor ligand binding domain
ARE Antioxidant responsive element
ATAD5 ATPase family AAA domain containing 5
AUC-ROC Area under the receiver operating characteristic curve
BA Balanced accuracy
BABM Biological activity-based modeling
BChE Butyrylcholinesterase
CAS Chemical abstract service
CERAPP Collaborative Estrogen Receptor Activity Prediction Project
CoMPARA Collaborative Modeling Project for Androgen Receptor Activity
COVID-19 Coronavirus disease 2019
CYP Cytochrome P450
ECFP4 Extended connectivity fingerprint 4
EPA Environmental Protection Agency
ER Estrogen receptor
FDA Food and Drug Administration
GPCR G-protein-coupled receptors
hERG Human Ether-à-go-go Related Gene
HSE Heat shock factor response element
HTS High-throughput screening
MACCS Molecular ACCess System
MCC Matthews correlation coefficient
MMP Mitochondrial membrane potential
NB Naive Bayes
NCATS National Center for Advancing Translational Sciences
NCCT National Center for Computational Toxicology
NIEHS National Institute of Environmental Health Sciences
NIH National Institutes of Health
NNET Neural networks
NTP National Toxicology Program
OPRs Opioid receptors
PLD Phospholipidosis
PP Pseudotyped particle
PPAR-gamma Peroxisome proliferator-activated receptor gamma
QC Quality control
QSAR Quantitative structure–activity relationship
RF Random forest
ROSE Random over sampling examples
SARS-CoV-2 Severe acute respiratory syndrome coronavirus 2
SD Standard deviation
SMOTE Synthetic minority over-sampling technique
SVM Support vector machine
Tox21 Toxicology in the 21st century
XGboost Extreme gradient boosting

Introduction

Currently, more than 200 million unique chemical substances, consisting of both organic and inorganic substances, are registered in the Chemical Abstract Service (CAS) database of the American Chemical Society (ACS) (https://www.cas.org/cas-data/cas-registry), and thousands of new chemical substances are added each day. These compounds may exhibit either harmful (e.g., toxins) or beneficial (e.g., drugs) properties to human health

https://doi.org/10.1016/B978-0-443-15339-6.00006-0

through interactions with specific molecular targets in vivo. For example, 2,3,7,8-tetrachlorodibenzo-p-dioxin, one of the unwanted environmental persistent organic pollutants, can bind to the aryl hydrocarbon receptor (AhR, a highly conserved nuclear receptor), leading to adverse biological effects (e.g., neurotoxicity) [1,2]. Rivastigmine, one of the US Food and Drug Administration (FDA)-approved clinical drugs, can relieve symptoms of human diseases related to acetylcholine (ACh) deficiency (e.g., Alzheimer's disease) by inhibiting the activities of acetylcholinesterase (AChE; EC 3.1.1.7) and butyrylcholinesterase (BChE; EC 3.1.1.8) [3]. However, information on known bioactive compounds is very limited. For instance, the latest release of the Comparative Toxicogenomics Database (accessed September 2022, http://ctdbase.org/) contained 14,346 unique compounds that were involved in 2,602,694 chemical–gene interactions. Moreover, the latest release of the DrugBank Database (version 5.1.9, accessed January 2022, https://go.drugbank.com/) contained 2724 approved small molecule drugs and over 6709 investigational drugs. Therefore, there is a strong need to conduct large-scale screening of chemical compound libraries in various biological assays to identify their potential biological activities.

High-throughput screening (HTS) has been successfully applied to identify thousands of bioactive compounds based on robust in vitro cell-based and cell-free assays. For example, the Toxicology in the 21st century (Tox21) program, a collaboration among multiple US federal agencies involving the intramural research division of the National Center for Advancing Translational Sciences (NCATS) at the National Institutes of Health (NIH), the National Center for Computational Toxicology (NCCT) at the Environmental Protection Agency (EPA), the National Toxicology Program (NTP) at the National Institute of Environmental Health Sciences (NIEHS), and the Food and Drug Administration (FDA), has screened a collection of ~10,000 compounds (including industrial and consumer products, food additives, chemical mixtures, and drugs (the NCATS Pharmaceutical Collection)) using >70 in vitro assays including nuclear receptor signaling, stress response, cytotoxicity, and other targets/pathways related to toxicity [4–12]. In response to the outbreak of coronavirus disease 2019 (COVID-19) caused by the novel severe acute respiratory syndrome coronavirus 2 (SARS-CoV-2), NCATS has screened >10,000 compounds, including approved drugs, and experimental therapeutics designed to modulate a wide range of cellular targets based on in vitro assays that cover a wide spectrum of the SARS-CoV-2 life cycle (e.g., viral entry and viral replication) [13]. The aforementioned databases are publicly available to the broad scientific community and accessible at https://tripod.nih.gov/tox21/pubdata and https://opendata.ncats.nih.gov/covid19/, respectively. Despite the advantages the HTS method offers during the drug discovery process, there remains a major challenge to test an enormous number of compounds using traditional HTS assays as this approach is time-consuming and costly.

The readily available HTS assay data, as a training dataset, provides an excellent resource for the development of machine learning models that can efficiently identify biologically active compounds from large-scale chemical compound libraries in a cost-efficient and timely manner (e.g., QSAR models that are developed by using the combination of various machine learning approaches and chemical descriptors). In this chapter, we will review the development and evaluation of QSAR models based on machine learning methods, and the application of these models in virtual screening of large, diverse compound collections to predict compound toxicity and accelerate drug discovery.

Overview of QSAR models based on machine learning methods

The most commonly used machine learning models can generally be classified into three main categories: supervised learning, unsupervised learning, and reinforcement learning. Supervised learning methods (e.g., classification algorithms) can predict the class of unknown data based on the labeled data features, whereas unsupervised learning methods (e.g., clustering) infer patterns directly based on the unlabeled data. Reinforcement learning methods (e.g., Q-learning) train an agent to choose better actions within a particular environment to maximize certain cumulative reward. In this chapter, an example case will be given to illustrate the workflow of the development and application of QSAR model-based supervised learning using various machine learning algorithms (Fig. 18.1). QSAR models are reliable and cost-effective alternatives to traditional experimental HTS assays, and they have been used to predict the biological properties of chemical compounds based on their physicochemical properties and molecular descriptors [14,15]. In vitro HTS assay data against a particular target (e.g., estrogen receptor alpha (ERα) and androgen receptor (AR) assays run in agonist or antagonist mode in the Tox21 program) in combination with chemical structural descriptors (e.g., ToxPrint, extended connectivity fingerprint 4 (ECFP4), Molecular ACCess System (MACCS)) were used as training datasets for QSAR model construction. Prior to model training, feature selection is performed to obtain the best combination of features that can improve computation efficiency and optimize model performance by removing irrelevant or redundant features [16]. Several strategies can be

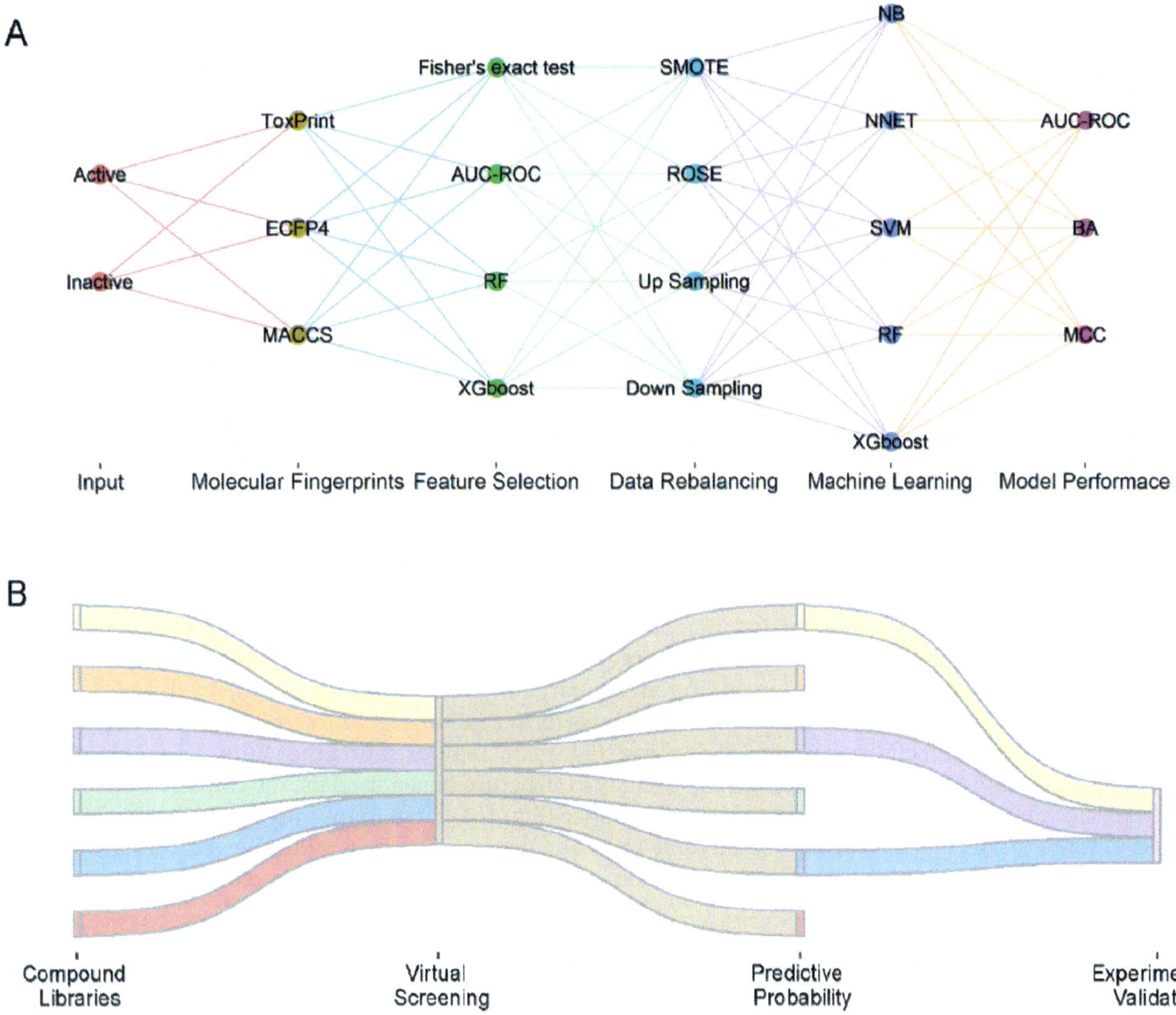

FIGURE 18.1 Workflow of quantitative structure-activity relationship (QSAR) models based on machine learning methods. (A) Model construction. (B) Virtual screening. *AUC-ROC*, area under the receiver operating characteristic curve; *BA*, balanced accuracy; *ECFP4*, extended connectivity fingerprint 4; *MACCS*, Molecular ACCess System; *MCC*, Matthews correlation coefficient; *NB*, Naive Bayes; *NNET*, neural networks; *RF*, random forest; *ROSE*, random over sampling examples; *SMOTE*, synthetic minority over-sampling technique; *SVM*, support vector machines; *XGboost*, extreme gradient boosting.

used to achieve this goal, such as Fisher's exact test with *P* value, the area under the receiver operating characteristic curve (AUC-ROC) value, and importance scores from the Random Forest (RF; Gini score) and the extreme gradient boosting (XGboost; Gain score) algorithms. The HTS assay results are often highly imbalanced (i.e., the number of active compounds is significantly smaller than the number of inactive compounds); therefore, such imbalanced datasets could lead to biased model performance (i.e., biased toward the majority labels) [17]. The training dataset can be rebalanced using four subsampling methods, such as Down sampling, Up sampling, random over sampling examples (ROSE), and synthetic minority over-sampling technique (SMOTE) via the "ROSE" and "DMwR" packages in R [18,19]. After data preparation and preprocessing, machine learning algorithms, such as Naive Bayes (NB), neural networks (NNET), RF, support vector machine (SVM), and XGboost, are performed for binary classification. Three metrics can be used to evaluate model performance: AUC-ROC, balanced accuracy (BA), and Matthews correlation coefficient (MCC). The combination of feature sets, rebalanced data, and machine learning algorithms yields the optimal models (e.g., maximum AUC-ROC value) that can be used to efficiently identify potential bioactive compounds (i.e., those with the highest predictive probability) through virtual screening of large compound libraries and further prioritize them for experimental validation (Fig. 18.1).

QSAR models for chemical risk assessment

Multispecies in vivo toxicity

During the production phase of the Tox21 program in 2015, ~10,000 chemical compounds were screened against a panel of 30 in vitro cell-based assays (including nuclear receptor and stress response pathway assays) in a quantitative HTS (qHTS) format, producing more than 50 million data points [20]. These data were used to build predictive models for different in vivo toxicity end points [20]. The in vivo toxicity data were derived from the Registry of Toxic Effects of Chemical Substances (RTECS) database compiled by Leadscope Inc., yielding 72 toxicity end points related to 13 different species, such as bird, cat, chicken, dog, duck, guinea pig, human, monkey, mouse, quail, rabbit, and rat (Fig. 18.2). The self-organizing map clustering approach was used to build three types of models that could predict these in vivo toxicity end points, namely, activity-based models, structure-based models, and combined models (i.e., the combination of activity data and compound structure). The AUC-ROC value was adopted as a measure of model performance.

The structure-based models generally exhibited better performance than the activity-based models and did not show any species selectivity (Fig. 18.2). Activity-based models for human in vivo toxicity end points performed markedly better than those that predicted rodent (mouse/rat/rabbit) in vivo toxicity endpoints (Fig. 18.2). The differences in model performance may be attributed to species difference, as all screening assays in this analysis were performed using human cells or cell lines [20]. For most in vivo toxicity end points, the models built with Tox21 HTS assay data showed reasonable but less than ideal performance. This may be explained by the fact that Tox21 assays run at the time were mainly focused on nuclear receptors and stress response pathways and did not fully capture all biological aspects involved in toxicity. The limited coverage of the assay panel then suggested the need to expand the scope of the biological response space by including assays targeting other pathways associated with toxicity [20]. The structure–activity combined models exhibited better performance for most in vivo endpoints when compared to structure-based models or activity-based models alone (Fig. 18.2), demonstrating the value of Tox21 in vitro assays for the prediction of in vivo toxicity endpoints [20]. Similar to structure-based models, the combined models did not show any species dependence in terms of model performance (Fig. 18.2). In a follow-up study, the Tox21 assay data were used to build models for predicting the adverse effects of drugs [21]. The results showed that the model performance was significantly improved with the addition of drug targets that represented the biological response space not covered by the Tox21 assays.

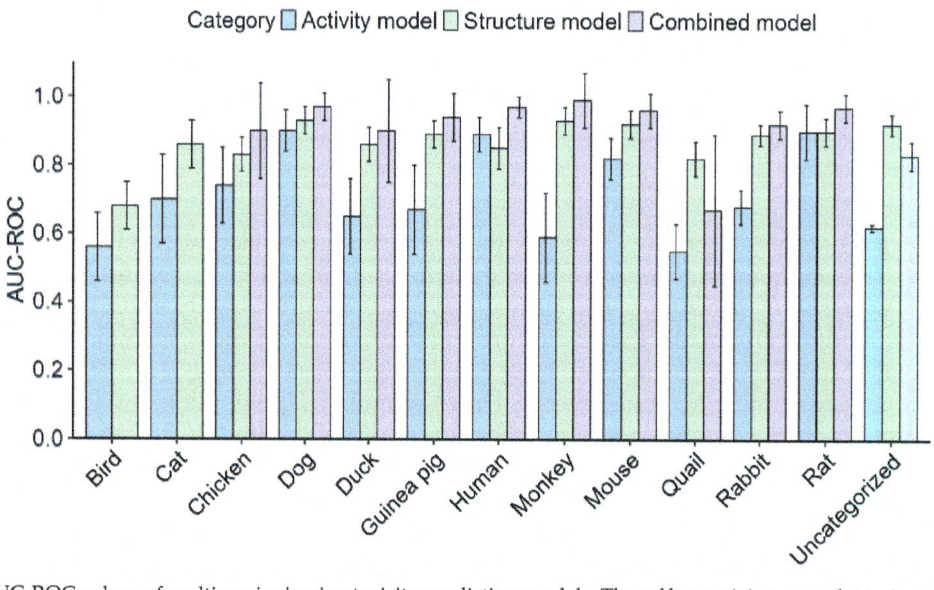

FIGURE 18.2 AUC-ROC values of multispecies in vivo toxicity prediction models. The self-organizing map clustering approach was used to build three types of models that could predict these in vivo toxicity end points based on either the Tox21 in vitro assay data (activity-based model), chemical structure (structure-based model), or both (combined model). Results are presented as mean ± standard deviation (SD). *AUC-ROC*, area under the receiver operating characteristic curve.

Human in vivo toxicity

In 2019, Tox21 in vitro assay data and two common structure-based fingerprints (ToxPrint or ECFP4) were used as input features to build machine learning classification models for predicting human in vivo toxicity [14]. For this study, the Tox21 in vitro assay data included qHTS data generated from screening the Tox21 compound library against 47 assays including 147 readouts. ToxPrint fingerprints (729-bit) were generated within the ChemoTyper application [22], while ECFP4 fingerprints (1024-bit) were generated using the CDK package in KNIME v. 4.0.2 [23]. In vivo human toxicity endpoints were taken from the ChemIDplus Advanced database of the U.S. National Library of Medicine (http://chem.sis.nlm.nih.gov/chemidplus/). A total of 14 human in vivo toxicity endpoints were collected: behavioral; blood; brain and coverings; cardiac; endocrine; gastrointestinal; kidney, ureter, and bladder; liver; lungs, thorax, or respiration; musculoskeletal; peripheral nerve and sensation; sense organs and special senses; skin and appendages skin; and vascular (Fig. 18.3) [14]. A total of 732 chemicals (identified by CAS numbers) was collected after merging the structure (ToxPrint and ECFP4) and in vitro assay data. To reduce data dimension and avoid overfitting, three feature selection methods, including Fisher's exact test with *P* value as well as the importance scores from the RF and XGboost algorithms, were used to remove redundant features for modeling. Five machine learning methods (NB, NNET, RF, SVM, and XGboost) were used to build models for predicting these human in vivo toxicity endpoints, and three metrics (AUC-ROC, BA, and MCC) were used to evaluate model performance.

The optimal AUC-ROC value, as well as BA and MCC for each model, are provided in Fig. 18.3, and the AUC-ROC value was adopted as the principal reference metric. Model performance for predicting each human in vivo toxicity endpoint was found to depend on the combination of the specific machine learning methods and the selected feature sets [14,24]. Models based on the ECFP4 fingerprints, either alone or in combination with in vitro assay data, showed better model performance in predicting the 14 human in vivo toxicity endpoints when compared to the ToxPrint-based models. The contribution of chemical structure and in vitro assay data to model performance varied across the different in vivo toxicity endpoints. Although the combination of in vitro assay data with chemical structure slightly improved the prediction accuracy for most in vivo toxicity endpoints (i.e., blood; endocrine;

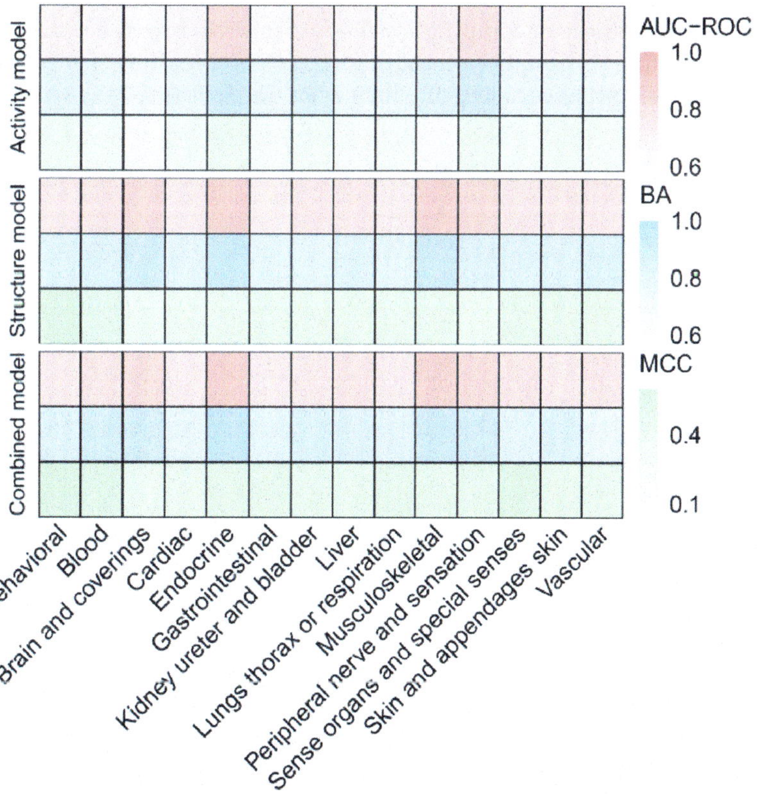

FIGURE 18.3 The performances of human in vivo toxicity prediction models. The machine learning models were built based on either the Tox21 in vitro assay data (activity-based models), chemical structure (ECFP4 structure-based models), or both (combined model). The optimal model was obtained from the best combination of five machine learning models and three feature selection methods. The darker the color, the better the model performance. *AUC-ROC*, area under the receiver operating characteristic curve; *BA*, balanced accuracy; *MCC*, Matthews correlation coefficient.

gastrointestinal; kidney, ureter, and bladder; liver; lungs, thorax, or respiration; musculoskeletal; peripheral nerve and sensation; sense organs and special senses; skin and appendages skin; and vascular), the performances of the combined models were nearly identical to that of the structure-only models, while the assay activity-only models performed relatively poorly in comparison. In principle, assay activity-only models can be applied to predict substances without a defined structure, such as mixtures and natural product extracts. The four best performing combined models were those predicting toxicity for endocrine (AUC-ROC=0.90), musculoskeletal (AUC-ROC=0.88), peripheral nerve and sensory (AUC-ROC=0.85), and brain and overlay (AUC-ROC=0.83), while the best models for the remaining 10 toxicity endpoints all had AUC-ROC values greater than 0.7 (Fig. 18.3).

In vitro toxicity

In addition to in vivo toxicity prediction, the Tox21 in vitro assay data can be used as a knowledge base to link chemical structures to their biological activities for the development of QSAR models. To encourage the mining and use of this publicly available dataset, the "Tox21 Data Challenge 2014" was initiated by NCATS to "crowd source" data analysis by independent researchers and reveal how well they could predict compound interference of cellular and biochemical pathways leading to potential toxicity using chemical structures [25]. Based on data quality and public interest, 12 in vitro assays were selected as training sets for this challenge, including seven nuclear receptor assays (i.e., aryl hydrocarbon receptor (AhR), androgen receptor ligand binding domain (LBD) (AR-LBD), androgen receptor full length (AR), aromatase, estrogen receptor alpha LBD (ER-LBD), estrogen receptor alpha full length (ER), and peroxisome proliferator-activated receptor gamma (PPAR-gamma)) and five stress response pathway assays (i.e., ATPase family AAA domain containing 5 (ATAD5), nuclear factor (erythroid-derived 2)-like 2/antioxidant responsive element (ARE), heat shock factor response element (HSE), mitochondrial membrane potential (MMP), and p53) (Fig. 18.4). This challenge attracted 125 participants from 40 teams representing 18 different countries and received nearly 400 computational models designed to better predict chemical toxicity. The winning models all had good model performance with AUC-ROC values greater than 0.8, some of which even exceeded 0.9 (Fig. 18.4). The winning models included both traditional machine learning methods (e.g., associative neural networks and RF) and deep learning method that used a cascade of many layers of nonlinear processing units to extract features, confirming the ability of computational approaches to provide meaningful predictions of toxicity responses using chemical structure information. In addition, the combination of the individual models produced consensus models with better performance, some of which even outperformed the winning models, demonstrating the wisdom of the crowd [25].

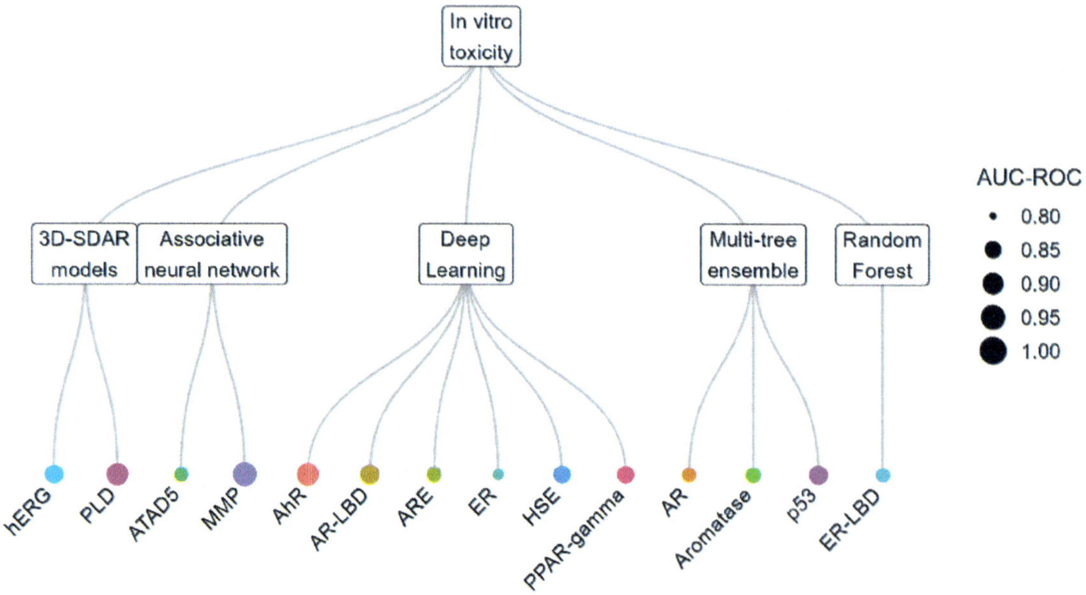

FIGURE 18.4 AUC-ROC values of in vitro toxicity prediction models. The size of the nodes is proportional to the performance of each model. *AhR*, aryl hydrocarbon receptor; *AR*, androgen receptor full length; *AR-LBD*, androgen receptor LBD; *ARE*, nuclear factor (erythroid-derived 2)-like 2/antioxidant responsive element; *ATAD5*, ATPase family AAA domain containing 5; *ER*, estrogen receptor alpha full length; *ER-LBD*, estrogen receptor alpha LBD; *hERG*, Human Ether-à-go-go Related Gene; *HSE*, heat shock factor response element; *MMP*, mitochondrial membrane potential; *PLD*, phospholipidosis; *PPAR-gamma*, peroxisome proliferator-activated receptor gamma.

In addition to the models developed for the Tox21 Challenge, Slavov et al. built 3D-SDAR models for predicting human Ether-à-go-go Related Gene (hERG) and phospholipidosis (PLD) activities of compounds using Tox21 data as the training sets, and both models showed excellent performance (AUC-ROC values >0.87) (Fig. 18.4) [26]. Wu et al. constructed >5000 models based on qHTS data of the Tox21 10K compound library [4] screened against 65 assays, and found that complex models (e.g., (LS-)SVM and RF) performed only slightly better than simple models (e.g., linear regression and k-nearest neighbors) [27]. Consistent with the findings in the "Tox21 Data Challenge 2014," Mansouri et al. used two consensus models, Collaborative Modeling Project for Androgen Receptor Activity (CoMPARA) and the Collaborative Estrogen Receptor Activity Prediction Project (CERAPP), to perform a virtual screening of chemicals for potential androgenic and estrogenic activity, showing that consensus models tend to outperform individual models [28,29]. In comparison to traditional consensus approaches, the Deep Learning Consensus Architecture that combines consensus and multitask deep learning approaches to generate large-scale QSAR models demonstrated improved prediction accuracies for both regression and classification tasks when predicting Tox21 assay activities [30].

QSAR models for drug discovery

Inhibitors and substrates of cytochrome P450 3A7 and 3A4

Inappropriate use of prescription drugs in neonates may be more harmful than their use in adults. The cytochrome P450 (CYP) subfamily CYP3A is one of the most abundant CYP enzymes in the human liver and exhibits a developmental switch in expression, whereby CYP3A7 is expressed in fetal tissues, followed by a transition to CYP3A4 expression during the first month of life [31,32]. Such a phenomenon may lead to differences in the effects of drugs on neonates and adults. In this section, two types of datasets, including CYP-binding qHTS data for ~5000 known drugs and bioactive molecules, as well as metabolic stability data for a subset of ~3300 of these compounds, were collected as training datasets for the construction of QSAR models [33]. To identify isozyme-specific activities, the CYP3A4 and CYP3A7 datasets were split into two categories: activity datasets and selectivity datasets (Fig. 18.5). The activity dataset consists of CYP3A4 inhibitors, CYP3A7 inhibitors, CYP3A4 substrates, and CYP3A7 substrates. By comparing compound activities in inhibition and metabolic stability assays, the compiled selectivity dataset consists of CYP3A7 selective inhibitor (vs. CYP3A4), CYP3A7 substrate (noninhibitor), CYP3A7 inhibitor (nonsubstrate), and CYP3A7 selective substrate (vs. CYP3A4). Five different classification algorithms (i.e., NB, NNET, RF, SVM, and XGboost), feature selection (i.e., Fisher's exact test), and multiple data rebalancing techniques (i.e., Down sampling, Up sampling, ROSE, and SMOTE) were used to build models. Model performance was evaluated by three metrics, namely AUC-ROC, BA, and MCC.

The performance of each CYP3A activity/selectivity prediction model varied depending on the combination of dataset type, model type, rebalancing method, and specific feature set (Fig. 18.5). For the activity datasets, the RF classifier produced the optimal performance for CYP3A4 inhibitor (AUC-ROC = 0.84 ± 0.01) and CYP3A7 inhibitor (AUC-ROC = 0.80 ± 0.01), while the SVM classifier produced the optimal performance for CYP3A4 substrate (AUC-ROC = 0.84 ± 0.02) and CYP3A7 substrate (AUC-ROC = 0.77 ± 0.01). For the selectivity models, the SVM classifier produced the optimal performance for CYP3A7 selective inhibitor (vs. CYP3A4) (AUC-ROC = 0.72 ± 0.02), CYP3A7 substrate (noninhibitor) (AUC-ROC = 0.74 ± 0.03), and CYP3A7 selective substrate (vs. CYP3A4) (AUC-ROC = 0.79 ± 0.04), while the NB classifier produced the optimal performance for CYP3A7 inhibitor (nonsubstrate) (AUC-ROC = 0.73 ± 0.02). The rebalancing approach was helpful to improve the performance of some prediction models, such as ROSE for CYP3A7 substrate (noninhibitor) and CYP3A7 selective substrate (vs. CYP3A4), and Up sampling for CYP3A7 selective inhibitor (vs. CYP3A4) (Fig. 18.5B). The number of ECFP4 features that generated the optimal models were: CYP3A4 inhibitor (415), CYP3A7 inhibitor (390), CYP3A4 substrate (343), CYP3A7 substrate (189), CYP3A7 selective inhibitor (vs. CYP3A4) (133), CYP3A7 substrate (noninhibitor) (33), CYP3A7 inhibitor (nonsubstrate) (116), and CYP3A7 selective substrate (vs. CYP3A4) (43) (Fig. 18.5B). These best-in-class models can be used to qualitatively assess the potential CYP3A activity of new drug candidates on fetuses/neonates prior to clinical trials.

Anti-SARS-CoV-2 compounds

Using compound screening to identify inhibitors that block the entry of SARS-CoV-2 into human host cells or the SARS-CoV-2 3-chymotrypsin-like (3CL) protease can reveal lead compounds that may be developed into

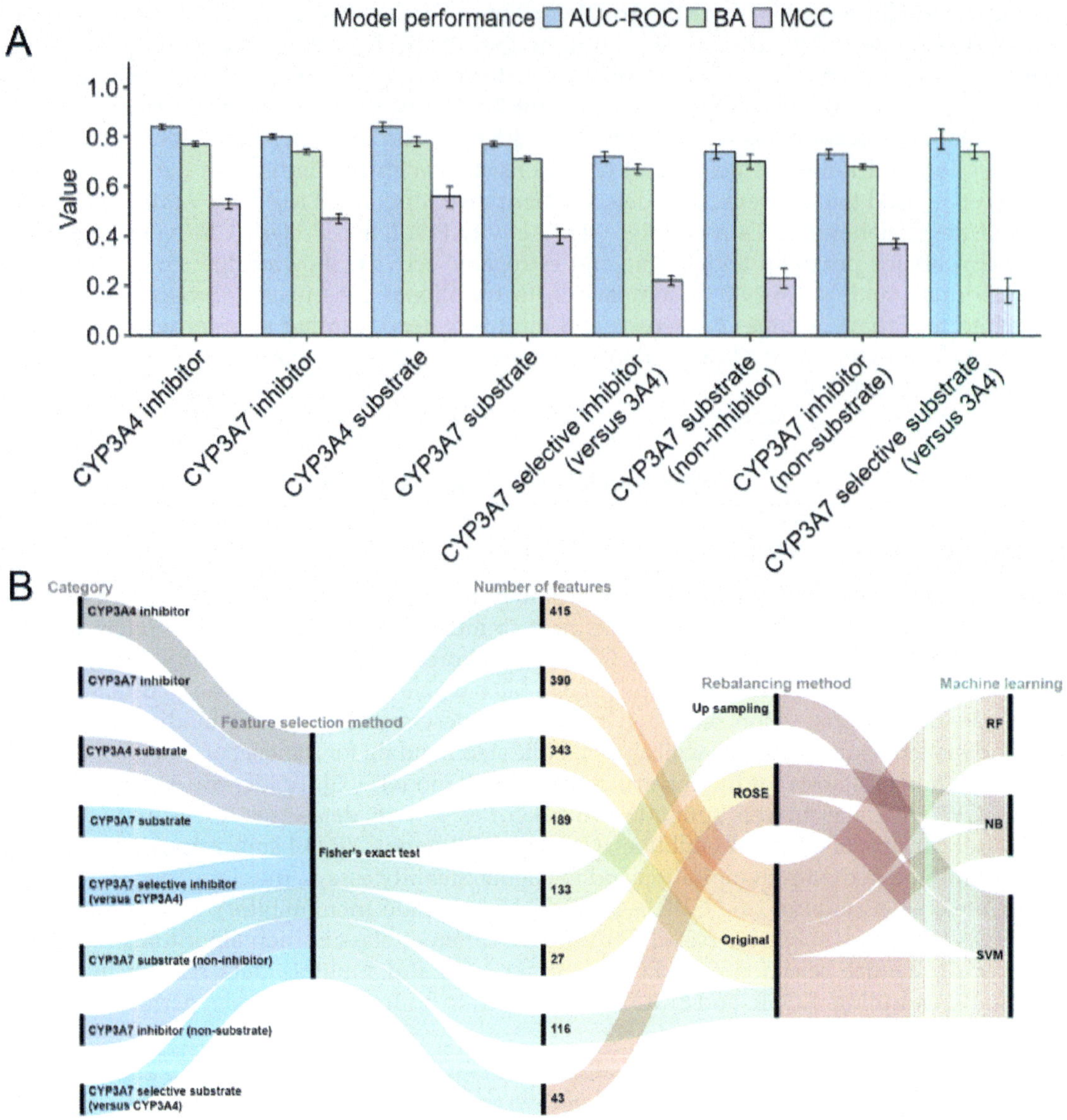

FIGURE 18.5 Optimal models for CYP3A7/3A4 activity and selectivity datasets. (A) The optimal AUC-ROC value together with BA and MCC for each model. Results are presented as mean ± standard deviation (SD). (B) Parameter sets for each optimal model. *AUC-ROC*, area under the receiver operating characteristic curve; *BA*, balanced accuracy; *MCC*, Matthews correlation coefficient; *NB*, Naive Bayes; *RF*, random forest; *ROSE*, random over sampling examples; *SVM*, support vector machine.

anti-COVID-19 therapeutics [34]. A total of 8149 compounds were screened against either the pseudotyped particle (PP) entry assay (identification of inhibitors that block the entry of SARS-CoV-2 into human host cells, including 2725 compounds) or the 3CL protease assay (identification of inhibitors of 3CL protease, including 8044 compounds) [13,35,36]. In this section, these two assays were applied to build machine learning models (e.g., QSAR and biological activity-based modeling [BABM] approaches) for predicting new inhibitors of PP entry or 3CL protease [15].

Multiple parameter combinations were performed on the training dataset to find the optimal QSAR model for each assay target. The combination of 157 ECFP4 features (Fisher's exact test, P value = 0.02) and RF (machine learning algorithm) resulted in the optimal QSAR model performance (AUC-ROC = 0.77 ± 0.02) for the prediction of PP entry inhibitors (Fig. 18.6A). The combination of 80 ECFP4 features (Fisher's exact test, P-value = 0.05), ROSE (rebalancing strategy), and SVM (machine learning algorithm) resulted in the optimal model performance (AUC-ROC = 0.90 ± 0.03) for the prediction of 3CL protease inhibitors (Fig. 18.6A), and this model outperformed previously reported models without applying any rebalancing strategy (AUC-ROC = 0.71) [37]. In addition, the structure–activity combined models (i.e., CM-M, structure (ToxPrint)-based model + BABM-M) produced the best performance for predicting PP entry inhibitors (AUC-ROC = 0.88 ± 0.01) and 3CL protease inhibitors

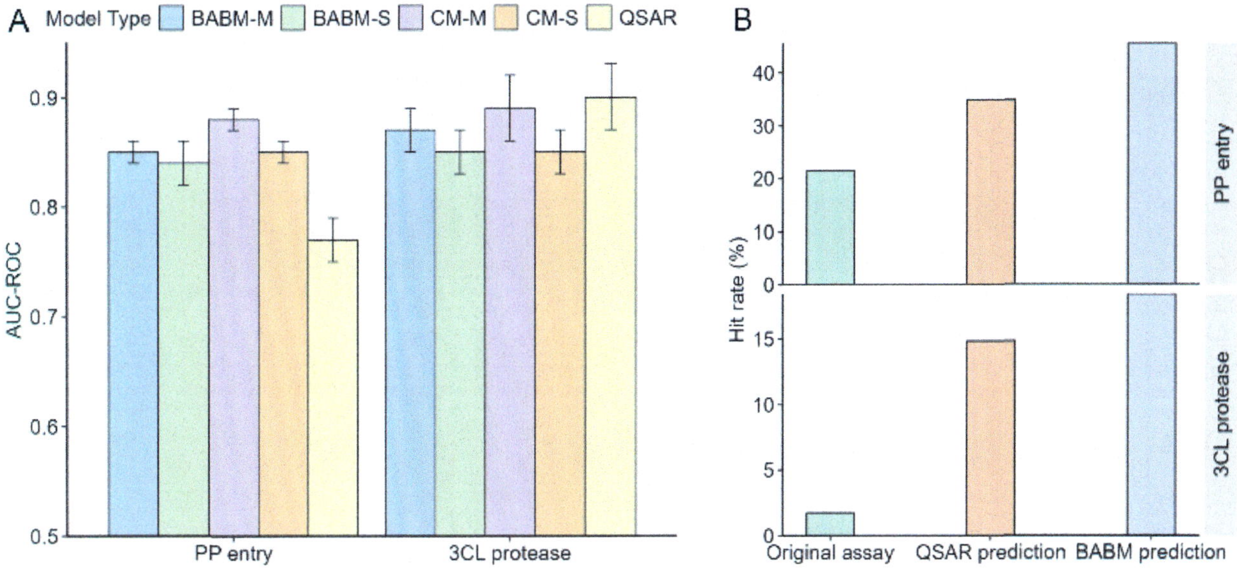

FIGURE 18.6 **Identification of anti-SARS-CoV-2 compounds using machine learning models.** (A) The model performance measured by AUC-ROC value. Results are presented as mean ± standard deviation (SD). (B) Comparison of hit rates between the original drug repurposing screens and model predictions. *BABM-M*, activity-based model (MLS); *BABM-S*, activity-based model (Sytravon); *CM-M*, combined model (SBM + BABM-M); *CM-S*, combined model (SBM + BABM-S); *QSAR*, quantitative structure (ECFP4)-activity relationship; *SBM*, structure (ToxPrint)-based model.

(AUC-ROC = 0.89 ± 0.03) (Fig. 18.6A). The combined models showed better performance than the QSAR models, further confirming the value of biological activity profiles in improving model prediction performance [38,39].

Subsequently, the optimal models (including QSAR and BABM) were applied to virtually screen a collection of ~360K diverse compounds to identify novel inhibitors of SARS-CoV-2 entry and 3CL protease. The optimal QSAR models increased the hit rate of the HTS assay by 1.6-fold (from 21.5% to 34.9%) for the PP entry inhibitors and 8.4-fold (from 1.76% to 14.8%) for the 3CL protease inhibitors (Fig. 18.6B). Likewise, the optimal BABM models increased the assay hit rate by 2.1-fold for the PP entry (from 21.5% to 45.6%), and 10.4-fold (from 1.76% to 18.4%) for the 3CL protease inhibitors (Fig. 18.6B). In summary, the use of machine learning classification models (including QSAR and BABM) as a complementary approach to HTS can improve the speed and efficiency of anti-SARS-CoV-2 drug discovery.

Inhibitors of AChE and BChE

Acetylcholinesterase (AChE; EC 3.1.1.7) and butyrylcholinesterase (BChE; EC 3.1.1.8) are responsible for ensuring normal cholinergic neurotransmission by catalyzing the hydrolysis of acetylcholine to choline and acetate. In patients with Alzheimer's disease, the reduction of acetylcholine (ACh) is further amplified by the hydrolysis of AChE/BChE [40,41]. AChE and BChE act as major ACh degrading enzymes at different stages of Alzheimer's disease progression, that is, early and advanced stages, respectively [42]. Therefore, the identification of AChE/BChE inhibitors would contribute to the expansion of the therapeutic options for Alzheimer's disease. A total of 12,092 compounds were screened for their inhibitory activity against AChE (6995 compounds) or BChE (7292 compounds) in a qHTS format [43,44]. The screening results showed that the number of inhibitors was much smaller than that of inactive compounds, leading to an activity rate of 6.8% for the AChE assay and 10.5% for the BChE assay. The compounds were divided into two subcategories based on their activity profile: AChE/BChE inhibitors (activity) and selective inhibitors (selectivity) (Fig. 18.7). Five different machine learning algorithms, that is, NB, NNET, RF, SVM, and XGboost, were applied to build predictive models based on ToxPrint chemotypes (729-bit).

The model performances varied depending on different combinations of feature selection methods, data rebalancing methods, and types of machine learning classification algorithms. The best classification model for each dataset was: XGboost (XGboost (feature selection method); original (data rebalancing approach)) for AChE activity dataset (AUC-ROC = 0.86 ± 0.02); RF (Fisher's exact test (feature selection method); original (data rebalancing approach)) for BChE activity dataset (AUC-ROC = 0.87 ± 0.01); SVM (XGboost (feature selection method); original

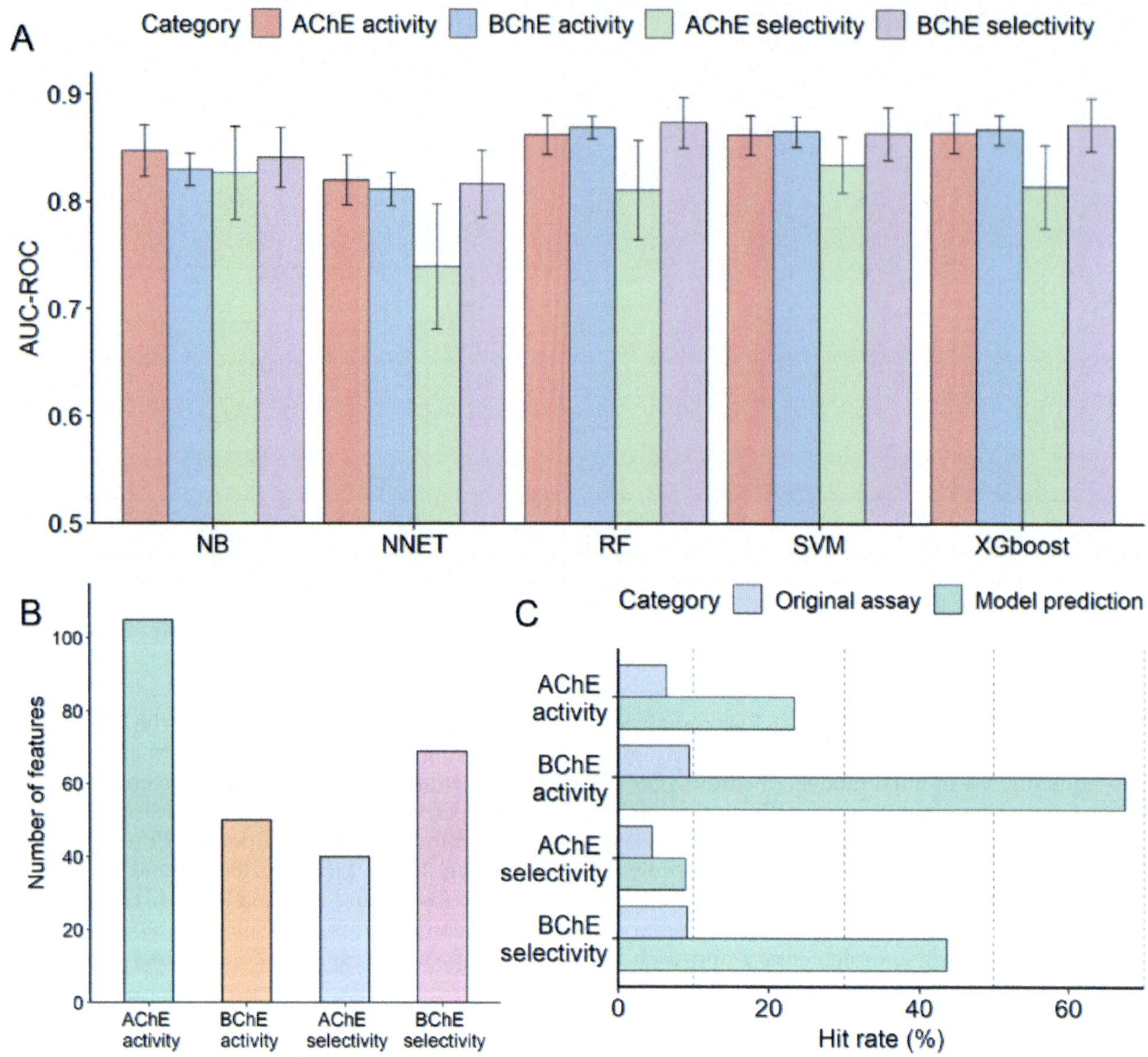

FIGURE 18.7 Identification of AChE/BChE inhibitors using machine learning models. (A) AUC-ROC value of the best machine learning model. Results are presented as mean ± standard deviation (SD). (B) The number of features for each optimal model. (C) Comparison of hit rates between the original drug repurposing screens and model predictions. *AUC-ROC*, area under the receiver operating characteristic curve; *NB*, Naive Bayes; *NNET*, neural networks; *RF*, random forest; *SVM*, support vector machine; *XGboost*, extreme gradient boosting.

(data rebalancing approach)) for AChE selectivity dataset (AUC-ROC = 0.83 ± 0.03); and RF (XGboost (feature selection method); original (data rebalancing approach)) for BChE selectivity dataset (AUC-ROC = 0.87 ± 0.02) (Fig. 18.7A). The number of features involved in the best models varied, ranging from 40 features in the AChE selectivity prediction model to 105 features in the AChE inhibitor prediction model (Fig. 18.7B). In total, 2762 structurally diverse compounds with a predicted probability >0.5 were selected for experimental validation. Compared to the original HTS assays, the optimal models significantly increased the assay hit rates, that is, 3.6-fold increase in the AChE inhibition assay (from 6.4% to 23.3%) and 7.3-fold increase in the BChE inhibition assay (from 9.5% to 68%), 2-fold increase in the identification of AChE selective inhibitors (from 4.5% to 8.9%), and 4.8-fold in the identification of BChE selective inhibitors (from 9.2% to 44%) (Fig. 18.7C). Experimental validation showed that the probability predicted by the model is a good indicator of AChE/BChE activity and selectivity for a test compound.

Activators and inhibitors of opioid receptors

Opioid receptors (OPRs) are members of the superfamily of G protein-coupled receptors (GPCR), and they are divided into three main types: delta (OPRD), kappa (OPRK), and mu (OPRM). These receptors play important roles in chronic pain transmission and are present in the central nervous system and peripheral tissues [45]. A collection of ~3000 approved drugs was screened against three cell-based assays (i.e., OPRD, OPRK, and OPRM) in both agonist and antagonist modes [46], and these results were used as training datasets to develop four QSAR models (i.e., NN, RF, SVM, and XGBoost) for the identification of new activators/inhibitors of the three OPRs based on chemical structure [47]. Furthermore, models with good performance were applied to virtually screen ~50K compounds to identify potential new OPR agonists and antagonists, which were subsequently experimentally confirmed.

After 5-fold cross-validation with 20 iterations on the training set [47], external validation was performed on models built with different combinations of structural fingerprints (i.e., PubChem), up sampling strategy, and machine learning algorithm (i.e., RF) (Fig. 18.8A). Model performance was evaluated by three metrics, including

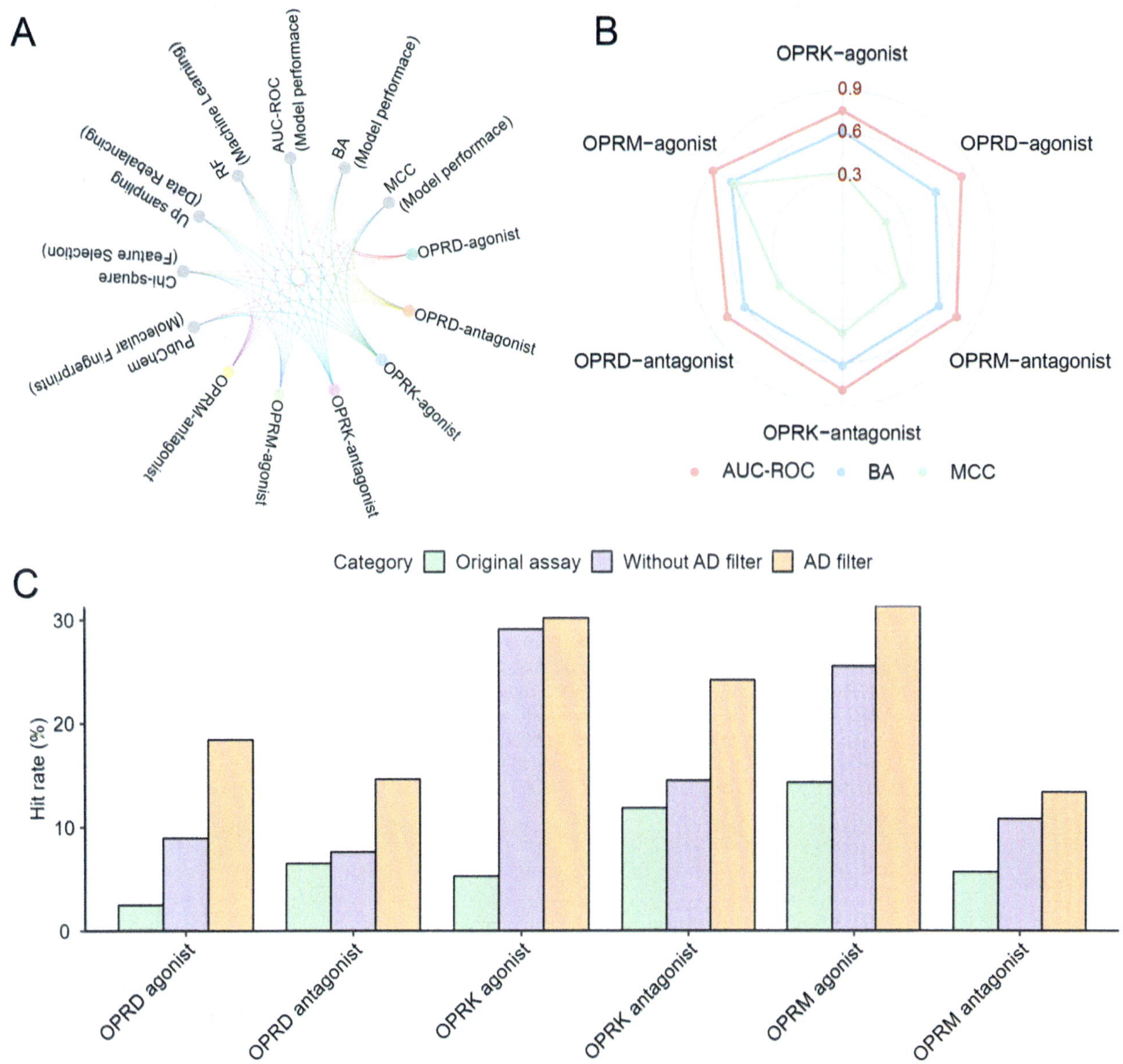

FIGURE 18.8 **Identification of agonists and antagonists of opioid receptors.** (A) Parameter sets for each optimal model. (B) The optimal AUC-ROC value together with BA and MCC for each model. (C) Comparison of hit rates between the original drug repurposing screens and model predictions. *AD*, applicability domain; *AUC-ROC*, area under the receiver operating characteristic curve; *BA*, balanced accuracy; *MCC*, Matthews correlation coefficient; *RF*, random forest.

AUC-ROC, BA, and MCC. The AUC-ROC values for all six models were >0.76, in which the models predicting OPRM-agonist and OPRD-agonist had AUC-ROC values >0.8 (Fig. 18.8B). The optimal models were further applied to make predictions on a library of ~50K diverse compounds that had no experimental data. Compounds with a predicted probability ≥0.5 were clustered using the k-modes algorithm, yielding 2450 clusters ranging in sizes from 1 to 112 compounds. For clusters with sizes <10, the compounds with the highest predicted probability scores were selected for experimental validation, and if the cluster size was >10, the top two compounds were selected. The highest hit rates were obtained from the prediction model for OPRM-agonist (31%), OPRK-agonist (30%) and OPRK-antagonist (24%), OPRD-agonist (18%) and OPRD-antagonist (15%), and OPRM-antagonist (14%) (Fig. 18.8C). The Tanimoto similarity between each predicted active compound and all active compounds in the training set was calculated to evaluate the applicability domain (AD) of the model. Predicted actives with Tanimoto scores >0.8 were selected to re-evaluate the hit rate of each model. In the original analysis (without the AD filter), only two models (i.e., OPRK-agonist and OPRD-agonist) increased the assay hit rate by ≥2-fold. When an AD was defined with a maximum Tanimoto score >0.8, all six models increased the hit rate by ≥2-fold, and the models for OPRK-agonist and OPRD-agonist even increased up to 6- and 7-fold, respectively. In summary, these robust OPR prediction models could be applied to prioritize compounds from large compound libraries for further experimental validation.

Conclusions and future directions

QSAR models based on machine learning methods could serve as efficient and effective alternatives to traditional experimental HTS assays, and they have been leveraged to virtually screen large chemical libraries for the identification of potential biological activities of compounds in the toxicology and drug discovery fields. In addition, the top-contributing structural features identified by the models could serve as structure alerts for toxicity and complement the structure–activity relationship analysis in the process of drug design and discovery. In contrast to QSAR, the BABM approach was developed based on the following assumptions: compounds that exhibit similar patterns of biological activity often share similar biological targets or mechanisms of action [20,38,48]. The combined use of these two methods (i.e., QSAR and BABM) can expand the scope of substances amenable for virtual screening, that is, their field of application is not limited to compounds with well-defined structures (e.g., QSAR) or existing biological activity (e.g., BABM). For the biological activity (e.g., assay activity)-based modeling approach to work more efficiently, it is important to establish a comprehensive set of activity profiles across a diverse range of biological targets and pathways. Lastly, consensus models were found to generally perform better than individual models, which may be a potential direction for future model improvements.

Acknowledgments

This work was supported by the Intramural Research Programs of the National Toxicology Program (Interagency agreement #Y2-ES-7020-01), National Institute of Environmental Health Sciences, the US Environmental Protection Agency (Interagency Agreement #Y3-HG-7026-03) and the Intramural Research Program of the National Center for Advancing Translational Sciences, National Institutes of Health. We also thank Srilatha Sakamuru, Jinghua Zhao, Caitlin Lynch, Li Zhang, Shuaizhang Li, Samuel Michael, Jameson Travers, and Carleen Klumpp-Thomas for assisting with the screens, Paul Shinn, Misha Itkin, Zina Itkin, Danielle Bougie, and Katlin Recabo for compound management and William Leister, Christopher LeClair, and Dingyin Tao for the Tox21 10K library quality control (QC). We also thank the NCATS OpenData Portal team for making the screening data publicly available.

References

[1] Xu T, Luo Y, Xie HQ, Xia Y, Li Y, Chen Y, et al. Systematic identification of molecular mechanisms for aryl hydrocarbon receptor mediated neuroblastoma cell migration. Environ Int 2022;168:107461.
[2] Kerr MA, Nasca PC, Mundt KA, Michalek AM, Baptiste MS, Mahoney MC. Parental occupational exposures and risk of neuroblastoma: a case–control study (United States). Cancer Causes Control 2000;11(7):635–43.
[3] Müller T. Rivastigmine in the treatment of patients with Alzheimer's disease. Neuropsychiatr Dis Treat 2007;3(2):211–8.
[4] Richard AM, Huang R, Waidyanatha S, Shinn P, Collins BJ, Thillainadarajah I, et al. The Tox21 10K compound library: collaborative chemistry advancing toxicology. Chem Res Toxicol 2020;34(2):189–216.
[5] NRC. Toxicity testing in the 21st century: a vision and a strategy. Washington, DC: The National Academies Press; 2007.
[6] Collins FS, Gray GM, Bucher JR. Toxicology. Transforming environmental health protection. Science 2008;319(5865):906–7.
[7] Kavlock RJ, Austin CP, Tice RR. Toxicity testing in the 21st century: implications for human health risk assessment. Risk Anal 2009;29(4): 485–7.

[8] Tice RR, Austin CP, Kavlock RJ, Bucher JR. Improving the human hazard characterization of chemicals: a Tox21 update. Environ Health Perspect 2013;121(7):756–65.
[9] Huang R, Southall N, Wang Y, Yasgar A, Shinn P, Jadhav A, et al. The NCGC pharmaceutical collection: a comprehensive resource of clinically approved drugs enabling repurposing and chemical genomics. Sci Transl Med 2011;3(80):80ps16.
[10] Huang R, Zhu H, Shinn P, Ngan D, Ye L, Thakur A, et al. The NCATS pharmaceutical collection: a 10-year update. Drug Discov Today 2019; 24(12):2341–9.
[11] NCATS. Tox21 Data Browser 2021. Available from: https://tripod.nih.gov/tox21/pubdata.
[12] PubChem. Tox21 phase II data 2021. Available from: http://www.ncbi.nlm.nih.gov/pcassay?term=tox21.
[13] Brimacombe KR, Zhao T, Eastman RT, Hu X, Wang K, Backus M, et al. An OpenData portal to share COVID-19 drug repurposing data in real time. bioRxiv 2020.
[14] Xu T, Ngan DK, Ye L, Xia M, Xie HQ, Zhao B, et al. Predictive models for human organ toxicity based on in vitro bioactivity data and chemical structure. Chem Res Toxicol 2020;33(3):731–41.
[15] Xu T, Xu M, Zhu W, Chen CZ, Zhang Q, Zheng W, et al. Efficient identification of anti-SARS-CoV-2 compounds using chemical structure-and biological activity-based modeling. J Med Chem 2022;65(6):4590–9.
[16] Saeys Y, Abeel T, Van de Peer Y. Robust feature selection using ensemble feature selection techniques. In: Joint European conference on machine learning and knowledge discovery in databases. Springer; 2008.
[17] Longadge R, Dongre S. Class imbalance problem in data mining review. 2013. arXiv preprint arXiv:1305.1707.
[18] Lunardon N, Menardi G, Torelli N. ROSE: a package for binary imbalanced learning. R J 2014;6(1).
[19] Torgo L, Torgo ML. Package 'DMwR'. Comprehensive R Archive Network; 2013.
[20] Huang R, Xia M, Sakamuru S, Zhao J, Shahane SA, Attene-Ramos M, et al. Modelling the Tox21 10 K chemical profiles for in vivo toxicity prediction and mechanism characterization. Nat Commun 2016;7:10425.
[21] Huang R, Xia M, Sakamuru S, Zhao J, Lynch C, Zhao T, et al. Expanding biological space coverage enhances the prediction of drug adverse effects in human using in vitro activity profiles. Sci Rep 2018;8(1):1–12.
[22] Yang C, Tarkhov A, Marusczyk J, Bienfait B, Gasteiger J, Kleinoeder T, et al. New publicly available chemical query language, CSRML, to support chemotype representations for application to data mining and modeling. J Chem Inf Model 2015;55(3):510–28.
[23] Beisken S, Meinl T, Wiswedel B, de Figueiredo LF, Berthold M, Steinbeck C. KNIME-CDK: workflow-driven cheminformatics. BMC Bioinf 2013;14:257.
[24] Ye L, Ngan DK, Xu T, Liu Z, Zhao J, Sakamuru S, et al. Prediction of drug-induced liver injury and cardiotoxicity using chemical structure and in vitro assay data. Toxicol Appl Pharmacol 2022;454:116250.
[25] Huang R, Xia M, Nguyen D-T, Zhao T, Sakamuru S, Zhao J, et al. Tox21Challenge to build predictive models of nuclear receptor and stress response pathways as mediated by exposure to environmental chemicals and drugs. Front Environ Sci 2016;3.
[26] Slavov S, Stoyanova-Slavova I, Li S, Zhao J, Huang R, Xia M, et al. Why are most phospholipidosis inducers also hERG blockers? Arch Toxicol 2017;91(12):3885–95.
[27] Wu L, Huang R, Tetko IV, Xia Z, Xu J, Tong W. Trade-off predictivity and explainability for machine-learning powered predictive toxicology: an in-depth investigation with Tox21 data sets. Chem Res Toxicol 2021;34(2):541–9.
[28] Mansouri K, Abdelaziz A, Rybacka A, Roncaglioni A, Tropsha A, Varnek A, et al. CERAPP: collaborative estrogen receptor activity prediction project. Environ Health Perspect 2016;124(7):1023–33.
[29] Mansouri K, Kleinstreuer N, Abdelaziz AM, Alberga D, Alves VM, Andersson PL, et al. CoMPARA: collaborative modeling project for androgen receptor activity. Environ Health Perspect 2020;128(2):027002.
[30] Zakharov AV, Zhao T, Nguyen D-T, Peryea T, Sheils T, Yasgar A, et al. Novel consensus architecture to improve performance of large-scale multitask deep learning QSAR models. J Chem Inf Model 2019;59(11):4613–24.
[31] Hines RN. Ontogeny of human hepatic cytochromes P450. J Biochem Mol Toxicol 2007;21(4):169–75.
[32] Gu X, Manautou JE. Molecular mechanisms underlying chemical liver injury. Expert Rev Mol Med 2012;14:e4–.
[33] Kabir M, Padilha EC, Shah P, Huang R, Sakamuru S, Gonzalez E, et al. Identification of selective CYP3A7 and CYP3A4 substrates and inhibitors using a high-throughput screening platform. Front Pharmacol 2022;2272.
[34] Xu T, Zheng W, Huang R. High-throughput screening assays for SARS-CoV-2 drug development: current status and future directions. Drug Discov Today 2021;26(10):2439–44.
[35] Chen CZ, Xu M, Pradhan M, Gorshkov K, Petersen JD, Straus MR, et al. Identifying SARS-CoV-2 entry inhibitors through drug repurposing screens of SARS-S and MERS-S pseudotyped particles. ACS Pharmacol Transl Sci 2020;3(6):1165–75.
[36] Zhu W, Xu M, Chen CZ, Guo H, Shen M, Hu X, et al. Identification of SARS-CoV-2 3CL protease inhibitors by a quantitative high-throughput screening. ACS Pharmacol Transl Sci 2020;3(5):1008–16.
[37] Kc GB, Bocci G, Verma S, Hassan MM, Holmes J, Yang JJ, et al. A machine learning platform to estimate anti-SARS-CoV-2 activities. Nat Mach Intell 2021;3(6):527–35.
[38] Huang R, Xu M, Zhu H, Chen CZ, Zhu W, Lee EM, et al. Biological activity-based modeling identifies antiviral leads against SARS-CoV-2. Nat Biotechnol 2021;39(6):747–53.
[39] Liu J, Mansouri K, Judson RS, Martin MT, Hong H, Chen M, et al. Predicting hepatotoxicity using ToxCast in vitro bioactivity and chemical structure. Chem Res Toxicol 2015;28(4):738–51.
[40] Bartus RT, Dean 3rd RL, Beer B, Lippa AS. The cholinergic hypothesis of geriatric memory dysfunction. Science 1982;217(4558):408–14.
[41] Bartus RT. On neurodegenerative diseases, models, and treatment strategies: lessons learned and lessons forgotten a generation following the cholinergic hypothesis. Exp Neurol 2000;163(2):495–529.
[42] Li Q, Yang H, Chen Y, Sun H. Recent progress in the identification of selective butyrylcholinesterase inhibitors for Alzheimer's disease. Eur J Med Chem 2017;132:294–309.
[43] Li S, Zhao J, Huang R, Travers J, Klumpp-Thomas C, Yu W, et al. Profiling the Tox21 chemical collection for acetylcholinesterase inhibition. Environ Health Perspect 2021;129(4):47008.
[44] Li S, Li AJ, Travers J, Xu T, Sakamuru S, Klumpp-Thomas C, et al. Identification of compounds for butyrylcholinesterase inhibition. SLAS Discov Adv Life Sci R & D 2021;26(10):1355–64.

[45] Hill S, Ganellin C, Timmerman H, Schwartz J, Shankley N, Young J, et al. International Union of Pharmacology. XIII. Classification of histamine receptors. Pharmacol Rev 1997;49(3):253–78.
[46] Inglese J, Auld DS, Jadhav A, Johnson RL, Simeonov A, Yasgar A, et al. Quantitative high-throughput screening: a titration-based approach that efficiently identifies biological activities in large chemical libraries. Proc Natl Acad Sci USA 2006;103(31):11473–8.
[47] Sakamuru S, Zhao J, Xia M, Hong H, Simeonov A, Vaisman I, et al. Predictive models to identify small molecule activators and inhibitors of opioid receptors. J Chem Inf Model 2021;61(6):2675–85.
[48] Xu T, Wu L, Xia M, Simeonov A, Huang R. Systematic identification of molecular targets and pathways related to human organ level toxicity. Chem Res Toxicol 2020;34(2):412–21.

CHAPTER

19

EADB—A database providing curated data for developing QSAR models to facilitate the assessment of endocrine activity

Fan Dong[1], Wenjing Guo[1], Jie Liu[1], Liang Xu[1], Myeongsang Lee[1], Meng Song[1], Zoe Li[1], Tucker A. Patterson[1,2] and Huixiao Hong[1]

[1]Division of Bioinformatics and Biostatistics, National Center for Toxicological Research, United States Food and Drug Administration, Jefferson, AR, United States [2]Office of Research, National Center for Toxicological Research, United States Food and Drug Administration, Jefferson, AR, United States

Abbreviations

AUC Area under the receiver operating characteristic curve
CERAPP Collaborative Estrogen Receptor Activity Prediction Project
ChEMBL Chemical Database of the European Molecular Biology Laboratory
EADB Estrogenic activity database
EDKB Endocrine disruptor knowledge base
EPA Environmental Protection Agency
ER Estrogen receptor
GPR Gaussian process regression
kNN k-nearest neighbors
MCC Mathews correlation coefficient
METI Ministry of Economy, Trade, and Industry
MLR Multiple linear regression
QSAR Quantitative structure–activity relationship
RA Relative activity
RBA Relative binding affinity
RMSE Root mean squared error
SVM Support vector machine
SVR Support vector regression

Introduction

The endocrine system, also known as the hormone system, is a complex system made up of glands, hormones, and receptors. Glands are located across the body to produce, store, and release different hormones. Major glands of the endocrine system are the hypothalamus, pituitary, thyroid, parathyroid, adrenals, pineal body, testes, and ovaries [1]. Hormones produced by glands are secreted into the blood and travel through tissues and organs. Hormones have different types of chemical structures to communicate between organs and tissues and coordinate body functions for behavioral activities and physiological regulation. Receptors located in various organs and tissues recognize and respond to hormones. Hormone receptors exist on the surface of the target cell or within the target cell depending on the type of hormone. When hormones are released into the bloodstream, those target cells with compatible receptors would bind and respond. The endocrine system is found in all mammals, fish, and birds, and it plays an important role in controlling and regulating all major body functions such as growth and development, behavior, immunity, energy control, and reproduction. For example, insulin is an essential hormone to control blood sugar by absorbing glucose in the bloodstream [2]. Estradiol [3] and testosterone [4] are the principal hormones that control the sexual characteristics of males and females, respectively, and impact the function of reproductive organs. Thyroid hormones [5], including triiodothyronine and thyroxine, control the body's metabolism to transfer the food you eat into energy. Problems such as diabetes, thyroid disorders, polycystic ovary syndrome,

QSAR in Safety Evaluation and Risk Assessment
https://doi.org/10.1016/B978-0-443-15339-6.00015-1

low testosterone, osteoporosis, excessive stress levels, weak bones, and improper development during growth may occur if the endocrine system is interrupted [6].

Endocrine disruptors are chemicals that mimic endogenous hormones in the body to interact with hormone receptors in the endocrine system, impairing endocrine system functions and disrupting the endocrine system. For example, the US Environmental Protection Agency (EPA) defines an endocrine disruptor as an exogenous agent that interferes with the production, release, transport, binding, action, or elimination of natural hormones in the body responsible for homeostatic maintenance and developmental regulation. The World Health Organization states that an endocrine disruptor is an exogenous substance or mixture that alters functions of the endocrine system and consequently causes an adverse effect on an organism or its progeny. Endocrine-disrupting chemicals can mimic hormones to induce inappropriate responses of the target receptors in the endocrine system or block the natural hormones from binding to their target receptors, which could cause developmental malformations, interference with reproduction, and disturbances in hormone and nervous system function. Some endocrine-disrupting chemicals can increase or decrease the level of hormones in blood by interfering with hormone production, storage, and transfer procedures [7,8], causing a variety of adverse effects to humans and animals. Therefore, endocrine disruptors have been a major research subject in the scientific community and have raised a huge concern to regulatory agencies over the past 20 years.

In 1996, two laws were passed in the US Congress. The Food Quality Protection Act of 1996 (https://www.congress.gov/bill/104th-congress/house-bill/1627) mandated that EPA develop a screening program with appropriate evaluation methods to determine whether certain chemicals may have adverse effects on the endocrine system in humans and wildlife. The Safe Drinking Water Act Amendments of 1996 (https://www.congress.gov/bill/104th-congress/senate-bill/1316) require actions to protect drinking water and its sources and authorize EPA to develop standards for drinking water to protect from contaminants such as endocrine-disrupting chemicals. Therefore, in 1998, EPA launched the Endocrine Disruptor Screening Program (https://www.epa.gov/endocrine-disruption/endocrine-disruptor-screening-program-edsp-overview) to screen chemicals, pesticides, and environmental contaminants for their potential effect on endocrine systems. Endocrine-disrupting chemicals are exogenous chemicals or mixtures of chemicals found in food sources, surrounding environments such as soil, water, and air, and personal care products that interfere with the normal function of the body's endocrine system.

Growing evidence in scientific research shows that adverse health consequences are found in humans, domestic animals, and wildlife after exposure to environmental chemicals that could interact with the endocrine system. Even though some health effects of endocrine-disrupting chemicals are not fully explored, and the cause-and-effect relationship between endocrine disrupting chemicals and adverse health outcomes is not yet fully proven, understanding the possible sources of common endocrine disrupting chemicals and risks would benefit public health. Endocrine-disrupting chemicals can enter the body through the air we breathe, the food we eat, the water we drink, and the environment we are exposed. They affect body endocrine functions such as psychological stress, metabolism, reproduction, growth, and development. Precautionary steps such as using a water filter, washing hands often for food preparation and eating, prohibiting smoking indoors, and so on could help to minimize the exposure to endocrine-disrupting chemicals, especially for children and pregnant women. Much research has been conducted to explore the relationship between endocrine-disrupting chemicals and adverse health outcomes in recent decades. For example, some endocrine-disrupting chemicals have been linked to obesity and type 2 diabetes, and some industrial chemicals can interfere with thyroid function [9]. Some endocrine-disrupting chemicals can affect reproductive health by mimicking the effect of male and female sex hormones [10]. Exposure to estrogen and androgen mimicking endocrine-disrupting chemicals can promote breast and prostate cancer and interfere with hormonal cancer therapy [11]. Endocrine-disrupting chemicals' effect that interfere with the body's endocrine system is a key research topic [12]. Endocrine-disrupting chemicals binding to hormone receptors is one of the main activities that can lead to disruption of hormonal function. Tier 1 assays of EPA's Endocrine Disruptor Screening Program were used to identify potential chemicals that interact with the major endocrine pathways and to test for direct interactions with the ER [13,14].

Estrogen is a sex steroid hormone that serves a vital role in sexual and reproductive development in women. Estrogen exhibits diverse biological functions through binding to its specific receptors: estrogen receptors including ER-α, ER-β, and G protein–coupled ER 1. ER-α and ER-β have a high degree of structural homology and belong to the nuclear receptor family [15,16]. ER-α and ER-β were identified and cloned in 1986 and 1996, respectively. The membrane ER family includes mostly G protein–coupled ERs. ER regulates endothelial functions, lipid profile, hemostasis, inflammatory markers, bone integrity, the growth of different tissues, and both prenatal and postnatal development.

ERs are target receptors that could be activated by estrogen hormones commonly found in breast cells, endometrium, and ovarian cells. When estrogens bind to their specific ERs, they could help the growth and function of breast cells. ER-α is mainly expressed in female sex organs such as the breast, ovaries, and uterus. ER-β is primarily expressed in different types of tissues and cells and to a higher degree in females compared with males. ER-α and ER-β have different effects on estrogen signaling and target gene regulation. Recent research related to ERs focuses on developing medicines for breast cancer, osteoporosis, and other endocrine female disorders [17,18]. Estrogen signaling pathways are selectively stimulated or inhibited based on a balance between the activities of ER-α and ER-β in target cells [19]. Most studies have focused on the biological role of ER-α because ER-α was the first receptor subtype identified in the breast. ER-α is considered a ligand-activated transcription factor. In breast cancer, ER-α promotes the growth of breast cancer cells by targeting the expression of signaling components of the insulin-like growth factor system. Another role of ER-α is to affect gene expression in the absence of direct DNA binding by associating with other transcription factors without binding directly to DNA [20,21]. Clinically, breast cancers have been classified into ER-α positive and ER-α negative categories. Patients with ER-α positive tumors have a better survival rate compared with patients with ER-α negative tumors.

Endocrine disruptors can bind to ERs and interfere with normal estrogen signaling through both genomic and nongenomic pathways. Estrogenic molecules such as 17b-estradiol can bind to ER to form ER–ligand complexes, regulating female reproductive functions [22]. Estrogenic molecules can also bind to membrane-bound ERs and interfere with the membrane ER-mediated activation [23].

Many environmental chemicals are subject to regulatory review for their potential to be endocrine disruptors. In the past 20 years, more and more chemicals have been assayed for estrogenic activity by research groups and government agencies [24–26]. Estrogenic activity data is a great asset to the scientific community and regulatory agencies. For most of the chemicals on the market and in the environment, estrogenic activity data is not available. Therefore, alternative methods such as computational modeling have been proposed to estimate estrogenic activity for chemicals without experimental data. For example, 58,000 environment chemicals were evaluated for their potential activity on ER using a computational method with integration of structural alerts and machine learning models [27].

Experimental estrogenic activity data has been reported in various sources such as public databases and the literature. Furthermore, estrogenic activity data is presented in a variety of formats, limiting utilization of the data for regulatory agencies and in the development of computational models, especially quantitative structure–activity relationship (QSAR) models, for predicting estrogenic activity of chemicals. Therefore, an urgent need exists to collect estrogenic activity data from all possible sources, standardize the collected data, and to present the data in a friendly way to the scientific community for more efficient use to facilitate QSAR models for safety evaluation of chemical products and risk assessment of environment chemicals. To meet this need, an estrogenic activity database (EADB) was implemented in the Instant JChem engine (https://chemaxon.com/products/jchem-engines) with browsing, querying, and exporting functions for users to fully utilize the estrogenic activity data comprehensively curated from diverse sources. The database contains estrogenic activity data gathered from in vitro and in vivo assays, which cover a wide range of six magnitudes. The chemicals in EADB are in diverse classes including pesticides, drugs, industrial chemicals, and new chemical entities, etc. The database contains 18,114 estrogenic activity data points for 8212 compounds tested in 1284 assays [5]. The estrogenic activity data were converted to standardized representations for comparability. EADB is publicly available from https://nctr-crs.fda.gov/eadb/home.

EADB has been widely used in the scientific community to develop QSAR models for predicting estrogenic activity of chemicals. This chapter will describe the data structure, database schema, and functions. We will also showcase some applications of EADB in QSAR model development and validation to illustrate its role for facilitating product safety evaluation and chemical risk assessment.

EADB schema

EADB is a database with comprehensive estrogenic activity data curated from a variety of data sources and is a component of the Endocrine Disruptor Knowledge Base (EDKB). Three data sources including published literature, the FDA's EDKB, and other public datasets are curated and managed in the database. Fig. 19.1 shows the EADB workflows of curation of estrogenic activity data from three heterogenous data sources.

Estrogenic activity data were curated from literature by searching the keywords "estrogen receptor" or "estrogenic." Estrogenic in vivo data include uterotrophic assay information that was merged from FDA's EDKB. Public datasets from the National Toxicology Program, and the Ministry of Economy, Trade, and Industry (METI) of Japan

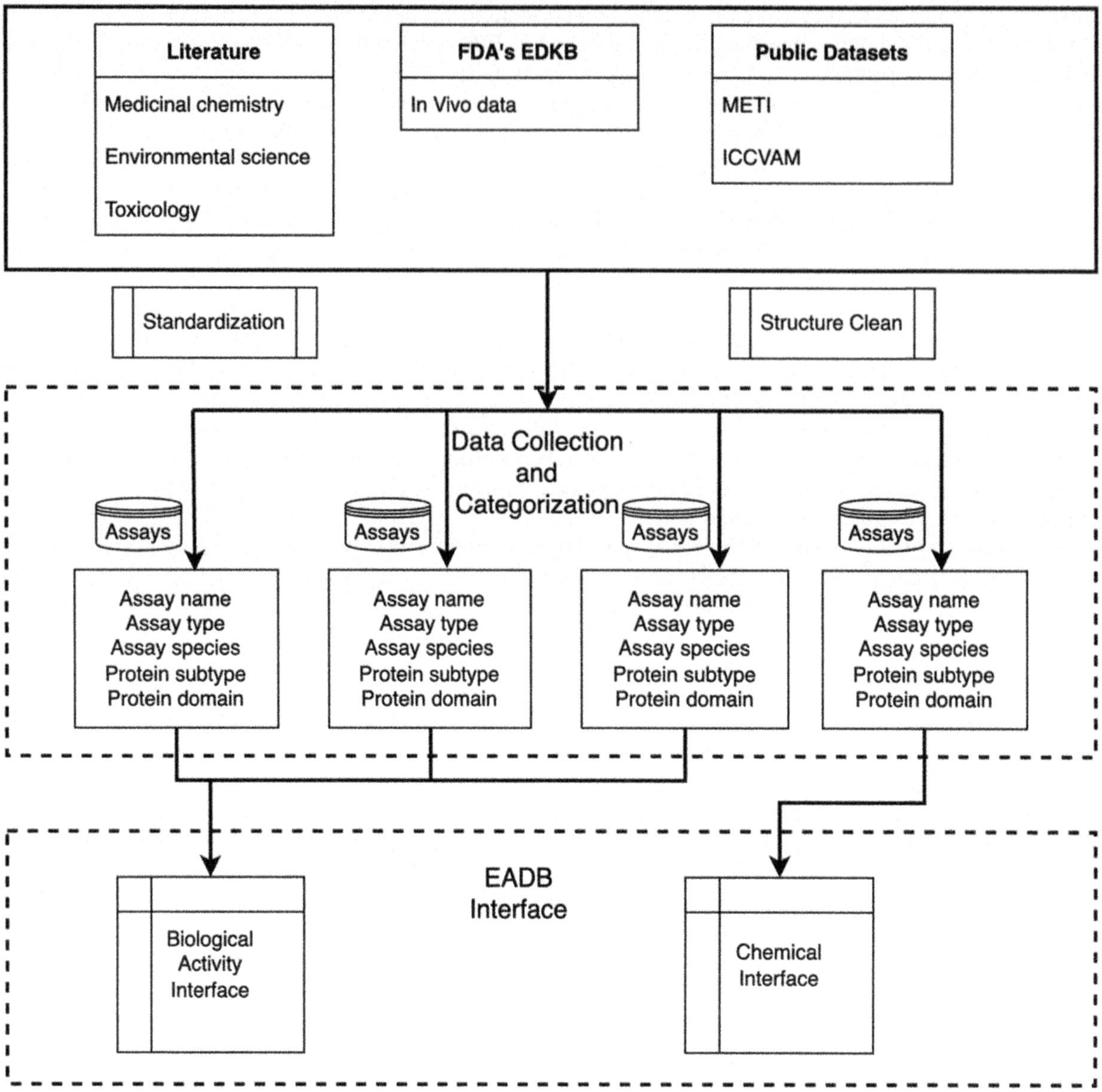

FIGURE 19.1 Overview of EADB data sources and data curation. Estrogenic activity data and related information are curated from the literature, FDA's EDKB, and public datasets. Assays, activity data, references, and chemicals are four internally related data tables. A biological activity interface based on estrogenic activity information and a chemical interface based on chemical information were designed and developed to support user interactions with the database.

were combined into EADB. After standardization and a structure cleaning procedure, the heterogenous estrogenic activity data were combined into four categories: assay descriptions, activity data, references, and chemicals. Assay metadata contain attributes such as assay name, assay type, assay species, protein subtype, and protein domain to describe the assay details. Assays were further organized into four categories: binding assays, reporter gene assays, cell proliferation assays, and in vivo assays. Activity data include detailed estrogenic activity information such as endpoint, activity value and unit, and relation. We generated internal identification numbers to link the activity information with assays, references, literature, and chemicals so that separately recorded activity data can be associated with all other data categories. These four data tables were linked using indexes, and EADB provides the query interface for managing the well-organized estrogenic activity data. The references data have attributes such as reference type, journal citation, PubMed ID, and URL so users have detailed information from the original source of the curated estrogenic activity data. The chemicals table consists of chemical structures, chemical abstracts service (CAS) registration numbers, names, and synonyms directly retrieved from the Chemical Database of the European

Molecular Biology Laboratory (ChEMBL) (https://www.ebi.ac.uk/chembl/) [28]. Chemical structure information was essential for the structure search functions in EADB. Marvin Sketch (https://chemaxon.com/products/marvin) open-source platform was adopted in the user interface design to implement the chemical structure view and query functions.

All datasets were collected from open-access databases and references. Four data tables were designed to store the metadata information of the four categorized data: assay, activity, reference, and chemical information. For each data table, unique internal identifiers were assigned to data records. The four data tables were linked together as the whole database using the identifiers.

The data were cleaned and standardized before they were organized into the four data tables. Because the data were generated using various platforms and reported in different formats, we parsed the curated data into chemical structures, assay descriptions, and activity data. We also checked the data quality and corrected typing errors by reading the original references. The chemical structures in EADB were standardized with JChem Standardizer (https://chemaxon.com/). The estrogenic activity data were standardized using the following approaches. If the endogenous estrogen E2 was tested in an assay, the estrogenic activity of other tested active compounds were normalized to the relative activity (RA) to E2. For instance, the binding affinity data of active compounds in binding assays were normalized to relative binding affinity (RBA) to E2 if E2 was assayed. Base-10 logarithmic calculation was applied to transfer the normalized activity data (RA and RBA) to logRBA and logRA, which were stored in EADB. The normalization enables integration of estrogenic activity data generated from different laboratories and using different assays, which, in turn, facilitates development of in silico prediction models based on multiple diverse data sources. In addition, some of the activity data only contains qualitative descriptions rather than quantitative values such as not active, not binding, and weak binding. Value −10,000 was used to represent not active and not binding, while value −5000 was used to indicate the weak binding activity in EADB.

Fig. 19.1 shows the back-end table schema with the inner associated four tables: assays, activity data, references, and chemicals. Table 19.1 shows the detailed number of entries in each table grouped by assay types. The majority of the estrogenic activity data in EADB were curated from 444 papers published in 21 journals. In addition, estrogenic activity data of 667 chemicals were obtained from the Interagency Coordinating Committee on the Validation of Alternative Methods report "Current Status of Test Methods for Detecting Endocrine Disruptors: In Vitro Estrogen Receptor Binding Assays," and estrogenic activity data of 938 chemicals were curated from Risk Assessment of Endocrine Disruptors, METI, Japan. The 1640 in vivo uterotrophic assay activity data (1604 from mice and 36 from rats) in FAD's EDKB were included in EADB. In total, 18,114 estrogenic activity data points from 8212 molecules tested in 1284 assays were curated from different data sources and merged into EADB after a data cleaning and standardized procedure. Table 19.2 shows the detailed endpoints and their corresponding data in different assay types. Species used in the assays are also important factors for assessing the potential estrogenic activity of chemicals. Most of the estrogenic activity assays in EADB are 11 species including human, rat, mouse, cattle, sheep, rabbit, trout, lizard, chicken, *Escherichia coli*, and monkey. Species information is unknown for 12% (2059) of the data in EADB.

We designed two hierarchical trees named DATA and MOLECULES to connect layer EADB and layer Data Collection and Categorization to support the complex query functions. The first tree is a joint table containing the data from three separate tables: assay descriptions, estrogenic activity data, and references using a one-to-many relationship with the root. The second data tree contains the molecular structures and other chemical information from the chemicals table. Two interfaces were implemented in the databases to enable browsing, searching, and exporting functions in EADB. The biological data interface can be used for examining chemical structures with a specific estrogenic activity. The chemical structure interface enables exploring estrogenic activity data for specific chemicals.

TABLE 19.1 Summary of data group by assay type contained in EADB.

Assay type	Chemicals	Data	Assays	References
Binding	5494	10,853	751	377
Reporter gene	1371	2633	234	80
Cell proliferation	1540	3039	297	107
In vivo	1351	1640	2	1
Total	8212	18,165	1284	447

TABLE 19.2 Summary of data endpoints statistics grouped by assay types.

Endpoint	Description	Data
Binding assay		10,853
logRBA	Log transfer of RBA compared with E2	8478
IC_{50}	50% inhibition concentration	1128
EC_{50}	50% effective concentration	63
Ki	Binding affinity	728
Kd	Dissociation constant	19
Ka	Association constant	41
INH	Inhibition	396
Reporter gene		2633
logRA	Log transfer of RA compared with E2	1212
logRA10	Log transfer of effective concentration equal to 10% of E2	518
REC10	Effective concentration equal to 10% of E2	13
EC_{50}	50% effective concentration	60
IC_{50}	50% inhibition concentration	496
Ki	Antagonist activity	25
INH	Inhibition percentage	167
Agonism	Agonistic estrogenic activities	71
Antagonism	Antiestrogenic antagonistic activities	71
Cell proliferation		3039
logRPE	Log transfer of relative proliferative effect compared with E2	57
logRPP	Log transfer of relative proliferative potency compared with E2	90
logRE	Log transfer of relative efficiency compared with E2	207
EC_{50}	50% effective concentration	18
IC_{50}	50% inhibition concentration	1649
ED_{50}	50% effective dose	44
GI_{50}	Concentration of drug that reduces cell growth by 50%	403
IC_{30}	30% inhibition concentration	10
Ki	Constant of cytotoxicity	22
INH	Inhibition percentage	480
Agonism	Agonistic activity	7
Antagonism	Antagonistic activity	52
In vivo		1640
logRP	Log transferred relative potency compared with E2	1640

Finally, a Java-based Web application is developed with a front-end client to interact with the back-end database through these two interfaces. Fig. 19.2 illustrates the Java Web Application User Interface to support functions such as text searching, chemical structure searching, data filtering, and exporting. The primary component of EADB is the activity data located in the button of the window to display the database query results. The filters button could help to sort and filter specific columns of the activity data to find chemicals and activities quickly. The Export

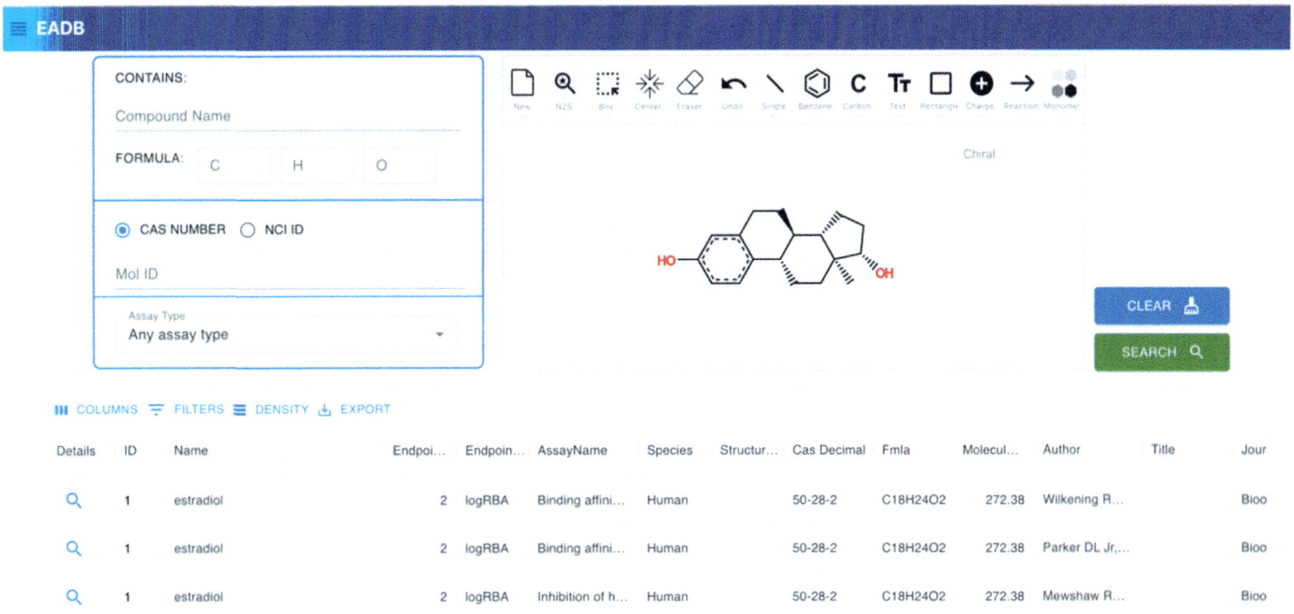

FIGURE 19.2 User interface of EADB. Top left panel supports text search and numeric search options. The top right panel supports the chemical structure view, edit, and search functions. The bottom panel supports estrogenic activity data display, filter, and export.

button could help to save the query result into CSV format for further analysis. The left search box provides text search options based on the chemical compound name, formula, and numerical search options based on CAS number. The right panel supports the chemical structure view, edit, and search operations.

In summary, EADB implements browsing, searching, updating, and exporting functions for users to fully utilize the data contained in the database. The browsing function supports users to view their search results easily in different ways such as compact, standard, and comfortable views. Filters and self-defined columns could be optimized to show specific information based on their interest. The searching function allows users to search estrogenic activity data based on multiple options such as text or numerical information input or chemical structures drawn. The updating function was designed for users to update their own database in their local computers by adding new chemicals or estrogenic activity data or editing the existing data. An exporting function was developed for users to download the query results in various table structures.

EADB applications

EADB is available for all researchers. An increasing number of research teams are using EADB as a scientific tool to assist their projects. Some teams have used EADB to develop ER binding activity prediction models, while other teams used EADB to evaluate their developed estrogenic activity prediction models. Table 19.3 summarizes research papers that used EADB for estrogenic activity prediction model development and validation. In this section, we will showcase some applications of EADB in development and validation of QSAR models for predicting estrogenic activity to demonstrate the usefulness of this database in facilitating product safety evaluation and chemical risk assessment.

Binding with ER is necessary for chemicals to exhibit estrogenic activity. Many QSAR models have been developed for predicting ER binding activity using a variety of datasets [27]. However, the small size of those datasets limited the performance and applications of the models. Therefore, to improve model performance and enhance applicability of models for predicting ER binding activity, a dataset that covers a large chemical space is needed for developing QSAR models. EADB contains ER binding activity data for a large number of chemicals from various sources and, thus, provides opportunities to integrate ER binding activity data to make a large dataset for developing reliable QSAR models for predicting ER binding activity. Shen et al. utilized EADB data and made a large ER binding activity dataset by integrating the diverse ER activity biding data, which contains 5394 chemicals (4719 binders and 675 nonbinders) [5]. They calculated Mold2 descriptors, which have been widely used in various

TABLE 19.3 Applications of EADB for model development and performance evaluation.

References	Develop	Validate	Remark
[5]	✓	✓	Model was built on the EADB data. Fivefold cross-validation was used for performance evaluation.
[29]		✓	EADB data was used as an external validation dataset.
[30]	✓	✓	Model was built on EADB data. Cross-validations were used for performance evaluation.
[31]		✓	Data from EADB, Tox21, METI, and ChEMBL were combined and used for model validation.
[32]	✓	✓	The models were built on EADB data. 10-fold cross-validation was conducted for performance evaluation.
[33]	✓	✓	Model was built on EADB data. Fivefold cross-validation was conducted for performance evaluation.
[34]		✓	Data from EADB, Tox21, METI, and ChEMBL were used for validation.
[35]		✓	Data from Nuclear Receptors Database including Negative Data and EADB were used for model validation.
[36]		✓	EADB were used in the validation dataset.

QSAR models [37]. After removal of descriptors with no or less information, the machine learning algorithm Decision Forest [38–40], which has been used in the development of various in silico predictive models [41–46], was used to construct the QSAR model for predicting ER binding activity based on this large dataset. The Decision Forest model was then evaluated using 10 iterations of fivefold cross-validations. The performance of the model is shown in Fig. 19.3. As can be seen in Fig. 19.3, the model had very good performance in the cross-validations with an average predictive accuracy of 93.84%, which is a big improvement compared with the previously reported models developed using much smaller datasets [27,38]. The findings demonstrate that EADB has a large collection of ER activity data to improve performance of QSAR models for predicting ER binding activity.

Estrogen receptor ER-α is the most investigated receptor and plays an essential role in the endocrine system. Understanding the binding activities of chemicals to ER-α could help to evaluate their endocrine activity and support

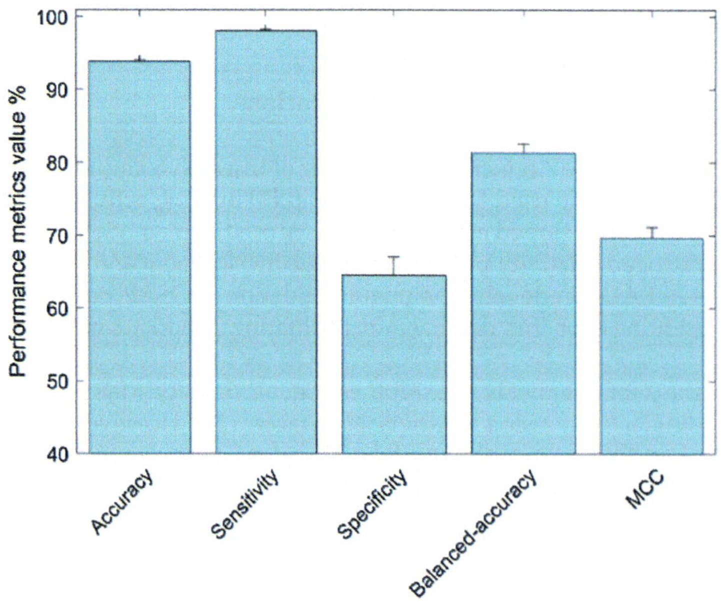

FIGURE 19.3 Results of the fivefold cross-validations. Average accuracy, sensitivity, specificity, balanced accuracy, and MCC are plotted as bars and shown on the x-axis. The sticks above the bars indicate standard deviations. The performance metrics values are indicated by the y-axis in percentage.

endocrine-disrupting risk assessment of chemicals. However, experimental ER-α binding activity data are not available for many chemicals. EADB contains a large set of chemicals with experimental ER-α binding activity data, enabling reliable QSAR model development for predicting ER-α binding activity. Ng et al. developed a decision forest classification model to predict chemical binding to ER-α with a training dataset of 3308 chemicals (2656 active and 652 inactive) from EADB [30]. First, 4161 ER-α binding data points for those 3308 chemicals were extracted from EADB. The dataset was then preprocessed. Molecular descriptors were calculated for the compounds using Mold2 software [37]. Descriptors with no or little information were removed before constructing the decision forest model, and the resulting 467 descriptors were used for developing the models. The prediction model was evaluated using fivefold cross-validations. Permutation test was conducted 1000 times. The fivefold cross-validation results are shown in Fig. 19.4. The model had a high prediction accuracy (91.98%). In addition, descriptors important to the ER-α binding activity prediction model were identified using their frequency values in the fivefold cross-validations. Eighteen descriptors were most often used in the fivefold cross-validations and were identified as informative descriptors for the ER-α binding activity prediction model. These 18 descriptors can be used for interpretation of the prediction model. The high prediction accuracy of the developed model demonstrated that a large dataset is necessary for building a prediction model for chemicals binding activity to ER-α with robust performance.

Most QSAR models used to predict ER-β binding activity were developed based on small-size datasets. More reliable QSAR models for ER-β binding activity prediction need to be developed using larger datasets. Sugunadevi et al. developed an ER-β binding activity prediction model based on a large dataset extracted from EADB [33]. The logRBA values of selected chemicals recorded in EADB were used to determine the chemicals as binders or nonbinders of ER-β. After preprocessing the ER-β binding activity data from EADB, 3349 experimental data points for 2492 compounds (2462 binders and 30 nonbinders) were used in this study. The Mold2 software was used to generate molecular descriptors for the 2492 chemicals to build the decision forest model. One thousand times of fivefold cross-validations were conducted to evaluate the model performance, which was measured using prediction accuracy, sensitivity, specificity, Mathews correlation coefficient (MCC), and balanced accuracy. One thousand permutation tests were also carried out for assessing the predictive capability of the developed decision forest model. As this dataset is very imbalanced with 99% binders, MCC and balanced accuracy are more appropriate for measuring model performance. The average MCC and balanced accuracy from the 1000 cross-validations are 0.2% and 75.5%, while the average MCC and balanced accuracy from 1000 permutation tests are 0% and 50%, indicating that the developed decision forest model for ER-β binding activity prediction has the capability to classify chemicals to ER-β binders or ER-β nonbinders and the model was not generated solely by chance.

Classification models are used to qualitatively predict chemicals as potential estrogenic actives or inactives. It is an important task to assess how active a chemical is if it is estrogenic active. Therefore, QSAR models for quantitative prediction of estrogenic activity play an important role in chemical risk assessment. Quantitative estrogenic data are needed for the development of QSAR models for quantitative prediction. EADB contains a large amount of quantitative ER binding data obtained using assays with different species such as human, mouse, and calf. These quantitative estrogenic activity data provide the scientific community a rich data source for developing quantitative

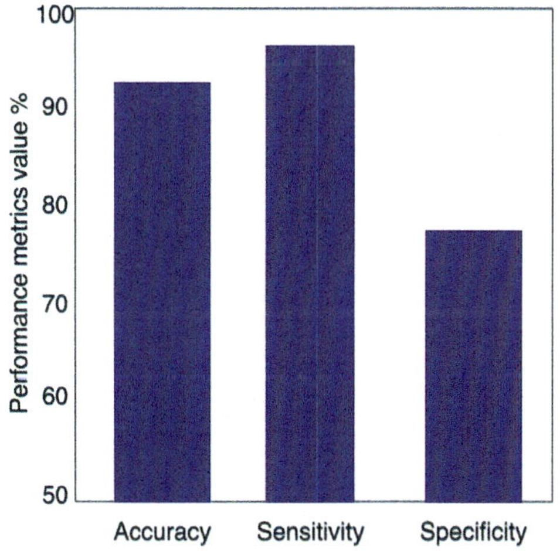

FIGURE 19.4 Results of the fivefold cross-validations. Accuracy, sensitivity, and specificity are plotted as blue bars and shown on the x-axis. The performance metrics values are indicated by the y-axis in percentage.

estrogenic activity prediction models. The machine learning models for quantitatively predicting ER binding affinity developed by Wrong et al. [32] illustrate applications of EADB. When 17β-estradiol (E2) was tested in an experiment, the ER binding affinity data of other chemicals tested in the same experiment was transformed to logRBA values in EADB using Eq. (19.1).

$$\text{logRBA} = \log\left(\frac{IC_{50}^{E2}}{IC_{50}^{\text{chemical}}} \times 100\right) \tag{19.1}$$

where IC_{50}^{E2} and $IC_{50}^{\text{chemical}}$ are the IC_{50} values of E2 and the chemical measured in the same experiment. Therefore, ER binding affinity data represented in logRBA values from different experiments can be comparable, enabling the combination of ER binding affinity data from different experiments to make large datasets for developing reliable quantitative prediction models. Of the 2119 chemicals with logRBA values in EADB, some chemicals have multiple logRBA values for the same chemical and the standard deviations in the logRBA values are large, indicating a large discrepancy among the experiments for the chemicals. After removal of such chemicals, 1589 chemicals with experimental logRBA values were used to build and validate models for quantitatively predicting ER binding affinity using machine learning algorithms multiple linear regression (MLR), support vector regression (SVR), and Gaussian process regression (GPR) [32]. After feature selection, 59 chemical descriptors including atomistic, spatial, physicochemical, charged partial surface area, and graph descriptors that were extracted using the three-dimensional molecular structures generated with SMILES codes in EADB were used to build the prediction models. The 1589 chemicals were classified into highly active (401), moderate (756), weak (402), and inactive (30) categories based on their logRBA values. A 10-fold cross-validation was conducted to evaluate the performance for each of the quantitative machine learning models. Root mean squared error (RMSE) was used to evaluate the prediction accuracy. The 10-fold cross-validation results are summarized in Table 19.4. Interestingly, chemicals in the category moderate have the smallest RMSE values for all the three machine learning algorithms, and the chemicals in the category inactive have the largest RMSE values. Comparative analysis on the performance of models built with different machine learning algorithms revealed that the models built with GPR achieved the best performance, while the models constructed using MLR had the worst prediction accuracy for predicting logRBA values of chemicals. In summary, this study demonstrated that EADB has a relatively large set of binding activity data in different species with logRBA values for developing reliable quantitative ER binding affinity prediction models.

External validation of existing models with a large dataset could help to assess generalizability of the predictive models to chemicals not used in training and internal validation. EADB can also serve as a powerful external database to help validate prediction models developed on other estrogenic activity datasets. These kinds of external validations could help to determine the generalizability and reproducibility for evaluation of the developed models.

TABLE 19.4 Models performance for chemicals in different activity ranges.

Activity range	Model	RMSE
Highly active	MLR	1.32
	SVR	1.01
	GPR	0.93
Moderate	MLR	0.67
	SVR	0.74
	GPR	0.66
Weak	MLR	1.50
	SVR	1.08
	GPR	1.03
Inactive	MLR	1.69
	SVR	1.33
	GPR	2.00

Some studies are introduced below to demonstrate how EADB was used as an external validation dataset for evaluating their models.

Most QSAR models for predicting estrogenic activity of chemicals were built with limited compounds and molecular descriptors. The underlying mechanism of estrogenic activity is hard to explain using few selected simple chemical descriptors. There is a need to explore new molecular features or mixed complex features to build models. Chen et al. developed four machine learning methods using k-nearest neighbors (kNN), decision tree, Naïve Bayes, and support vector machine (SVM) to predict ER binders using molecular seven fingerprints that were calculated using a PaDEL-Descriptor tool [29]. They curated ER binding activity data of 333 chemicals (188 binders and 145 nonbinders) from the literature. This dataset was used to build machine learning models for predicting ER binding activity. The ER binding activity data were extracted from EADB for 4648 chemicals (4070 binders and 578 nonbinders), which were used as an external dataset to evaluate the developed machine learning models in addition to the internal fivefold cross-validations on the training dataset. Classification accuracy, specificity, sensitivity, and area under the receiver operating characteristic curve (AUC) were used for the measurements for model performance. The model built with PubChem fingerprint using SVM performed best in the fivefold cross-validation, while the model built with CDK Extended Fingerprint using SVM performed best in the external validation. The performance metrics values for the best models are given in Fig. 19.5. As shown in Fig. 19.5, even though the model training dataset is quite small and the external validation dataset from EADB is large, the external validation results of the best model (the cyan bars) are close to the fivefold cross-validation results of the best model (the red bars), indicating that the developed ER binding activity prediction models had a good generalization capability. Examination of the developed models identified a direct connection between chemical structures and their ER binding activity and showed that the estate fingerprint did not perform well overall when using the same machine learning algorithms, which may be due to the short length of the estate fingerprint.

EPA organized the Collaborative Estrogen Receptor Activity Prediction Project (CERAPP) to develop predictive models based on the estrogenic activity data obtained from ToxCast high-throughput screening assays for predicting potential estrogenic activity of chemicals in the environment [31]. Predictive models were developed by 17 US and European groups. Those models were then used to produce a consensus model. The models were evaluated using a dataset with experimental estrogenic activity data in which the data from EADB is the main component before they were used to predict estrogenic active of 32,464 chemicals that are regulated by EPA. The estrogenic activity data in EADB included in the evaluation dataset in CERAPP were used in a case study of the integrated approaches for a testing and assessment program [34]. This study identified 4-tert-butyl-phenol as a potential endocrine disruptor among the three assessed substituted phenols. Utilization of EADB data for evaluation of the developed models in such large international programs demonstrates the usefulness of EADB data in chemical risk assessment.

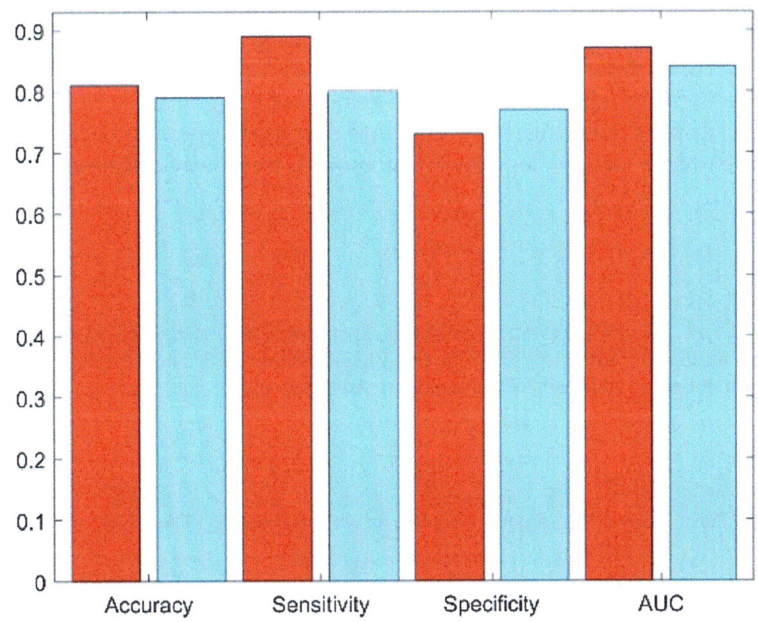

FIGURE 19.5 Performance metrics values of the best models in the fivefold cross-validation (red bars) and in the external validation (cyan bars). Performance metrics are shown on the x-axis.

ER binding is an important endpoint for assessing endocrine-disrupting potential of chemicals. QSAR models have been developed to predict ER binding activity of chemicals. Extensively evaluating QSAR models for ER binding activity prediction is vital for application of QSAR models in endocrine disruption risk assessment. Sellami et al. designed a pipeline to combine structure-based (docking and pharmacophore models) and ligand-based (pharmacophore models) methods to predict ER-α binding activity [35]. A dataset of 2442 chemicals (2219 nonbinding and 223 binding compounds) with ER-α binding activity data was curated from the EPA database and used to build the prediction model. The chemicals with ER binding activity data in EADB were used as an external evaluation dataset for evaluation of the performance of the developed model. Ligand-based, structure-based, and combination of ligand-based and structure-based pharmacophores models were developed and validated. The external validation results showed that the combination of models achieved better performance than the individual models. This study presented an optimized protocol for docking studies and a combined structure-based and ligand-based pharmacophore model to predict the potential compounds binding to ER-α.

Machine learning and deep learning techniques have been attracting attention in predictive toxicology. Ciallella et al. explored classic machine learning and deep learning algorithms for predicting estrogenic activity of chemicals. They built machine learning and deep learning models based on 7500 chemicals tested with the 18 ER-related assays in the ToxCast program to predict ER agonism, antagonism, and binding activities [36]. Three binary fingerprint data including molecular access system, extended connectivity fingerprint, and functional connectivity fingerprint were calculated with RDkit python package and were used for developing prediction models. Four machine learning algorithms (Bernoulli Naive Bayes, kNN, random forest, and SVM) and two deep learning neural networks were used to develop the models. The 966 compounds with in vivo rodent uterotrophic agonist data in EADB were used as an external validation dataset to evaluate the performance of the machine learning and deep learning models. The deep neural network models developed using the functional connectivity fingerprint achieved the best performance in the external validation with the EADB dataset. This study further demonstrated that EADB data can be used for external validation of models for predicting estrogenic activity.

Conclusions

Endocrine-disrupting chemicals can be found in our surrounding environment. They bind to various receptors such as ER in the endocrine system to activate, block, or alter the natural endocrine system. Though the estrogenic activity of a large number of chemicals has been experimentally tested, the data are in many sources and in different formats, limiting the use of this asset in the scientific community and regulatory science. A database for managing such large and diverse estrogenic activity data is needed to fully utilize the data. EADB fills this important gap by curating chemicals containing ER activity data from comprehensive sources including the literature and public databases. Data cleaning and standardization procedures were conducted to guarantee data quality before importing them to EADB. Many applications have been conducted using data from EADB for model development and validation in recent years, illustrating that EADB is a large and reliable estrogenic activity data source. EADB provides a user-friendly web client to support querying, browsing, and exporting operations on estrogenic activity data. This open public resource can be utilized to develop prediction models to rapidly estimate the potential estrogenic activity of new chemicals.

Acknowledgments

This work was supported in part by an appointment to the Research Participation Program at the National Center for Toxicological Research (Fan Dong, Myeongsang Lee, Meng Song, Zoe Li), administered by the Oak Ridge Institute for Science and Education through an interagency agreement between the US Department of Energy and the US Food and Drug Administration.

Disclaimer

This chapter reflects the views of the authors and does not necessarily reflect those of the US Food and Drug Administration.

References

[1] Hiller-Sturmhöfel S, Bartke A. The endocrine system: an overview. Alcohol Health Res World 1998;22(3):153–64.
[2] Baltaci AK, Mogulkoc R, Baltaci SB. The role of zinc in the endocrine system. Pak J Pharm Sci 2019;32(1):231–9.

[3] Russell N, Grossmann M. Mechanisms in endocrinology: estradiol as a male hormone. Eur J Endocrinol 2019;181(1):R23–43.
[4] Eisenegger C, Haushofer J, Fehr E. The role of testosterone in social interaction. Trends Cognit Sci 2011;15(6):263–71.
[5] Shen J, Xu L, Fang H, Richard AM, Bray JD, Judson RS, et al. EADB: an estrogenic activity database for assessing potential endocrine activity. Toxicol Sci 2013;135(2):277–91.
[6] De Leo V, Musacchio MC, Cappelli V, Massaro MG, Morgante G, Petraglia FJ. Genetic, hormonal and metabolic aspects of PCOS: an update. Reprod Biol Endocrinol 2016;14(1):38.
[7] Zoeller RT, Brown TR, Doan LL, Gore AC, Skakkebaek NE, Soto AM, et al. Endocrine-disrupting chemicals and public health protection: a statement of principles from the Endocrine Society. Endocrinology 2012;153(9):4097–110.
[8] Ford AT, LeBlanc GA. Endocrine disruption in invertebrates: a survey of research progress. Environ Sci Technol 2020;54(21):13365–9.
[9] Gore AC, Chappell VA, Fenton SE, Flaws JA, Nadal A, et al. Executive summary to EDC-2: the Endocrine Society's second scientific statement on endocrine-disrupting chemicals. Endocr Rev 2015;36(6):593–602.
[10] Caserta D, Mantovani A, Marci R, Fazi A, Ciardo F, La Rocca C, et al. Environment and women's reproductive health. Hum Reprod Update 2011;17(3):418–33.
[11] Gibson DA, Saunders PT. Endocrine disruption of oestrogen action and female reproductive tract cancers. Endocr Relat Cancer 2014;21(2): T13–31.
[12] Monneret C. What is an endocrine disruptor? Comptes Rendus Biol 2017;340(9–10):403–5.
[13] Willett CE, Bishop PL, Sullivan KM. Application of an integrated testing strategy to the US EPA endocrine disruptor screening program. Toxicol Sci 2011;123(1):15–25.
[14] Shanle EK, Xu W. Endocrine disrupting chemicals targeting estrogen receptor signaling: identification and mechanisms of action. Chem Res Toxicol 2011;24(1):6–19.
[15] Warner M, Nilsson S, Gustafsson JA. The estrogen receptor family. Curr Opin Obstet Gynecol 1999;11(3):249–54.
[16] Hall JM, McDonnell DP. Coregulators in nuclear estrogen receptor action: from concept to therapeutic targeting. Mol Interv 2005;5(6):343–57.
[17] Duffy MJ. Estrogen receptors: role in breast cancer. Crit Rev Clin Lab Sci 2006;43(4):325–47.
[18] Pettersson K, Gustafsson JA. Role of estrogen receptor beta in estrogen action. Annu Rev Physiol 2001;63(1):165–92.
[19] Matthews J, Gustafsson JA. Estrogen signaling: a subtle balance between ERα and ERβ. Mol Interv 2003;3(5):281.
[20] Cavalieri EL, Rogan EG. Depurinating estrogen-DNA adducts, generators of cancer initiation: their minimization leads to cancer prevention. Clin Transl Med 2016;5(1):1–15.
[21] Alluri PG, Speers C, Chinnaiyan AM. Estrogen receptor mutations and their role in breast cancer progression. Breast Cancer Res 2014;16(6): 1–8.
[22] Safe S, Kim K. Nonclassical genomic ER/Sp and ER/AP-1 signaling pathways. J Mol Endocrinol 2008;41(5):263.
[23] Yager JD, Davidson NE. Estrogen carcinogenesis in breast cancer. N Engl J Med 2016;354(3):270–82.
[24] Blair RM, Fang H, Branham WS, Hass BS, Dial SL, Moland CL, et al. The estrogen receptor relative binding affinities of 188 natural and xenochemicals: structural diversity of ligands. Toxicol Sci 2000;54(1):138–53.
[25] Hong H, Branham WS, Dial SL, Moland CL, Fang H, Shen J, et al. Rat α-Fetoprotein binding affinities of a large set of structurally diverse chemicals elucidated the relationships between structures and binding affinities. Chem Res Toxicol 2012;25(11):2553–66.
[26] Shen J, Jiang J, Kuang G, Tan C, Liu G, Huang J, et al. Discovery and structure–activity analysis of selective estrogen receptor modulators via similarity-based virtual screening. Eur J Med Chem 2012;54:188–96.
[27] Hong H, Tong W, Fang H, Shi L, Xie Q, Wu J, et al. Prediction of estrogen receptor binding for 58,000 chemicals using an integrated system of a tree-based model with structural alerts. Environ Health Perspect 2002;110(1):29–36.
[28] Gaulton A, Bellis LJ, Bento AP, Chambers J, Davies M, Hersey A, et al. ChEMBL: a large-scale bioactivity database for drug discovery. Nucleic Acids Res 2012;40(D1):D1100–7.
[29] Chen Y, Cheng F, Sun L, Li W, Liu G, Tang Y. Computational models to predict endocrine-disrupting chemical binding with androgen or oestrogen receptors. Ecotoxicol Environ Saf 2014;110:280–7.
[30] Ng HW, Doughty SW, Luo H, Ye H, Ge W, Tong W, et al. Development and validation of decision forest model for estrogen receptor binding prediction of chemicals using large data sets. Chem Res Toxicol 2015;28(12):2343–51.
[31] Mansouri K, Abdelaziz A, Rybacka A, Roncaglioni A, Tropsha A, Varnek A, et al. CERAPP: collaborative estrogen receptor activity prediction project. Environ Health Perspect 2016;124(7):1023–33.
[32] Wong JC, Zidar J, Ho J, Wang Y, Lee KK, Zheng J, et al. Assessment of several machine learning methods towards reliable prediction of hormone receptor binding affinity. Chemical Data Collections 2017;9:114–24.
[33] Sakkiah S, Selvaraj C, Gong P, Zhang C, Tong W, Hong H. Development of estrogen receptor beta binding prediction model using large sets of chemicals. Oncotarget 2017;8(54):92989–3000.
[34] Webster F, Gagné M, Patlewicz G, Pradeep P, Trefiak N, Judson RS, et al. Predicting estrogen receptor activation by a group of substituted phenols: an integrated approach to testing and assessment case study. Regul Toxicol Pharmacol 2019;106:278–91.
[35] Sellami A, Montes M, Lagarde N. Predicting potential endocrine disrupting chemicals binding to estrogen receptor α (ERα) using a pipeline combining structure-based and ligand-based in silico methods. Int J Mol Sci 2021;22(6):2846.
[36] Ciallella HL, Russo DP, Aleksunes LM, Grimm FA, Zhu H. Predictive modeling of estrogen receptor agonism, antagonism, and binding activities using machine-and deep-learning approaches. Lab Invest 2021;101(4):490–502.
[37] Hong H, Xie Q, Ge W, Qian F, Fang H, Shi L, et al. Mold2, molecular descriptors from 2D structures for chemoinformatics and toxicoinformatics. J Chem Inf Model 2008;48(7):1337–44.
[38] Tong W, Hong H, Fang H, Xie Q, Perkins R. Decision Forest: combining the predictions of multiple independent decision tree models. J Chem Inf Comput Sci 2003;43(2):525–31.
[39] Hong H, Tong W, Perkins R, Fang H, Xie Q, Shi L. Multiclass decision forest–a novel pattern recognition method for multiclass classification in microarray data analysis. DNA Cell Biol 2004;23(10):685–94.
[40] Hong H, Tong W, Xie Q, Fang H, Perkins R. An in silico ensemble method for lead discovery: decision Forest. SAR QSAR Environ Res 2005;16: 339–47.

[41] Sakkiah S, Leggett C, Pan B, Guo W, Valerio Jr LG, Hong H. Development of a nicotinic acetylcholine receptor nAChR α7 binding activity prediction model. J Chem Inf Model 2020;60(4):2396–404.
[42] Hong H, Thakkar S, Chen M, Tong W. Development of decision forest models for prediction of drug-induced liver injury in humans using a large set of FDA-approved drugs. Sci Rep 2017;7(1):17311.
[43] Xie Q, Ratnasinghe LD, Hong H, Perkins R, Tang ZZ, Hu N, et al. Decision Forest analysis of 61 single nucleotide polymorphisms in a case-control study of esophageal cancer; a novel method. BMC Bioinf 2005;6(Suppl. 2):S4.
[44] Hong H, Rua D, Sakkiah S, Selvaraj C, Ge W, Tong W. Consensus modeling for prediction of estrogenic activity of ingredients commonly used in sunscreen products. Int J Environ Res Publ Health 2016;13(10):958.
[45] Hong H, Shen J, Ng HW, Sakkiah S, Ye H, Ge W, et al. A rat α-fetoprotein binding activity prediction model to facilitate assessment of the endocrine disruption potential of environmental chemicals. Int J Environ Res Publ Health 2016;13(4):372.
[46] Hong H, Harvey BG, Palmese GR, Stanzione JF, Ng HW, Sakkiah S, et al. Experimental data extraction and in silico prediction of the estrogenic activity of renewable replacements for bisphenol A. Int J Environ Res Publ Health 2016;13(7):705.

CHAPTER

20

Centralized data sources and QSAR methods for the prediction of idiosyncratic adverse drug reaction

Yuan Liu[1,2], *Menglong Li*[1] *and Zhining Wen*[1]

[1]College of Chemistry, Sichuan University, Chengdu, China [2]University of California, Department of Bioengineering, San Diego, CA, United States

Abbreviations

1D 1 dimensional
2D 2 dimensional
3D 3 dimensional
ADR Adverse drug reaction
ASMs Antiseizure medications
ATC Anatomical therapeutic chemical
AUPRC The area under the precision-recall curve
AUROC The area under the receiver operating characteristic curve
BACC Balance accuracy
CK Creatine kinase
DIR Drug-induced rhabdomyolysis
DIRA Drug-induced rhabdomyolysis atlas
DIT Drug-induced teratogenesis
DTs Decision trees
FAERS FDA adverse event reporting system
FDA Food and Drug Administration
FN False negatives
FP False positives
FPR False-positive rate
IADR Idiosyncratic adverse drug reaction
LR Logistic regression
MCC Matthews correlation coefficient
MCMs Major congenital malformations
ML Machine learning
OHDSI Observational health data sciences and informatics
QSAR Quantitative structure–activity relationship
RF Random forest
ROR Reporting odds ratio
RXCUIs RxNorm clinical drug concepts
SAR Structure–activity relationship
SAs Structural alerts
SDFs Structure data files
SNOMED-CT Systematized nomenclature of medicine clinical terms
SVM Support vector machine
TN True negatives
TP True positives
TPR True-positive rate
UNII Unique ingredient identifier
XGBoost Extreme gradient boosting

Introduction

Adverse drug reactions (ADRs) are harmful events attributed to the use of medicines, which can be described as an appreciably harmful or unpleasant reaction resulting from an intervention related to the use of a medicinal product [1]. With the increasing use of new regimens and medications, ADRs become an important complication and can increase the risk of drug use. Better understanding of ADRs' molecular mechanisms and clinical relevance with therapies can be beneficial for preventing the occurrence of adverse events. In addition, the evidence of drug-induced side effects is also a considerable concern in new drug research and development for the pharmaceutical industry. Currently, machine learning (ML) has been widely used in various ADR prediction studies and has achieved good results [2–8]. On the basis of the information on the biomarkers from genomics and proteomics and the chemical structures of the drugs, ML methods are introduced to predict ADRs [7,8]. Recently, Ietswaart et al. [4] used random forest (RF) models to predict the relationship between drug targets and ADRs. Liu et al. [9] used postmodified

https://doi.org/10.1016/B978-0-443-15339-6.00030-8

nonnegative matrix factorization [10] to deconvolute the toxicological data for identifying drug-related factors to facilitate the understanding of drug using, and Zhu et al. used ML methods for predicting chemical reactions [11]. He et al. [3] used the quantitative structure–activity relationship (QSAR) model to predict drugs that induce damage to the liver. Galeano et al. [2] used ML algorithms to predict the frequency of multiple ADRs. The feasibility of using ML algorithms to study ADRs has been confirmed by these studies. Investigations on different ADRs using QSAR and ML models would significantly boost the relevant research of the safety evaluation and risk assessment for certain types of drugs.

Although substantial efforts have been made on applying proper models to investigate different ADRs of drugs, some major obstacles remain. First, for some ADRs that happen rarely, like some idiosyncratic ADRs (IADRs) such as rhabdomyolysis and torsades de pointes, due to the lack of treatment experience and evidence, it is hard to apply any comprehensive investigation due to the lack of a centralized data source. Second, with numerous QSAR models and ML models, generating appropriate descriptors of structure–activity information, choosing compatible models by proper evaluation, and optimizing parameters based on specific drugs and cases are all delicate procedures that need to be careful to deal with. Third, the potential risk cannot be effectively assessed unless a valid model is utilized; hence, it is hard to carry out the early detection and intervention of ADRs and prevent the occurrence of ADRs.

Thus, in this chapter, the constriction of centralized data sources was demonstrated, and a strategy of selecting proper structure–activity relationship (SAR) models along with ML models was illustrated. To address the problems that mentioned above, two projects on two different ADRs, drug-induced rhabdomyolysis (DIR) and drug-induced teratogenesis (DIT), and the corresponding drugs were introduced. The strategy for constructing SAR models of ADRs consists of following 5 steps (Fig. 20.1).

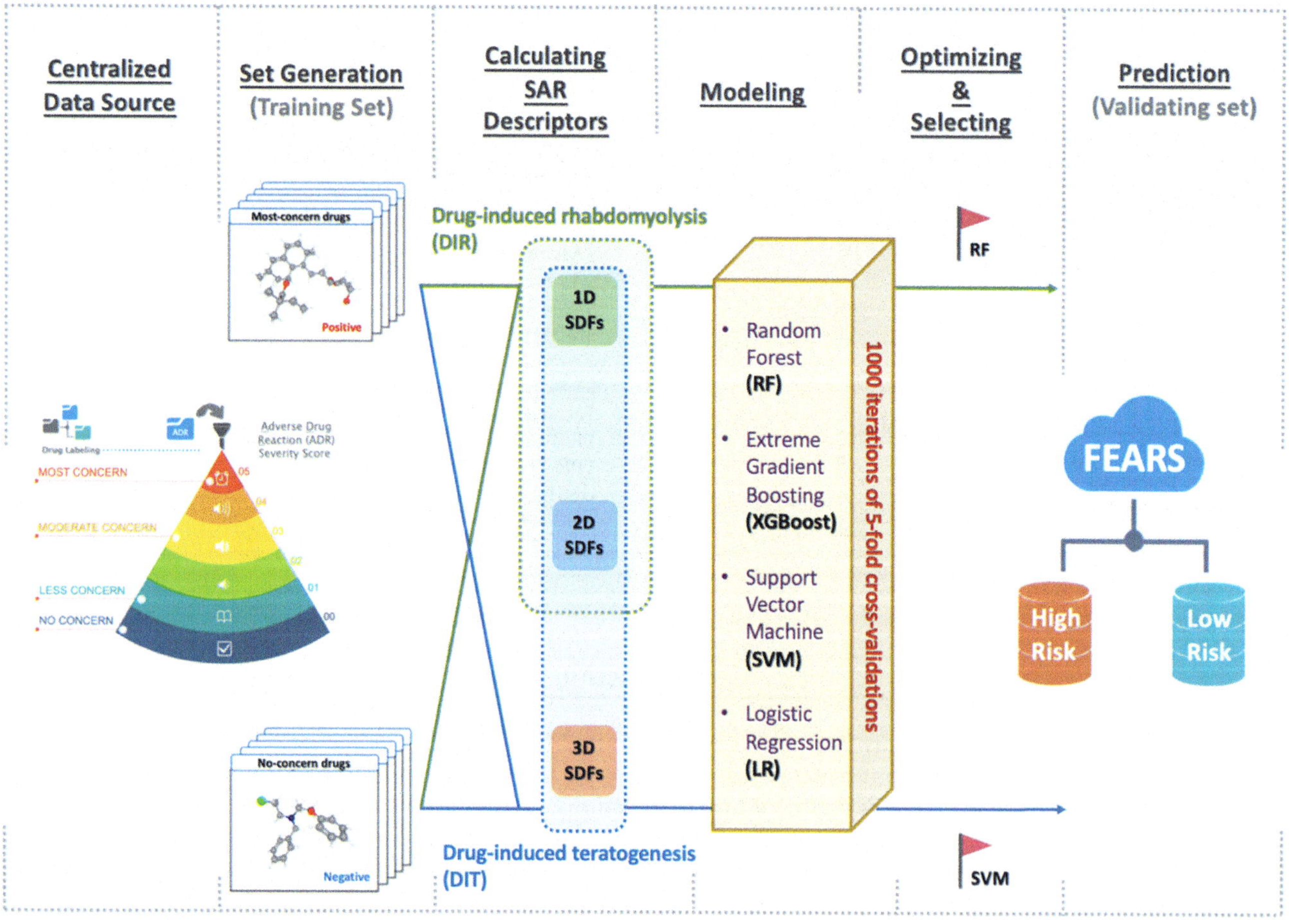

FIGURE 20.1 The workflow of strategy that constructs the SAR model of ADRs by using centralized data sources. *ADRs*, adverse drug reactions; *DIR*, drug-induced rhabdomyolysis; *DIT*, drug-induced teratogenesis; *FAERS*, US Food and Drug Administration (FDA) spontaneous adverse events reporting system; *LR*, logistic regression; *RF*, random forest; *SAR*, structure–activity relationship; *SDFs*, structure data files; *SVM*, support vector machine; *XGBoost*, extreme gradient boosting.

1) Obtaining the centralized dataset for targeted ADRs with a schema for stratifying the risk of marketed drugs and collecting training sets based on the drugs in the centralized data source.
2) Generating the validation dataset based on the drugs extracted from FAERS.
3) Calculating 1D/2D/3D molecular descriptors of drugs.
4) Developing, optimizing, and selecting valid model among multiple models.
5) Evaluating the performance on both training and validation datasets.

Drug-induced rhabdomyolysis is a potentially life-threatening muscle injury that leads to the damage of skeletal muscle, resulting in the release of cellular contents into the blood and a variety of clinical symptoms ranging from nonspecific fatigue and muscle pain to acute kidney failure [12–14]. The specific pathogenesis of rhabdomyolysis is still unknown due to many interfering factors leading to rhabdomyolysis [14,15]. The timely identification and early intervention of rhabdomyolysis are critical because it causes approximately 10%~25% cases of acute renal failure [15–17]. A full recovery can be expected through early diagnosis and the treatment of the complications that may arise [18]. At present, the main diagnostic method in clinics is to monitor the levels of serum and urine myoglobin as well as the level of serum creatine kinase (CK) which is 5–10 times higher than the normal serum levels in patients [15,19]. However, detected CK values are usually inconsistent, limiting the diagnostic performance [20]. It is still an arduous task in clinical practice to carry out the early detection and intervention of DIR. Therefore, the urgent problem to be solved is to develop effective in silico methods to predict the DIR concerns of the drugs. To date, only a few ML models were proposed to predict DIR using chemical structures of drugs [21,22]. However, because of the scarcity of DIR cases and the lack of negative drugs, to our best knowledge, the performance of the models was still suboptimal and can be improved by using a well-defined DIR dataset.

For the DIT project, the antiseizure medications (ASMs) were mainly involved with the data of congenital malformations in utero. Congenital malformations are defined as the malformations of organs or body parts during development in utero, mainly attributed to hereditary, maternal, external environmental, or some unknown factors. Major congenital malformations (MCMs) are defined as structural abnormalities with surgical, medical, functional, and/or cosmetic importance [23]. It is not only an irreparable blow to the family but also a burden on the society. As a potential teratogen, a number of drugs can pass through the placental barrier and affect the growth and development of the fetus in utero. It is commonly accepted that the most sensitive period to teratogens is during active organogenesis, which is from three to 8 weeks after fertilization. Some organs, such as the brain, will continue to be very developmentally active after active organogenesis and may still be affected by teratogens [24]. ASMs are used to control various types of convulsive disorders or as mood stabilizers. About 0.3%–0.7% of pregnant women are diagnosed with epilepsy [25], most of which receive ASMs monotherapy (74%) [26]. Although the incidence of MCMs in offspring of women with epilepsy receiving ASMs monotherapy during pregnancy is very low (4%), the rate is still higher than that of MCMs in the general population [26]. In clinical practice, the lowest effective dose is usually recommended for seizure control for the safety of both the mother and child. Furthermore, women with epilepsy should avoid unexpected pregnancies because ASMs take time to metabolize to a safe level for embryos. Therefore, the teratogenic risk evaluation of ASMs is crucial for the clinical medications. During the last decade, data from large prospective epilepsy and pregnancy registries have been used to compare the incidence and the risk of MCMs in offspring exposed to different ASMs. When exposed to different ASMs, the incidence of teratogenesis exhibits significant difference. The incidence of MCMs in offspring exposed to valproate and phenobarbital is 5%–11% and 6%–7%, respectively, while it is only 2%–3% for lamotrigine and levetiracetam [27–29]. Fetal exposure to ASMs is associated with an increased risk of MCMs. Although there is a variation in the rate, the evidence in pregnancy registries implies that valproate is associated with the highest risk of MCMs among ASMs. The postmarketing surveillance data in US Food and Drug Administration (FDA) Adverse Events Reporting System (FAERS) also indicates the high prevalence of congenital malformations induced by valproate. In addition, carbamazepine, phenobarbital, phenytoin, and topiramate are also associated with the increased risk of MCMs in offspring, while lamotrigine and levetiracetam are of relatively low teratogenic risk [29,30]. In preclinical studies, the zebrafish embryo model and the animal models have been established to evaluate the developmental toxicity and teratogenic activity of the drugs [31–34]. In addition, a few in silico models have been proposed to predict the reproductive toxicity of chemical compounds [35,36], but none of them is developed for the purpose of teratogenic risk prediction. Although large-scale epilepsy pregnancy registries played an important role in evaluating the teratogenic risk of ASMs, the teratogenic risks of a large proportion of ASMs, especially the newly approved ones, are still unknown because of the limited postmarketing surveillance data. Thus, the selected model was utilized to predict the potential teratogenic risk of drugs based on their chemical structural information.

To wipe out the obstacles of selecting a proper SAR model for studying ADRs and their corresponding drugs, two projects were investigated by using the 5-step strategy that mentioned above. The results of the two related projects were demonstrated to show the applicability of the whole strategy and the selected models. The detailed workflow of this strategy was described in Fig. 20.1.

Methods and materials

Constructing centralized dataset and obtaining training set

In previous studies, a systematic workflow for constructing databases with centralized data source by classifying the FDA-approved drug labeling was designed. By using this workflow, the data source could be generated by integrating the information of postmarketing surveillance data, the physicochemical properties, and key biomarkers of drugs. The drugs related to a certain type of ADRs related were evaluated and stratified into five levels, and the drugs were scored from one to five according to the labeling content. The highest severity score (severity score = 5) was assigned to drugs that were withdrawn from the market or have a warning in the "Boxed warning" in their drug labels because they can induce fatal ADRs. The lowest severity score (severity score = 1) was assigned to drugs without any significant side effect. Subsequently, with the severity score, the drugs were categorized into four concern levels: most-DIR concern drugs (severity score = 4 or 5), moderate-DIR concern drugs (severity score = 2 or 3), less-DIR concern drugs (severity score = 1), and no-DIR concern drugs (severity score = 0).

For DIR project, to generate the SAR model, the related drugs were schemed by using the workflow that mentioned above. The Drug-Induced Rhabdomyolysis Atlas (DIRA) dataset was curated [37]. In DIRA, 213 DIR drugs were collected and categorized into five severity levels according to the FDA drug labeling content. The most-DIR concern drugs cause serious clinical outcomes, that is, acute renal failure. The moderate-DIR concern drugs mainly cause DIR by the concomitant use with other drugs or overdose. The less-DIR concern and no-DIR concern drugs do not have severe clinical outcomes and any muscular-injury-related ADRs, respectively. To obtain the training set for establishing an SAR model of DIR, the 55 most-DIR concern drugs and 40 no-DIR concern drugs were used as positive and negative samples, respectively. After removing the protein drugs, 49 most-DIR concern drugs and 40 no-DIR drugs remained in the training set.

For DIT project, a similar procedure was applied. By using the workflow of generating centralized data sources of teratogenicity, drugs related to ASMs and teratogenicity were collected and classified for constructing SAR models. FDA-approved drug labeling contains the essential information for the safe and effective use of the drug, which is a relatively accurate and stable data source to evaluate the teratogenic risk of a drug [38]. Drug labeling with keywords "teratogenicity" were extracted from the FDA Label public version (https://nctr-crs.fda.gov/fdalabel/ui/search) (data acquired on April 26, 2021) by using full text searching. In total, 7926 drug labels with keywords "teratogenicity" were obtained. Then, 7059 drugs labeled with single active ingredient based on Unique Ingredient Identifier (UNII) were kept. Subsequently, only the drug labels for drugs administered through oral or parenteral route were kept (6180 drug labeling). If multiple drug labels contain the same UNII, the latest version of the drug label was kept. At last, a dataset containing 286 drugs with potential teratogenic risk were generated, and the drugs were classified into five severity scores and four concern levels. Based on the FDA use-in-pregnancy ratings, the most-concern drugs and the no-concern drugs were kept in the training dataset. That is, 112 drugs were divided into high teratogenic risk (positives, 67 drugs) group and low teratogenic risk (negatives, 45 drugs) group based on the drug labeling information. A drug was categorized into high teratogenic risk if it was indicated in drug labeling that adequate and well-controlled studies or animal studies had shown the teratogenic risk to the fetus, while a drug was categorized into low teratogenic risk if both animal and human studies showed no risk to the fetus.

To generate the structure data files (SDFs) of the drugs in training sets, the drugs were mapped to the PubChem [39] database to obtain the 2D SDFs. The 3D SDFs were generated by using E-Dragon webserver [40]. Molecular descriptors were calculated by Mold2 for drugs using the SDFs [41]. The details of calculating the descriptors are given in Calculating 1D/2D/3D molecular descriptors section.

Collection of FAERS data as validation dataset

FAERS database contains the postmarketing surveillance data of human prescription drugs and is commonly used for drug safety surveillance (https://www.fda.gov/drugs/drug-approvals-and-databases/fda-adverse-event-reporting-system-faers) [42]. Several hurdles of the FAERS database exist, including nonstandard usage of

TABLE 20.1 Contingency table for the calculation of reported odds ratio.

	Cases with current ADR	Cases without current ADR
Cases with current drugs	a	b
Cases with other drugs	c	d

terms, duplicate records, and mapping of drugs and adverse events inaccurately [43]. Therefore, the data in the FAERS database need to be preprocessed. The preprocessing procedures proposed by Banda et al. [43] were used. All the reported cases from 2004 to 2020 in FAERS were compiled by following steps: (1) removing the duplicated cases based on four demographic data fields (event date, age, sex, and reporter country); (2) mapping the drug names into RxNorm clinical drug concepts (RXCUIs) and Observational Health Data Sciences and Informatics (OHDSI) standard vocabulary identifiers; (3) mapping the drug indications and reactions from MedDRA to the Systematized Nomenclature of Medicine Clinical Terms (SNOMED-CT) standard codes by using OHDSI vocabulary; and (4) establishing the drug-event pairs by linking each drug in each case with the associated outcome and calculating the reporting odds ratio (ROR) for each drug by the equation:

$$\mathrm{ROR} = \frac{a/c}{b/d}$$

where a, b, c, and d are the numbers of cases defined in the contingency table (Table 20.1).

For DIR project, a total of 131,491,140 drug—event pairs were generated from the FAERS dataset. Subsequently, we extracted the drug—event pairs related to rhabdomyolysis, resulting a total of 2878 drugs that were reported to cause DIR. To indicate the ADR severity of the drugs, the reporting odds ratio (ROR) [44] was calculated for each drug. Then, the 2878 DIR drugs were mapped to the DrugBank database (https://www.drugbank.ca) [45] and extracted 1441 DIR drugs with the chemical structures. After removing 39 drugs that were contained in the training set, 1402 drugs were finally kept for validating the SAR model. The drugs were ranked by their RORs in descending order.

For DIT project, the related cases were retrieved, and the relevant drugs were obtained with teratogenic risk according to the following steps. Firstly, all the preferred terms related to congenital malformation under the subject word "Congenital Abnormalities" in MeSH were extracted. Secondly, by using preferred terms as keywords to search the FAERS database, the cases were kept if the preferred terms were reported as adverse events. Finally, the drugs reported as the primary suspected drug in each case were considered as the teratogenic risk drug. After removing the drugs that existed in the training set, the drugs that are administered through the oral or parenteral route and have more than three reported cases and the lower limit of 95% CI > 1 were kept. In addition, teratogenic risk of the drugs in the low teratogenic risk group was validated by using the postmarketing surveillance data in FAERS. Eventually, the drugs were uploaded into the PubChem database to generate 2D and 3D SDFs. Similar to the DIR project, the Anatomical Therapeutic Chemical (ATC) code for each drug was extracted from the Drugbank database [45].

Calculating 1D/2D/3D molecular descriptors

In order to have a relatively comprehensive description of chemical structures and molecular activities, three levels of molecular descriptors were used in these two projects, namely, 1D, 2D, and 3D molecular descriptors. Molecular descriptors were generated for the drugs in both the training and validation sets.

For the DIR project, the molecular descriptors of the drugs were generated by using Mold2 (https://www.fda.gov/science-research/bioinformatics-tools/mold2), which is a free software package for calculating 777 1D and 2D molecular descriptors from chemical structures [46]. The 777 descriptors can be grouped into 20 categories. Compared with other software that calculates molecular descriptors, Mold2 has the advantage of a faster speed of calculation. Mold2 is able to calculate the amount of information, but the other software is only based on the 2D structure of a chemical [41]. Mold2 was first applied to calculate the molecular descriptors for each drug in the training set and then removed the descriptors with a variance less than 0.001. Eventually, 572 molecular descriptors were kept as the features of model construction. The input to Mold2 is the SDFs of the drugs' 2D chemical structures. After the preprocessing of Mold2, the 777 descriptors were calculated for the drugs. The preprocessing is a structural standardization process that corrects the input file and converts it into Mold2 operable form.

The drugs in the training dataset were mapped to the PubChem (https://pubchem.ncbi.nlm.nih.gov/) [39] database, and 1D and 2D SDFs were obtained for the 89 drugs for SAR model construction, among which 49 mostDIR concern drugs were positives and 40 no-DIR concern drugs were negatives. For the drugs in the validation dataset, 2878 DIR drugs were subsequently mapped to the DrugBank database [45], and 1441 DIR drugs were extracted with the chemical structures. After removing 39 drugs from the 1441 DIR drugs that overlapped with the training set, 1402 drugs were finally kept for validating the SAR model from FAERS. The drugs were ranked by their RORs in descending order.

For DIT project, similar to DIR, the 1D and 2D molecular descriptors of the drugs were calculated by Mold2. Additionally, 3D molecular descriptors were used. The 3D molecular descriptors were calculated by E-Dragon web server (http://www.vcclab.org/lab/edragon/) [40] with 3D SDFs of the drugs.

To generate the SDFs of the drugs for the training dataset of teratogenicity, the drugs were mapped to the PubChem [39] database to obtain the 2D and 3D SDFs of 112 drugs (67 positives and 45 negatives) for SAR model construction. When processing the validation dataset that was obtained from the FAERS database, candidate drugs were uploaded into the PubChem database to generate 2D and 3D SDFs. In addition, the ATC code for each drug was extracted from the Drugbank database [45].

SAR model development

To develop, optimize, and select the valid model among multiple models, for both project, four state-of-the-art ML methods were used to predict the severity of corresponding ADR for comparison, including RF, extreme gradient boosting (XGBoost), support vector machine (SVM), and logistic regression (LR). All the parameters of XGBoost, SVM, and LR were optimized by the grid search approach. All the model construction and prediction were conducted in Python 3.7.5 with the scikit-learn 0.24.1.

RF

RF algorithm [47] integrates a number of decision trees (DTs) to form an ensemble classifier and classified the samples on the basis of a majority voting scheme [48]. In these projects, three parameters, namely, the number of trees, the number of features, and the maximum depth of a tree were optimized accordingly.

XGboost

XGBoost is also an ensemble classifier and integrates multiple gradient boosted decision trees [49]. Different from RF, XGBoost establishes the association between the decision trees and can be considered as a scalable end-to-end tree boosting system that was proposed by Chen et al. in 2016. In these two projects, two parameters, namely, minimum loss reduction parameter (γ) and the maximum depth of a tree, were determined by using the training set and were optimized.

SVM

SVM is a classical ML method based on a statistical learning theory [50], that maps vectors nonlinearly to a high-dimensional space and constructs decision surfaces in the high-dimensional space. In both projects, the radial basic function was used as the kernel function and the regularization parameter (c) and kernel width (γ) were optimized in the modeling procedure.

LR

LR is a linear regression approach for classification by using a sigmoid active function [51]. LR outputs the probability of a sample that belonged to positive or negative categories. The inverse of regularization strength (C) was optimized in model training procedure.

After applying all methods, RF showed the best performance among these four models in the DIR project, and SVM outperformed the other three models in the teratogenicity project. With the corresponding outperformed medals of each project, further investigations and optimizations were conducted. To ensure the stability of the model, both selected models involved 1000 iterations of 5-fold cross-validations.

For DIR project, the RF model was constructed with 572 molecular descriptors as the features in the 5-fold cross-validation. Two parameters (the number of trees and the maximum depth) were optimized by a grid search

TABLE 20.2 The optimized parameters of each model.

Model	Optimized parameters
RF	'max_depth': 3, 'min_samples_split': 10, 'n_estimators': 20
XGBoost	'eta': 0.01, 'gamma': 0.01, 'learning_rate': 0.3, 'max_depth': 4
SVM	'svc__C': 0.5, 'svc__gamma': 0. 01
LR	'LR__C': 2.0, 'LR__solver': 'newton-cg'

approach. A drug with a predicted probability higher than 0.5 was considered a high-risk DIR drug. After selecting the best parameters, to obtain a statistically robust estimation of model performance, the modeling procedures were conducted 1000 times. At the same time, we conducted two permutation tests. One is that the DIR labels for the drugs were randomly permuted to generate a permutated labels dataset, and the other is that the features of the drugs that were randomly permuted to generate a permutated features dataset. Using the permutated labels dataset and the permutated features dataset, the RF models were constructed by 1000 iterations of 5-fold cross-validations. Compared with the performance of the models in the two types of permutation tests, the performance of the models with the real dataset was really better than that of the random models. The optimal parameters selected by all models are in Table 20.2.

Similarly, for the DIT project, SVM was proposed for teratogenic risk prediction. The radial basic function was used as the kernel function, and two parameters, regularization parameter (c) and kernel width (γ), were optimized by grid searching and 5-fold cross-validation. The optimal regularization parameter (c) and kernel width (γ) were 2 and 0.001, respectively. After determining the optimal parameters, we performed 1000 iterations of 5-fold cross-validation to investigate the performance of the models.

Evaluation of model performance

To evaluate the performance of the models in both training dataset and validation dataset, nine performance metrics (accuracy, recall rate, precision, specificity, Matthews correlation coefficient (MCC), balance accuracy (BACC), F1 score, the area under the receiver operating characteristic curve (AUROC), and the area under the precision-recall curve (AUPRC)) were applied in both projects. The equations of the nine metrics were given as follow.

$$\text{accuracy} = \frac{\text{TP} + \text{TN}}{\text{TP} + \text{TN} + \text{FP} + \text{FN}}$$

$$\text{recall rate} = \frac{TP}{TP + FN}$$

$$\text{precision} = \frac{\text{TP}}{\text{TP} + \text{FP}}$$

$$\text{specificity} = \frac{\text{TN}}{\text{TN} + \text{FP}}$$

$$\text{MCC} = \frac{\text{TP} \times \text{TN} - \text{FP} \times \text{FN}}{\sqrt{(\text{TP}+\text{FP})(\text{TP}+\text{FN})(\text{TN}+\text{FP})(\text{TN}+\text{FN})}}$$

$$\text{BACC} = \frac{(\text{TPR} + \text{TNR})}{2}$$

$$\text{F1 score} = \frac{2 \times (\text{precision} \times \text{recall rate})}{\text{precision} + \text{recall rate}}$$

$$\text{AUROC} = \int_{x=0}^{1} \text{TPR}\left(\text{FPR}^{-1}(x)\right)dx$$

$$\text{AUPRC} = \int_{-\infty}^{\infty} \text{precision } (x)dP\left[Y \leq x\right]$$

Where TP, TN, FP, and FN indicate true positives, true negatives, false positives, and false negatives, respectively. The average metrics of 1000 iterations of 5-fold cross-validations were used to evaluate the model performance.

Results

Study design

The SAR models based on two curated datasets of two ADRs (DIR and teratogenicity) were constructed. The workflow of our study design is shown in Fig. 20.1.

In DIR project, the SAR model was based on the curated DIR dataset named DIRA (https://www.adratlas.com) to predict the DIR severity of the marketed drugs. The 49 most-DIR concern drugs and 40 no-DIR concern drugs were used as the positive and negative samples, respectively. The 1D and 2D chemical descriptors of the drugs were calculated. In the study of a real application, the RF model was applied to predicting the potentially serious DIR of 1402 drugs, which were reported to cause DIR by FAERS.

Similarly, in the DIT project, to facilitate the prevention and early detection of DIT in pregnancy, a dataset was curated. An SAR model was developed to predict the potential teratogenic risk of a drug. In addition to 1D and 2D chemical descriptors, 3D chemical descriptors were calculated in this project. The model was subsequently applied to evaluating the teratogenic risk of ASMs. The SVM model was validated by using 196 drugs that were reported to induce congenital malformations in the FDA FAERS database.

Furthermore, four models (RF, XG-Boost, SVM, LR) were constructed. The selected model was compared with the other three state-of-the-art ML methods. To test the reliability of the models, 1000 iterations of 5-fold cross-validations were applied to the construction procedures of all models. Nine performance metrics, namely, accuracy, recall rate, precision, MCC, BACC, F1 score, AUROC, AUPRC, and specificity, were used for evaluating the models.

Performance of SAR models

Performance of RF model for DIR

To evaluate the performance of the RF model, 1000 iterations of 5-fold cross-validations were applied in the modeling procedure. The average accuracy, recall rate, precision, MCC, BACC, F1, AUROC, AUPRC, and specificity scores were 0.73, 0.81, 0.73, 0.46, 0.73, 0.77, 0.73, 0.70, and 0.64, respectively, indicating that the RF model performed well in predicting the most-DIR concern drugs. Specifically, the RF model can better identify the positive samples in current data set (recall rate = 0.81). The average metrics and the corresponding standard deviations of the RF model are listed in Table 20.2. For comparison, we randomly permuted the DIR labels and features of the drugs and constructed RF models. The RF models were evaluated by 1000 iterations of 5-fold cross-validations in the permutation test. For the dataset of permutated labels, the average accuracy, recall rate, precision, MCC, BACC, F1 score, AUROC, AUPRC, and specificity were 0.51, 0.62, 0.55, 0.00, 0.50, 0.58, 0.50, 0.55, and 0.38, respectively, indicating that the RF model in this permutation test randomly predicted the DIR labels as expected. Similar results were achieved for the RF model in the permutation test with the dataset of permutated features (the average accuracy, recall rate, precision, MCC, BACC, F1 score, AUROC, AUPRC, and specificity were 0.52, 0.69, 0.55, −0.01, 0.50, 0.61, 0.50, 0.55, and 0.31, respectively). The average MCC achieved by the RF model using the actual dataset (MCC = 0.46) was significantly higher than those of the RF models using the permutation data sets, demonstrating that the RF model built with the actual data can predict the DIR severity using the chemical structure of drugs.

The performance of the RF model was compared with those of the state-of-the-art methods. The frequency distributions of the MCCs and recall rates from 1000 iterations of 5-fold cross-validations for different methods are shown in Fig. 20.2A and B, respectively. The results of MCCs and recall rates showed that the performance of the models with the actual dataset is significantly superior to those of the models constructed using the permutation dataset. The RF model outperformed the XGBoost, SVM, and LR models (Table 20.3). We used the Mann−Whiney U-test to indicate the significance of the difference in prediction results between the RF model and the other models. Except for specificity between RF and XGBoost (P-value = .094), the P-values for all performance metrics between RF and other models were less than 0.0001 (see Table 20.4), indicating that the performance of the RF model was significantly better than those of the other models.

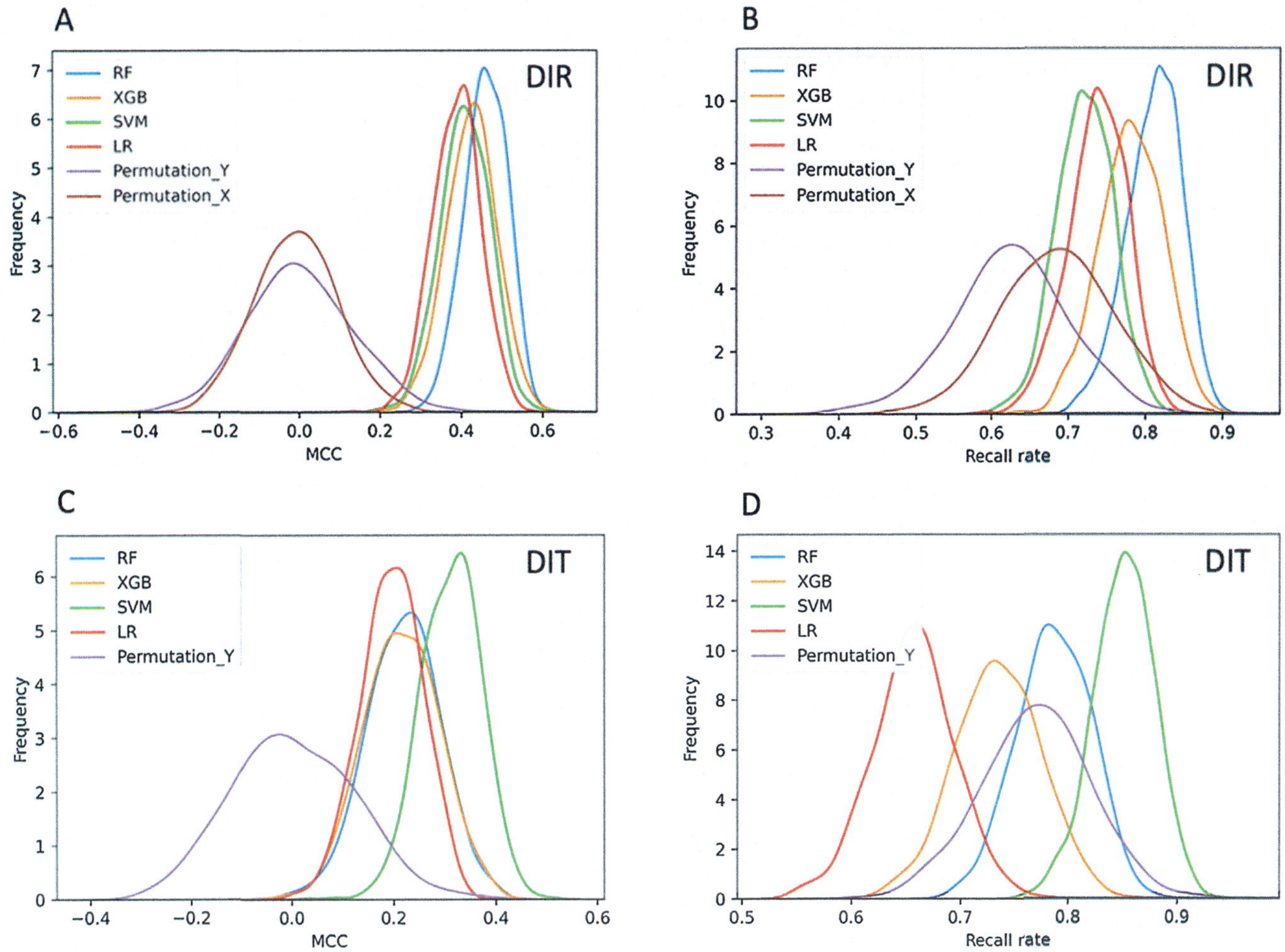

FIGURE 20.2 The distribution of Matthews correlation coefficients (MCCs) and recall rates achieved by random forest (RF), XGBoost, support vector machine (SVM), logistic regression (LR), Permutation_X, and Permutation_Y. (A) MCC distributions of DIR. (B) Recall rates distributions of drug-induced rhabdomyolysis (DIR). (C) MCC distributions of drug-induced teratogenesis (DIT). (D) Recall rates distributions of DIT. For (A and C) the Y-axis represents the frequency of the predictive model MCC, and the X-axis represents the MCCs achieved by different methods. For (B and D) the Y-axis represents the frequency of the predictive model recall rates, and the X-axis represents the recall rates achieved by different methods. Permutation_X and Y indicated that the models were constructed by using the samples with randomly permutated labels.

TABLE 20.3 Prediction results of RF, XGBoost, SVM, LR, Permuatation_X, and Permutation_Y[a] of DIR.

	RF	XGBoost	SVM	LR	Permutation X	Permutation Y
Accuracy	0.73 ± 0.03	0.72 ± 0.03	0.71 ± 0.03	0.70 ± 0.03	0.52 ± 0.05	0.51 ± 0.06
Recall rate	0.81 ± 0.04	0.78 ± 0.04	0.72 ± 0.04	0.74 ± 0.04	0.69 ± 0.07	0.62 ± 0.08
Precision	0.73 ± 0.02	0.73 ± 0.03	0.74 ± 0.03	0.72 ± 0.03	0.55 ± 0.03	0.55 ± 0.05
MCC	0.46 ± 0.05	0.42 ± 0.06	0.41 ± 0.06	0.39 ± 0.06	−0.01 ± 0.10	0.00 ± 0.10
BACC	0.72 ± 0.03	0.71 ± 0.03	0.71 ± 0.03	0.69 ± 0.03	0.50 ± 0.05	0.50 ± 0.06
F1 score	0.77 ± 0.02	0.75 ± 0.03	0.73 ± 0.03	0.73 ± 0.03	0.61 ± 0.05	0.58 ± 0.06
AUROC	0.73 ± 0.03	0.71 ± 0.03	0.71 ± 0.03	0.69 ± 0.03	0.50 ± 0.05	0.50 ± 0.06
AUPRC	0.70 ± 0.02	0.69 ± 0.03	0.69 ± 0.03	0.68 ± 0.02	0.55 ± 0.02	0.55 ± 0.03
Specificity	0.64 ± 0.04	0.64 ± 0.05	0.69 ± 0.05	0.65 ± 0.05	0.31 ± 0.08	0.38 ± 0.09

[a]*Permutation_X, the RF model constructed by using the samples with permutated features; Permutation_Y, the RF model constructed by using the samples with permutated labels.*

TABLE 20.4 The *P*-values for the predicted results of RF models and other models of DIR.

	RF versus XGBoost	RF versus LR	RF versus SVM	RF versus Permutation_X	RF versus Permutation_Y
Accuracy	<0.0001	<0.0001	<0.0001	<0.0001	<0.0001
Recall rate	<0.0001	<0.0001	<0.0001	<0.0001	<0.0001
Precision	<0.0001	<0.0001	<0.0001	<0.0001	<0.0001
MCC	<0.0001	<0.0001	<0.0001	<0.0001	<0.0001
BACC	<0.0001	<0.0001	<0.0001	<0.0001	<0.0001
F1 score	<0.0001	<0.0001	<0.0001	<0.0001	<0.0001
AUROC	<0.0001	<0.0001	<0.0001	<0.0001	<0.0001
AUPRC	<0.0001	<0.0001	<0.0001	<0.0001	<0.0001
Specificity	0.09389	<0.0001	<0.0001	<0.0001	<0.0001

Permutation_X: the RF model constructed by using the samples with permutated features; Permutation_Y: the RF model constructed by using the samples with permutated labels.

Performance of SVM model for DIT

We generated 777 1D and 2D molecular descriptors and 1666 3D molecular descriptors. After filtering the descriptors with the standard deviation across all samples less than 0.001, 2083 1D/2D/3D descriptors remained as features for model construction. Of the 2083 molecular descriptors, 585 were 1D and 2D molecular descriptors and 1498 were 3D molecular descriptors.

The frequency distributions of MCCs and recall rates from the 1000 iterations of 5-fold cross-validations for the four models and the random model were shown in Fig. 20.2C and D, respectively. The performance of four models was much better than that of the random model. The SVM model had higher MCCs and recall rates than LR, RF, and XGBoost models. Table 20.5 showed the average values of performance metrics and the corresponding standard deviations from the 1000 iterations of cross validations. Among the four modeling algorithms, SVM exhibited the best performance on predicting the teratogenic risk. The average MCC, BACC, F1 score, AUROC, AUPRC, specificity, and recall rate of SVM model were 0.312, 0.640, 0.762, 0.640, 0.676, 0.428, and 0.851, respectively. Specifically, the average recall rate was 0.851 for the SVM model, significantly higher than those from the LR, RF, and XGBoost models (0.657, 0.787, and 0.735 for LR, RF, and XGBoost, respectively), indicating that the SVM model performed better on detecting the drugs with high teratogenic risk than the other three models.

TABLE 20.5 The prediction results of SVM, SVM, LR, RF, XGBoost, and Permutation_Y[a] of DIT.

	RF	XGBoost	SVM	LR	Permutation Y
Accuracy	0.638 ± 0.001	0.630 ± 0.001	0.681 ± 0.001	0.610 ± 0.001	0.554 ± 0.002
Recall rate	0.787 ± 0.001	0.735 ± 0.002	0.851 ± 0.001	0.657 ± 0.001	0.770 ± 0.003
Precision	0.668 ± 0.001	0.676 ± 0.001	0.690 ± 0.000	0.680 ± 0.001	0.599 ± 0.001
MCC	0.219 ± 0.005	0.215 ± 0.005	0.312 ± 0.004	0.195 ± 0.004	0.001 ± 0.015
BACC	0.602 ± 0.001	0.604 ± 0.001	0.640 ± 0.001	0.598 ± 0.001	0.501 ± 0.003
F1 score	0.722 ± 0.001	0.704 ± 0.001	0.762 ± 0.000	0.668 ± 0.001	0.673 ± 0.001
AUROC	0.602 ± 0.001	0.604 ± 0.001	0.640 ± 0.001	0.598 ± 0.001	0.501 ± 0.003
AUPRC	0.653 ± 0.000	0.656 ± 0.000	0.676 ± 0.000	0.653 ± 0.000	0.600 ± 0.001
Specificity	0.416 ± 0.004	0.474 ± 0.004	0.428 ± 0.003	0.539 ± 0.002	0.233 ± 0.006

[a]*Permutation_Y, the SVM model constructed by using the samples with permutated labels.*

Prediction based on validation dataset from FAERS

Predicting serious DIR drugs

To verify the prediction ability of the RF model in detecting the DIR severity for the marketed drugs, we extracted 1402 drugs from the FAERS database, which had been reported to cause DIR to varying degrees. We then applied the RF model to detect the drugs with serious DIR. The 1402 drugs were first ranked by the RORs in a descending order and then were predicted by the RF model. The 1402 drugs were related to DIR and their default label is 1. Of the top 10 ranked drugs, seven (70.0%) were correctly predicted by the RF model (Table 20.6). In addition, from the ranked drugs, the recall rates of the top ranked drugs with a sliding window of 150 drugs are shown in Fig. 20.3B. As the number of drugs increased from 150 to 600, the recall rate decreased rapidly from 62.7% to 55.3%. After that, the recall rate remained almost unchanged. The results indicate that the RF model was more sensitive to the drugs with serious DIR. Since DIR is mainly caused by drugs taken orally or parenterally, we extracted 1043 oral or injected drugs from 1402 drugs in FAERS for the DIR prediction by using the RF model. The results of 1043 drugs were similar to those of 1402 drugs (Fig. 20.4A).

We selected the top 150 DIR drugs and 150 drugs without DIR from FAERS to construct a test set. In addition, to validate the prediction performance of the RF model on FAERS, we separately used the datasets of randomly shuffled labels and randomly shuffled features as the negative controls. The prediction results are listed in Table 20.7, and the ROC curves of the model prediction results are shown in Fig. 20.4B. It can be seen from the prediction results that the prediction performance of the RF model with a real dataset was significantly higher than those of models with permutation datasets. To illustrate the predictive ability of the RF model in FAERS, we separately used the dataset of randomly shuffled labels and randomly shuffled features as the negative controls to validate our model. We selected the top 150 drugs ranked by ROR values in a descending order among 1402 DIR drugs as positive samples and randomly selected 150 drugs without reported DIR from FAERS as negative samples. The curves of predictive AUROC of the models were shown in Fig. 20.4B.

Of the 1402 DIR drugs, a total of 774 were predicted as the positive drugs by the RF model. We extracted the ATC codes of these positive drugs from the DrugBank database. The numbers of drugs in different therapeutic categories are shown in Fig. 20.4. The 774 positive drugs fell into 14 therapeutic categories. Four 482 (62.3%) drugs belong to four categories including anti-infectives for systemic use (J, 128 drugs), cardiovascular system (C, 127 drugs), alimentary tract and metabolism (A, 115 drugs), and nervous system (N, 112 drugs) (Fig. 20.3A). From the distribution of drugs in the training set, it can be seen that 31 of the 45 drugs (68.9%) were in the therapeutic categories of nervous system (N, 22 drugs) and cardiovascular system (C, nine drugs) (Fig. 20.4C). The drugs in these two categories accounted for a large proportion of the DIR drugs, which should be the main reason why our model can predict more drugs in the categories of nervous system and cardiovascular system. We summarized the recall rate of the RF model for each therapeutic category. According to the recall rate, the top three therapeutic categories were systemic hormonal preparations excluding reproductive hormones and insulins (H), anti-infectives for systemic use (J), and genitourinary system and reproductive hormones (G), for which the recall rates were 87.5% (21/24), 75.6% (127/168), and 65.0% (65/100), respectively, suggesting that our model was more sensitive to detect the DIR drugs in these categories.

TABLE 20.6 Information of the top 10 drugs of the DIR drug list.

Drug name	Odds ratio	ATC code
Acepromazine[a]	44.978	N05
Rapacuronium[a]	38.996	Withdrawn
Oxyphenbutazone[a]	32.902	M01; M02; S01
Naringenin[a]	31.905	
Ferrous ascorbate	28.716	B03
Etofibrate[a]	27.707	C10
Perhexiline	25.406	C08
Dextrothyroxine[a]	23.397	C10
Kojic acid	21.27	
Podofilox[a]	19.829	D06

[a]*Drugs that were predicted as positives by the RF model.*

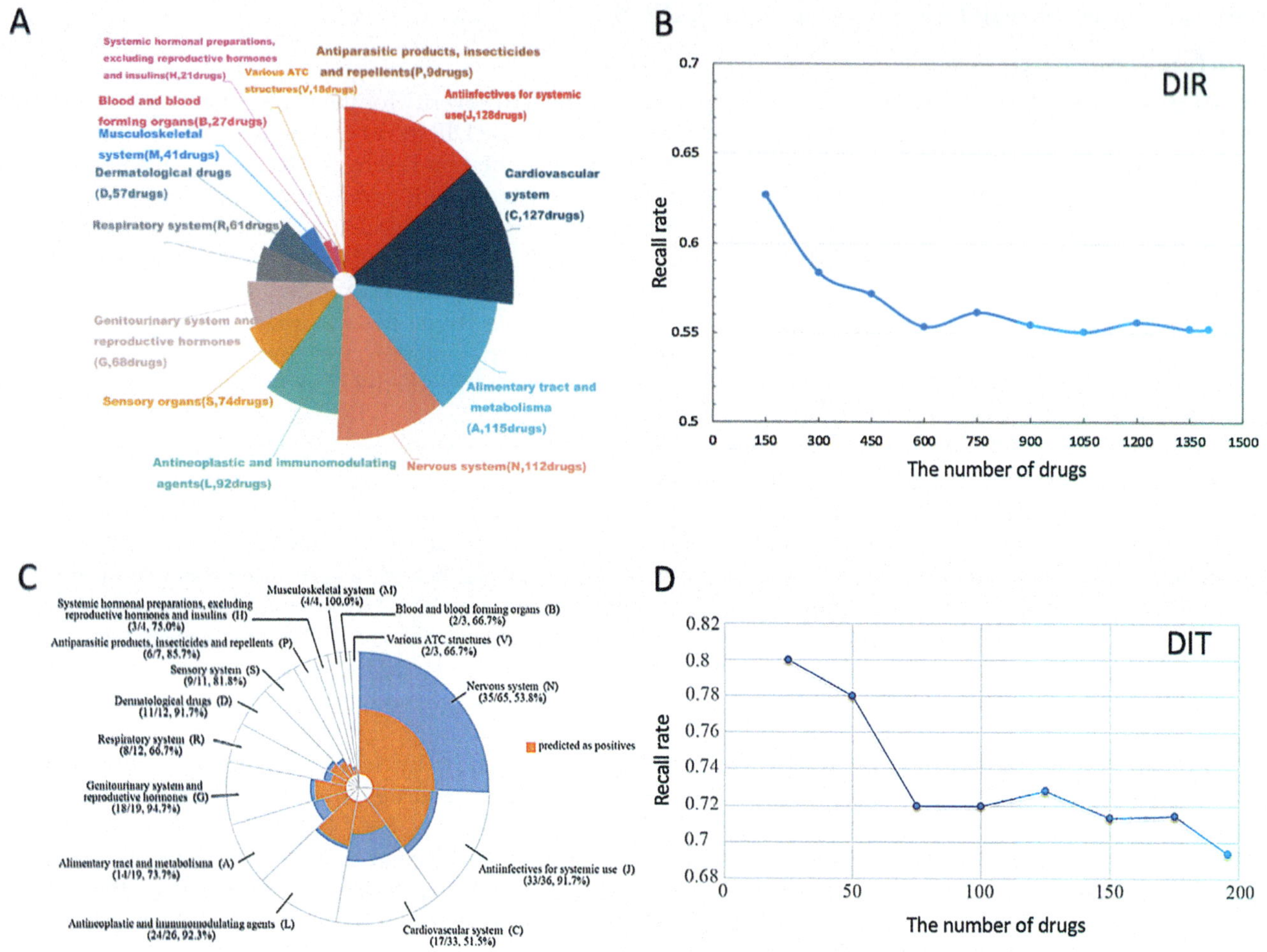

FIGURE 20.3 (A) Distribution of different therapeutic categories of the drugs predicted to be positive in FAERS for drug-induced rhabdomyolysis (DIR). (B) The trend of recall rate changed with the numbers of drugs in the DIR prediction. (C) The distribution of the drugs predicted as positives by support vector machine (SVM) in different therapeutic categories from FAERS in drug-induced teratogenesis (DIT); (D) the trend of recall rate changed with the number of drugs increased in DIT. For (B and D) the Y-axis indicates the recall rate, and the X-axis is the number of drugs used for the prediction.

To indicate the significance of the therapeutic categories that can be well predicted by our model, we calculated the Fisher's exact test P-value for each category. According to the Fisher's exact test P-values, the drugs in the categories of A ($P = .007$), L ($P = .002$), N ($P = .008$), and P ($P < .0001$) can be significantly identified by our model. The specific steps of establishing the confusion matrix and calculating the Fisher's exact test P-values are described in Table 20.8.

Predicting teratogenic risk drugs

The SVM model was validated by using 196 drugs that were reported to induce congenital malformations in the FDA FAERS database. The distribution of drugs in different therapeutic categories was shown in Fig. 20.3C. The ATC code for each drug was extracted from the Drugbank database by searching drug names. There is no therapeutic category associated with two of the 196 drugs. Thirty-seven drugs are associated with more than one therapeutic category. For each of the 37 drugs, we counted it once in each of the categories when calculating the recall rate. Among the 196 drugs, over half belonged to four therapeutic categories including nervous system (N, 65 drugs, 53.8%), anti-infectives for systemic use (J, 36 drugs, 91.7%), cardiovascular system (C, 33 drugs, 51.5%), and antineoplastic and immunomodulating agents (L, 26 drugs, 92.3%). A total of 136 drugs (69.4%) were predicted to be high teratogenic risk by the SVM model. For each therapeutic category, the SVM model can detect more than 50% drugs (Fig. 20.3C). Specifically, for the therapeutic category of antineoplastic and immunomodulating agents (L), 92.3% drugs were correctly predicted with potential teratogenic risk.

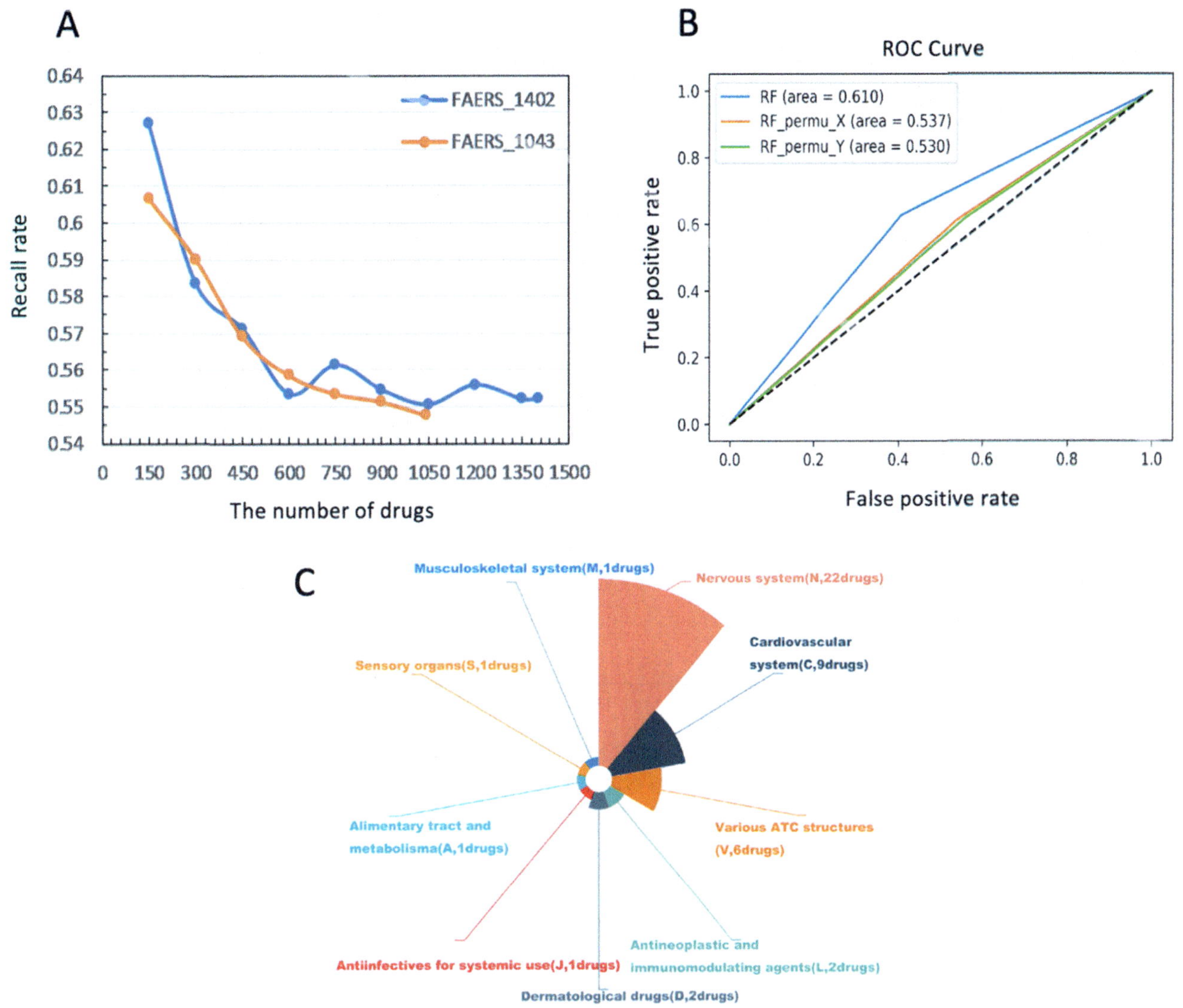

FIGURE 20.4 (A) The trend of recall rate changed with the numbers of drugs in prediction. The Y-axis indicates the recall rate and the X-axis is the numbers of drugs used for prediction. The blue line indicated the recall rates calculated by using all the drug-induced rhabdomyolysis (DIR) drugs in FAERS. The orange line indicates the recall rates calculated by using the DIR drugs administered through an oral or parenteral route in FAERS. (B) Prediction AUROC curves of RF model, Permutation_X, and Permutation_Y models in data set (150 DIR-drugs and 150 no DIR drugs) RF_permu_X: The model constructed by using the samples with permutated features; PF permu Y: The model constructed by using the samples with permutated labels. (C) Distribution of the drugs in different therapeutic categories of Most-DIR concern drugs in the training set.

TABLE 20.7 The prediction results of RF, permutation_X, and permutation_Y in the data set (150 DIR-drugs and 150 no DIR drugs).

	RF	Permutation_X	Permutation_Y
Accuracy	0.61	0.537	0.53
Recall rate	0.606	0.532	0.525
Precision	0.627	0.613	0.62
MCC	0.22	0.074	0.061
BACC	0.616	0.57	0.569
F1 score	0.61	0.537	0.53
AUROC	0.61	0.537	0.53
AUPRC	0.567	0.519	0.516
Specificity	0.593	0.46	0.44

Permutation_X: the RF model constructed by using the samples with permutated features; Permutation_Y: the RF model constructed by using the samples with permutated labels.

TABLE 20.8 The therapeutic category and the Fisher's exact test *P* value of the drugs predicted to have a risk of DIR.

ATC code	Num. of predicted drugs	*P*-value
Alimentary tract and metabolism (A)	115	0.007
Blood and blood forming organs (B)	27	0.247
Cardiovascular system (C)	127	0.066
Dermatological drugs (D)	57	0.142
Genitourinary system and reproductive hormones (G)	68	0.115
Systemic hormonal preparations, excluding reproductive hormones and insulins (H)	21	1
Anti-infectives for systemic use (J)	128	0.139
Antineoplastic and immunomodulating agents (L)	92	0.002
Musculoskeletal system (M)	41	0.921
Nervous system (N)	112	0.008
Antiparasitic products, insecticides and repellents (P)	9	0
Respiratory system (R)	61	0.235
Sensory organs (S)	74	0.223
Various ATC structures (V)	18	1.762

Subsequently, the 196 drugs were ranked by their RORs in a descending order and were used to investigate the performance of the SVM model on detecting high teratogenic risk drugs. The recall rates were calculated for the top ranked drugs with a sliding widow of 25 drugs and are shown in Fig. 20.3D. It can be seen that the SVM model can detect 80.0% drugs from the top 25 drugs. The recall rate decreased with the increase of the number of top ranked drugs. Only 69.4% drugs can be detected in the top 196 drugs, indicating that the SVM model was more sensitive to detecting the drugs with a high teratogenic risk. In addition, we investigated the prediction results of the top 10 drugs and found that eight were correctly predicted (Table 20.9).

For the teratogenic risk prediction of ASMs, the predicted probabilities of ASMs were shown in Fig. 20.4. We took 0.5 as the cut-off predicted probability to classify the drugs. The drugs with predicted probability larger than 0.5 were considered as high-teratogenic-risk drugs. It can be seen from the figure that 21 ASMs (21/39, 53.8%) including four commonly used ASMs, namely, valproate, topiramate, phenobarbital, and phenytoin, were predicted to be of a

TABLE 20.9 The information of top 10 drugs with high DIT teratogenic risk in FAERS.

Drug name	Odds ratio	ATC code
Cisapride[a]	364.7	Withdrawn
Levamisole	284.7	Withdrawn
Bedaquiline[a]	146.7	J04AK05
Ibutilide[a]	139.3	C01BD05
Ondansetron[a]	95.0	A04AA01
Clofazimine[a]	78.5	J04BA01
Nelfinavir	77.0	J05AE04
Folic acid[a]	70.7	V04CX02; B03BB01
Procainamide	64.1	C01BA02
Vandetanib[a]	63.8	L01XE12

[a]*Drugs that were predicted as positives by the SVM model.*

high teratogenic risk. In addition, five commonly used ASMs including lamotrigine, levetiracetam, gabapentin, carbamazepine, and oxcarbazepine were predicted to be low teratogenic risk.

Discussions

Drug-induced rhabdomyolysis-DIR

As an idiosyncratic ADR, DIR is rare but lethal in clinics. Limited methods in clinics, such as monitoring CK values, make it difficult to diagnose DIR effectively, which makes the early detection of DIR very arduous [15,19,20]. A few in silico methods have been proposed to predict the DIR of the drugs on the basis of their chemical structures, but the performance of the models were still suboptimal because of the scarcity of DIR cases and the lack of negative drugs [21]. The previous study established a curated DIR drug dataset (DIRA), in which the drugs were categorized into four severity levels (i.e., most-, moderate-, less-, and no-DIR concern) according to their DIR severity indicated by FDA drug labeling content. On the basis of this dataset, in this study, we endeavored to develop an SAR model for predicting the potentially serious DIR drugs from the marketed drugs. Therefore, we trained an RF model by using the most-DIR concern and no-DIR concern drugs and subsequently validated the model with the FAERS dataset. In our prediction, we specified the positive and negative samples as most-DIR and non-DIR, respectively, so that we predicted the serious DIR drugs, which refer to the drugs that have the high-risk DIR attribute.

The RF model performed the best (mean MCC = 0.46) when compared with the state-of-the-art methods including XGBoost, SVM, and LR (mean MCCs = 0.42, 0.41, and 0.39 for XGBoost, SVM, and LR, respectively) (Fig. 20.2A and B, Table 20.3). According to the predicting results of the models based on the dataset from DIRA, the RF model performed the best. To further validate the performance of the RF model and distinguish it from that of the random model, we also compared the prediction results of 1000 iterations of 5-fold cross-validations by using the actual dataset with that of the random model which was developed by using the permutated data set [52]. The mean MCC achieved by the actual dataset was 0.46 which was significantly higher than that of the models using the permutated labels dataset (mean MCC = 0.00) and the models using the permutated features dataset (mean MCC = −0.01), indicating that our proposed RF model was the most suitable for predicting the severity DIR of the drugs.

When applying the RF model to the FAERS dataset, it tended to find the drugs with higher RORs. After ranking the DIR drugs by their RORs in a descending order, the top four drugs (ROR > 30.00) and seven out of top 10 drugs (ROR >19.00) were correctly predicted. Of first 600 DIR drugs, recall rates were calculated for every 150 drugs and dropped from 62.7% to 55.3% with the threshold value of ROR decreasing from 4.64 to 1.45 (Fig. 20.3). The results illustrated that our proposed RF model can identify the drugs with serious DIR in practical application and should be a powerful tool for the early detection of most-DIR concern drugs. Among the drugs predicted as positives in the FAERS dataset, over half (482/774) belonged to four therapeutic categories, namely, anti-infectives for systemic use (J, 128 drugs), cardiovascular system (C, 127 drugs), alimentary tract and metabolism (A, 115 drugs), and nervous system (N, 112 drugs), indicating that our proposed model was more inclined to detect the serious DIR caused by the drugs in these four categories. The main reason might be that 68.9% (31/45) of most-DIR concern drugs in the DIRA were distributed in the therapeutic categories of nervous system (N, 22 drugs) and cardiovascular system (C, 9 drugs) (Fig. 20.4). More most-DIR concern drugs in other categories to train the model should be helpful for the model performance to detect more serious DIR drugs in other therapeutic categories. According to the significance analysis, the drugs in the four therapeutic categories of A ($P = .007$), L ($P = .002$), N ($P = .008$), and P ($P < .001$) can be significantly identified by the RF model.

DIT

Although the incidence of drug-induced malformations accounts for only 1% or less of the cases of birth defects with known causes [53], the teratogenicity of drugs has always been a widespread concern in clinical medication. Since the first case report indicated that ASMs were associated with congenital malformations in the 1960s [54,55], a number of efforts had been made for assessing the teratogenic risk of ASMs, so as to help pregnant women avoid the harm of drugs to the offspring while treating epilepsy [56,57]. Case-controlled studies and animal studies are currently reliable approaches to evaluate the teratogenic risk of ASMs. The teratogenicity of commonly used ASMs including valproate, carbamazepine, and lamotrigine had been systematically investigated by reviewing the prevalence of major congenital malformations in epilepsy pregnancy registries [29]. However, for a large proportion of ASMs, especially the newly approved ones, the lack of evidence is a big hurdle to effectively estimate the teratogenic risk. To fill this gap, we proposed an SAR model to detect the drugs with high teratogenic risk and applied it to assess the teratogenic risk of ASMs. The model was constructed by using SVM algorithm with the chemical structural descriptors of the drugs as features.

The SVM model performed the best when compared to the state-of-the-art ML approaches (Fig. 20.1). When validated by the FAERS dataset, the SVM model detected 136 of 196 drugs with teratogenic risk (recall rate = 0.694). The distribution of the predicted probability of the teratogenic risk of the drugs was shown in Fig. 20.5. For the subsets of top 10 and top 25 drugs ranked by their RORs, the recall rates achieved by the SVM model were 0.700 and 0.800, respectively, indicating that the model was more sensitive to detecting the high teratogenic risk drugs (Fig. 20.3B). Interestingly, the model seems to be sensitive to detecting the teratogenic risk drugs in two therapeutic categories, namely anti-infectives for systemic use and antineoplastic and immunomodulating agents, for which the recall rates were 91.7% and 92.3% (Fig. 20.3A).

For the risk prediction of ASMs, valproate, topiramate, phenobarbital, and phenytoin were predicted to be of a high teratogenic risk, which were consistent with drug labeling information and reported prevalence in pregnancy registries. The FDA-approved drug labeling of these four drugs included a clear statement of the teratogenic risk of use during pregnancy. Valproate labeling contains a BOXED WARNING section warning that it can cause major congenital malformations, for example, neural tube defects, and should avoid treating women with epilepsy during pregnancy. The teratogenic severity of topiramate, phenobarbital, and phenytoin was indicated in the WARNINGS AND PRECAUSIONS labeling section indicating that infants exposed to these drugs in utero had an increased risk for congenital malformations and other developmental outcomes. Especially for topiramate and phenobarbital, case-controlled studies and the data in pregnancy registries indicated that the use of these two drugs might increase the risk of fetal abnormalities. Furthermore, relatively high prevalence of major congenital malformations of valproate, topiramate, phenobarbital, and phenytoin is reported in the pregnancy registries. In addition, five commonly used ASMs including lamotrigine, levetiracetam, gabapentin, carbamazepine, and oxcarbazepine were predicted to be low teratogenic risk, which was also consistent with the low reported prevalence of major congenital malformations in pregnancy registries.

The chemical structures of drugs containing the toxicophores (or structural alerts (SAs)) were considered as one of the important factors of DIT [58]. For instance, the aromatic ring in the chemical structure of a drug, for exmaple, lamotrigine, may produce a highly reactive epoxide or arene oxide in the metabolism through cytochrome P-450 family of enzymes and generate reactive metabolites, which may contribute to teratogenesis [58]. To further

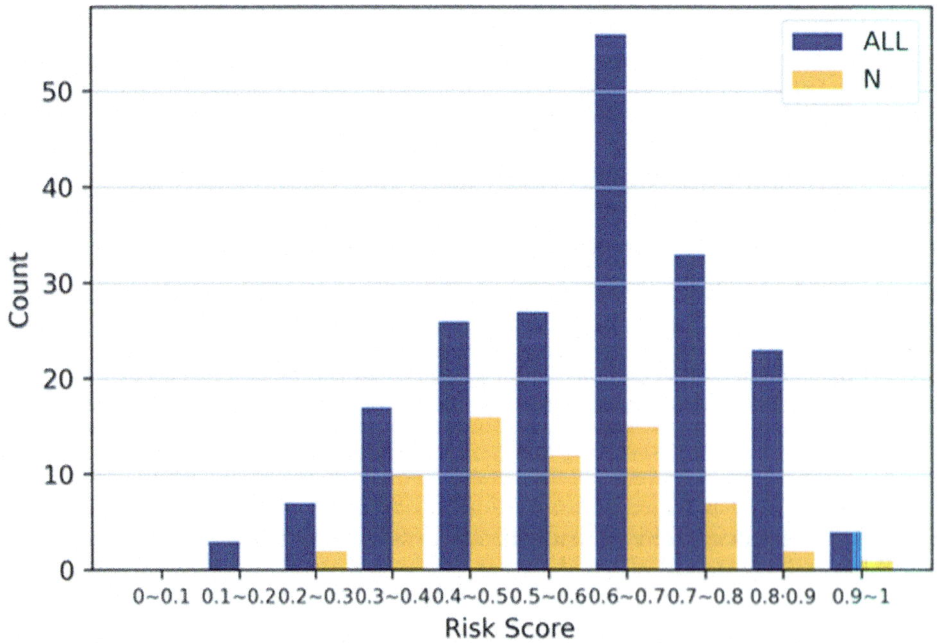

FIGURE 20.5 The distribution of the predicted probability of the teratogenic risk of the drugs in FAERS data set. ALL represents all drugs and N represents the nervous system drugs among them.

investigate the structural features of ASMs and help better understand the mechanisms of teratogenicity, the SAs associated with known toxic properties were extracted from the structures of 39 ASMs. We uploaded the SDFs of the ASMs into the Online Chemical Modeling Environment (OCHEM, https://ochem.eu/home/) and used a web server named ToxAlerts [59] for mapping the toxicological SAs. Consequently, a total of 23 SAs that might be associated with teratogenicity were extracted, which belonged to four end points, namely, genotoxic carcinogenicity and mutagenicity, idiosyncratic adverse reaction, potential electrophilic agents, and endocrine disruption (Fig. 20.6). The chemical structure of the common antiepileptic drug valproate contained the carboxylic acids, which might lead to the formation of reactive metabolites and then induce the dose-dependent idiosyncratic adverse reaction [60]. Another drug phenytoin contained a toxicological alert ethane-1,1-diyldibenzene, which was reported to be of potential estrogenic and androgenic activities [60,61]. In addition, ezogabine was predicted to be a high teratogenic risk drug by the SVM model with the highest predicted probability (0.858), for which the chemical structure contained three SAs. One SA was anilines which might form the reactive metabolites by bioactivation and was linked to the idiosyncratic adverse reaction [60]. The other two SAs were aromatic amines and derivatives of urethane, which were reported to be associated with genotoxic carcinogenicity and mutagenicity [62].

Three third-generation antiepileptic drugs (perampanel, rufinamide, and brivaracetam) were used in clinics, and there is insufficient data to support the risk assessment of teratogenicity for these three drugs. The predicted probability of teratogenicity was 0.80, 0.75, and 0.50 for perampanel, rufinamide, and brivaracetam, respectively, indicating a high risk of perampanel and rufinamide and a moderate risk of brivaracetam. In the FDA drug labeling, the evidence provided by animal studies showed that these three ASMs had developmental toxicity in pregnant rats and rabbits at clinically relevant doses or at plasma exposures greater than clinical exposures, including visceral abnormalities, skeletal abnormalities, decreased fetal weight, and embryo-fetal death for perampanel exposure at dose of 1, 3, or 10 mg/kg/day, skeletal abnormalities, decreased fetal weight, and embryo-fetal death for rufinamide

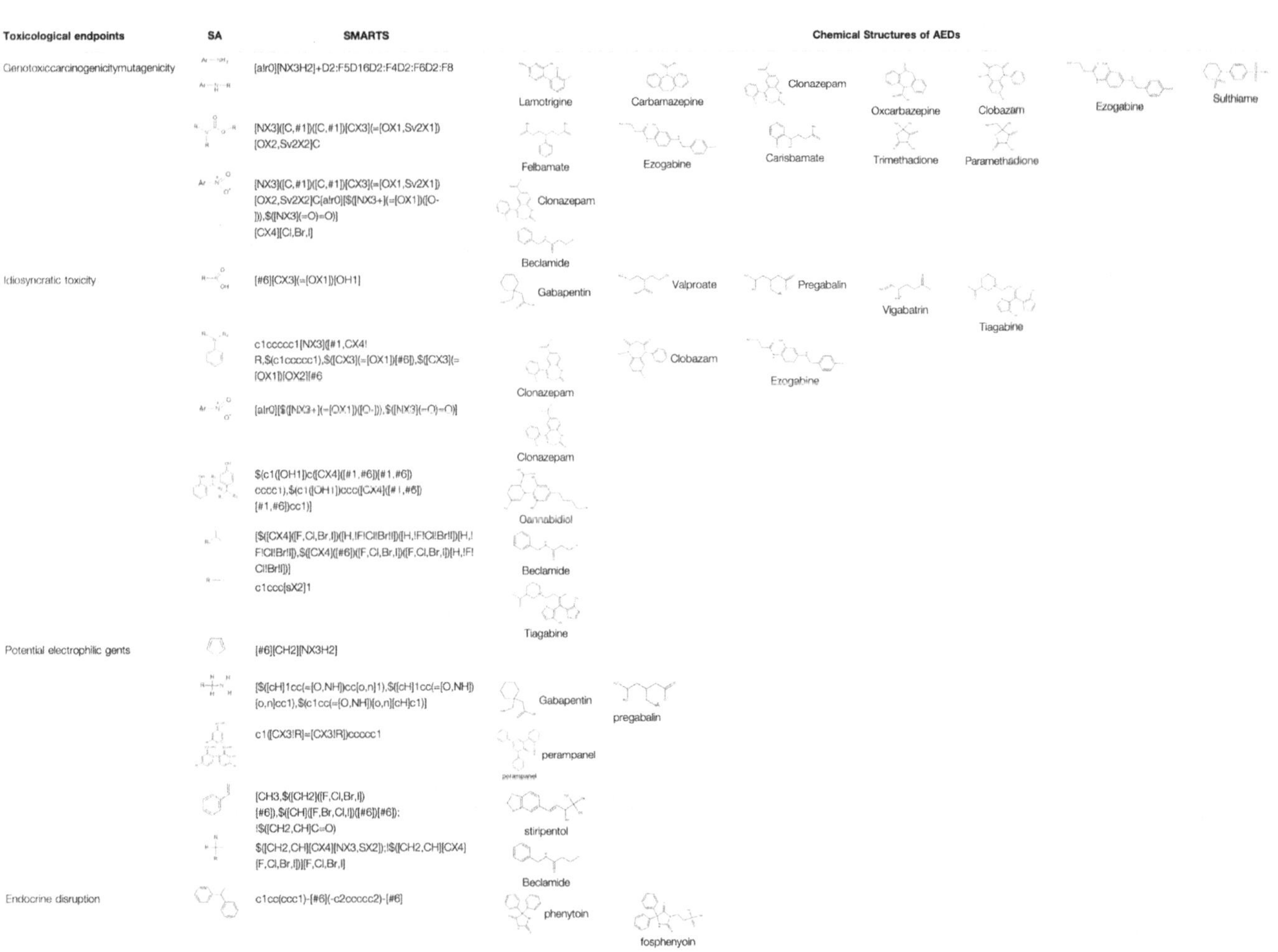

FIGURE 20.6 The structural alerts extracted from the antiseizure medication.

exposure at dose of 20, 100, and 300 mg/kg/day, and embryo-fetal death and decreased fetal weight only for highest dose brivaracetam exposure. One case report reported two cases of minor congenital malformations in the offspring exposed to brivaracetam during pregnancy [63]. Therefore, our model can be used as a tool to alert new drugs that lack clinical evidence for teratogenicity.

Perspectives

In this chapter, on the basis of the centralized data source and corresponding curated dataset, two SAR models for predicting the potential risks of certain ADRs were proposed based on the chemical structures of the drugs, the potential DIR risk, and the potential teratogenic risk of the ASMs. The RF model can well-predict the DIR severity of the marketed drugs using chemical structures. The current model was sensitive to detect the drugs in the specific therapeutic categories. The SVM model was sensitive to detecting the high teratogenic risk drugs. The SAR models can be good complementary tools in drug safety evaluation. The selected ML model can be helpful for the prevention and early detection of DIR and DIT.

In addition, a few caveats or challenges were still worthy of further discussion. For DIR project, the RF model discriminated the serious DIR drugs from others only based on chemical structures. The model could be used for early screening of drug candidates during the development process to identify those with the greatest likelihood of causing DIR and to eliminate them. Although the model is simple, the limited information makes it difficult to further improve the prediction performance of the model. Because drug-induced rhabdomyolysis is a very complex ADR affected by many factors, additional information, for example, the energy homeostasis in cells [64], involved in the model construction procedures should be helpful to improve the model performance. Considering the fact that the concomitant use of drugs or overdose can also cause the DIR, a more comprehensive evaluation method is needed to be developed in a future study. For DIT project, it is still a big challenge to develop a reproducible procedure to assess the teratogenic risk for the drugs. A set of attributes such as severity, expectedness, and causality should be considered in annotating the teratogenic risk for drugs. In this project, the teratogenic risk for drugs were annotated based on the FDA-approved drug labeling because drug labeling contains a compilation of up-to-date drug safety information and is a relatively accurate and stable data source. Thus, 67 drugs with teratogenic risk (positive samples) and 45 drugs with low teratogenic risk (negative samples) were annotated. The average precision achieved by our model in 1000 iterations of 5-fold cross-validation was 0.690. When validating the model by using the FAERS data, we can only determine the drugs with teratogenic risk when the teratogenicity is reported. As a result, the recall rate of relevant models is the only indicator that has been tested. The teratogenic risk of the drugs used in the modeling is determined by the FDA drug labeling. In the drug list, all of the 26 drugs in the therapeutic category of antineoplastic and immunomodulating agents (L) were assigned to the high-teratogenic-risk group (positive samples). It may be one of the main reasons that 92.3% of drugs were predicted to have teratogenic risk when using the FAERS dataset to validate the predictive model. In comparison, the number of drugs in the therapeutic category N (nervous system) in the high-teratogenic-risk group (positive samples) and the low-teratogenic-risk group (negative samples) is relatively balanced (7 positive and 10 negative samples). As a result, 53.8% of the drugs in the FAERS data-set were predicted to be teratogenic.

For future studies, it is foreseeable that when more balanced data can be obtained with the update of the FDA drug labeling, the predictive performance of our model can be further improved. For all ADRs studies, note that the annotation was based on the US FDA-approved drug labeling. The classification of the drugs might be different when using the drug labeling of other countries. Additionally, the current proposed models are only based on the 1D/2D/3D chemical structures of the drugs. Although it is simple and easy to use, the limited information restricted the predictive performance of the model. Drug-induced ADRs are usually dose-dependent and affected by many factors. Including additional information such as the cell-based in vitro data [61,65], toxicogenomic data [66–68], the data from animal studies [67], and the pharmacogenomics data [69] in the modeling procedure may improve the model performance.

References

[1] Aronson JK, Ferner RE. Clarification of terminology in drug safety. Drug Saf 2005;28(10):851–70.

[2] Galeano D, Li S, Gerstein M, Paccanaro A. Predicting the frequencies of drug side effects. Nat Commun 2020;11(1):1–14.

[3] He S, Ye T, Wang R, Zhang C, Zhang X, Sun G, et al. An in silico model for predicting drug-induced hepatotoxicity. Int J Mol Sci 2019;20(8):1897.

[4] Ietswaart R, Arat S, Chen AX, Farahmand S, Kim B, DuMouchel W, et al. Machine learning guided association of adverse drug reactions with in vitro target-based pharmacology. EBioMedicine 2020;57:102837.

[5] Jamal S, Ali W, Nagpal P, Grover S, Grover A. Computational models for the prediction of adverse cardiovascular drug reactions. J Transl Med 2019;17(1):1–13.

[6] Yamanishi Y, Pauwels E, Kotera M. Drug side-effect prediction based on the integration of chemical and biological spaces. J Chem Inf Model 2012;52(12):3284–92.

[7] Vo AH, Van Vleet TR, Gupta RR, Liguori MJ, Rao MS. An overview of machine learning and big data for drug toxicity evaluation. Chem Res Toxicol 2019;33(1):20–37.

[8] Yang H, Sun L, Li W, Liu G, Tang Y. Corrigendum: in silico prediction of chemical toxicity for drug design using machine learning methods and structural alerts. Front Chem 2018;6:129.

[9] Liu Y, Liang Y, Kuang Q, Xie F, Hao Y, Wen Z, et al. Post-modified non-negative matrix factorization for deconvoluting the gene expression profiles of specific cell types from heterogeneous clinical samples based on RNA-sequencing data. J Chemom 2018;32(11):e2929.

[10] Liu Y, Wen Z, Li M. The power of matrix factorization: methods for deconvoluting genetic heterogeneous data at expression level. Curr Bioinf 2020;15(8):841–53.

[11] Zhu XY, Ran CK, Wen M, Guo GL, Liu Y, Liao LL, et al. Prediction of multicomponent reaction yields using machine learning. Chin J Chem 2021;39(12):3231–7.

[12] Stahl K, Rastelli E, Schoser B. A systematic review on the definition of rhabdomyolysis. J Neurol 2020;267(4):877–82.

[13] Warren JD, Blumbergs PC, Thompson PD. Rhabdomyolysis: a review. Muscle Nerve 2002;25(3):332–47.

[14] Vanholder R, Sever MS, Erek E, Lameire N. Rhabdomyolysis. J Am Soc Nephrol 2000;11(8):1553–61.

[15] Nance JR, Mammen AL. Diagnostic evaluation of rhabdomyolysis. Muscle Nerve 2015;51(6):793–810.

[16] Chavez LO, Leon M, Einav S, Varon J. Beyond muscle destruction: a systematic review of rhabdomyolysis for clinical practice. Crit Care 2016; 20(1):1–11.

[17] Larbi E. Drug-induced rhabdomyolysis. Ann Saudi Med 1998;18(6):525–30.

[18] Sauret JM, Marinides G, Wang GK. Rhabdomyolysis. Am Fam Physician 2002;65(5):907.

[19] Bagley W, Yang H, Shah K. Rhabdomyolysis. Intern Emerg Med 2007;2(3):210–8.

[20] Keltz E, Khan FY, Rhabdomyolysis MG. The role of diagnostic and prognostic factors. Muscles Ligaments Tendons J 2013;3(4):303–12.

[21] Cui X, Liu J, Zhang J, Wu Q, Li X. In silico prediction of drug-induced rhabdomyolysis with machine-learning models and structural alerts. J Appl Toxicol 2019;39(8):1224–32.

[22] Hu X, Yan A. In silico prediction of rhabdomyolysis of compounds by self-organizing map and support vector machine. Toxicol Vitro 2011; 25(8):2017–24.

[23] Tomson T, Battino D, Bonizzoni E, Craig J, Lindhout D, Perucca E, et al. Declining malformation rates with changed antiepileptic drug prescribing: an observational study. Neurology 2019;93(9):e831–40.

[24] Ornoy A. Valproic acid in pregnancy: how much are we endangering the embryo and fetus? Reprod Toxicol 2009;28(1):1–10.

[25] Viinikainen K, Heinonen S, Eriksson K, Kälviäinen R. Community-based, prospective, controlled study of obstetric and neonatal outcome of 179 pregnancies in women with epilepsy. Epilepsia 2006;47(1):186–92.

[26] Bosak M, Cyranka K, Dudek D, Kowalik M, Mołek P, Słowik A. Psychiatric comedication in patients with epilepsy. Epilepsy Behav 2018;83: 207–11.

[27] Brodie M. Major congenital malformations and antiepileptic drugs: prospective observations. J Neurol Neurosurg Psychiatry 2006;77(2):145.

[28] Bromley RL, Weston J, Marson AG. Maternal use of antiepileptic agents during pregnancy and major congenital malformations in children. JAMA 2017;318(17):1700–1.

[29] Tomson T, Battino D, Bonizzoni E, Craig J, Lindhout D, Perucca E, et al. Comparative risk of major congenital malformations with eight different antiepileptic drugs: a prospective cohort study of the EURAP registry. Lancet Neurol 2018;17(6):530–8.

[30] Campbell E, Kennedy F, Russell A, Smithson W, Parsons L, Morrison P, et al. Malformation risks of antiepileptic drug monotherapies in pregnancy: updated results from the UK and Ireland Epilepsy and Pregnancy Registers. J Neurol Neurosurg Psychiatry 2014;85(9):1029–34.

[31] Murayama N, Suemizu H, Uehara S, Kusama T, Mitsui M, Kamiya Y, et al. Association of pharmacokinetic profiles of lenalidomide in human plasma simulated using pharmacokinetic data in humanized liver mice with liver toxicity detected by human serum albumin RNA. J Toxicol Sci 2018;43(6):369–75.

[32] Chetot T, Taufana S, Benoit E, Lattard V. Vitamin K antagonist rodenticides display different teratogenic activity. Reprod Toxicol 2020;93: 131–6.

[33] Jarque S, Rubio-Brotons M, Ibarra J, Ordoñez V, Dyballa S, Miñana R, et al. Morphometric analysis of developing zebrafish embryos allows predicting teratogenicity modes of action in higher vertebrates. Reprod Toxicol 2020;96:337–48.

[34] Tandon N, Yadav SS. Safety and clinical effectiveness of *Withania somnifera* (Linn.) Dunal root in human ailments. J Ethnopharmacol 2020;255: 112768.

[35] Basant N, Gupta S, Singh KP. In silico prediction of the developmental toxicity of diverse organic chemicals in rodents for regulatory purposes. Toxicol Res 2016;5(3):773–87.

[36] Jiang C, Yang H, Di P, Li W, Tang Y, Liu G. In silico prediction of chemical reproductive toxicity using machine learning. J Appl Toxicol 2019; 39(6):844–54.

[37] Wen Z, Liang Y, Hao Y, Delavan B, Huang R, Mikailov M, et al. Drug-Induced Rhabdomyolysis Atlas (DIRA) for idiosyncratic adverse drug reaction management. Drug Discov Today 2019;24(1):9–15.

[38] Wu L, Ingle T, Liu Z, Zhao-Wong A, Harris S, Thakkar S, et al. Study of serious adverse drug reactions using FDA-approved drug labeling and MedDRA. BMC Bioinf 2019;20(2):129–39.

[39] Kim S, Thiessen PA, Bolton EE, Chen J, Fu G, Gindulyte A, et al. PubChem substance and compound databases. Nucleic Acids Res 2016;44: D1202–13.

[40] Mauri A, Consonni V, Pavan M, Todeschini R. Dragon software: an easy approach to molecular descriptor calculations. Match 2006;56(2): 237–48.

[41] Hong H, Xie Q, Ge W, Qian F, Fang H, Shi L, et al. Mold2, molecular descriptors from 2D structures for chemoinformatics and toxicoinformatics. J Chem Inf Model 2008;48(7):1337–44.

[42] Vilar S, Harpaz R, Chase HS, Costanzi S, Rabadan R, Friedman C. Facilitating adverse drug event detection in pharmacovigilance databases using molecular structure similarity: application to rhabdomyolysis. J Am Med Inf Assoc 2011;18(Suppl. 1):i73–80.
[43] Banda JM, Evans L, Vanguri RS, Tatonetti NP, Ryan PB, Shah NH. A curated and standardized adverse drug event resource to accelerate drug safety research. Sci Data 2016;3(1):1–11.
[44] van Puijenbroek EP, Bate A, Leufkens HG, Lindquist M, Orre R, Egberts AC. A comparison of measures of disproportionality for signal detection in spontaneous reporting systems for adverse drug reactions. Pharmacoepidemiol Drug Saf 2002;11(1):3–10.
[45] Wishart DS, Knox C, Guo AC, Cheng D, Shrivastava S, Tzur D, et al. DrugBank: a knowledgebase for drugs, drug actions and drug targets. Nucleic Acids Res 2008;36(Suppl. 1_1):D901–6.
[46] Hong H, Rua D, Sakkiah S, Selvaraj C, Ge W, Tong W. Consensus modeling for prediction of estrogenic activity of ingredients commonly used in sunscreen products. Int J Environ Res Publ Health 2016;13(10):958.
[47] Zhang L, Ai H, Chen W, Yin Z, Hu H, Zhu J, et al. CarcinoPred-EL: novel models for predicting the carcinogenicity of chemicals using molecular fingerprints and ensemble learning methods. Sci Rep 2017;7(1):1–14.
[48] Qi Y. Random forest for bioinformatics. In: Ensemble machine learning. Springer; 2012. p. 307–23.
[49] Guo X, Li W, Iorio F. Convolutional neural networks for steady flow approximation. In: Proceedings of the 22nd ACM SIGKDD international conference on knowledge discovery and data mining; 2016.
[50] Cortes C, Vapnik V. Support-vector networks. Mach Learn 1995;20(3):273–97.
[51] Park H-A. An introduction to logistic regression: from basic concepts to interpretation with particular attention to nursing domain. J Korean Acad Nurs 2013;43(2):154–64.
[52] He Y, Huang Z, Chen B, Tsutsui M, Shui Miao X, Taniguchi M. Electrokinetic analysis of energy harvest from natural salt gradients in nanochannels. Sci Rep 2017;7(1):1–15.
[53] De Santis M, Carducci B, Cavaliere AF, De Santis L, Straface G, Caruso A. Drug—induced congenital defects. Drug Saf 2001;24(12):889–901.
[54] Janz D. Are antiepileptic drugs harmful when given during pregnancy? German Med Mon 1964;9:20–2.
[55] Meadow S. Anticonvulsant drugs and congenital abnormalities. In: Problems of birth defects. Springer; 1968. p. 224.
[56] Tomson T, Battino D, Bonizzoni E, Craig J, Lindhout D, Sabers A, et al. Dose-dependent risk of malformations with antiepileptic drugs: an analysis of data from the EURAP epilepsy and pregnancy registry. Lancet Neurol 2011;10(7):609–17.
[57] Hernández-Díaz S, Smith C, Shen A, Mittendorf R, Hauser W, Yerby M, et al., North American AED Pregnancy Registry. Comparative safety of antiepileptic drugs during pregnancy. Neurology 2012;78(21):1692–9.
[58] Sankar R, Lerner JT. Teratogenicity of antiepileptic drugs: role of pharmacogenomics. Int Rev Neurobiol 2008;83:215–25.
[59] Sushko I, Salmina E, Potemkin VA, Poda G, Tetko IV. ToxAlerts: a web server of structural alerts for toxic chemicals and compounds with potential adverse reactions. ACS Publications; 2012.
[60] Kalgutkar AS, Soglia JR. Minimising the potential for metabolic activation in drug discovery. Expert Opin Drug Metabol Toxicol 2005;1(1): 91–142.
[61] Liu Y, Jing R, Wen Z, Li M. Narrowing the gap between in vitro and in vivo genetic profiles by deconvoluting toxicogenomic data in silico. Front Pharmacol 2020;10:1489.
[62] Corwin D, LeMert R. Construction and evaluation of an inexpensive weighing lysimeter for studying contaminant transport. J Contam Hydrol 1994;15(1–2):107–23.
[63] Paolini SL, Pilato M, Rajasekaran V, Waters JF, Bagic A, Urban A. Outcomes in three cases after brivaracetam treatment during pregnancy. Acta Neurol Scand 2020;141(5):438–41.
[64] Wagner BK, Kitami T, Gilbert TJ, Peck D, Ramanathan A, Schreiber SL, et al. Large-scale chemical dissection of mitochondrial function. Nat Biotechnol 2008;26(3):343–51.
[65] Liu Z, Kelly R, Fang H, Ding D, Tong W. Comparative analysis of predictive models for nongenotoxic hepatocarcinogenicity using both toxicogenomics and quantitative structure–activity relationships. Chem Res Toxicol 2011;24(7):1062–70.
[66] Liu Y, Jing R, Xu J, Liu K, Xue J, Wen Z, et al. Comparative analysis of oncogenes identified by microarray and RNA-sequencing as biomarkers for clinical prognosis. Biomark Med 2015;9(11):1067–78.
[67] Chen M, Zhang M, Borlak J, Tong W. A decade of toxicogenomic research and its contribution to toxicological science. Toxicol Sci 2012;130(2): 217–28.
[68] Liu Z, Huang R, Roberts R, Tong W. Toxicogenomics: a 2020 vision. Trends Pharmacol Sci 2019;40(2):92–103.
[69] Phillips KA, Van Bebber SL. Measuring the value of pharmacogenomics. Nat Rev Drug Discov 2005;4(6):500–9.

PART III

QSAR models for safety evaluation of drugs and consumer products

CHAPTER

21

QSAR modeling for predicting drug-induced liver injury

Tsung-Jen Liao[1], *Jingwen Zhao*[2] *and Minjun Chen*[1]

[1]Division of Bioinformatics and Biostatistics, National Center for Toxicological Research, United States Food and Drug Administration, Jefferson, AR, United States; [2]Bioinformatics Program, Department of Information Science, University of Arkansas at Little Rock, Little Rock, AR, United States

Abbreviations

ALP Alkaline phosphatase
ALT Alanine aminotransferase
ANN Artificial neural network
AST Aspartate aminotransferase
CNN Convolution neural network
DILI Drug-induced liver injury
DL Deep learning
DNN Deep neural network
EV External validation
ML Machine learning
QSAR Quantitative structure–activity relationship
SVM Support vector machine
TBL Total bilirubin
UGRNN Undirected graph recursive neural network

Introduction

Drug-induced liver injury (DILI) can cause acute and chronic liver diseases, which deteriorate liver function and may result in liver failure [1]. In the United States, DILI is the most frequent cause of acute liver failure, and over 40,000 individuals experience DILI annually [2]. DILI is classified as intrinsic, indirect, and idiosyncratic [3]. Intrinsic DILI is caused by an overdose of hepatoxic drugs, frequently acetaminophen, which account for most DILI cases [4]. Indirect DILI is unexpected liver injury that is associated with known actions of a drug [3]. Idiosyncratic DILI is unpredictable and may occur without preexisting liver disease(s) [5], with the causes remaining unclear. Idiosyncratic DILI is a major contributor to drug failure in clinical trials and market withdrawal after approval [6]. Up to 18% of safety-related drug withdrawal from the market between 1953 and 2013 were due to hepatoxicity [7]. Early identification of potential DILI risk will improve drug discovery and development.

The liver is the main organ of drug metabolism. The polymorphic cytochrome P450 enzymes in the liver are responsible for oxidizing drugs into their metabolites, which are largely excreted via the bile, urine, and feces [8]. Drug metabolites can be toxic and considered reactive metabolites because of their ability to bind to hepatocyte proteins [8]. Binding of drug metabolites to hepatocyte proteins may trigger cell stress, mitochondrial impairment, and specific immune reactions that lead to cell apoptosis and necrosis, subsequently resulting in DILI [9].

DILI can be clinically diagnosed by the elevation of the liver enzymes alanine aminotransferase (ALT) and/or aspartate aminotransferase (AST) greater than threefold the upper limit of normal (ULN) [10], or a twofold increase of alkaline phosphatase (ALP) and/or serum total bilirubin (TBL) above ULN accompanied with any rise of ALT and AST [11]. According to different biochemical patterns, DILI could be categorized as hepatocellular, cholestatic, and mixed types of injury using the R value, a ratio of (ALT/ULN)/(ALP/ULN) [12]. Hepatocellular injury is indicated by an $ALT/ULN > 3$ and $R > 5$, cholestatic injury as an $ALP/ULN > 2$ and $R < 2$, and the mixed type of injury as an $ALT/ULN > 3$ and $2 < R < 5$ [13]. DILI clinical manifestations include hepatitis, cholestasis, cirrhosis, and steatosis [14].

QSAR in Safety Evaluation and Risk Assessment
https://doi.org/10.1016/B978-0-443-15339-6.00009-6

Predicting DILI risk caused by drugs is challenging and has been investigated for decades. Multiple methodologies including in vitro and in vivo experiments, as well as computational techniques, have been used to identify the risk of DILI [15]. Among these methods, quantitative structure–activity relationship (QSAR) modeling was viewed as one of the most useful tools for screening potentially hepatotoxic drugs at the early stage of drug discovery, without requiring the physical substances and allowing for a quick return of results [16]. However, the chemical structure–based QSAR approach may not embody sufficient predictive performance [17]. A recent investigation indicated that a high daily dosage and high lipophilicity were associated with significant risk of DILI [18]. Subsequently, a DILI predictive model based on dosage, lipophilicity, and formation of reactive metabolites was developed to grade the severity of DILI [19].

The basic concepts of QSAR modeling for DILI were comprehensively summarized in several reviews [20,21]. Here, we focus on three issues: (1) how QSAR was applied to DILI prediction, (2) how deep learning (DL) was used to assist DILI QSAR modeling, and (3) DILI prediction performance of current QASR models. By reviewing the recent developments of the past 3 years, we attempt to provide a comprehensive view of QSAR modeling for predicting DILI in humans.

How does QSAR apply to DILI prediction?

A QSAR model is a regression or classification, which can construct a mathematical relationship between the compounds' chemical structures and the bioactivities measured. In DILI prediction, QSAR modeling usually employs chemical structures as predictors and the corresponding DILI annotation as the independent variable, with the association made through machine learning (ML) algorithms [22,23]. DILI annotations can be retrieved from multiple DILI databases [24–32] and were also discussed in Chapter 19. A drug's chemical structure is described by molecular descriptors, which are quantitative representations of the molecule's properties. Molecular descriptors are mainly divided into two categories: (1) theoretical and (2) experimental measurements such as partition coefficient and molar refractivity. The theoretical molecular descriptors include 1D (list of fragments), 2D (graphical), and 3D (steric) descriptors, which were widely used for predicting hepatotoxicity by QSAR modeling [28]. Recent developments of QSAR modeling fused molecular descriptors with in vivo or in vitro assays for improving DILI prediction [33].

A recent DILI predictive model correlated the chemical structure–based descriptors and the predicted protein targets, elucidating the potential protein targets involved in DILI [17]. The 2D molecular descriptors were generated from chemical fingerprints [34], the molecular descriptor calculator Mordred [35], and the protein targets were predicted by PIDGIN [36]. The drug hepatotoxicity annotation was retrieved from DILIrank [37] and SIDER [38] databases. The model was trained and tested by using ML algorithms, including random forest and support vector machine (SVM), and evaluated by an FDA validation set, which contained 49 compounds, yielding an accuracy of 0.759. This study identified that most proteins were consistent with mechanisms of action and toxicity of nonsteroidal antiinflammatory drugs, which are viewed as a common cause of DILI.

A predictive model of DILI was developed by combining molecular descriptors and genetic algorithms [39]. One genetic algorithm used principal component analysis to flip the feature space and extracted toxic fingerprints. A combination of genetic algorithms and ensemble learning was used to improve the accuracy of this model, which was developed using 2D molecular descriptors. The molecular descriptors were generated from Mold2 [40] and PaDEL [41] software, while the DILI annotation was obtained from Xu et al. [42]. This genetic algorithm–based method outcompeted the conventional ML algorithms as well as ensemble learning and graph neural network methods.

How does deep learning assist QSAR for DILI prediction?

DL is a subfield of ML based on an artificial neural network (ANN). An ANN is made of multiple layers that can be grouped by input, hidden, and output layers. Each layer can contain different numbers of neurons as well as different activation functions (turn on/off the neuron). DL usually contains more than two hidden layers. Training a DL model would minimize the error of prediction. DL could provide stronger predictive power than ML but requires larger computing power and amount of data. Common DL methods include convolution neural network

(CNN) and deep neural network (DNN) models. These methods have been applied to DILI prediction in recent years [42–44].

A CNN model using molecular fingerprint-embedded features was proposed for predicting DILI [44]. This study utilized 2048 molecular fingerprints from Morgan [34] to describe chemical structures of compounds. The molecular fingerprint algorithm read the compound structures and assigned 1 for the substructure presence and 0 for the absence. The binary vectors of molecular fingerprints formed an embedded matrix, which was then trained and tested by the CNN method. In total, 1919 compounds of DILI data were retrieved from literature and public databases [20,37,45–47]. The 1919 compounds were separated into a training set and an independent test set that include 1597 samples and 322 samples, respectively. This study yielded an average accuracy of 0.89, Matthews's correlation coefficient of 0.80, and an area under the recursive operating curve (AUC) of 0.96 for the independent test set.

A DNN-based predictive model was also developed for predicting DILI risk [48]. This model used four fingerprints (ECFP4, ECFP6, FCFP4, and FCFP6) to describe compounds by determining the existence of substructures. This DL model consisted of one input layer, five hidden layers, and one output layer. The DILI data from DILIrank [37] and LiverTox [49] databases were used as the training set, and the datasets from Xu et al. [50] and Greene et al. [27] were used for validation. The best model was the ECFP4-based DNN model, with an accuracy of 0.73, a sensitivity of 0.71, and a specificity of 0.75 on the validation set. ECFP4 is a topological fingerprint that contains 1024 binary bits as molecular descriptors for each compound. The model was further evaluated by 15 drugs with reported DILI cases after 2019, and DILI risk was correctly predicted for 13 drugs.

Another DNN-based model was proposed in 2022 and showed better performance than previously published DILI predictive models [51]. This model used an 18-layer residual neural network (ResNet18) in addition to 5-layer blocks (ResNet18DNN) to improve DILI prediction by vectorizing molecular structure images. The molecular structure images were generated from SMILES and used as the feature matrix learned by the CNN model ResNet18, vectorizing 512 features. The features were then used as the input for the ResNet18DNN model. This study included 1446 chemical compounds from various DILI databases [22,29,37,38,49,50,52] and resulted in a test set with AUC of 0.958, accuracy of 0.976, recall of 0.935, precision of 0.947, F1-score of 0.926, and specificity of 0.913. However, the extraordinary model performance was questioned by Liu et al. regarding data augmentation strategy, detail of model development, and reproducibility [53]. As this study had an imbalanced number of DILI positives and negatives (the number of DILI positives is fivefold that of DILI negatives), the predictive performance may be biased. One-time random splitting may also result in chance correlation that boosts the model's performance. Chen et al. explained their inability to release their data and codes, but used an external validation (EV) set to validate ResNet18DNN [54].

Prediction performance of current DILI QSAR models

Conventional QSAR modeling using molecular descriptors and/or fingerprints for DILI prediction generally possesses an accuracy of 60%–70% (Table 21.1). Ekins et al. used a Bayesian modeling method with ECFC6 fingerprints and obtained an accuracy of 0.6 for the EV set [28]. Liew et al. built an ensemble model of 617 classifiers with PaDEL descriptors with an accuracy of 0.75 for EV [45]. Chen et al. used decision forest algorithm with Mold2 molecular descriptors with accuracies of 0.62–0.69 for the 3 EV sets [22]. Zhang et al. used 5 ML methods based on MACCS and FP4 fingerprints, with SVM being the best model with an accuracy of 0.75 for EV [46]. Ai et al. developed an ensemble model using 3 ML methods and 12 molecular fingerprints, resulting in an accuracy of 0.711 for the EV [47]. As mentioned previously, Liu et al. used SVM methods with ECFP4 fingerprints, predictive protein targets, and Mordred descriptors as features and obtained an accuracy of 0.759 for EV [17]. Yan et al. utilized a genetic algorithm based on the molecular fingerprints obtained from Rdkit2D, PubChem, MACCS, and ECFP2 and obtained an accuracy of 0.77 for EV [39].

The DILI predictive models based on DL may display superior performance compared to ML-assisted models (Table 21.2). Xu et al. developed a DL architecture, undirected graph recursive neural network (UGRNN), which utilized the SMILES strings of chemical structures as the input features for learning, resulting in an accuracy of 0.869 for EV set [42]. Li et al. developed a DNN-based model, DeepDILI, which combined model-level representation generated from conventional ML algorithms and used Mold2 descriptors as features, generating an accuracy of 0.687 for the EV [43]. The recent DL-assisted model by Nguyen-Vo et al. used a nature language processing inspired framework for describing chemical structures by the presence of substructures and the CNN method for learning, and had an accuracy of 0.89 for EV [44]. Kang et al. also developed a DNN-based model using ECFP4 fingerprints to describe chemical structures, and obtained an accuracy of 0.731 for EV [48].

TABLE 21.1 Selected literature of ML-assisted DILI predictions.

Study	Year	Models	Input features	Size of training	Size of EV	Accuracy of EV	AUC of EV
Ekins et al. [28]	2010	Bayesian	ECFC6 fingerprints	295	237	0.60	0.62
Liew et al. [45]	2011	Ensemble of mixed learning	PaDEL descriptors	1087	120	0.75	–
Chen et al. [22]	2013	Decision forest	Mold2 descriptors	197	3 datasets from 190 to 328	0.62–0.69	–
Zhang et al. [46]	2016	SVM	MACCS and FP4 fingerprints	1317	88	0.75	0.61
Ai et al. [47]	2018	Ensemble-Top5	12 types of molecular fingerprints	1241	286	0.711	0.764
Liu et al. [17]	2021	SVM	ECFP4, protein targets, Mordred descriptors	401	49	0.759	–
Yan et al. [39]	2022	R-E-GA	Molecular fingerprints from Rdkit2D, PubChem, MACCS, and ECFP2	2931	673	0.77	0.842

AUC, area under the recursive operating curve; *EV*, external validation; *R-E-GA*, rotation–ensemble–genetic algorithm; *SVM*, support vector machine.

TABLE 21.2 Selected literature of DL-assisted DILI predictions.

Study	Year	Models	Input features	Size of training	Size of EV	Accuracy of EV	AUC of EV
Xu et al. [42]	2015	UGRNN	SMILES string	475	198	0.869	0.955
Li et al. [43]	2021	DeepDILI	Mold2 descriptors	753	249	0.687	0.659
Nguyen-Vo et al. [44]	2021	CNN	Molecular fingerprints converted from SMILES	1597	322	0.89	0.96
Kang et al. [48]	2021	DNN	ECFP4 fingerprints	940	179	0.731	–

AUC, area under the recursive operating curve; *CNN*, convolutional neural network; *DNN*, deep neural network; *EV*, external validation; *SMILES*, simplified molecular-input line-entry system; *UGRNN*, undirected graph recursive neural network.

DL utilizes a more complex structure of algorithms and enables the processing of unstructured data, possibly providing superior performance than ML methods. However, a robust predictive artificially intelligent model relies on the quality of input data. A recent study reported that models based on curated data for predicting skin sensitization and irritation showed a 7%–24% higher correct classification rate than the models using uncurated data [55]. Therefore, the quality of data source(s) is as important as the computational modeling methods and is essential for generating reliable predictions.

Perspectives

Computational approaches designed for DILI QSAR modeling remain challenging due to the limited sample size and quality of DILI databases. Many efforts have been devoted to the development of QSAR modelings by using ML and DL algorithms with various molecular descriptors for chemical compounds. While DL methods usually outperformed the conventional ML methods, ML-based models can rank the importance of molecular descriptors, which may help identify potential risk factors of DILI.

High quality and large sizes of DILI databases are essential for the development of reliable DILI predictive models. When applying ML methods, the number of data should be 10 times more than the number of features. Similarly, DL usually requires at least 5000 samples for each classification for acceptable performance. However, most of the published DILI databases have limited drugs annotated, and even though more additional databases have been released, the risk classification based on DILI severity is not thoroughly standardized. This results in difficulties when combining multiple DILI databases for training and testing. Therefore, standardization of risk classification will be necessary to improve future predictive model development. As more drug data become available, we anticipate that future predictive models will become a powerful screening tool for preventing DILI.

Acknowledgments

This research was supported in part by an appointment to the Research Participation Program at the National Center for Toxicological Research (Tsung-Jen Liao), administered by the Oak Ridge Institute for Science and Education through an interagency agreement between the US Department of Energy and the US Food and Drug Administration. This chapter reflects the views of the authors and does not necessarily reflect those of the US Food and Drug Administration. Any mention of commercial products is for clarification only and is not intended as approval, endorsement, or recommendation.

References

[1] Li M, Luo Q, Tao Y, Sun X, Liu C. Pharmacotherapies for drug-induced liver injury: a current literature review. Front Pharmacol 2021;12:806249.
[2] Lisi DM. Drug-induced liver injury: an overview. US Pharm 2016;41(12):30–4.
[3] Hoofnagle JH, Bjornsson ES. Drug-induced liver injury - types and phenotypes. N Engl J Med 2019;381(3):264–73.
[4] Moreno-Torres M, Quintas G, Castell JV. The potential role of metabolomics in drug-induced liver injury (DILI) assessment. Metabolites 2022;12(6):564.
[5] Katarey D, Verma S. Drug-induced liver injury. Clin Med 2016;16(Suppl. 6):s104–9.
[6] Babai S, Auclert L, Le-Louet H. Safety data and withdrawal of hepatotoxic drugs. Therapie 2021;76(6):715–23.
[7] Onakpoya IJ, Heneghan CJ, Aronson JK. Post-marketing withdrawal of anti-obesity medicinal products because of adverse drug reactions: a systematic review. BMC Med 2016;14(1):191.
[8] Chen J, Zhao KN, Chen C. The role of CYP3A4 in the biotransformation of bile acids and therapeutic implication for cholestasis. Ann Transl Med 2014;2(1):7.
[9] Zhou Y, Wang J, Zhang D, Liu J, Wu Q, Chen J, et al. Mechanism of drug-induced liver injury and hepatoprotective effects of natural drugs. Chin Med 2021;16(1):135.
[10] Mosedale M, Watkins PB. Drug-induced liver injury: advances in mechanistic understanding that will inform risk management. Clin Pharmacol Ther 2017;101(4):469–80.
[11] Fontana RJ, Seeff LB, Andrade RJ, Bjornsson E, Day CP, Serrano J, et al. Standardization of nomenclature and causality assessment in drug-induced liver injury: summary of a clinical research workshop. Hepatology 2010;52(2):730–42.
[12] Chalasani NP, Maddur H, Russo MW, Wong RJ, Reddy KR, Practice Parameters Committee of the American College of G. ACG clinical guideline: diagnosis and management of idiosyncratic drug-induced liver injury. Am J Gastroenterol 2021;116(5):878–98.
[13] Kullak-Ublick GA, Andrade RJ, Merz M, End P, Benesic A, Gerbes AL, et al. Drug-induced liver injury: recent advances in diagnosis and risk assessment. Gut 2017;66(6):1154–64.
[14] Devarbhavi H. An update on drug-induced liver injury. J Clin Exp Hepatol 2012;2(3):247–59.
[15] Hamburg MA. Advancing regulatory science. Science 2011;331(6020):987.
[16] Merlot C. Computational toxicology–a tool for early safety evaluation. Drug Discov Today 2010;15(1–2):16–22.
[17] Liu A, Walter M, Wright P, Bartosik A, Dolciami D, Elbasir A, et al. Prediction and mechanistic analysis of drug-induced liver injury (DILI) based on chemical structure. Biol Direct 2021;16(1):6.
[18] Chen M, Borlak J, Tong W. High lipophilicity and high daily dose of oral medications are associated with significant risk for drug-induced liver injury. Hepatology 2013;58(1):388–96.
[19] Chen M, Borlak J, Tong W. A Model to predict severity of drug-induced liver injury in humans. Hepatology 2016;64(3):931–40.
[20] Kotsampasakou E, Montanari F, Ecker GF. Predicting drug-induced liver injury: the importance of data curation. Toxicology 2017;389:139–45.
[21] Vall A, Sabnis Y, Shi J, Class R, Hochreiter S, Klambauer G. The promise of AI for DILI prediction. Front Artif Intell 2021;4:638410.
[22] Chen M, Hong H, Fang H, Kelly R, Zhou G, Borlak J, et al. Quantitative structure-activity relationship models for predicting drug-induced liver injury based on FDA-approved drug labeling annotation and using a large collection of drugs. Toxicol Sci 2013;136(1):242–9.
[23] Hong H, Thakkar S, Chen M, Tong W. Development of decision forest models for prediction of drug-induced liver injury in humans using a large set of FDA-approved drugs. Sci Rep 2017;7(1):17311.
[24] O'Brien PJ, Irwin W, Diaz D, Howard-Cofield E, Krejsa CM, Slaughter MR, et al. High concordance of drug-induced human hepatotoxicity with in vitro cytotoxicity measured in a novel cell-based model using high content screening. Arch Toxicol 2006;80(9):580–604.
[25] Rodgers AD, Zhu H, Fourches D, Rusyn I, Tropsha A. Modeling liver-related adverse effects of drugs using knearest neighbor quantitative structure-activity relationship method. Chem Res Toxicol 2010;23(4):724–32.
[26] Fourches D, Barnes JC, Day NC, Bradley P, Reed JZ, Tropsha A. Cheminformatics analysis of assertions mined from literature that describe drug-induced liver injury in different species. Chem Res Toxicol 2010;23(1):171–83.
[27] Greene N, Fisk L, Naven RT, Note RR, Patel ML, Pelletier DJ. Developing structure-activity relationships for the prediction of hepatotoxicity. Chem Res Toxicol 2010;23(7):1215–22.
[28] Ekins S, Williams AJ, Xu JJ. A predictive ligand-based Bayesian model for human drug-induced liver injury. Drug Metab Dispos 2010;38(12):2302–8.
[29] Chen M, Vijay V, Shi Q, Liu Z, Fang H, Tong W. FDA-approved drug labeling for the study of drug-induced liver injury. Drug Discov Today 2011;16(15–16):697–703.
[30] Liu Z, Shi Q, Ding D, Kelly R, Fang H, Tong W. Translating clinical findings into knowledge in drug safety evaluation–drug induced liver injury prediction system (DILIps). PLoS Comput Biol 2011;7(12):e1002310.

[31] Zhu X, Kruhlak NL. Construction and analysis of a human hepatotoxicity database suitable for QSAR modeling using post-market safety data. Toxicology 2014;321:62–72.
[32] Liu R, Yu X, Wallqvist A. Data-driven identification of structural alerts for mitigating the risk of drug-induced human liver injuries. J Cheminf 2015;7:4.
[33] Zhu XW, Sedykh A, Liu SS. Hybrid in silico models for drug-induced liver injury using chemical descriptors and in vitro cell-imaging information. J Appl Toxicol 2014;34(3):281–8.
[34] Rogers D, Hahn M. Extended-connectivity fingerprints. J Chem Inf Model 2010;50(5):742–54.
[35] Moriwaki H, Tian YS, Kawashita N, Takagi T. Mordred: a molecular descriptor calculator. J Cheminf 2018;10(1):4.
[36] Mervin LH, Afzal AM, Drakakis G, Lewis R, Engkvist O, Bender A. Target prediction utilising negative bioactivity data covering large chemical space. J Cheminf 2015;7:51.
[37] Chen M, Suzuki A, Thakkar S, Yu K, Hu C, Tong W. DILIrank: the largest reference drug list ranked by the risk for developing drug-induced liver injury in humans. Drug Discov Today 2016;21(4):648–53.
[38] Kuhn M, Letunic I, Jensen LJ, Bork P. The SIDER database of drugs and side effects. Nucleic Acids Res 2016;44(D1):D1075–9.
[39] Yan B, Ye X, Wang J, Han J, Wu L, He S, et al. An algorithm framework for drug-induced liver injury prediction based on genetic algorithm and ensemble learning. Molecules 2022;27(10).
[40] Hong H, Xie Q, Ge W, Qian F, Fang H, Shi L, et al. Mold(2), molecular descriptors from 2D structures for chemoinformatics and toxicoinformatics. J Chem Inf Model 2008;48(7):1337–44.
[41] Yap CW. PaDEL-descriptor: an open source software to calculate molecular descriptors and fingerprints. J Comput Chem 2011;32(7):1466–74.
[42] Xu Y, Dai Z, Chen F, Gao S, Pei J, Lai L. Deep learning for drug-induced liver injury. J Chem Inf Model 2015;55(10):2085–93.
[43] Li T, Tong W, Roberts R, Liu Z, Thakkar S. DeepDILI: deep learning-powered drug-induced liver injury prediction using model-level representation. Chem Res Toxicol 2021;34(2):550–65.
[44] Nguyen-Vo TH, Nguyen L, Do N, Le PH, Nguyen TN, Nguyen BP, et al. Predicting drug-induced liver injury using convolutional neural network and molecular fingerprint-embedded features. ACS Omega 2020;5(39):25432–9.
[45] Liew CY, Lim YC, Yap CW. Mixed learning algorithms and features ensemble in hepatotoxicity prediction. J Comput Aided Mol Des 2011; 25(9):855–71.
[46] Zhang C, Cheng F, Li W, Liu G, Lee PW, Tang Y. In silico prediction of drug induced liver toxicity using substructure pattern recognition method. Mol Inform 2016;35(3–4):136–44.
[47] Ai H, Chen W, Zhang L, Huang L, Yin Z, Hu H, et al. Predicting drug-induced liver injury using ensemble learning methods and molecular fingerprints. Toxicol Sci 2018;165(1):100–7.
[48] Kang MG, Kang NS. Predictive model for drug-induced liver injury using deep neural networks based on substructure space. Molecules 2021;26(24).
[49] Hoofnagle JH, Serrano J, Knoben JE, Navarro VJ. LiverTox: a website on drug-induced liver injury. Hepatology 2013;57(3):873–4.
[50] Xu JJ, Henstock PV, Dunn MC, Smith AR, Chabot JR, de Graaf D. Cellular imaging predictions of clinical drug-induced liver injury. Toxicol Sci 2008;105(1):97–105.
[51] Chen Z, Jiang Y, Zhang X, Zheng R, Qiu R, Sun Y, et al. ResNet18DNN: prediction approach of drug-induced liver injury by deep neural network with ResNet18. Briefings Bioinf 2022;23(1):bbab503.
[52] Quinton A, Latry P, Biour M. Hepatox: database on hepatotoxic drugs. Gastroenterol Clin Biol 1993;17(5 Pt 2):H116–20.
[53] Liu Z, Li T, Connor S, Thakkar S, Roberts R, Tong W. Best practice and reproducible science are required to advance artificial intelligence in real-world applications. Briefings Bioinf 2022;23(4):bbac237.
[54] Chen Z, Jiang Y, Zhang X, Zheng R, Qiu R, Sun Y, et al. The prediction approach of drug-induced liver injury: response to the issues of reproducible science of artificial intelligence in real-world applications. Briefings Bioinf 2022;23(4).
[55] Alves VM, Auerbach SS, Kleinstreuer N, Rooney JP, Muratov EN, Rusyn I, et al. Curated data in - trustworthy in silico models out: the impact of data quality on the reliability of artificial intelligence models as alternatives to animal testing. Altern Lab Anim 2021;49(3):73–82.

CHAPTER

22

The need of QSAR methods to assess safety of chemicals in food contact materials

Serena Manganelli, Sander Koster and Benoit Schilter

Nestlé Research, Lausanne, Switzerland

Abbreviations

DT Decision tree
EFSA European Food Safety Authority
FCM Food contact material
FCS Food contact substance
GC Gas chromatography
HBGV Health-Based Guidance Value
IAS Intentionally added substances
LC Liquid chromatography
LOAEL Lowest observed adverse effect level
MoE Margin of exposure
MS Mass spectrometry
NIAS Non-intentionally added substances
NOAEL No observed adverse effect level
PoD point of departure
QSAR Quantitative structure–activity relationship
REACH Registration, Evaluation, Authorization and Restriction of Chemicals
RMSE Root mean square error
TD_{50} Tumorigenic dose for 50% of test species
TTC Threshold of toxicological concern

Introduction

Food contact materials: Background

Food contact materials (FCMs) such as plastics or coatings are manufactured by polymerization of monomers in the presence of various additives. FCMs may contain numerous chemicals that have the potential to migrate into foodstuffs (see Fig. 22.1) and result in undesirable consumer exposures [1,2].

Some of these chemicals having technological functions in the production of FCMs (e.g., monomers, antioxidants, plasticizers) are called intentionally added substances (IAS). From a food safety perspective, they are characterized and can be managed accordingly, for example, using specific migration limits when established. However, it is also well documented that FCMs can contain other substances, the so called non-intentionally added substances (NIAS), which are not directly added. NIAS may be numerous and originate from various sources such as impurities in starting materials and chemical transformations during manufacturing processes. The scope of this chapter is to highlight the current challenges regarding risk assessment of FCM chemicals and to outline solutions such as the application of the threshold of toxicological concern (TTC) and QSAR models.

Regulatory considerations

In Europe, FCMs are generally regulated under the EU Framework Regulation EC 1935/2004 stating that no food contact materials shall transfer constituents into food at levels that endanger human health [3]. Only for some FCMs such as plastics, a specific regulation at European level exists, whereas other materials such as papers or coatings are regulated at national level. NIAS are mentioned in the European Regulation on plastic materials and articles intended to come into contact with food [4], as impurities and reaction and degradation products formed during manufacturing of the plastic material and article. The EU Regulation 10/2011 specifies that the presence of NIAS

QSAR in Safety Evaluation and Risk Assessment
https://doi.org/10.1016/B978-0-443-15339-6.00023-0

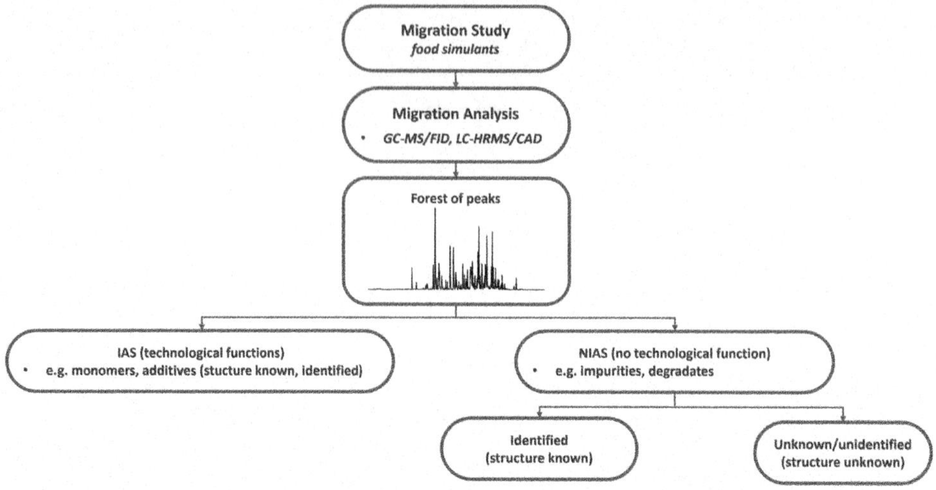

FIGURE 22.1 Chemicals associated with food contact materials.

in the final plastic article is acceptable but should be assessed by the manufacturer in accordance with internationally recognized scientific principles of risk assessment [4]. A 10 μg/kg (10 ppb) limit was set as a pragmatic analytical limit of detection for migration from behind a functional barrier in Commission Regulation (EU) 10/2011, with exclusions of substances that are classified as carcinogenic, mutagenic, toxic to reproduction, or nanomaterials [5]. In summary, to comply with the European legislation, manufacturers and operators are obliged to document safety, not only for IAS, but also for NIAS [1,4]. Demonstrating that the presence of NIAS is without any health concern for consumers is the main challenge in FCMs safety evaluation.

In the US, according to 21 CFR 170.3 [1,6], any food contact substance (FCS) expected to migrate into food through the Food Contact Articles must conform with the legal requirements. An FCS is defined as "any substance that is intended for use as a component of materials used in manufacturing, packing, packaging, transporting, or holding food if such use is not intended to have any technical effect in such food." Like the concept of IAS in EU, the definition of FCS does not cover substances which are not added intentionally, and the term NIAS, introduced originally in the European FCM plastic regulation, is not mentioned and applied in a legal context in the US. However, there are several provisions concerning substances which could fit to the NIAS definition, for example, the safety assessment shall also include "any substance formed in or on food because of its use." Under 21 CFR 174.5, any FCS "shall be of a purity suitable for its intended use," and also in the case of polymers, the submission of information on the significant impurities and side reactions is recommended [1].

Assessing safety of FCMs: The standard approach

Theoretically, the safety evaluation of FCM substances could be done according to standard food chemical risk assessment principles [2,7,8], which include the four following steps:

- *Hazard identification* to determine the intrinsic toxicological properties of chemicals. This is usually achieved with in vivo animal toxicological studies addressing various end points including chronic, carcinogenicity, developmental, and reproductive end points. Mechanistic data such as genotoxic (e.g., micronucleus test) and mutagenic (e.g., Ames test) potentials are also generated in order to decide on the hazard characterization strategy to be applied.
- *Hazard characterization* determining the most appropriate toxicological data in order to establish a safe level of human exposure. Classically, a dose level without any effect in animal, the No Observed Adverse Effect Level (NOAEL), is defined and used as a point of departure (PoD). The Health-Based Guidance Value (HBGV), safe level of human exposure, is then derived by applying uncertainty factors to the PoD. In general, the uncertainty factors cover potential interspecies and interindividual differences. NOAELs are single-dose estimates. They are largely dependent on study design. Therefore, they are highly uncertain and consequently increasingly replaced by the use of benchmark dose, which better accounts for the full dose–response curve. For chemicals which are carcinogenic and mutagenic (no threshold of effect assumed), a determined level of excess cancer risk is

calculated, for example, the benchmark dose of 10% or the TD_{50}, a daily dose inducing cancer in 10% and 50% of the treated animals respectively [9–13].

- *Exposure assessment* combining occurrence levels with food intake. In case of FCMs, migrating levels in food simulants are often used as food occurrence surrogates and food intake scenarios may be chosen to reflect the amount of food anticipated to be in contact with the FCM under investigation.
- *Risk characterization* provides the level of safety concern through the calculation of the margin of safety calculated as the ratio between the HBGV and calculated exposure. The size of margin of safety determines the safety concern. A margin of safety >1 would indicates a low safety concern or reasonable evidence of safety. For a mutagenic chemical, a margin of exposure (MoE) corresponding to the ratio between determined level of risk and exposure is calculated. The size of MoE determines the level of safety concern [11,12].

For FCM chemicals, the application of the standard risk assessment approach as described above has been limited to a relatively small number of chemicals, mostly IAS such as monomers and additives. IAS are well characterized from chemical structure and purity perspectives and depending on the functionality can be present in high quantities. Therefore, these substances can be easily tested in animal regulatory toxicology studies to obtain toxicological information adequate for standard risk assessment. However, only a restricted number of identified FCM chemicals have been extensively tested toxicologically and assigned with official HBGVs. Many FCM chemicals have no, or very limited, toxicological information available and would require additional information for comprehensive and standard risk assessments. Some chemicals having a defined function as IAS in specific applications can be identified as NIAS in others. For these substances, availability of toxicological data may still be adequate. Conversely, for the majority of chemicals regularly identified as NIAS, typically no toxicological data are available, and consequently, their safety significance cannot be assessed [2,14].

This situation is raising questions among scientific, consumer, regulatory, and industrial communities. It is now widely acknowledged that the toxicological information available on FCMs is in general insufficient or inadequate to demonstrate confidently that an FCM is compliant with the European regulatory framework, emphasizing that migration of any chemicals should not endanger human health. Generating additional toxicological information on FCM substances, especially on NIAS, has been strongly warranted [2,14–16]. The US Food and Drug Administration (FDA) is also facing growing pressure because of presence of impurities in FCM, with congressional representatives requesting to initiate a review of FDA procedures [1,17]. Despite the widely recognized importance to address health risks of NIAS, official harmonized guidance on how they should be assessed is still missing.

Assessing the safety of NIAS: The challenges

Chromatographic analyses of FCM extracts and/or migrates generally reveal the presence of forests-of-peaks corresponding to large numbers (dozens to thousands) of individual NIAS present in various amounts ranging over several orders of magnitude. Importantly, some of these chemicals may be structurally known, while others may only be partially characterized [2,14]. Therefore, many of these substances remain unidentified and must be considered as unknowns. In addition, for most of the identified NIAS, toxicological information is missing, thus preventing any hazard identification and characterization. Consequently, to address the safety of NIAS through a traditional risk assessment, toxicological data need to be generated.

Keeping standard risk assessment in mind, two possibilities may be envisaged to obtain appropriate toxicological information. The first one refers to the direct testing of migrates in animal regulatory toxicology studies. However, the production of sufficiently high amounts of relevant and chemically characterized materials to be tested in animals is not feasible in practice. Additionally, regulatory toxicology animal studies are very limited to detect potential toxic effects resulting from the presence of relatively low but still toxicologically relevant levels of substances (e.g., impurities) in complex mixtures such as FCM migrates. The statistical designs of animal studies focus only on the identification of effects at high doses.

Another possibility would ideally require detecting and quantifying all migrating substances and testing them individually in animal toxicological studies. Obviously, this would not be possible in practice because of excessive need for resources. In addition, experience suggests that many NIAS would be expected to present only low safety concern, meaning that testing all of them in toxicological studies would result in unnecessary use of laboratory animals.

The considerations outlined above clearly speak for the development of a novel approach relying less on standard toxicity testing and more on nontesting strategy. In this context, QSAR models have the potential to play a significant role to obtain cost-efficiently toxicological information necessary to establish levels of safety concern of NIAS.

The application of these techniques would certainly help to prioritize the substances on which additional investigations should be focused.

Non-testing approaches for hazard identification and characterization

To address the toxicology challenges highlighted above, developing, applying, and integrating fit-for-purpose nontesting methods are key. The minimum set of data required for conducting a risk assessment with reasonable confidence includes information on the mutagenic potential of the substances under investigation and a conservative PoD suitable for hazard and risk characterization. Since, in general, the main purpose of risk assessment is to protect the whole human population from long-term chemical exposures, initial priority has been to predict PoDs related to chronic toxicity which is often the most conservative. It has been recognized that for some chemicals, the pivotal effect (in general the most sensitive one) may be on reproduction or development resulting from nonchronic exposures. Therefore, efforts should also be invested to identify such chemicals.

There have been initiatives to screen food contact chemicals using in silico toxicology, but many of them use qualitative approaches, mainly mutagenicity predictions, suitable for hazard identification [18]. They do not provide any information about hazard characterization such as prediction of chronic toxicity and even less about health risks. Schilter et al. [11] developed a decision tree (DT) reflecting the risk assessment paradigm to rapidly and cost-efficiently integrate exposure information with predicted toxicological values. It results in a conceptual scheme, which adopts a sequential screening on mutagenicity, followed by the quantitative evaluation of chronic toxicity (LOAEL, NOAEL) and/or carcinogenic potency (TD_{50}) when mutagenic is identified. The next sections provide a review of models and approaches currently available and applicable in the DT. A simplified version of the DT is provided in Fig. 22.2.

Predicting mutagenicity

The Ames test is the most recommended test to identify gene mutation-inducing chemicals [14]. In absence of experimental information, several databases and QSAR models predicting Ames mutagenicity have been consolidated over years, integrating knowledge-based and statistic-based models. The guideline for the assessment and

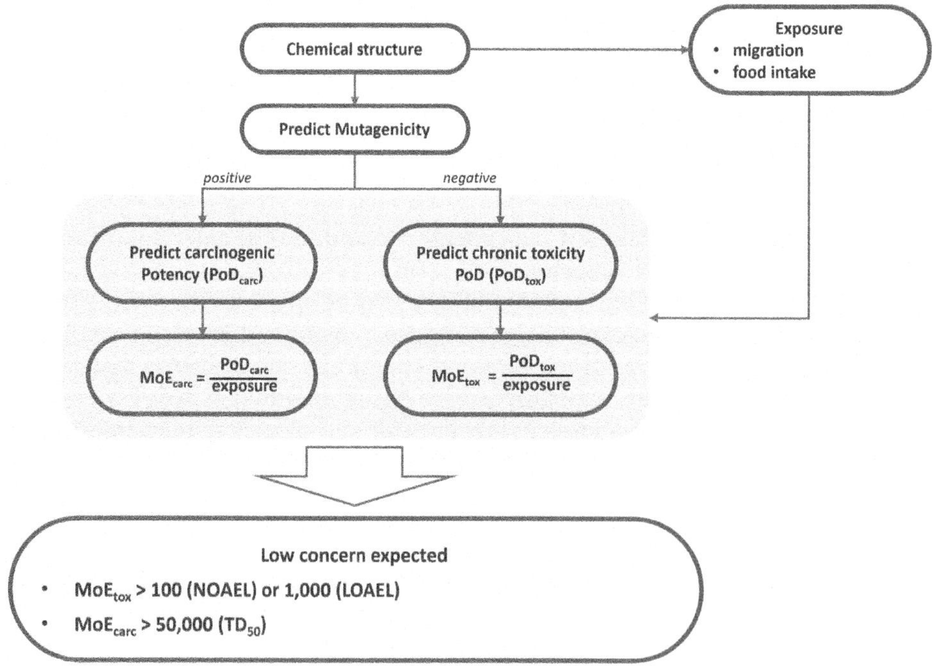

FIGURE 22.2 Application of quantitative structure–activity relationship (QSAR) to establish level of safety concern of exposures to toxicologically nontested chemicals. *MoE*, margin of exposure; *PoD_{carc}*, prediction of TD_{50}; *PoD_{tox}*, prediction of chronic toxicity. *Adapted from Schilter et al. [11].*

control of mutagenic impurities in pharmaceuticals to limit potential carcinogenic risk permits the use of QSAR tools for the initial mutagenicity assessment. This guideline was established during the International Conference on Harmonization (ICH) M7 [13] and was the first international guideline that addresses the use of QSAR models instead of experimental toxicological study in the frame of human health assessment. Thus, QSAR is no longer merely a prediction or screening tool but can be regarded as a mutagenicity test under ICH M7. The guideline requires two QSAR prediction methodologies, one rule-based and one statistical-based to maximize sensitivity [19]. It is also recommended by the European Food Safety Authority (EFSA) that at least two independent QSAR models based on different training sets or algorithms (knowledge-based and statistically based models) are used in order to obtain sufficient confidence [20]. REACH and ICH M7 [21] guidelines on the use of in silico models for regulatory purposes ask for predictions within the applicability domain of the model [22]. Based on this consideration, a wide range of commercial and free publicly available QSAR models can be applied.

The Division of Genetics and Mutagenesis, National Institute of Health Sciences (DGM/NIHS) of Japan recently established a unique proprietary Ames mutagenicity database containing about 12,000 new chemicals not previously used for developing QSAR models. The DGM/NIHS provided this Ames database to QSAR developers/suppliers to validate and improve their QSAR tools. In total, 12 QSAR vendors using 17 different tools participated in this project representing the most used models. Overall, this initiative is a good example of integration of complementary tools. The detailed features of each QSAR tool, either commercial or freely available, were reviewed and described [19].

All QSAR tools demonstrated improved predictive power after the study compared to the original versions, indicating that this project successfully fulfilled the principal aim of enhancing QSAR performance. Some QSAR tools demonstrated greater than 80% accuracy for predicting Ames mutagenicity, which is almost equivalent to interlaboratory reproducibility of experimental Ames test results [19].

TTC for hazard characterization

The TTC concept is not a chemical-specific approach but is driven by the assignment of structures to classes to which a specific TTC is allocated. It is based on a statistical and probabilistic analysis of animal toxicological data from which generic human chronic exposure thresholds were derived such that the probability of adverse effects at human exposures below these threshold values is considered to be low [23]. It is applicable to substances with known structure but unknown toxicity [20,23] but is not applicable to several exclusion categories such as polyhalogenated-dibenzodioxins and N-nitroso compounds. For substances that bear alert of direct DNA-reactive mutagenicity, a TTC value of 0.0025 μg/kg bw/day can be used. For organophosphates and carbamates, the TTC value is 0.3 μg/kg bw/day. All other substances can be classified in one of the three Cramer classes, with the Cramer class III-TTC (1.5 μg/kg bw/day) being the most conservative to be applied to chemicals with presumption of high toxicity.

EFSA (2019) [20] acknowledged and defined the conditions of applicability of TTCs to conduct food chemical risk assessment in absence of toxicological data and promotes the use of QSAR to exclude genotoxicity through DNA reactivity. The TTC approach is currently widely accepted for use such as in European member states and the United States [1], as well as by EFSA and WHO [24].

The concept of prioritization of large numbers of chemicals based on TTC has been recently explored by Nelms et al. [25]; the authors established new TTC values using the EPA's Toxicity Values database (ToxValDB), which were similar though not identical to the original TTC values. Even though the original 5th percentile NOAEL values for Cramer class I–III were lower than those using the data from ToxValDB, only the Cramer class III values were significantly different between the two datasets [26].

QSAR to predict repeated-dose toxicity

General considerations. As mentioned above, chronic toxicity and carcinogenicity studies are providing the key data for hazard and risk assessment. Consequently, predicting such end points appears very valuable. Results from QSAR models are independent from the TTC approach and support and confirm the decisions taken using the TTC concept. In addition, it would provide an opportunity to address the safety of toxicologically uncharacterized NIAS, occurring in food at levels resulting in exposures higher than the TTC. As already suggested, to be applicable in a risk assessment framework, models should provide not only qualitative information on hazard (e.g., mutagenicity) but also quantitative predictions such as chronic toxicity PoDs allowing comparisons with exposure estimates [11,27]. The models should be based on reliable data from studies complying with official guidelines such as from OECD. They should be validated according to recognized standards. NIAS and food chemicals in general present a high structural diversity. Consequently, models should be global rather than referring to particular chemical classes (local). Since manual expert review will always be instrumental to interpret the predictions, models should be transparent, and all relevant information should be available to the users.

Data availability. There are several data sources which can be used for batch processing of long lists of chemicals. Most of them have been used as a basis to develop QSAR models. Experimental NOAELs and LOAELs from (sub) chronic rat oral toxicity studies generally complying with the OECD guidelines are the most available. Many of them are implemented and/or easily accessible:

- the Munro and Hazard Evaluation Support System (HESS) databases, both of which are available from the OECD QSAR Toolbox (https://qsartoolbox.org/) [9],
- the Integrated Risk Information System (IRIS) database from the US Environmental Protection Agency (EPA) (https://www.epa.gov/iris) [9],
- the COSMOS database [28] and the US EPA's ToxRefDB, v. 2.0 [29],
- the EFSA OpenFoodTox database [30,31],
- the ToxValDB [32],
- the Carcinogenic Potency Database (CPDB) [33].

Most of freely available QSAR models rely on these databases.

Several in silico models have been developed in the past predicting different effect levels for repeated dose toxicity using relatively small datasets (<500 chemicals) with limited coverage and/or performance. As reviewed by Pradeep et al. [32], several different modeling approaches for small datasets were unable to achieve a coefficient of determination (R^2) greater than 43%–70% and 10-fold cross-validated root mean square error (RMSE) values in log10-mg/kg/day of less than approximately 0.5 log10-mg/kg/day [34–38]. Only a small number of models assessed underlying variability in the experimental data coming from a variety of in vivo studies and incorporated that variability to characterize the uncertainty in model predictions.

Mazzatorta et al. [39] reported a predictive in silico study of more than 400 compounds based on multivariate analysis. The predictive model error of 0.73 (in a logarithmic scale) on a leave-one-out cross-validation was close to the estimated variability of experimental values (0.64). Aside from incorporating underlying variability in the experimental data, both Pradeep et al. [32] and Helma et al. [40] derived confidence intervals for each prediction. Helma et al. [40] developed QSAR models predicting oral chronic rodents' toxicity based on lazy structure–activity relationships (lazar) algorithm, consisting of local random forests regression model with similar compounds (neighbors). The RMSE of bootstrapped local model predictions was used to construct 95% prediction intervals at 1.96*RMSE. The width of the prediction interval indicates the expected prediction accuracy. The "true" value of a prediction should be with 95% probability within the prediction interval. They demonstrated that predictions within the applicability domain of the training data had the same variability as the experimental training data. Predictions with a larger distance from the applicability domain (lower similarity threshold) were also significantly better than random guessing, but the errors to be expected were higher, and a manual inspection of prediction results was highly recommended by the authors.

Pradeep et al. [32] presented a set of QSAR models developed using chemical structural and physicochemical properties to provide a robust prediction for repeat dose in vivo toxicity PoDs using a large dataset of 3592 chemicals. A bootstrap resampling technique was applied to incorporate typical study-to-study variability (~0.5 log10-mg/kg/day) in the PoD values for deriving confidence intervals in model predictions. In this work, the 95% confidence interval (±2 standard deviations) around each PoD prediction of repeat dose toxicity approached ±1 log10-mg/kg/day, based on the variability in the training set values. Quantitative predictions of PoDs within ±1 log10-mg/kg/day where the training set of PoD values typically ranges over approximately 1–2 log10-mg/kg/day per chemical indicated that variability in the PoD values clearly limits QSAR accuracy. Moreover, the majority of experimental effect levels span about 4 orders of magnitude.

A few attempts have been made to apply QSAR methods to simultaneously model both NOAEL and LOAEL. Gadaleta et al. [41] developed two regression models (NOAEL and LOAEL) based on the optimal descriptors calculated by the Monte Carlo method. These models were built using toxicity data from 327 compounds from the Fraunhofer RepDose database, a nonpublic database. Regression models were based on quantitative relationships between the endpoints and SMILES-based attributes. Sub-chronic 90 days feeding rat studies were included. For the NOAEL model, the validation set in Gadaleta et al. [41] consisted of 38 compounds reaching $R^2 = 0.65$. When excluding the compounds outside the applicability domain, the predictive capacity reached $R^2 = 0.69$, with 33 compounds covered. NOAEL and LOAEL models were integrated based on differences between NOAEL and LOAEL. This strategy improved the performance of single models, leading to R^2 greater than 0.70 and RMSE lower than 0.60 (for regression models) on the validation set, based on the end point and the threshold applied for selecting predictions.

Selvestrel et al. [9] provided models for NOAEL and LOAEL predictions. Focus was on systemic and oral organ-specific toxicity in rats. Data were obtained from different publicly available databases. They were compliant with the OECD guideline 408 dealing with effects on liver, kidney, and brain. The authors developed and validated eight QSAR models based on the optimal descriptors calculated by the Monte Carlo method, using the CORAL software. The predictive ability R^2 in the validation sets reached 0.55 and 0.53 for NOAEL and LOAEL, respectively.

A few highly relevant models predicting sub-chronic NOAELs have been implemented and are or will be made freely available at the VEGA platform (VEGA HUB—Virtual models for property Evaluation of chemicals within a Global Architecture). Some refer to systemic toxicity, other to organ toxicity [9].

Only a few attempts aimed at modeling carcinogenic potency. This would be valuable to address safety concern of mutagenic chemicals which are acting through a nonthreshold direct DNA damage mechanism. The CPDB database (also called "Gold database") contains the results of 6529 chronic, long-term animal cancer studies on 1547 chemicals. It provides a numerical value of carcinogenic potency (TD_{50}) calculated for each set of tumor incidence data available. This database has been used to build QSAR and read across like models [42–44] including also commercial models (developed by Simulation Plus). Lhasa Limited calculated its own TD_{50} values based on the CPDB database. Values are available in the Lhasa Limited Carcinogenicity database [45]. The prediction errors observed in models by Lo Piparo et al. [44] were similar in size to the experimental variability for chemicals falling in the applicability domain. This suggests that the availability and quality of experimental data constitute the main source of limitations of the models.

Use of QSAR predictions for establishing level of safety concern

Predicting PoDs provides the opportunity to integrate toxicology and exposure. However, predicted PoDs are not considered suitable to derive HBGVs. Consequently, to establish the level of safety concern, an MoE approach (MoE = PoD/exposure) similar to the one developed earlier for genotoxic carcinogens [12]. MoEs can be interpreted as described previously [11]. In brief, for nonmutagenic chemicals, an MoE covering the standard factors used to deal with potential interspecies and interindividual differences, as well as other uncertainties triggered by the data available for the MoE calculation, for example, to convert LOAEL into NOAEL or incompleteness of available toxicological data, would indicate a low safety concern. For mutagenic chemicals, an MoE greater than 50,000 using TD_{50} as PoD [13,14] would suggest a potential excess risk $<10^5$. This is considered of low priority for further investigation [14].

Safety assessment protocol for FCM chemicals

In absence of clear advice given by authorities, different stakeholders have developed scientific approaches and guidance documents dealing with some aspects of the FCM safety assessment. They describe how to fill specific gaps and address challenges highlighted above [1,2,14]. Fig. 22.3 presents an integration of these approaches in the format of a practical protocol/procedure to be followed for chemical safety assessment of FCM chemicals. This protocol favors the use of alternative and nontesting methods such as QSAR for hazard identification and characterization.

Migration studies

The assessment is initiated by migration studies using simulants reflecting the food to be packed in the FCM under investigation. For that purpose, regulatory guidelines are available [4] specifying the migration solvents and the time/temperature to be applied. For example, for an acidic (pH < 4.5) food with hydrophilic character and a contact with the packaging of more than 30 days, a migration study in acetic acid 3% (w/v) at 60°C over 10 days would be recommended [4].

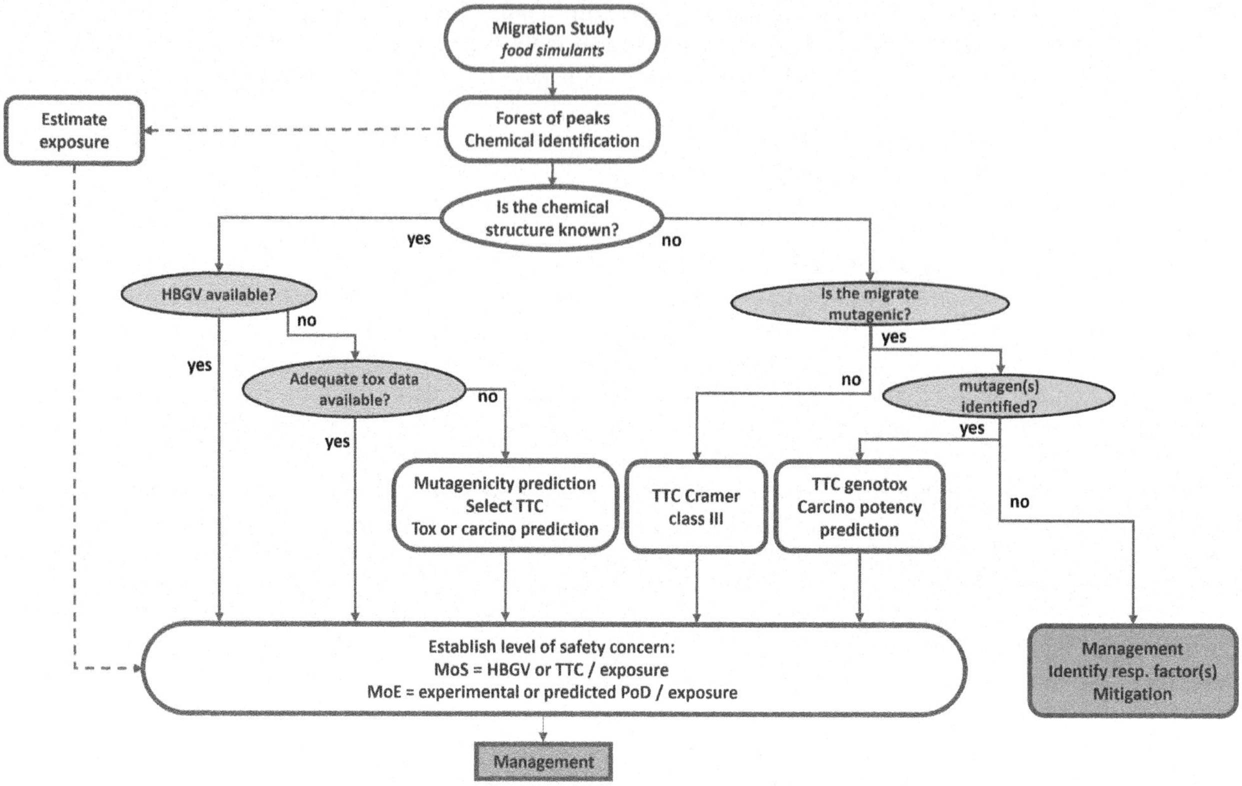

FIGURE 22.3 Safety assessment of chemicals migrating from food contact materials.

Migrate analysis

The migration samples (the migrates) undergo a sample treatment if necessary and are submitted to various analytical methods based on gas chromatography (GC) and liquid chromatography (LC) coupled with mass spectrometry (MS) for identification, for example, flame ionization detector for semi-quantification [2,14,46]. For IAS and known NIAS, targeted analytical methods can be applied. Nontargeted methods are applied to analyze known and unknown NIAS. Advances in analytical techniques and growing databases steadily facilitate the detection and identification of NIAS, but a comprehensive analysis of all NIAS is not available yet. Many FCM migrants have known identity but still a large number of unexpected and unidentified NIAS are also being detected [2,14]. A significant challenge is the quantification of unidentified NIAS in order to prioritize them with respect to safety. For that purpose, methods allowing to get rapidly quantitative information on substances of unknown structures have been developed using universal detectors, for example, flame ionization detector in GC-high-resolution mass spectrometry and charged aerosol detector in LC-high-resolution mass spectrometry, together with appropriate standard molecules [2,14]. Analytical data obtained from migrates are most important since they can be translated directly into exposure estimates, though other sources of exposure than the packaging material investigated should also be considered. However, the analytical techniques described above can also be applied on other relevant materials such as FCM extracts or starting compounds (e.g., resins, biopolymers, monomers, adhesives). This may target the generation and facilitate the interpretation of data obtained with the migrates [2,14].

In parallel to chemical analysis, a part of the concentrated migrate is transferred in a solvent compatible with in vitro, cell culture-based bioassays (generally dimethyl sulfoxide, DMSO). The use of bioassays in the context of FCM safety assessment has been a lot debated [14]. A recent research proposed that FCM migrates should be tested in bioassays in order to identify the potential presence of directly acting mutagenic substances [14]. To document an absence of mutagenicity in migrates would allow to apply the threshold of Toxicological Concern Cramer (TTC) class III to facilitate the prioritization of unknown NIAS (see below hazard identification/characterization) once specific chemicals not covered by the TTC. This can be done through either measurements or expert judgment [47]. For that purpose, it has been recognized that the best test currently available to exclude mutagenicity is the Ames test, although its capacity to detect mutagens at sufficiently low, safety relevant levels are still limited [14]. New approaches based on high-performance thin layer chromatography coupled with bioassay (high-performance thin layer chromatography bioassay) may adequately cope with the challenge of health driven limits of mutagen analytical detection [48,49].

Exposure assessment

Exposures are calculated by combining analytical data with specific information on intended usages of the FCM under investigation for packing foods. Generic food intakes for broad categories of packaged food products have been developed by EFSA, taking into account specific target human populations (infants, toddlers, adults) [50].

Hazard assessment and risk characterization

Quantitative toxicology values are necessary to evaluate the significance of the calculated exposures. The hazard assessment approaches must adjust to the information brought by the analytical investigation of migrates. Three scenarios may have to be envisaged to address substances of different structural characterization:

- *Identified substances toxicologically characterized.* For those chemicals, a standard risk assessment as already described above may be applied.
- *Identified substances without toxicological data.* As highlighted earlier, different methods are available to get useable information for hazard characterization. The concept of the TTC is currently the most widely applied and accepted. Exposures lower than the appropriate TTC are considered as safe [20,23]. To confirm the assessment based on the TTC or to deal with exposures higher than the appropriate TTC, the use of QSAR predicted PoDs for calculation of MoEs is recommended.
- *Substances detected, but not identified.* In such cases, a cut-off value of 10 mg/kg food (or simulant) has been often promoted for early management purposes [14]. However, a new and more scientifically driven way of dealing with unknowns may now be recommended [14,46]. It consists of exploiting the TTC Cramer class III. This was considered sufficiently protective for consumer health, provided that evidence is available indicating that the migrate does not contain any mutagenic chemicals [14]. As mentioned above, this should be supported by experimental data obtained with the migrate directly tested in an Ames test. In addition, the understanding of the chemistry involved in the manufacturing of the investigated FCM may also bring useful and highly relevant data to exclude the presence of mutagens in migrates. Unknowns present at levels lower than the Cramer class III TTC would be considered of low priority for further structural characterization and risk assessment.

Interpretation in context of risk management

The new strategy outlined above is expected to bring a more complete, scientific, evidence-based, and accurate picture on the potential safety concern associated with NIAS in specific FCMs. Because of the uncertainties involved, it should not be considered as a formal risk assessment procedure triggering automatically straightforward operational management decisions and actions. It should be seen as a tool or a reasoning guide, contributing to the identification and prioritization of NIAS which would deserve additional investigations, either on the assessment (e.g., improve exposure or hazard assessment) or on the management sides (e.g., study the possibility to reduce formation over manufacturing). Application may be particularly valuable in the context of new FCM development, in a context of safety by design to identify and manage as early as possible substances of potential safety concern.

Conclusion and perspectives

FCMs illustrate the limitations of a standard risk assessment procedure based on animal toxicological studies to address safety of complex materials containing large number of chemicals. In Europe, the difficulties of the task have been handled by a wide application of either a cut-off threshold value of 10 ppb in food products (for chemicals not classified as carcinogenic, mutagenic, toxic to reproduction), and/or the concept of the TTC. This approach has been considered as reasonably conservative and protective, although many NIAS may occur at higher levels and are often ignored. In addition, for many NIAS, 10 ppb could be overconservative, triggering unproportionate management actions. In this context, nontesting in silico toxicology methods appear very promising since they may provide rapidly and cost-efficiently highly relevant toxicological information to improve hazard assessment of toxicologically uncharacterized chemicals and then support decision-making. In that regards, QSAR models have been identified as promising. Validated models with reasonable accuracy and performance are available to reliably predict parameters necessary to conduct risk assessment, such as mutagenicity and quantitative PoDs with particular caution to be given to out of domain predictions. When available, error estimates of the models were found to be close to the variability of the experimental dataset. Although most useful for safety by design in new packaging development, QSAR predictions may bring other advantages, for example, to handle emergency chemical contamination situations requiring rapid decisions and actions. Importantly, QSAR methods are fast and in batch mode are applicable to large numbers of chemicals. This potential could be exploited with databases of FCM specific chemicals to further study and confirm the level of protection associated with the use of the 10 ppb and/or the TTC.

As already highlighted, the proposed nontesting strategy is mainly based on predictions of mutagenicity and PoDs for chronic toxicity and carcinogenic potency. Although considered reasonably conservative, this approach is not fully reflecting the toxicological database required for standard risk assessment. Further research to address additional end points is therefore necessary. Currently, attempts to predict other genotoxic mechanisms than direct DNA-reactivity are being made. This appears significantly more challenging. In the EFSA evaluation of the applicability of existing QSAR models for predicting the genotoxicity of pesticides it was concluded that the reliability of the QSAR models for assays/end points different from in vitro bacterial mutagenicity (Ames) appears to be still quite far from optimality [22].

QSAR and read-across models predicting micronucleus genotoxicity [51,52] have been recently made freely available in the VEGA hub platform [53]. In silico initiatives have been proposed to integrate different genotoxicity end points encompassing gene mutation, aneugenicity, and clastogenicity effects associated to confidence levels [54]. Similarly, predictions of developmental and reproductive toxicities have been shown to be associated with several limitations [55]. The integration of these additional predictions into the framework as depicted in Figs. 22.2 and 22.3 is not yet clear and could involve the applications of specific uncertainty factors still to be defined.

Chronic toxicity, including carcinogenicity, is highly complex and multifactorial phenomena embracing a plethora of different biological effects and mechanisms of action making QSAR studies intrinsically very challenging. There are known limitations and sometimes reluctance in using QSAR models for repeated dose toxicity predictions. This is mainly because of limited applicability domain, high experimental variability impacting model building and validation (introducing noise in the training data and on the prediction data), as well as lack of endpoint specificity [9,40,41]. These weaknesses originate from the standard experimental design of animal studies (e.g., dose regimen), the classical end points measured, and the complexity of the vastly unknown toxicity mechanisms and processes behind the observed adverse effects.

Solutions to incorporate more systematically available knowledge on experimental variability have been proposed with the aim of better characterizing model performance. QSAR predictions associated with a confidence interval may be useful in understanding model performance but may also facilitate the prediction of a range of doses for a given PoD [32,40]. The size of the confidence interval should provide insight on the performance of the model. Some models predicting PoDs have explicitly considered the experimental variability in the underlying data, so that the performance metrics may more accurately reflect the reality of the variable experimental data. Some tools exploit all data available to predict only one generic value per chemical and not individual discrete PoDs (e.g., NOAEL or LOAEL) bound to study design or type of toxicity [32]. The need to choose one type of PoDs (e.g., chronic LOAEL instead of NOAEL) for calculating MoEs may introduce bias in the interpretation of the results. Importantly, parts of these issues may be addressed manually by expert reviews of the QSAR studies. For example, for the models of the VEGA platform, an evaluation of the reliability of the predictions is provided together with the six most similar substances used by models. This transparency facilitates the expert reviews by end-users [53].

It is recognized that using independent approaches and modeling techniques together with thorough manual expert review may significantly improve the reliability of toxicity predictions and increase confidence for in silico driven decision-making [56]. Consequently, an approach focusing solely on QSAR-based methods predicting highly complex end points such as standard quantitative chronic toxicity values and carcinogenic potency is still unsatisfactory and, in isolation, may not be assumed as sufficiently robust for formal safety assessment. Many relevant parameters such as absorption, distribution, metabolism, excretion, organ toxicity, and mechanistic properties (e.g., receptor mediated activity) are not identified and transparently considered. However, some of these parameters may now be modeled and possibly used to better incorporate the basis of toxicity into the final in silico hazard assessment. Interestingly, the availability of mechanistic information could be instrumental to determine the need to apply mixture toxicology principles, for example, dose-addition, for chemicals expected to act through the same mode of action. Additionally, they may serve building groups of toxicologically populated compounds similar to the substances of interest and finally support comparisons across chemicals, facilitating toxicity predictions through read across. This technique is less formalized and more resource intensive than QSAR and requires significant expert knowledge and judgment, despite broader similarity concepts and improved strategies and tools [11,57–59]. Moreover, whereas during the years QSARs have undergone many performance evaluations, with special emphasis on comparative prospective exercises, nothing analogous exists yet for read across. Nevertheless, to integrate QSAR modeling with read across constitutes the best avenue to improve the overall performance of in silico toxicology.

In conclusion, considering the difficulty to obtain robust predictions of complex end points relevant for risk assessment, QSAR models and other techniques currently available are quite promising. Although not very suited to obtain precise predictions of human toxicity, the applications of the approaches are considered to be protective [11]. This is encouraging, with a view to a future integrated approach, where the predictions of PoDs will support the paradigm shift toward animal-free next generation risk assessment methodologies currently under development.

Acknowledgments

We acknowledge the European Union's Horizon 2020 research and innovation program under the Marie Sklodowska-Curie grant agreement No 799288. We also acknowledge Gina Montoya for her careful reading.

References

[1] Kato LS, Conte-Junior CA. Safety of plastic food packaging: the challenges about non-intentionally added substances (NIAS) discovery, identification and risk assessment. Polymers 2021;13(13):2077.
[2] Koster S, Bani-Estivals M, Bonuomo M, Bradley E, Chagnon M-C, Garcia M, et al. Guidance on best practices on the risk assessment of non-intentionally added substances (NIAS) in food contact materials and articles. 2015.
[3] Regulation EC. No 1935/2004 of the European Parliament and of the Council of 27 October 2004 on materials and articles intended to come into contact with food and repealing Directives 80/590/EEC and 89/109/EEC. Off J Eur Union 2004;338:4–16.
[4] European Commission Regulation (EU). No 10/2011 of 14 January 2011 on plastic materials and articles intended to come into contact with food off. Off J Eur Union January 14, 2011;12:1–89.
[5] Nerín C, Bourdoux S, Faust B, Gude T, Lesueur C, Simat T, et al. Guidance in selecting analytical techniques for identification and quantification of non-intentionally added substances (NIAS) in food contact materials (FCMS). Food Addit Contam 2022;39(3):620–43.
[6] US Food and Drug Administration. CFR-code of federal regulations title 21. 2017.
[7] FAO/WHO. Principles and methods for the risk assessment of chemicals in food. Environmental Health Criteria.; 2009. p. 240.
[8] World Health Organization. WHO human health risk assessment toolkit: chemical hazards. 2021.
[9] Selvestrel G, Lavado GJ, Toropova AP, Toropov AA, Gadaleta D, Marzo M, et al. Monte Carlo models for sub-chronic repeated-dose toxicity: systemic and organ-specific toxicity. Int J Mol Sci 2022;23(12):6615.
[10] EFSA. Update: use of the benchmark dose approach in risk assessment. EFSA J 2017;15(1):e04658.
[11] Schilter B, Benigni R, Boobis A, Chiodini A, Cockburn A, Cronin MT, et al. Establishing the level of safety concern for chemicals in food without the need for toxicity testing. Regul Toxicol Pharmacol 2014;68(2):275–96.
[12] EFSA. Statement on the applicability of the margin of Exposure approach for the safety assessment of impurities which are both genotoxic and carcinogenic in substances added to food/feed. EFSA J 2012;10(3):2578.
[13] EMA. ICH guideline M7(R1) on assessment and control of DNA reactive (mutagenic) impurities in pharmaceuticals to limit potential carcinogenic risk. European Medicines Agency; 2018. https://www.ema.europa.eu/en/ich-m7-assessment-control-dna-reactive-mutagenic-impurities-pharmaceuticals-limit-potential. [Accessed 1 November 2022].
[14] Schilter B, Burnett K, Eskes C, Geurts L, Jacquet M, Kirchnawy C, et al. Value and limitation of in vitro bioassays to support the application of the threshold of toxicological concern to prioritise unidentified chemicals in food contact materials. Food Addit Contam 2019;36(12):1903–36.
[15] Muncke J. Exposure to endocrine disrupting compounds via the food chain: is packaging a relevant source? Sci Total Environ 2009;407(16):4549–59.
[16] Grob K, Biedermann M, Scherbaum E, Roth M, Rieger K. Food contamination with organic materials in perspective: packaging materials as the largest and least controlled source? A view focusing on the European situation. Crit Rev Food Sci Nutr 2006;46(7):529–35.
[17] Karmaus AL, Osborn R, Krishan M. Scientific advances and challenges in safety evaluation of food packaging materials: workshop proceedings. Regul Toxicol Pharmacol 2018;98:80–7.
[18] Van Bossuyt M, Van Hoeck E, Raitano G, Manganelli S, Braeken E, Ates G, et al. Q) SAR tools for priority setting: a case study with printed paper and board food contact material substances. Food Chem Toxicol 2017;102:109–19.
[19] Honma M, Kitazawa A, Cayley A, Williams RV, Barber C, Hanser T, et al. Improvement of quantitative structure–activity relationship (QSAR) tools for predicting Ames mutagenicity: outcomes of the Ames/QSAR International Challenge Project. Mutagenesis 2019;34(1):3–16.
[20] EFSA. Guidance on the use of the threshold of toxicological concern approach in food safety assessment. EFSA J 2019;17(6):e05708.
[21] Fioravanzo E, Bassan A, Pavan M, Mostrag-Szlichtyng A, Worth A. Role of in silico genotoxicity tools in the regulatory assessment of pharmaceutical impurities. SAR QSAR Environ Res 2012;23(3–4):257–77.
[22] Benigni R, Laura Battistelli C, Bossa C, Giuliani A, Fioravanzo E, Bassan A, et al. Evaluation of the applicability of existing (Q) SAR models for predicting the genotoxicity of pesticides and similarity analysis related with genotoxicity of pesticides for facilitating of grouping and read across. EFSA Support Publ 2019;16(3):1598E.
[23] Kroes R, Renwick A, Cheeseman M, Kleiner J, Mangelsdorf I, Piersma A, et al. Structure-based thresholds of toxicological concern (TTC): guidance for application to substances present at low levels in the diet. Food Chem Toxicol 2004;42(1):65–83.

[24] EFSA and WHO. Review of the threshold of toxicological concern (TTC) approach and development of new TTC decision tree. EFSA Supporting Publication 2016: EN-1006; 2016. p. 50.

[25] Nelms MD, Pradeep P, Patlewicz G. Evaluating potential refinements to existing threshold of toxicological concern (TTC) values for environmentally-relevant compounds. Regul Toxicol Pharmacol 2019;109:104505.

[26] Patlewicz G, Worth A, Yang C, Zhu T. Advances and refinements in the development and application of threshold of toxicological concern. Front Toxicol 2022;4.

[27] Lo Piparo E, Worth A, Manibusan M, Yang C, Schilter B, Mazzatorta P, et al. Use of computational tools in the field of food safety. Regul Toxicol Pharmacol 2011;60(3):354–62.

[28] Yang C, Barlow SM, Jacobs KLM, Vitcheva V, Boobis AR, Felter SP, et al. Thresholds of toxicological concern for cosmetics-related substances: new database, thresholds, and enrichment of chemical space. Food Chem Toxicol 2017;109:170–93.

[29] Watford S, Pham LL, Wignall J, Shin R, Martin MT, Friedman KP. ToxRefDB version 2.0: improved utility for predictive and retrospective toxicology analyses. Reprod Toxicol 2019;89:145–58.

[30] Kovarich S, Ceriani L, Ciacci A, Baldin R, Perez Miguel M, Gibin D, et al. OpenFoodTox: EFSA's chemical hazards database. Palma, Italy: EFSA; 2020.

[31] Dorne J, Richardson J, Livaniou A, Carnesecchi E, Ceriani L, Baldin R, et al. EFSA's OpenFoodTox: an open source toxicological database on chemicals in food and feed and its future developments. Environ Int 2021;146:106293.

[32] Pradeep P, Friedman KP, Judson R. Structure-based QSAR models to predict repeat dose toxicity points of departure. Comput Toxicol 2020;16: 100139.

[33] Gold LS, Sawyer CB, Magaw R, Backman GM, De Veciana M, Levinson R, et al. A carcinogenic potency database of the standardized results of animal bioassays. Environ Health Perspect 1984;58:9–319.

[34] Mumtaz M, Knauf L, Reisman D, Peirano W, DeRosa C, Gombar V, et al. Assessment of effect levels of chemicals from quantitative structure-activity relationship (QSAR) models. I. Chronic lowest-observed-adverse-effect level (LOAEL). Toxicol Lett 1995;79(1–3):131–43.

[35] Hisaki T, née Kaneko MA, Yamaguchi M, Sasa H, Kouzuki H. Development of QSAR models using artificial neural network analysis for risk assessment of repeated-dose, reproductive, and developmental toxicities of cosmetic ingredients. J Toxicol Sci 2015;40(2):163–80.

[36] Veselinović JB, Veselinović AM, Toropova AP, Toropov AA. The Monte Carlo technique as a tool to predict LOAEL. Eur J Med Chem 2016;116: 71–5.

[37] Toropova AP, Toropov AA, Veselinović JB, Veselinović AM. QSAR as a random event: a case of NOAEL. Environ Sci Pollut Control Ser 2015; 22(11):8264–71.

[38] Truong L, Ouedraogo G, Pham L, Clouzeau J, Loisel-Joubert S, Blanchet D, et al. Predicting in vivo effect levels for repeat-dose systemic toxicity using chemical, biological, kinetic and study covariates. Arch Toxicol 2018;92(2):587–600.

[39] Mazzatorta P, Estevez MD, Coulet M, Schilter B. Modeling oral rat chronic toxicity. J Chem Inf Model 2008;48(10):1949–54.

[40] Helma C, Vorgrimmler D, Gebele D, Gütlein M, Engeli B, Zarn J, et al. Modeling chronic toxicity: a comparison of experimental variability with (Q) SAR/read-across predictions. Front Pharmacol 2018;9:413.

[41] Gadaleta D, Marzo M, Toropov A, Toropova A, Lavado GJ, Escher SE, et al. Integrated in silico models for the prediction of no-observed-(adverse)-effect levels and lowest-observed-(adverse)-effect levels in rats for sub-chronic repeated-dose toxicity. Chem Res Toxicol 2020; 34(2):247–57.

[42] Bercu JP, Morton SM, Deahl JT, Gombar VK, Callis CM, van Lier RB. In silico approaches to predicting cancer potency for risk assessment of genotoxic impurities in drug substances. Regul Toxicol Pharmacol 2010;57(2–3):300–6.

[43] Contrera JF. Improved in silico prediction of carcinogenic potency (TD50) and the risk specific dose (RSD) adjusted threshold of toxicological concern (TTC) for genotoxic chemicals and pharmaceutical impurities. Regul Toxicol Pharmacol 2011;59(1):133–41.

[44] Lo P, Maunz A, Helma C, Vorgrimmler D, Schilter B. Automated and reproducible read-across like models for predicting carcinogenic potency. Regul Toxicol Pharmacol 2014;70(1):370–8.

[45] Thresher A, Gosling JP, Williams R. Generation of TD50 values for carcinogenicity study data. Toxicol Res 2019;8(5):696–703.

[46] Koster S, Rennen M, Leeman W, Houben G, Muilwijk B, van Acker F, et al. A novel safety assessment strategy for non-intentionally added substances (NIAS) in carton food contact materials. Food Addit Contam 2014;31(3):422–43.

[47] Koster S, Boobis AR, Cubberley R, Hollnagel HM, Richling E, Wildemann T, et al. Application of the TTC concept to unknown substances found in analysis of foods. Food Chem Toxicol 2011;49(8):1643–60.

[48] Debon E, Rogeboz P, Latado H, Morlock GE, Meyer D, Cottet-Fontannaz C, et al. Incorporation of metabolic activation in the HPTLC-SOS-Umu-C bioassay to detect low levels of genotoxic chemicals in food contact materials. Toxics 2022;10(9):501.

[49] Meyer D, Marin-Kuan M, Debon E, Serrant P, Cottet-Fontannaz C, Schilter B, et al. Detection of low levels of genotoxic compounds in food contact materials using an alternative HPTLC-SOS-Umu-C assay. ALTEX-Altern Anim Exp 2021;38(3):387–97.

[50] EFSA. Recent developments in the risk assessment of chemicals in food and their potential impact on the safety assessment of substances used in food contact materials. EFSA J 2016;14(1):4357.

[51] Van Bossuyt M, Raitano G, Honma M, Van Hoeck E, Vanhaecke T, Rogiers V, et al. New QSAR models to predict chromosome damaging potential based on the in vivo micronucleus test. Toxicol Lett 2020;329:80–4.

[52] Baderna D, Gadaleta D, Lostaglio E, Selvestrel G, Raitano G, Golbamaki A, et al. New in silico models to predict in vitro micronucleus induction as marker of genotoxicity. J Hazard Mater 2020;385:121638.

[53] Manganelli S, Gamba A, Colombo E, Benfenati E. Using VEGAHUB within a weight-of-evidence strategy. In: In silico methods for predicting drug toxicity. Springer; 2022. p. 479–95.

[54] Hasselgren C, Ahlberg E, Akahori Y, Amberg A, Anger LT, Atienzar F, et al. Genetic toxicology in silico protocol. Regul Toxicol Pharmacol 2019;107:104403.

[55] Manganelli S, Schilter B, Scholz G, Benfenati E, Lo Piparo E. Value and limitation of structure-based profilers to characterize developmental and reproductive toxicity potential. Arch Toxicol 2020;94(3):939–54.

[56] Myatt GJ, Ahlberg E, Akahori Y, Allen D, Amberg A, Anger LT, et al. In silico toxicology protocols. Regul Toxicol Pharmacol 2018;96:1–17.

[57] Schultz T, Amcoff P, Berggren E, Gautier F, Klaric M, Knight D, et al. A strategy for structuring and reporting a read-across prediction of toxicity. Regul Toxicol Pharmacol 2015;72(3):586–601.

[58] Patlewicz G, Helman G, Pradeep P, Shah I. Navigating through the minefield of read-across tools: a review of in silico tools for grouping. Comput Toxicol 2017;3:1–18.
[59] Patlewicz G, Cronin MT, Helman G, Lambert JC, Lizarraga LE, Shah I. Navigating through the minefield of read-across frameworks: a commentary perspective. Comput Toxicol 2018;6:39–54.

C H A P T E R

23

QSAR models for predicting in vivo reproductive toxicity

Jie Liu[1], *Fan Dong*[1], *Wenjing Guo*[1], *Zoe Li*[1], *Liang Xu*[1], *Meng Song*[1], *Tucker A. Patterson*[2] *and Huixiao Hong*[1]

[1]Division of Bioinformatics and Biostatistics, National Center for Toxicological Research, United States Food and Drug Administration, Jefferson, AR, United States [2]Office of Research, National Center for Toxicological Research, United States Food and Drug Administration, Jefferson, AR, United States

Abbreviations

ALogP Ghose-Crippen-Viswanadhan octanol–water partition coefficient
AUC Area under the receiver operating characteristic curve
C&L Classification and Labelling
CDK Chemistry Development Kit
DF Decision forest
DT Decision tree
ECFC Atom-type extended-connectivity fingerprint counts
ECFP Atom-type extended-connectivity fingerprint
ECHA European Chemicals Agency
EPA Environmental Protection Agency
EPFC Atom-type connectivity fingerprint counts
EPFP Atom-type daylight path-based fingerprints
FCFC Functional class extended-connectivity fingerprint counts
FCFP Functional class extended-connectivity fingerprints
FPFC Functional class daylight path-based fingerprint counts
FPFP Functional class daylight path-based fingerprints
kNN k-nearest neighbors
LCFC ALogP-type extended-connectivity fingerprint counts
LCFP ALogP extended-connectivity fingerprint
LDA Linear discriminant analysis
LOAEL Lowest observed adverse effect level
LR Logistic regression
MCC Matthews correlation coefficient
NB Naïve Bayes
NPR Negative predictive rate
OECD Organisation for Economic Co-operation and Development
PND Postnatal day
PPR Positive predictive rate
QSAR Quantitative structure–activity relationship
RF Random forest
SVM Support vector machine
ToxRefDB Toxicity reference database
XGBoost Extreme gradient boosting

Introduction

Reproductive toxicity is one of the 14 categories of toxicity that has caused drug withdrawal and attrition [1]. Therefore, assessing reproductive toxicity is required for safety evaluation by regulatory agencies and industry before products enter the market. Reproductive toxicity can be measured with various endpoints related to the reproductive systems, including the alternations in sexual behavior, fertility, estrous cycle, and other adverse effects in males and females [2,3]. The most used method for assessment of reproductive toxicity is in vivo animal testing. The Organisation for Economic Co-operation and Development (OECD) has made animal study guidelines for several reproductive toxicity tests to standardize and regulate these studies [4].

The study design of reproductive toxicity testing is very complicated. As an example, the general dosing schedule of a multigeneration reproductive toxicity study is outlined in Fig. 23.1 [5,6]. The dosing of test compounds starts on postnatal day (PND) n of the first generation of male and female animals. The first generation of animals are observed for adverse effects and sacrificed on PND m. The second, third, and fourth generations of animals are

QSAR in Safety Evaluation and Risk Assessment
https://doi.org/10.1016/B978-0-443-15339-6.00013-8

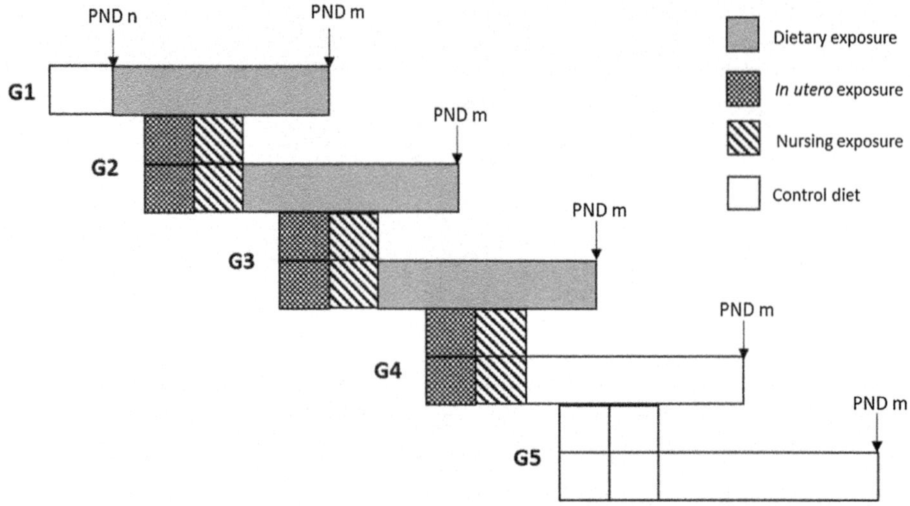

FIGURE 23.1 Design of chemical dosing for multigeneration reproductive toxicity study. G1 indicates the first-generation (parent) animals. G2, G3, G4, and G5 generations are separated into two groups to observe the parental and offspring toxicity endpoints. Dietary exposure started on postnatal day n (PND-n) and the adult animals were sacrificed on postnatal day m (PND-m).

separated into two groups. One group is used to observe toxic effects on their parental animals and to produce the next-generation animals. The other group is used for observing the adverse effects in themselves as the offspring. With multiple generations, male and female groups, and dosing groups, in vivo reproductive toxicity testing requires many animals and may take several years for multiple generations, which is extremely expensive and time-consuming [7,8]. Moreover, it is difficult to interpret the complicated effects and mechanisms, and to extrapolate the results from animals to humans [9]. Therefore, it is difficult to perform in vivo reproductive toxicity testing on such a large number of compounds found in products on the market and in the environment. There are also increasing ethical concerns with the use of animals in research. The 3R's (replacement, reduction, and refinement) principle was proposed to instruct the usage of animals. Replacement is to apply nonanimal approaches to substitute for animal use. Reduction is to minimize animal use through optimizing the study design. Refinement is to reduce the pain on experimental animals [10–12]. Therefore, the need exists for regulatory agencies and industry to seek nonanimal alternative methods for reproductive toxicity testing.

In vitro and in silico approaches have been proposed as alternatives for reproductive toxicity testing in animals. There are many in vitro models established for reproductive toxicity endpoints, such as the testis-on-a-chip [13], the ex vivo human placenta model [14], and the in vitro ovary and follicle culture [15]. However, it is impossible to build the in vitro models for the entire reproductive system due to its complex mechanisms and involvement in multiple organs [4,12,16]. Therefore, computational models have become an attractive alternative method for in vivo reproductive toxicity testing. Quantitative structure–activity relationship (QSAR) models are based on the principle that compounds having similar structural features have similar biological activities. QSAR models are constructed using numerical molecular descriptors, which are transformed from chemical structures [17,18]. Several QSAR models have been built for potential reproductive toxicity prediction [19–22]. Moreover, there are guidelines for the regulatory use of QSAR models. Comparative analysis on QSAR models for reproductive prediction from some tools using pesticides revealed that most of the testing pesticides are out of the chemical space of the models, the definitions of reproductive toxicity vary in the data in training the models, and structure similarity is not sufficiently defined, indicating the need for improvement of the QSAR models [23].

This chapter will review the QSAR models developed for in vivo reproductive toxicity prediction, including data curation, data processing, feature selection, machine learning algorithms, and model validation used in the models. The applications of available tools for QSAR models in reproductive toxicity prediction will also be discussed.

QSAR models based on ECHA-C&L inventory

Dataset

Chemicals with reproductive toxicity data were curated from the European Chemicals Agency (ECHA)-Classification and Labelling (C&L) Inventory (ECHA-C&L) (https://echa.europa.eu/information-on-chemicals/cl-inventory-database) and OECD-eChemPortal (https://www.echemportal.org/echemportal/). ECHA-C&L is a publicly available database with registered substances and their hazard category information. The toxicants cause multiple reproductive toxicity endpoints. Substances with reproductive toxicity are assigned to two categories, category 1 and category 2. Category 1 includes two subcategories. Category 1A contains known reproductive toxicants with adverse effects on humans. Category 1B represents presumed reproductive toxicants with adverse effects based on animal studies. Category 2 indicates the suspected reproductive toxicants with the adverse effects from humans or animal studies.

After removing mixtures, inorganic salts, metals, and duplicate compounds, 1823 compounds were used for model construction of reproductive toxicity in multiple studies [20,21,24]. This dataset has a large number of compounds with significant diversity and contains a series of reproductive toxicity endpoints. However, the quality of this dataset needs to be assessed because the accuracy of the classification and labeling information submitted by industry to the ECHA-C&L are not verified by ECHA. Moreover, this dataset was curated from data mining and has not been reviewed by experts for quality control.

Models based on fingerprints

Jiang et al. [20] constructed binary classification models for reproductive toxicity prediction based on this dataset. Nine molecular fingerprints were calculated for the 1823 chemicals using PaDEL-Descriptor software [25], including MACCS fingerprint, the Chemistry Development Kit (CDK) fingerprint, CDK extended fingerprint, CDK graph only fingerprint, Atom Pairs 2D fingerprint, PubChem fingerprint, Substructure fingerprint, Estate fingerprint, and Klekota-Roth fingerprint. The 1823 chemicals were randomly split into the training (80%) and external validation set (20%). Classification models were built using six machine learning algorithms including support vector machine (SVM), random forest (RF), artificial neural network (ANN), k-nearest neighbors (kNN), decision tree (DT), and naïve Bayes (NB). Performance of the developed models was evaluated by 10-fold cross-validation and external validation.

Prediction accuracies of the 54 classification models generated using the six machine learning algorithms based on the nine datasets with different molecular descriptors in the cross-validations and external validations are shown in Fig. 23.2. As shown, there were large variations in the performance of the 54 machine learning models for both cross-validations with accuracy 65.7%−86.9% and the external validations with accuracy of 67.9%−83.8%. The SVM model built with PubChem fingerprint best performed in the cross-validations with the highest accuracy of 86.9%, while the SVM model constructed with Klekota-Roth fingerprint achieved the best performance (83.8% accuracy) in the external validation. Furthermore, the performance variation in the machine learning algorithms is larger than the performance variation in the fingerprints. SVM outperformed the other algorithms in both cross-validations and external validations, while the NB models performed worse in both cross-validations and external validations. Therefore, selection of machine learning algorithms such as SVM is vital for constructing highly performing models for predicting reproductive toxicity. Interestingly, for the same machine leaning algorithm, the predictive models built with different fingerprints performed similarly in both cross-validations and external validations except for the models generated with Atom Pairs 2D fingerprint and CDK graph-only fingerprint, which performed much worse than the other models. Therefore, selection of fingerprints may not be a concern in the development of machine learning models for predicting reproductive toxicity of chemicals. However, caution is advised when using Atom Pairs 2D fingerprint and CDK graph-only fingerprint as the representation of chemicals.

Models based on combination of molecular descriptors and fingerprints

Zhang et al. [24] used the same ECHA-C&L Inventory dataset. From the 1823 compounds, metal−organic chemicals, inorganic compounds, and compounds with less than three carbon atoms were removed. The remaining 1685 compounds were used for developing machine learning models. The dataset was first randomly split into a training set (75%) and validation set (25%). An external validation set of 94 chemicals was curated from the toxicity reference database (ToxRefDB) version 1 [26]. Molecular descriptors and fingerprints were calculated for all chemicals in the training set, validation set, and external validation set using Discovery Studio. Eleven maximum diameters were used for each fingerprint. The maximum diameters of 0, 1, 2, 3, 4, 5, 6, 7, 8, 9, and 10 were applied to the fingerprints, atom-type connectivity fingerprint counts (EPFC), functional class daylight path-based fingerprints (FPFP), functional class daylight path-based fingerprint counts (FPFC), and atom-type daylight path-based fingerprints (EPFP). The other fingerprints, Ghose-Crippen-Viswanadhan octanol−water partition coefficient (ALogP) types extended-connectivity fingerprint counts (LCFC), atom-type extended-connectivity fingerprint counts (ECFC), functional class extended-connectivity fingerprints (FCFP), atom-type extended-connectivity fingerprints (ECFP), ALogP extended-connectivity fingerprint (LCFP), and functional class extended-connectivity fingerprint counts (FCFC) used the maximum diameters of 0, 2, 4, 6, 8, 10, 12, 14, 16, 18, 20. After feature selection, 6 molecular

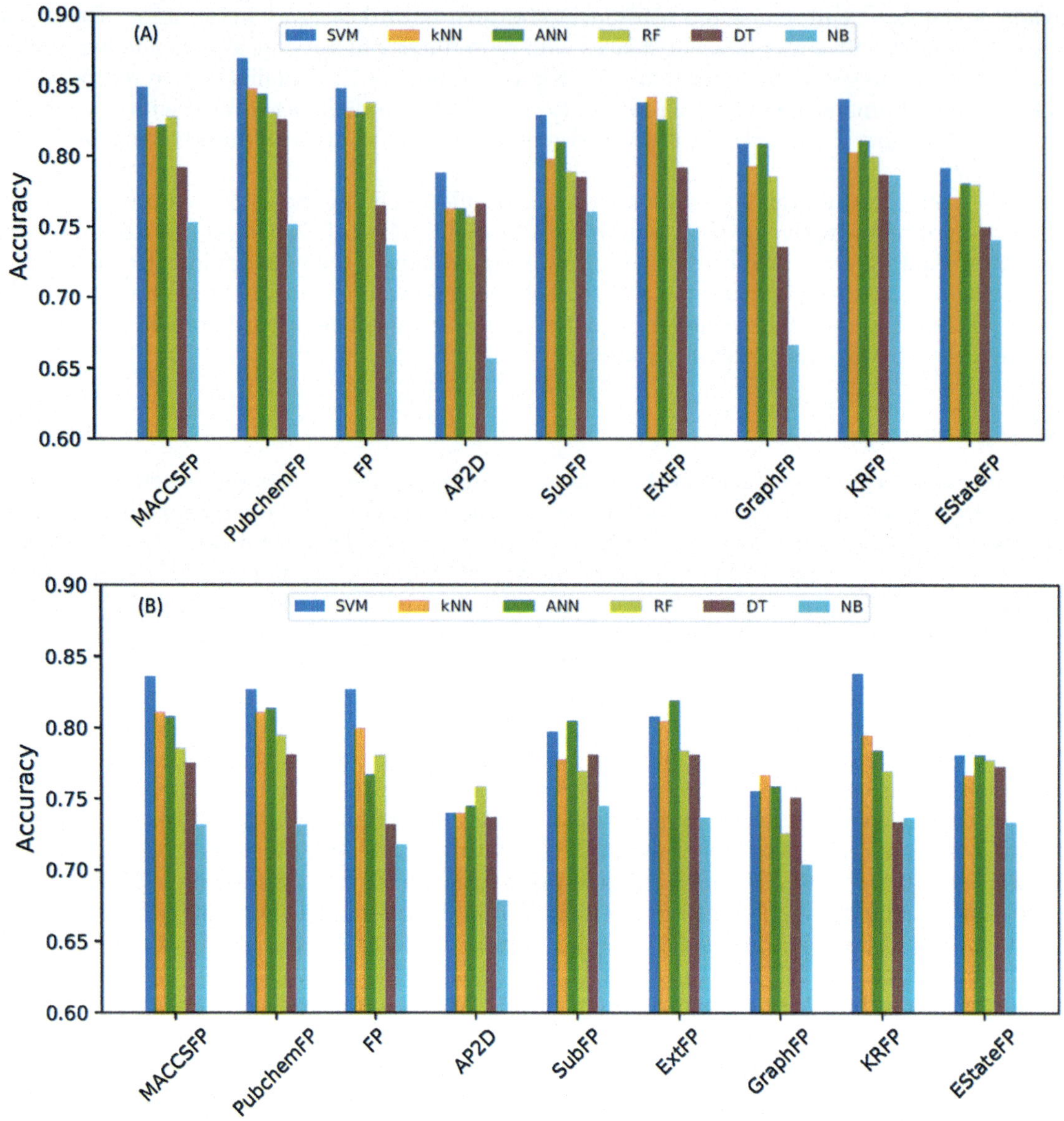

FIGURE 23.2 Accuracy of cross-validations (A) and external validations (B). The x-axis represents 9 types of molecular fingerprints. The bars represent models from different machine learning algorithms. *ANN*, artificial neural network; *DT*, decision tree; *kNN*, k-nearest neighbors; *NB*, naïve Bayes; *RF*, random forest; *SVM*, support vector machine.

descriptors and 10 types of fingerprints were kept for model development. The naïve Bayes (NB) models were constructed using six molecular descriptors and one of the 10 types of molecular fingerprints with 11 maximum diameters, respectively. The fivefold cross-validation was conducted using the training set to evaluate performance. The NB model constructed based on the training set was evaluated using the validation set and the external validation set. Accuracy, sensitivity, specificity, balanced accuracy, positive predictive rate (PPR), negative predictive rate (NPR), and Matthews correlation coefficient (MCC) were calculated for performance evaluation. Among the 110 constructed models, the NB model generated using the six molecular descriptors and fingerprint LCFC_20 achieved the best performance with accuracy values 0.918, 0.83, and 0.851 from the fivefold cross-validations, predicting the validation set and predicting the external validation set, respectively.

MCC was the metrics used to examine performance difference across the fingerprints and maximum diameters. MCC values from the 110 NB models constructed with combinations of the six molecular descriptors and 10 fingerprints at 11 diameters are summarized in Fig. 23.3. The average MCC values and standard deviations of the 11 NB models for each of the 10 fingerprints at 11 maximum diameters were plotted in Fig. 23.3A. It can be seen from Fig. 23.3A that the NB models built with different fingerprints vary in performance at a quite large range (average

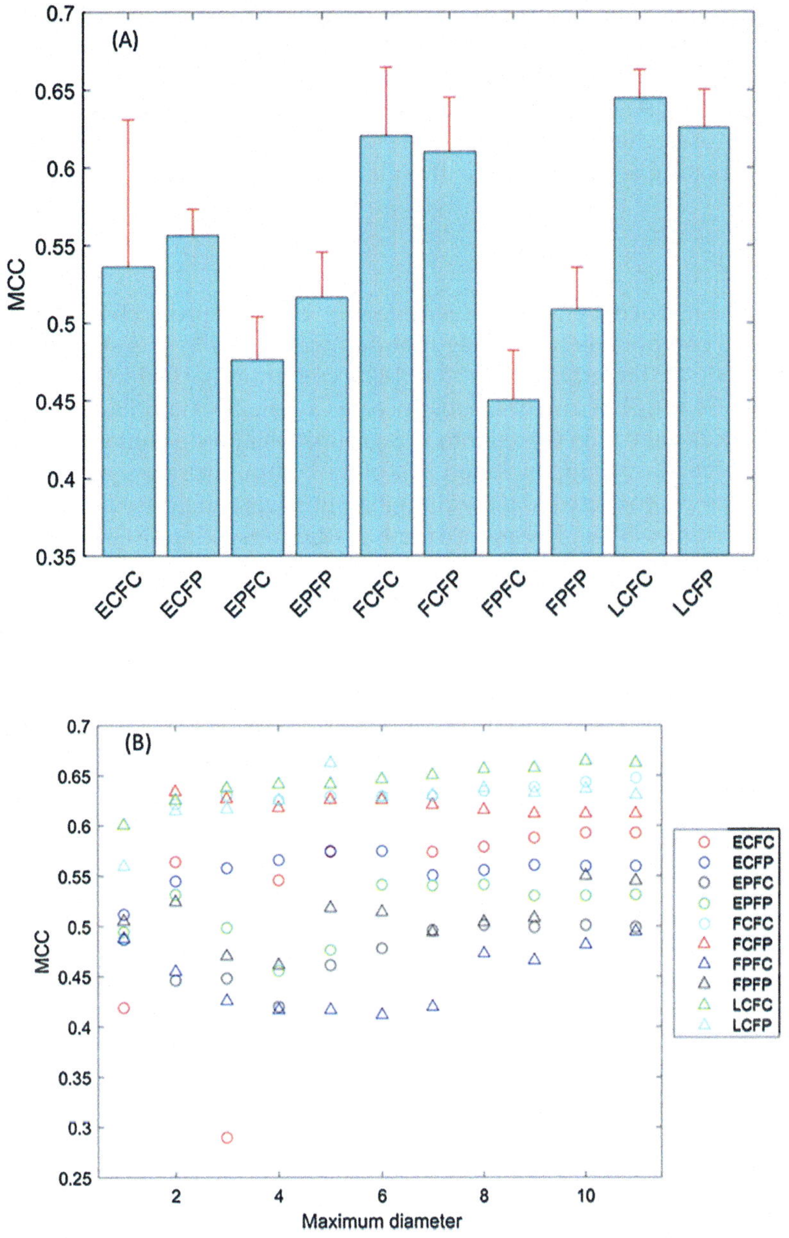

FIGURE 23.3 Matthews correlation coefficient (MCC) values from models built with 10 fingerprints (A) and with 11 maximum diameters of each fingerprint (B).

MCC from 0.441 to 0.635). Though the models built with fingerprints LCFC, LCFP, FCFC, and FCFP outperformed the models built with the other 6 fingerprints, the model built with fingerprint LCFC reached the highest average MCC, while the model constructed with fingerprint FPFC yielded the lowest average MCC. Comparative analysis on the MCC values from each pair of fingerprints and maximum diameter found that the variation in diameters is smaller than that in fingerprints (Fig. 23.3B), indicating that selection of fingerprints is more crucial than determination of diameters. For each fingerprint, an upward trend in MCC values with the increasing maximum diameters was observed. Moreover, the models built with fingerprint LCFC (the green up triangles in Fig. 23.3B) outperformed other models, and the model generated with fingerprint LCFC at diameter 20 achieved the highest MCC value. To better understand the model for predicting reproductive toxicity of chemicals, informative substructures were identified from fingerprint LCFC at diameter 20 for the reproductive toxicants and nontoxicants. 20 structural features were identified associated with reproductive toxicants, while another 20 structural features were found to be related to reproductive nontoxicants. The results indicate that the substructures from fingerprint LCFC at diameter 20 had good coverage on the common features for the prediction of reproductive toxicity.

Chemical diversity in training compounds is crucial for building a robust model since the prediction on a new chemical is based on its structural similarity with the compounds in the training set. The OECD principle on QSAR model evaluation requires that a QSAR model needs a predefined applicability domain [27]. This study defined the chemical space using ALogP and molecular weight. The training set and the two test sets were found to have a high similarity in this chemical space. The results indicated that the reproductive toxicity prediction models built based on the training set have good extrapolations to chemicals in the test sets.

Ensemble models

Recently, Feng et al. [21] constructed ensemble learning models for reproductive toxicity prediction. The reproductive toxicity data of 1823 compounds previously reported were used in this study. PaDEL Descriptor [25] was used to calculate nine types of fingerprints for the 1823 compounds, including PubChem, MACCS, Estate, Klekota-Roth, Klekota-Roth Count, Substructure, Substructure Count, 2D Atom Pairs, and 2D Atom Pairs Count fingerprints. The 1823 compounds were then placed into two groups: one group contained 80% of the compounds used as training compounds, and the other group included 20% of the compounds used as the validation compounds. The training compounds and their reproductive data and fingerprints were then used to train predictive models using three machine learning algorithms, SVM, RF, and extreme gradient boosting (XGBoost). The 27 individual models were combined to construct 27 ensemble models by combining a part of the individual models that produced more accurate predictions. Performance of those models was evaluated using100 iterations of fivefold cross-validations and external validations.

For the cross-validations, the ranges of accuracy, sensitivity, specificity, and area under the receiver operating characteristic curve (AUC) from the individual models were 0.771−0.859, 0.692−0.833, 0.804−0.900, and 0.831−0.929, respectively. The best ensemble model achieved 0.863 accuracy, 0.820 sensitivity, 0.902 specificity, and AUC 0.937. The performance metrics values of the best individual models from different machine learning algorithms and the top ensemble model in the fivefold cross-validations were plotted in Fig. 23.4A. More specifically, the model built using SVM based on the PubChem fingerprint reached the highest accuracy of 0.859 with 0.833 sensitivity, 0.880 specificity, and AUC 0.920. Among the RF models, the model constructed using PubChem fingerprint performed the best with an accuracy of 0.858. Comparing the performance of the developed XGBoost models revealed that the model generated with PubChem fingerprint also produced the highest accuracy of 0.845. The performance metrics values of models in the external validations were plotted in Fig. 23.4B. Interestingly, though the XGBoost models did not perform well in the fivefold cross-validations (Fig. 23.4A), they outperformed the models built using SVM and RF (Fig. 23.4B). Moreover, the ensemble models showed better prediction performance with higher metrics values.

Comparing with the previously reported models using the same dataset [20,24], the models from this study achieved better performance. Though good prediction performance was reported, the reproductive data reliability should be assessed since there were no quality control and expert review of this dataset. As an example of the quality issues of this dataset, ethanol was assigned as a reproductive toxicant in this dataset based on a study with inappropriate design [28], which was to compare the reproductive toxicity of ethanol and ginger extract. The study used a lower dose 1 g/kg for ginger extracts and an extremely high dose of 4 g/kg ethanol and concluded that ethanol caused reproductive toxicity compared with the control group and ginger extracts. Though adverse effects were observed in this study, the conclusion was not reliable. It is well known that everything has potential toxicity and the dose determines if it is poison or not [29,30].

QSAR models based on ToxRefDB

ToxRefDB

ToxRefDB is a public database developed by the US Environmental Protection Agency (EPA). It is a major resource of legacy in vivo animal toxicity study data. Guideline studies in data evaluation records submitted by industry to EPA's Office of Pesticide Programs for review were extracted and manually entered into the database [31−33]. ToxRefDB contains many types of toxicity testing studies (subchronic testing, chronic testing, multigeneration reproductive toxicity testing, etc.), toxicity endpoints, toxicity target organs, etc. To establish QSAR models for reproductive toxicity prediction, multigeneration reproductive toxicity data were extracted. The quality of reproductive toxicity data from guideline animal studies are believed to be high; thus, models established based on such data

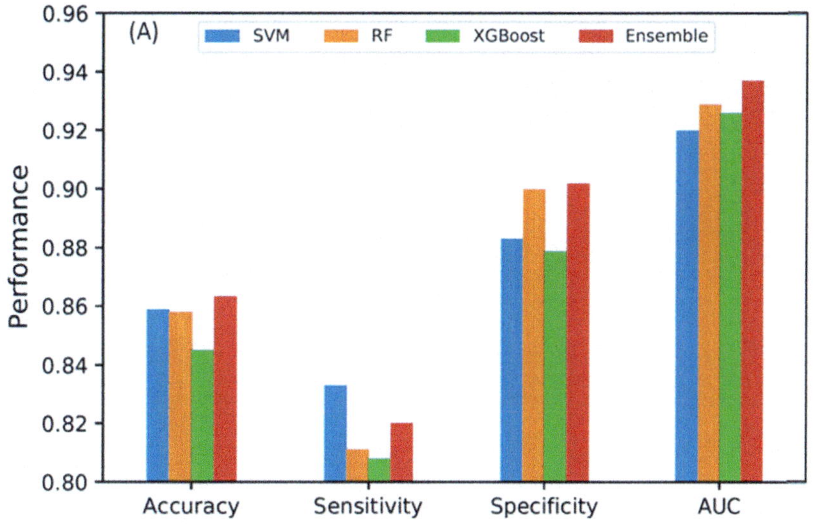

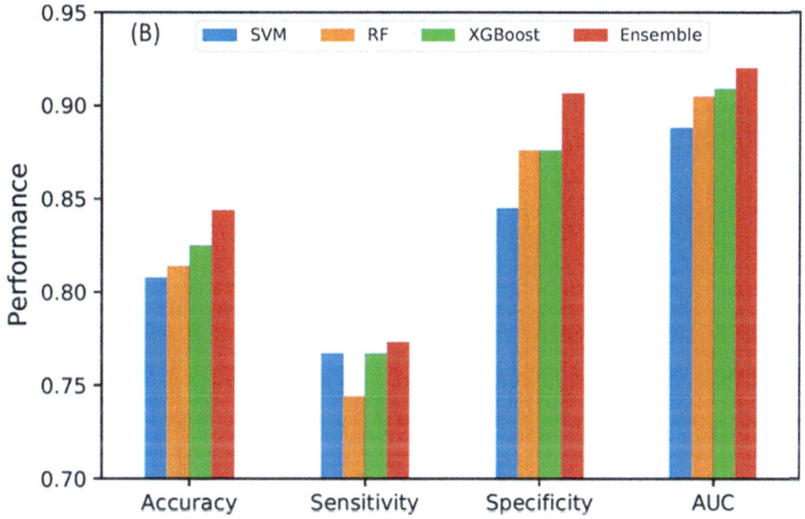

FIGURE 23.4 Top performance of individual and ensemble models from cross-validations (A) and external validations (B). The x-axis labels the performance metrics. The bars represent models from different machine learning algorithms. *AUC*, area under the receiver operating characteristic curve; *RF*, random forest; *SVM*, support vector machine; *XGBoost*, extreme gradient boosting.

should be reliable. However, most compounds in ToxRefDB are pesticides. Therefore, the relatively small chemical space of the compounds that were used to train models may limit the prediction power of the developed models in reproductive toxicity prediction on other types of chemicals.

Regression models

To develop reliable QSAR models following OECD principles for predicting potential lowest observed adverse effect level (LOAEL), Basant et al. [34] performed the first attempt to develop regression models for reproductive toxicity prediction. Compounds with LOAEL values at rat multigeneration reproductive toxicity endpoints were collected from ToxRefDB. After removing duplicates, mixtures, and salts, 334 chemicals remained for model development. For chemicals that have more than one LOAEL value in ToxRefDB, the LOAEL value was used in this study. The distribution of the LOAEL values of the 334 chemicals is shown in Fig. 23.5A. This distribution is very different from a normal distribution, and the LOAEL values are not suitable for developing regression models. Therefore, the LOAEL values were transformed by logarithm of base 10 before the dataset was used for model development. The distribution of the logarithm transformed data is given in Fig. 23.5B, which is close to a normal distribution and

FIGURE 23.5 Distribution of LOAEL values (A) and log10(LOAEL) values (B). The x-axis represents the LOAEL or log10(LOAEL) values. The y-axis indicates the number of chemicals.

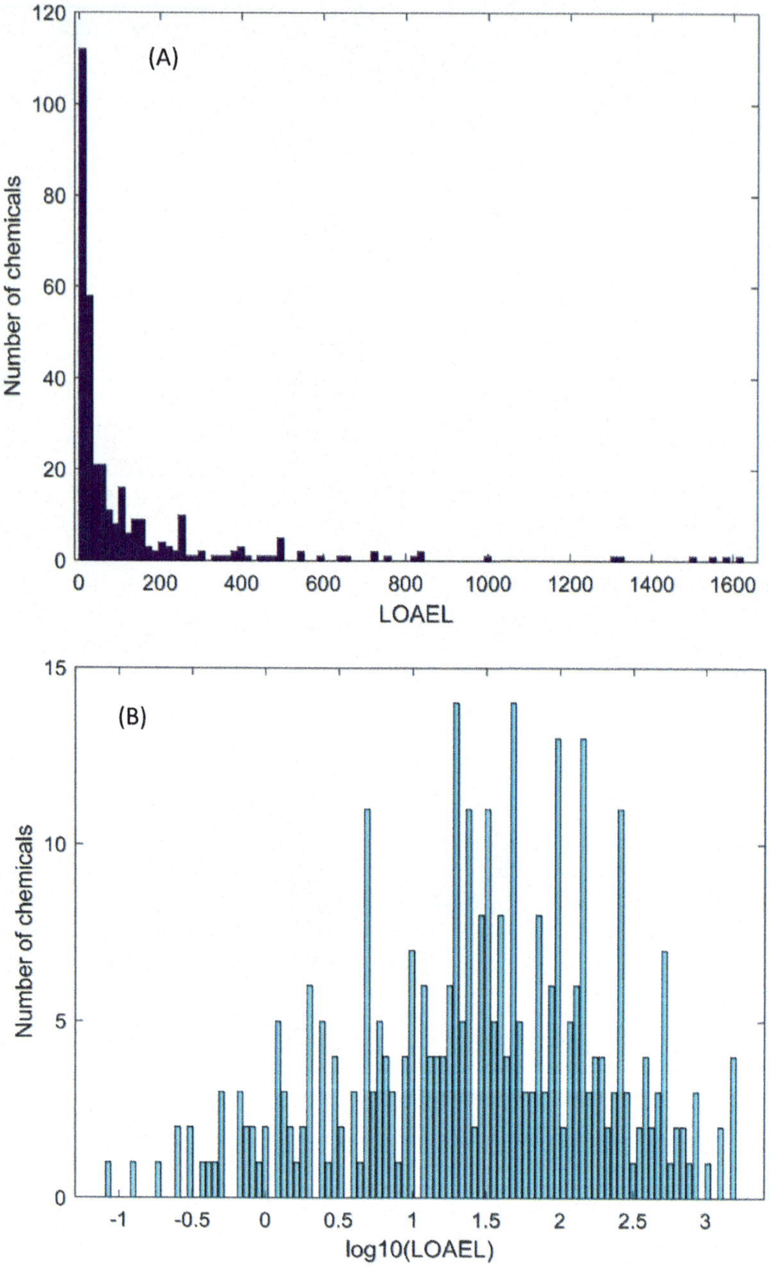

should be good for developing regression models. Chemopy [35] was used to calculate 633 molecular descriptors for each of the 334 chemicals. The 334 chemicals were divided into two groups with an 8:2 ratio. The group of 267 chemicals was used as the training dataset, and the other group of 67 chemicals served as the external validation data set. Regression models were constructed using decision tree forest and decision tree boost. The models were evaluated using 10-fold cross-validations and external validation. The R^2 values from the external validations were 0.856 for the decision tree forest model and 0.945 for the decision tree boost model. The good performance of the developed models in the external validations suggested the models could be used for screening chemicals with potential multi-generation reproductive toxicity.

The quality and representation of input data influence the performance of in silico models developed based on the data. Therefore, data preprocessing such as data cleaning, data integration, and transformation is a required step and critical for model development [36]. In this study, the distribution of original LOAEL values was extremely skewed from a normal distribution as shown in Fig. 23.5A. With a base-10 logarithmic transformation, the reproductive toxicity data represented by the transformed values were closed to a normal distribution as shown Fig. 23.5B, which should contribute to the good performance of the models in the external validation.

Classification models

Classification models were generated by Liu et al. [22] to provide an approach for screening and prioritizing chemicals with potential rat multigeneration reproductive toxicity. High-quality (acceptable and complete) rat multigeneration reproductive toxicity data were collected from ToxRefDB. The LOAEL values were observed on the three effect categories, parental, offspring, and reproductive [37]. "Parental" category contained systemic toxicity, such as body weight, organ weight, and relative weight to liver. "Offspring" category included most of the developmental toxicity endpoints, such as offspring weight. "Reproductive" category had reproductive toxicity endpoints, such as fertility, sperm morphology, and ovary weight. Therefore, this study only curated the data in the reproductive category as the reproductive toxicity effects and the LOAEL values based on the reproductive category used to classify chemicals as the reproductive toxicants and nontoxicants. The rat multigeneration reproductive toxicity data of 275 compounds were curated as the training dataset. Additionally, the rat multigeneration reproductive toxicity data of 29 chemicals were curated from COSMetics to Optimise Safety (COSMOS) database (https://archive.cosmostox.org/home/welcome/) and literature as the external validation set. Then, 777 molecular descriptors were calculated using Mold2 [38,39] software for all the chemicals in the training and external testing datasets. Finally, 34 informative descriptors were selected for model construction using Shannon entropy [40] analysis and importance of descriptors in the decision tree–based models.

The models were constructed using seven machine learning algorithms, including decision forest (DF), DT, SVM, linear discriminant analysis (LDA), RF, kNN, and logistic regression (LR). Moreover, a consensus model was constructed from the seven machine learning models using the majority voting strategy. The individual machine learning models and the consensus model were validated using fivefold cross-validations and external validation. Prediction accuracy, sensitivity, specificity, balanced accuracy, PPR, NPR, and MCC from cross-validations were plotted in Fig. 23.6. The validation results indicated the predictive power of the developed models. It is worth noting that a small variation in validation performance among the models built using different machine learning algorithms suggested that the algorithms had no big influence on model performance. Moreover, the consensus model achieved better performance than the seven individual models. Not surprising, the performance of cross-validations was slightly better than the performance of external validation. The analysis of chemical space between chemicals in the training dataset and chemicals in the external validation set provided a possible interpretation for the differences in prediction performances. Moreover, most of the chemicals in the training set are pesticides, while the chemicals in the external validation set are primarily cosmetics and drugs.

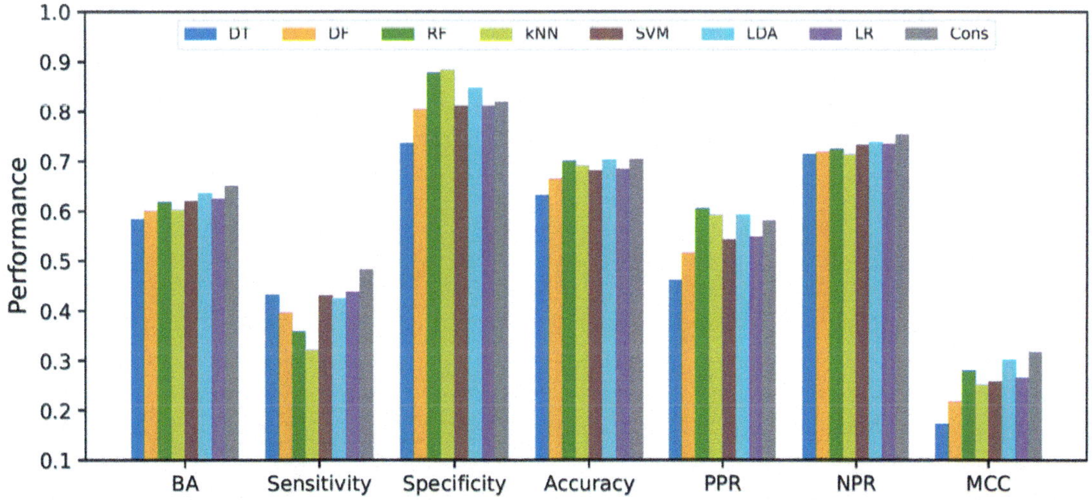

FIGURE 23.6 Performance of fivefold cross-validations. The x-axis indicates seven performance metrics. Each bar represents the average performance metrics from the 500 iterations of fivefold cross-validations. *BA*, balanced accuracy; *Cons*, consensus model; *DF*, decision forest; *DT*, decision tree; *kNN*, k-nearest neighbors; *LDA*, linear discriminant analysis; *LR*, logistic regression; *MCC*, Matthews correlation coefficient; *NPR*, negative predictive rate; *PPR*, positive predictive rate; *RF*, random forest; *SVM*, support vector machine.

QSAR models based on P&G and leadscope

Marzo et al. [19] developed naïve Bayesian models for reproductive toxicity prediction using two datasets from Procter and Gamble (P&G) [41] and the Leadscope database [42]. After removing mixtures, 283 chemicals (positive/negative = 228/55) were collected from the P&G dataset and 1101 chemicals (positive/negative = 328/773) were taken from the Leadscope database. These two datasets were combined, and the chemicals with the opposite reproductive toxicity labels were removed. Finally, the dataset for model development contains 1172 chemicals (positive/negative = 434/738) after removing chemicals with only one carbon atom, inorganic, and outlier compounds. For model development, the dataset was randomly split into three data sets: training dataset (752 chemicals), test dataset (187 chemicals), and validation set (233 chemicals). The extended-connectivity fingerprint of diameter 2 was calculated for all compounds using CDK [43]. The KNIME platform was then used for model development and for calculating the probability scores of positive and negative predictions for each compound. The chemical was predicted to class positive or class negative if the probability score to class positive or class negative is higher. The prediction performances on the training dataset, test dataset, and validation set resulted in balanced accuracy values of 0.72, 0.62, and 0.61, respectively. Moreover, for all three datasets, the sensitivity was lower than 0.5 and the specificity was higher than 0.9, which was likely due to the imbalance in the distribution of positives and negatives of the datasets.

Analysis was performed to determine the best probability score threshold that can be used to accurately classify the compounds to positive or negative classes. Different thresholds were explored to find the best one that resulted in the maximum MCC, balanced accuracy, or true predictions from training or test datasets. With the best threshold, the AUC values from the best three models reached 0.87–0.89 and 0.74–0.76 for the training dataset and test dataset, respectively. As a comparison, without using the threshold, these three models had AUC values of 0.84–0.85 and 0.72 for the training data set and test data set, respectively. As shown in Fig. 23.7, the models with the best threshold for probability score outperformed the models without the best threshold. Therefore, using the best threshold for probability score was important for the prediction performance, especially for the imbalanced data set.

QSAR model application

OECD has collected QSAR models on chemical toxicity prediction since 2000 [4] and proposed OECD guidelines of QSAR models for regulatory applications. The OECD guidelines suggest that a QSAR model needs to have a defined applicability domain, a defined endpoint, a clear algorithm for model development, performance metrics for evaluating the predictions, and an interpretation for the established model [27,44]. The US FDA and EPA also have taken great efforts to explore approaches for utilizing in silico methods for screening potential toxicants.

Software tools are available for building QSAR models to predict reproductive toxicity. Weyrich et al. [23] evaluated the available models from the software tools using the guideline study data from 310 pesticides. The available

FIGURE 23.7 AUC values of the three best models with and without threshold of probability score. The threshold was determined by balanced accuracy for Model-1 and by MCC for Model-2 and Model-3.

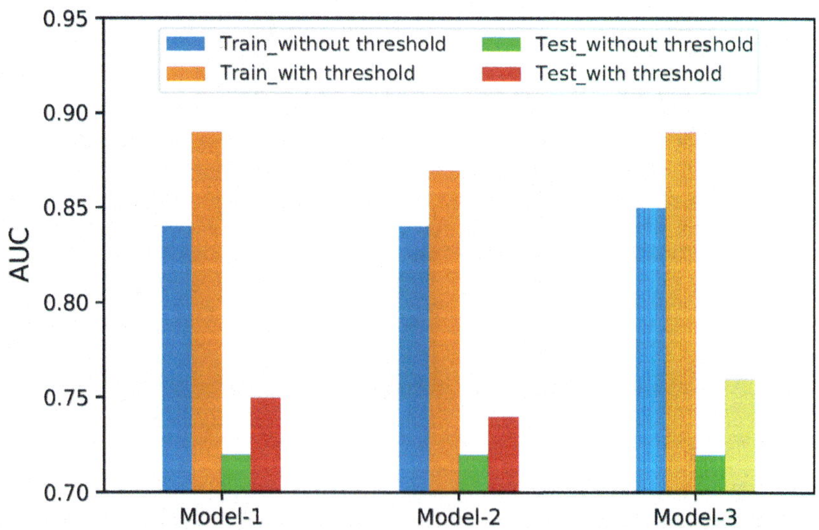

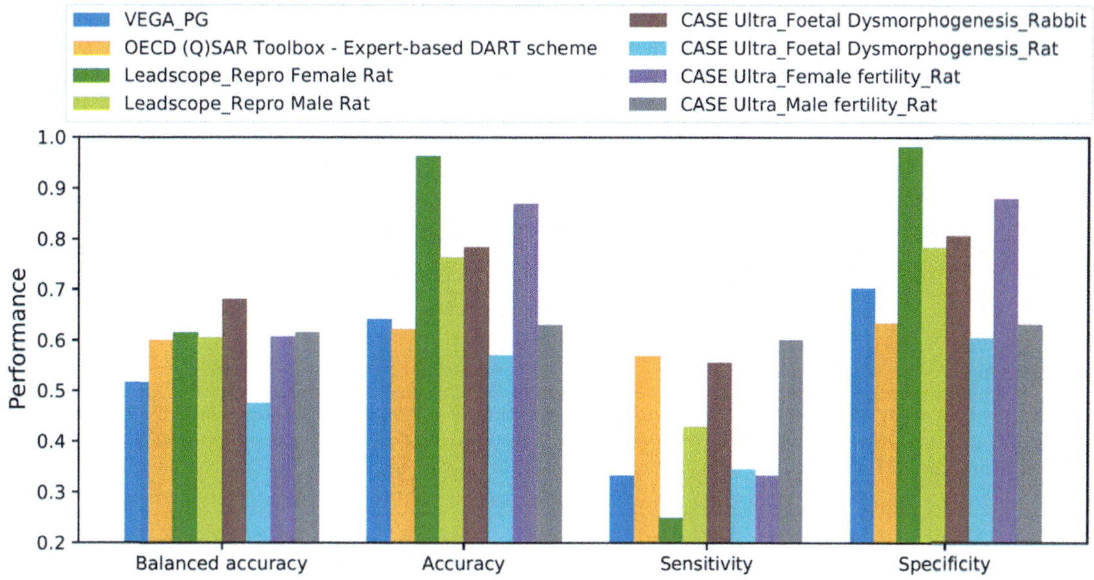

FIGURE 23.8 Performance of reproductive toxicity prediction from the QSAR models provided in the software tools.

QSAR models for reproductive toxicity prediction include VEGA_PG, OECD (Q)SAR Toolbox_Expert-based developmental and reproductive toxicity scheme, Leadscope_Repro Female Rat, Leadscope_Repro Male Rat, and CASE Ultra models (DYSM_Rabbit, FDYSM_Rat, FFERT_Rat, and MFERT_Rat). VEGA_PG is an expert rule-based model built with DT algorithm. This model classified two-thirds of the reproductive toxicants as nontoxicants since the chemical space of pesticides is different from the chemical space of compounds used in training the model. OECD (Q)SAR Toolbox_Expert-based developmental and reproductive toxicity scheme is also an expert rule-based model [41]. Leadscope_Repro Female Rat is an ensemble QSAR model for predicting toxicity of chemicals to female reproductive organs and fertility. Leadscope_Repro Male Rat is a QSAR model for predicting toxicity of chemicals to male reproductive organs and fertility. CASE Ultra models are structure alert models. The performance metrics values of the tested models in the reproductive toxicity prediction on the 310 pesticides were plotted in Fig. 23.8. The balanced accuracy, accuracy, sensitivity, and specificity are in ranges of 0.52–0.68, 0.57–0.96, 0.25–0.6, and 0.61–0.98, respectively. The results showed that the models performed quite different on predicting reproductive toxicity of pesticides, and it is difficult to suggest which model is suitable for predicting reproductive toxicity of pesticides. It is noted that the models could not be compared directly since they were different types (expert rule-based or statistical model), had a different training data set, and had different sets of reproductive toxicity endpoints.

In this case study, the majority of the pesticides fell outside of the chemical spaces of the models. Moreover, the rules to define reproductive toxicants and nontoxicants were different among the various models. Additionally, different molecular descriptors were used in the models, which might affect prediction performances. Therefore, prediction performances of the models were not good and consistent. However, the integration of different models might improve prediction performance since all models showed some predictive capability.

Conclusions

Reproductive toxicity plays an important role in safety evaluation of drugs, cosmetics, and other products as well as in risk assessment of environmental chemicals. Evaluation of reproductive toxicity of chemicals is crucial for protecting public health. Traditional reproductive toxicity testing in animals is costly and time-consuming, which makes it impossible to perform animal testing on a large amount of chemicals. In vitro and in silico methods were explored as nonanimal alternative approaches for assessing reproductive toxicity. It is not feasible to establish in vitro models for the entire reproductive system for all chemicals used in products and present in the environment. Therefore, machine learning is a promising approach. QSAR models have been established using limited reproductive toxicity data and showed prediction power in potential reproductive toxicity prediction. The developed QSAR models could be used to screen and prioritize potential reproductive toxicants for in vitro and animal testing. However, comparing with QSAR models for predicting other toxicity endpoints [45–53], there are some challenges and limitations for the developed reproductive toxicity prediction QSAR models. First, only a small amount of reproductive toxicity data is available for training reliable and accurate models. Second, it is difficult to interpret and ensure the results on the complicated reproductive endpoints and mechanisms. Third, a big chemical space formed by a large number of structurally diverse chemicals is important for constructing an accurate and robust model for predicting reproductive toxicity. Therefore, more efforts are needed to improve in silico methods for reproductive toxicity prediction. One way could be to integrate different models and data resources for improving the performance of reproductive toxicity prediction models [54,55].

Acknowledgments

This work was supported in part by an appointment to the Research Participation Program at the National Center for Toxicological Research (Fan Dong, Zoe Li, Meng Song), administered by the Oak Ridge Institute for Science and Education through an interagency agreement between the US Department of Energy and the US Food and Drug Administration.

Disclaimer

This chapter reflects the views of the authors and does not necessarily reflect those of the U.S. Food and Drug Administration.

References

[1] Siramshetty VB, Nickel J, Omieczynski C, Gohlke BO, Drwal MN, Preissner R. WITHDRAWN—a resource for withdrawn and discontinued drugs. Nucleic Acids Res 2016;44(D1):D1080–6.
[2] Novic M, Vracko M. QSAR models for reproductive toxicity and endocrine disruption activity. Molecules 2010;15(3):1987–99.
[3] Matthews EJ, Kruhlak NL, Daniel Benz R, Contrera JF. A comprehensive model for reproductive and developmental toxicity hazard identification: I. Development of a weight of evidence QSAR database. Regul Toxicol Pharmacol 2007;47(2):115–35.
[4] Hofer T, Gerner I, Gundert-Remy U, Liebsch M, Schulte A, Spielmann H, et al. Animal testing and alternative approaches for the human health risk assessment under the proposed new European chemicals regulation. Arch Toxicol 2004;78(10):549–64.
[5] National Toxicology Program. Multigenerational reproductive study of genistein (Cas No. 446-72-0) in Sprague-Dawley rats (feed study). Natl Toxicol Program Tech Rep Ser 2008;539:1–266.
[6] National Toxicology Program. Multigenerational reproductive toxicology study of ethinyl estradiol (CAS No. 57-63-6) in Sprague-Dawley rats. Natl Toxicol Program Tech Rep Ser 2010;547:1–312.
[7] Rorije E, Muller A, Beekhuijzen ME, Hass U, Heinrich-Hirsch B, Paparella M, et al. On the impact of second generation mating and offspring in multi-generation reproductive toxicity studies on classification and labelling of substances in Europe. Regul Toxicol Pharmacol 2011;61(2): 251–60.
[8] Beekhuijzen M. The era of 3Rs implementation in developmental and reproductive toxicity (DART) testing: current overview and future perspectives. Reprod Toxicol 2017;72:86–96.
[9] Vo AH, Van Vleet TR, Gupta RR, Liguori MJ, Rao MS. An overview of machine learning and big data for drug toxicity evaluation. Chem Res Toxicol 2020;33(1):20–37.
[10] Fischer I, Milton C, Wallace H. Toxicity testing is evolving. Toxicol Res 2020;9(2):67–80.
[11] Tannenbaum J, Bennett BT. Russell and Burch's 3Rs then and now: the need for clarity in definition and purpose. J Am Assoc Lab Anim Sci 2015;54(2):120–32.
[12] Brannen KC, Chapin RE, Jacobs AC, Green ML. Alternative models of developmental and reproductive toxicity in pharmaceutical risk assessment and the 3Rs. ILAR J 2016;57(2):144–56.
[13] Baert Y, Ruetschle I, Cools W, Oehme A, Lorenz A, Marx U, et al. A multi-organ-chip co-culture of liver and testis equivalents: a first step toward a systemic male reprotoxicity model. Hum Reprod 2020;35(5):1029–44.
[14] Aengenheister L, Favaro RR, Morales-Prieto DM, Furer LA, Gruber M, Wadsack C, et al. Research on nanoparticles in human perfused placenta: state of the art and perspectives. Placenta 2021;104:199–207.
[15] Stefansdottir A, Fowler PA, Powles-Glover N, Anderson RA, Spears N. Use of ovary culture techniques in reproductive toxicology. Reprod Toxicol 2014;49:117–35.
[16] Nikolaidis E. Relevance of animal testing and sensitivity of end points in reproductive and developmental toxicity. In: Reproductive and developmental toxicology. Academic Press; 2017. p. 211–24.
[17] Hong H, Tong W, Fang H, Shi L, Xie Q, Wu J, et al. Prediction of estrogen receptor binding for 58,000 chemicals using an integrated system of a tree-based model with structural alerts. Environ Health Perspect 2002;110(1):29–36.
[18] Ng HW, Doughty SW, Luo H, Ye H, Ge W, Tong W, et al. Development and validation of decision forest model for estrogen receptor binding prediction of chemicals using large data sets. Chem Res Toxicol 2015;28(12):2343–51.
[19] Marzo M, Benfenati E. Classification of a Naive Bayesian Fingerprint model to predict reproductive toxicity. SAR QSAR Environ Res 2018; 29(8):631–45.
[20] Jiang C, Yang H, Di P, Li W, Tang Y, Liu G. In silico prediction of chemical reproductive toxicity using machine learning. J Appl Toxicol 2019; 39(6):844–54.
[21] Feng H, Zhang L, Li S, Liu L, Yang T, Yang P, et al. Predicting the reproductive toxicity of chemicals using ensemble learning methods and molecular fingerprints. Toxicol Lett 2021;340:4–14.

[22] Liu J, Guo W, Dong F, Aungst J, Fitzpatrick S, Patterson TA, et al. Machine learning models for rat multigeneration reproductive toxicity prediction. Front Pharmacol 2022.

[23] Weyrich A, Joel M, Lewin G, Hofmann T, Frericks M. Review of the state of science and evaluation of currently available in silico prediction models for reproductive and developmental toxicity: a case study on pesticides. Birth Defects Res 2022;114(14):812–42.

[24] Zhang H, Shen C, Liu RZ, Mao J, Liu CT, Mu B. Developing novel in silico prediction models for assessing chemical reproductive toxicity using the naive Bayes classifier method. J Appl Toxicol 2020;40(9):1198–209.

[25] Yap CW. PaDEL-descriptor: an open source software to calculate molecular descriptors and fingerprints. J Comput Chem 2011;32(7):1466–74.

[26] Matthews EJ, Kruhlak NL, Daniel Benz R, Ivanov J, Klopman G, Contrera JF. A comprehensive model for reproductive and developmental toxicity hazard identification: II. Construction of QSAR models to predict activities of untested chemicals. Regul Toxicol Pharmacol 2007; 47(2):136–55.

[27] Sahigara F, Mansouri K, Ballabio D, Mauri A, Consonni V, Todeschini R. Comparison of different approaches to define the applicability domain of QSAR models. Molecules 2012;17(5):4791–810.

[28] Akbari A, Nasiri K, Heydari M, Mosavat SH, Iraji A. The protective effect of hydroalcoholic extract of *Zingiber officinale* Roscoe (ginger) on ethanol-induced reproductive toxicity in male rats. J Evid Based Complement Altern Med 2017;22(4):609–17.

[29] Della Pasqua O. On the role of concentration-effect relationships in safety pharmacology: only the dose makes a drug not to be poison. Br J Clin Pharmacol 2015;79(6):1042–3.

[30] Stumpf WE. The dose makes the medicine. Drug Discov Today 2006;11(11–12):550–5.

[31] Martin MT, Mendez E, Corum DG, Judson RS, Kavlock RJ, Rotroff DM, et al. Profiling the reproductive toxicity of chemicals from multigeneration studies in the toxicity reference database. Toxicol Sci 2009;110(1):181–90.

[32] Watford S, Ly Pham L, Wignall J, Shin R, Martin MT, Friedman KP. ToxRefDB version 2.0: improved utility for predictive and retrospective toxicology analyses. Reprod Toxicol 2019;89:145–58.

[33] Knudsen TB, Martin MT, Kavlock RJ, Judson RS, Dix DJ, Singh AV. Profiling the activity of environmental chemicals in prenatal developmental toxicity studies using the U.S. EPA's ToxRefDB. Reprod Toxicol 2009;28(2):209–19.

[34] Basant N, Gupta S, Singh KP. QSAR modeling for predicting reproductive toxicity of chemicals in rats for regulatory purposes. Toxicol Res 2016;5(4):1029–38.

[35] Cao DS, Xu QS, Hu QN, Liang YZ. ChemoPy: freely available python package for computational biology and chemoinformatics. Bioinformatics 2013;29(8):1092–4.

[36] Idri A, Benhar H, Fernandez-Aleman JL, Kadi I. A systematic map of medical data preprocessing in knowledge discovery. Comput Methods Programs Biomed 2018;162:69–85.

[37] Martin MT, Knudsen TB, Reif DM, Houck KA, Judson RS, Kavlock RJ, et al. Predictive model of rat reproductive toxicity from ToxCast high throughput screening. Biol Reprod 2011;85(2):327–39.

[38] Hong H, Thakkar S, Chen M, Tong W. Development of decision forest models for prediction of drug-induced liver injury in humans using a large set of FDA-approved drugs. Sci Rep 2017;7(1):17311.

[39] Hong H, Xie Q, Ge W, Qian F, Fang H, Shi L, et al. Mold(2), molecular descriptors from 2D structures for chemoinformatics and toxicoinformatics. J Chem Inf Model 2008;48(7):1337–44.

[40] Godden JW, Stahura FL, Bajorath J. Variability of molecular descriptors in compound databases revealed by Shannon entropy calculations. J Chem Inf Comput Sci 2000;40(3):796–800.

[41] Wu S, Fisher J, Naciff J, Laufersweiler M, Lester C, Daston G, et al. Framework for identifying chemicals with structural features associated with the potential to act as developmental or reproductive toxicants. Chem Res Toxicol 2013;26(12):1840–61.

[42] Leadscope. Leadscope toxicity database. 2014. Data set available at: http://www.leadscope.com/.

[43] Willighagen EL, Mayfield JW, Alvarsson J, Berg A, Carlsson L, Jeliazkova N, et al. The Chemistry Development Kit (CDK) v2.0: atom typing, depiction, molecular formulas, and substructure searching. J Cheminform 2017;9(1):33.

[44] OECD. Report on the regulatory uses and applications in OECD member countries of (quantitative) structure-activity relationship [(Q)SAR] models in the assessment of new and existing chemicals. OECD papers 6/11; 2006.

[45] Ng HW, Shu M, Luo H, Ye H, Ge W, Perkins R, et al. Estrogenic activity data extraction and in silico prediction show the endocrine disruption potential of bisphenol A replacement compounds. Chem Res Toxicol 2015;28(9):1784–95.

[46] Shi L, Tong W, Fang H, Xie Q, et al. An integrated "4-phase" approach for setting endocrine disruption screening priorities—phase I and II predictions of estrogen receptor binding affinity. SAR QSAR Environ Res 2002;13:69–88.

[47] Idakwo G, Luttrell J, Chen M, Hong H, Zhou Z, Gong P, Zhang C. A review on machine learning methods for in silico toxicity prediction. J Environ Sci Health C Environ Carcinog Ecotoxicol Rev 2018;36(4):169–91.

[48] Selvaraj C, Sakkiah S, Tong W, Hong H. Molecular dynamics simulations and applications in computational toxicology and nanotoxicology. Food Chem Toxicol 2018;112:495–506.

[49] Ng HW, Zhang W, Shu M, Luo H, Ge W, Perkins R, et al. Competitive molecular docking approach for predicting estrogen receptor subtype α agonists and antagonists. BMC Bioinf 2014;15(Suppl. 11):S4.

[50] Tang W, Chen J, Wang Z, Xie H, Hong H. Deep learning for predicting toxicity of chemicals: a mini review. J Environ Sci Health C Environ Carcinog Ecotoxicol Rev 2018;36(4):252–71.

[51] Wang Z, Chen J, Hong H. Applicability domains enhance application of PPARγ agonist classifiers trained by drug-like compounds to environmental chemicals. Chem Res Toxicol 2020;33(6):1382–8.

[52] Advances in computational toxicology. In: Hong H, editor. Methodologies and applications in regulatory science. vol 30. Cham: Springer International Publishing; 2019.

[53] Tang W, Liu W, Wang Z, Hong H, Chen J. Machine learning models on chemical inhibitors of mitochondrial electron transport chain. J Hazard Mater 2022;426:128067.

[54] Marzo M, Kulkarni S, Manganaro A, Roncaglioni A, Wu S, Barton-Maclaren TS, et al. Integrating in silico models to enhance predictivity for developmental toxicity. Toxicology 2016;370:127–37.

[55] Hougaard KS. Next generation reproductive and developmental toxicology: crosstalk into the future. Front Toxicol 2021;3:652571.

C H A P T E R

24

Aryl hydrocarbon receptors and their ligands in human health management

Ana-Maria Udrea[1], *Catalina Sogor*[2], *Catalin Buiu*[3] *and Speranta Avram*[2]

[1]Laser Department, National Institute for Laser, Plasma and Radiation Physics, Magurele, Romania [2]Department of Anatomy, Animal Physiology and Biophysics, Faculty of Biology, University of Bucharest, Bucharest, Romania [3]Department of Automatic Control and Systems Engineering, Politehnica University of Bucharest, Bucharest, Romania

Abbreviations

2,7-DBCZ 2,7-Dibromocarbazole
3,6-DBCZ 3,6-Dibromocarbazole
3,6-DCCZ 3,6-Dichlorocarbazole
3-BCZ 3-Bromocarbazole
3-CCZ 3-Chlorocarbazole
3-ICZ 3-Iodocarbzole
AHR Aryl hydrocarbon receptor
ARNT AHR nuclear transporter
AUC Area under the curve
MAX Maximum fold-change value of AhR reporter activity
MCC Matthews' correlation coefficient
PDB Protein Data Bank
PHCZ Polyhalogenated carbazoles
QSAR Quantitative structure–activity relationship

Introduction

The aryl hydrocarbon receptor (AHR) is a ligand-dependent transcription factor identified by Poland et al. in his in vitro experiments on [3H]TCDD binding to liver cytosol from C57BL/6J mice [1].

This receptor is normally inactive, but upon ligand binding, cytosolic AHR translocates to the nucleus, dissociating from its chaperone proteins and dimerizing with the AHR nuclear transporter (ARNT). The AHR/ARNT complex thus formed binds to the dioxin response elements located in the genome, disrupting gene expression [2].

AHR binds to exogenous and endogenous chemicals and can regulate the expression of some genes, depending on the ligand [3]. Numerous studies tried to elucidate the basic mechanisms by which AHR produces biological and toxicological effects [4]. The structural aspects of molecular complexes are essential for predicting biological activity by combining experimental with in silico studies [5].

Initially, AHR was considered as having a biochemical adaptive mediator role to xenobiotics, functioning as an environmental sensor for pollutants [6]. However, studies on mice showed a high expression in several tissues, such as the liver [7], heart [8], placenta [9], lung [10], and heart, but also in periods of uterine development [9].

ARH is an important modulator in host–environment interactions, having a unique role in immune and inflammatory responses [11]. However, the importance of AHR is also given by its ability to control the severity of autoimmune disease and can improve the prognosis of autoimmune inflammatory diseases [12].

AHR ligands come from different sources, from endogenous to exogenous metabolisms such as ingested food, drugs, or intestinal microbiota. These ligands can cause liver microlesions [13]. Depending on their chemical structure and quantity, AHR activation ligands can increase the formation of toxic metabolites.

Physiological AHR ligands in the diet were identified by in vitro tests. Flavonoids apigenin, luteolin, galangin, quercetin, kaempferol, and naringenin are present in fruits and vegetables [14], and indoles from the cruciferous class such as indole-3-acetonitrile, indole-3-carbinol, 3,3′-diindolylmethane, and indolo(3,4)bicarbazole represent the source of natural AHR ligands [15]. Natural compounds represent a possible therapeutic approach in

QSAR in Safety Evaluation and Risk Assessment
https://doi.org/10.1016/B978-0-443-15339-6.00037-0

pathologies determined by AHR activation, being involved in numerous studies. However, they have also been studied for their beneficial effects on the nervous system [16,17] and as antidiabetic potential [18,19].

Lactobacillus species can use tryptophan as an energy source, thus producing AHR ligands such as indole-3-acetic acid, tryptamine, and 3-methyl indole [20]. These bacteria can degrade tryptophan into indole so that it can later be metabolized in liver cells to the AHR ligands, indoxyl-3-sulfate, and indole-3-propionic acid.

AHR ligands can have both agonistic and antagonistic activities, with some compounds having tissue selectivity [20]. The classification of a compound as an AHR agonist or antagonist depends on the gene that is targeted in the tissue [14].

De Juan and Segura's study showed the importance of AHR ligands from food in maintaining homeostasis. They analyzed intestinal inflammation symptoms in mice whose diet did not contain any AHR ligands. The results showed more pronounced symptoms in mice with a synthetic diet without ligands, with weight loss and tissue damage [21].

Dietary or microbial AHR ligands can regulate type I beta interferon signaling in inflammatory cells of the central nervous system [22].

Quantitative structure–activity relationships (QSARs) are in silico studies that predict the biological activity of the compounds using high-precision statistical models [23,24]. Several QSAR models were made to predict the interaction or hepatotoxicity of several ligands [25–27].

Molecular docking studies use computational methods to predict the interaction site between a protein and a ligand and their estimated binding free energy [28]. For molecular docking studies, the 3D structures of the protein and ligand are required; the ligand structure is flexible, and the protein structure can be either flexible or rigid [29]. Molecular docking studies are a fast and cheap alternative to in vitro and in vivo studies and, consequently, are used in various biological fields, including microbiology, cancer therapy, and depression [28–31]. Moreover, during the COVID-19 pandemic, molecular docking approaches were extensively used to find possible treatments [32,33].

Protein 3D structures can be predicted or determined experimentally. A global database including around 200,000 experimentally determined biological macromolecules is the RCSB Protein Data Bank (PDB) [34]. In 2017, Schulte et al. reported the crystal structure of a heterodimeric AHR-ARNT complex with the heterodimeric basic helix-loop-helix-Per-ARNT-Sim domain attached to its target DNA (Fig. 24.1) [35].

Over 200 million protein structure predictions are made by an AI system called the Alpha Fold Protein Structure Database, which is openly accessible to scientists. Alfa Fold predicted the AHR structure (AF-P35869-F1) with a very high or confident per-residues confidence score (pLDDT) in several regions (Fig. 24.2) [36,37].

We superimpose the structures to compare the experimentally determined 3D AHR structure with the one predicted by the Alpha Fold (Fig. 24.3).

Studies on AHR-generated hepatotoxicity

AHR influences hepatotoxicity because the receptor is linked to enhanced inflammatory cell infiltration. However, the toxic manifestation is influenced by the ligand that produces liver microlesions [38]. AHR activation in hepatocytes causes hepatotoxicity, thus explaining the individual sensitivity to liver failure [39].

AHR mediates the toxic effects of dioxin in the environment. Exposure to dioxin can produce a series of toxic effects from exhaustion, suppression of immune function, hepatotoxicity, tumor promotion, and even collapse. Toxic effects disturb the normal physiological functioning of the body [40]. 2,3,7,8-Tetrachlorodibenzo-p-dioxin induces hepatotoxicity by interacting with AHR, to which it binds with high affinity mediating its toxicity and promoting carcinogenesis. AHR can induce the transcription of cytochromes P450 [41,42].

The bioactive alkaloids from *Euodia rutaecarpa* (rutaecarpine, evodiamine, and dehydroevadiamine) activated AHR, but the dose administered in the study by Zhang et al. in vivo did not cause liver damage. This was achieved because the alkaloids were not orally absorbed orally in wild-type AHR mice in the experimental model. However, although these compounds did not cause hepatotoxicity, they disrupted bile acid flow, depending on the degree of AHR activation [43].

The study by Ozeki et al. shows that CYP1A enzymes induced by AHR are necessary against dioxin toxicity because they protect against acute hepatotoxicity. The accumulation of hepatic bile acid, the increase in the level of inflammation with the help of cytokines, and the necrosis of hepatocytes determine severe liver toxicity in mice [44].

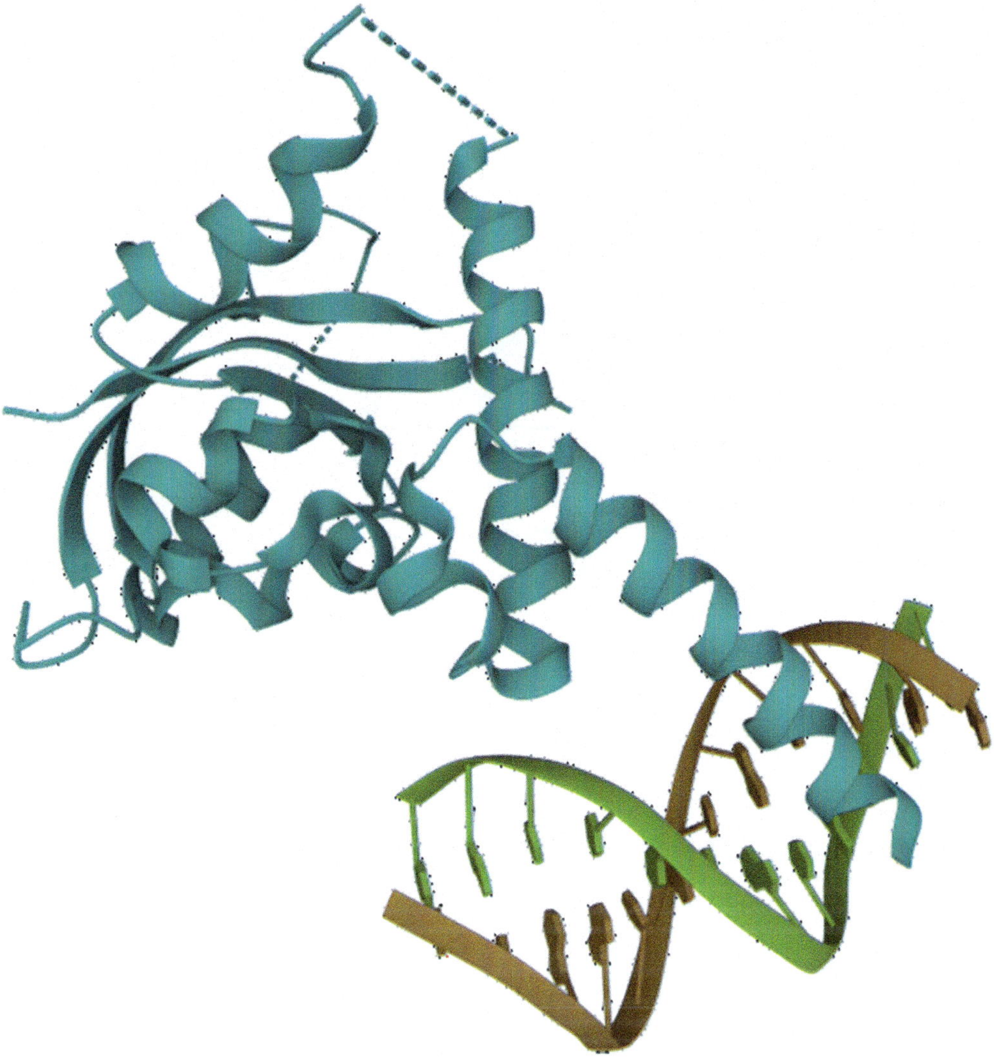

FIGURE 24.1 3D experimentally determined structure of AHR chain A (*green*) interacting with DNA (*light green* and *brown*). *Image from the RCSB PDB [34] of PDB ID 5NJ8 [35].*

In studies on HepG2 cells and mice, Zhang et al. showed the induction of CYP1A2 expression and activity through AHR activation based on psoralen and isopsoralen treatment. Following treatment with these natural medicinal compounds that were isolated from *Psoraleae fructus*, an increase in CYP1A2 transcription was observed [45].

Matsuzaka et al. conducted a study that predicts AHR activation using a QSAR model based on a DeepSnap–Deep Learning method [25]. DeepSNAP is a tool that uses a ball-and-stick model to illustrate the steric conformation of a chemical structure. Images might be generated automatically depending on the x-, y-, and z-axis directions. A deep learning method is employed in this model to identify and produce the prediction model based on using as inputs the previously generated images. As a consequence, the model is known as DeepSNAP-deep learning [25]. The investigation was carried out using a database of 201 molecules with in vitro-determined AHR inhibition, randomly split into three sets: training, validation, and testing. They used as indicators of performance the percentage of accurate responses based on the results obtained from the validation dataset and its labeled dataset and the error rate between the results received from the validation dataset and its labeled dataset; the set used for those indicators is the validation one. Future investigations will focus on the image collection angle, assessing teen angles on the x, y, and z coordinates ranging from 360 degrees to 38 degrees on two datasets of 201 chemicals divided into training, validation, and testing sets. At an angle of 176 degrees, the authors obtained the highest accuracy. When evaluating AHR-induced hepatotoxicity, this in silico model outperforms standard machine-learning approaches

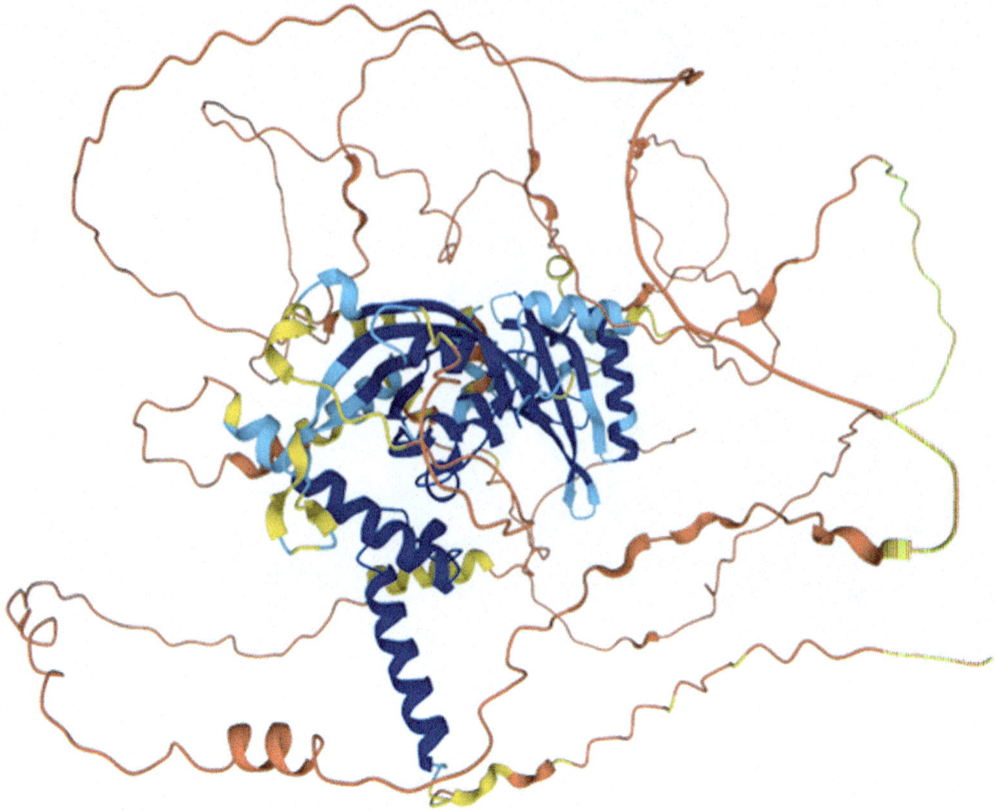

FIGURE 24.2 Alfa Fold predicted AHR 3D structure. Regions colored in *navy blue* have very high model confidence, with pLDDT higher than 90; *light blue*-colored areas have a pLDDT between 90 and 70; *yellow*-colored regions have low confidence, with a pLDDT between 70 and 50; the *orange*-colored areas have very low confidence with a pLDDT lower than 50 [36,37].

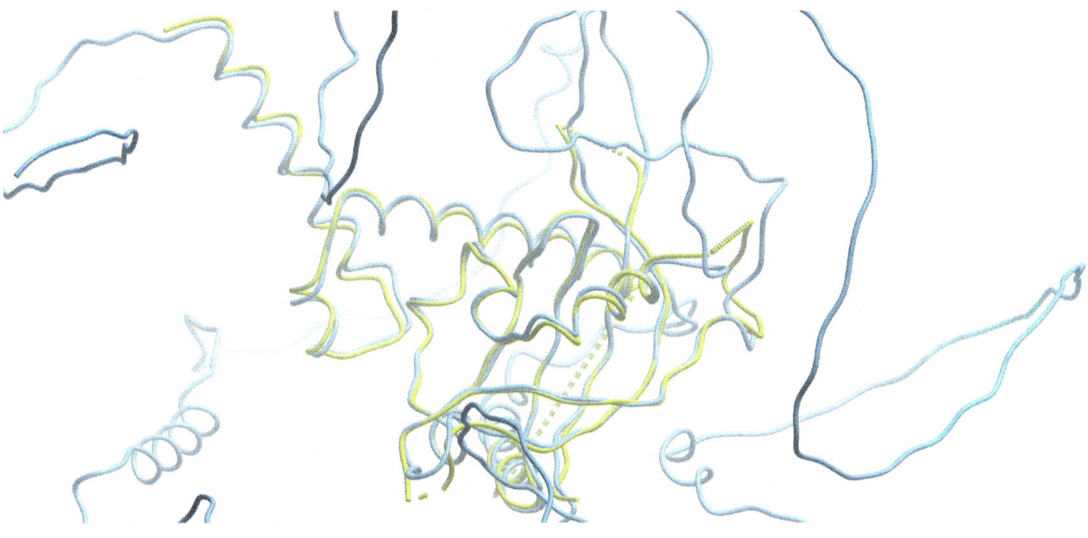

FIGURE 24.3 Superimpose of Alfa Fold predicted AHR (*cyan*) structure with the experimentally determined structure (*yellow*, PDB: 5NJ8).

such as the random forest, XGBoost, LightGBM, and CatBoost [25]. The accuracy (Acc), Matthews correlation coefficient (MCC), area under the curve (AUC), and maximum fold-change value of AhR reporter activity (MAX) were evaluated to predict the performance of the QSAR models. The authors classified the compound datasets using the top 10%, 15%, 20%, 30%, 35%, 40%, 45%, 50%, and 55% of the MAX values of AhR activation and evaluated how the AhR activation threshold influenced the DeepSnap-DL approach's ability to make predictions. The model with MAX

TABLE 24.1 Performance evaluations of the QSAR models computed by Šala et al. and Matsuzaka et al.

Šala et al. QSAR models [46]			
	r^2	q^2	s
Polychlorobiphenyls	0.73	0.64	0.40
Polychlorodibenzodioxins	0.92	0.88	0.43
Polychlorodibenzofurans	0.89	0.86	0.48
Matsuzaka et al. QSAR models			
MAX scores	Average AUC ± SD	Average Acc (test) ± SD	Average MCC ± SD
0.30	0.920 ± 0.029	0.940 ± 0.032	0.814 ± 0.024
0.40	0.959 ± 0.025	0.922 ± 0.036	0.845 ± 0.075

score of 0.30 has the highest performances of prediction on average Acc (Test) (Table 24.1), and the model with the MAX score of 0.40 has the highest performances on average AUC and average MCC (Table 24.1).

Šala et al. created three multiple linear regression models to predict AHR binding affinity: one for polychlorobiphenyls, one for polychlorodibenzodioxins, and one for polychlorodibenzofurans [46]. CODESSA software was used to compute the molecular descriptors required for the QSAR models and to choose an optimal subset of structural descriptors by lowering prediction errors using the multiple linear regression techniques. The leave-one-out cross-validation approach was used to examine the models' prediction capabilities throughout the descriptor selection process [46].

For each multiple linear regression model, the squared correlation coefficients (r^2, q^2) and standard error (s) were determined. The polychlorodibenzodioxin model had the best statistical parameters, whereas the polychlorobiphenyl model had the worst (Table 24.1) [46].

Studies of AHR antagonist

Medicines and compounds in the environment can act on the AHR either agonistically, activating the receptor, which performs the function of gene transcription, or antagonistically bound to the receptor and blocking its activity [47,48].

Spices and essential oils contain chemical compounds with an agonistic and antagonistic role in the human AHR. Bartoňková and Dvořák's study on spices and essential oils included some of the foods usually used as AHR: (1) agonists, mentioning, among others, cumin, jasmine, and vanilla; (2) partial agonists such as cloves, nutmeg, and oregano, and (3) antagonists mentioning tarragon, cumin, anise, and mint [47].

The usefulness of AHR as a drug target depends on its expression level in specific organs and the function it performs in pathological tissue. If AHR activity can inhibit the evolution of the disease, therapies with specific ligands can be used that can have an agonistic or antagonistic effect [49]. Studies on AHR in several types of cancer cells have shown a decrease in invasion and migration following treatment with AHR agonists or antagonists [50].

The constitutive androstane receptor is a gene differentially regulated by AHR, its transcription being higher following AHR activation by exogenous and endogenous ligands [51].

Sorafenib is a tyrosine kinase inhibitor that induces autophagy and can suppress several types of cancer, including hepatocellular carcinoma. The drug can suppress the transcription activity determined by endogenous factors by suppressing the protein expression of β-NF and ITE genes. Sorafenib shows the antagonistic activity of AHR by suppressing the activity depending on the dose and is a helpful drug in diseases that have AHR inhibition as a therapeutic strategy [52].

Aromatic polycyclic compounds can have such a beneficial role in the body being able to be administered as medicines, but also they can have a series of adverse effects on the body. These effects can be highlighted through bioinformatics analysis methods of the compounds' physicochemical properties, based on the chemical structure of the compounds that can suggest the function performed by them in the body [53].

High levels of AHR can cause malignant transformation, and exposure to exogenous agonists such as halogenated aromatic hydrocarbons, polycyclic aromatic hydrocarbons, and other dioxin-like compounds can lead to activation of AHR in the mammary gland [54]. Papoutsis et al. showed that resveratrol, a phytoalexin identified in grapes and red wine, can be an AHR antagonist in cell lines treated with dioxin. At the same time, resveratrol has a protective role in MCF-7 cell lines [55].

Epigallocatechin gallate, genistein [56], α-naphthoflavone [57], and quercetin [58] have an antagonistic role on AHR in studies on breast cancer cells. These compounds have an antagonistic role in AHR-dependent processes involved in breast tumorigenesis [54].

Goya-Jorge et al. estimated the AHR antagonist using QSAR models based on experimental in vivo studies. The dataset contains 116 compounds reported as AHR antagonists, including organohalogenated, polyaromatic hydrocarbon, flavonoid, quinone, and pyrethroid chemical classes (68 in vitro evaluated in the same study and 48 from various literature sources) [26].

The QSAR model descriptors were generated using the DRAGON and PADEL software packages. After removing constant variables and chemical descriptors with pair correlations larger than 0.9, 401 descriptors remained [26].

The collection of variables and 10 main components were subjected to principal components analysis using MATLAB software [26].

Five QSAR models were built using linear discriminant analysis or logistic regression as model-building classification. The metrics that evaluate the performance are accuracy, precision, sensitivity, and specificity. The QSAR validation was also performed using internal leave-group-out cross-validation to check for possible overfitting. These five QSAR models have adequate robustness and predictive power based on the quality of the obtained statistical validation parameters [26].

Studies on AHR activation

Transcription factors regulate physiological and pathophysiological processes. AHR targets the cytochrome P450 gene family. The high-affinity ligands are dioxins and similar compounds. Considering the stability of dioxin-type pollutants, they are resistant to degradation, leading to long-term activation of AHR. At the same time, the ligands with lower affinity to AHR are flavonoids, bilirubin, and other endogenous molecules [59,60].

Norisoboldine is a natural AHR agonist that promotes the differentiation of Treg cells in hypoxia and can alleviate ulcerative colitis. Lv et al., in their study on mice with induced colitis, showed the importance of the regulation between glycolysis and AHR [61].

Pelargonidin has an agonistic, protective effect on the AHR, which blocks the translocation of NF-κB in the nucleus, thus inhibiting the expression of ACE2 and, indirectly, the binding of the SARS-CoV2 virus [62].

The study by Jin et al. on some drugs described as agonists of AHR can determine several responses depending on the cell lines on which they were tested. Shown as 4-hydroxytamoxifen, mexiletine, and tranilast, although they are classified as AHR agonist drugs, their effects are different depending on the cell line in which the test is performed. On breast cancer cell lines BT474 and MDA-MB-468, mexiletine showed the most potent agonist effect for AHR, followed by tranilast. Sulindac and leflunomide were less active for AHR than omeprazole [63].

Cinnabaric acid, the tryptophan metabolite, is an endogenous AHR agonist ligand that stimulates the differentiation of T cells and is involved in a broad spectrum of biological functions for the immune system [64].

To predict the activator capability of a set of 80,086 chemical structures, Klimenko et al. developed two QSAR models (a rational model and a random model). The authors developed the QSAR models starting from an experimental dataset of 324,858 chemicals with a PubChem AHR activity score ranging from 0 to 100 [27].

The two QSAR models were developed using Leadscope Predictive Data Miner, a feature of Leadscope Enterprise, on a complete dataset of 925 active compounds and 204,513 inactive compounds. This program performs a substructural analysis by employing 27,000 structural keys and nine molecular descriptors. Utilizing the software's automated approach and a partial logistic regression (PLS) model, the structural descriptors linked to AhR activation were found [27].

High specificity (rational 87 $\pm$ 1.6; random 91.5 $\pm$ 0.8), high sensitivity (rational 74.5 $\pm$ 4.7; random 86.4 $\pm$ 2.6), and balanced accuracy (rational 80.8 $\pm$ 2.5; random 88.9 $\pm$ 0.8) characterize both QSAR models [27].

The prediction performance of the models is validated using cross-validation and external validation, where both models exhibit comparable or superior specificity, sensitivity, and balanced accuracy to their respective predecessors by cross-validation. Both models also showed better robustness (lower standard deviations) than their predecessors. The random-final model is the most effective model developed in the study. According to the screening results, 1256 of the 80,086 chemicals tested are predicted by the rational model to be AHR active, whereas 3214 are predicted by the random model [27]. Both QSAR models are available in the Danish (Q)SAR Database [65].

Molecular modeling provides a highly potent collection of tools for adapting fundamental scientific concepts from quantum mechanics and statistical mechanics to acquire a high-resolution understanding of structure–function interactions in biomolecules [66]. Using a 3D structure of AHR determined via a homology model, Song et al. predict the binding of D3 vitamin derivatives in interaction with AHR. For the molecular docking predictions, they used Glide, implemented in Schrödinger software [67]. The authors predicted the binding affinity of indirubin, indole acetic acid, vitamin D3, and several vitamin D3 derivatives. According to their results, vitamin D3 and its derivatives have lower binding energy than indirubin and indole acetic acid (Table 24.2) [67].

Chaubah et al. studied the binding affinity of three polycyclic aromatic hydrocarbons, benzo[a] anthracene (BaA), benzo[a] pyrene (BaP), and phenanthrene (PA), bound to AHR to predict the carcinogenic activity of these three compounds. The predictions were made using the crystal structure of AHR (PDB 5NJ8) with HEX 8.0.0 software, and the outcomes revealed that BaA had the lowest expected energy (Table 24.2) [68].

Products such as plastics and textiles use flame retardants known as polybrominated diphenyl ethers (PBDEs). Using the AHR ligand binding domain obtained via molecular modeling, Gu et al. evaluated the AHR-induced toxicity of those molecules through molecular docking. The results show that several PBDEs interacted with AHR with low interaction energies (Table 24.2) [69].

Garcia et al. used the molecular docking approach to predict the activation probability of the AHR: ARNT heterodimer by several flavonoids, including genistein and resveratrol. The authors used the crystal structure of ARNT (PDB: 4zp4) from *Mus musculus*. The SwissModel Web Server was used for the molecular docking of AHR and ARNT monomers, and MOE 2019.01 was utilized for the visualization of the interactions between the model and the investigated flavonoids. The investigated flavonoids frequently interact with the bHLH, PAS-A, and bHLH/PAS-A domains at this contact interface. The compounds investigated by Garcia et al. have a high affinity for AHR, acting antagonistically by preventing the interaction between AHR and ARNT. The AA residues Asp161, Leu164, Lys165, Ile168, Leu176, and Tyr188 are essential for the binding of flavonoids to ARNT. Quercetin, taxifolin, and robinetin are described in the literature as polyphenolic compounds capable of activating AHR, while luteolin, genistein, and resveratrol have been observed by molecular modeling to act as AHR blockers. Flavone, galangin, 5-hydroxy-2-(4′-hydroxy)-7-methoxy-flavonone, and 7-hydroxy-5-methoxy-flavonone show the highest probabilities of blocking the formation of the AHR [71].

TABLE 24.2 The lowest binding energies in molecular docking studies presented.

Compound	Predicted energy of binding	Software used	References
Indirubin	−9.555 (kcal/mol)	Glide	[67]
Indole acetic acid	−9.027 (kcal/mol)		
Vitamin D3	−12.162 (kcal/mol)		
BaA	−232.9 (kcal/mol)	HEX 8.0.0	[68]
BaP	−230.6 (kcal/mol)		
PA	−191.93 (kcal/mol)		
PBDE, with bromination site 2,2′,4,4′,6	−12.16 [−lg(kd)]	Surflex-Dock	[69]
PBDE, with bromination site 2,2′,4,4′,5,6′	−11.85 [−lg(kd)]		
PBDE, with bromination site 2,3′,4,4′,6	−11.11 [−lg(kd)]		
3,6-DCCZ	−9.6 (kcal/mol)	AutoDock	[70]
3,6-DBCZ	−9.4 (kcal/mol)		
2,7-DBCZ	−9.2 (kcal/mol)		
3-ICZ	−9.1 (kcal/mol)		
3-BCZ	−9 (kcal/mol)		
3-CCZ	−9 (kcal/mol)		

Polyhalogenated carbazoles (PHCZs) are environmental pollutants with a chemical structure like dioxin. Ji et al. tried to highlight the agonistic activity of AHR of PHCZ derivatives on Danio rerio, H. Molecular docking studies were performed using the AutoDock software to identify the bioactive conformations of the compounds derived from PHCZ. The amino acid sequence of the AHR binding domain is found in UniProt (entry ID: P35869), and the model used was built based on sequence homology. The most active compounds are selected according to the binding score obtained after docking. PHCZ derivatives interacted with the AHR binding pocket, having a good interaction, observing the agonistic effect of the compounds with an affinity score lower than −7. Derivatives with values below −8.8 such as 3,6-dichlorocarbazole (3,6-DCCZ), 3,6-dibromocarbazole (3,6-DBCZ), 2,7-dibromocarbazole (2,7-DBCZ), 3-iodocarbazole (3-ICZ), and 3-bromocarbazole (3-BCZ, 3-chlorocarbazole (3-CCZ) presented in Table 24.2 represent the AHR agonist effect [70]. Through molecular docking studies, the interactions between PHCZ and AHR were observed, having binding affinity comparable with that of dioxin. 3,6-DCCZ, 3,6-DBCZ, and 2,7-DBCZ obtained the highest binding affinity to AHR [70].

Studies that determine the AHR inhibitors in food

Lo Piparo et al. developed four 3D-QSAR models based on the experimentally determined binding affinity of 93 AHR ligands (25 dibenzo-p-dioxins, 38 dibenzofurans, 14 biphenyls, 5 naphthalenes, 7 indolocarbazoles, and 4 indolocarbazole derivatives). CoMFA (comparative molecular field analysis), VolSurf (a method that is based on computed molecular interaction fields and multivariate statistics), HQSAR (Hologram QSAR), and a hybrid algorithm were the four main methods used to create the QSAR models. Relevant descriptors included Log P, Hologram 257, CoMFA (electrostatic and steric), and strain energy [72].

According to the statistical evaluations, the CoMFA method's R^2 value is more significant (R^2 0.91) than those of the HQSAR (R^2 0.85), VolSurf (R^2 0.79), and hybrid (R^2 0.73) approaches. Additionally, leave-one-out (LOO) and cross-validation (CV) q^2 were used to confirm the model's prediction performance. The findings show that LOO q^2 is the same on the CoMFA and HQSAR models (q^2 0.62), lower on the VolSurf model (q^2 0.58) and higher on the hybrid model (q^2 0.7). The CVq2 has the highest predicted value in the HQSAR model (q^2 0.66), while the CoMFA (q^2 0.62) and VolSurf model (q^2 0.50) are lower [72].

The study conducted by Lo Piparo et al. was designed to predict the binding affinity of compounds that had not yet been examined as AHR ligands in food. The study employed specific guidelines for the QSAR models to guarantee their applicability and reliability [72].

Conclusions

In this chapter, we highlighted the importance of endogenous (e.g., bilirubin) and exogenous chemical compounds (pollution compounds) that can regulate the expression of the cytochrome P450 1A1 and AHR genes. Although AHR was initially considered to be a mediator of xenobiotics, its role as a modulator of interactions with the immune system and the response to inflammation was observed. AHR can have agonistic and antagonistic ligands, their selectivity being given by the targeted gene in the tissue. QSAR and molecular docking studies highlighted the ability of nontoxic compounds such as apigenin, luteolin, galangin, quercetin, kaempferol, and naringenin to bind selectively to AHR. Hepatotoxicity caused by agonistic compounds in the environment can cause effects that determine the physiological disturbance of the functionality of the organs.

Flavonoids (quercetin, taxifolin, and resveratrol) are one of the significant classes of diet-derived AHR ligands found widely distributed in fruits and vegetables and can be agonistic ligands of AHR. However, some ligands can have antagonistic effects on AHR, such as 25 dibenzo-p-dioxins, 38 dibenzofurans, 14 biphenyls, 5 naphthalenes, 7 indolocarbazoles, luteolin, and 4 indolocarbazoles and their derivatives.

References

[1] Poland A, Glover E, Kende AS. Stereospecific, high affinity binding of 2,3,7,8-tetrachlorodibenzo-p-dioxin by hepatic cytosol. Evidence that the binding species is receptor for induction of aryl hydrocarbon hydroxylase. J Biol Chem 1976;251:4936–46.

[2] Rothhammer V, Quintana FJ. The aryl hydrocarbon receptor: an environmental sensor integrating immune responses in health and disease. Nat Rev Immunol 2019;19:184–97.

[3] Denison MS, Soshilov AA, He G, DeGroot DE, Zhao B. Exactly the same but different: promiscuity and diversity in the molecular mechanisms of action of the aryl hydrocarbon (dioxin) receptor. Toxicol Sci 2011;124:1–22.

[4] Nguyen LP, Bradfield CA. The search for endogenous activators of the aryl hydrocarbon receptor. Chem Res Toxicol 2008;21:102–16.
[5] Avram S, Borcan F, Borcan L-C, Milac AL, Mihailescu D. QSAR approaches applied to antidepressants induced neurogenesis—in vivo and in silico applications. Mini Rev Med Chem 2015;16:230–40.
[6] Larigot L, Juricek L, Dairou J, Coumoul X. AhR signaling pathways and regulatory functions. Biochim Open 2018;7:1–9.
[7] Carambia A, Schuran FA. The aryl hydrocarbon receptor in liver inflammation. Semin Immunopathol 2021;43:563–75.
[8] Ansari MA, Maayah ZH, Bakheet SA, El-Kadi AO, Korashy HM. The role of aryl hydrocarbon receptor signaling pathway in cardiotoxicity of acute lead intoxication in vivo and in vitro rat model. Toxicology 2013;306:40–9.
[9] Jiang Y, Wang K, Fang R, Zheng J. Expression of aryl hydrocarbon receptor in human placentas and fetal tissues. J Histochem Cytochem 2010; 58:679–85.
[10] Guerrina N, Traboulsi H, Eidelman DH, Baglole CJ. The aryl hydrocarbon receptor and the maintenance of lung health. Int J Mol Sci 2018;19: E3882.
[11] Kawajiri K, Fujii-Kuriyama Y. The aryl hydrocarbon receptor: a multifunctional chemical sensor for host defense and homeostatic maintenance. Exp Anim 2017;66:75–89.
[12] Quintana FJ, Basso AS, Iglesias AH, Korn T, Farez MF, Bettelli E, et al. Control of T(reg) and T(H)17 cell differentiation by the aryl hydrocarbon receptor. Nature 2008;453:65–71.
[13] Yoon E, Babar A, Choudhary M, Kutner M, Pyrsopoulos N. Acetaminophen-induced hepatotoxicity: a comprehensive update. J Clin Transl Hepatol 2016;4:131–42.
[14] Jin U-H, Park H, Li X, Davidson AL, Allred C, Patil B, et al. Structure-dependent modulation of aryl hydrocarbon receptor-mediated activities by flavonoids. Toxicol Sci 2018;164:205.
[15] Bjeldanes LF, Kim JY, Grose KR, Bartholomew JC, Bradfield CA. Aromatic hydrocarbon responsiveness-receptor agonists generated from indole-3-carbinol in vitro and in vivo: comparisons with 2,3,7,8-tetrachlorodibenzo-p-dioxin. Proc Natl Acad Sci USA 1991;88:9543–7.
[16] Avram S, Stan MS, Udrea AM, Buiu C, Boboc AA, Mernea M. 3D-ALMOND-QSAR models to predict the antidepressant effect of some natural compounds. Pharmaceutics 2021;13:1449.
[17] Avram S, Milac A-L, Mihailescu D. 3D-QSAR study indicates an enhancing effect of membrane ions on psychiatric drugs targeting serotonin receptor 5-HT1A. Mol Biosyst 2012;8:1418–25.
[18] Buiu C, Avram S, Duda-Seiman D, Milac AL, Duda-Seiman C, Pacureanu L, Borcan F. More effective DPP4 inhibitors as antidiabetics based on sitagliptin applied QSAR and clinical methods. Curr Comput Aided Drug Des 2014;10:237–49.
[19] Udrea AM, Gradisteanu Pircalabioru G, Boboc AA, Mares C, Dinache A, Mernea M, Avram S. Advanced bioinformatics tools in the pharmacokinetic profiles of natural and synthetic compounds with anti-diabetic activity. Biomolecules 2021;11:1692.
[20] Jin U-H, Lee S-O, Sridharan G, Lee K, Davidson LA, Jayaraman A, et al. Microbiome-derived tryptophan metabolites and their aryl hydrocarbon receptor-dependent agonist and antagonist activities. Mol Pharmacol 2014;85:777–88.
[21] De Juan A, Segura E. Modulation of immune responses by nutritional ligands of aryl hydrocarbon receptor. Front Immunol 2021;12:645168.
[22] Quintana FJ, Sherr DH. Aryl hydrocarbon receptor control of adaptive immunity. Pharmacol Rev 2013;65:1148–61.
[23] Avram S, Mernea M, Borcan F, Mihailescu D. Evaluation of the therapeutic properties of mastoparan- and sifuvirtide-derivative antimicrobial peptides using chemical structure-function relationship—in vivo and in silico approaches. Curr Drug Deliv 2016;13:202–10.
[24] Avram S, Udrea AM, Negrea A, Ciopec M, Duteanu N, Postolache C, et al. Prevention of deficit in neuropsychiatric disorders through monitoring of arsenic and its derivatives as well as through bioinformatics and cheminformatics. Int J Mol Sci 2019;20:E1804.
[25] Matsuzaka Y, Hosaka T, Ogaito A, Yoshinari K, Uesawa Y. Prediction model of aryl hydrocarbon receptor activation by a novel QSAR approach, deepsnap—deep learning. Molecules 2020;25:1317.
[26] Goya-Jorge E, Doan TQ, Scippo ML, Muller M, Giner RM, Barigye SJ, Gozalbes R. Elucidating the aryl hydrocarbon receptor antagonism from a chemical-structural perspective. SAR QSAR Environ Res 2020;31:209–26.
[27] Klimenko K, Rosenberg SA, Dybdahl M, Wedebye EB, Nikolov NG. QSAR modelling of a large imbalanced aryl hydrocarbon activation dataset by rational and random sampling and screening of 80,086 REACH pre-registered and/or registered substances. PLoS One 2019;14: e0213848.
[28] Udrea A-M, Dinache A, Pagès J-M, Pirvulescu RA. Quinazoline derivatives designed as efflux pump inhibitors: molecular modeling and spectroscopic studies. Molecules 2021;26:2374.
[29] Nistorescu S, Gradisteanu Pircalabioru G, Udrea A-M, Simon A, Pascu ML, Chifiriuc M-C. Laser-irradiated chlorpromazine as a potent antibiofilm agent for coating of biomedical devices. Coatings 2020;10:1230.
[30] Tozar T, Santos Costa S, Udrea A-M, Nastasa V, Couto I, Viveiros M, et al. Anti-staphylococcal activity and mode of action of thioridazine photoproducts. Sci Rep 2020;10:18043.
[31] Nistorescu S, Udrea A-M, Badea MA, Lungu I, Boni M, Tozar T, et al. Low blue dose photodynamic therapy with porphyrin-iron oxide nanoparticles complexes: in vitro study on human melanoma cells. Pharmaceutics 2021;13:2130.
[32] Udrea A-M, Mernea M, Buiu C, Avram S. Scutellaria baicalensis flavones as potent drugs against acute respiratory injury during SARS-CoV-2 infection: structural biology approaches. Processes 2020;8:1468.
[33] Udrea A-M, Avram S, Nistorescu S, Pascu M-L, Romanitan MO. Laser irradiated phenothiazines: new potential treatment for COVID-19 explored by molecular docking. J Photochem Photobiol B 2020;211:111997.
[34] Berman HM, Westbrook J, Feng Z, Gilliland G, Bhat TN, Weissig H, et al. The protein data bank. Nucleic Acids Res 2000;28:235–42.
[35] Schulte KW, Green E, Wilz A, Platten M, Daumke O. Structural basis for aryl hydrocarbon receptor-mediated gene activation. Structure 2017; 25:1025–1033.e3.
[36] Jumper J, Evans R, Pritzel A, Green T, Figurnov M, Ronneberger O, et al. Highly accurate protein structure prediction with AlphaFold. Nature 2021;596:583–9.
[37] Varadi M, Anyango S, Deshpande M, Nair S, Natassia C, Yordanova G, et al. AlphaFold Protein Structure Database: massively expanding the structural coverage of protein-sequence space with high-accuracy models. Nucleic Acids Res 2022;50:D439–44.
[38] Yoda T, Tochitani T, Usui T, Kouchi M, Inada H, Hosaka T, et al. Involvement of the CYP1A1 inhibition-mediated activation of aryl hydrocarbon receptor in drug-induced hepatotoxicity. J Toxicol Sci 2022;47:359–73.

[39] Schuran FA, Lommetz C, Steudter A, Ghallab A, Wieschendorf B, Schwinge D, et al. Aryl hydrocarbon receptor activity in hepatocytes sensitizes to hyperacute acetaminophen-induced hepatotoxicity in mice. Cell Mol Gastroenterol Hepatol 2020;11:371–88.
[40] Furness SGB, Whelan F. The pleiotropy of dioxin toxicity—xenobiotic misappropriation of the aryl hydrocarbon receptor's alternative physiological roles. Pharmacol Ther 2009;124:336–53.
[41] Murray IA, Patterson AD, Perdew GH. AH receptor ligands in cancer: friend and foe. Nat Rev Cancer 2014;14:801–14.
[42] Ovando BJ, Ellison CA, Vezina CM, Olson JR. Toxicogenomic analysis of exposure to TCDD, PCB126 and PCB153: identification of genomic biomarkers of exposure to AhR ligands. BMC Genom 2010;11:583.
[43] Zhang Y, Yan T, Sun D, Xie C, Zheng Y, Zhang L, et al. Structure-activity relationships of the main bioactive constituents of *Euodia rutaecarpa* on aryl hydrocarbon receptor activation and associated bile acid homeostasis. Drug Metab Dispos Biol Fate Chem 2018;46:1030–40.
[44] Ozeki J, Uno S, Ogura M, Choi M, Maeda T, Sakurai K, et al. Aryl hydrocarbon receptor ligand 2,3,7,8-tetrachlorodibenzo-p-dioxin enhances liver damage in bile duct-ligated mice. Toxicology 2011;280:10–7.
[45] Zhang C, Zhao J-Q, Sun J-X, Li H-J. Psoralen and isopsoralen from Psoraleae Fructus aroused hepatotoxicity via induction of aryl hydrocarbon receptor-mediated CYP1A2 expression. J Ethnopharmacol 2022;297:115577.
[46] Šala M, Balaban A, Veber M, Pompe M. QSAR models for estimating aryl hydrocarbon receptor binding affinity of polychlorobiphenyls, polychlorodibenzodioxins, and polychlorodibenzofurans. MATCH Commun Math Comput Chem 2016;75:559–82.
[47] Bartoňková I, Dvořák Z. Essential oils of culinary herbs and spices display agonist and antagonist activities at human aryl hydrocarbon receptor AhR. Food Chem Toxicol 2018;111:374–84.
[48] Hoskin PJ, Hanks GW. Opioid agonist-antagonist drugs in acute and chronic pain states. Drugs 1991;41:326–44.
[49] Safe S, Jin U, Park H, Chapkin RS, Jayaraman A. Aryl hydrocarbon receptor (AHR) ligands as selective AHR modulators (SAhRMs). Int J Mol Sci 2020;21:6654.
[50] Opitz CA, Litzenburger UM, Sahm F, Ott M, Tritschler I, Trump S, et al. An endogenous tumour-promoting ligand of the human aryl hydrocarbon receptor. Nature 2011;478:197–203.
[51] Patel RD, Hollingshead BD, Omiecinski CJ, Perdew GH. Aryl-hydrocarbon receptor activation regulates constitutive androstane receptor levels in murine and human liver. Hepatology 2007;46:209–18.
[52] Wei K-L, Gao G-L, Chou Y-T, Lin C-Y, Chen S-C, Chen Y-L, et al. Sorafenib is an antagonist of the aryl hydrocarbon receptor. Toxicology 2022; 470:153118.
[53] Avram S, Mernea M, Mihailescu D, Duda-Seiman D, Duda-Seiman C. Advanced QSAR methods evaluated polycyclic aromatic compounds duality as drugs and inductors in psychiatric disorders. Curr Org Chem 2013;17:2880–90.
[54] Donovan MG, Selmin OI, Romagnolo DF. Aryl hydrocarbon receptor diet and breast cancer risk. Yale J Biol Med 2018;91:105–27.
[55] Papoutsis AJ, Borg JL, Selmin OI, Romagnolo DF. BRCA-1 promoter hypermethylation and silencing induced by the aromatic hydrocarbon receptor-ligand TCDD are prevented by resveratrol in MCF-7 cells. J Nutr Biochem 2012;23:1324–32.
[56] Romagnolo DF, Donovan MG, Papoutsis AJ, Doetschman TC, Selmin OI. Genistein prevents BRCA1 CpG methylation and proliferation in human breast cancer cells with activated aromatic hydrocarbon receptor. Curr Dev Nutr 2017;1:e000562.
[57] Miller ME, Holloway AC, Foster WG. Benzo-[a]-pyrene increases invasion in MDA-MB-231 breast cancer cells via increased COX-II expression and prostaglandin E2 (PGE2) output. Clin Exp Metastasis 2005;22:149–56.
[58] Mense SM, Chhabra J, Bhat HK. Preferential induction of cytochrome P450 1A1 over cytochrome P450 1B1 in human breast epithelial cells following exposure to quercetin. J Steroid Biochem Mol Biol 2008;110:157–62.
[59] Jin J, Wahlang B, Thapa M, Head KZ, Hardesty JE, Srivastava S, et al. Proteomics and metabolic phenotyping define principal roles for the aryl hydrocarbon receptor in mouse liver. Acta Pharm Sin B 2021;11:3806–19.
[60] Esser C, Rannug A. The aryl hydrocarbon receptor in barrier organ physiology, immunology, and toxicology. Pharmacol Rev 2015;67:259–79.
[61] Lv Q, Wang K, Qiao S, Yang L, Xin Y, Dai Y, Wei Z. Norisoboldine, a natural AhR agonist, promotes Treg differentiation and attenuates colitis via targeting glycolysis and subsequent NAD+/SIRT1/SUV39H1/H3K9me3 signaling pathway. Cell Death Dis 2018;9:258.
[62] Biagioli M, Marchianò S, Roselli R, Di Giorgio C, Bellini R, Bordoni M, et al. Discovery of a AHR pelargonidin agonist that counter-regulates Ace2 expression and attenuates ACE2-SARS-CoV-2 interaction. Biochem Pharmacol 2021;188:114564.
[63] Jin U-H, Lee S, Safe S. Aryl hydrocarbon receptor (AHR)-active pharmaceuticals are selective AHR modulators in MDA-MB-468 and BT474 breast cancer cells. J Pharmacol Exp Ther 2012;343:333–41.
[64] Lowe MM, Mold JE, Kanwar B, Huang Y, Louie A, Pollastri MP, et al. Identification of cinnabarinic acid as a novel endogenous aryl hydrocarbon receptor ligand that drives IL-22 production. PLoS One 2014;9:e87877.
[65] Danish (Q)SAR Database. https://qsar.food.dtu.dk/. (Accessed 19 October 2022).
[66] Aminpour M, Montemagno C, Tuszynski JA. An overview of molecular modeling for drug discovery with specific illustrative examples of applications. Molecules 2019;24:1693.
[67] Song Y, Slominski RM, Qayyum S, Kim T-K, Janjetovic Z, Raman C, et al. Molecular and structural basis of interactions of vitamin D3 hydroxyderivatives with aryl hydrocarbon receptor (AhR): an integrated experimental and computational study. Int J Biol Macromol 2022;209: 1111–23.
[68] Lukman Chaubah MA, Wihelmina Bontes B. Molecular docking of polycyclic aromatic hydrocarbons as potentially carcinogenic molecules through binding with aryl hydrocarbon receptor. JSMARTech 2019;1:12–5.
[69] Gu C, Cai J, Fan X, Bian Y, Yang X, Xia Q, et al. Theoretical investigation of AhR binding property with relevant structural requirements for AhR-mediated toxicity of polybrominated diphenyl ethers. Chemosphere 2020;249:126554.
[70] Ji C, Shen C, Zhou Y, Zhu K, Sun Z, Zuo Z, Zhao M. AhR agonist activity confirmation of polyhalogenated carbazoles (PHCZs) using an integration of in vitro, in vivo, and in silico models. Environ Sci Technol 2019;53:14716–23.
[71] García OC, De Winter H, Cos P, Matos MJ, Uriarte E, Maury GL, et al. Molecular docking study of flavonoids to block the aryl hydrocarbon receptor. Chem Proc 2021;8:77.
[72] Lo Piparo E, Koehler K, Chana A, Benfenati E. Virtual screening for aryl hydrocarbon receptor binding prediction. J Med Chem 2006;49: 5702–9.

CHAPTER

25

Use of in silico protocols to evaluate drug safety

Kevin P. Cross, Glenn J. Myatt and Candice Johnson

Instem, Columbus, OH, United States

Abbreviations

3Rs Replacement, reduction, and refinement of animal use
ADME Absorption, distribution, metabolism, and excretion
AI Acceptable limits
AOPs Adverse Outcome Pathways
GLP Good laboratory practices
IATAs Integrated Approaches to Testing and Assessment
ICH International Council for Harmonisation of Technical Requirements for Pharmaceuticals for Human Use
IST In silico toxicology
KE Key events
NAMs New Approach Methodologies
OECD Organisation for Economic Co-operation and Development
R&D Research and development
WoE Weight of evidence

Introduction

The development of new drugs requires the simultaneous evaluation and optimization of (1) potency, (2) absorption, distribution, metabolism, and excretion (ADME) properties along with (3) adverse effects. Computational approaches to assessing safety or in silico toxicology (IST) have experienced significant growth as a stand-alone method and as part of weight of evidence (WoE) considerations. This increased acceptance is driven by its recognition within specific regulatory guidelines (such as the ICH M7 pharmaceutical impurities guideline [1]) as well as a general need to (1) fill data gaps for chemicals with limited toxicology information to support lead identification and optimization, (2) improve R&D efficiency and avoid late-stage attrition, and (3) support the replacement, reduction, and refinement of animal use (3Rs). This expansion has been helped by advancements in IST science and documentation of best practices such as the establishment of validation principles [2] and guidelines for expert review of in silico results [3].

At the same time, our understanding of the complex mechanistic basis of different toxicological endpoints has been progressing. Initiatives, such as the Adverse Outcome Pathways (AOPs) [4], Integrated Approaches to Testing and Assessment (IATAs) [5,6], and New Approach Methodologies (NAMs) [7] have supported the documentation of the effects or mechanisms that need to be considered in a hazard assessment. These WoE hazard assessments incorporate information from a series of experimental and in silico results predictive of these defined effects or mechanisms. Such approaches, therefore, enable a more transparent assessment of the mechanistic and often structural basis that is supportive of the 3Rs and increased research efficiency.

Further position papers alongside the development and documentation of fit-for-purpose in silico models that incorporate advancements in our understanding of structure–activity (toxicity) relationships will support the continued expansion in the use of IST methods. However, one impediment restricting the deployment of in silico approaches is the limited availability of IST protocols, equivalent to in vivo or in vitro test guidelines. These protocols would provide in silico experts with a clear procedure to (1) select suitable in silico models, (2) run the models and perform an expert review, (3) derive an overall assessment and score the reliability of the information, (4) establish an overall assessment of the hazard endpoints (based on the WoE) and an associated confidence, and (5)

QSAR in Safety Evaluation and Risk Assessment
https://doi.org/10.1016/B978-0-443-15339-6.00028-X

document the results. Such protocols will support defending the use of in silico methods with peers and regulators and provide a procedure resulting in a consistent and reproducible hazard assessment. They build on the IST scientific progress and best practices as well as the mechanistic understanding.

This following book chapter describes the core IST protocol concepts alongside how published protocols can be executed within an interactive visual tool and incorporated across drug discovery.

In silico toxicology protocol concepts

Hazard assessment framework

IST protocols describe how a series of effects or mechanisms are combined to support a hazard assessment for a toxicological endpoint [8]. In silico models, alongside experimental data, can be used as part of the assessment of these effects or mechanism, which are outlined in the IST protocol. This is illustrated in Fig. 25.1, where experimental data and in silico results are collected to support an assessment of a series of effects or mechanisms ("bacterial mutation" and "mouse lymphoma"). This information is then used to assess an endpoint ("gene mutation"). The following section outlines how to:

(1) identify experimental data,
(2) select suitable in silico models,
(3) make an assessment and score the reliability of all information for any effect or mechanism using experimental data, in silico model results and an expert review,
(4) establish an overall assessment of the hazard endpoints (based on the WoE) and derive its associated confidence, and
(5) document the results.

Identifying experimental data

High-quality structure-searchable databases, which include the underlying study information, will support the assessment of different effects or mechanisms. This would ideally include access to both public and proprietary study data. Searching the databases for study information that maps onto effects or mechanisms outlined in the IST protocol is one of the first steps in making a hazard assessment. This is usually performed using an exact match structure search and assessing the availability and reliability of the information collected. For example, a matching chemical in the database may include an acceptable test result, but a summary of the results or any underlying study report should be examined to determine whether the data was generated in concordance with an accepted protocol. Other factors, such as whether it was performed according to GLP quality standard, may also be important to assess. In Fig. 25.1, even though a series of in silico models had been generated, the overall positive assessment for bacterial mutation is associated with a reliability score of RS2 reflecting the availability of reliable experimental data. The underlying information was then inspected, and the reliability documented (described in Assessing effects or mechanisms section).

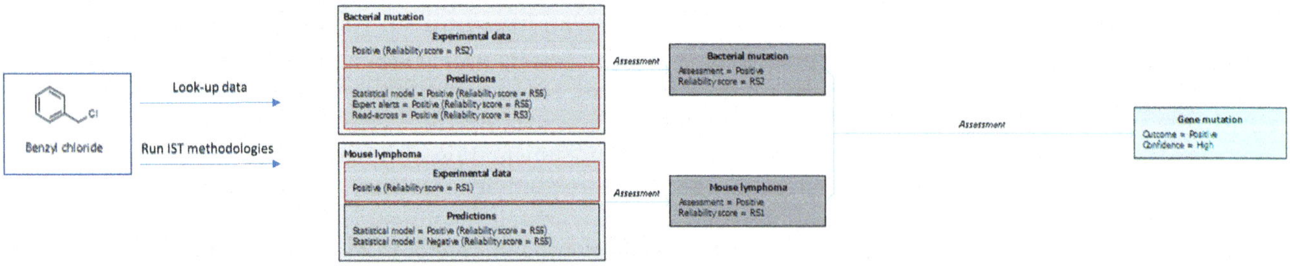

FIGURE 25.1 Collecting experimental data and predictions to support an assessment of hazard.

Selection of models

Any in silico approaches should be predictive of an effect or mechanism described in the hazard assessment framework. The performance and scope of any selected models are ideally validated (e.g., using OECD validation principles [2]). Another consideration is whether the models are sufficiently transparent to support an expert review of the results, for example, does a statistical method provide access to the features and weightings used to make a prediction as well as access to the underlying training set chemicals and studies used.

When multiple computational models are used to assess an effect or mechanism, it is important to consider if they are complementary methodologies. Three commonly used complementary methodologies are statistical-based, expert rule-based, and read-across. It is also possible to use other predictions methodologies to predict the effect or mechanism, such as quantum mechanical models. A summary of some of these methodologies is provided:

- **Statistical-based**: This methodology uses a training set of chemical structures each with an associated measured property (biological response). A series of calculated descriptors (such as the presence or absence of predefined chemical fragments or a series of physicochemical properties) are generated for all chemicals in the training set, and a mathematical model is derived and optimized to encode the relationship between these calculated descriptors and the biological response. This model can be now used to make predictions on chemicals with no experimental data. As part of the prediction process, a step to assess whether the test chemical is within the applicability domain of the model is performed. Next, the same calculated descriptors that are used as part of the model are computed for the test chemical(s) and are then used with the model to compute a prediction for the biological (toxicology) response.
- **Expert rule-based**: This methodology is based on a series of structural alerts that are encoded from the literature or derived from reference databases that contain experimental results. These alerts detail the structure–activity (toxicity) relationships, specifically features that both activate and deactivate the toxicity. A test chemical is usually compared against the alert's reference database to determine whether it is within the applicability domain of the alert system. The alerts are then matched against the test chemical which is flagged as potentially toxic if one or more alerts fire. The biological basis and database analogs are usually presented as supporting information to enable an expert review.
- **Read-across**: This is a semiautomated methodology where, for a test chemical (referred to as the target), one or more analogs are identified with experimental data (referred to as the source). The analogs are usually selected through a combination of chemical structural similarity, physicochemical similarity, and biological or toxicological similarity (using experimental data or expert systems that profile mechanistic domains). It is important to justify the why the source chemicals are meaningfully similar. Experiment data on these analogs is then used to "read-across" onto the target to generate a prediction, such as selecting the most conservative value from the analogs.

Assessing effects or mechanisms

Fig. 25.1 presents a hazard assessment framework that includes an evaluation of bacterial mutagenicity for the chemical benzyl chloride using both experimental data and in silico model results. In this example, positive experimental data is identified for benzyl chloride, and three prediction methodologies are also performed on the chemical: (1) a statistical-based, (2) an expert rule-based, and (3) a read-across.

It is considered the best practice is to use multiple complementary in silico methodologies coupled with an expert review of the information. This review will increase the reliability of the overall assessment. Although an expert review of the information is inherently subjective, a series of guidelines have been established to structure such a review [8]. For example, Table 25.1 illustrates different considerations as part of an expert review of the different methodologies.

For an individual effect or mechanism, an overall assessment is made based on the information collected and a score is assigned that reflects its reliability. This score ranges from RS5 (least reliable) to RS1 (most reliable). A single in silico model result is considered the least reliable and is assigned a reliability score of RS5. Multiple concurring in silico result would increase the overall reliability assessment to RS4. An expert review of the in silico result(s) will further increase the reliability score to RS3, as long as the review concludes there is sufficient supporting evidence. Any experimental data is also considered as part of the reliability score based on the criteria established in the Klimisch score [9], where Klimisch score 3 (not reliable) or 4 (not assignable) maps onto RS5, Klimisch score 2 (reliable with restrictions) maps onto RS2, and Klimisch score 1 (reliable without restrictions) maps onto RS1. An expert

TABLE 25.1 Expert review consideration for the different in silico methodologies.

Methodology	Example guideline elements to consider
Statistical-based	- Inspect the rationale behind the applicability domain assessment - Review information supporting the model's confidence score - Assess whether the structural features highlighted in the prediction are biologically meaningly and not just statistical anomalies - Evaluate the quality and diversity of the underlying experimental data on analogs - Review additional literature or other experimental data supporting the prediction
Expert rule-based	- Inspect the information supporting any confidence scores associated with alerts that fire - Qualify that the absence of an alert can justify a lack of toxicity - Assess and justify any deactivation of an alert - Evaluate the environment around the alert to determine whether there are additional factors that may influence potential activation or deactivation of the toxicity - Look for additional support from chemical analogs - Review additional literature or other experimental data supporting the prediction
Read-across	- Justify the analogs based on the structural similarity, physicochemical property similarity, and biological or toxicological relevance - Support the source chemical(s) data quality - Assess any bias in the analog selection process - Review the overall read-across prediction

review of the experimental data is required in deriving the Klimisch score. An overall assessment in then based on the highest reliability score from the information collected. In Fig. 25.1, even though a series of in silico models had been generated, the overall positive assessment is associated with a reliability score of RS2 reflecting the availability of reliable experimental data. In situations where there is conflicting evidence, an expert review of the information is prudent.

Assessing hazard endpoints

An assessment of one or more effects or mechanisms is used to assess one or more toxicological endpoints. Fig. 25.1 illustrates how the results from two effects or mechanisms ("bacterial mutation" and "mouse lymphoma") are combined to generate an assessment of a toxicological hazard endpoint ("gene mutation"). An overall assessment and confidence score are determined using information on the associated effects or mechanisms. The rules or principles for calculating both the overall outcome and the confidence score are enumerated within the IST protocol. In general, the confidence score takes into consideration the reliability of the information generated (as codified in the reliability score), the completeness of the information collected (e.g., missing information on critical effects or mechanisms would lower the confidence), and the relevance of the effects or mechanisms in determining the hazard endpoint. In general, this confidence score is a quantitative output using values such as "high," "medium," and "low." Although this assessment and confidence score can be calculated directly using the rules and principles outlined in the IST protocol, additional consideration may also be considered as part of an expert review, and such a review may adjust the overall assessment or the confidence level with sufficient supporting evidence.

Documentation

The information used in determining the overall hazard assessment should be ideally clearly documented including (1) information on the overall decision content as well as when the analysis was performed and by whom, (2) an executive summary of the results and conclusions, (3) an explanation of the purpose of the assessment, (4) materials and methods outlining the systems, models, and databases used and their associated version numbers, (5) detailed information on the results and expert review performed within the context of the hazard assessment framework, and (6) conclusions that summarize the approach and overall findings within the decision context. For completeness, it is often helpful to also include an appendix with study and in silico reports generated by the software application.

Current status

The development of an IST protocols requires input from a variety of stakeholders, and as such, a series of cross-industry working groups were established to determine whether it is feasible to generate an IST protocol for a given endpoint. In some situations when the state of the science was evolving or when there was a lack of availability or suitable in silico models, it was not possible to generate an IST protocol. These findings were documented as position papers summarizing a roadmap toward an IST protocol. Protocols for genetic toxicology [9], skin sensitization [10], and endocrine activity have been completed, whereas position papers around acute lethality [11], liver toxicity [12], heart, kidney, and lung toxicities [13]; carcinogenicity [14]; and neurotoxicity [15] describe the current state of the science and in silico technology development.

Applying in silico toxicology concepts and protocols to assess genotoxic impurities

The IST protocol concepts can be applied as part of different applications where effects or mechanisms have been defined and principles for evaluating the overall hazard assessment have been described. For example, the concepts have been applied to the assessment of pharmaceutical impurities [1]. The application of these concepts to the assessment of DNA-reactive mutagenicity of pharmaceutical impurities is described in this section.

The International Committee for Harmonization (ICH) M7 guideline "Assessment and Control of DNA Reactive (Mutagenic) Impurities in Pharmaceuticals to Limit Potential Carcinogenic Risk" [1] includes an in silico option to predict the bacterial reverse mutation assay [16] when no adequate experimental data is available. Experimental data on carcinogenicity can also be used as part of the assessment.

In Fig. 25.2, a hazard assessment framework has been put together that is aligned with the ICH M7 guideline. Two effects—bacterial mutation and carcinogenicity—are included, and it is possible to use adequate experimental data to assess both effects. In addition, in the absence of sufficient bacterial mutagenicity data, the results from two in silico methodologies can be used. No in silico approaches may be used to access carcinogenicity; however, knowing if the impurity belongs to a cohort of concern, as defined in the ICH M7 guideline, is important in the overall assessment.

The required output from this analysis of the effects (shown in the blue "ICH M7 Protocol" box) is for each chemical to be assigned to one of five classes. Each class number has a specific meaning and determines whether the impurities need to be further controlled or whether additional experimental testing is needed. The rules and principles for how this information is combined to establish an overall class is described in the guidelines [1], as well as in supporting Q&A documents and other papers [17]. Adequate experimental data can be used to assign an impurity to classes 1 (known mutagenic carcinogen), 2 (known mutagen with unknown carcinogenic potential), and 5 (sufficient data to demonstrate lack of mutagenicity or carcinogenicity). In the absence of experimental data, the in silico results can be used to assign the impurity to class 3 (alerting structure, unrelated to the structure of the drug substance, no

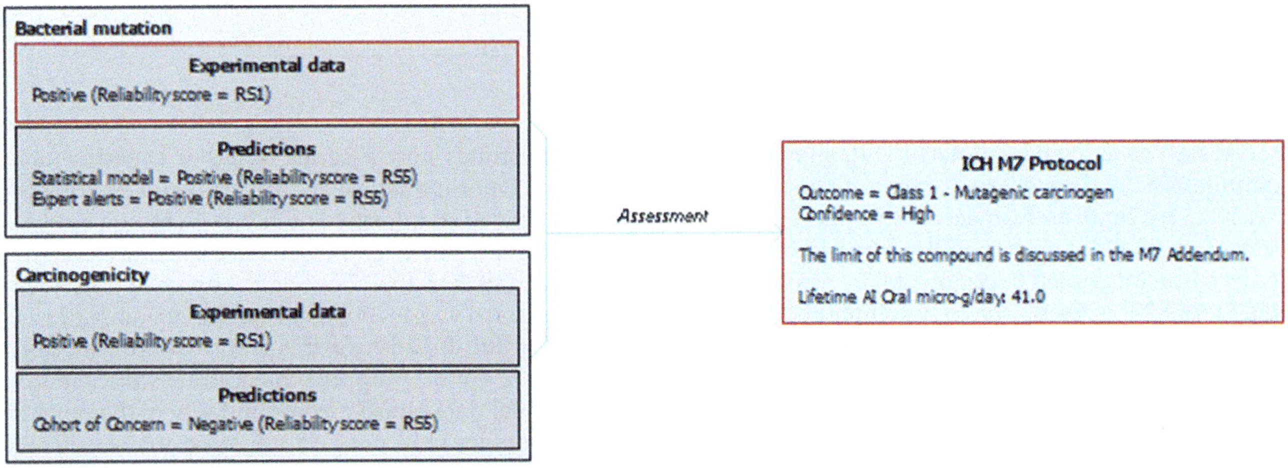

FIGURE 25.2 ICH M7 hazard assessment framework.

mutagenic data), class 4 (alerting structure, same alert in drug substance which has been tested and is nonmutagenic), and class 5 (no structural alerts).

In Fig. 25.2, experimental and in silico information is collected and included within the hazard assessment framework. For this chemical (benzyl chloride), experimental data was identified from an exact structure database search. An inspection of the studies performed on this chemical identifies a number of positive studies that, after an expert review, could be assigned to the reliability score of RS1, i.e., they were performed according to the OECD 471 protocol [16] and GLP quality standards. Two complementary in silico models that predict the results of the bacterial mutagenicity test were also run and the results concurred with the experimental data. It should be noted that in the absence of adequate experimental data, an expert review of the in silico results would be prudent. In addition, positive carcinogenicity data was identified from a database search for this chemical, and after an expert review of the information, it was assigned to a reliability score of RS1. The cohorts of concern profiler were also run against the chemicals, and no concerning chemical class was flagged. This information on bacterial mutagenicity and carcinogenicity was then combined to assign the chemical to an ICH M7 class (class 1) based on the principles outlined before. Since this analysis made use of high reliable experimental data (RS1), the confidence in this assessment was set to high. Additional supporting information on the chemical's acceptable intake (AI) further supports how this impurity should be controlled.

Genetic toxicology in silico toxicology protocol

Understanding the potential for a compound to induce genetic alterations helps to determine whether there is a possible carcinogenic, reproductive, or other safety issue. Genetic toxicology generally evaluates the following endpoints: (1) gene mutation (DNA adducts, insertions, inversions, and deletions), (2) clastogenicity (duplication, deletion, and rearrangement of the chromosome), and (3) aneugenicity (generation of an abnormal number of chromosomes). In addition, general DNA damage may also be considered as part of the battery of tests.

The genetic toxicology in silico protocol [18] can be used to assess whether a chemical is or is not genotoxic, based on these major endpoints. Fig. 25.3 presents the hazard assessment framework for genetic toxicology in which the overall assessment of "genetic toxicity" is based on the subendpoints "gene mutation" and "clastogenicity/aneugenicity" (which is further separated into in vitro and in vivo "clastogenicity/aneugenicity"). For each of the endpoints, a series of effects or mechanisms are assessed using available experimental data and/or in silico model(s) results. For example, "bacterial mutation" and "mouse lymphoma" are used to assess "gene mutation," and experimental data and in silico model results are collected and organized. Table 25.2 summarizes the different Organisation for Economic Co-operation and Development (OECD) test guidelines used to assess the different endpoints under consideration.

After having collected the experimental data and in silico results for each of the effects or mechanisms, an overall call and reliability score is then computed, as described in Assessing effects or mechanisms section. This information is then used to assess the endpoints (and their associated confidence values) based on both the general concepts outlined in Assessing hazard endpoints section as well as specific rules or principles outlined in the genetic toxicology in silico protocol [9]. An expert review of all endpoints is prudent and may take into consideration other related information related to DNA damage, for example.

Skin sensitization IST protocol

The skin sensitization protocol [19], shown in Fig. 25.4, demonstrates how a series of either experimental or computationally derived assessments for mechanisms or effects relevant for skin sensitization could be combined to derive an overall assessment. The protocol includes an assessment of AOP-informed methods including an assessment of covalent interaction with proteins (KE 1), events in keratinocytes (KE 2), and events in dendritic cells (KE 3). To derive an integrated experimental in vitro call, a defined approach such as the "2 out of 3" approach described in OECD 497 [20] is used. The local lymph node assay (LLNA) is used to assess KE 4. Guinea pig maximization test (GPMT) and human studies inform on the adverse outcome. The full breadth of this information may not be required for a reliable assessment. As indicated by OECD 497, the "2 out of 3" approach could be sufficient given a high confidence in the assessment. Likewise, where sufficient animal data already exists and where the reliability of that assessment is sufficient for purpose, then additional information will not add value on the sensitization assessment. Where data gaps exist, or where the reliability of the data is not sufficient for purpose, in silico models

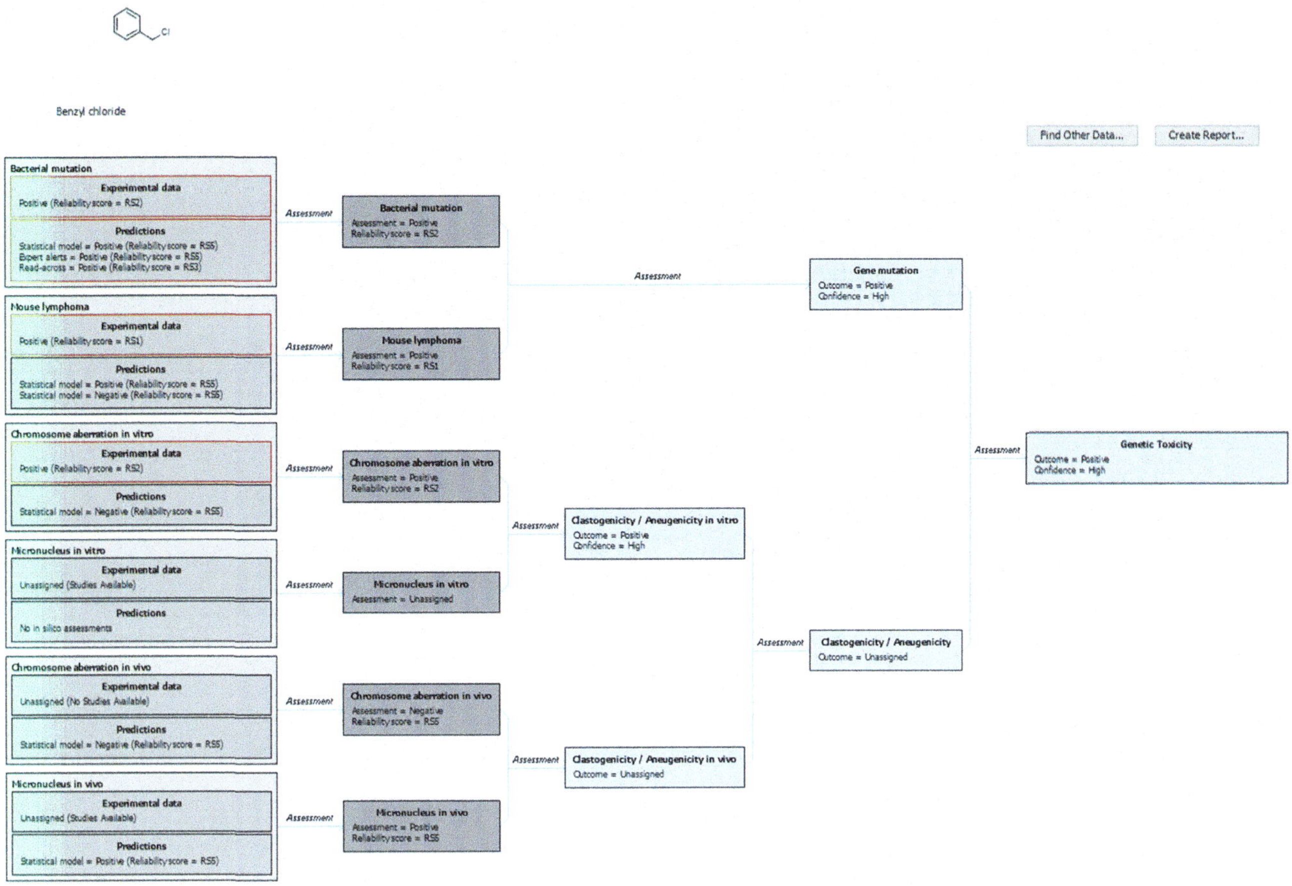

FIGURE 25.3 Hazard assessment framework for genetic toxicology.

TABLE 25.2 Examples of OECD test guidelines.

Endpoint	Examples of OECD experimental test guideline
Gene mutation	471, 476
Clastogenicity/aneugenicity in vitro	487, 490
Clastogenicity/aneugenicity in vivo	474, 475, 478, 483

are used to add value. The reliability of the in silico predictions is lower than experimental data conducted from guideline adherent studies, and this factor is taken into consideration when assigning and assessing the confidence in a prediction. Johnson [19] demonstrates this integrative approach using a series of case studies for a data-rich and data-poor compound, which is assessed primarily based on computational approaches.

Interactive visual hazard assessment framework

To complete an assessment aligned with any IST published protocol requires a series of operations to be performed including (1) database searches to identify adequate experimental data for each defined effect or mechanisms, (2) running the selected in silico models, (3) performing an expert review of the available information, (4)

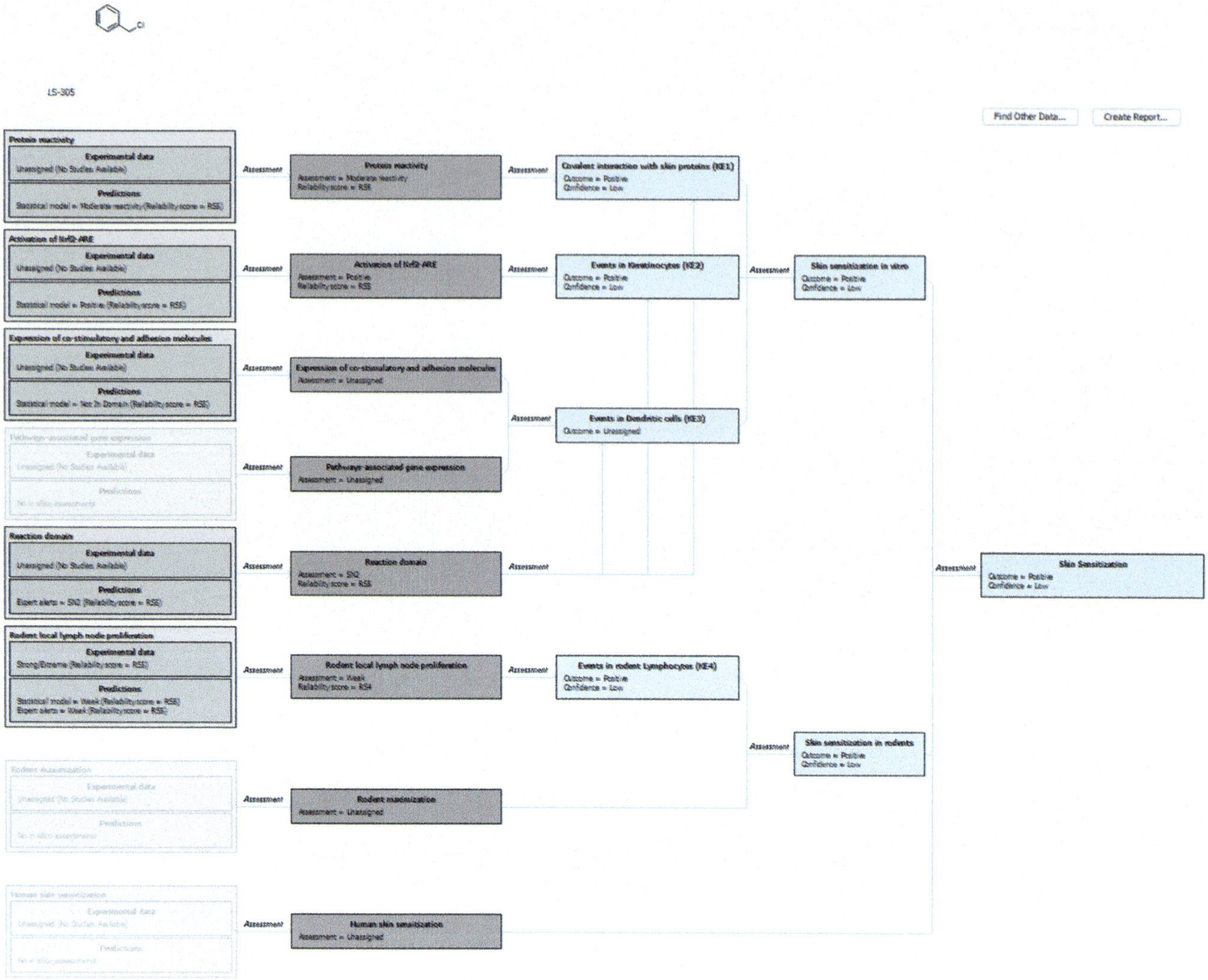

FIGURE 25.4 Skin sensitization in silico protocol.

determining and reviewing an overall assessment and reliability score for each effect or mechanism, (5) combining and reviewing the information based on the IST protocol rules and principles (and expert review) to establish an overall assessment and confidence score for each hazard endpoint, and (6) documenting the results and expert review. Although an expert review of the information requires manual intervention, most of the other operations can be automated enabling a faster analysis that is less susceptible to errors.

Fig. 25.5 illustrates the complete hazard assessment framework for genetic toxicology encoded in a visual and interactive graphical user interface. Here, a series of effects mechanisms (shown in the left column as light gray boxes) are assessed, and reliability score is assigned. Each effect or mechanism is linked to a database search to identify any relevant experimental data. It is also linked directly to multiple complementary computational models. Based on a series of general and protocol-specific rules, an overall assessment and reliability score is calculated for each effect or mechanism. This information is in turn used to derive a series of hazard endpoints (shown in light blue) and confidence scores.

An exact structure database search was performed for benzyl chloride to identify bacterial mutagenicity study results. A number of studies were returned that can be viewed by clicking on the experimental data subbox, including the study summarized in Fig. 25.5(A). A summary of the experimental design and overall results is provided, with a link to the full study report. This information is presented to support an expert review to assess the

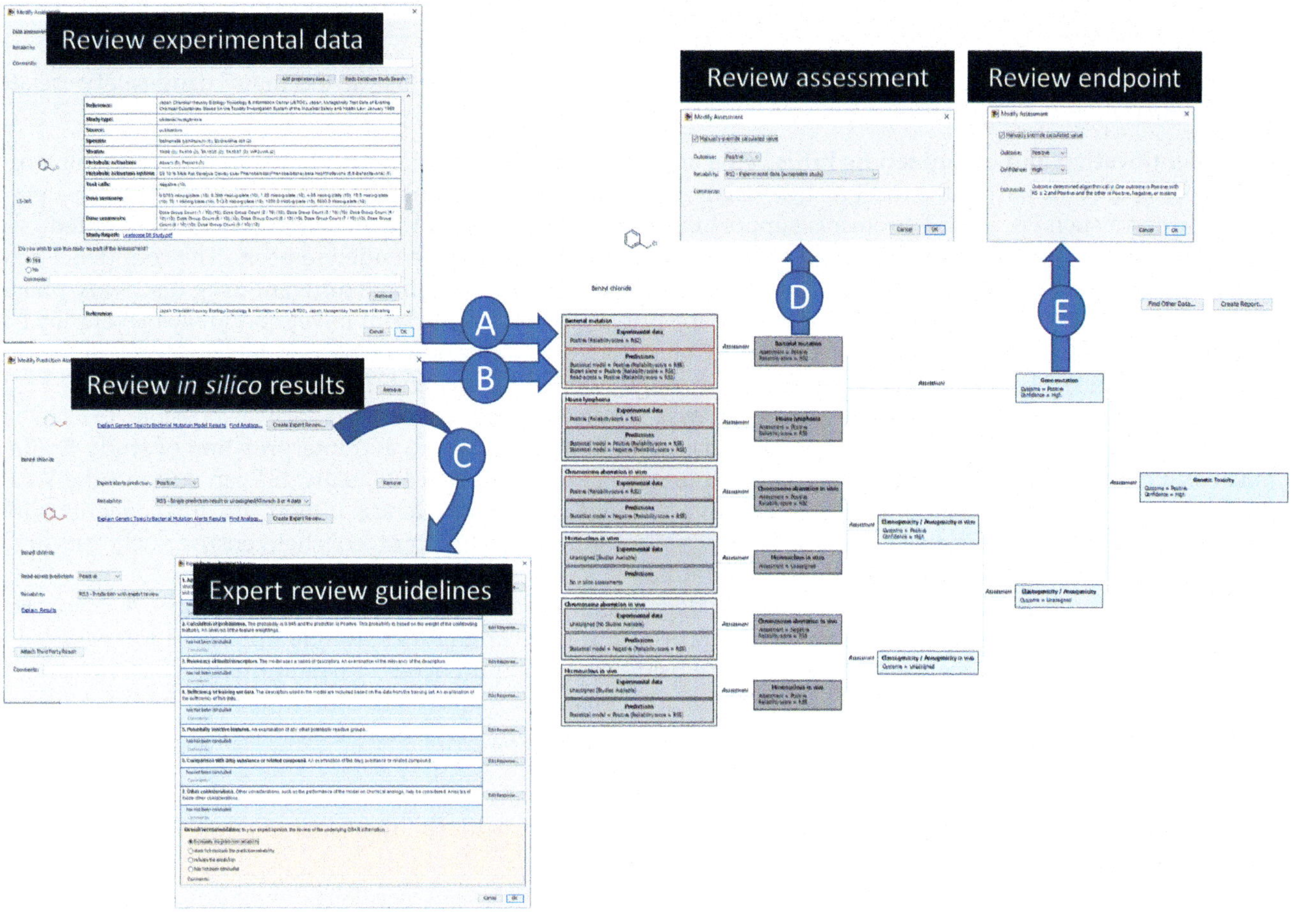

FIGURE 25.5 Hazard assessment framework for genetic toxicology.

reliability of the information provided. In this example, based on an examination of the results, it was concluded that the default reliability score (RS5) should be raised to RS2.

A series of complementary in silico methodologies are also directly linked to the bacterial mutation box, in the area shown as "Predictions". Further information on how these predictions were calculated can be explored by clicking on the predictions subbox shown in Fig. 25.5(B) where an explanation of the different in silico methodologies results is available. Each prediction is also linked to a "Create Expert Review" that provides a set of guidelines for conducting such a review along with the capability to document each one (Fig. 25.5(C)). Such a review may increase the reliability of a prediction result when there is sufficient supporting information. The review may even overturn or refute a prediction, again with sufficient and documented supporting evidence.

Each effect or mechanism is accompanied by an overall assessment and a reliability score. These values are determined based on the experimental data and in silico results, either the original results or the modified results based on an expert review. This information is in turn then used to derive an overall assessment for the hazard endpoint and a confidence score. Similarly, both the overall effect or mechanism assessment (Fig. 25.5(D)) and the overall hazard endpoint (Fig. 25.5(E)) can be inspected and potentially modified with sufficient information, which is ideally documented.

Once an inspection of the results and any necessary expert review has been completed, then the results can be automatically documented. This automation both saves time and avoids transcription errors that may result when generating the documentation by hand. It also provides a completed audit trail of the evaluation process, including any expert review of the information and justification of any changes that deviate from the automated procedures outlined in the protocol.

Discussion and conclusions

The concepts outlined as part of the IST protocols framework can be applied across many drug safety–related decisions including:

- **Lead identification and optimization**: IST is being applied to the identification of drug leads as a screening tool. It is also being used as part of the design–test–analyze lead optimization that incorporates predictions of safety to design-around such issues. IST protocols support the use of computational models through a documented standardized execution and interpretation of the results that can be easily defended to peers. The confidence score generated as part of the protocol implementation can be strategically used in the different drug discovery phases. For example, at an early phase where IST is used for prioritization, results with lower confidence level may be acceptable, whereas higher confidence scores may be required as the candidates progress.
- **Nonclinical safety**: As chemicals are progressed through nonclinical safety, IST can provide supportive information into the sequencing of these tests to derisk predicted safety concerns as early as possible. IST can also support experiment design considerations (such as setting initial test doses) and interpretation of study findings (such as a particular nonclinical finding predicted to have clinical relevance). Finally, IST can support the WoE assessment, in conjunction with in vitro and in vivo findings, and help to assess if there are any concerns with metabolites. In a similar manner to in vitro or in vivo test guidelines, IST protocols help ensure such tests are performed according to appropriate quality standards.
- **Quality**: IST is routinely used to assess impurities, degradants as well as extractable and leachables. A major focus is the assessment of DNA-reactive mutagenicity; however, it is becoming increasingly important to assess sensitization, irritation, as well as systemic toxicity using in silico approaches. This information also needs to be submitted to regulatory authorities in a transparent and consistent manner which is supported by IST protocols.
- **Manufacturing**: IST is being used to support occupational toxicology, classification and labeling, green chemistry, and emergency response situations, and IST protocols support how to apply computational models in these situations. It is important to predict acute effects, such as acute lethality, sensitization, and irritation/corrosion, and IST protocols provide a rapid and defendable approach that is particularly important to addressing emergency response situations.
- **Regulatory submissions**: In addition to the assessment of impurities as well as extractables and leachables, a few additional regulatory submissions also permit the use of in silico approaches, including pharmaceuticals in the environment and assessment of mechanism-based inhibition of cytochrome P450. There are also several emerging regulatory areas where in silico approaches may play an important role such as the ICH S1B guideline to support the recently published WoE factors [21] and the assessment of nongenotoxic impurities [22].

In summary, the use of IST protocol concepts provides a transparent conceptual framework for assessing a variety of toxicological endpoints in a consistent and repeatable manner. Where the science and technology permit, peer-reviewed IST protocols further support the acceptance of in silico methods across drug discovery and development, manufacturing, as well as for regulatory submissions. Although the use of IST protocols is critical for making and defending predictions, the continual development of position papers and guidelines documenting the evolving state of the science and technology is also important. In addition, academic programs provide training to effectively apply these IST protocol concepts along with technology solutions that allow for the rapid application of IST protocols that incorporate new models and in silico approaches, as illustrated in this book chapter.

References

[1] ICH. ICH guideline M7 (R1) on assessment and control of DNA reactive (mutagenic) impurities in pharmaceuticals to limit potential carcinogenic risk. European Medicines Agency; 2017. https://database.ich.org/sites/default/files/M7_R1_Guideline.pdf.

[2] OECD. Guidance document on the validation of (quantitative) structure-activity relationship [(Q)SAR] models. Paris: OECD Environment, Health and Safety Publications; 2007.

[3] Amberg A, et al. Principles and procedures for implementation of ICH M7 recommended (Q)SAR analyses. Regul Toxicol Pharmacol 2016;77: 13–24. https://doi.org/10.1016/j.yrtph.2016.02.004.

[4] Ellison CM, Piechota P, Madden JC, Enoch SJ, Cronin MTD. Adverse Outcome Pathway (AOP) informed modeling of aquatic toxicology: QSARs, read-across, and interspecies verification of modes of action. Environ Sci Technol 2016;50:3995–4007. https://doi.org/10.1021/acs.est.5b05918.

[5] OECD. Guidance document on the reporting of defined approaches to be used within integrated approaches to testing and assessment. Paris: OECD Environment, Health and Safety Publications; 2016.

[6] OECD. Guidance document on the reporting of defined approaches and individual information sources to be used within integrated approaches to testing and assessment (IATA) for skin sensitisation. Paris: OECD Environment, Health and Safety Publications; 2016.

[7] US EPA. Integrated risk information system. US EPA; 2021. https://www.epa.gov/assessing-and-managing-chemicals-under-tsca/strategic-plan-reduce-use-vertebrate-animals-chemical. [Accessed 9 April 2021].
[8] Myatt GJ, et al. In silico toxicology protocols. Regul Toxicol Pharmacol 2018;96:1–17. https://doi.org/10.1016/j.yrtph.2018.04.014.
[9] Klimisch H-J, Andreae M, Tillmann U. A systematic approach for evaluating the quality of experimental toxicological and ecotoxicological data. Regul Toxicol Pharmacol 1997;25:1–5. https://doi.org/10.1006/rtph.1996.1076.
[10] Johnson C. Skin sensitization in silico protocol. Regul Toxicol Pharmacol 2020;116:104688. https://doi.org/10.1016/j.yrtph.2020.104688.
[11] Zwickl C, et al. Principles and procedures for assessment of acute toxicity incorporating in silico methods. Comput Toxicol 2022.
[12] Bassan A, et al. In silico approaches in organ toxicity hazard assessment: current status and future needs in predicting liver toxicity. Comput Toxicol 2021;20:100187.
[13] Bassan A, et al. In silico approaches in organ toxicity hazard assessment: current status and future needs for predicting heart, kidney and lung toxicities. Comput Toxicol 2022;20:100188.
[14] Tice R, et al. In silico approaches in carcinogenicity hazard assessment: current status and future needs. Comput Toxicol 2021;20:100191.
[15] Crofton KM, et al. Current status and future directions for a neurotoxicity hazard assessment framework that integrates in silico approaches. Comput Toxicol 2022;22:100223. https://doi.org/10.1016/j.comtox.2022.100223.
[16] OECD. Guideline no. 471: bacterial reverse mutation test. OECD; 2020. https://www.oecd-ilibrary.org/docserver/9789264071247-en.pdf?expires=1671212867&id=id&accname=guest&checksum=513DC78D3B6FA4191D8F3BFA5FF8FDB3. [Accessed 15 December 2022].
[17] Myatt GJ, et al. Implementation of in silico toxicology protocols within a visual and interactive hazard assessment platform. Comput Toxicol 2021:100201.
[18] Hasselgren C, et al. Genetic toxicology in silico protocol. Regul Toxicol Pharmacol 2019;107:104403. https://doi.org/10.1016/j.yrtph.2019.104403.
[19] Johnson C, et al. Evaluating confidence in toxicity assessments based on experimental data and in silico predictions. Comput Toxicol 2022;21:100204.
[20] OECD. Guideline no. 497: defined approaches on skin sensitisation. OECD; 2021. https://www.oecd-ilibrary.org/environment/guideline-no-497-defined-approaches-on-skin-sensitisation_b92879a4-en. [Accessed 21 September 2021].
[21] ICH. International council for harmonisation of technical requirements for pharmaceuticals for human use. In: ICH harmonized guideline—addendum to the guideline on testing for carcinogenicity of pharmaceuticals S1B(R1); 2021. https://database.ich.org/sites/default/files/ICH_S1BR1_Step2_DraftGuideline_2021_0510.pdf.
[22] EMA. Reflection paper on the qualification of non-genotoxic impurities. EMA/CHMP/SWP/545588/2017; 2017. https://www.ema.europa.eu/en/documents/scientific-guideline/reflection-paper-qualification-non-genotoxic-impurities_en.pdf.

C H A P T E R

26

QSAR models for predicting cardiac toxicity of drugs

Igor H. Sanches[1], Sabrina S. Mendonca[1], Vinicius M. Alves[2], Rodolpho C. Braga[3] and Carolina H. Andrade[1]

[1]Laboratory for Molecular Modeling and Design, Faculty of Pharmacy, Universidade Federal de Goiás, Goiania, GO, Brazil
[2]Laboratory for Molecular Modeling, University of North Carolina at Chapel Hill, Chapel Hill, NC, United States
[3]InsilicAll Ltda, São Paulo, SP, Brazil

Abbreviations

ACC Accuracy
AUC Area under the curve
CCR Correct classification ratio
CHO Chinese hamster ovary
DNN Deep neural networks
FDA Food and Drug Administration
HEK Human embryonic kidney
LBDD Ligand-based drug design
LQTS Long QT syndrome
MCC Matthews correlation coefficient
OECD Organisation for Economic Co-operation and Development
QSAR/QSPR Quantitative structure—activity/property relationship
RF Random forest
SBDD Structure-based drug design
SEN Sensitivity
SPE Specificity
SVM Support vector machine
XO *Xenopus laevis* oocytes

Introduction

Toxicity is any harmful effect on a living organism caused by a chemical compound, e.g., cleaning supplies, pesticides, pharmaceuticals, pollutants, or toxins. Furthermore, the stricter definition relates to the use of drugs in concentrations higher than recommended. According to the National Cancer Institute of the United States, cardiotoxicity is any toxic effect that affects the heart [1].

There are several mechanisms through which cardiotoxicity can occur, including the direct toxic effect or accumulation of the agent on the heart tissue, the generation of reactive oxygen species, and the induction of apoptosis or necrosis in heart cells. In addition, some agents can alter the expression of genes [2] involved in cardiac function and lead to cardiac dysfunction. Adverse drug effects via disruption of electrophysiology, contractibility, mitochondrial toxicity, growth factor, and cytokine regulation are also often associated with the mechanisms of cardiotoxicity, which are often multifactorial and diverse [3].

Detecting cardiotoxicity is challenging. Clinical signs and symptoms may not appear until the drug has been approved and is in widespread use. Regulatory agencies, such as the US Food and Drug Administration (FDA), established requirements to anticipate the elucidation of cardiotoxicity in pharmacovigilance [4]. The management of cardiotoxicity often involves the discontinuation of the offending agent and the use of supportive therapies to address the underlying cardiac dysfunction.

The mechanisms of cardiotoxicity are complex and multifactorial (see Fig. 26.1). They can be divided into four types, drug-induced disruption of electrophysiology, which can be caused by the blockage of transport ion channels, e.g., human ether-a-go-go-related gene (hERG), L-type calcium channel (I_{CaL}) inward sodium current (I_{Na}) and late

QSAR in Safety Evaluation and Risk Assessment
https://doi.org/10.1016/B978-0-443-15339-6.00039-4

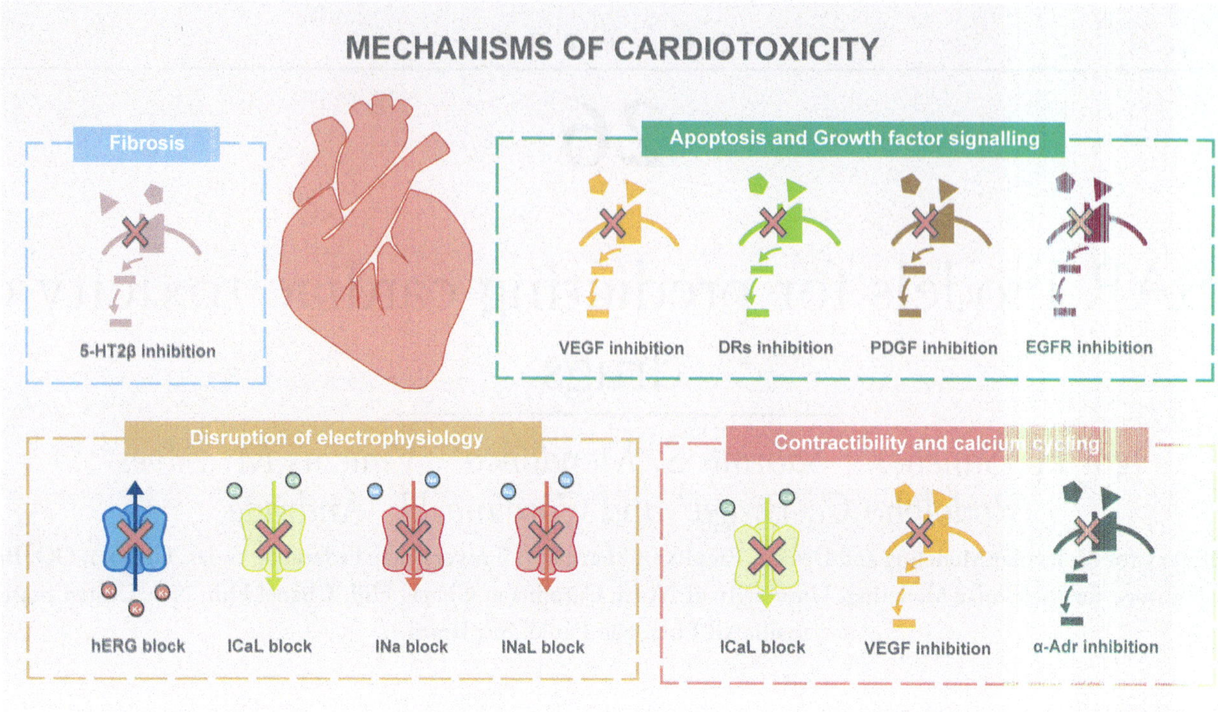

FIGURE 26.1 Mechanisms of cardiotoxicity.

sodium current (I_{NaL}), disruption of the normal states of contractibility, and calcium cycling (blockage of the ion channel I_{CaL}). There is also the disruption of the vascular endothelial growth factor (VEGF) and alpha-adrenergic receptor (α-Adr) linked to contractibility and calcium cycling pathways. Another mechanism is the disruption of 5-hydroxytryptamine receptor (5-HT2β) pathway linked to cardiac fibrosis induction, and VEGF, death receptors (DRs), platelet-derived growth factor (PDGF), and epidermal growth factor receptor (EGFR) pathways linked to apoptosis and growth factor signaling disruption (Fig. 26.1).

Drugs were withdrawn from the market

Cardiotoxicity is a particular concern due to the increase in postmarket drug withdrawals, representing 14% of 462 medicinal products withdrawn from the market between 1953 and 2013 [5]. Cardiac disorders represent 18.8% of 133 withdrawn drugs worldwide between 1990 and 2010 [6]. Naturally, this is a broad topic and can be caused by various medications and other related agents, attracting significant attention from scientists and clinicians.

Most drugs causing cardiotoxicity are classified as central nervous system inhibitors, followed by cardiovascular drugs, antiinflammatory, gastrointestinal, antiallergic, antiinfective, genitourinary, and others [3]. Kalyaanamoorthy and Barakat [7] summarized the FDA-approved drugs that were withdrawn from the market due to the prolongation of QT interval, while Mamoshina et al. in 2021 [3] reviewed all classes of known drugs and cardiotoxicity mechanisms.

For example, the FDA approved the drug astemizole in the United States in 1988, and Johnson & Johnson withdrew it from the global market voluntarily in 1999. The studies revealed that astemizole and its metabolite demethylated astemizole effectively blocked the hERG channel (see the next section), reducing the conductivity of the delayed rectifier potassium channel (I_{kr} channel encoded by hERG) [8]. In vitro studies showed that inhibiting the hERG potassium channel preferentially in a concentration-dependent manner may prolong the cardiac action potential duration associated with long QT syndrome (LQTS) [9]. As a result, an increase in drug concentration may influence the occurrence of LQTS. The mutation on the subunit gene, *KCNQ1* (c793t), on the other hand, was found to prolong cardiac membrane depolarization and increase susceptibility to LQTS [10,11].

hERG channel

The hERG K^+ channel, also known as the human ether-a-go-go-related gene channel (*KCNHA* gene), is a potassium channel commonly expressed in the heart, but also in various brain regions, smooth muscle cells, endocrine cells, and a wide range of tumor cell lineages [12]. These channels may also be referred to as I_{kr} (delayed rectifier K^+), when discussing the native channel, Kv11.1, when referring to the channel expressed in heterologous systems, and hERG, when referring to both. hERG is part of a group of three subfamilies of the large ether-a-go-go family of *voltage-gated potassium* channels, namely, Kv10.x, Kv11.x and Kv12.x. Within the Kv11.x subfamily, there are three members, Kv11.1 (hERG), Kv11.2 (hERG2), and Kv11.3 (hERG3) [13]. The hERG channel is specific for K^+ and sensitive to voltage changes in cell membrane potentials—having their function well elucidated in the heart, and is responsible for mediating the repolarization current I_{kr}, in the cardiac action potential, contributing to phase 3 repolarization, thus coordinating the heartbeat.

Genetic defects in the major pore-forming subunit hERG are associated with congenital (LQT2 and LQT6) and acquired long-QT syndrome, which carries an increased risk of life-threatening cardiac arrhythmias [14–17]. LQT2 is caused by dysregulation of the K^+ hERG channel. It is a condition in which the repolarization of the heart after a heartbeat is affected due to the prolongation of the interval between Q and T waves [18]. This increases the risk of irregular heartbeat (cardiac arrhythmia), which can lead to fainting, drowning, seizures, or sudden death. These episodes can be triggered by exercise or stress. Some rare forms of long QT are associated with other symptoms and signs, including deafness and periods of muscle weakness [19].

Cryo-EM structure—Possibilities of structure-based drug design

Until 2021, homology modeling techniques have been applied to predict the hERG structure. Many homology models were developed to further the knowledge of potassium channel functioning and drug-binding interactions rather than being used as a predictive tool [21].

The crystal structure of the hERG channel (Fig. 26.2) has been recently resolved by cryogenic electron microscopy (cryo-EM) [20]. The channel is formed by four identical alpha subunits (Fig. 26.2A and B), which form the transmembrane pore and the K^+ selectivity filter (Fig. 26.2C). Each hERG subunit consists of six transmembrane alpha helices, numbered S1–S6. The S4 helix contains a positively charged arginine or lysine amino acid residue at every 3rd position. It is believed to act as a voltage-sensitive sensor, which allows the channel to respond to changes in voltage by changing conformations between open, inactive, and closed [20]. A selectivity filter is present in the pore domain (Fig. 26.2D), formed by S5 and S6 segments, allowing potassium to cross the membrane [21].

The K^+ selectivity filter is sensitive to the binding of many compounds with no apparent structural correlation, as well as K^+ at deficient extracellular levels, which can result in impaired channel function and induced (acquired) long QT syndrome. Among the types of drugs that can cause QT prolongation, the most common include antiarrhythmics, antipsychotic agents, and certain antibiotics (including quinolones and macrolides) [22]. Although there are other potential targets for cardiac toxicity, it is known that most compounds associated with acquired QT prolongation interact with the hERG potassium channel. One of the main reasons for this phenomenon is its large hydrophobic intracellular binding pocket (selectivity filter, Fig. 26.2D) compared with other K^+ channels in the same family, thus providing more space for many different classes of compounds to bind and block this potassium channel [23,24].

hERG inhibitors often interact with the selectivity filter. Usually, most inhibitors interact with T623, S624, and V625 at the P1, and residues G648, Y652, and F656, present on the S6 helix. A high-affinity aromatic binding determinant for blockers is also located in helix S5, F557 [21].

Therefore, the expanded knowledge on the hERG channel structure can fuel the structure-based drug design (SBDD) techniques to aid detecting the risk of cardiac toxicity in early stages of drug discovery.

In vivo and in vitro approaches for the evaluation of hERG safety

In vivo telemetry experiments in nonrodents is the definitive preclinical test for cardiotoxicity, with data generated under physiological conditions and related to the pharmacokinetic profile of the compound [25]. These assays are often performed in dogs and monkeys in later stages of drug development, as the difference in the mechanisms of repolarization in adult rats and mice makes them unsuited for evaluating hERG blockage [4]. The conventional patch-clamp electrophysiology in primary cardiac tissue, such as Purkinje fibers, is the preferred in vitro method,

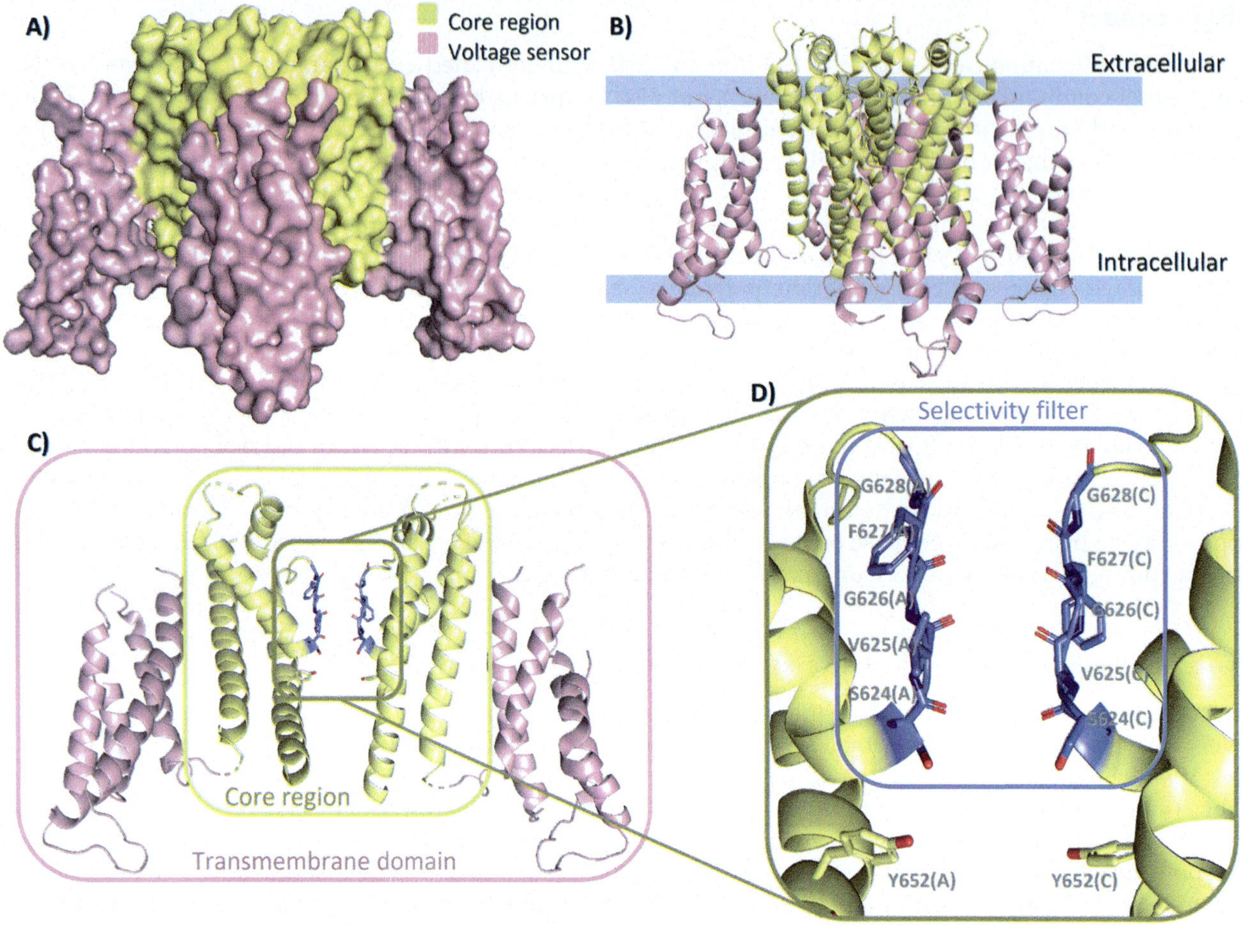

FIGURE 26.2 Structure of the hERG channel. (A and B) Core region coupled with four voltage sensors. (C and D) Selectivity filter formed by G628(A), F627(A), G626(A), V625(A), S624(A), G628(B), F627(B), G626(B), V625(B), S624(B), and two Y652(A) and Y652(C) at the core region.

which allows the evaluation of the electric current passing through hERG channels expressed in cells and provides real-time mechanistic information about ion channels. The experiments are commonly performed in mammalian cells, e.g., Chinese hamster ovary cells (CHO), HEK 293 cells, or *Xenopus laevis* oocytes (XO) cells transfected with the gene for hERG [26]. Disadvantages are the cost and time required to carry out such studies. More recently, various in vitro assays were developed to assess hERG safety, such as the transgenic zebrafish models have been validated as high-throughput animal models for hERG safety assessment [27]. Several studies aiming to understand the mechanism of action of hERG and identify the structural–toxicity relationships have been also published in the literature [28–33].

Computational approaches

In the past decade, automated hERG assays [34] have fairly contributed to the expansion of publicly available data for hERG in databases such as ChEMBL [35]. The increase in data has allowed the development of both quantitative structure–activity/property relationship (QSAR/QSPR) and pharmacophore models. When this chapter was written, the publicly available hERG data increased to 14,831 molecules [36] from 11,958 in 2014, when our group published our first hERG paper [37]. As already mentioned, a complete Cryo-EM structure of hERG is now available, allowing the use of SBDD methodologies, such as molecular docking and SBDD pharmacophore models, aiming to understand the mechanism of hERG inhibition and virtually evaluate compounds lacking experimental data.

QSAR approaches

Computational methods to predict hERG blockage can reduce the number of compounds to be tested in vitro and in vivo experiments, thus minimizing animal testing and reducing costs and saving time [38]. The most common ligand-based drug design (LBDD) approach to predict cardiotoxicity is the QSAR/QSPR. Modern QSAR models are statistical/machine learning models that use various algorithms to develop models capable of correctly predicting the biological property of compounds based on their chemical structures [21].

The bases of the QSAR emerged in 1962, proposed by Hansch et al. [39]. In this work, the authors studied the influence of fragment substitution in aromatic compounds and developed a linear equation describing the relationship between the chemical structures and the biological activity. The contemporary QSAR models use diverse machine learning algorithms (machine learning and deep learning) to learn from the available data and predict the data for untested compounds. These machine learning methods calculate different weights for the molecular descriptors and thus adjust the relationship equation between structure and activity or biological property, making the model increasingly predictive [40].

Since 2014, several papers have been published using QSAR/QSPR models to predict hERG blockers and help in the research and development of new drugs. Here, we update our group's comprehensive analysis of QSAR models for predicting hERG blockage [21], commenting on the most exciting works published from 2016 to 2022. Table 26.1 contains all the publications in this period. These studies used the data derived from the patch-clamp technique to analyze the inhibition of hERG K^+ channels on various cells.

Several tools are freely available to predict hERG blockage, mainly from academic groups; however, as shown in Table 26.1, the majority of the published QSAR models do not adhere to the standard validation protocols and statistical criteria outlined in the best practices of QSAR modeling nor are they compliant with the Organisation for Economic Co-operation and Development (OECD) guidance on QSAR model development and validation [38]. For instance, (1) most models lack proof of passing the Y-randomization test [29–31,33,36,41–47], (2) no evidence is provided for applicability domain (AD) estimation [29,31,33,41,44–47], (3) model predictivity is unacceptable [29,30,41,45], and (4) implementing the best practices to curate the datasets is not described, as proposed by Fourches and collaborators [48,49]. As a result, despite the availability of many QSAR models for hERG blockage in the literature, only few models have real validity to be used in practice. Furthermore, most models and associated datasets used to develop them are unavailable to the scientific community.

A critical examination of recent models reveals that most of them exhibit the same failures as those reported in the previous study [21]. These significant drawbacks jeopardize the practical application of these models for the reliable assessment of drug-induced QT syndrome. A few enhancements are worth mentioning. Some models used more sophisticated algorithms to predict the inhibition of hERG K^+ channels, such as implementing deep neural networks (DNNs), convolution neural networks, and optimized machine learning algorithms [29,31,33,41,44,46,50]. Also, diverse types of molecular descriptors were applied to obtain a better chemical representation from molecules and add protein–ligand information to teach the algorithm to differentiate minor differences between the molecules [29,30,36,43,45,46,50].

The two papers published by Siramshetty and colleagues [36,42] clearly show the application of DNNs and latent descriptors to improve model performance. The results show that machine learning algorithms outperform DNN, but using latent descriptors appears to be better at distinguishing blockers and nonblockers of hERG K^+ channels. Wang and colleagues [44], on the other hand, proposed a promising DNN alternative to improve performance by employing convolution–capsule networks and restricted Boltzmann capsule networks. Except for Zang and colleagues, these approaches showed an increase in prediction while outperforming machine learning developed in other scientific papers. Even though these models perform well, the authors did not report Y-randomization results or the estimation of the applicability domain in their papers. Only Wacker and Noskov [50] presented these preconized methodologies, which unfortunately do not have high performance, implying that these models can be overcome if these approaches are used.

Applications to predict hERG blockage

Several web servers implementing computational models to predict hERG inhibition are freely available, such as HergSpred [32], ToxTree [33], CardioTox net [29], AMED Cardiotoxicity [41], ADMETsar [51,52], ADMETlab [53,54], CardPred [55], and Pred-hERG [28,37]. StarDrop—ADME QSAR models (Optibrium Ltd., http://www.optibrium.com/stardrop/) and QuikProp (Schrödinger, LLC, http://www.schrodinger.com/) are commercial tools that can

TABLE 26.1 hERG blockage prediction studies published between 2016 and 2022.

Nº compounds (training)	Nº compounds (external)	Dataset (training/ external)	Curation	Activity	Activity threshold	Performance (external test set unless otherwise specified)		AD	Y-rand	Features	Availability	Year	References
421	External test = 309 In silico test = 7361	Misc	No	Ki	Inactive 5–54 μM Moderate 50 nM –5 μM Potent ≤ 50 nM	QSAR-3D	Training set = Q^2 0.80 External set = r^2 0.86 In silico test = ROC 0.96	No	No	Physicochemical properties	No	2017	[47]
700	283	ChEMBL v23	No	pIC50	>5 active	Lasso SVM RF XGBM	MAE 0.80 R^2 0.61 RMSE 0.98	Yes	Yes	Pharmacophore features (FS2)	No	2017	[50]
8380	499	ChEMBL v22/ misc	Yes	pIC50	≥6 strong blocker 6 > pIC50 ≥ 5 moderate blocker 5 > pIC50 ≥ 4.5 weak blocker Pic50 < 4.5 nonblocker	DNN Kernelized-SVM RF	Q4 0.75 Q2 0.94 SE 0.99 SP 0.75 MCC 0.80 CCR 0.87	No	No	PaDEL-Descriptor v2.21	Yes	2022	[33]
14,831	700	ChEMBL v23/ misc	No	IC50/ Ki	≤1 μM—blocker >10 μM—nonblocker	RF	AUC 0.97 ACC 0.93 SE 0.87 SP 0.97 MCC 0.85 BA 0.92	Yes	No	PubChem and Morgan fingerprints	No	2018	[36]
12,620	924	Misc	Yes	IC50	>10 μM nonblocker	DNN	MCC 0.599 NVP 0.688 ACC 0.810 PPV 0.893 SPE 0.786 SEN 0.833	No	No	Node feature and adjacency vector	Yes	2021	[29]
8337	100 (random f/training set)	ChEMBL v25	No	IC50	≤1 μM blocker	Lasso SVM	ACC 0.79 NPV 0.80 AUC 0.86	No	No	Interaction fingerprint (i.e., MOLPRINT2D)	No	2021	[46]
6561	4630	Misc/randomly from train set	Yes	IC50	Blocker if ≤10 μM or ≥50% inhibition at 10 μM	SVR RF DNN	R 0.72 R^2 0.28 RMSE 0.87 R 0.73 R^2 0.53 RMSE 0.67 R 0.69 R^2 0.48 RMSE 0.68	No	No	Morgan, Mordred, and RDKit fingerprints	Yes	2021	[41]

233	27	Misc	No	pIC50	NA	PLS regression	R^3 0.72 RMSD 0.7	No	No	Protein–ligand interaction fingerprint	No	2021	[45]
9215	1146	ChEMBL v29/ misc	No	pIC50	NA	SVR	PCC 0.765 RMSE 0.58 Spearman's ρ 0.757	Yes	No	Morgan, MACCS, and Estate fingerprints	No	2021	[30]
2644	60	Misc	No	IC50	<10 μM blocker >30 μM nonblocker	Capsule networks	SE 0.89 SP 0.89 MCC 0.774 SD 0.0109 Q 0.89	No	No	MACCS fingerprints	No	2020	[44]
14,440	44	Misc	Yes	IC50	<10 μM blocker ≥10 μM nonblocker	DNN	ACC 0.773 MCC 0.476 SE 0.833 NPV 0.643	No	No	Mordred, Morgan, and Pubchem fingerprints	Yes	2020	[31]
1428	4095	Misc	Yes	IC50	IC50 ≤ 40 μM blocker	Pharmacophore modeling	TP: 119/177 TN: 68/177 FN: 24/32 FP: 6/32 MCC = 0.72	Yes	No	Grid independent molecular descriptor (GRIND)	No	2018	[43]
8154	1016	Misc	No	IC50	Between 1 and 10 μM are removed	Consensus (RF, XGBoost, DNN, LSTM)	BACC 0.80 SEN 0.74 SP 0.86	No	No	RDKit and Morgan fingerprints	No	2020	[42]

predict pIC50 values for inhibition of hERG K^+ channels expressed in mammalian cells. Nevertheless, these projects were not updated in scientific journals since 2016. Thus, we can only discuss the web servers freely available.

HergSpred

HergSpred [32] has an hERG model that predicts the inhibition of hERG channels. This model uses a raw dataset of 12,850 (6907 positives and 5943 negatives) compounds from the ChEMBL database, using a threshold of $IC_{50} \leq 10$ μM of hERG blockers, and the molecular fingerprints used were MACCS and Morgan. The model was trained using five different training types and test sets with random forest (RF), DNN, XGBoost, and consensus. The quality of the models during the validation achieved an AUC of 0.908, MCC of 0.681, and ACC of 0.840.

ToxTree

ToxTree [33] is a web-based application trained to predict the pIC_{50} values of hERG inhibitors and voltage-gated sodium channels (Nav1.5). They used the ChEMBL v.25 and PubChem BioAssay, using only the IC_{50} values to train classification and regression models. The algorithms used were kernelized SVM, DNN, and RF with PaDEL descriptor and 90% of data for the training set and 10% for the test set; they used different thresholds ($pIC_{50} = 4.5$, 5, and 6). It is composed of RF models trained with a dataset of 8380 unique molecular compounds for hERG blockage. SVM models were introduced with a manually curated dataset of 1550 compounds from ChEMBL and PubChem for Nav1.5. The hERG model yielded a multiclass accuracy of Q4 = 74.5%, MCC = 80.3%, and correct classification ratio (CCR) = 86.8%, while the Nav1.5 model yielded a binary classification of Q2 = 86.7%, MCC = 71.2%, and F1 = 89.7%.

POEM

POEM [56] is a supervised machine learning algorithm developed by Cyclica that is based on chemical similarity. Unlike other machine learning approaches, POEM simultaneously uses multiple chemical fingerprints to build property prediction models resulting in less information loss as compared with using a single fingerprint for compound representation. Further, POEM models do not require hyperparameter optimization, allowing for the rapid creation of predictive models that can be enhanced easily as new data becomes available. The models were generated against a balanced dataset of active and inactive, the hERG channel POEM model leave-one-out cross-validation yielded both high sensitivity (0.9380) and specificity (0.9772). Applying an external dataset of 10,606 compounds to hERG channel POEM model yielded a sensitivity and specificity of 0.83 and 0.93, respectively.

CardioTox net

CardioTox net [29] is a deep learning framework based on stepwise training to predict hERG blockage. They implemented a methodology based on five individual deep learning base models trained on different base features and a sixth deep learning model to combine the output. They used three various datasets containing IC_{50} activity information and a threshold of 10 μM, achieving a performance of ACC = 84%, MCC = 68.2%, SPE = 86.8%, and SEN = 84.3%.

AMED cardiotoxicity

AMED cardiotoxicity [41] generated a regression model with 6561 compounds, using IC50 and Ki values extracted from the AMED cardiotoxicity database and randomly divided into training (70%) and test (30%) sets for model evaluation. They used NSGA-II, a variation of a genetic algorithm for multiple objective optimizations, to maximize Q2 and minimize RMSE in fivefold cross-validation and reduce the number of descriptors used. For the test set, the SVR model produced an R^2 of 0.594 and an RMSE of 0.604.

admetSAR

admetSAR [51,52] is a comprehensive tool with 47 computational models as of 2022. The training data were collected from several databases such as DrugBank, ChEMBL, and CPDB and peer-reviewed scientific papers and high-throughput screening projects such as Tox21 and CYP450. The raw data were submitted to a series of curation procedures, such as salt removal, structural standardization, and duplicate removal. The molecular fingerprints used were MACCS, ECFP, and AtomParis implemented with RDkit. The models were trained using SVM and kNN algorithms and yielded binary models with AUC ranging from 0.65 to 0.992 and an average AUC value of 0.842. Regression models yielded an R^2 of 0.668 for PPB, 0.810 for water solubility, 0.822 for pyriformis toxicity, and 0.522 for acute oral toxicity.

ADMETlab

ADMETlab [53,54] is an integrated online platform for ADMET property prediction. It implemented a multitask graph attention framework to model data retrieved from ChEMBL, PubChem, OCHEM, and peer-reviewed literature. The raw data was submitted to a series of curation procedures, such as salt removal, structural standardization, and duplicate removal. A total of 53 prediction models were implemented, including 40 classification models and 13 regression models, yielding a classification accuracy of 0.889, AUC = 0.943, MCC = 0.778, Se = 0.869, and SP = .909.

CardPred

CardPred [55] generated models to predict hERG-related cardiotoxicity, with a total of 2130 compounds. It used a neural network approach and five classical machine learning approaches, namely, regression algorithms, ridge regression, logistic regression, naïve Bayes, and RF. The models yielded an accuracy of 90.1%, an MCC of 0.368, an SEN of 0.321, and an SPE of 0.967. The model was further evaluated in vivo showing an accuracy rate of 80.0% and an MCC of 0.655.

Pred-hERG

Pred-hERG is a web-based application that helps identify potential hERG blockers and nonblockers in chemical libraries using binary and multiclass QSAR models [28,37]. An enhanced version was last published in 2016 based on ChEMBL v21 and other public datasets with 16,932 chemical records in raw data. The application was created using models trained with 8489 compounds, which were available for modeling after curation, including 4437 nonblockers (activity $\geq$ 10 μM), 2753 weak/moderate blockers (1 μM $\leq$ activity $\leq$ 10 μM), and 1299 strong blockers (activity $\leq$ 1 μM). Both binary (CCR $\approx$ 0.82) and multiclass models (accuracy $\approx$ 0.7) are robust and externally predictive. These models are available through the Pred-hERG web app, freely accessible to the public at http://predherg.labmol.com.br/. Three results are available to users: the probability maps of atomic contribution, binary model prediction, and multiclass model prediction. The probability map allows users to see which fragments contribute to the prediction (Fig. 26.3), which can help reduce the perception of QSAR as a "black box" method.

Other approaches

Several other in silico methods have been developed to predict the toxicity of chemical compounds, such as models based on structural alerts, which use substructures of a molecule that indicate or are associated with toxicity [57,58]. Read-across is another method in which compounds lacking biological data are grouped according to their chemical similarity to compounds with known toxicity [38]. Other approaches have also been described in the literature, such as molecular docking, to improve the prediction of hERG inhibition. For instance, Creanza and collaborators united Cryo-EM proteins and homology models to perform docking simulations (induced-fit and standard docking) to generate protein–ligand interaction fingerprints to apply in SVM and LASSO models [46].

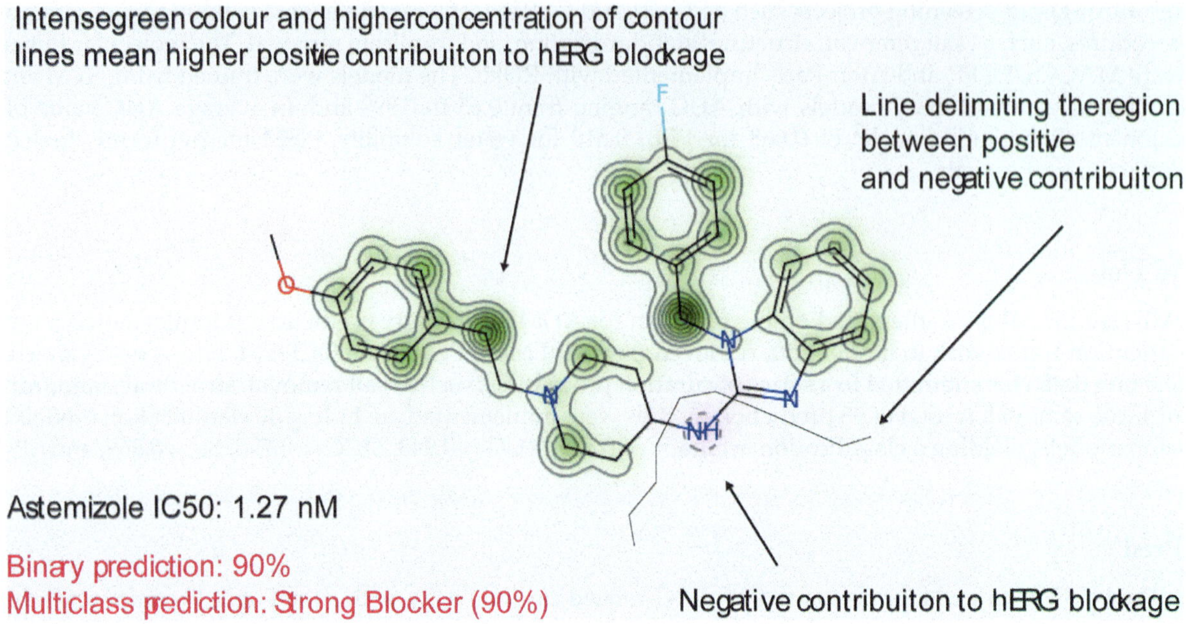

FIGURE 26.3 Pred-hERG result interpretation. (A) Negative contribution to hERG blockage (*red*). (B) Positive contribution to hERG blockage (*green*). (C) More contour lines can mean a higher contribution of an atom or fragment to hERG blockage. (D) Binary classification probability is given in %. (E) Multiclass classification probability is given in % and three possible classes, nonblocker, weak/moderate, and strong blocker.

Conclusions and future directions

In the past decade, there has been some progress in predicting hERG blockage, with the introduction of new methodologies, e.g., QSAR/QSPR, models based on structural alerts, identification, and analysis of structural trends and read-across. The lack of an elucidated 3D structure has led to current shortcomings for years. Considering the chemical diversity of hERG blockers, there is still a need to improve computational models to ensure hERG safety. Therefore, for the next few years, we expect that using the newly available Cryo-EM structure, combined with LBDD methods, will show significant potential for developing new and more predictive models for understanding and identifying hERG blockers in the early stages of drug discovery.

References

[1] Mladenka P, Applova L, Patocka J, Costa VM, Remiao F, Pourova J, et al. Comprehensive review of cardiovascular toxicity of drugs and related agents. Med Res Rev 2018;38(4):1332–403.
[2] Papazoglou P, Peng L, Sachinidis A. Epigenetic mechanisms involved in the cardiovascular toxicity of anticancer drugs. Front Cardiovasc Med 2021;8:658900.
[3] Mamoshina P, Rodriguez B, Bueno-Orovio A. Toward a broader view of mechanisms of drug cardiotoxicity. Cell Rep Med 2021;2(3):100216.
[4] Group IEW. S7B nonclinical evaluation of the potential for delayed ventricular repolarization (QT interval prolongation) by human pharmaceuticals. In: International conference on harmonisation of technical requirements for registration of pharmaceuticals for human use; 2005.
[5] Onakpoya IJ, Heneghan CJ, Aronson JK. Post-marketing withdrawal of 462 medicinal products because of adverse drug reactions: a systematic review of the world literature. BMC Med 2016;14:10.
[6] Craveiro NS, Lopes BS, Tomas L, Almeida SF. Drug withdrawal due to safety: a review of the data supporting withdrawal decision. Curr Drug Saf 2020;15(1):4–12.
[7] Kalyaanamoorthy S, Barakat KH. Development of safe drugs: the hERG challenge. Med Res Rev 2018;38(2):525–55.
[8] Zhou Z, Vorperian VR, Gong Q, Zhang S, January CT. Block of HERG potassium channels by the antihistamine astemizole and its metabolites desmethylastemizole and norastemizole. J Cardiovasc Electrophysiol 1999;10(6):836–43.
[9] Kim KS, Kim EJ, Lee HA, Park SJ. Effect of sibutramine HCl on cardiac hERG K^+ channel. Mol Cell Biochem 2009;320(1–2):125–31.
[10] Harrison-Woolrych M, Clark DW, Hill GR, Rees MI, Skinner JR. QT interval prolongation associated with sibutramine treatment. Br J Clin Pharmacol 2006;61(4):464–9.
[11] Zhang W, Roederer MW, Chen WQ, Fan L, Zhou HH. Pharmacogenetics of drugs withdrawn from the market. Pharmacogenomics 2012;13(2):223–31.
[12] Abbott GW, Sesti F, Splawski I, Buck ME, Lehmann MH, Timothy KW, et al. MiRP1 forms IKr potassium channels with HERG and is associated with cardiac arrhythmia. Cell 1999;97(2):175–87.

[13] Vandenberg JI, Perry MD, Perrin MJ, Mann SA, Ke Y, Hill AP. hERG K(+) channels: structure, function, and clinical significance. Physiol Rev 2012;92(3):1393–478.
[14] Chiesa N, Rosati B, Arcangeli A, Olivotto M, Wanke E. A novel role for HERG K^+ channels: spike-frequency adaptation. J Physiol 1997;501(Pt 2):313–8.
[15] Farrelly AM, Ro S, Callaghan BP, Khoyi MA, Fleming N, Horowitz B, et al. Expression and function of KCNH2 (HERG) in the human jejunum. Am J Physiol Gastrointest Liver Physiol 2003;284(6):G883–95.
[16] Smith GA, Tsui HW, Newell EW, Jiang X, Zhu XP, Tsui FW, et al. Functional up-regulation of HERG K^+ channels in neoplastic hematopoietic cells. J Biol Chem 2002;277(21):18528–34.
[17] Warmke JW, Ganetzky B. A family of potassium channel genes related to eag in Drosophila and mammals. Proc Natl Acad Sci USA 1994;91(8): 3438–42.
[18] Fermini B, Fossa AA. The impact of drug-induced QT interval prolongation on drug discovery and development. Nat Rev Drug Discov 2003; 2(6):439–47.
[19] Levine E, Rosero SZ, Budzikowski AS, Moss AJ, Zareba W, Daubert JP. Congenital long QT syndrome: considerations for primary care physicians. Cleve Clin J Med 2008;75(8):591–600.
[20] Asai T, Adachi N, Moriya T, Oki H, Maru T, Kawasaki M, et al. Cryo-EM structure of K(+)-bound hERG channel complexed with the blocker astemizole. Structure 2021;29(3):203–212 e4.
[21] Alves VM, Braga RC, Andrade CH. Computational approaches for predicting hERG activity. In: Computational toxicology: risk assessment for chemicals; 2018. p. 69–91.
[22] Sanguinetti MC, Tristani-Firouzi M. hERG potassium channels and cardiac arrhythmia. Nature 2006;440(7083):463–9.
[23] Milnes JT, Crociani O, Arcangeli A, Hancox JC, Witchel HJ. Blockade of HERG potassium currents by fluvoxamine: incomplete attenuation by S6 mutations at F656 or Y652. Br J Pharmacol 2003;139(5):887–98.
[24] Mitcheson JS, Chen J, Lin M, Culberson C, Sanguinetti MC. A structural basis for drug-induced long QT syndrome. Proc Natl Acad Sci USA 2000;97(22):12329–33.
[25] Aronov AM. Predictive in silico modeling for hERG channel blockers. Drug Discov Today 2005;10(2):149–55.
[26] Wan H, Selvaggio G, Pearlstein RA. Toward in vivo-relevant hERG safety assessment and mitigation strategies based on relationships between non-equilibrium blocker binding, three-dimensional channel-blocker interactions, dynamic occupancy, dynamic exposure, and cellular arrhythmia. PLoS One 2020;15(11):e0234946.
[27] Wen D, Liu A, Chen F, Yang J, Dai R. Validation of visualized transgenic zebrafish as a high throughput model to assay bradycardia related cardio toxicity risk candidates. J Appl Toxicol 2012;32(10):834–42.
[28] Alves V, Braga R, Muratov E, Andrade C. Development of web and mobile applications for chemical toxicity prediction. J Braz Chem Soc 2018.
[29] Karim A, Lee M, Balle T, Sattar A. CardioTox net: a robust predictor for hERG channel blockade based on deep learning meta-feature ensembles. J Cheminform 2021;13(1):60.
[30] Meng J, Zhang L, Wang L, Li S, Xie D, Zhang Y, et al. TSSF-hERG: a machine-learning-based hERG potassium channel-specific scoring function for chemical cardiotoxicity prediction. Toxicology 2021;464:153018.
[31] Ryu JY, Lee MY, Lee JH, Lee BH, Oh KS. DeepHIT: a deep learning framework for prediction of hERG-induced cardiotoxicity. Bioinformatics 2020;36(10):3049–55.
[32] Zhang X, Mao J, Wei M, Qi Y, Zhang JZH. HergSPred: accurate classification of hERG blockers/nonblockers with machine-learning models. J Chem Inf Model 2022;62(8):1830–9.
[33] Arab I, Wonneberger D, Barakat K. ToxTree: descriptor-based machine learning models to predict hERG and Nav1. 5 cardiotoxicity. 2022.
[34] Li X, Zhang R, Zhao B, Lossin C, Cao Z. Cardiotoxicity screening: a review of rapid-throughput in vitro approaches. Arch Toxicol 2016;90(8): 1803–16.
[35] Gaulton A, Hersey A, Nowotka M, Bento AP, Chambers J, Mendez D, et al. The ChEMBL database in 2017. Nucleic Acids Res 2017;45(D1): D945–54.
[36] Siramshetty VB, Chen Q, Devarakonda P, Preissner R. The catch-22 of predicting hERG blockade using publicly accessible bioactivity data. J Chem Inf Model 2018;58(6):1224–33.
[37] Braga RC, Alves VM, Silva MF, Muratov E, Fourches D, Tropsha A, et al. Tuning HERG out: antitarget QSAR models for drug development. Curr Top Med Chem 2014;14(11):1399–415.
[38] Raies AB, Bajic VB. Silico toxicology: computational methods for the prediction of chemical toxicity. Wiley Interdiscip Rev Comput Mol Sci 2016;6(2):147–72.
[39] Hansch C, Maloney PP, Fujita T, Muir RM. Correlation of biological activity of phenoxyacetic acids with Hammett substituent constants and partition coefficients. Nature 1962;194(4824):178–80.
[40] Rodríguez-Pérez R, Miljković F, Bajorath J. Machine learning in chemoinformatics and medicinal chemistry. Ann Rev Biomed Data Sci 2022;5.
[41] Sato T, Yuki H, Honma T. Quantitative prediction of hERG inhibitory activities using support vector regression and the integrated hERG dataset in AMED cardiotoxicity database. Chem Bio Inform J 2021;21:70–80.
[42] Siramshetty VB, Nguyen DT, Martinez NJ, Southall NT, Simeonov A, Zakharov AV. Critical assessment of artificial intelligence methods for prediction of hERG channel inhibition in the "Big Data" Era. J Chem Inf Model 2020;60(12):6007–19.
[43] Munawar S, Windley MJ, Tse EG, Todd MH, Hill AP, Vandenberg JI, et al. Experimentally validated pharmacoinformatics approach to predict hERG inhibition potential of new chemical entities. Front Pharmacol 2018;9:1035.
[44] Wang Y, Huang L, Jiang S, Wang Y, Zou J, Fu H, et al. Capsule networks showed excellent performance in the classification of hERG blockers/ nonblockers. Front Pharmacol 2019;10:1631.
[45] Sekhar Pagadala N. Computational prediction of hERG blockers using homology modelling, molecular docking and QuaSAR studies. Results Chem 2021;3.
[46] Creanza TM, Delre P, Ancona N, Lentini G, Saviano M, Mangiatordi GF. Structure-based prediction of hERG-related cardiotoxicity: a benchmark study. J Chem Inf Model 2021;61(9):4758–70.

[47] Chemi G, Gemma S, Campiani G, Brogi S, Butini S, Brindisi M. Computational tool for fast in silico evaluation of hERG K(+) channel affinity. Front Chem 2017;5:7.
[48] Fourches D, Muratov E, Tropsha A. Trust, but verify II: a practical guide to chemogenomics data curation. J Chem Inf Model 2016;56(7): 1243–52.
[49] Fourches D, Muratov E, Tropsha A. Trust, but verify: on the importance of chemical structure curation in cheminformatics and QSAR modeling research. J Chem Inf Model 2010;50(7):1189–204.
[50] Wacker S, Noskov SY. Performance of machine learning algorithms for qualitative and quantitative prediction drug blockade of hERG1 channel. Comput Toxicol 2018;6:55–63.
[51] Yang H, Lou C, Sun L, Li J, Cai Y, Wang Z, et al. admetSAR 2.0: web-service for prediction and optimization of chemical ADMET properties. Bioinformatics 2019;35(6):1067–9.
[52] Cheng F, Li W, Zhou Y, Shen J, Wu Z, Liu G, et al. admetSAR: a comprehensive source and free tool for assessment of chemical ADMET properties. J Chem Inf Model 2012;52(11):3099–105.
[53] Xiong G, Wu Z, Yi J, Fu L, Yang Z, Hsieh C, et al. ADMETlab 2.0: an integrated online platform for accurate and comprehensive predictions of ADMET properties. Nucleic Acids Res 2021;49(W1):W5–14.
[54] Dong J, Wang NN, Yao ZJ, Zhang L, Cheng Y, Ouyang D, et al. ADMETlab: a platform for systematic ADMET evaluation based on a comprehensively collected ADMET database. J Cheminform 2018;10(1):29.
[55] Lee HM, Yu MS, Kazmi SR, Oh SY, Rhee KH, Bae MA, et al. Computational determination of hERG-related cardiotoxicity of drug candidates. BMC Bioinform 2019;20(Suppl. 10):250.
[56] Brereton AE, MacKinnon S, Safikhani Z, Reeves S, Alwash S, Shahani V, et al. Predicting drug properties with parameter-free machine learning: pareto-optimal embedded modeling (POEM). Mach Learn Sci Technol 2020;1(2).
[57] Valerio Jr LG. In silico toxicology for the pharmaceutical sciences. Toxicol Appl Pharmacol 2009;241(3):356–70.
[58] Venkatapathy R, Wang NC. Developmental toxicity prediction. Methods Mol Biol 2013;930:305–40.

PART IV

QSAR models for risk assessment of chemicals

CHAPTER

27

Curation of more than 10,000 Ames test data used in the Ames/QSAR International Challenge Projects

Ayako Furuhama[1], Toshio Kasamatsu[1], Kei-ichi Sugiyama[1] and Masamitsu Honma[2]

[1]**Division of Genetics and Mutagenesis, National Institute of Health Sciences, Kawasaki, Kanagawa, Japan** [2]**Division of General Affairs, National Institute of Health Sciences, Kawasaki, Kanagawa, Japan**

Abbreviations

DMSO Dimethyl sulfoxide
GLP Good Laboratory Practice
H_2O Water
ISHA Industrial Safety and Health Act
IWGT International Workshops on Genotoxicity Testing
OECD Organisation for Economic Co-operation and Development
QSAR Quantitative structure–activity relationship
SMILES Simplified molecular-input line-entry system

Introduction

In silico prediction models for in vitro Ames mutagenicity are important tools evaluating chemical substance genotoxicity. Quantitative structure–activity relationship (QSAR) models are specifically used for regulations. Therefore, QSAR model improvement becomes vital for practical use. Ames test data are important for the improvement of QSAR models for mutagenicity knowledge of chemicals and technique of modeling, as well as quantity and quality. However, detailed test information such as solvent and strains were not always accessible to QSAR model developers and users. In this chapter, the curated Ames test data were described and analyzed. The Ames data were composed of more than 10,000 chemicals and were tested under good laboratory practice (GLP) compliance.

Data curation procedure of 12,140 ANEI-HOU chemicals

The Division of Genetics and Mutagenesis, National Institute of Health Sciences (DGM/NIHS), organized the first Ames/QSAR International Challenge Project in 2014–2017 [1]. Twelve QSAR vendors' 17 QSAR tools challenged to predict the Ames mutagenicity of 12,140 chemicals were based on the submitted Ames test reports of new chemical substances [2] to the Ministry of Health Labor and Welfare (MHLW), Japan, and were subject to the Industrial Safety and Health Act (ISHA) in Japan [3] since 1979. ISHA was expressed as "ANEI-HOU" in the project [1]. The Ames tests were conducted according to the OECD TG 471 [4] and ANEI-HOU guidance under GLP-compliant conditions. ANEI-HOU Ames data have been undisclosed and are only available as a list for the strongly positive (strongly mutagenic; Class A) chemical substances, under the definition of ANEI-HOU, which generally induce more than 1000 revertant colonies per mg of at least one Ames test strain in the presence or absence of metabolic activation. The list of Class A chemicals used for the first projects is available via https://www.nihs.go.jp/dgm/amesqsar.

QSAR in Safety Evaluation and Risk Assessment
https://doi.org/10.1016/B978-0-443-15339-6.00022-9

html (accessed on October 30, 2022). The list of positive (mutagenic; Class B) and negative (non-mutagenic; Class C) chemicals are not open to the public. According to ANEI-HOU, positive (Class B) chemicals induce at least two-fold increases in revertant colonies (but <1000) compared with the negative control in at least one Ames strain in the presence or absence of metabolic activation; negative (Class C) chemicals induce no revertant increases or less than two-fold increases.

DGM/NIHS started the second Ames/QSAR International Challenge Projects in February 2020 (https://www.nihs.go.jp/dgm/2nd_amesqsar.html accessed on October 30, 2022). After completing the first project, new "ANEI-HOU" Ames tests results (about 1600 chemicals) had been accumulated. DGM/NIHS used these chemicals as challenge chemicals in the second project. DGM/NIHS also provided the 12,140 chemicals' Ames test results used in the first project as a training dataset. These 12,140 chemicals' Ames test study reports were reviewed in detail because some test results appeared skeptical [1]. Unfortunately, 1461 study reports were not available, some of which were probably discarded because of old ones (before 1985). The remaining 10,679 chemicals' Ames test study reports were reviewed (Table 27.1).

Five genotoxicity test experts in DGM/NIHS scrutinized not only the accuracy of the test report descriptions but also the validity of the positive/negative test results. Obvious description errors and overestimation of test results due to special test methods (e.g., gas exposure method) were immediately corrected. Six studies could not be assigned positive or negative because the studies were not properly conducted. The positive/negative boundary judgment results were carefully discussed. Considering the reproducibility of the preliminary and main studies, dose related responses, non-specific reactions due to strong toxicity and precipitation, etc., positive/negative test results were changed for about 110 studies. The mutagenicity of 63 chemicals judged negative (Class C) in the study reports was changed to positive (Class B) (Table 27.1). These chemicals did not induce mutagenicity more than two-fold in any experiment, but the nearly two-fold inducibility was reproducibly observed in more than one experiment (e.g., between preliminary and main studies, among multiple strains, with and without metabolic activation). The experts judged that these weak responses were biologically relevant. 53 among the 63 chemicals were predicted as positive by some QSAR tools.

Information of the tested chemicals' purity, the solvent used in the Ames test, and each strain's mutagenicity with and/or without metabolic activation were collected from the 10,673 study reports and added to the new database.

Strain information

All the collected 10,673 Ames tests were performed with and without metabolic activation with *Salmonella typhimurium* TA100 (base-pair substitution type), TA1535 (base-pair substitution type), and TA98 (frameshift type) strains. The 10,601 sets (99.3%) of Ames test were performed with the combination of "TA100, TA1535, TA98" and *S. typhimurium* TA1537 (frameshift type) plus *Escherichia coli* WP2*uvrA* (base-pair substitution type); the 19 sets (0.18%) were TA100, TA1535, TA98, TA1537, and *S. typhimurium* TA102 (base-pair substitution type); the four sets (<0.1%) were TA100, TA1535, TA98, *S. typhimurium* TA97a (frameshift type), and TA102; the six sets (<0.1%) were TA100, TA1535, TA98, TA97a, and WP2*uvrA* (Fig. 27.1). These strain combinations, i.e., TA100, TA1535,

TABLE 27.1 Ames test database in DGM/NIHS.

	Class A			Class B			Class C			NA	Total
Original Ames database [1]	672			1085			10,383			0	12,140
No available Ames study reports	44			164			1253			0	1461
New Ames database before expert reviews	628			921			9130			0	10,679
Change the class by expert reviews	to B: 20	to C: 3	to NA: 3	to A: 0	to C: 15	to NA: 3	to A: 3	to B: 63	to NA: 0		
New Ames database after expert reviews	605			986			9082			6	10,679

Class A: strongly positive (strongly mutagenic), Class B: positive (mutagenic). Class C: negative (non-mutagenic), NA: not assigned.

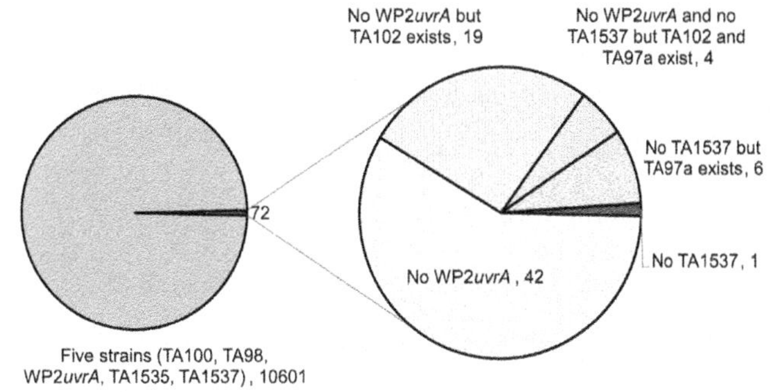

FIGURE 27.1 Strain information for the available 10,673 Ames test data.

TA98, either TA1537 or TA97a, and either WP2*uvrA* or TA102, followed the recommendation of the OECD TG 471 [4] and covered 99.6% of the Ames test set. Therefore, only 0.4% sets of data did not follow the sets of the standard strains.

Solvent and purity of the ames tests

The curated data informed us that dimethyl sulfoxide (DMSO) was the most frequently chosen solvent and was used in 8068 of the 10,673 Ames tests, and water (H_2O) and acetone were the second and the third most frequently chosen solvents, respectively, according to the current analysis (Fig. 27.2A). These three solvents were used in 97% of the 10,673 Ames tests. For the Ames tests by using DMSO, 1290 (16.0%) chemicals were positive and 6778 (84.0%) chemicals were negative. For the Ames test by using H_2O, 188 (13.6%) chemicals were positive and 1192 (86.4%) chemicals were negative. For the Ames test by using acetone, 77 (8.8%) chemicals were positive and 796 (91.2%) chemicals were negative. The percentage of positive chemicals was higher for the Ames tests using DMSO solvent than those using H_2O and especially acetone solvents. Fig. 27.2B indicated that 81% positive results were from the Ames tests using solvent DMSO, while less negative results (75%) were from the test using solvent DMSO as shown in Fig. 27.2C. On the contrary, 12% and 5% tests had positive results when using solvent H_2O and acetone, respectively, which were lower than the corresponding percentages of the tests having negative results (13% and 9%).

Furthermore, let us focus on the test data obtained by using the 95% and more than 95% (≥95%) chemical purity. Of the 10,673 tests, the number of test data with ≥95% purity is 9723. The tests using H_2O and acetone solvents were 1090 and 816 tests, respectively. Still, DMSO was the most frequently used solvent in the Ames tests as well as H_2O and acetone were the second and third chosen ones (see Figs. 27.2A and 27.3A). Comparing the tests with (Fig. 27.3) and without (Fig. 27.2) considering the ≥95% purity revealed that the ratio of positive to negative results from tests with ≥95% purity (Fig. 27.3) is similar as that from all tests for the Ames tests using DMSO, H_2O, and acetone.

In section Relationship with mutagenicity and strain, let us focus on the relationship between in vitro Ames mutagenicity and strain for the 10,673 Ames data and restricted with the tests by "≥95% purity chemicals and DMSO solvent" (i.e., for the 7490 Ames data).

Relationship with mutagenicity and strain

Ames test reported data

The 1591 [= 10,673 − 9082] Ames mutagenic (positive) data can be categorized by the presence or absence of metabolic activation as shown in Table 27.2. About 61% of the positive data (967 of the 1591 Ames-positive compounds) was positive with both the presence and absence of metabolic activation. Only 27% and 12% of positive data (431 and 193 of the 1591 compounds) were positive with only presence and absence of metabolic activation, respectively. Metabolic activation worked to detect the 431 positive chemicals, which did not become positive under the absence of metabolic activation for the 10,673 chemicals.

The Ames test analysis using the TA100 and/or TA98 combination and the TA100, TA98, and/or TA1535 combination were listed in Tables 27.3 and 27.4, respectively, because all the 10,673 tests were performed using these

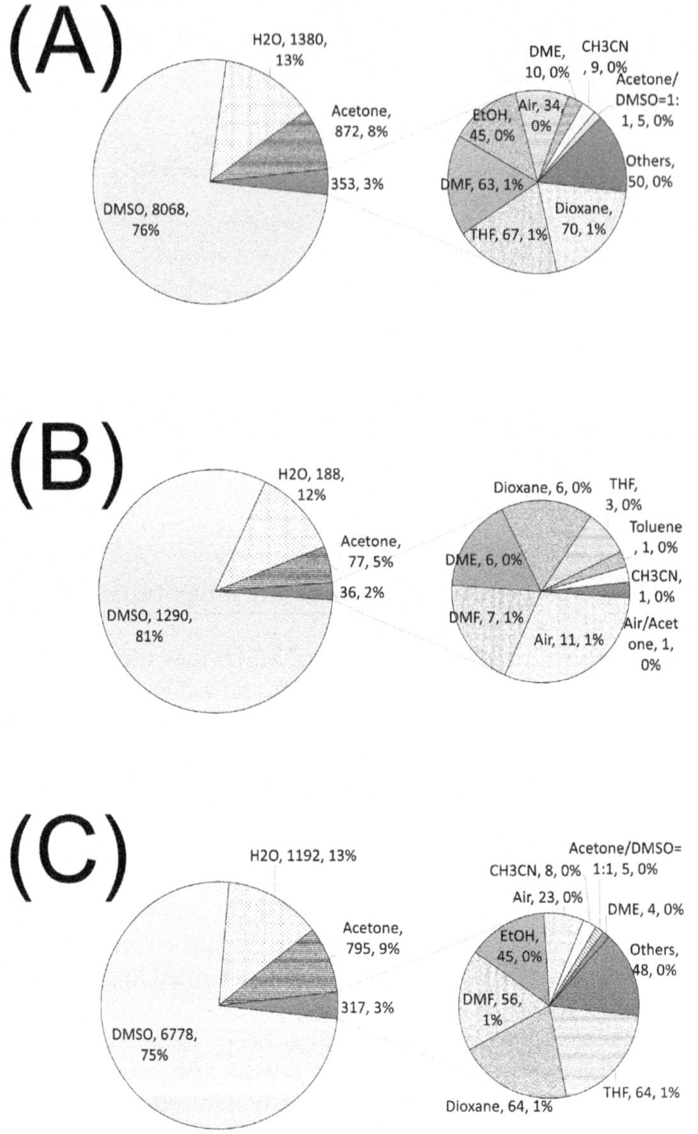

FIGURE 27.2 Solvent information obtained (A) from the 10,673 Ames test reports, (B) from 1591 positive data (mutagenic; class A and class B), and (C) from 9082 negative data (non-mutagenic; class C).

strains. The tables indicated how TA100, TA98, and TA1535 played important roles that detect Ames positive. By using a single strain, 68%, 51%, and 45% positive chemicals (i.e., 1086, 806, and 712 of the 1591 Ames-positive compounds) were detected by the Ames tests by TA100, TA98, and TA1535, respectively, under the presence and/or absence of metabolic activation. Although TA100 was the most effective to detect mutagenic chemicals, 32% positive data still remained to be detected by other strain conditions.

If combined with a base-pair substitution type TA100 and a frameshift type TA98, 18% positive data (293 of the 1591 compounds in Table 27.3) remained to be detected by the rest strains. Additionally, if combined with two base-pair substitution types TA100 and TA1535 and one frameshift type TA98, 9% positive data (147 of the 1591 compounds in Table 27.4) remained to be detected by the other strains. Table 27.4 analysis also included another combination of one frameshift and base-paired substitution types, that is, TA98 and TA1535, which had less positive predictive powers (72%, 1149 [= 1591 − 295 − 147] of the 1591 compounds) than the combination of TA100 and TA98 (82%, 1298 [= 1591 − 293] of the 1591 compounds).

According to the 2017 IWGT discussion [5], the chemicals in the dataset of Ames tests identified that 93% was mutagenic in either TA98 and/or TA100. Additionally, from the analysis for the mutagen sensitivity for TA1535 comparing with TA100, there were 43 of 1167 compounds (3.7%), which were positive in TA1535 and not in

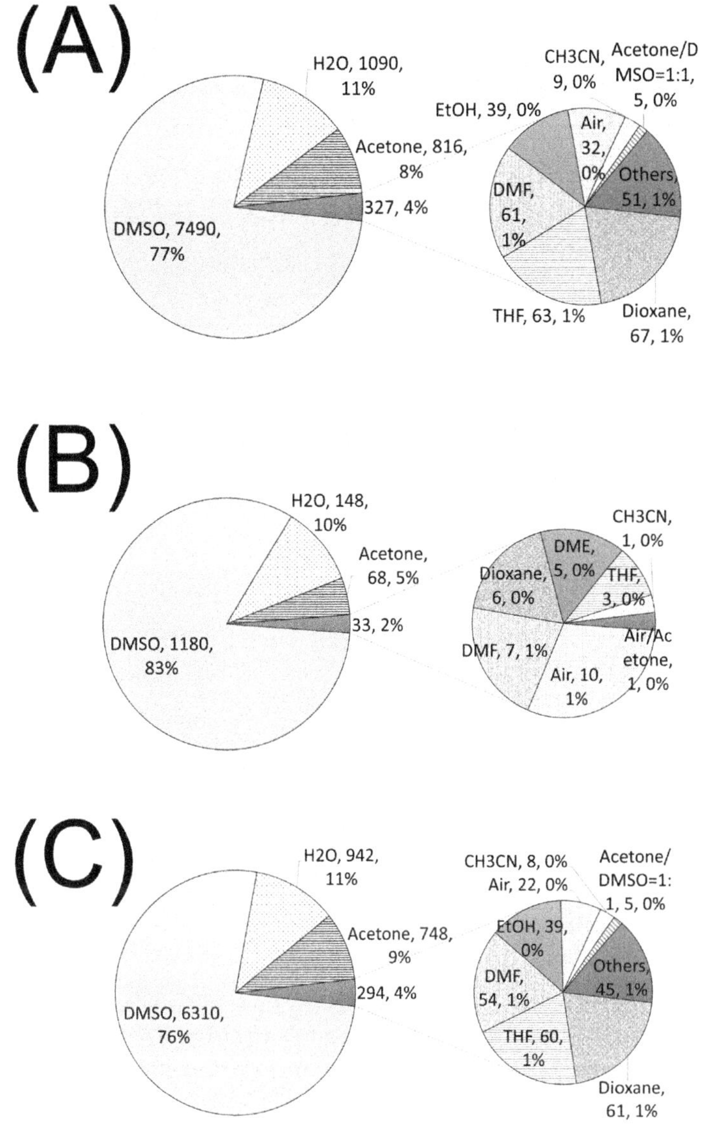

FIGURE 27.3 Solvent information obtained (A) from the 9723 Ames test reports of chemicals with purity ≥95%, (B) from 1429 positive data (mutagenic; class A and class B) of chemicals with purity ≥95%, and (C) from 8294 negative data (non-mutagenic; class C) of chemicals with purity ≥95%.

TABLE 27.2 The relationship between metabolic activation and mutagenicity for the 10,673 Ames test data.[a]

+S9	−S9	Total
Positive	Positive	967
Positive	Negative	431
Negative	Positive	193
Negative	Negative	9082

[a]The + S9 and−S9 means with and without metabolic activation. Positive indicates Class A and B, and Negative indicates Class C.

TABLE 27.3 The relationship between two strains (*Salmonella typhimurium* TA100 and TA98) and mutagenicity for the 10,673 Ames test data.[a]

TA100	TA98	Other strains	Total
Positive	Positive	Positive or Negative	594
Positive	Negative	Positive or Negative	492
Negative	Positive	Positive or Negative	212
Negative	Negative	Positive	293
Negative	Negative	Negative	9082

[a]*Positive indicates Class A and B, and Negative indicates Class C.*

TABLE 27.4 The relationship between three strains (*Salmonella typhimurium* TA100, TA98, and TA1535) and mutagenicity for the 10,673 Ames test data.[a]

TA100	TA98	TA1535	Other strains	Total
Positive	Positive	Positive	Positive or Negative	350
Positive	Positive	Negative	Positive or Negative	244
Positive	Negative	Positive	Positive or Negative	197
Negative	Positive	Positive	Positive or Negative	19
Positive	Negative	Negative	Positive or Negative	295
Negative	Positive	Negative	Positive or Negative	193
Negative	Negative	Positive	Positive or Negative	146
Negative	Negative	Negative	Positive	147
Negative	Negative	Negative	Negative	9082

[a]*Positive indicates Class A and B, and Negative indicates Class C.*

TA100 on the basis of the two-fold rule. Our curation and analysis were not equivalent to the results in this discussion [5]. Nevertheless, our analysis indicated 82% was mutagenic in either TA98 and/or TA100 and there were 146 of the 1591 compounds (9.2%), which were positive in TA1535 and neither in TA100 nor in TA98 (Table 27.4).

Data tested with ≥95% purity chemicals with DMSO solvent

The 1180 [= 7490 − 6310] Ames-positive data can be categorized by the presence or absence of metabolic activation as shown in Table 27.5. A total of 65% positive data (770 of the 1180 compounds) was positive in both presence and absence of metabolic activation. The 25% and 9% positive data (300 and 110 of the 1180 compounds) were

TABLE 27.5 The relationship between metabolic activation and mutagenicity for the Ames test data using the DMSO solvent with the ≥95% purity.[a]

+S9	−S9	Total
Positive	Positive	770
Positive	Negative	300
Negative	Positive	110
Negative	Negative	6310

[a]*The + S9 and−S9 means with and without metabolic activation. Positive indicates Class A and B, and Negative indicates Class C.*

positive only in the presence and absence of metabolic activation, respectively. Metabolic activation worked to effectively detect the 300 positive chemicals, which were not active without metabolic activation in the 7490 chemicals. Compared with Tables 27.2 and 27.5, the purity and solvent restrictions are not significantly influenced by the ratio of positives to negatives with or without metabolic activation.

Similar to the discussion in Ames test reported data section, the Ames test analysis using the TA100 and/or TA98 combination and the combination of the tests by TA100, TA98, and/or TA1535 were listed in Tables 27.6 and 27.7, respectively. By using a single strain with the tests with ≥95% purity chemicals by the DMSO solvent, 70%, 50%, and 43% positive chemicals (i.e., 830, 586, and 502 of the 1180 Ames-positive compounds) were detected by the Ames tests by TA100, TA98, and TA1535, respectively, under the presence and/or absence of metabolic activation. TA100 was still the most effective in detecting mutagenic chemicals, but 30% positive data were remained to be detected by other strain conditions.

If combined with a base-pair substitution type TA100 and a frameshift type TA98, 17% positive data (197 of the 1180 compounds in Table 27.6) remained to be detected by the rest strains. Additionally, if combined with two base-pair substitution types TA100 and TA1535 and one frameshift type TA98, 9% positive data (106 of the 1180 compounds in Table 27.7) remained to be detected by the other strains. Table 27.7 analysis also provided information that another combination of one frameshift and base-paired substitution types, that is, TA98 and TA1535, had less positive predictive powers (71%, 832 [= 1180 − 242 − 106] of the 1180 compounds) than the combination of TA100 and TA98 (83%, 983 [= 1180 − 197] of the 1180 compounds). Overall, TA100, TA98, and/or TA1535 for the Ames test data with ≥95% purity chemicals by the DMSO solvent seemed to have almost equivalent detectivity powers to the 10,673 Ames data as discussed in Ames test reported data section.

TABLE 27.6 The relationship between two strains (*Salmonella typhimurium* TA100 and TA98) and mutagenicity for the Ames test data using the DMSO solvent with the ≥95% purity.[a]

TA100	TA98	Other strains	Total
Positive	Positive	Positive or Negative	433
Positive	Negative	Positive or Negative	397
Negative	Positive	Positive or Negative	153
Negative	Negative	Positive	197
Negative	Negative	Negative	6310

[a]*Positive indicates Class A and B, and Negative indicates Class C.*

TABLE 27.7 The relationship between three strains (*Salmonella typhimurium* TA100, TA98, and TA1535) and mutagenicity for the Ames test data using the DMSO solvent with the ≥95% purity.[a]

TA100	TA98	TA1535	Other strains	Total
Positive	Positive	Positive	Positive or Negative	245
Positive	Positive	Negative	Positive or Negative	188
Positive	Negative	Positive	Positive or Negative	155
Negative	Positive	Positive	Positive or Negative	11
Positive	Negative	Negative	Positive or Negative	242
Negative	Positive	Negative	Positive or Negative	142
Negative	Negative	Positive	Positive or Negative	91
Negative	Negative	Negative	Positive	106
Negative	Negative	Negative	Negative	6310

[a]*Positive indicates Class A and B, and Negative indicates Class C.*

Summary

The data of all 12,140 chemicals used in the Ames/QSAR International Challenge Projects were curated, and their detailed data were collected. These data were confidential, and the test results could not be released even for the Ames mutagenic and non-mutagenic (positive and negative) information except for the strongly positive (i.e., Class A) chemicals. Nevertheless, the overview of the curated Ames test data and the mutagenic dependencies of metabolic activation and strain, as described here, may become the basic information for developing QSAR models at Ames mutagenicity. The curated and detailed data were provided to the second project participants for supporting their QSAR model developments.

Acknowledgments and fundings

We greatly acknowledge the participants in the Ames/QSAR International Challenge Projects, as well as Ms. Airi Kitazawa and Dr. Takayoshi Suzuki NIHS Japan. The Chemical Hazards Control Division, Industrial Safety and Health Department, MHLW Japan for providing the ANEI-HOU Ames dataset and allowing us to use the data in the projects. The studies were performed by the Ministry of Health, Labour, and Welfare under Grants (H30-Chemistry-Destination-005, 21KD2005, and 21KA1001).

References

[1] Honma M, Kitazawa A, Cayley A, Williams RV, Barber C, Hanser T, et al. Improvement of quantitative structure–activity relationship (QSAR) tools for predicting Ames mutagenicity: outcomes of the Ames/QSAR International Challenge Project. Mutagenesis 2019;34(1):3–16.
[2] ISHA: Existing Chemical Substances/Newly Announced Chemical Substances Available from: https://www.nite.go.jp/en/chem/chrip/chrip_search/html/hazardHelp.html#RJ_04_002.
[3] Industrial safety and Health Act of Japan. 1972. Information available from: https://www.mhlw.go.jp/stf/english/labour_standards_index.html#m-naAncTarget3.
[4] OECD. OECD Guidelines for Testing of Chemicals. Test No. 471: Bacterial Reverse Mutation Test. Paris: OECD; 2020. https://doi.org/10.1787/9789264071247-en.
[5] Williams RV, DeMarini DM, Stankowski LF, Escobar PA, Zeiger E, Howe J, et al. Are all bacterial strains required by OECD mutagenicity test guideline TG471 needed? Mutat Res, Genet Toxicol Environ Mutagen 2019;848:503081.

CHAPTER

28

QSAR model of photolysis kinetic parameters in aquatic environment

Jiale He and Zhongyu Wang

Key Laboratory of Industrial Ecology and Environmental Engineering (MOE), School of Environmental Science and Technology, Dalian University of Technology, Dalian, China

Abbreviations

BTHMs Brominated trihalomethanes
CA Cluster analysis
ClPAHs Chlorinated polycyclic aromatic hydrocarbons
DBPs Disinfection by-products
DOM Dissolved organic matter
FA Factor analysis
GCM Group contribution method
HOF Heat of formation values
ITHMs Iodinated trihalomethanes
MLR Multiple linear regression
MRA Multiple regression analysis
OACs Oxidated aromatic compounds
OPEs Organophosphate esters
PAHs Polycyclic aromatic hydrocarbons
PBDEs Polybrominated diphenyl ethers
PCA Principal component analysis
PCBs Polychlorinated biphenyls
PCDD/Fs Polychlorinated dibenzo-*p*-dioxins and dibenzofurans
PLS Partial least squares
PPRIs Photochemically produced reactive intermediates
QSAR Quantitative structure–activity relationship

Introduction

Chemicals play an important role in promoting the development of human society and improving the quality of human life. However, chemicals can enter the environment via unorganized release or organized emission throughout the life cycle of chemicals-containing products, become pollutants, and pose risk to human and ecosystem health. Exposure is a key determinant of the risk. In general, exposure is quantitatively characterized by the concentration, which further depends on the rate of release into the environment and the rate of degradation of chemicals inside the environment.

Transformation of chemicals triggered by light is called *photolysis*. Sunlight is the driving force behind the material cycles and energy flows of the earth's ecosystem. Meanwhile, the earth's surface is 70% covered by water. Therefore, photolysis in the aquatic environment makes an important part in the degradation of chemicals.

Photolysis of chemicals in the aquatic environment involves several processes (Fig. 28.1). A chemical could absorb light and consequently undergoes transformation, which is commonly referred as *direct photolysis*. Chemicals can also be transformed by photochemically produced reactive intermediates (PPRIs) [1], which is known as *indirect photolysis*.

The photolysis processes are determined by the photolysis kinetic parameters. However, the photolysis kinetic parameters of chemicals are typically obtained via experiments, which can be low-throughput and time-consuming. To fill the data gap in the photolysis kinetic parameters, quantitative structure–activity relationship (QSAR) models could serve as an efficient tool. In this chapter, a brief introduction to the photolysis kinetic parameters is presented. History and progress of related research are archived. Existing QSAR models for predicting photolysis kinetic parameters of chemicals in the aquatic environment are summarized.

https://doi.org/10.1016/B978-0-443-15339-6.00020-5

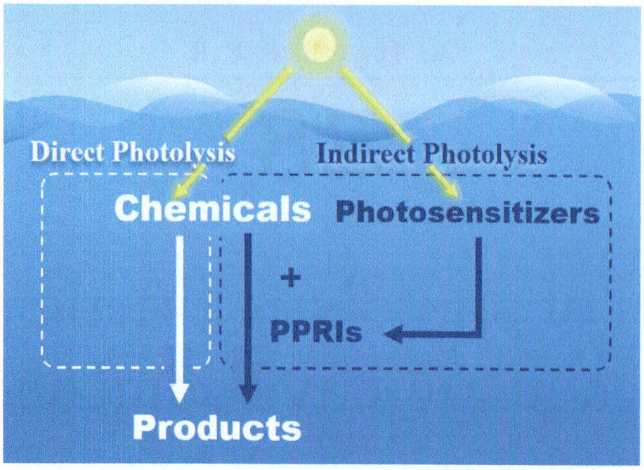

FIGURE 28.1 Photolysis pathways of chemicals in the aquatic environment.

Direct photolysis

The natural water environment contains a large number of light-absorbing substances that gradually attenuate the intensity of the sunlight in water [2]. Therefore, direct photolysis of chemicals occurs mainly in the surface water column. When a photon emitted by the sun approaches a molecule in the aquatic environment, the electromagnetic fields of both the molecule and the light radiation interact with each other. The radiation is considered "effective" in initiating a photochemical reaction of the molecule if and only if the radiation is absorbed by the molecule as a result of the interaction (Grotthus-Draper law).

The absorption of light by a molecule is a one-photon process. Normally, each molecule absorbs only one photon to reach its excited state (Stark–Einstein law). Excited-state molecules are very unstable and short-lived, and some would return to the ground state via intra- or intermolecular physical processes, while others shall undergo photochemical processes. These physical and chemical processes are known as the *primary steps* in the photolytic transformation. The relative efficiency with which a primary step occurs can be expressed in terms of *quantum yields*, defined by Eq. (28.1).

$$\Phi_i = \frac{n_i}{n_p} \tag{28.1}$$

where Φ_i is the quantum yield of a single primary step i; n_i is the number of excited state species consumed by the step i; and n_p is the total number of photons absorbed by the system owing to the presence of the compound. If both photophysical and photochemical processes occur, the sum of the quantum yields of all primary steps must be equal to 1.

Unlike photochemical processes, photophysical processes do not change the constitution of molecules. For photochemical reactions, in addition to the primary quantum yield, the so-called *apparent quantum yield* (Φ) defined by Eq. (28.2) is also taken into account.

$$\Phi = \frac{n_r}{n_p} \tag{28.2}$$

where n_r is the number of reactant molecules consumed or products produced in a photochemical process. Φ may exceed 1 in cases where the absorption of a photon by a given molecule causes a chain reaction that consumes more than one molecule.

Process during which molecules undergo photochemical processes is referred as direct photolysis. Direct photolysis generally follows a primary reaction kinetic with an apparent photolysis rate constant (k, s^{-1}) obtained from Eqs. (28.3) and (28.4).

$$-\mathrm{d}c / \mathrm{d}t = \Phi \cdot \left(\sum I_\lambda \cdot \varepsilon_\lambda\right) \cdot c = k \cdot c \tag{28.3}$$

$$c_t = c_0 \cdot \mathrm{e}^{-k \cdot t} \tag{28.4}$$

where I_λ is the average light intensity the chemical exposed in the wavelength $\Delta\lambda$ interval, Einstein/(cm^2·s·10^3); ε_λ is the average molar absorption coefficient of the chemical in the wavelength $\Delta\lambda$ interval, L/(mol·cm); $\sum I_\lambda \cdot \varepsilon_\lambda$ represents the cumulative light absorption of the chemical over the full wavelength range (accumulated at fixed $\Delta\lambda$ intervals); c_t, c_0 are the concentration (mol/L) of the chemical at moment t and the initial moment, respectively. Accordingly, the half-life of direct photolysis of the chemical ($t_{1/2}$) can be obtained using Eq. (28.5).

$$t_{1/2} = \ln 2/k \tag{28.5}$$

Direct photolysis tests on chemical substances in water can be carried out according to OECD 316 Standard—"Test Method for Direct Photolysis of Substances in Water" [3]. The direct photolysis kinetic parameters (Φ, k, $t_{1/2}$) can be obtained by measuring the degradation rate of chemical substances under natural or simulated sunlight. However, direct photolysis tests are low-throughput, time-consuming, and demanding on equipment. The data obtained are far from sufficient to support the huge demand for chemical risk assessment. Modeling methods enabling high-throughput prediction are thus necessary.

QSAR models of direct photolysis kinetic parameters

The Φ of direct photolysis of a chemical is determined by the structure of the chemical itself [4]. Theoretically, it is possible to develop QSAR models for Φ based on the structural descriptors of molecules.

In 1992, Peijenburg et al. first reported a QSAR model on Φ of selected meta-substituted halobenzenes under irradiation of artificial light sources (250—350nm) [5]. In this model, four empirical structural descriptors were considered: (1) the bond strength (*BS*) of the carbon—halogen bond to be broken; (2) the summation of the Hammett σ constants of the additional substituents to the benzene ring; (3) the inductive constant of the additional substituents σ_I; and (4) the summation of the steric factors, E_s, of all substituents to the aromatic ring. The best model was constructed by multiple linear regression (MLR) using two descriptors as given in Eq. (28.6).

$$\log \Phi = -0.010\, BS + 0.665\, E_s + 1.031 \tag{28.6}$$

$$n = 12;\ R^2 = 0.94$$

where n is the number of compounds and R^2 is the squared regression coefficient.

With the first QSAR model of Φ reported, more related studies have been progressively advanced. Chen et al. used a series of algorithms to build QSAR models for logΦ of substituted aromatic halides [6—8]. In these studies, the authors used the PM3 method (implemented in the MOPAC 6.0 software) to optimize the molecular structures and first proposed and calculated 18 quantum chemical descriptors. Six of the descriptors reflect the overall properties of the molecules (Table 28.1).

The other descriptors reflect the nature of the carbon—halogen bond. For example, the bond order of the carbon—halogen bond (*BO*) reflects the magnitude of the strength of the carbon—halogen bond. If a compound has more than one carbon—halogen bond, only the carbon—halogen bond with the lowest *BO* value is selected (usually the difference in bond order between the same type of carbon—halogen bond is very small), as this carbon—halogen bond is the weakest and most likely to be broken. The following 11 parameters, q_x, q_C, E_{EE1}, E_{EN1}, E_J, E_K, E_{EE2}, E_{EN2}, E_{NN2}, E_C, E_{TE}, all correspond to this carbon—halogen bond (Table 28.1).

Subsequently, a series of QSAR models were obtained using simple regression analysis, multiple regression analysis (MRA), stepwise regression analysis, principal component analysis (PCA), factor analysis (FA), cluster analysis (CA), and partial least squares (PLS) analysis (Fig. 28.2).

Based on the experience of previous studies, Chen et al. further developed a PLS model for the Φ of polycyclic aromatic hydrocarbons (PAHs) [9]. The authors concluded that PAHs with large α, molecular weight, and heat of formation values (HOF) tend to have small Φ. PAHs with high E_{LUMO}, low E_{LUMO}, and large E_{LUMO}−E_{HOMO} values tend to have large Φ.

Subsequently, Chen et al. developed a QSAR model to predict the Φ of polychlorinated dibenzo-*p*-dioxins (PCDDs) [10]. In this study, PLS algorithm was used, and 27 quantum chemical descriptors were computed by the PM3 Hamiltonian. The authors showed that Φ of the PCDDs are dependent on the number of chlorine atoms bonded with the parent structures, the number of carbon—oxygen bonds, and molecular polarity. Increasing bulkiness and polarity of PCDDs lead to a decrease of the logΦ values. Meanwhile, increasing E_{LUMO}, E_{HOMO}, and HOF lead to an increase of the logΦ values.

TABLE 28.1 Quantum chemical descriptors used in QSAR models.

Descriptor	Definition
μ	Dipole moment
α	Average molecular polarizability
E_{HOMO}	Energy of the highest occupied molecular orbital
E_{LUMO}	Energy of the lowest unoccupied molecular orbital
q^-	The largest negative atomic charge on an atom
q^+	The most positive net atomic charges on an atom
BO	The bond order of the carbon–halogen bond
q_x	The net atomic charges on the halogen atoms
q_C	The net atomic charges on the carbon atoms in the benzene ring that are connected to the halogen atoms
E_{EE1}	The electron–electron repulsion energy of the one-center term for the halogen atoms
E_{EN1}	The electron–nuclear attraction energy of the one-center terra for the halogen atoms
E_J	The resonance energy of the two-center term for the carbon–halogen bonds
E_K	The exchange energy of the two-center term for the carbon–halogen bonds
E_{EE2}	The electron–electron repulsion energy of the two-center term for the carbon–halogen bonds
E_{EN2}	The electron–nuclear attraction energy of the two-center term for the carbon–halogen bonds
E_{NN2}	The nuclear–nuclear repulsion energy of the two-center term for the carbon–halogen bonds
E_C	The coulombic interaction energy of the two-center term for the carbon–halogen bonds ($E_{Cs} = E_{EE2} + E_{EN2} + E_{NN2}$)
E_{TE}	The total of electron and nuclear energies of the two-center term for the carbon–halogen bonds

Although many QSAR models for Φ of chemicals have been established, Φ itself is not a straightforward parameter for describing the rate of degradation of chemicals. Many simulated experiments have shown that photolysis of chemicals in the environment mostly follows first-order (or pseudo–first-order) kinetics [11]. Eq. (28.5) indicates that the direct photolysis rate constant (k) of a chemical is a more intuitive parameter than Φ. In 1996, Chen et al. first reported a QSAR model for k of selected PAHs under the irradiation of artificial light sources [12]. In this study, the authors found a parabolic relationship between log$k$ and $E_{LUMO} - E_{HOMO}$ described by Eq. (28.7).

$$\log k = 9.770(E_{LUMO} - E_{HOMO}) - 0.715(E_{LUMO} - E_{HOMO})^2 - 32.738 \tag{28.7}$$

$$n = 11, R = 0.921, SE = 0.322, F = 38.911, P = 0.0000$$

where n is the number of compounds; R is the correlation coefficient; SE is the standard error of estimates for the full regression; F is the F value for analysis of variance; P is the significance level.

In recent years, in addition to the aforementioned typical pollutants, photolysis of disinfection by-products (DBPs) in drinking water has also received attention. In 2010, Chen et al. developed a model for predicting k of DBPs with Eq. (28.8) [13].

$$\log k = \delta E_S + \rho\sigma + c' \tag{28.8}$$

where δ is the sensitivity coefficient of the steric effect of a substituent; E_S is the steric effect constant of the substituent; ρ is the sensitivity coefficient of the electronic effect of the substituent; σ is the electronic effect constant of the substituent; and c' is an empirical constant. This study was among the first investigations into the solar photolysis effect on a wide variety of carbonaceous and nitrogenous DBPs and related compounds ($n = 170$) under a set of common exposure conditions.

Subsequently, more researchers have focused on the photolysis of DBPs, and several QSAR models have been built to facilitate the understanding of the removal process of DBPs in aquatic environment [14–16]. Xiao et al.

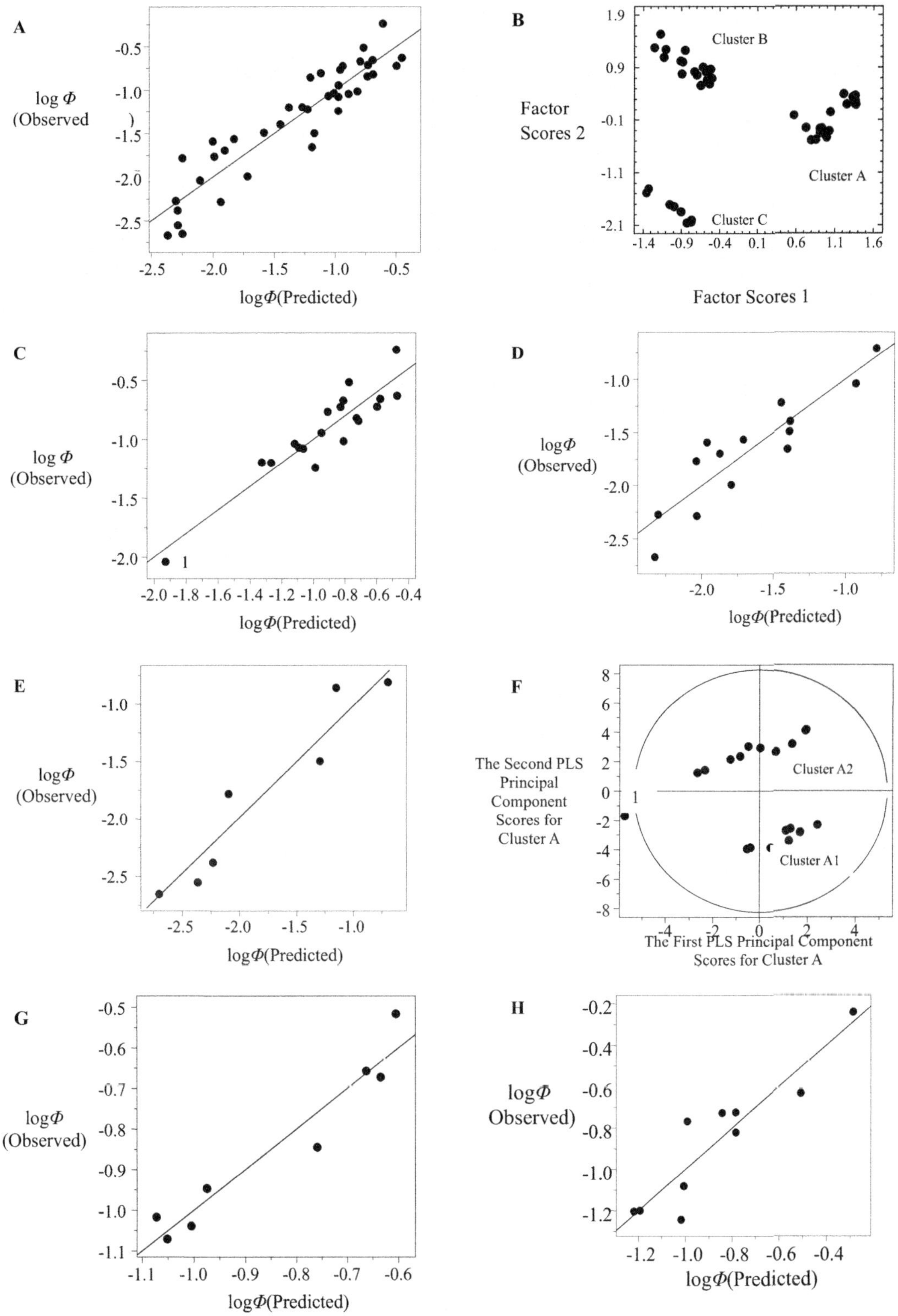

FIGURE 28.2 (A) plot of observed quantum yields (log*Φ*) versus those predicted by PLS model for 40 substituted aromatic halides; (B) grouping of 45 substituted aromatic halides based on factor scores; (C) plot of observed log*Φ* versus those predicted by a PLS model for cluster A; (D) plot of observed log*Φ* versus those predicted by a PLS model for cluster B; (E) plot of observed log*Φ* versus that predicted by a PLS model for cluster C; (F) plot of the first and second PLS principal component scores for the compounds in cluster A (the circle indicates the 95% confidence area); (G) plot of observed log*Φ* versus those predicted by a PLS model for cluster A1; (H) plot of observed log*Φ* versus that predicted by a PLS model for cluster A2 [8].

determined the k values of six iodinated trihalomethanes (ITHMs) and three brominated trihalomethanes (BTHMs) and built QSAR models for log$k$ based on Eq. (28.8) [16]. Zhang et al. determined the k values of 40 halogenated DBPs and built a QSAR model using quantum chemical descriptor ($E_{\text{LUMO}}-E_{\text{HOMO}}$) and Dragon descriptor (*Eta_C*) by means of MLR [15]. $E_{\text{LUMO}}-E_{\text{HOMO}}$ indicates the energy gap between the highest occupied molecular orbital to the lowest unfilled molecular orbital, which encodes information on electronic effect. Since the energy gap directly relates to the ionization potential and exhibits high correlation with the chemical reactivity, it can influence the photolysis rate of DBPs. *Eta_C* refers to the composite extended topochemical atom descriptor calculated by Dragon software. This descriptor indicates the size, shape, branching, and functionality contributions of a molecular graph and includes bond dissociation energy and steric effect of DBPs.

As chemicals continue to grow in the variety and functions, such as various types of dyes, pharmaceuticals, and personal care products, QSAR models for photolysis kinetic parameters of these "new chemicals" are gradually developed [17–20]. Selected QSAR models are summarized in Table 28.2.

Discussion of QSAR models on direct photolysis parameters

For Φ of direct photolysis of chemicals, mature QSAR modeling methods have been developed. However, it is difficult to obtain experimental data of Φ, thus making it difficult to build models with large applicability domains. Indeed, the existing models are only applicable to a relatively small number of classes of chemicals.

In addition, for some hydrophobic chemicals such as PBDEs, photolysis kinetic parameters cannot be obtained experimentally in water but only in organic solvents or in aqueous solution supplemented with organic cosolvent. However, due to differences in the photochemical properties of these chemicals in different solvents, it may be irrational to use these experimental data for modeling the photolysis kinetic parameters of chemicals in the aquatic environment. As experimental photochemistry data on chemicals are so scarce, it is worth to consider experimental data in nonaqueous solvents for modeling the kinetics of photolysis in an aquatic environment. In recent years, some researchers have used deep learning methods to characterize the effects of solvents on the optical and photophysical properties of organic compounds using graph convolutional networks [34], which may provide new ideas for modeling the Φ of chemicals.

Finally, the photolysis processes are de facto complex, including hydrogen atom abstraction, photoisomerization, and photodehalogenation. Different reaction pathways may be associated with different molecule structural features, which makes the existing structural descriptors insufficient for modeling the Φ of different classes of chemicals. This is one of the reasons why QSAR models of Φ cannot be built with large applicability domains.

On the other hand, although many QSAR studies on k and $t_{1/2}$ for chemicals have been reported, they also face challenges of both scientific and practical validity. From the previous equations, the k is dependent on both internal factors, which are governed by the molecular structure and external environmental factors. If the environmental conditions for training compounds are the same or similar, the effects of environmental conditions on the photolysis might be omitted. Therefore, in principle, it is possible to develop QSARs on k or $t_{1/2}$ of organic chemicals.

When the experimental conditions (e.g., light source) are changed, discrepant k and $t_{1/2}$ values may be obtained for the same chemical with the same Φ [35,36]. This makes QSAR models only applicable for k and $t_{1/2}$ values under certain environmental conditions. To address this problem, it is necessary to introduce features that have an impact on the k and $t_{1/2}$ other than the molecule's own structure (such as the information on light sources) into the modeling process. In this way, all these multimodal data (descriptors, light sources irradiance and spectrum, solvent information) that characterize the features of photolysis processes could be used for developing multimodal photolysis models. However, it is difficult to fuse multimodal data of photolysis processes into a multimodal model. In recent years, a network architecture, the transformer, has been developed to change this situation and may provide new ideas for modeling photolysis [37].

Indirect photolysis

Chemicals in aquatic environment can undergo indirect photolysis by reacting with PPRIs. PPRIs are generated by the photosensitizers such as dissolved organic matter (DOM), nitrate, and nitrite. Some chemicals can also act as photosensitizers, generating PPRIs that react with the chemicals themselves.

TABLE 28.2 Established QSAR models on direct photolysis kinetics of organic pollutants in different media.

Compounds	Number of compounds	Photolysis kinetic parameters	Media	Light sources	Algorithms	References
PAHs	n = 17	*K*	Methanol/water (1:1, v:v)	500W medium pressure mercury lamp	MLR	[12]
	n = 13	*Φ*	Methanol/water (1:1, v:v); water	450W medium pressure mercury lamp, UV light (313 nm, 366 nm, 436 nm)	PLS	[9]
	n = 13	$t_{1/2}$	Near-surface water body	Sunlight, latitude 40 degrees N, midday and midsummer	PLS	[21]
	n = 13	$t_{1/2}$	Near-surface water body	sunlight, latitude 40 degrees N, midday and midsummer	PLS	[22]
Substituted aromatic halides	n = 12	*Φ*	Water (for compounds having a very low aqueous solubility, 10% acetonitrile (v/v) was added)	RUL 3000 Å lamps	MLR	[5]
	n = 45	*Φ*	Water	RUL 3000 Å lamps	FA, CA, stepwise variable selection regression analysis	[6]
	n = 45	*Φ*	Water	RUL 3000 Å lamps	MRA, FA	[7]
	n = 45	*Φ*	Water	RUL 3000 Å lamps	PLS	[8]
PCDD/Fs	n = 7	*Φ*	Acetonitrile/water (3:2, v:v; 1:1, v:v)	Rayonet RPR 3000 Å lamps (313 nm)	PLS	[10]
	n = 9	*Φ*	Acetonitrile/water (3:2, v:v; 1:1, v:v)	Rayonet RPR 3000 Å lamps (313 nm, 300 nm)	PLS	[23]
PCBs	n = 22	$t_{1/2}$	*n*-Hexane	15W UV lamp (254 nm)	PLS	[24]
	n = 16	*K*	*n*-Hexane	500-W high-pressure mercury lamp	PLS	[25]
	n = 7	$t_{1/2}$	Water	15W UV lamp (254 nm)	PLS	[26]
PBDEs	n = 11	*Φ*	Methanol/water (4:1, v:v)	UV light in the sun light region	PLS	[27]
	n = 15	*K*	Methanol/water (4:1, v:v)	UV light in the sun light region	PLS	
	n = 9	*K*	Methanol	UV light in the sun light region	PLS	
	n = 11	*Φ*	Methanol/water (4:1, v:v)	UV light in the sun light region	PLS	[28]
	n = 15	*K*	Methanol/water (4:1, v:v)	UV light in the sun light region	PLS	
	n = 9	*K*	Methanol	UV light in the sun light region	PLS	
	n = 18	*K*	Hexane	Rayonet RPR 3000 Å lamps	PLS	[29]

Continued

TABLE 28.2 Established QSAR models on direct photolysis kinetics of organic pollutants in different media.—cont'd

Compounds	Number of compounds	Photolysis kinetic parameters	Media	Light sources	Algorithms	References
	n = 18	K	Methanol	Rayonet RPR 3000 Å lamps	PLS	
	n = 18	Φ	Hexane	Rayonet RPR 3000 Å lamps	PLS	
	n = 18	Φ	Methanol	Rayonet RPR 3000 Å lamps	PLS	
ClPAHs[a] and the corresponding 5 parent PAHs	n = 16	$t_{1/2}$, Φ	Cyclohexane	450W high-pressure mercury lamp(>290 nm)	MLR	[30]
PBDEs, OH-PBDEs, and selected organochlorines	n = 30	$t_{1/2}$	Methanol/water (4:1, v:v)	UV light in the sun light region	PCA, PLS, self-organizing map	[31]
1,4-Dihydropyridine antihypertensives	n = 9	$t_{0.33}$[b]	Water	Xenon lamp	PLS	[17]
Aromatic pollutants	n = 30	Φ	Water	Low-pressure mer cury lamp	Genetic algorithm, MLR	[32]
Triazine dyes	n = 9	Φ	Water	5.4W low-pressure mercury lamp	Genetic algorithm, MLR	[18]
ITHMs and BTHMs[c]	n = 9	K	Water	5W low-pressure mercury lamp	Simple linear regression (SLRs), MLR	[16]
Halogenated DBPs	n = 40	K	Water	5W low-pressure mercury lamp	MLR	[15]
Dihydroindolizines	n = 28	$t_{1/2}$,	Organic solvents	Laser flash photolysis	Genetic Algorithm-MLR	[33]
Azo dyes	n = 28	K	Water	300 W medium-pressure mercury lamp	Correlation analysis, PCA, MLR	[20]
Antibiotics	n = 77	K	Water	Xenon lamp(>290 nm)	SMLR	[19]

[a]*ClPAHs: chlorinated polycyclic aromatic hydrocarbons.*
[b]$t_{0.33}$: *the time necessary to reach the concentration which equivalent to a third of the starting concentration.*

PPRIs in aquatic environment include hydroxyl radicals (·OH), singlet oxygen ($^1O_2$), hydrated electrons (e^-_{aq}), triplet-excited DOM ($^3DOM^*$), etc. Indirect photolysis reactions of chemicals initiated by PPRIs generally follow second-order reaction kinetics expressed in Eq. (28.9).

$$-dc/dt = \sum k_{PPRI} \cdot [PPRI] \cdot c \tag{28.9}$$

where c is the concentration of chemicals, mol/L; k_{PPRI} is the second-order reaction rate constant for the reaction of chemicals with one of the PPRIs, L/(mol·s), characterizing the reactivity of chemicals with PPRIs; and [PPRI] is the concentration of PPRIs, mol/L. Given that [PPRI] in natural aquatic environment is approximately in a steady state, indirect photolysis of chemicals approximately follows the pseudo−first-order kinetics described using Eqs. (28.10) and (28.11).

$$-dc/dt = \sum k_p \cdot c \tag{28.10}$$

$$k_p = k_{PPRI} \cdot [PPRI]_{ss} \tag{28.11}$$

where k_p is the pseudo−first-order reaction rate constant, s^{-1}, and $[PPRI]_{ss}$ is the steady-state concentration of PPRIs, mol/L. The $t_{1/2}$ for the reaction of chemicals with PPRIs can be calculated by Eq. (28.12).

$$t_{1/2} = \ln 2 / k_p \tag{28.12}$$

In aquatic environment, k_{PPRI} can be obtained using either direct or indirect assays.

The aforementioned equations show that the factors affecting the indirect photolysis rate of chemicals are mainly two parts: (1) the steady-state concentration of PPRIs in the aquatic environment and (2) the rate constants for chemicals reacting with respective PPRIs. The two parts correspond to two processes of indirect photolysis: (1) the absorption of light by photosensitizers in water to generate PPRIs and (2) the reaction of PPRIs with the chemicals.

The first process and relevant QSAR models

In the first process, photosensitizers (mainly DOM) in aquatic environment absorb sunlight and produce various types of PPRIs. For example, $^1O_2$ is formed mainly through the energy transfer between $^3DOM^*$ and ground-state O_2. These PPRIs have multiple quenching pathways and rapid quenching rate in aquatic environment, which makes them generally very short-lived.

While continuously produced and quenched, PPRIs in the aquatic environment would eventually exist in a steady-state concentration (or in a narrow range of concentrations). For example, the steady-state concentrations of $^1O_2$ in natural waters under solar light are approximately 10^{-14}–10^{-12} mol/L [38]. Noticeably, the steady-state concentrations could vary spatially and temporally, depending on meteorological conditions (e.g., intensity and duration of sunlight irradiation) and the composition of the aquatic environment. The steady-state concentration of OH in natural aquatic environment could be in the range of 10^{-19}–10^{-16} mol/L [39], considering that generation of OH in natural aquatic environment could be facilitated with a variety of components in water, including NO_3^-, NO_2^-, H_2O_2, and DOM [39,40].

The production rate of PPRIs depends on the quantum yield of PPRIs generation from the photosensitizer (Φ_{PPRI}). Previous studies have shown that the Φ_{PPRI} of DOM is closely related to the structural composition of the DOM and that the Φ_{PPRI} of DOM in different water bodies varies considerably [41]. However, as DOM is mixture, the descriptors previously employed to characterize the structure of chemicals are not applicable.

Given that the functional groups of DOM such as aromatic ketones and quinones play a key role in the production of PPRIs, organic compounds that are chemically similar to DOM (i.e., DOM analogs) were proposed as alternative models for DOM. Such strategy allows the traditional descriptors applicable to the DOM analogs. In 2020, Zhao et al. selected 16 DOM analogs as the training set to build the QSAR model for predicting $^1O_2$-generating quantum yields (Φ_{1O2}) using quantum chemical descriptors and Dragon descriptors [42]. The QSAR model was built by means of MLR and is given in Eq. (28.13).

$$\log \Phi_{1O2} = -5.861 + 0.658\, CIC1 + 4.480\, \text{DLS\_cons} - 1.327\, \text{Mor27}i \tag{28.13}$$

$$n = 16, R^2 = 0.901, \text{RMSE} = 0.194, F = 34.447, P < 0.0001$$

where n is the number of the analogs in the training set and *RMSE* is the root mean squared error. The model contains three structural molecular descriptors: the complementary information content index (neighborhood symmetry of 1-order, *CIC*1), the Dragon consensus drug-like score (DLS_cons), and a 3D-MoRSE (molecule representation of structures based on electron diffraction) descriptor weighted by ionization potential (signal 27/weighted by ionization potential, Mor27i). The simulation results were validated with Suwannee River fulvic acid: the predicted value ($\log\Phi_{1O2} = -1.76$) was in good agreement with the experimental value ($\log\Phi_{1O2} = -1.79$).

In addition to the studies of the DOM analogs, a number of QSAR studies on the Φ_{PPRI} of specific classes of organics in aquatic environment are available. Wang et al. measured the Φ_{1O2} from a series of oxidated aromatic compounds (OACs) in water, and they selected 11 of them to calculate quantum chemical descriptors [43]. PLS regression algorithm was used to build the QSAR model. Their results showed that E_{LUMO} and the most positive net charge of hydrogen atoms had the greatest effect on the generation of $^1O_2$ by OACs.

The second process and relevant QSAR models

In the second process of indirect photolysis, PPRIs in aquatic environment react with chemicals to transform them. The kinetics of this process is determined by second-order reaction rate constants Eq. (28.9): k_{1O2}, k_{OH}, k_{eaq-}, k_{3DOM^*}, etc. The most QSAR studies of PPRIs available are for $^1O_2$ and OH.

QSAR models for k_{OH} in aquatic-phase include single-parameter and multiparameter models [44–54]. Earlier single-parameter models mostly established a linear relationship between k_{OH} and specific experimentally derived descriptors (e.g., Hammett's constant), such as the model reported by Einschlag et al. [44]. Herrmann et al. [45] built single-parameter models using the C—H bond dissociation energy based on the Evans—Polanyi relationship, which says that at a given temperature condition, there is a linear relationship between the rate constant of the reaction and its activation energy [55]. However, earlier single-parameter models of k_{OH} have limited accuracy in describing molecular structural features. In contrast, multiparameter models, combining with experimental descriptors and theoretically calculated descriptors, provide a more comprehensive and accurate description of the relationship between the molecular structures of chemicals and their ·OH reactivity. However, some early multiparameter models focused more on revealing the mathematical relationships between descriptors and k_{OH} values of small data sets, and less on model validation and applicability domain characterization.

With the increasing understanding of the ·OH reaction mechanism and the development of computational methods, the development of multiparameter models has been greatly enhanced in terms of dataset, performance, and applicability. Following the promulgation of the OECD guidelines in 2007 [56], a number of prediction models based on multiple theoretical computational descriptors and linear algorithms began to dominate. These models are represented by the PLS and MLR models, most of which are based on the requirements of the OECD guidelines. Cheng et al. [53] built a PLS model of k_{OH} based on 116 k_{OH} values for 24 organic compounds and their quantum chemical parameters, some basic information of molecules as well as the temperature of the reaction system. Luo et al. [51] calculated Dragon descriptors, 21 quantum chemical descriptors such as E_{HOMO}, and built a QSAR model of organic pollutant k_{OH} by a stepwise MLR approach. The training set of the model had 526 compounds, including olefins, alcohols, ethers, phenols, ketones, aldehydes, aromatic hydrocarbons, aromatic heterocycles, organic acids, nitrogen compounds, sulfur compounds, phosphorus compounds, and organohalogen compounds. Li et al. [52] calculated Dragon descriptors and quantum chemical descriptors (*e.g.*, E_{HOMO}, the most negative charge of C atoms in an organic molecule) for organophosphate esters (OPEs) and used the MLR algorithm to build a QSAR model for prediction of k_{OH} of OPEs. The chemical potential $[(E_{LUMO} - E_{HOMO})/2]$ of the molecules was found to be the main factor affecting the magnitude of k_{OH} in the aquatic phase of OPEs. The model had good fitting, robustness, and predictive power ($R^2_{tra} = 0.877$, $R^2_{ext} = 0.862$, $RMSE_{ext} = 0.235$; see Fig. 28.3).

In addition to the MLR and PLS models, group contribution method (GCM)—based models have been developed in some studies based on the ·OH reaction mechanisms. Minakata et al. [47] built a GCM model using 66 group rate constants and 80 group contribution factors. Furthermore, some improved GCM methods such as Atkinson group addition method were used to build the QSAR models for k_{OH} [48]. Most of the existing linear models such as MLR, PLS, and GCM have good fitting and prediction performance, but some of them cover a limited variety and number of compounds and have a relatively narrow applicability domain. Moreover, some models lack necessary internal and external validation and applicability domain characterization, and some GCM models have too many descriptors for application.

Besides linear models, some models based on nonlinear machine learning algorithms have also emerged with the development of big data and artificial intelligence in recent years. Some machine learning models such as the artificial neural network model built by Dutot et al. [46], the feed-forward neural network model built by Borhani et al. [49], and a series of neural network models built by Zhong et al. [54] have been reported in succession. Compared with traditional linear models, these machine learning models with larger datasets generally have better fitting and prediction capabilities.

It is worth mentioning that, in aquatic environment, pH and acid/base dissociation can significantly affect the reaction kinetics of many organic chemicals with ·OH. Particularly for ionizable chemicals, they often coexist in multiple dissociated forms in the aquatic environment, and the different dissociated forms have different kinetic parameters. However, building models that can be used to predict the k_{OH} of the ionic forms of chemical is limited by experimental data and other factors [57]. Therefore, existing models for ionizable chemicals generally cover only the molecular/neutral forms. In 2019, Luo et al. [58] optimized the structures of antibiotic (fluoroquinolones and sulfonamides) molecules based on the density functional theory method and calculated Dragon descriptors and quantum chemical descriptors to build a QSAR model of k_{OH} described in Eq. (28.14) for different dissociation forms of antibiotics with stepwise MLR algorithm:

$$\log k_{OH} = 9.724 + 0.2180\ C-006 - 0.6650\ q_C^+ + 0.01300\mu \tag{28.14}$$

$$n_{tra} = 45, R^2_{adj} = 0.805, RMSE_{tra} = 0.228,$$

$$n_{ext} = 15, R^2_{ext} = 0.820, Q^2_{ext} = 0.816, RMSE_{ext} = 0.197$$

where n_{tra} and n_{ext} are the number of the data points in the training set and validation set, respectively; R^2_{adj} and R^2_{ext} are adjusted and external determination coefficients, respectively; Q^2_{ext} is external validation coefficient; $RMSE_{tra}$ and $RMSE_{ext}$ are root mean square errors for the training set and validation set, respectively. The model uses three structural molecular descriptors: the Dragon descriptor *C-006* (the number of CH_2RX fragments), the most positive net atomic charge on C atoms (q_C^+), and the molecular dipole moment (μ). The QSAR model has high goodness-of-fit, robustness, and predictive performance (Fig. 28.4).

Similar to k_{OH} models, QSAR models for k_{1O2} in aquatic phase include single-parameter and multi-parameter models [59–63]. Of the single-parameter models, some studies built linear relationships based on experimentally descriptors (e.g., Hammett's constant and half-wave potential), such as the SLR models reported by and Tratnyek et al. [59] and Mill [61]. In other studies, theoretically computational descriptors (which are more convenient to obtain and apply) are used to build QSAR models. For instance, Rorije et al. [60] and Arnold et al. [62] built SLR models using descriptors including E_{HOMO} and single-electron oxidation potential.

Most multiparameter models are built based on MLR and PLS using theoretical computational descriptors including quantum chemical descriptors and Dragon descriptors [63]. Li et al. [63] developed three QSAR models (i.e., one model for all forms, and two models for neutral and deprotonated molecules) to predict k_{1O2} values of dissociating organic compounds with the MLR algorithm. The large training set (180 $\log k_{1O2}$ values including 109 $\log k_{1O2}$ values for neutral states and 71 $\log k_{1O2}$ values for deprotonated states) allows the models to have larger applicability domains.

Except for k_{OH} and k_{1O2}, other k_{PPRI} are also investigated [64]. In 1964, Anbar's study on the reactivity of organic compounds with e^-_{aq} showed that the constitutional properties of molecules are related to their reactivity [65],

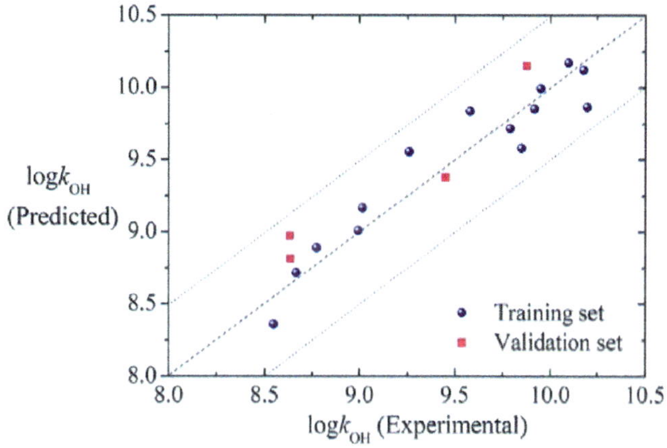

FIGURE 28.3 Plot of predicted versus experimental $\log k_{OH}$ values for OPEs in the training and validation sets [52].

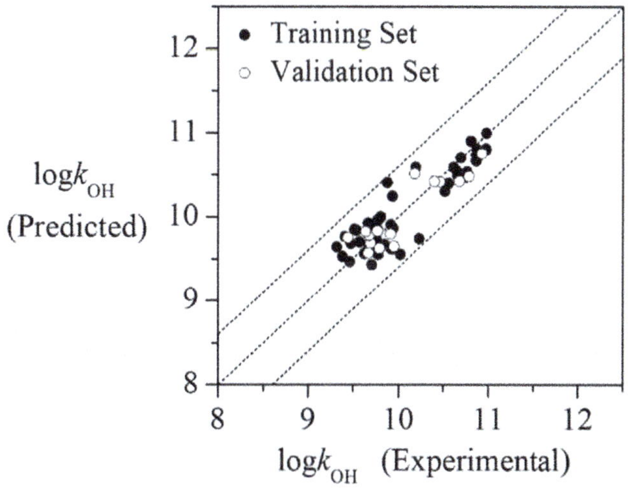

FIGURE 28.4 Plot of the predicted versus experimental $\log k_{OH}$ values for the training and the validation set [58].

indicating that there is an intrinsic correlation between the k_{eaq^-} and chemical structures. Based on this, Li et al. [64] collected a large dataset of k_{eaq^-} for aliphatic phenyl-based compounds (e.g., benzene and its derivatives, polybrominated diphenyl ethers, PCBs, fluorenone, quinolone, atropine, and polycyclic aromatic hydrocarbons) and developed QSAR models with MLR and support vector machine regression techniques. Both QSAR models exhibit satisfactory goodness-of-fit, robustness, and good predictability. E_{LUMO}, one-electron reduction potential, and α are the most important molecular structural parameters governing the reactivity of organic compounds with e^-_{aq}. Compounds with electron-withdrawing groups tend to possess higher second-order rate constants. Halogen radicals play an important role in photochemical reactions in sunlit estuarine and coastal waters rich in halogen ions [66]. In recent years, Luo et al. [67,68] developed a series of QSAR models for the reaction rate constants of $Cl^·$, Cl^{-2}, $Br^·$, and Br^{-2} with organic chemicals using Hammett constants.

Discussion of QSAR models on indirect photolysis parameters

Nowadays, QSAR modeling studies on indirect photolysis kinetic parameters have been much explored, but many shortcomings remain. For the first process, only a few studies have been conducted to predict the Φ of PPRIs in the aquatic environment. As the existing QSAR models cover only a portion of DOM or DOM analogs in water, these models do not reflect the Φ_{PPRI} of DOM as a whole. With the development of DOM analysis techniques, the photochemically active components of the DOM are continually being identified. If increasing understanding of the DOM can combine with developed modeling algorithms to calculate descriptors that more accurately describe the composition of DOM, the problems mentioned before can potentially be solved. In addition, the production of PPRIs by photosensitizer in aquatic environment is influenced by various environmental factors. Previous studies have shown that changes in temperature and light wavelength can alter Φ_{PPRI} of the photosensitizers in water [69]. It is also necessary to consider the influential environmental factors when modeling Φ_{PPRI}.

For the second process, although second-order reaction rate constant of PPRIs with chemicals has been widely modeled, some PPRIs (such as $CO\cdot - 3$) still have not received much attention. In specific water bodies, or for specific chemicals, these "neglected PPRIs" may play a significant role in the indirect photolysis kinetics of the chemicals [70]. Therefore, the reactivity of various PPRIs with different chemicals is suggested to be explored by photochemical experiments in future studies, and QSAR models should be developed to predict their kinetics.

Perspectives

QSAR models on photolysis kinetic parameters in aquatic environment have improved the knowledge of the photochemistry and provided tools for chemical risk assessment. However, there is still a lot of development to be done in this area of research. While it is unrealistic to think that a solution will be obtained soon, positive steps forward could include the following:

(1) Generation of meaningful and precise data. Data are the basis for modeling and are the first factor to be considered in conducting QSAR studies. Different amounts and forms of data determine the modeling algorithms we use. Moreover, data of larger quantity and higher quality can help researchers to build more effective and robust QSAR models with larger applicability domains. For this reason, researchers need to standardize photochemical experimental procedures and follow existing guidelines (such as OECD 316 [3]) for the determination of photolysis kinetic parameters of chemicals. At the same time, the photolysis parameters are supposed to be given along with as much experimental information as possible, such as light source and solvent information. Hence, a database of chemical photochemical conversion kinetic parameters with standardized data forms and uniform data labels needs to be established.

(2) Building a more comprehensive and multimodal model. First, there are multiple processes involved in the photolysis of chemicals, which makes it impossible in some cases to assess the actual photolysis kinetics of chemicals in aquatic environment by considering only single pathway-endpoint QSAR models. In addition, the photolysis of chemicals is influenced by multiple factors, which make the photolysis kinetics of the same chemical vary considerably under different environmental conditions. Therefore, researchers need to develop more comprehensive models that take into account the photolysis pathways and influencing factors of chemicals in a holistic manner. Researchers should integrate existing models and experimental data, incorporate models for other environmental factors such as the attenuation of underwater light intensity and the temperature dependence of Φ of PPRIs, and build multimodal models by multimodal data.

(3) Striving for more supports. Data on the photolysis kinetic parameters of chemicals require a large number of experiments and many specialist/researchers to carry out the tests. Support in terms of funding and specialist training are major contributions to the development of photochemical QSAR models. In addition, the support by researchers from other fields such as computational science and geosciences will also enable a rapid development of photochemical QSAR studies.

References

[1] Vione D, Minella M, Maurino V, Minero C. Indirect photochemistry in sunlit surface waters: photoinduced production of reactive transient species. Chem Eur J 2014;20(34):10590–606.
[2] Kirk JTO. Attenuation of light in natural waters. Mar Freshw Res 1977;28:497–508.
[3] OECD. Test No. 316: phototransformation of chemicals in water–direct photolysis. Paris: OECD Publishing; 2008.
[4] Hites RA. Environmental organic chemistry. 2nd ed. Easton: American Chemical Society; 2003.
[5] Peijnenburg WJGM, De Beer KGM, De Haan MWA, Den Hollander HA, Stegeman MHL, Verboom H. Development of a structure-reactivity relationship for the photohydrolysis of substituted aromatic halides. Environ Sci Technol 1992;26(11):2116–21.
[6] Chen JW, Peijnenburg WJGM, Quan X, Zhao YZ, Xue DM, Yang FL. The application of quantum chemical and statistical technique in developing quantitative structure-property relationships for the photohydrolysis quantum yields of substituted aromatic halides. Chemosphere 1998;37(6):1169–86.
[7] Chen JW, Peijnenburg WJGM, Wang LS. Using Pm3 Hamiltonian, factor analysis and regression analysis in developing quantitative structure-property relationships for photohydrolysis quantum yields of substituted aromatic halides. Chemosphere 1998;36(13):2833–53.
[8] Chen JW, Peijnenburg WJGM, Quan X, Chen S, Zhao YZ, Yang FL. The use of pls algorithms and quantum chemical parameters derived from Pm3 Hamiltonian in QSPR studies on direct photolysis quantum yields of substituted aromatic halides. Chemosphere 2000;40(12):1319–26.
[9] Chen JW, Peijnenburg WJGM, Quan X, Yang FL. Quantitative structure-property relationships for direct photolysis quantum yields of selected polycyclic aromatic hydrocarbons. Sci Total Environ 2000;246(1):11–20.
[10] Chen JW, Quan X, Peijnenburg WJGM, Yang FL. Quantitative structure–property relationships (QSPRs) on direct photolysis quantum yields of pcds. Chemosphere 2001;43(2):235–41.
[11] Zhang YN, Wang JQ, Chen JW, Zhou CZ, Xie Q. Phototransformation of 2,3-dibromopropyl-2,4,6-tribromophenyl ether (DPTE) in natural waters: important roles of dissolved organic matter and chloride ion. Environ Sci Technol 2018;52(18):10490–9.
[12] Chen JW, Kong LR, Zhu CM, Huang QG, Wang LS. Correlation between photolysis rate constants of polycyclic aromatic hydrocarbons and frontier molecular orbital energy. Chemosphere 1996;33(6):1143–50.
[13] Chen BY, Lee W, Westerhoff PK, Krasner SW, Herckes P. Solar photolysis kinetics of disinfection byproducts. Water Res 2010;44(11):3401–9.
[14] Chen BY, Zhang T, Bond T, Gan YQ. Development of quantitative structure activity relationship (QSAR) model for disinfection byproduct (DBP) research: a review of methods and resources. J Hazard Mater 2015;299:260–79.
[15] Zhang YQ, Xiao YJ, Zhang YC, Lim TT. UV direct photolysis of halogenated disinfection byproducts: experimental study and Qsar modeling. Chemosphere 2019;235:719–25.
[16] Xiao YJ, Fan RL, Zhang LF, Yue JQ, Webster RD, Lim TT. Photodegradation of iodinated trihalomethanes in aqueous solution by UV 254 irradiation. Water Res 2014;49:275–85.
[17] Ioele G, De Luca M, Oliverio F, Ragno G. Prediction of photosensitivity of 1,4-dihydropyridine antihypertensives by quantitative structure-property relationship. Talanta 2009;79(5):1418–24.
[18] Kusic H, Koprivanac N, Bozic AL. Environmental aspects on the photodegradation of reactive triazine dyes in aqueous media. J Photochem Photobiol Chem 2013;252:131–44.
[19] Lyu YT, Huang TB, Liu W, Sun WL. Unveil the quantum chemical descriptors determining direct photodegradation of antibiotics under simulated sunlight: Batch experiments and model development. J Environ Chem Eng 2022;10(3):108086.
[20] Zhang GY, Zhang SJ. Quantitative structure-activity relationship in the photodegradation of azo dyes. J Environ Sci 2020;90:41–50.
[21] Chen JW, Peijnenburg WJGM, Quan X, Chen S, Martens D, Schramm KW, et al. Is it possible to develop a QSPR model for direct photolysis half-lives of pahs under irradiation of sunlight? Environ Pollut 2001;114(1):137–43.
[22] Lu GN, Dang Z, Tao XQ, Peng PA, Zhang DC. QSPR study on direct photolysis half-lives of pahs in water surface. J Theor Comput Chem 2005;4(3).
[23] Chen JW, Quan X, Schramm KW, Kettrup A, Yang FL. Quantitative structure-property relationships (QSPRS) on direct photolysis of pcdds. Chemosphere 2001;45(2):151–9.
[24] Niu JF, Yang ZF, Shen ZY, Wang LL. QSPRs for the prediction of photodegradation half-life of PCBs in *n*-hexane. SAR QSAR Environ Res 2006;17(2):173–82.
[25] Li X, Fang L, Huang J, Yu G. Photolysis of mono-through deca-chlorinated biphenyls by ultraviolet irradiation in *n*-hexane and quantitative structure-property relationship analysis. J Environ Sci 2008;20(6):753–9.
[26] Bao YP, Huang QY, Wang WL, Xu JJ, Jiang F, Feng CH. Application of quantum chemical descriptors into quantitative structure-property relationship models for prediction of the photolysis half-life of PCBs in water. Front Environ Sci Eng China 2011;5(4):505–11.
[27] Niu JF, Shen ZY, Yang ZF, Long XX, Yu G. Quantitative structure–property relationships on photodegradation of polybrominated diphenyl ethers. Chemosphere 2006;64(4):658–65.
[28] Chen JW, Wang DG, Wang SL, Qiao XL, Huang LP. Quantitative structure–property relationships for direct photolysis of polybrominated diphenyl ethers. Ecotoxicol Environ Saf 2007;66(3):348–52.
[29] Fang L, Huang J, Yu G, Li X. Quantitative structure–property relationship studies for direct photolysis rate constants and quantum yields of polybrominated diphenyl ethers in hexane and methanol. Ecotoxicol Environ Saf 2009;72(5):1587–93.
[30] Ohura T, Amagai T, Makino M. Behavior and prediction of photochemical degradation of chlorinated polycyclic aromatic hydrocarbons in cyclohexane. Chemosphere 2008;70(11):2110–7.

[31] Heimstad ES, Bastos PM, Eriksson J, Bergman Å, Harju M. Quantitative structure—photodegradation relationships of polybrominated diphenyl ethers, phenoxyphenols and selected organochlorines. Chemosphere 2009;77(7):914—21.
[32] Juretic D, Kusic H, Papic A, Smidt M, Jezovita O, Peternel I, et al. Modeling of photodegradation kinetics of aromatic pollutants in water matrix. J Photochem Photobiol Chem 2013;271:65—76.
[33] Beiknejad D, Chaichi MJ, Fatemi MH. Prediction of photolysis half-lives of dihydroindolizines by genetic algorithm-multiple linear regression (Ga-Mlr). J Phys Org Chem 2016;29(6):312—20.
[34] Ksenofontov AA, Lukanov MM, Bocharov PS. Can machine learning methods accurately predict the molar absorption coefficient of different classes of dyes? Spectrochim Acta Mol Biomol Spectrosc 2022;279:121442.
[35] Kim M, O'Keefe PW. Photodegradation of polychlorinated dibenzo-P-dioxins and dibenzofurans in aqueous solutions and in organic solvents. Chemosphere 2000;41(6):793—800.
[36] Choudhry G, Webster G. Environmental photochemistry of PCDDs.2.[1] Quantum yields of the direct phototransformation of 1,2,3,7-Tetra-, 1,3,6,8-Tetra-, 1,2,3,4,6,7,8-Hepta-, and 1,2,3,4,6,7,8,9-Octachlorodibenzo-*p*-dioxin in aqueous acetonitrile and their sunlight half-lives[2]. J Agric Food Chem 1989;37(1):254—61.
[37] Vaswani A, Shazeer N, Parmar N, Uszkoreit J, Jones L, Gomez AN, et al. Attention is all you need. Adv Neural Inf Process Syst 2017;(30).
[38] Peterson BM, McNally AM, Cory RM, Thoemke JD, Cotner JB, McNeill K. Spatial and temporal distribution of singlet oxygen in lake superior. Environ Sci Technol 2012;46(13):7222—9.
[39] Timko SA, Romera-Castillo C, Jaffé R, Cooper WJ. Photo-reactivity of natural dissolved organic matter from fresh to marine waters in the Florida Everglades, USA. Environ Sci: Process Impacts 2014;16(4):866—78.
[40] Nosaka Y, Nosaka AY. Generation and detection of reactive oxygen species in photocatalysis. Chem Rev 2017;117(17):11302—36.
[41] Wasswa J, Driscoll CT, Zeng T. Photochemical characterization of surface waters from lakes in the Adirondack region of New York. Environ Sci Technol 2020;54(17):10654—67.
[42] Zhao JC, Zhou YJ, Li C, Xie Q, Chen JW, Chen GC, et al. Development of a quantitative structure-activity relationship model for mechanistic interpretation and quantum yield prediction of singlet oxygen generation from dissolved organic matter. Sci Total Environ 2020;712:136450.
[43] Wang MJ, Xiang XY, Zuo YG, Peng JB, Lu K, Dempsey C, et al. Singlet oxygen production abilities of oxidated aromatic compounds in natural water. Chemosphere 2020;258:127308.
[44] Garcia Einschlag FS, Carlos L, Capparelli AL. Competition kinetics using the UV/H_2O_2 process: a structure reactivity correlation for the rate constants of hydroxyl radicals toward nitroaromatic compounds. Chemosphere 2003;53(1):1—7.
[45] Herrmann H. Kinetics of aqueous phase reactions relevant for atmospheric chemistry. Chem Rev 2003;103(12):4691—716.
[46] Dutot A, Rude J, Aumont B. Neural network method to estimate the aqueous rate constants for the oh reactions with organic compounds. Atmos Environ 2003;37(2):269—76. 1994.
[47] Minakata D, Li K, Westerhoff P, Crittenden J. Development of a group contribution method to predict aqueous phase hydroxyl radical (HO•) reaction rate constants. Environ Sci Technol 2009;43(16):6220—7.
[48] Doussin JF, Monod A. Structure—activity relationship for the estimation of OH-oxidation rate constants of carbonyl compounds in the aqueous phase. Atmos Chem Phys 2013;13(23):11625—41.
[49] Borhani TNG, Saniedanesh M, Bagheri M, Lim JS. QSPR prediction of the hydroxyl radical rate constant of water contaminants. Water Res 2016;98:344—53.
[50] Ortiz EV, Bennardi DO, Bacelo DE, Fioressi SE, Duchowicz PR. The conformation-independent QSPR approach for predicting the oxidation rate constant of water micropollutants. Environ Sci Pollut Res 2017;24(35):27366—75.
[51] Luo X, Yang XH, Qiao XL, Wang Y, Chen JW, Wei XX, et al. Development of a QSAR model for predicting aqueous reaction rate constants of organic chemicals with hydroxyl radicals. Environ Sci J Integr Environ Res: Process Impacts 2017;19(3):350—6.
[52] Li C, Wei GL, Chen JW, Zhao YH, Zhang YN, Su LM, et al. Aqueous oh radical reaction rate constants for organophosphorus flame retardants and plasticizers: experimental and modeling studies. Environ Sci Technol 2018;52(5):2790—9.
[53] Cheng ZW, Yang BW, Chen QC, Shen ZM, Yuan T. Quantitative relationships between molecular parameters and reaction rate of organic chemicals in fenton process in temperature range of 15.8 °C—60 °C. Chem Eng J 2018;350:534—40.
[54] Zhong SF, Hu JJ, Yu X, Zhang HC. Molecular image-convolutional neural network (CNN) assisted QSAR models for predicting contaminant reactivity toward OH radicals: Transfer learning, data augmentation and model interpretation. Chem Eng J 2021;408:127998.
[55] Evans MG, Polanyi M. Inertia and driving force of chemical reactions. Trans Faraday Soc 1938;34:11.
[56] OECD. Guidance document on the validation of (quantitative) structure-activity relationship [(Q)SAR] models. Paris: OECD Publishing; 2014.
[57] Nolte TM, Ragas AMJ. A review of quantitative structure—property relationships for the fate of ionizable organic chemicals in water matrices and identification of knowledge gaps. Environ Sci J Integr Environ Res: Process Impacts 2017;19(3):221—46.
[58] Luo X, Wei XX, Chen JW, Xie Q, Yang XH, Peijnenburg WJGM. Rate constants of hydroxyl radicals reaction with different dissociation species of fluoroquinolones and sulfonamides: combined experimental and QSAR studies. Water Res 2019;166:115083.
[59] Tratnyek PG, Hoigne J. Oxidation of substituted phenols in the environment: a QSAR analysis of rate constants for reaction with singlet oxygen. Environ Sci Technol 1991;25(9):1596—604.
[60] Rorije E, Peijnenburg WJGM. QSARs for oxidation of phenols in the aqueous environment, suitable for risk assessment. J Chemometr 1996; 10(2):79—93.
[61] Mill T. Predicting photoreaction rates in surface waters. Chemosphere 1999;38(6):1379—90.
[62] Arnold WA, Oueis Y, O'Connor M, Rinaman JE, Taggart MG, McCarthy RE, et al. QSARs for phenols and phenolates: oxidation potential as a predictor of reaction rate constants with photochemically produced oxidants. Environ Sci J Integr Environ Res: Process Impacts 2017;19(3): 324—38.
[63] Li TT, Huang Y, Wei GL, Zhang YN, Zhao YH, Crittenden JC, et al. Quantitative structure-activity relationship models for predicting singlet oxygen reaction rate constants of dissociating organic compounds. Sci Total Environ 2020;735:139498.
[64] Li C, Zheng SS, Li TT, Chen JW, Zhou JH, Su LM, et al. Quantitative structure-activity relationship models for predicting reaction rate constants of organic contaminants with hydrated electrons and their mechanistic pathways. Water Res 2019;151:468—77.
[65] Anbar M, Hart EJ. The reactivity of aromatic compounds toward hydrated electrons. J Am Chem Soc 1964;86(24):5633—7.

[66] Parker KM, Mitch WA. Halogen radicals contribute to photooxidation in coastal and estuarine waters. Proc Natl Acad Sci USA 2016;113(21): 5868–73.
[67] Lei Y, Cheng SS, Luo N, Yang X, An TC. Rate constants and mechanisms of the reactions of Cl• and Cl2•$^{-}$ with trace organic contaminants. Environ Sci Technol 2019;53(19):11170–82.
[68] Lei Y, Lei X, Yu YF, Li KZ, Li Z, Cheng SS, et al. Rate constants and mechanisms for reactions of bromine radicals with trace organic contaminants. Environ Sci Technol 2021;55(15):10502–13.
[69] McKay G, Rosario-Ortiz FL. Temperature dependence of the photochemical formation of hydroxyl radical from dissolved organic matter. Environ Sci Technol 2015;49(7):4147–54.
[70] Ye ZM, Guo ZY, Wang JQ, Zhang LL, Guo YC, Yoshimura C, et al. Photodegradation of acebutolol in natural waters: important roles of carbonate radical and hydroxyl radical. Chemosphere 2022;287:132318.

C H A P T E R

29

(Q)SAR models on transthyretin disrupting effects of chemicals

Xianhai Yang[1], Huihui Liu[1] and Jingwen Chen[2]

[1]Jiangsu Key Laboratory of Chemical Pollution Control and Resources Reuse, School of Environmental and Biological Engineering, Nanjing University of Science and Technology, Nanjing, China [2]Key Laboratory of Industrial Ecology and Environmental Engineering (Ministry of Education), Dalian Key Laboratory on Chemicals Risk Control and Pollution Prevention Technology, School of Environmental Science and Technology, Dalian University of Technology, Dalian, China

Abbreviations

1-Br-2-NA 1-Bromo-2-naphthoic acid
2,4,6-TBP 2,4,6-Tribromophenol
2,4,6-TCP 2,4,6-Trichlorophenol
2,4,6-TCTP 2,4,6-Trichlorothiophenol
2,4,6-TFP 2,4,6-Trifluorophenol
2,4,6-TIP 2,4,6-Triiodophenol
2,4-DBP 2,4-Dibromophenol
2,4-DCP 2,3-Dichlorophenol
2,4-DCTP 2,4-Dichlorothiophenol
2-CP 2-Chlorophenol
2-CTP 2-Chlorothiophenol
2-I-4-MBA 2-Iodo-4-methylbenzoic acid
3-HO-BDE 47 3-HO-2,2′,4,4′-tetrabrominated diphenyl ethers
4-hexyl-BA 4-Hexylbenzoic acid
5-HO-BDE 47 5-HO-2,2′,4,4′-Tetrabrominated diphenyl ethers
6-HO-BDE 47 6-HO-2,2′,4,4′-Tetrabrominated diphenyl ethers
ACF Atom-centered fragment
AD Applicability domain
ALB Albumin
ANSA 8-Anilino-1-naphthalenesulfonic acid ammonium salt
AOP Adverse outcome pathway
BA Benzoic acid
BFRs Brominated flame retardants
C-024 R–CH–R
C-025 R–CR–R
C-026 R–CX–R
CALUX Chemically activated LUciferase eXpression
CCC Concordance correlation coefficient
Cl-089 Cl attached to C1(sp2)
CoMSIA Comparative molecular similarity indices analysis
DBPs Disinfection by-products
FGC Functional group counts
FITC-T4 Fluorescein isothiocyanate–thyroxine
GFA-MLR Genetic function approximation–multiple linear regression
H-047 H attached to C1(sp3)/C0(sp2)
H-050 H attached to heteroatom
HBCD Hexabromocyclododecane
HO-PBDEs Hydroxylated polybrominated diphenyl ethers
HO-PCBs Hydroxylated polychlorinated biphenyls
HO-PCDDs Hydroxylated chlorodibenzo-p-dioxins
HO-PCDFs Hydroxylated chlorodibenzofurans
HPLC High-performance liquid chromatography
hTTR Human transthyretin
IC_{50} Concentration of inhibitor resulting in 50% inhibition
ITC Isothermal titration calorimetry
k Number of nearest neighbor
K_d Dissociation constant
K_i Affinity constant
kNN *k*-nearest neighbor
LDA Linear discriminant analysis
LIB-SVM Library for support vector machine
logD_{OW} Logarithm of the n-octanol/water distribution coefficient
log*K* Logarithm of the binding constant
log*RP* Logarithm of the relative competitive potency
MIEs Molecular initiating events
MLR Multiple linear regression
n Number of total compounds
n_{ArCOOH} Number of carboxylic acids (aromatic)
nArOH Number of aromatic hydroxyls
nArX Number of X on aromatic ring
nCar Number of aromatic C(sp2)
nCb- Number of substituted benzene C(sp2)
nCbH Number of unsubstituted benzene C(sp2)
n_{EXT} Numbers of model compounds in validation set
nHAcc Number of acceptor atoms for H-bonds (N,O,F)
nHDon Number of donor atoms for H-bonds (N and O)
n_N Number of nitrogen atoms
n_{RCOOH} Number of carboxylic acids (aliphatic)
n_{ROH} Number of hydroxyl groups
n_{SO2OH} Number of sulfonic (thio-/dithio-) acids
n_{SOOH} Number of sulfinic (thio-/dithio-) acids
n_{train} Numbers of model compounds in training set
n_X Number of halogen atoms
O-057 Phenol/enol/carboxyl OH
PBDEs Polybrominated diphenyl ethers
PBP Pentabromophenol

https://doi.org/10.1016/B978-0-443-15339-6.00024-2

PCBs Polychlorinated biphenyls
PCP Pentachlorophenol
PCTP Pentachlorothiophenol
PDB Protein Data Bank
PFCs Poly-/perfluorinated chemicals
PFTP Pentafluorothiophenol
pK_a Acid dissociation constants
PLS Partial least squares regression
PLS-DA Partial least squares–discriminant analysis
Q^2_{BOOT} Bootstrapping coefficient
Q^2_{CUM} Total variation of the dependent variables that can be predicted by all of the extracted PLS components
Q^2_{EXT} Externally explained variance
Q^2_{LOO} Leave-one out cross-validation Q^2
Q_{EXT} Predictive accuracy of validation set
Q_{Train} Predictive accuracy of training set
(Q)SAR (Quantitative) structure–activity relationship model
R^2_{train} Squared correlation coefficient of observed and predicted values for the training set
RLDM Radiolabeled ligand displacement method
$RMSE_{EXT}$ Root mean square errors for the validation set
$RMSE_{train}$ Root mean square errors for the training set
$S_{n\text{-}EXT}$ Sensitivity of validation set
$S_{n\text{-}Train}$ Sensitivity of training set
$S_{p\text{-}Train}$ Specificity of training set
$S_{p\text{-}EXT}$ Specificity of validation set
SPRB Surface plasmon resonance biosensor-based method
T_3 3,3′,5-Tetraiodo-L-thyronine
T_4 3,3′,5,5′-Tetraiodo-L-thyronine
TBBPA Tetrabromobisphenol A
TBG Thyroxine-binding globulin
TDCs Thyroid-disrupting chemicals
THDPs Thyroid hormones distributor proteins
THs Thyroid hormones
T_S Tanimoto similarity index
TTR Transthyretin
TTR-TRβ CALUX TTR-TRβ-based chemically activated LUciferase eXpression bioassay
δ_I Ionized species
δ_M Neutral species

Introduction

Thyroid hormones (THs) that included 3,3′,5,5′-tetraiodo-L-thyronine (T_4) and 3,3′,5-triiodo-L-thyronine (T_3) (Fig. 29.1) were usually synthesized in the thyroid gland (e.g., mammals, birds, amphibians, reptiles) or follicular cells (e.g., fish without discrete thyroid gland) and played a vital physiological role in vertebrates, such as growth, differentiation, energy metabolism, and so on [3]. Normally, the THs should be delivered from the site of synthesis to their sites of action throughout the body by three thyroid hormones distributor proteins (THDPs), i.e., albumin (ALB), transthyretin (TTR), and thyroxine-binding globulin (TBG) [4]. Among the three THDPs, not all of them were synthesized in the all vertebrates and in the whole life cycle of vertebrates. For example, only ALB and TTR were involved in transporting THs in fish, amphibians, and reptile [5]. In previous research, TTR received more attention than that of ALB or TBG because TTR was not only the critical THDPs in the serum of vertebrates, but also the major THs deliverer across the blood–brain and placental barrier [6,7]. Unfortunately, hundreds of anthropogenic chemicals or unintentionally produced chemicals referred as thyroid-disrupting chemicals (TDCs) could disturb the TH's delivery process through competing the binding sites of TTR with THs [8,9]. In this regard, the normal physiological functions of thyroid system and a variety of thyroid-related harmful effects in both humans and wildlife related to this molecular initiating events (MIEs), i.e., TTR binding, may be perturbed and evoked in the view of adverse outcome pathway (AOP) [10–13]. Thus, the first purpose of this chapter was to introduce the research progress of the disrupting effects between TDCs and human/other species TTR.

Under normal physiological conditions, TTR is a homotetramer composed of four identical subunits for all the vertebrates (Fig. 29.1). The function of TTR is phylogenetically conserved across different vertebrate [14]. However, some amino acid residues were changed during vertebrate evolution [15]. For example, the No. 117 amino acid residue in the binding site of TTR was threonine in seabream (*Sparus aurata*, PDB ID: 1SN0 [16]) and serine in human (*Homo sapiens*, PDB ID: 1ICT [1]), mouse (*Mus musculus*, PDB ID: 2QPF [17]), chicken (*Gallus*, PDB ID: 1TFP [18]), respectively (Fig. 29.2). The change of amino acid residues during vertebrate evolution may result in the structural differences of TTR among different vertebrate. For instance, the side chain of threonine was larger than that of serine, which resulted in the volume of the binding site for seabream TTR smaller than that of human transthyretin (hTTR) [19]. The structural difference may further lead to the distinct TTR binding affinity to T_3 and T_4 in different vertebrate. Indeed, the results from previous studies proved that T_3 binds TTR from fish, birds, reptiles, and amphibians with higher affinity, whereas T_4 binds TTR from eutherians and marsupials with higher affinity (Fig. 29.3) [25]. Whether the TTR from different vertebrate may exhibit distinct binding affinity with TDCs remains unclear. Therefore, the next purpose of this chapter was to analyze the interspecies variation of potential TTR disrupting effects among different species.

Foresaid, only hundreds of substances tested their potential TTR disrupting effects. However, more than 350,000 compounds had been used in the current global market [26,27], which implies that there exists a big data gap on the

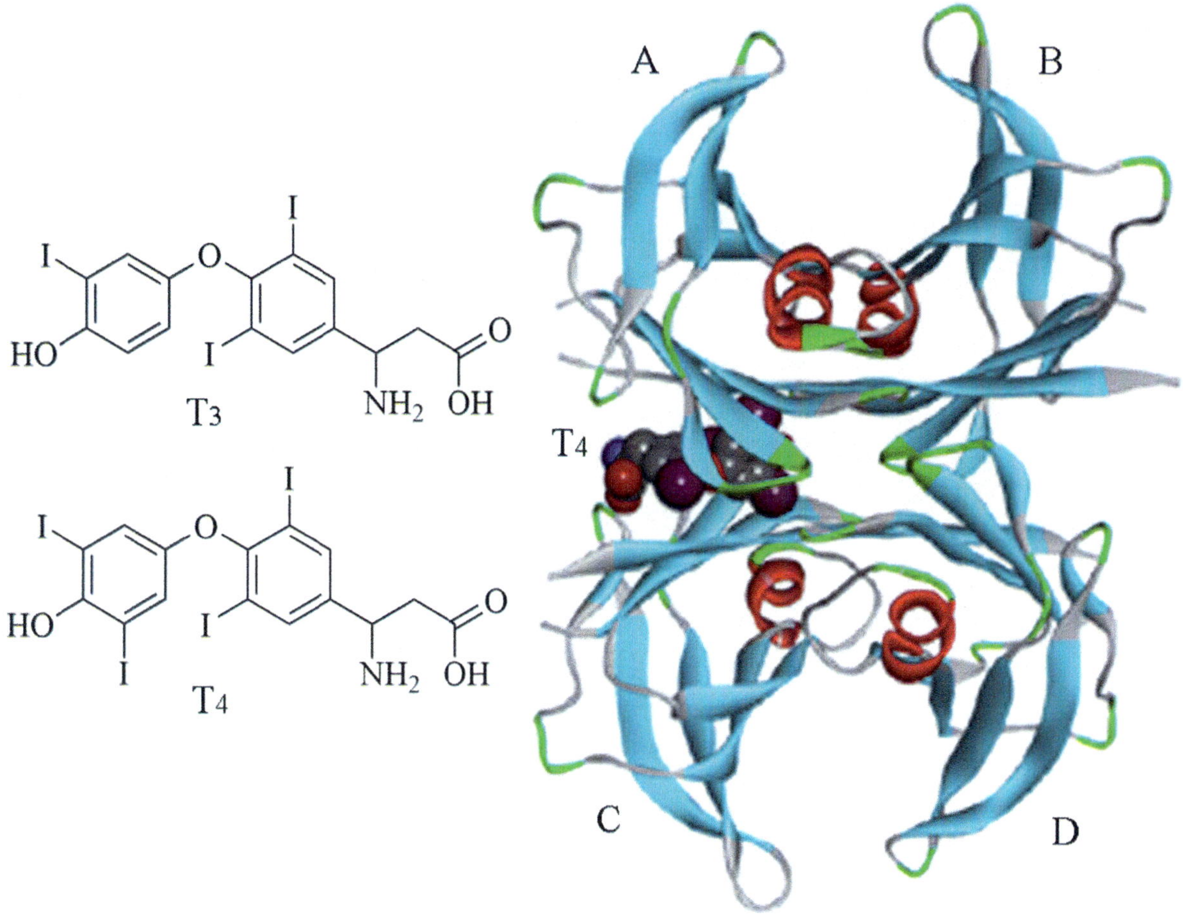

FIGURE 29.1 Molecular structure of thyroxine (T_4), triiodo-L-thyronine (T_3), and human transthyretin with four identical subunits (PDB ID: 1ICT [1], all the TTR structures used in this chapter were downloaded from RCSB Protein Data Bank (RCSB PDB) [2], https://www.rcsb.org/). TTR structure was saved from Discovery Studio Visualizer v17.2.0.16349 (Dassault Systèmes Biovia Co.).

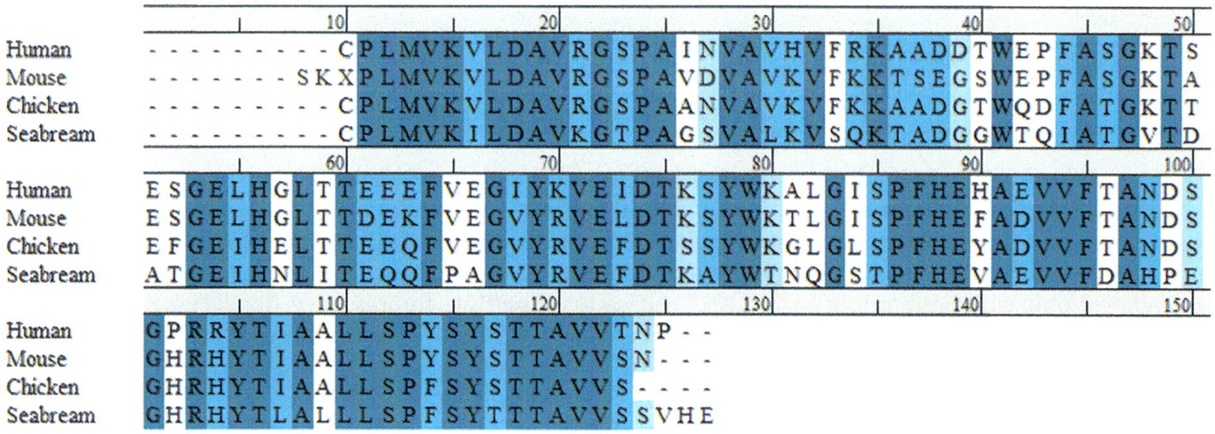

FIGURE 29.2 Multiple amino acid sequence alignment of human (*Homo sapiens*, PDB ID: 1ICT [1]), mouse (*Mus musculus*, PDB ID: 2QPF [17]), Chicken (*Gallus gallus*, PDB ID: 1TFP [18]), and seabream (*Sparus aurata*, PDB ID: 1SN0 [16]). Sequence was aligned at Discovery Studio Visualizer v17.2.0.16349 (Dassault Systèmes Biovia Co.).

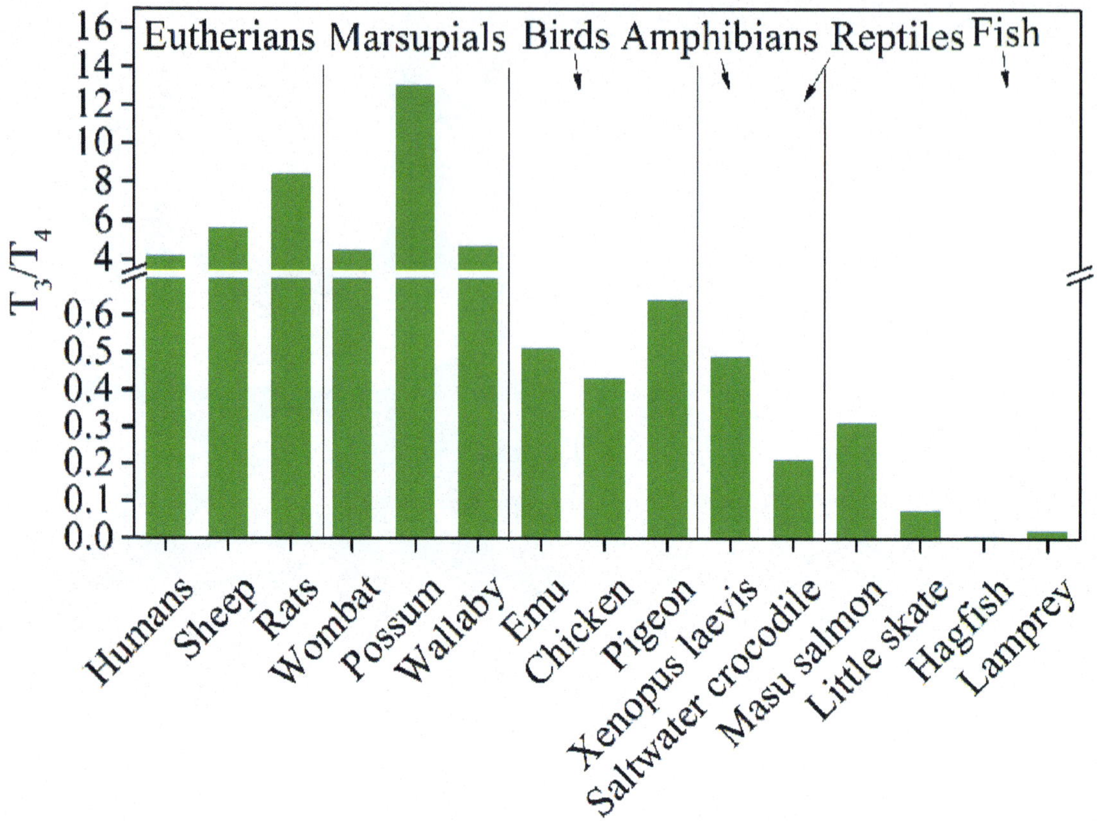

FIGURE 29.3 Ratio of TTR binding affinity between T_3 and T_4 for different vertebrates. TTR binding affinity data of T_3 and T_4 for different vertebrates were collected from Schreiber [20], Yamauchi et al. [21], Suzuki et al. [22,23] and Kasai et al. [24].

potential TTR disrupting effects for most of the used substance. To fill the data gap effectively, high-throughput experimental assays and predictive methods should be developed and employed. Thus, the third purpose of this chapter was to summarize the available in vitro assay used to test the potential TTR disrupting effects of TDCs, and the (quantitative) structure–relationship models (Q)SAR as well as software could be employed to fill the data gap.

Profile of transthyretin disrupting effects

To date, various in vitro assays had been developed to identify the potential TTR disruptors from commercially used substances and unintentional production compounds (e.g., disinfection by-products (DBPs)). The reported in vitro assays included radiolabeled ligand displacement method (RLDM) [8,22,28–38], high-performance liquid chromatography (HPLC) [39], surface plasmon resonance biosensor-based method (SPRB) [40], 8-anilino-1-naphthalenesulfonic acid ammonium salt (ANSA)–based competitive fluorescence displacement assay (referred as ANSA) [41–45], fluorescein isothiocyanate (FITC)-thyroxine (FITC-T4)–based fluorescence polarization competitive binding assays (referred as FITC-T4) [46–49], triple bioaffinity mass spectrometry (BioMS) [50], isothermal titration calorimetry (ITC) [6,51], and TTR-TRβ-based chemically activated LUciferase eXpression (CALUX) bioassay (TTR-TRβ CALUX) [52,53]. Among them, radiolabeled ligand displacement method was the first assay employed to investigate the binding interactions of TDCs with TTR since 1990, and also the most used screening assay up to now. For example, 327 out of 584 (56%) hTTR binding affinity data from 445 substance were tested by RLDM according to the summarized results of our recent study [9]. Considering that the RLDM depended on ^{125}I labeled T_4 as the radiotracer, the handling and disposal should be subjected to stringent regulations. Nonradioactivity-based assay was desired to be developed and used.

As elaborated before, several nonradioactivity-based assay had been developed and reported since 2004. Thereinto, the fluorescence displacement method, i.e., ANSA and ITC-T4 assay, had attracted increasing attention. For instance, 158 out of 584 (27%) hTTR binding affinity data from 445 substance was tested by ANSA (98 data, 17%) and ITC-T4 (60 data, 10%), which indicated that fluorescence displacement method was also the most used screening assay second only to RLDM. In addition, the high-throughput screening methods were also developed to screening potential TTR binders based on ANSA and FITC-T4 assay [54,55].

On the basis of aforementioned in vitro assays, hundreds of environmental contaminants had assessed their potential disrupting effects with TTR from human, birds, amphibians, and fish. In the subsequent section, we would analyze the most common chemical structure characteristics of human and other species (birds, amphibians, and fish) TTR disruptors, and the interspecies variation of transthyretin disrupting effects for different species.

Human transthyretin disrupting effects of chemicals

Totally, more than 445 substances with 260 quantitative and 584 qualitative hTTR binding affinity data were available up to now. Those data and corresponding chemicals were presented in our recent study [9]. The quantitative hTTR binding affinity data was expressed as the logarithm of the relative competitive potency (log*RP*), which is defined as follows:

$$\log RP = \log \frac{IC_{50,\mathrm{T}_4}}{IC_{50,\mathrm{disruptor}}} \tag{29.1}$$

where $IC_{50,T4}$ and $IC_{50,\mathrm{disruptor}}$ are the half-maximal inhibitory concentrations of T_4 and TTR disruptor, respectively. Those 445 compounds could be divided into five groups by following the criteria for classification proposed by Hamers et al. [31]; the substance with response <20% of control at 10 μM is referred as inactive hTTR binder, with $\log RP < -2.26$, and/or response 20%–40% of control at 10 μM is a low-potency hTTR binder, with $-2.26 \leq \log RP < -1.26$, and/or response 40%–70% of control at 10 μM is a moderate-potency hTTR binder, with $-1.26 \leq \log RP \leq 0$, and/or response 70%–100% of control at 10 μM is a high-potency hTTR binder, and that with $\log RP > 0$ is a very-high-potency hTTR binder. There were 216, 63, 45, 64, and 57 compounds in inactive, low-, moderate-, high-, very-high-potency hTTR binder groups, respectively.

To identify the most common chemical structure characteristics, we firstly calculated the functional group counts and atom-centered fragments of the compounds from aforementioned five groups by DRAGON 6.0 software, respectively [56]. Then, the occurrence rate for each functional group counts and atom-centered fragments in each group were analyzed. The functional group count or atom-centered fragment in each group with occurrence rate >50% were extracted, and 15 functional group counts (7) and atom-centered fragments (8) were selected finally (Fig. 29.4). As shown in Fig. 29.4, there were eight functional group counts, and atom-centered fragments with occurrence rate >50% were contained in all groups, which indicated that those eight descriptors could not be used to distinguish the active and inactive hTTR binders. In addition, we found that there were six functional group counts, and atom-centered fragments with occurrence rate >50% were only in active hTTR disruptors groups, which implied that those descriptors would help in revealing which structure characteristics were critical in the molecular recognition process between TDCs and hTTR.

For the very-high-potency hTTR binder group, the occurrence rate of *nHDon* (number of donor atoms for H-bonds (N and O)), *nArX* (number of X on aromatic ring), *O-057* (phenol/enol/carboxyl OH), *nArOH* (number of aromatic hydroxyls), and *Cl-089* (Cl attached to C1(sp2)) were 98.2%, 100%, 98.2%, 98.2%, and 57.9%, respectively. The reason why occurrence rate of *nHDon*, *O-057*, and *nArOH* were 98.2% was that the pentachlorothiophenol ($\log RP = .133 \pm 0.0393$) contained aromatic sulfhydryl group (*nArSH*) in its molecular structure, not the aromatic hydroxyls (*nArOH*). In this regard, the occurrence rate of all the three descriptors should be 100% if aromatic sulfhydryl group was counted. Considering that the occurrence rate of number of halogen atom (X) on aromatic ring (*nArX*) was 100%, those data proved that all the very-high-potency hTTR binders contained three critical chemical structure characteristics, i.e., aromaticity, presence of hydroxyl or sulfydryl group, and halogenation.

For the high-potency hTTR binder group, the occurrence rate of *nHDon*, *nArX*, *O-057*, *nArOH*, and *Cl-089* were 95.3%, 89.1%, 79.7%, 65.6%, and 54.7%, respectively. This data indicated that the aromaticity, presence of hydroxyl or sulfhydryl group, and halogenation were still the three critical chemical structure characteristics in this group. Thus, it is may be reasonable to assume that the substance with aromatic ring, halogen atom, and hydroxyl or sulfhydryl on aromatic ring in its molecular structure may be high- or very-high-potency hTTR disruptors. In addition, it worth mentioned that a few compounds without aromatic ring substituent, but with halogen and hydrogen bond donor substituents, may also exhibit high hTTR binding affinity. Why do the compounds with aromatic ring, halogen atom, and hydroxyl or sulfhydryl on aromatic ring in its molecular structure would exhibit distinct binding

		Very high	High	Moderate	Low	Inactive
FGC	*nCar*	100	95.3	75.6	73.0	79.2
FGC	*nCb-*	100	95.3	73.3	69.8	78.2
ACF	*C-026*	100	95.3	68.9	73.0	70.4
ACF	*H-050*	100	98.4	80.0	69.8	50.0
FGC	*nHAcc*	98.2	95.3	82.2	87.3	76.4
FGC	*nCbH*	89.5	92.2	73.3	68.3	75.0
ACF	*C-024*	89.5	92.2	75.6	68.3	75.5
ACF	*H-047*	89.5	93.8	84.4	79.4	88.4
ACF	*C-025*	59.6	54.7	-	-	50.5
FGC	*nHDon*	98.2	95.3	71.1	68.3	-
FGC	*nArX*	100	89.1	-	57.1	-
ACF	*O-057*	98.2	79.7	53.3	54.0	-
FGC	*nArOH*	98.2	65.6	-	-	-
ACF	*Cl-089*	57.9	54.7	-	-	-
ACF	*O-058*	-	-	51.1	60.3	-

FIGURE 29.4 Most important 15 functional group counts and atom-centered fragments used to characterize the hTTR disruptors. The number was occurrence rate (%). Functional group counts (FGC): *nCar* (number of aromatic C(sp2)), *nCb-* (number of substituted benzene C(sp2)), *nHAcc* (number of acceptor atoms for H-bonds (N,O,F)), *nCbH* (number of unsubstituted benzene C(sp2)), *nHDon* (number of donor atoms for H-bonds (N and O)), *nArX* (number of X on aromatic ring), *nArOH* (number of aromatic hydroxyls); atom-centered fragments (ACF): *C-026* (R–CX–R), *H-050* (H attached to heteroatom), *C-024* (R–CH–R), *H-047* (H attached to C1(sp3)/C0(sp2)), *C-025* (R–CR–R), *O-057* (phenol/enol/carboxyl OH), *Cl-089* (Cl attached to C1(sp2)), *O-058* (#NOME?).

affinity with hTTR? The different position of hydroxyl or sulfhydryl on aromatic ring, type of halogen atom, and number of halogen atom may be the reasons.

The position of hydroxyl or sulfhydryl on aromatic ring could influence the hTTR binding affinity of substances. For example, the log*RP* values were 0.602, 0.477, and −0.490 for 3-HO-2,2′,4,4′-tetrabrominated diphenyl ethers (3-HO-BDE 47), 5-HO-2,2′,4,4′-tetrabrominated diphenyl ethers (5-HO-BDE 47), and 6-HO-2,2′,4,4′-tetrabrominated diphenyl ethers (6-HO-BDE 47), respectively (Fig. 29.5).

Usually, the hTTR binding affinity of compounds with iodine atom substitution > that of bromine atom substitution > that of chlorine atom substitution > that of fluorine atom substitution. For example, the log*RP* values were inactive, −0.314, 0.267, and 0.421 for 2,4,6-trifluorophenol (2,4,6-TFP), 2,4,6-trichlorophenol(2,4,6-TCP), 2,4,6-tribromophenol(2,4,6-TBP), and 2,4,6-triiodophenol (2,4,6-TIP), respectively (Fig. 29.6).

The hTTR binding potency would increase with the increase of halogen atom number in the molecular structure. For instance, the log*RP* values were −1.78, −1.45, −0.882, and 0.133 for 2-chlorothiophenol (2-CTP), 2,4-dichlorothiophenol (2,4-DCTP), 2,4,6-trichlorothiophenol (2,4,6-TCTP), and pentachlorothiophenol (PCTP), respectively (Fig. 29.7).

What is the role of aromatic ring, halogen atom, and hydroxyl or sulfhydryl on aromatic ring in the interaction between substances and hTTR? Our previous molecular modeling results had proved that those structural alerts could form dominant and orientational noncovalent interactions, which had critical role in the molecular recognition process of compounds with hTTR. For example, the cation−π interaction would form between the aromatic ring of ligands and the $-NH_3^+$ group of Lys15 in hTTR [57]. The halogen bonds and halogen−hydrogen bonds would form between halogen atom in compounds and the residues in hTTR directly [43,45,58]. Furthermore, we also found that the halogen atom in substances would influence the molecular recognition process through hydrophobic effects and inductive effects [58]. For the hydroxyl or sulfhydryl on aromatic ring, they may ionize in the normal physiological and experimental pH conditions, and both dissociated and undissociated forms may coexist. The fractions of the neutral (δ_M) and ionized species (δ_I) at a given pH condition for monoacid and monobase are defined as follows [59]:

$$\delta_M = \frac{1}{1 + 10^{(pH - pK_a)\cdot I_{ab}}} \tag{29.2}$$

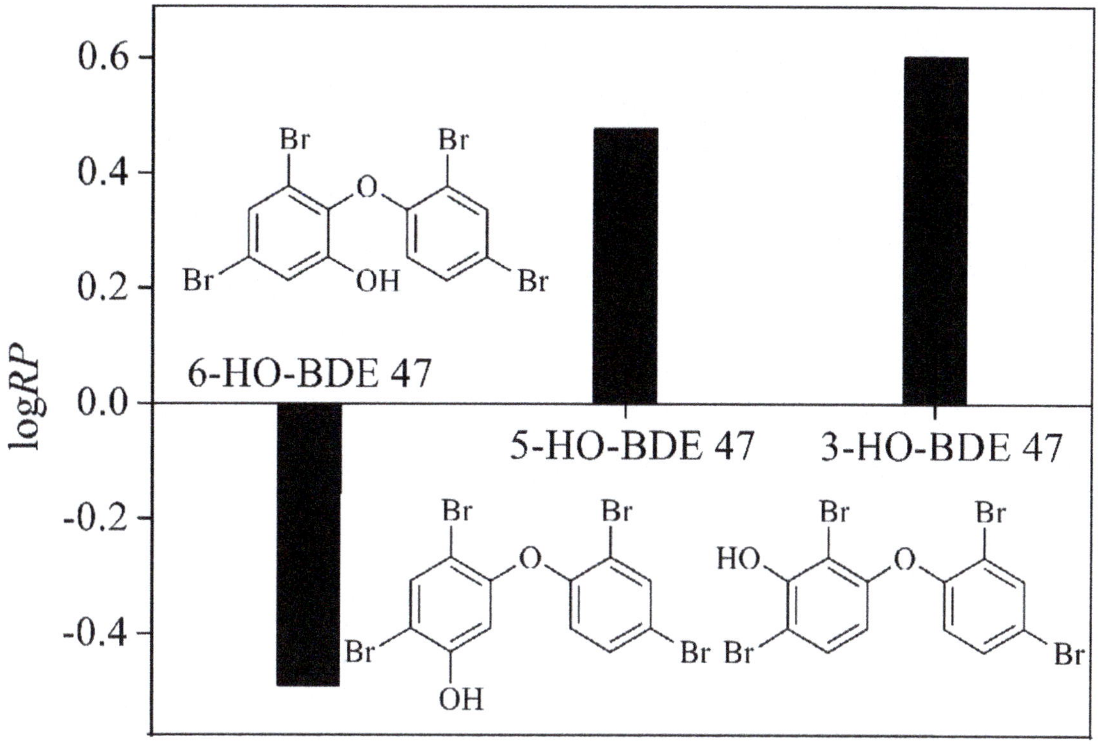

FIGURE 29.5 Influence of hydroxyl position on the binding potency of compounds with hTTR. Molecular structures of 6-HO-2,2′,4,4′-tetrabrominated diphenyl ethers (6-HO-BDE 47), 5-HO-2,2′,4,4′-tetrabrominated diphenyl ethers (5-HO-BDE 47), 3-HO-2,2′,4,4′-tetrabrominated diphenyl ethers (3-HO-BDE 47).

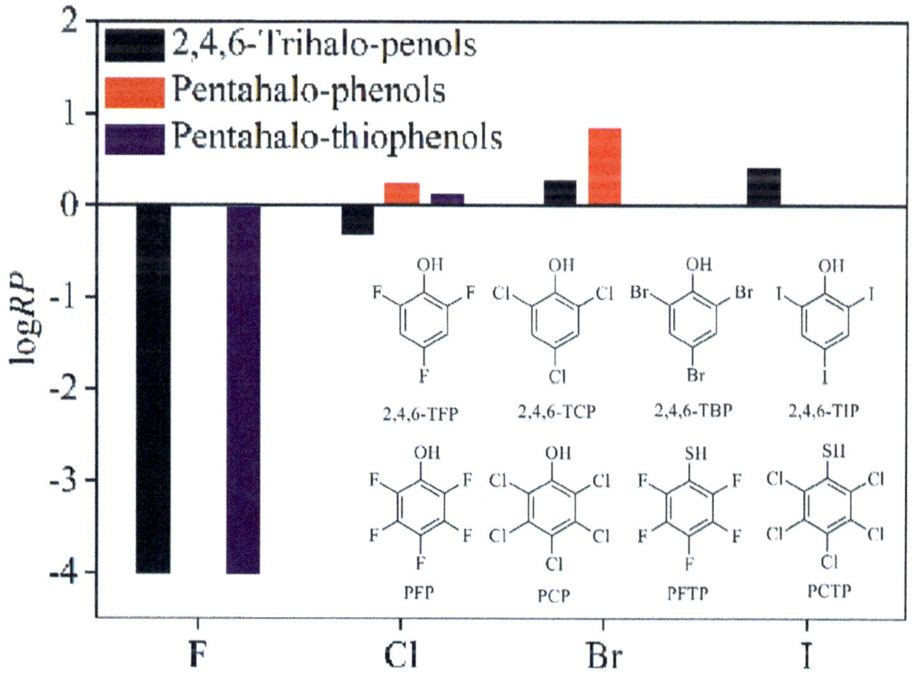

FIGURE 29.6 Influence of different halogen atom on the binding potency of compounds with hTTR. Molecular structure of 2,4,6-trifluorophenol (2,4,6-TFP), 2,4,6-trichlorophenol (2,4,6-TCP), 2,4,6-tribromophenol (2,4,6-TBP), 2,4,6-triiodophenol (2,4,6-TIP), pentachlorophenol (PCP), pentabromophenol (PBP), pentafluorothiophenol (PFTP), and pentachlorothiophenol (PCTP).

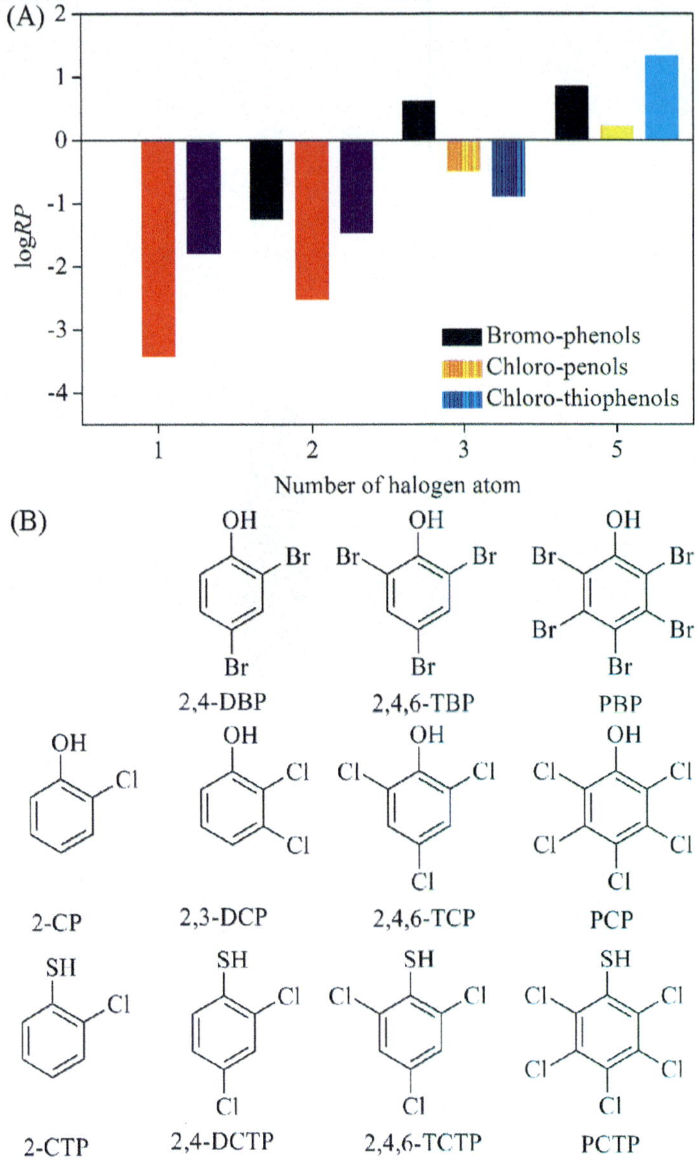

FIGURE 29.7 Influence of halogen atom number on the binding potency of compounds with hTTR. Molecular structure of 2,4-dibromophenol (2,4-DBP), 2,4,6-tribromophenol (2,4,6-TBP), pentabromophenol (PBP), 2-chlorophenol (2-CP), 2,3-dichlorophenol (2,4-DCP), 2,4,6-trichlorophenol (2,4,6-TCP), pentachlorophenol (PCP), 2-chlorothiophenol (2-CTP), 2,4-dichlorothiophenol (2,4-DCTP), 2,4,6-trichlorothiophenol (2,4,6-TCTP), and pentachlorothiophenol (PCTP).

$$\delta_I = \frac{10^{(pH - pK_a) \cdot I_{ab}}}{1 + 10^{(pH - pK_a) \cdot I_{ab}}} \tag{29.3}$$

where pK_a is acid dissociation constant. The I_{ab} value for monoacid and monobase is 1 and −1, respectively. Our results suggested that the anionic form of halogenated phenols and halogenated thiophenols binds stronger to hTTR than that of corresponding neutral form. The anionic form of hydroxyl or sulfhydryl group on aromatic ring would form ionic interaction and hydrogen bond interaction with hTTR [43,45,57,58].

The degree of dissociation for the compounds with −COOH, $-SO_3H$ as end groups usually were higher than that of substances with −OH in experimental and normal physiological pH conditions because the pK_a values of compounds with −COOH, $-SO_3H$ groups were smaller than that with −OH [60]. For example, under pH = 7.40, the fractions of the anionic form for 2,4-dibromobenzoic acid ($pK_a = 3.22$) and 2,4-dibromophenol ($pK_a = 7.71$) were 100% and 33%, respectively (Fig. 29.8). In the view of molecular structures, the halo-benzoic acids and halo-benzenesulfonic acids should be potent hTTR disruptors. However, our recent study documented that halo-benzoic acids and halo-benzenesulfonic acids/sulfates only exhibited moderate and low binding affinity to hTTR; even some of them was inactive hTTR binder [44]. For instance, the log*RP* values of 2,4,6-trichlorobenzoic acid and 2,4,5-trichlorobenzenesulfonic acid potassium salt were −2.95 and −2.26, respectively.

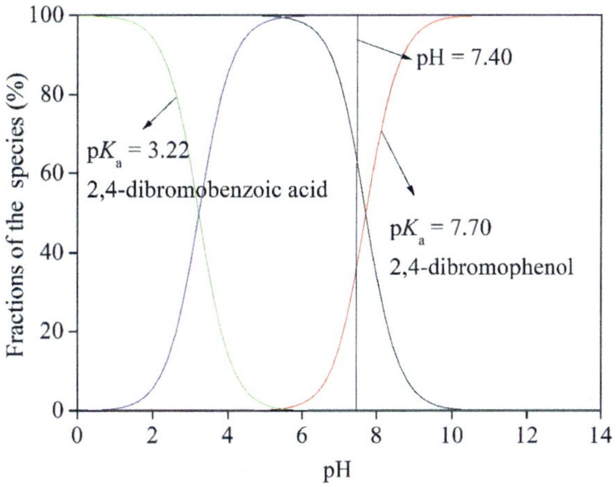

FIGURE 29.8 Species distribution of 2,4-dibromobenzoic acid (pK_a = 3.22) and 2,4-dibromophenol (pK_a = 7.71). The pK_a values were obtained from the MarvinSketch 15.6.29.0, 2015, ChemAxon (http://www.chemaxon.com).

The reason why the hTTR binding affinity of halo-benzoic acids and halo-benzenesulfonic acids/sulfates was lower than that of halo-phenols with similar structures was analyzed. The experimental observed hTTR binding affinity of a given substance may be controlled by two processes [61]. The first step was the distribution of compound from water to hTTR. Then, the compound interacted with the amino acid residues of hTTR. As the halo-benzoic acids and halo-benzenesulfonic acids/sulfates ionized more completely than that of halo-phenols in experimental pH conditions, it was reasonable to conclude that halo-benzoic acids and halo-benzenesulfonic acids/sulfates were more water-soluble and less proteinophilic than that of halo-phenols. The distribution ability of halo-benzoic acids and halo-benzenesulfonic acids/sulfates from water to hTTR may be the limiting step. In this case, the hTTR binding affinity of halo-benzoic acids and halo-benzenesulfonic acids/sulfates would expect to increase with the increase of their molecular hydrophobicity. We indeed observed that the hTTR binding potency of benzoic acid could be increased after introducing hydrophobic functional groups into its molecular structure. For example, the logarithm of the n-octanol/water distribution coefficient (logD_{OW}) values of 1-bromo-2-naphthoic acid (logD_{OW} = −0.06) and 4-hexylbenzoic acid (logD_{OW} = 1.36) was higher than that of benzoic acid (log logD_{OW} = −1.48), and the higher logRP values of 1-bromo-2-naphthoic acid (logRP = −2.47) and 4-hexylbenzoic acid (logRP = −1.37) than that of benzoic acid (inactive) were observed (Fig. 29.9) [44].

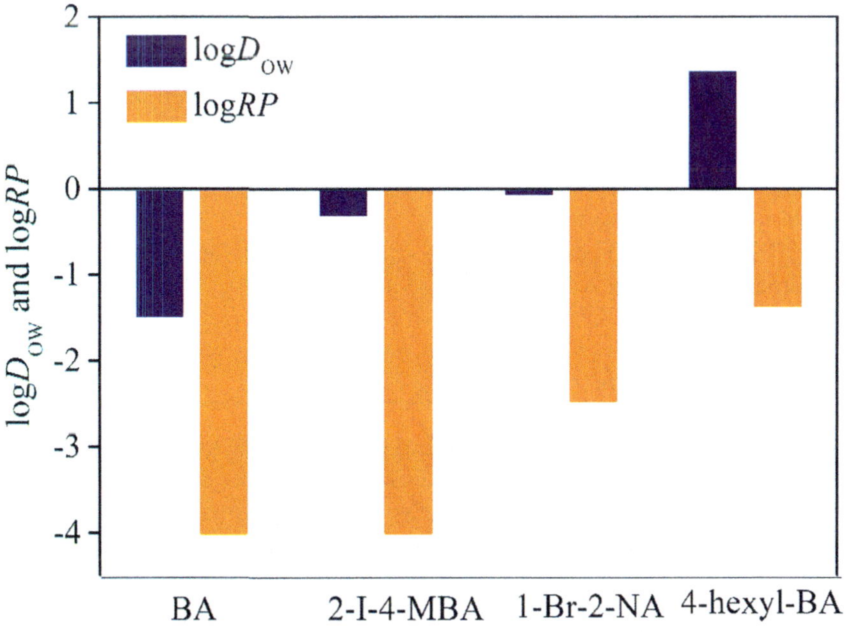

FIGURE 29.9 Relationship of logD_{OW} and logRP for benzoic acid (BA), 2-iodo-4-methylbenzoic acid (2-I-4-MBA), 1-bromo-2-naphthoic acid (1-Br-2-NA), 4-hexylbenzoic acid (4-hexyl-BA). The logD_{OW} and logRP values were collected from Xi et al. [44]. The logRP value of inactive substance was set as −4 [58].

Nonhuman species transthyretin disrupting effects of chemicals

Besides hTTR binding affinity data, the data from other species including gulls (*Larus hyperboreus* and *Larus argentatus*), chickens (*Gallus gallus*), Japanese quails (*Coturnix japonica*), bullfrog tadpoles (*Rana catesbeiana*), little skate (*Leucoraja erinacea*), and seabream (*Sparus aurata*) were also collected and analyzed (Table 29.1) [22,29,30,33–35,51]. As shown in Table 29.1, the numbers of available data for gulls, chickens, Japanese quails, bullfrog tadpoles, little skate, and seabream were 7, 9, 7, 9, 36 and 35, respectively. The most common chemical structure characteristics of the TTR disruptors for each specie were also attempted to identify by the following aforementioned method. However, no most common chemical structure characteristics were selected except for the TTR disruptors of little skate.

For little skate, there were 14 functional group counts, and atom-centered fragments with occurrence rate >50% were contained in both active and inactive TTR binder groups, respectively. Three functional group counts named *nArOH*, *nArX*, and *nHDon* with occurrence rate = 100% were only selected in active little skate TTR disruptors group, which implied that the aromaticity, presence of hydroxyl, and halogenation were still the three critical chemical structure characteristics for the potential little skate TTR disruptors.

For gulls, chickens, Japanese quails, and bullfrog tadpoles, no inactive TTR binders as reference may be the reason why no most common chemical structure characteristics were selected. According to their TTR binding affinity data, however, we also could found that the aromatic ring, halogen atom, and hydroxyl on aromatic ring were also the critical chemical structure characteristics for their potential TTR disruptors. For example, the gulls TTR binding affinity of 6-HO-2,2′,4,4′-tetrabrominated diphenyl ethers and 4-HO-2,2′,3,4′,5,5′,6-heptachlorobiphenyl were higher than that of their parent compound, i.e., 2,2′,4,4′-tetrabrominated diphenyl ethers and 2,2′,3,4′,5,5′,6-heptachlorobiphenyl, respectively. The chickens and bullfrog tadpoles TTR binding affinity of bisphenol A and 4-nonylphenol chlorinated derivatives were higher than that of bisphenol A and 4-nonylphenol, respectively. We also observed that their TTR binding potency would increase with the increasing number of chlorine atoms. For instance, the chickens TTR log*RP* values of 4-nonylphenol, 2-chloro-4-nonylphenol, and 2,6-dichloro-4-nonylphenol were −1.77, −0.865, and 0.479, respectively.

Interspecies variation of transthyretin disrupting effects for different species

According to the log*RP* values from different species for same compounds, there was not an evident trend in the TTR binding affinity. For example, the log*RP* values of 3,3′,5,5′-tetrabromobisphenol A with hTTR, little skate TTR, and seabream TTR were 0.351, 0.424, and 0.998, respectively, indicating that 3,3′,5,5′-tetrabromobisphenol A was very-high-potency TTR disruptors for the three species. Pentachlorophenol (log*RP* = .24) and 2,2′,4,4′-tetrabromodiphenyl ether (log*RP* = −2.77) were very-high- and low-potency hTTR binder, respectively. However, pentachlorophenol is a moderate-potency seabream TTR binder (log*RP* = −2.04). 2,2′,4,4′-tetrabromodiphenyl ether was a high-potency gulls TTR binder and a very-high-potency seabream TTR binder, respectively. The correlation analysis was performed by employing the log*RP* values from human, gulls, chickens, bullfrog tadpoles, Japanese quails, little skate, and seabream. As shown in Fig. 29.10, the significant correlations were only observed between chickens and bullfrog tadpoles ($n = 9$, $r = 0.821$, $P < .01$), Japanese quails, and little skate ($n = 5$, $r = 0.954$, $P < .02$), respectively.

Those results highlighted that the interspecies variation of transthyretin disrupting effects for different species was unneglectable in the screening of potential TTR disruptors [62]. Considering that the most of available TTR binding affinity data was related to human, it may be overrated or underrated the potential TTR binding potency of nonhuman species by extrapolating from hTTR data. More testing efforts should be taken to determine the potential TTR disrupting effects of TDCs on nonhuman species.

(Q)SARs models of transthyretin disrupting effects

To date, more than 26 ((Q)SAR) models have been developed to predict the potential TTR binding potency of TDCs, and all of those models were constructed based on hTTR binding affinity data (Table 29.2) [9,37,41,44,45,57,58,63–69]. As shown in Table 29.2, most of models were derived based on dozens of substances, and the applicability domain (AD) of those models only covered one or several groups of compounds that included poly-/perfluorinated chemicals (PFCs) [58,66–68], brominated flame retardants (BFRs) [63,65,67], such as polybrominated diphenyl ethers (PBDEs) and their hydroxylated derivatives (HO-PBDEs) [41,64], tetrabromobisphenol A (TBBPA) and hexabromocyclododecane (HBCD), hydroxylated chlorodibenzo-p-dioxins (HO-PCDDs) and

TABLE 29.1 The available binding affinities data of compounds with TTR for different species.

Species	CAS NO.	Chemical name	IC_{50} nM	Assay method	log*RP*	Reference
Gulls	000051-48-9	3,3′,5,5′-tetraiodo-L-thyronine	15.4[a]	RLDM[b]	–	[34,35]
	052663-68-0	2,2′,3,4′,5,5′,6-heptachlorobiphenyl	92.6	RLDM	−0.779	
	169295-23-2	4-MeO-2,2′,3,4′,5,5′,6-heptachlorobiphenyl	972	RLDM	−1.80	
	158076-68-7	4-HO-2,2′,3,4′,5,5′,6-heptachlorobiphenyl	19.3	RLDM	−0.0980	
	005436-43-1	2,2′,4,4′-tetrabrominated diphenyl ethers	91.1	RLDM	−0.772	
	079755-43-4	6-HO-2,2′,4,4′-tetrabrominated diphenyl ethers	10.8	RLDM	0.154	
	602326-23-8	4′-HO-2,2′,4,5′-tetrabrominated diphenyl ethers	8.62	RLDM	0.252	
	102739-99-1	6-MeO-2,2′,4,4′-tetrabrominated diphenyl ethers	56.4	RLDM	−0.564	
Chickens (*Gallus gallus*)	006893-02-3	3,3′,5-triiodo-L-thyronine	22.5	RLDM	–	[29]
	014151-65-6	3,5-Dichlorobisphenol A	19	RLDM	0.0734	
	000079-95-8	3,3′,5,5′-tetrachlorobisphenol A	22.7	RLDM	−0.00384	
	040346-55-2	3,3′,5-trichlorobisphenol A	33.7	RLDM	−0.175	
	000079-98-1	3,3′-dichlorobisphenol A	86.3	RLDM	−0.584	
	074192-35-1	3-Chlorobisphenol A	367	RLDM	−1.21	
	000080-05-7	Bisphenol A	9070	RLDM	−2.61	
	102127-41-3	2,6-Dichloro-4-nonylphenol	7.47	RLDM	0.479	
	060044-33-9	2-Chloro-4-nonylphenol	165	RLDM	−0.865	
	000104-40-5	4-Nonylphenol	1330	RLDM	−1.77	
Bullfrog tadpoles (*Rana catesbeiana*)	006893-02-3	3,3′,5-triiodo-L-thyronine	1.43	RLDM	–	[29]
	014151-65-6	3,5-Dichlorobisphenol A	44.7	RLDM	−1.49	
	000079-95-8	3,3′,5,5′-tetrachlorobisphenol A	18.7	RLDM	−1.12	
	040346-55-2	3,3′,5-trichlorobisphenol A	10.8	RLDM	−0.878	
	000079-98-1	3,3′-dichlorobisphenol A	18.3	RLDM	−1.11	
	074192-35-1	3-Chlorobisphenol A	173	RLDM	−2.08	
	000080-05-7	Bisphenol A	1530	RLDM	−3.03	
	102127-41-3	2,6-Dichloro-4-nonylphenol	48	RLDM	−1.53	
	060044-33-9	2-Chloro-4-nonylphenol	597	RLDM	−2.62	
	000104-40-5	4-Nonylphenol	2730	RLDM	−3.28	
Japanese quails (*Coturnix japonica*)	006893-02-3	3,3′,5-Triiodo-L-thyronine	9.70	RLDM	–	[30]
	006898-97-1	Diethylstilbestrol	<0.4	RLDM	1.38	
	001689-83-4	Ioxynil	1.1	RLDM	0.945	
	000087-86-5	Pentachlorophenol	6.3	RLDM	0.187	
	000104-40-5	4-Nonylphenol	1000	RLDM	−2.01	

Continued

TABLE 29.1 The available binding affinities data of compounds with TTR for different species.—cont'd

Species	CAS NO.	Chemical name	IC_{50} nM	Assay method	log*RP*	Reference
	000080-05-7	Bisphenol A	1500	RLDM	−2.19	
	000121-75-5	Malathion	1400	RLDM	−2.16	
	000085-68-7	Butylbenzyl phthalate	2400	RLDM	−2.39	
Little skate (*Leucoraja erinacea*)	000051-48-9	3,3′,5,5′-Tetraiodo-L-thyronine	26	RLDM	–	[22]
	006893-02-3	3,3′,5-Triiodo-L-thyronine	2	RLDM	–	
	066701-15-3	3,3′,5-Triiodo-D-thyronine	4.4	RLDM	0.772	
	005817-39-0	3,3′,5′-Triiodo-L-thyronine	81	RLDM	−0.494	
	000051-24-1	3,3′,5-Triiodothyroacetic acid	1	RLDM	1.41	
	000067-30-1	3,3′,5,5′-Tetraiodothyroacetic acid	9.7	RLDM	0.428	
	000300-39-0	3,5-Diiodo-L-tyrosine	>1000	RLDM	I	
	000070-78-0	3-Iodo-L-tyrosine	>1000	RLDM	I	
	000087-86-5	Pentachlorophenol	5.1	RLDM	0.707	
	000609-23-4	2,4,6-Triiodophenol	2.2	RLDM	1.07	
	000118-79-6	2,4,6-Tribromophenol	49	RLDM	−0.275	
	000088-06-2	2,4,6-Trichlorophenol	44	RLDM	−0.228	
	000608-71-9	Pentabromophenol	1.4	RLDM	1.27	
	000079-94-7	3,3′,5,5′-Tetrabromobisphenol A	9.8	RLDM	0.424	
	000079-95-8	3,3′,5,5′-Tetrachlorobisphenol A	27	RLDM	−0.0164	
	001041-01-6	3,5-Diiodo-L-thyronine	250	RLDM	−0.983	
	001689-83-4	Ioxynil	33	RLDM	−0.104	
	000085-68-7	Butylbenzyl phthalate	–	RLDM	I	
	000117-81-7	Bis(2-ethylhexyl) phthalate	–	RLDM	I	
	000084-61-7	Dicyclohexyl phthalate	–	RLDM	I	
	000084-74-2	Di-n-butyl phthalate	–	RLDM	I	
	000084-66-2	Diethyl phthalate	–	RLDM	I	
	002385-85-5	Mirex	–	RLDM	I	
	000121-75-5	Malathion	–	RLDM	I	
	000115-32-2	Dicofol	–	RLDM	I	
	000094-75-7	2,4-Dichlorophenoxyacetic acid	–	RLDM	I	
	040596-69-8	Methoprene	–	RLDM	I	
	034256-82-1	Acetochlor	–	RLDM	I	
	000119-61-9	Benzophenone	–	RLDM	I	
	000103-23-1	Bis(2-ethylhexyl) adipate	–	RLDM	I	
	001461-22-9	Tributyltin (IV) chloride	–	RLDM	I	
	000115-29-7	Endosulfan	–	RLDM	I	
	000050-32-8	Benzo(a)pyrene	–	RLDM	I	
	000104-40-5	4-Nonylphenol	–	RLDM	I	

TABLE 29.1 The available binding affinities data of compounds with TTR for different species.—cont'd

Species	CAS NO.	Chemical name	IC_{50} nM	Assay method	log*RP*	Reference
	000080-05-7	Bisphenol A	–	RLDM	I	
	000098-54-4	4-Tert-butylphenol	–	RLDM	I	
	000088-69-7	2-Isopropylphenol	–	RLDM	I	
	000051-28-5	2,4-Dinitrophenol	–	RLDM	I	
Seabream (*Sparus aurata*)	000051-48-9	3,3′,5,5′-Tetraiodo-L-thyronine	230	ITC	–	[51]
	000131-55-5	2,2′,4, 4′-Tetrahydroxybenzophenone	54,610	ITC	−2.38	
	017737-65-4	2-(3-Chloro-2-methylanilino) pyridine-3-carboxylic acid	3610	ITC	−1.20	
	000335-67-1	Pentadecafluorooctanoic acid	2160	ITC	−0.973	
	000079-94-7	3,3′,5,5′-Tetrabromobisphenol A	180	ITC	0.106	
	006515-38-4	3,5,6-Trichloro-2-pyridinol	28,410	ITC	−2.09	
	000093-76-5	2,4,5-Trichlorophenoxyacetic acid	10,200	ITC	−1.65	
	000051-48-9	3,3′,5,5′-Tetraiodo-L-thyronine	20.9	RLDM	–	[33]
	006893-02-3	3,3′,5-Triiodo-L-thyronine	23.8	RLDM	–	
	001689-83-4	Ioxynil	38.9	RLDM	−0.270	
	000087-86-5	Pentachlorophenol	2480	RLDM	−2.07	
	006898-97-1	Diethylstilbestrol	4849	RLDM	−2.37	
	000104-40-5	4-Nonylphenol	17,035	RLDM	−2.91	
	243982-82-3	2,2′,4,5′-Tetrabrominated diphenyl ethers	0.5	RLDM	1.62	
	000079-94-7	3,3′,5,5′-Tetrabromobisphenol A	2.1	RLDM	0.998	
	005436-43-1	2,2′,4,4′-Tetrabrominated diphenyl ethers	5.25	RLDM	0.600	
	060348-60-9	2,2′,4,4′,5-Pentabrominated diphenyl ethers	6.7	RLDM	0.494	
	407606-57-9	3,4′,5 Tribrominated diphenyl ethers	9.5	RLDM	0.342	
	041318-75-6	2,4,4′-Tribrominated diphenyl ethers	14.9	RLDM	0.147	
	000118-79-6	2,4,6-Tribromophenol	30	RLDM	−0.157	
	147217-73-0	2,2′,6-Tribrominated diphenyl ethers	43.9	RLDM	−0.322	
	189084-64-8	2,2′,4,4′,6-Pentabrominated diphenyl ethers	55.5	RLDM	−0.424	
	337513-54-9	3,4,5-Tribrominated diphenyl ethers	65.1	RLDM	−0.493	
	079755-43-4	6-HO-2,2′,4,4′-Tetrabrominated diphenyl ethers	700.6	RLDM	−1.53	
	000080-05-7	Bisphenol A	–	RLDM	I	
	068631-49-2	2,2′,4,4′,5,5′-Hexabrominated diphenyl ethers	–	RLDM	I	

Continued

TABLE 29.1 The available binding affinities data of compounds with TTR for different species.—cont'd

Species	CAS NO.	Chemical name	IC_{50} nM	Assay method	log*RP*	Reference
	446254-86-0	3,3′,4,5,5′-Pentabrominated diphenyl ethers	–	RLDM	I	
	035854-94-5	2,2′,4,4′,6,6′-Hexabrominated diphenyl ethers	–	RLDM	I	
	446255-18-1	3,3′,4,4′,5,5′-Hexabrominated diphenyl ethers	–	RLDM	I	
	189084-67-1	2,2′,3,4,4′,5,6-Heptabrominated diphenyl ethers	–	RLDM	I	
	207122-16-5	2,2′,3,4,4′,5′,6-Heptabrominated diphenyl ethers	–	RLDM	I	
	405237-86-7	2,2′,3,4,5,5′,6-Heptabrominated diphenyl ethers	–	RLDM	I	
	189084-68-2	2,3,3′,4,4′,5,6-Heptabrominated diphenyl ethers	–	RLDM	I	
	063387-28-0	2,2′,3,3′,4,4′,5,5′,6-Nonabrominated diphenyl ethers	–	RLDM	I	
	001163-19-5	2,2′,3,3′,4,4′,5,5′,6,6′-Decabrominated diphenyl ethers	–	RLDM	I	
	021850-44-2	Benzene, 1,1′-(1-methylethylidene) bis[3,5-dibromo-4-(2,3-dibromopropoxy)-	–	RLDM	I	
	138257-19-9	Hexabromocyclododecane alpha	–	RLDM	I	
	138257-18-8	Hexabromocyclododecane beta	–	RLDM	I	
	169102-57-2	Hexabromocyclododecane gamma	–	RLDM	I	

[a]*The concentration of inhibitor resulting in 50% inhibition (IC_{50}) value of a given compound was defined as follows:*

$$IC_{50} = \frac{K_i}{1 + \frac{[radioligand]}{K_d}} \quad (29.4)$$

where K_i is affinity constant; K_d is dissociation constant of inhibitor–binder reaction; [radioligand] is the radiolabeled ligand concentration.

[b]ANSA, *ANSA-competitive fluorescence displacement assay;* FITC-T4, *FITC-T4-fluorescence polarization–based competitive binding assays;* ITC, *isothermal titration calorimetry;* RLDM, *radiolabeled ligand displacement method;* SPRB, *surface plasmon resonance biosensor–based method.*

hydroxylated chlorodibenzofurans (HO-PCDFs), halo-phenols, polychlorinated biphenyls (PCBs) and their hydroxylated successors (e.g., HO-PCBs) [57], and so on. Recently, Zhang et al. [37] and Yang et al. [9,69] developed several global models by employing more than 200 and 400 compounds with structural diversity, respectively.

Zhang et al. [37] retrieved the hTTR binding affinity data from 222 unique substances (88 hTTR binders and 134 hTTR nonbinders) in 2015. Three classification models were developed by library for support vector machines (LIBSVM), partial least squares–discriminant analysis (PLS-DA), and *k*-nearest neighbor (kNN) algorithms. The kNN model exhibited the best predictive performance with correct classification rate of 88% and 82% for training and validation set compounds, respectively.

Considering that the researchers had tested potential hTTR disrupting effects of some new groups of compounds, e.g., tetrabromobisphenol A monoether structural analogs, halosalicylic acids, 2,6-dihalo-4-nitrophenols, halobenzoic acids, 3,5-dihalo-4-hydroxybenzaldehydes, halo-benzenesulfonic and halo-phenylboronic acids, bisphenols, halogenated thiophenols since 2015, Yang et al. [9] carried out an in-depth literature survey and collected the qualitative and quantitative hTTR binding affinity data from 445 unique compounds (229 hTTR binders and

	Human	Gulls	Chickens	Bullfrog tadpoles	Japanese quails	Little skate	Seabream
Human		$n=4$ $r=0.867$ $p>0.05$	$n=1$	$n=1$	$n=1$	$n=6$ $r=0.601$ $p>0.05$	$n_1=6$ $r_1=0.470$ $p_1>0.05$ $n_2=11$ $r_2=-0.232$ $p_2>0.05$
Gulls	$n=4$ $r=0.867$ $p>0.05$		$n=0$	$n=0$	$n=0$	$n=0$	$n=2$
Chickens	$n=1$	$n=0$		$n=9$ $r=0.821$ $p<0.01$	-	$n=3$ $r=0.949$ $p>0.05$	-
Bullfrog tadpoles	$n=1$	$n=0$	$n=9$ $r=0.821$ $p<0.01$		$n=2$	$n=3$ $r=0.994$ $p>0.05$	$n=2$
Japanese quails	$n=1$	$n=0$	$n=2$	$n=2$		$n=5$ $r=0.954$ $p<0.02$	$n=5$ $r=0.746$ $p>0.05$
Little skate	$n=6$ $r=0.601$ $p>0.05$	$n=0$	$n=3$ $r=0.949$ $p>0.05$	$n=3$ $r=0.994$ $p>0.05$	$n=5$ $r=0.954$ $p<0.02$		$n=6$ $r=0.793$ $p>0.05$
Seabream	$n_1=6$ $r_1=0.470$ $p_1>0.05$ $n_2=11$ $r_2=-0.232$ $p_2>0.05$	$n=2$	$n=2$	$n=2$	$n=5$ $r=0.746$ $p>0.05$	$n=6$ $r=0.793$ $p>0.05$	

FIGURE 29.10 Correlation analysis results of log*RP* from human, gulls, chickens, bullfrog tadpoles, Japanese quails, little skate, and seabream. n is the number substance with available TTR binding affinity data from two species; r, r_1, r_2 are Pearson correlation coefficient; p, p_1, p_2 are significance level.

216 hTTR nonbinders). Then a two-tiered approach that contained both classification and quantitative models was developed. In this tiered approach, the classification model was employed to distinguish whether a given compound was a potential hTTR disruptor or not. For the classified potential hTTR binder, its quantitative hTTR binding affinity values would be predicted by quantitative models. All the classification models had good classification performance, while the quantitative models had good internal and external predictive performance (Table 29.2) [9,69].

Software could be used to screen potential transthyretin disruptors

Recently, we summarized the available software that could be employed to predict the potential endocrine disrupting effects of compounds and found 12 software named VEGA (https://www.vegahub.eu), ToxProfiler (https://toxpro.bhsai.org), ENDOCRINE DISRUPTOME (http://endocrinedisruptome.ki.si), Danish (Q)SAR Models (https://qsarmodels.food.dtu.dk), QSAR Toolbox (https://qsartoolbox.org), VirtualToxLab (www.virtualtoxlab.org), ProTox-II (https://tox-new.charite.de/protox_II), admetSAR (http://lmmd.ecust.edu.cn/admetsar2/), ChemBioSim (https://zenodo.org/record/4761226#.YQTZ7eSP6Uk), SepPCNET (https://pubs.acs.org/doi/10.1021/acs.est.1c01228), CPTP (http://cptp.dlut.edu.cn/), and ED Profiler (http://jszy.njust.edu.cn/hjsw/yxh_en/list.psp) [69]. Among those software, only CPTP and ED Profiler integrated the predictive models related to hTTR disrupting effects.

TABLE 29.2 (Q)SAR models of transthyretin disrupting effects.

Method[a]	*Endpoint*[b]	m[c]	n[d]	Training set[e]	Validation set[f]	AD[g]	Year	References
PLS	log*RP*	2	17	$n_{\text{Train}} = 9$, $R^2_{\text{Train}} = 0.960$, $Q^2_{\text{LOO}} = 0.960$, $Q^2_{\text{BOOT}} = 0.870$, $RMSE_{\text{Train}} = 0.290$	$n_{\text{EXT}} = 8$, $R^2_{\text{EXT}} = 0.930$ $Q^2_{\text{EXT}} = 0.900$, $RMSE_{\text{EXT}} = 0.470$	Yes	2010	[63]
MLR	log*K*	1	14	$n_{\text{Train}} = 14$, $R^2_{\text{Train}} = 0.880$	-	NO	2010	[41]
CoMSIA	log*RP*	4	28	$n_{\text{Train}} = 22$, $Q^2_{\text{LOO}} = 0.752$	$n_{\text{EXT}} = 6$, $Q^2_{\text{EXT}} = 0.928$	NO	2011	[64]
kNN	Classification	2	29	$n_{\text{Train}} = 20$, $Q_{\text{Train}} = 0.950$, $S_{\text{n-Train}} = 1$, $S_{\text{p-Train}} = 0.870$	$n_{\text{EXT}} = 9$, $Q_{\text{EXT}} = 0.889$, $S_{\text{n-EXT}} = 0.800$, $S_{\text{p-EXT}} = 1$	Yes	2011	[65]
kNN	Classification	2 (3)	19	$n_{\text{Train}} = 10$, $Q_{\text{Train}} = 0.900–1$, $S_{\text{n-Train}} = 0.830–1$, $S_{\text{p-Train}} = 0.750–1$	$n_{\text{EXT}} = 9$, $Q_{\text{EXT}} = 0.900–1$, $S_{\text{n-EXT}} = 1$, $S_{\text{p-EXT}} = 0.750–1$	Yes	2012	[66]
kNN	Classification	3 (5)	53	$n_{\text{Train}} = 37$, $Q_{\text{Train}} = 0.840$, $S_{\text{n-Train}} = 0.910$, $S_{\text{p- Train}} = 0.790$	$n_{\text{EXT}} = 16$, $Q_{\text{EXT}} = 0.810$, $S_{\text{n-EXT}} = 0.890$, $S_{\text{p-EXT}} = 0.860$	Yes	2013	[67]
MLR	log*RP*	3	32	$n_{\text{Train}} = 23$, $Q^2_{\text{LOO}} = 0.810$, $Q^2_{\text{BOOT}} = 0.780$	$n_{\text{EXT}} = 9$, $CCC = 0.940$			
PLS	log*RP*	3	47	$n_{\text{Train}} = 38$, $R^2_{\text{Train}} = 0.860$, $Q^2_{\text{CUM}} = 0.840$, $RMSE_{\text{Train}} = 0.510$	$n_{\text{EXT}} = 9$, $R^2_{\text{EXT}} = 0.950$ $Q^2_{\text{EXT}} = 0.930$, $RMSE_{\text{EXT}} = 0.320$	Yes	2013	[57]
kNN	Classification	-(4)	222	$n_{\text{Train}} = 178$, $Q_{\text{Train}} = 0.880$, $S_{\text{n-Train}} = 0.890$, $S_{\text{p- Train}} = 0.870$	$n_{\text{EXT}} = 44$, $Q_{\text{EXT}} = 0.820$, $S_{\text{n-EXT}} = 0.780$, $S_{\text{p-EXT}} = 0.850$	Yes	2015	[37]
LIB-SVM	Classification	-	222	$n_{\text{Train}} = 178$, $Q_{\text{Train}} = 0.820$, $S_{\text{n-Train}} = 0.950$, $S_{\text{p- Train}} = 0.780$	$n_{\text{EXT}} = 44$, $Q_{\text{EXT}} = 0.770$, $S_{\text{n-EXT}} = 0.900$, $S_{\text{p-EXT}} = 0.730$			
PLS-DA	Classification	-	222	$n_{\text{Train}} = 178$, $Q_{\text{Train}} = 0.800$, $S_{\text{n-Train}} = 0.930$, $S_{\text{p- Train}} = 0.720$	$n_{\text{EXT}} = 44$, $Q_{\text{EXT}} = 0.790$, $S_{\text{n-EXT}} = 0.940$, $S_{\text{p-EXT}} = 0.690$			
PLS	log*RP*	3	29	$n_{\text{Train}} = 20$, $R^2_{\text{Train}} = 0.934$, $Q^2_{\text{CUM}} = 0.903$, $RMSE_{\text{Train}} = 0.258$	$n_{\text{EXT}} = 9$, $R^2_{\text{EXT}} = 0.674$ $Q^2_{\text{EXT}} = 0.654$, $RMSE_{\text{EXT}} = 0.874$	Yes	2017	[58]
PLS	log*RP*	3	29	$n_{\text{Train}} = 20$, $R^2_{\text{Train}} = 0.922$, $Q^2_{\text{CUM}} = 0.900$, $RMSE_{\text{Train}} = 0.281$	$n_{\text{EXT}} = 9$, $R^2_{\text{EXT}} = 0.670$ $Q^2_{\text{EXT}} = 0.649$, $RMSE_{\text{EXT}} = 0.879$			
LDA	Classification	3	24	$n_{\text{Train}} = 16$, $Q_{\text{Train}} = 0.934$, $S_{\text{n-Train}} = 1$, $S_{\text{p- Train}} = 0.833$	$n_{\text{EXT}} = 8$, $Q_{\text{EXT}} = 1$, $S_{\text{n-EXT}} = 1$, $S_{\text{p-EXT}} = 1$	NO	2017	[68]
GFA-MLR	p*IC*$_{50}$	2	15	$n_{\text{Train}} = 10$, $R^2_{\text{Train}} = 0.860$, $Q^2_{\text{LOO}} = 0.730$	$n_{\text{EXT}} = 5$, $Q^2_{\text{EXT}} = 0.640$			
MLR	log*RP*	2	23	$n_{\text{Train}} = 18$, $R^2_{\text{Train}} = 0.828$, $Q^2_{\text{LOO}} = 0.729$, $Q^2_{\text{BOOT}} = 0.776$,$RMSE_{\text{Train}} = 0.418$	$n_{\text{EXT}} = 5$, $Q^2_{\text{EXT}} = 0.902$, $CCC = 0.936$, $RMSE_{\text{EXT}} = 0.316$	Yes	2019	[43]
kNN	Classification	3 (3)	94	$n_{\text{Train}} = 78$, $Q_{\text{Train}} = 0.923$, $S_{\text{n-Train}} = 0.979$, $S_{\text{p- Train}} = 0.839$	$n_{\text{EXT}} = 26$, $Q_{\text{EXT}} = 0.923$, $S_{\text{n-EXT}} = 0.938$, $S_{\text{p-EXT}} = 0.900$	Yes	2020	[44]
MLR	log*RP*	3	29	$n_{\text{Train}} = 22$, $R^2_{\text{Train}} = 0.850$, $Q^2_{\text{LOO}} = 0.782$, $Q^2_{\text{BOOT}} = 0.736$,$RMSE_{\text{Train}} = 0.376$	$n_{\text{EXT}} = 7$, $Q^2_{\text{EXT}} = 0.795$, $CCC = 0.887$, $RMSE_{\text{EXT}} = 0.399$	Yes	2021	[45]
kNN	Classification	3 (3)	407	$n_{\text{Train}} = 305$, $Q_{\text{Train}} = 0.856–0.872$, $S_{\text{n-Train}} = 0.854–0.888$, $S_{\text{p- Train}} = 0.854–0.859$	$n_{\text{EXT}} = 102$, $Q_{\text{EXT}} = 0.853–0.873$, $S_{\text{n-EXT}} = 0.818–0.907$, $S_{\text{p-EXT}} = 0.833–0.930$	Yes	2021	[9]
kNN	log*RP*	4 (3)	88	$n_{\text{Train}} = 70$, $R^2_{\text{Train}} = 0.886$, $Q^2_{\text{LOO}} = 0.784$, $Q^2_{\text{BOOT}} = 0.639$,$RMSE_{\text{Train}} = 0.442$	$n_{\text{EXT}} = 18$, $Q^2_{\text{EXT}} = 0.897$, $CCC = 0.941$, $RMSE_{\text{EXT}} = 0.466$			
MLR	log*RP*	6	88	$n_{\text{Train}} = 66$, $R^2_{\text{Train}} = 0.812$, $Q^2_{\text{LOO}} = 0.757$, $Q^2_{\text{BOOT}} = 0.755$,$RMSE_{\text{Train}} = 0.567$	$n_{\text{EXT}} = 22$, $Q^2_{\text{EXT}} = 0.852$, $CCC = 0.906$, $RMSE_{\text{EXT}} = 0.561$			

TABLE 29.2 (Q)SAR models of transthyretin disrupting effects.—cont'd

Method[a]	*Endpoint*[b]	*m*[c]	*n*[d]	Training set[e]	Validation set[f]	AD[g]	Year	References
kNN	log*RP*	2 (3)	41	$n_{Train} = 32$, $R^2_{Train} = 0.960$, $Q^2_{LOO} = 0.902$, $Q^2_{BOOT} = 0.800$, $RMSE_{Train} = 0.218$	$n_{EXT} = 9$, $Q^2_{EXT} = 0.925$, $CCC = 0.963$, $RMSE_{EXT} = 0.294$			
MLR	log*RP*	4	41	$n_{Train} = 32$, $R^2_{Train} = 0.909$, $Q^2_{LOO} = 0.855$, $Q^2_{BOOT} = 0.770$, $RMSE_{Train} = 0.331$	$n_{EXT} = 9$, $Q^2_{EXT} = 0.908$, $CCC = 0.959$, $RMSE_{EXT} = 0.326$			
kNN	Classification	3 (3)	445	$n_{Train} = 333$, $Q_{Train} = 0.868$, $S_{n\text{-}Train} = 0.867$, $S_{p\text{-}Train} = 0.868$	$n_{EXT} = 112$, $Q_{EXT} = 866$, $S_{n\text{-}EXT} = 0.841$, $S_{p\text{-}EXT} = 0.898$	Yes	2022	[69]
kNN	log*RP*	2 (3)	88	$n_{Train} = 66$, $R^2_{Train} = 0.927$, $Q^2_{LOO} = 0.839$, $Q^2_{BOOT} = 0.810$, $RMSE_{Train} = 0.354$	$n_{EXT} = 18$, $Q^2_{EXT} = 0.833$, $CCC = 0.910$, $RMSE_{EXT} = 0.560$			
kNN	log*RP*	2 (3)	41	$n_{Train} = 30$, $R^2_{Train} = 0.910$, $Q^2_{LOO} = 0.852$, $Q^2_{BOOT} = 0.760$, $RMSE_{Train} = 0.336$	$n_{EXT} = 9$, $Q^2_{EXT} = 0.870$, $CCC = 0.930$, $RMSE_{EXT} = 0.390$			

[a]CoMSIA, *comparative molecular similarity indices analysis;* GFA-MLR, *genetic function approximation—multiple linear regression;* kNN, *k-nearest neighbor;* LDA, *linear discriminant analysis;* LIB-SVM, *library for support vector machine;* MLR, *multiple linear regression;* PLS, *partial least squares regression;* PLS-DA, *partial least squares—discriminant analysis.*
[b]*logK, logarithm of the binding constant; logRP, logarithm of the relative competitive potency.*
[c]*m, number of predictive variables; numbers in parentheses, value of* k *(number of nearest neighbor); —, number of predictive variables is not available.*
[d]*n, number of total compounds.*
[e]n_{train}, *numbers of model compounds in training set;* Q^2_{BOOT}, *bootstrapping coefficient;* Q^2_{CUM}, *total variation of the dependent variables that can be predicted by all of the extracted PLS components;* Q^2_{LOO}, *leave-one out cross-validation* Q^2; $RMSE_{train}$, *root mean square errors for the training set;* Q_{Train}, *predictive accuracy of training set;* R^2_{train}, *squared correlation coefficient of observed and predicted values for the training set;* $S_{n\text{-}Train}$, *sensitivity of training set;* $S_{p\text{-}Train}$, *specificity of training set.*
[f]*CCC, concordance correlation coefficient;* n_{EXT}, *numbers of model compounds in validation set;* Q_{EXT}, *predictive accuracy of validation set;* Q^2_{EXT}, *externally explained variance;* $RMSE_{EXT}$, *root mean square errors for the validation set;* $S_{n\text{-}EXT}$, *sensitivity of validation set;* $S_{p\text{-}EXT}$, *specificity of validation set.*
[g]AD, *applicability domain.*

In CPTP, the valid inputted substance was firstly classified into one of the two major categories and eight subcategories by 10 functional group counts (i.e., named number of nitrogen atoms (n_N), number of aromatic C(sp2) (n_{Car}), number of aromatic hydroxyls (n_{ArOH}), number of carboxylic acids (aromatic) (n_{ArCOOH}), number of hydroxyl groups (n_{ROH}), number of carboxylic acids (aliphatic) (n_{RCOOH}), number of sulfonic (thio-/dithio-) acids (n_{SO2OH}), number of sulfinic (thio-/dithio-) acids (n_{SOOH}), number of X on aromatic ring (n_{ArX}), number of halogen atoms (n_X). The two major categories and eight subcategories were aromatic compounds (ionizable halogenated aromatic compounds, ionizable nonhalogenated aromatic compounds, nonionizable halogenated aromatic compounds, nonionizable nonhalogenated aromatic compounds) and aliphatic compounds (ionizable halogenated aliphatic compounds, ionizable nonhalogenated aliphatic compounds, nonionizable halogenated aliphatic compounds, nonionizable nonhalogenated aliphatic compounds). Then, the quantitative hTTR binding affinity value of classified aromatic compounds or aliphatic compounds would be predicted by the aromatic QSAR model or aliphatic QSAR model, respectively. In the final report, the predicted value and the AD assessment result based on Euclidean distance method were presented.

In ED Profiler, the valid inputted compound was firstly grouped into active or inactive hTTR binder by the classification model. Secondly, the missing quantitative hTTR binding affinity value of potential active hTTR disruptor would be predicted by the QSAR models. Besides reporting the Euclidean distance-based AD evaluation result, the reliability assessment result of the predicted values from classification model and/or QSAR models for the valid inputted substance was also listed in the final report. The Tanimoto similarity index (T_S) was used to evaluate the reliability of predicted value. Eq. (29.5) gives the definition of T_S [9]:

$$T_S = \frac{\sum_{i=1}^{n} Y_{X-i} \cap Z_{X-i}}{\sum_{i=1}^{n} Y_{X-i} + \sum_{i=1}^{n} Z_{X-i} - \sum_{i=1}^{n} Y_{X-i} \cap Z_{X-i}} \tag{29.5}$$

where $Y_{X\text{-}i}$ and $Z_{X\text{-}i}$ were the *i*-th descriptors of compound Y and Z, respectively. The Pubchem Fingerprints descriptors of training set compounds and inputted substance were employed to calculate the T_S value. The magnitude of the T_S value reflects the level of similarity between training set compounds and inputted substance. A higher value of T_S meant that the inputted substance had higher level of similarity with at least one training set compound. In this tool, if the T_S value of the inputted compound was ≥ 0.900, the predicted value for this compound was referred as high reliability; $0.750 \leq T_S < 0.900$, the predicted value was labeled as moderate reliability; otherwise was low reliability. The reliability assessment result could be useful to determine whether the predicted value of inputted compound is receivable.

Conclusions and future directions

In this contribution, we have described the research progress for the disrupting effects of TDCs on human, gulls, chickens, Japanese quails, bullfrog tadpoles, little skate, and seabream TTR and also identified the most vital chemical structure characteristics of high- and very-high-potency TTR binder. We found that aromaticity, presence of hydroxyl or sulfhydryl group, and halogenation were the three critical chemical structure characteristics, and the interspecies variation of TTR disrupting effects for different species was unneglectable. We also summarized the available in silico models based on the hTTR binding affinity data, and the software could be used to screening potential hTTR disruptors.

Some future directions are proposed as follows:

Attention to the transthyretin disrupting effects of nonhuman species. As there was unneglectable interspecies variation of TTR disrupting effects for different species and the available TTR binding affinity data was mainly related to human up to now, more efforts should be paid to test the potential disrupting effects of TDCs with nonhuman species TTR.

Developing more predictive models and software for transthyretin disrupting effects. Firstly, more modeling action should be taken to derive the toxicant-based qualitative and quantitative models for the nonhuman species TTR disrupting effects. Secondly, target-based predictive models also suggested to be constructed. Finally, the new constructed toxicant-based and/or target-based could be integrated into the existing software or new software.

Acknowledgments

This research was funded by the National Natural Science Foundation of China (No.22176097).

References

[1] Wojtczak A, Neumann P, Cody V. Structure of a new polymorphic monoclinic form of human transthyretin at 3 A resolution reveals a mixed complex between unliganded and T4-bound tetramers of TTR. Acta Crystallogr D Biol Crystallogr 2001;57(Pt 7):957–67.

[2] Berman HM, Westbrook J, Feng Z, Gilliland G, Bhat TN, Weissig H, et al. The protein data bank. Nucleic Acids Res 2000;28(1):235–42.

[3] Power DM, Llewellyn L, Faustino M, Nowell MA, Björnsson BT, Einarsdottir IE, et al. Thyroid hormones in growth and development of fish. Comp Biochem Physiol C Toxicol Pharmacol 2016;130(4):447–59.

[4] Yamauchi K. Evolution of thyroid hormone distributor proteins in fish. Gen Comp Endocrinol 2021;305:113735.

[5] Rabah SA, Gowan IL, Pagnin M, Osman N, Richardson SJ. Thyroid hormone distributor proteins during development in vertebrates. Front Endocrinol 2019;10:506.

[6] Zhang J, Begum A, Brännström K, Grundström C, Iakovleva I, Olofsson A, et al. Structure-based virtual screening protocol for in silico identification of potential thyroid disrupting chemicals targeting transthyretin. Environ Sci Technol 2016;50(21):11984–93.

[7] Richardson SJ. Evolutionary changes to transthyretin: evolution of transthyretin biosynthesis. FEBS J 2009;276(19):5342–56.

[8] Weiss JM, Andersson PL, Zhang J, Simon E, Leonards PE, Hamers T, et al. Tracing thyroid hormone-disrupting compounds: database compilation and structure-activity evaluation for an effect-directed analysis of sediment. Anal Bioanal Chem 2015;407(19):5625–34.

[9] Yang XH, Ou W, Zhao SH, Xi Y, Wang LJ, Liu HH. Rapid screening of human transthyretin disruptors through a tiered in silico approach. ACS Sustainable Chem Eng 2021;9:5661–72.

[10] Miller MD, Crofton KM, Rice DC, Zoeller RT. Thyroid-disrupting chemicals: interpreting upstream biomarkers of adverse outcomes. Environ Health Perspect 2009;117(7):1033–41.

[11] Noyes PD, Friedman KP, Browne P, Haselman JT, Gilbert ME, Hornung MW, et al. Evaluating chemicals for thyroid disruption: opportunities and challenges with in vitro testing and adverse outcome pathway approaches. Environ Health Perspect 2019;127(9):95001.

[12] Zhang X, Cui S, Pan L, Dong W, Ma M, Liu W, et al. The molecular mechanism of the antagonistic activity of hydroxylated polybrominated biphenyl (OH-BB80) toward thyroid receptor β. Sci Total Environ 2019;697:134040.

[13] Lu L, Zhan T, Ma M, Xu C, Wang J, Zhang C, et al. Thyroid disruption by bisphenol s analogues via thyroid hormone receptor β: in vitro, in vivo, and molecular dynamics simulation study. Environ Sci Technol 2018;52(11):6617–25.

[14] Richardson SJ. Cell and molecular biology of transthyretin and thyroid hormones. Int Rev Cytol 2007;258:137–93.

[15] Funkenstein B, Perrot V, Brown CL. Cloning of putative piscine (*Sparus aurata*) transthyretin: developmental expression and tissue distribution. Mol Cell Endocrinol 1999;157(1–2):67–73.

[16] Eneqvist T, Lundberg E, Karlsson A, Huang S, Santos CR, Power DM, et al. High resolution crystal structures of piscine transthyretin reveal different binding modes for triiodothyronine and thyroxine. J Biol Chem 2004;279(25):26411–6.

[17] Reixach N, Foss TR, Santelli E, Pascual J, Kelly JW, Buxbaum JN. Human-murine transthyretin heterotetramers are kinetically stable and non-amyloidogenic. A lesson in the generation of transgenic models of diseases involving oligomeric proteins. J Biol Chem 2008;283(4):2098–107.

[18] Sunde M, Richardson SJ, Chang L, Pettersson TM, Schreiber G, Blake CC. The crystal structure of transthyretin from chicken. Eur J Biochem 1996;236(2):491–9.

[19] Folli C, Pasquato N, Ramazzina I, Battistutta R, Zanotti G, Berni R. Distinctive binding and structural properties of piscine transthyretin. FEBS Lett 2003;555(2):279–84.

[20] Schreiber G. The evolutionary and integrative roles of transthyretin in thyroid hormone homeostasis. J Endocrinol 2002;175:61–73.

[21] Yamauchi K, Nakajima J, Hayashi H, Hara A, Yamauchi K. Purification and characterization of thyroid-hormone-binding protein from masu salmon serum: a homolog of higher-vertebrate transthyretin. Eur J Biochem 1999;265:944–9.

[22] Suzuki S, Kasai K, Yamauchi K. Characterization of little skate (*Leucoraja erinacea*) recombinant transthyretin: zinc-dependent 3, 3′, 5-triiodo-l-thyronine binding. Gen Comp Endocrinol 2015;217:43–53.

[23] Suzuki S, Kasai K, Nishiyama N, Ishihara A, Yamauchi K. Characteristics of the brown hagfish Paramyxine atami transthyretin: metal ion-dependent thyroid hormone binding. Gen Comp Endocrinol 2017;249:1–14.

[24] Kasai K, Nishiyama N, Yamauchi K. Molecular and thyroid hormone binding properties of lamprey transthyretins: the role of an N-terminal histidine-rich segment in hormone binding with high affinity. Mol Cell Endocrinol 2018;474:74–88.

[25] Prapunpoj P, Leelawatwatana L, Schreiber G, Richardson SJ. Change in structure of the N-terminal region of transthyretin produces change in affinity of transthyretin to T4 and T3. FEBS J 2006;273(17):4013–23.

[26] Wang Z, Walker GW, Muir DCG, Nagatani-Yoshida K. Toward a global understanding of chemical pollution: a first comprehensive analysis of national and regional chemical inventories. Environ Sci Technol 2020;54(5):2575–84.

[27] Tang WH, Chen JW, Hong HX. Development of classification models for predicting inhibition of mitochondrial fusion and fission using machine learning methods. Chemosphere 2020;273:128567.

[28] van den Berg KJ. Interaction of chlorinated phenols with thyroxine binding sites of human transthyretin, albumin and thyroid binding globulin. Chem Biol Interact 1990;76:63–75.

[29] Yamauchi K, Ishihara A, Fukazawa H, Terao Y. Competitive interactions of chlorinated phenol compounds with 3,3′,5-triiodothyronine binding to transthyretin: detection of possible thyroid-disrupting chemicals in environmental waste water. Toxicol Appl Pharmacol 2003;187: 110–7.

[30] Ishihara A, Nishiyama N, Sugiyama S, Yamauchi K. The effect of endocrine disrupting chemicals on thyroid hormone binding to Japanese quail transthyretin and thyroid hormone receptor. Gen Comp Endocrinol 2003;134:36–43.

[31] Hamers T, Kamstra JH, Sonneveld E, Murk AJ, Kester MH, Andersson PL, et al. In vitro profiling of the endocrine-disrupting potency of brominated flame retardants. Toxicol Sci 2006;92:157–73.

[32] Harju M, Hamers T, Kamstra JH, Sonneveld E, Boon JP, Tysklind M, et al. Quantitative structure-activity relationship modeling on in vitro endocrine effects and metabolic stability involving 26 selected brominated flame retardants. Environ Toxicol Chem 2007;26(4):816–26.

[33] Morgado I, Hamers T, Van der Ven L, Power DM. Disruption of thyroid hormone binding to sea bream recombinant transthyretin by ioxinyl and polybrominated diphenyl ethers. Chemosphere 2007;69:155–63.

[34] Ucán-Marín F, Arukwe A, Mortensen A, Gabrielsen GW, Fox GA, Letcher RJ. Recombinant transthyretin purification and competitive binding with organohalogen compounds in two gull species (*Larus argentatus* and *Larus hyperboreus*). Toxicol Sci 2009;107:440–50.

[35] Ucán-Marín F, Arukwe A, Mortensen AS, Gabrielsen GW, Letcher RJ. Recombinant albumin and transthyretin transport proteins from two gull species and human: chlorinated and brominated contaminant binding and thyroid hormones. Environ Sci Technol 2010;44:497–504.

[36] Simon E, Bytingsvik J, Jonker W, Leonards PE, de Boer J, Jenssen BM, et al. Blood plasma sample preparation method for the assessment of thyroid hormone-disrupting potency in effect-directed analysis. Environ Sci Technol 2011;45(18):7936–44.

[37] Zhang J, Kamstra JH, Ghorbanzadeh M, Weiss JM, Hamers T, Andersson PL. In silico approach to identify potential thyroid hormone disruptors among currently known dust contaminants and their metabolites. Environ Sci Technol 2015;49(16):10099–107.

[38] Hill KL, Mortensen ÅK, Teclechiel D, Willmore WG, Sylte I, Jenssen BM, et al. In vitro and in silico competitive binding of brominated poly phenyl ether contaminants with human and gull thyroid hormone transport proteins. Environ Sci Technol 2018;52(3):1533–41.

[39] Purkey HE, Palaninathan SK, Kent KC, Smith C, Safe SH, Sacchettini JC, et al. Hydroxylated polychlorinated biphenyls selectively bind transthyretin in blood and inhibit amyloidogenesis: rationalizing rodent PCB toxicity. Chem Biol 2004;11(12):1719–28.

[40] Marchesini GR, Meimaridou A, Haasnoot W, Meulenberg E, Albertus F, Mizuguchi M, et al. Biosensor discovery of thyroxine transport disrupting chemicals. Toxicol Appl Pharmacol 2008;232(1):150–60.

[41] Cao J, Lin Y, Guo LH, Zhang AQ, Wei Y, Yang Y. Structure-based investigation on the binding interaction of hydroxylated polybrominated diphenyl ethers with thyroxine transport proteins. Toxicology 2010;277(1–3):20–8.

[42] Grimm FA, Lehmler HJ, He X, Robertson LW, Duffel MW. Sulfated metabolites of polychlorinated biphenyls are high-affinity ligands for the thyroid hormone transport protein transthyretin. Environ Health Perspect 2013;121(6):657–62.

[43] Yang X, Ou W, Xi Y, Chen J, Liu H. Emerging polar phenolic disinfection byproducts are high-affinity human transthyretin disruptors: an in vitro and in silico study. Environ Sci Technol 2019;53(12):7019–28.

[44] Xi Y, Yang X, Zhang H, Liu H, Watson P, Yang F. Binding interactions of halo-benzoic acids, halo-benzenesulfonic acids and halo-phenylboronic acids with human transthyretin. Chemosphere 2020;242:125135.

[45] Yang X, Ou W, Zhao S, Wang L, Chen J, Kusko R, et al. Human transthyretin binding affinity of halogenated thiophenols and halogenated phenols: an in vitro and in silico study. Chemosphere 2021;280:130627.

[46] Ren XM, Guo LH. Assessment of the binding of hydroxylated polybrominated diphenyl ethers to thyroid hormone transport proteins using a site-specific fluorescence probe. Environ Sci Technol 2012;46:4633–40.

[47] Xin Y, Ren XM, Ruan T, Li CH, Guo LH, Jiang G. Chlorinated polyfluoroalkylether sulfonates exhibit similar binding potency and activity to thyroid hormone transport proteins and nuclear receptors as perfluorooctanesulfonate. Environ Sci Technol 2018;52(16):9412–8.

[48] Ren XM, Yao L, Xue Q, Shi J, Zhang Q, Wang P, et al. Binding and activity of tetrabromobisphenol A mono-ether structural analogs to thyroid hormone transport proteins and receptors. Environ Health Perspect 2020;128(10):107008.
[49] Huang K, Wang X, Zhang H, Zeng L, Zhang X, Wang B, et al. Structure-directed screening and analysis of thyroid-disrupting chemicals targeting transthyretin based on molecular recognition and chromatographic separation. Environ Sci Technol 2020;54(9):5437–45.
[50] Aqai P, Fryganas C, Mizuguchi M, Haasnoot W, Nielen MW. Triple bioaffinity mass spectrometry concept for thyroid transporter ligands. Anal Chem 2012;84(15):6488–93.
[51] Zhang J, Grundström C, Brännström K, Iakovleva I, Lindberg M, Olofsson A, et al. Interspecies variation between fish and human transthyretins in their binding of thyroid-disrupting chemicals. Environ Sci Technol 2018;52:11865–74.
[52] Collet B, Simon E, van der Linden S, El Abdellaoui N, Naderman M, Man HY, et al. Evaluation of a panel of in vitro methods for assessing thyroid receptor β and transthyretin transporter disrupting activities. Reprod Toxicol 2020;96:432–44.
[53] Šauer P, Švecová H, Grabicová K, Gönül Aydın F, Mackuľak T, Kodeš V, et al. Bisphenols emerging in Norwegian and Czech aquatic environments show transthyretin binding potency and other less-studied endocrine-disrupting activities. Sci Total Environ 2021;751:141801.
[54] Montaño M, Cocco E, Guignard C, Marsh G, Hoffmann L, Bergman A, et al. New approaches to assess the transthyretin binding capacity of bioactivated thyroid hormone disruptors. Toxicol Sci 2012;130(1):94–105.
[55] Ouyang X, Froment J, Leonards PEG, Christensen G, Tollefsen KE, de Boer J, et al. Miniaturization of a transthyretin binding assay using a fluorescent probe for high throughput screening of thyroid hormone disruption in environmental samples. Chemosphere 2017;171:722–8.
[56] DRAGON. Software for molecular descriptor calculation, version 6.0. Milano, Italy: Talete srl; 2012.
[57] Yang X, Xie H, Chen J, Li X. Anionic phenolic compounds bind stronger with transthyretin than their neutral forms: nonnegligible mechanisms in virtual screening of endocrine disrupting chemicals. Chem Res Toxicol 2013;26(9):1340–7.
[58] Yang X, Lyakurwa F, Xie H, Chen J, Li X, Qiao X, et al. Different binding mechanisms of neutral and anionic poly-/perfluorinated chemicals to human transthyretin revealed by in silico models. Chemosphere 2017;182:574–83.
[59] Scherrer RA, Howard SM. Use of distribution coefficients in quantitative structure-activity relationships. J Med Chem 1977;20:53–8.
[60] Franco A, Ferranti A, Davidsen C, Trapp S. An unexpected challenge: ionizable compounds in the REACH chemical space. Int J Life Cycle Assess 2010;15:321–5.
[61] Li F, Xie Q, Li X, Li N, Chi P, Chen J, et al. Hormone activity of hydroxylated polybrominated diphenyl ethers on human thyroid receptor-beta: in vitro and in silico investigations. Environ Health Perspect 2010;118(5):602–6.
[62] Akiyoshi S, Sai G, Yamauchi K. Species-dependent effects of the phenolic herbicide ioxynil with potential thyroid hormone disrupting activity: modulation of its cellular uptake and activity by interaction with serum thyroid hormone-binding proteins. J Environ Sci (China) 2012; 24(5):949–55.
[63] Papa E, Kovarich S, Gramatica P. QSAR modeling and prediction of the endocrine-disrupting potencies of brominated flame retardants. Chem Res Toxicol 2010;23(5):946–54.
[64] Yang W, Shen S, Mu L, Yu H. Structure-activity relationship study on the binding of PBDEs with thyroxine transport proteins. Environ Toxicol Chem 2011;30(11):2431–9.
[65] Kovarich S, Papa E, Gramatica P. QSAR classification models for the prediction of endocrine disrupting activity of brominated flame retardants. J Hazard Mater 2011;190(1–3):106–12.
[66] Kovarich S Papa E, Li J, Gramatica P. QSAR classification models for the screening of the endocrine-disrupting activity of perfluorinated compounds. SAR QSAR Environ Res 2012;23(3–4):207–20.
[67] Papa E, Kovarich S, Gramatica P. QSAR prediction of the competitive interaction of emerging halogenated pollutants with human transthyretin. SAR QSAR Environ Res 2013;24(4):333–49.
[68] Kar S, Sepúlveda MS, Roy K, Leszczynski J. Endocrine-disrupting activity of per- and polyfluoroalkyl substances: exploring combined approaches of ligand and structure based modeling. Chemosphere 2017;184:514–23.
[69] Yang XH, Liu HH, Kusko R, Hong HX. Chapter 10 ED profiler: machine learning tool for screen-ing potential endocrine disrupting chemicals. In: Hong HX, editor. Machine learning and deep learning in computational toxicology. Cham, Switzerland: Springer International Publishing; 2023. p. 243–62.

CHAPTER

30

QSAR models for toxicity assessment of multicomponent systems

Laura Lomba and Beatriz Giner

Facultad de Ciencias de la Salud, Universidad San Jorge, Campus Universitario, Villanueva de Gállego, Zaragoza, Spain

Abbreviations

A. baumannii *Acinetobacter baumannii*
A. fischeri *Allivibrio fischeri*
B. subtilis *Bacillus subtilis*
C. albicans *Candida albicans*
C. elegans *Caenorhabditis elegans*
D. magna *Daphnia magna*
DES Deep eutectic solvents
DFT Density functional theory
E. coli *Escherichia coli*
EC50 Half maximal effective concentration
ELM Extreme learning machine
ELUMO Energy of the lowest unoccupied molecular orbital
ETA Extended topochemical atom
GC-COSMO Group contribution-based conductor-like screening model
HBA Hydrogen bond acceptor
HBD Hydrogen bond donors
HQSAR Hologram quantitative structure activity relationship
IC50 Half maximal inhibitory concentration
LC50 Median lethal concentration
LFER Linear free energy relationship
MBC Minimal bactericidal concentration
MIC Minimal inhibitory concentration
OECD Organization for Economic Cooperation and Development
P. aeruginosa *Pseudomonas aeruginosa*
PLS Partial least squares
QSTR Quantitative structure toxicity relationship
QTMS Quantum topological molecular similarity
S. aureus *Staphylococcus aureus*
S. vacuolatus *Scenedesmus vacuolatus*
SiRMS Simplex representation of molecular structure
SMILES Simplified molecular input line entry specification

Introduction

A quantitative structure–activity relationship (QSAR) is a mathematical model used for the prediction or estimation of the properties or biologic activity of a substance from its molecular structure. According to QSAR, the activity of a certain chemical is given as a function of one or more properties or molecular characteristics of the chemical, including molecular descriptors.

QSAR assumes that the macroscopic behavior of any chemical substance is a consequence of its chemical structure and therefore, of its nanoscopic or molecular behavior. The complex interactions between molecules that form a certain substance that determines the global behavior too can also be included in the model, by means of macroscopic descriptors.

These methods are very useful, on the one hand, allow the prediction of a certain biological activity (also called end point) or property without the need to measure it experimentally. On the other hand, and inform in which sense the structural modifications in a molecule affect the biological activity that can be produced due to the exposition of that chemical. The latter is of great importance when it comes to rationally designing new chemical substances that maintain certain properties, minimizing, for example, their toxic risks. For those reasons, QSAR is a very robust method that allows reducing costs and time in any case, as well as animal experimentation in certain applications.

In order to apply QSAR, some precepts must be met: first of all, a dataset is needed; a property (dependent variable) generally measured or determined experimentally for each of the substances that are part of the group of

QSAR in Safety Evaluation and Risk Assessment
https://doi.org/10.1016/B978-0-443-15339-6.00010-2

chemicals to be studied. Secondly, the independent variable(s) are needed, which must be related to the structure for this same group of chemical substances. There are several types of independent variables that are used. One of the most common are molecular descriptors (1D, 2D, or 3D type), but it al also possible to find thermodynamic, electric, or other type of physicochemical properties, topological descriptors, or group contribution [1]. Finally, it must be possible to correlate, through a mathematical method or statistical analysis, these two data matrices, always fulfilling certain requirements that ensure the validity of the method.

The methodology that allows obtaining the mathematical model is well described and is included in various Organization for Economic Cooperation and Development (OECD) standards, as main guide documents.

The scientific literature has collected information related to QSAR since the 1970s, although the first OECD standard that regulates QSAR dates back to 1992, referring to the evaluation of aquatic environmental toxicity produced by chemical substances [2]. This first OECD recommendation for the development of QSAR methods contemplated only molecular organic substances as study matrices.

Since 2000, there has been a steady growth in QSAR-related studies. Fig. 30.1 shows the historical data, from 2000 to 2022, of the scientific publications on QSAR collected on the Web of Science, exceeding one and a half 1000 annual publications, in recent years. Among the end points that can be predicted through QSAR, the most popular are those that allow estimating the toxicity produced by the substance. Fig. 30.1 also includes this fact, showing those scientific works that explicitly use QSAR to estimate toxicity. In any case, this number could be increased if we take into account of all the end points that are related to toxicity, even if not explicitly (for example, cytotoxicity or mutagenicity, among others).

The very nature of the method means that the models are developed around groups of molecules that maintain a correlation criterion with each other. That is why it is very common to find QSAR methods for organic chemical substances with a certain structural homogeneity [3], drugs (or a certain type of them) [4,5], metals [6,7], pesticides [8,9], nanomaterials [10,11], etc … The vast majority of QSAR models are developed for molecular substances, which are not part of a more complex mixture or system such as ionic liquids or deep eutectic solvents (Fig. 30.1).

In fact, only a very small percentage of QSAR studies address multicomponent systems (Fig. 30.2). One possible reason to explain this fact lies in the complexity of the definition of the matrix itself, allowing for multiple composition and combinations of the components of the system, that make difficult to manage and approach the uniformity, homogeneity and universality of the results obtained [12].

In this work, we focus precisely on the approach to the development of QSAR methods for complex systems, formed by mixtures of two or more components, paying attention to ionic liquids, deep eutectic solvents (DES), and other complex mixtures.

Ionic liquids are molten salts, described as liquid systems that show ionic-covalent crystal structures. The ions in ionic liquids generally have a high degree of asymmetry that makes packing difficult and thus, the crystallization process. The ions can be pure inorganic compounds, organic compounds, or organo-mineral combinations. In addition, these ions are usually considerably large, poorly coordinated, and diffusely charged; factors that also favor a relatively poor ordering and therefore justify the liquid state in which they are shown [13]. The interactions present in ionic liquids are electrostatic and strong. This gives mixtures low volatility/flammability and high chemical and

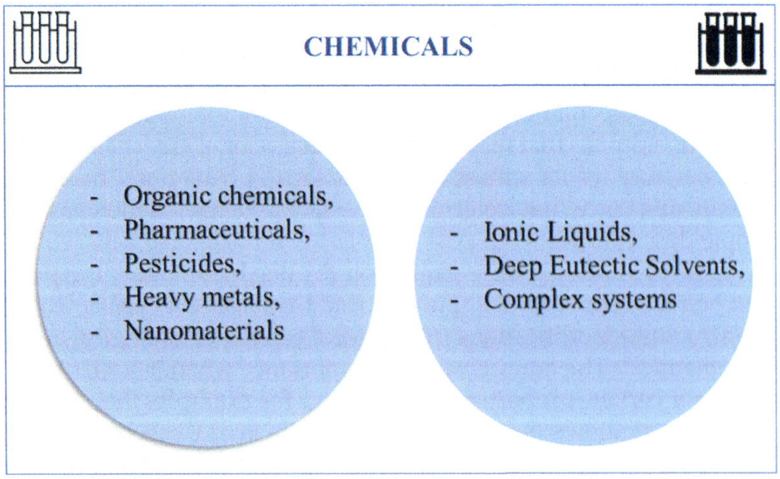

FIGURE 30.1 Classification of types of chemicals or substances used for developing quantitative structural −activity relationship (QSAR) methods.

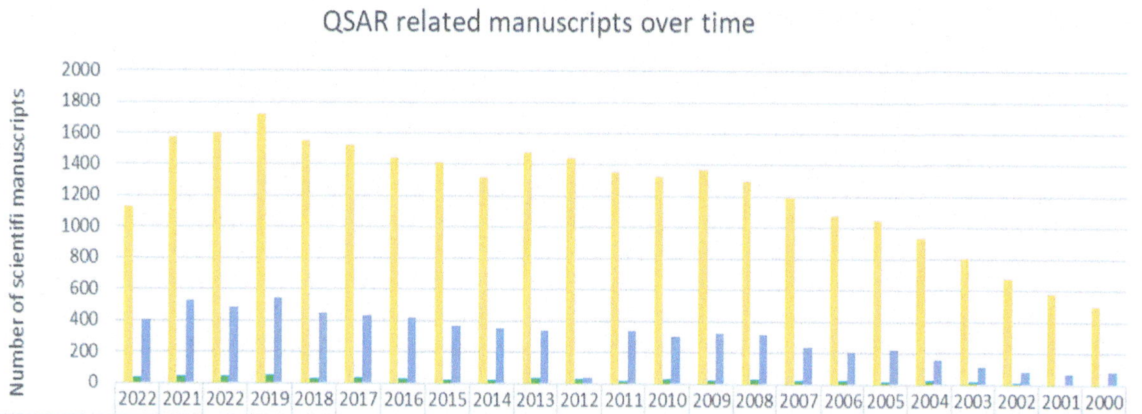

	2022	2021	2022	2019	2018	2017	2016	2015	2014	2013	2012	2011	2010	2009	2008	2007	2006	2005	2004	2003	2002	2001	2000
Total	1135	1579	1604	1723	1555	1530	1446	1417	1325	1481	1444	1359	1334	1376	1298	1202	1083	1051	946	813	684	589	504
Multicomponent	39	48	46	58	38	42	33	24	30	42	36	21	32	25	34	25	25	20	26	21	10	6	8
Toxicity	407	529	488	545	452	433	421	372	356	337	42	337	303	326	319	237	209	221	161	115	87	69	82

FIGURE 30.2 Quantitative structural–activity relationship (QSAR)-related scientific manuscripts over time. *Source Web of Science, October 16, 2022.*

electrochemical stability [14]. The study of ionic liquids has been very intensive during the last 2 decades precisely because of its great stability and versatility. There are different families of ionic liquids depending on the structure of the cation (based on the pyridine ring, imidazolium, tetraalkylammonium, for example). This cation can be linked in turn to different anions (halides, tetrafluoroborate, dicyanamide, triflate). The result of all this is a wide variety of ionic liquids formed by combining the different ions, which translates into a wide range of possibilities in terms of the properties that can be obtained. The combination possibilities are practically infinite, making them deserve the category of designer solvents that can be easily tuned by the suitable combination of selected ions. That is why ionic liquids have been widely applied and industrialized in various fields, such as batteries, extraction, chromatography, biomass conversion, adsorption, catalysis, CO_2 capture, adsorbents, and (bio)medical or pharmaceutical applications [15]. The toxicity of ionic liquids, however, is the cause for concern. From very early on, they were labeled as green solvents and alternatives to volatile organic compounds, mainly due to their low vapor pressure compared to traditional solvents. However, the toxic properties, both for health and for the environment, had not been previously assessed when classifying them as green. Afterward, great effort has been made in determining their toxic properties, trying to rationalize behavior, in order to facilitate the long-awaited directed and sustainable synthesis (interesting ionic liquids for a certain application but safe and with low toxic profiles). In this sense, the possibility of having predictive methods that allow this design to be carried out on demand is very interesting and necessary for this type of solvents.

As concerns about ionic liquids and their toxicity increased, DES emerged strongly as alternative solvents, sharing some of their good properties with ionic liquids (high solubility, low volatility) and apparently better toxicity. This issue is also subject to debate and numerous studies are being carried out, both for health and for the environment, to ensure their safety [16–20]. In 2003, DES were defined as mixtures whose melting temperature is lower than that of the substances that form it by more than 100°C, and a classification based on the mixture composition was proposed [21]. Types I, II, and IV contain metal salts or hydrated metal salts (as hydrogen bond acceptor (HBA)) combined with organic salts or hydrogen bond donors (HBD). Type III includes mixtures of organic salts (typically an ammonium halide) with metabolites; they are often called natural deep eutectic solvents because they come from renewable sources of raw material [22].

In the category of complex systems, we can include polymers which usually present polydispersity and polymolecularity phenomena, colloidal systems, which can be formed by molecular aggregates or clusters and other types of complex systems such as high-molecular-weight biological substances (proteins, nucleic acids, etc.)

Multicomponent systems or mixtures

As mentioned before, the first OECD recommendations about the development of QSAR methods did not contemplate the possibility of using mixtures or multicomponent systems to create the matrix dataset. In fact, in the OECD monograph No. 67 [23], it is specifically said that it was not yet achievable to deal with complex or undefined mixtures. Mixtures of organic chemicals are explicitly excluded from the method development recommendation. It is not until 2007, when OECD recommendations for the grouping of mixtures are included [24]. Concretely, it is contemplated that substances may contain chemicals of distinct chain lengths at varying amounts, being this category in itself an example of a complex system, which can present polymolecularity or even polydispersion (polymer-like substance). Mixtures of isomers are also mentioned, resolving their treatment as a whole. This document also addresses the case of complex systems (defined as "substances of unknown or variable composition, complex reaction products or biological mater") in order to define a category for the development of the method. In these complex cases, the definition of the composition of the mixture is considered to be one of the main blocks to construct the category.

One of the premises of QSAR is the similarity in chemical structure or mode of action of the substances studied to develop the project. This concept is undoubtedly a challenge when the system to model is multicomponent. For example, if it is an ionic system, the variety of inorganic-type counterions that can be included, even maintaining clear molecular correlation between the organic structures that can be part of it, makes it difficult to establish structural patterns. In the case of DES, there is an almost unlimited number of substances that can act either as donors or acceptors of hydrogen bonds, ranging from sugars, or alcohols to amino acids, including a wide variety of substances of natural origin, like choline chloride. It is evident that the mechanisms of action that produce the toxicity are very different given the chemical nature of the substances that are part of the eutectics.

Ionic liquids

The very nature of ionic liquids (molten salts at room temperature) means that the development of QSAR methods must consider the presence of two components that make up each substance. For this, the way to represent the chemical structure through the ions properly is essential for the subsequent development of the model. Normally, the two ions (cation and anion) that make up the ionic liquid are described separately. However, there are other possibilities, such as obtaining the independent variables of the ion pair instead of the individual ions. Following, a review of the strategies used to approach the modeling of these ionic mixtures through QSAR-type methods is shown. The different studies have been classified in biological studies (cells, antibacterial studies, enzymatic reactions, etc.) and environmental predictions (Table 30.1).

In relation to the prediction of cytotoxicity, there are a limited number of studies. In most cases, ions and cations are described separately. This is the case, for example, of the work of Das et al. In this study, to construct the model, 289 ionic liquids and 2D descriptors (topochemical ETA and non-ETA parameters) indices, Quantum Topological Molecular Similarity (QTMS) indices and lipophilicity were used. The dependant variable or endpoint was the toxicity of ionic liquids on rat leukemia cell line (IPC-81). The R^2 value obtained was 0.869, which is pretty good [25].

However, in the work of Peng et al. [26], the GC-COSMO method was used to describe the ionic liquids, that were decomposed into three parts: on the one hand, the anions with a single contribution and the cations on the other through two contributions (the cationic skeleton and the substituents attached to this main structure). A total of 127 ionic liquids were used as dataset (93 as training set, 15 as validation set and 19 as test set). A total of 4 models were obtained, with R^2 ranging from 0.87 to 0.98 using between 7 and 9 descriptors. These results are also very good, considering the disparate nature of the starting mixtures.

The antibacterial or bactericidal capacity of ionic liquids has been studied more in depth. Once again, most of the studies, the ions are described separately by means of 2D descriptors obtained in turn, from other predictive models such as ALogPS, Chemaxon or DRAGON. This is the case of the work of Hodyna et al. who developed a QSAR model for predicting the antibacterial activity of imidazolium-based ionic liquids for *Bacillus subtilis* (*B. subtilis*) and *Pseudomonas aeruginosa* (*P. aeruginosa*) [27]. For this, 20 synthesized ionic liquids based on 1,3-dialkylimidazolium as a cation and different anions were prepared. Six descriptor packages were used: E-State indices 2D (atom/bond type, atom, and bonds counts, topological environment), ALogPS (lipophilicity and solubility), Adriana Code 3D (molecular dipole moment, number of hydrogen bonds, molecular weight, number of rotatable bonds, topological polar surface area, molecular and ring complexity, number of atoms, special properties among others), DRAGON descriptors (topological indices, functional group counts, path counts, information

TABLE 30.1 Summary and main characteristics of QSAR studies for ionic liquids.

Ionic liquid	QSAR strategy	Endpoint	Independent variables	Author	References
Biological studies					
Cations: Imidazolium, pyridinium, ammonium, pyrrolidinium, morpholinium, piperidinium, phosphonium, quinolinium, purinium, guanidinium Anions: Organic and inorganic anions	Description of ions separately	Toxicity toward rat leukemia cell line (IPC-81)	2D descriptors (topochemical ETA and non-ETA parameters) indices, quantum topological molecular *similarity* (QTMS) indices, lipophilicity	Das et al.	[25]
Cations: Piperidinium, pyridinium, imidazolium, morpholinium, pyrrolidinium, ammonium	Description of ions separately (cation skeleton + substitutes and anion)	Leukemia rat cell line (ICP-81)	Descriptors from GC-COSMO: Cation skeleton, substitutes in the cation skeleton and anion	Peng et al.	[26]
Anions: Halides, thiocyanate, bis(trifluoromethylsulfonyl)amide, tetrafluoroborate, sulfates, sulfonates, dicyanamide, trifluoroacetate, hexafluorophosphate, bis(trifluoromethylsulfonyl)imide	Description of ionic pair	Predicted the minimal inhibitory concentration (MIC), *Pseudomonas aeruginosa*	E-state indices, ALogPS, Adriana Code, DRAGON descriptors, chemaxon descriptors, inductive descriptors	Hodyna et al.	[27]
Cations: Ammonium, phosphonium, imidazolium, pyridinium Anions: Lactate, histidinate, aspartate, asparaginate, glutaminate, methioninate, serinate, isoleucinate, threoninate, glycinate, leucinate, valinate, alaninate, lysinate, argininate, glutaminate, proline, tryptophan, tetrafluoroborate, decanoate, bis(trifluoromethylsulfonyl)amide, dicyanide, hexafluorophosphate, iodine, chloride, bromide, methylsulfate, p-toluenesulfonate, octylsulfate, trifluoromethansulfonate, nitrate.	Description of ions separately	Predicted the minimal inhibitory concentration (MIC) and minimal bactericidal concentration (MBC) in *Escherichia coli*, *Staphylococcus aureus* and *Candida albicans*	Linear free energy relationship (LFER) descriptors: Excess molar refraction, polarizability/dipolarity, H-bonding acidity, H-bonding basicity, McGowan volume, cationic interaction, anionic interaction.	Cho et al.	[28]
Cations: Imidazole Anions: Bromide, chloride, acetate, nitrate, tetrafluoroborateion and thiocyanate	Description of ions separately	Toxicity towards *Caenorhabditis elegans*, logEC50	Molecular descriptors, including conformational, electrotopological, electronic, information-content, quantum mechanical, spatial, structural, thermodynamic, and topological information	Lu et al.	[29]

Continued

TABLE 30.1 Summary and main characteristics of QSAR studies for ionic liquids.—cont'd

Ionic liquid	QSAR strategy	Endpoint	Independent variables	Author	References
Cations: (Ethoxymethyl) ethyldimethylammonium, 1-(2-ethoxyethyl)-1-methylpiperidinium, 1-(2-hydroxyethyl)pyridinium, 1-(2-methoxyethyl)-1-methylpiperidinium, 1-(2-methoxyethyl)pyridinium, 1-(3-hydroxypropyl)-1-methylpiperidinium, 1-(3-hydroxypropyl)pyridinium, 1-(3-methoxypropyl)-1-methylpiperidinium, 1-(3-methoxypropyl)pyridinium, 1-(ethoxymethyl)pyridinium), 1,3-diethylimidazolium, 1-butyl-3-methylimidazolium, 1-butylpyridinium, 1-decyl-3-methylimidazolium, 1-ethylpyridinium, 1-heptyl-3-methylimidazolium, 1-hexylpyridinium, 1-octyl-4-methylpyridinium, 3-methyl-1-octadecylimidazolium Anions: Chloride, bis(trifluoromethylsulfonyl)amide, bromide, iodide, bis(trifluoromethylsulfonyl)amide, dicyanamide, tetrafluoroborate, trifluoromethanesulfonate	Description of ions separately	Toxicity toward acetylcholinesterasa	Charge density distribution area, electrostatic potential surface area.	Zhu et al.	[30]
Cations: 2-hydroxy ethylammonium, 2-hydroxy diethylammonium, Anions: Formic, lactic, acetic, propionic, adipic, citric, benzoic, maleic, salicylic acid	Description of ions separately	Toxicity toward HaCat and HepG2, EC50	Group contribution method with Boolean distribution	Varela et al.	[31]
Cations: Imidazolium, pyridinium, pyrrolidinium, piperidinium. Anions: Bromide, thiocyanate, bis(trifluoromethylsulfonyl)amide.	Description of ionic pair	Toxicity towards *Escherichia coli*, logEC50	Molecular descriptors DFT (B3LYP/6-311 + G*)	Rybińska-Fryca et al.	[32]
Multiple combinations		Predicted the minimal inhibitory concentration (MIC) and minimal bactericidal concentration (MBC) in *Staphylococcus aureus*	Hybrid optimal descriptors, hydrogen suppressed molecular graph and hydrogen filled graph	Lofti et al.	[33]
Cations: Imidazolium and pyridinium Anions: Inorganic	Description of ions separately	Toxicity toward *Acinetobacter baumannii* and *Staphylococcus aureus*, MIC	Electro-topological state indices, AlogPS, CDK2, and DRAGON descriptors	Semenyuta et al.	[34]
Multiple combinations	Description of ions separately	Citoxicity toward Hep-2 cells, IC50	DRAGON descriptors	Hodyna et al.	[35]

Multiple combinations	Description of ionic pair	Citoxicity toward IPC-81 cells, EC50	RDKit molecular descriptors (number of atoms, heavy atoms, carbon atoms, oxygen atoms, nitrogen atoms, chlorine atoms, topological polar surface area, molecular weight, number of valence electrons and number of heteroatoms). Other features using Mol2vec	Chipofya et al.	[36]
Environmental studies					
Cations: 1-Methylimidazolium, imidazolium, 1,1,3,3-tetramethylguanidinium, 1-butylimidazolium and melamine (4,6-diamino-1,3,5-triazin-2-aminium) Anions: Trifluoroacetate, triflate (trifluoromethane sulfonate), formate, acetate and caprylate.	Description of ions separately	Toxicity toward *A. fischeri*, logEC50	Group contribution methods in Boolean manner (anions, cations and substitutions). Type of anions, type of cations, number of carbons in short and long chain	Luis et al.	[37]
Cations: Imidazolium, pyridinium, pyrrolidinium, quinolinium, quaternary ammonium, piperidinium, morpholinium, choline Anions: Bromide, chloride	Description of ions separately	Toxicity toward *A. fischeri*, logEC50	Constitutional, topological, geometrical, electrostatic and quantum chemical descriptors	Bruzzone et al.	[38]
Multiple combinations	Description of ions separately	Toxicity toward IPC-81, *A. fischeri* and *Scenedesmus vacuolatus*, log EC50	Linear free energy relationship descriptors (excess molar refraction, dipolarity/polarizability, hydrogen bonding acidity, hydrogen bonding basicity, McGowan volume)	Cho et al.	[39]
Cations: 1-n-Butyl-3-methylimidazolium, 1-n-octyl-3-methylimidazolium bromide, 1-methyl-3-[(trimethylsilyl)methyl]-imidazolium, 1-n-dodecyl-3-methylimidazolium, 1-methyl-1,1,1-tri-n-octylammonium bis(trifluoromethanesulfonyl)imide, 1-(2-chloroethyl)imidazolium, 1-2-[2-(2-methoxyethoxy)ethoxy]ethyl]-3-methylimidazolium, N-methyl-N,N,N-	Description of ions separately	Toxicity toward *D. magna*, EC49	Extended topochemical atom indices (ETA) with topological non-ETA thermodynamic parameters. DRAGON and PaDEL descriptors.	Roy et al.	[40]

Continued

TABLE 30.1 Summary and main characteristics of QSAR studies for ionic liquids.—cont'd

Ionic liquid	QSAR strategy	Endpoint	Independent variables	Author	References
triethylammonium, ethyltetrahydrothiophenium, 1,1,1-triethylsulfonium, among others Anions: Bromide, methylsulfate, methylsulfonate, methylsulfonate, chloride, chloride, among others					
Cations: 1-n-Butyl-3-methylimidazolium, 1-n-octyl-3-methylimidazolium bromide, 1-methyl-3-[(trimethylsilyl)methyl]-imidazolium, 1-n-dodecyl-3-methylimidazolium, 1-methyl-1,1,1-tri-n-octylammonium bis(trifluoromethanesulfonyl)imide, 1-(2-chloroethyl)imidazolium, 1-[2-[2-(2-methoxyethoxy)ethoxy]ethyl]-3-methylimidazolium, N-methyl-N,N,N-triethylammonium, ethyltetrahydrothiophenium, 1,1,1-triethylsulfonium among others Anions: Bromide, methylsulfate, methylsulfonate, methylsulfonate, chloride, chloride, among others	Description of ions separately	Toxicity toward *D. magna*, EC50	Alog P, ETA descriptors, quantum chemical topology descriptors, lipophilicity	Roy et al.	[41]
Cations: Pyrrolidinium, piperidinium, pyridinium, imidazolium Anion: Bromide	Description of ions separately	Toxicity toward *A. fischeri* and *D. magna*, logEC50	Unoccupied molecular orbitals (ELUMO), dipole moment, total energy, molecular volume.	Wang et al.	[42]
Cations: Ammonium, cholinium, imidazolium, morpholinium, melanimium, phosphonium, tropinium, piperidinium, pyridinium, pyrrolidinium, quinuclidinium, sulfonium Anions: Organic and inorganic	Description of ions separately	Toxicity toward *A. fischeri*, logEC50	Constitutional and connectivity parameters, extended topochemical atom (ETA) indices, group contribution, and quantum topological molecular similarity (QTMS) parameters, DRAGON descriptors	Das et al.	[43]
Multiple combinations	Description of ions separately	Toxicity toward *A. fischeri*, logEC50	Number of carbon, hydrogen, nitrogen, and oxygen atoms in cation and in anion. Increasing and decreasing factor of moietures	Jafari et al.	[44]

Cations: Benzalkonium, benzyldimethyldodecylammonium, imidazolium, octylsulfate, 1-alkyl-3-methylimidazolium Anions: Chloride, bromide, tetrafluoroborate, dicyanamide, octylsulfate, bistriflamide	Description of ions separately and ionic pair	Ecotoxicity on leukemia rat cell lines (IPC-81)	Euclidean spatial distance, number of a outermost electrons, electronegativity, number of protons, atom weight, number of electron shells, enthalpy-vapourization of organic compounds	Yan et al.	[45]

indices, drug like indices, atom centered fragments, or molecular properties), Chemaxon descriptors (elemental analysis, geometry, charge, physicochemical properties), and inductive descriptors (intermolecular distances, atomic electronegativities, covalent radii and charge) The R^2 values were 0.77−0.92 and the accuracy obtained was 83% −88% [27].

In the work of Cho et al., a QSAR model was developed to predict the minimal inhibitory concentration (MIC) and minimal bactericidal concentration (MBC) for ionic liquids against *Escherichia coli* (*E. coli*), *Staphylococcus aureus* (*S. aureus*), and *Candida albicans* (*C. albicans*). In this case, the ions were described individually thorough several linear free energy relationship (LFER) descriptors such as anionic interaction, McGowan volume, excess molar refraction, polarizability/dipolarity, H-bonding acidity, H-bonding basicity, and cationic interaction were used. The R^2 values were 0.900 (MIC) and 0.934 (MBC) for *E. coli*, 0.910 (MIC) and 0.947 (MBC) for *S. aureus*, and finally, 0.892 (MIC) and 0.803 (MBC) for *C. albicans* [28].

Other models such as *Caenorhabditis elegans* (*C. elegans*) has been used by Lu et al. to obtain a QSAR model, which can predict the viscosity, melting point, and toxicity (LC50). To carry out this study 30 imidazolium-based ionic liquids were used. They optimized the geometry structures of ionic liquids using density functional theory calculations. Molecular descriptors such as electronic, electrotopological, information-content, quantum mechanical, spatial, structural, topological, and thermodynamic information. The R^2 value in this QSAR was 0.938 [29].

Trush et al. developed an effective QSAR model to search for new *Staphylococcus aureus* (*S. aureus*) inhibitors using quaternary phosphonium salts (ionic liquids). Additionally, some ionic liquids were synthetized and their antibacterial activity and their toxicity using *Danio rerio* and acetylcholinesterase enzyme as biological models was tested. The obtained QSAR model used artificial neural network approach [46].

Sosnowska et al. published a chemoinformatic method using 13 ionic liquids to obtain a QSAR model for predicting the cytotoxicity in leukemia cell line (IPC-81) through EC50 values and the acetylcholinesterase inhibition. Several programs such as Chemsketch or Avogadro were used to obtain the SMILES of chemicals. Additionally, Dragon or Padel software was used to calculate molecular descriptors; however, in the case of 3D descriptor or geometry, MOPAC software were used instead [47].

Another QSAR model was developed by Zhu et al. for predicting the toxicity of ionic liquids in acetylcholinesterase enzymes. They used 160 ionic mixtures and a multiple linear regression and extreme learning machine (ELM). Some of the used descriptors are charge density distribution area and electrostatic potential surface area [30].

Varela et al. presented a paper where they synthetized and evaluated the toxic profile of several ionic mixtures in 2 cell lines (HapG2 and HaCat). Additionally, they developed a QSAR in order to predict the cytotoxicity. The model was obtained using the group contribution method, which consists of obtaining the different molecular properties as the sum of the contributions of each of the atoms or fragments of the molecules. The correlation coefficient for this study was 0.9260 [31].

Unlike the previous ones, in the Rybińska-Fryca et al. study, the ionic pair for a set of 24 ionic liquids was also modeled, and the results were compared with those obtained by the most common process (modeling of the ions independently). The independent variables were both 2D and 3D molecular descriptors, obtained by DFT methods (B3LYP/6-311 + G*). The structure of the ionic pair is constructed using an additive approach; the molecular descriptors of each ion are obtained, and the contribution of the ionic pair is calculated, taking into account the molar composition of each ion, so that the presence of each ion in the moiety is weighted [32]. A total of 6 models were developed and for each of them, with no more than 3 uncorrelated descriptors were used. Results obtained indicated that 3D molecular descriptors did not improve the model compared to 2D type. However, the models that contemplated the ionic pair gave better results than those that used the ions of the ionic liquid separately.

Other study related to QSAR of ionic liquids toward *S. aureus* was published by Lofti et al. They predicted the minimal inhibitory concentration (MIC) and minimal bactericidal concentration (MBC) using 204 and 114 ionic liquids, respectively. Their descriptors were hybrid optimal descriptors, hydrogen suppressed molecular graph, and hydrogen filled graph. The correlation coefficient was 0.8036−0.8903 for MIC and 0.8171−0.8901 for MBC [33].

In the field of pharmaceutical applications of ionic liquids, their antibiotic capacity has been also evaluated, in order to combat bacterial resistance. In the work of Semenyuta et al. an important number of ionic liquids were tested against the biocide activity toward *Acinetobacter baumannii* (*A. baumannii*), and *S. aureus*. The dataset consisted of 1078 ionic liquids, of which 862 were used for the dataset and 216 for the test set [34].

Hodyna et al. described a QSAR formed by 1195 compounds against Hep-2 cell line. They obtained the values to construct the dataset from ChEMBL database. Molecules were modeled using ChemAxon and descriptors were obtained through DRAGON package obtaining topological, geometrical and constitutional properties, atom-centered fragments among others. The data analysis was carried out using Batch Pruning Algorithm (combination between

Associative Neural Network method, Self-Organized Map of Kohonen and descriptor selection methods). The correlation coefficient was 0.76 for training set and 0.75 for external test [35].

As traditional approaches to predict the toxicity of substances are often costly, Chipofya et al. decided to use computational methods. In their recently published article, a probabilistic model that is built thanks to deep kernel learning and that allows to determine the toxicity of ionic liquids in the IPC-81 cell line was presented. In their study, they used different tools such as RDKiet and Mol2vec; additionally, they have only used data on the chemical structure of ILs. The value of the correlation coefficient has been very good, 0.943 [36].

As mentioned before, there have also been studies related to environmental toxicity modeling and prediction. This is the case of the work of Mutalib et al. They published a review chapter where the influence of the changes in ionic liquids structure such as: cation and functional group substituent, length of alkyl chain and anion nature on the toxicity of these compounds in *Allivibrio fischeri* (*A. fiseheri*) was investigated. Additionally, they presented a comparative of several QSAR published in literature by different authors [1]. For instance, in this review, they presented the first published by Couling et al. [48]. They used 25 ionic liquids to develop a QSAR model in order to determine log EC50 values using four descriptors (E-state indices, surface change density, surface area, and shadow parameter), the R^2 obtained value was 0.78.

Luis et al. published two studies using QSAR models. In the first case, they use 43 ionic liquids and 9 molecular descriptors and, in the second one, 96 ionic liquids and 15 molecular descriptors to obtain the ecotoxicity (EC50). In both cases, they used multiple linear regression models with good regression statistics [37,49].

Bruzzone et al. developed a QSAR model based in descriptors of cations in order to predict the toxicity of ionic liquids in *A. fischeri*. They assumed that the play of the anions is not relevant for the toxicity. The structure of cations was optimized obtaining electronic, geometric and energetic parameters (gas phase and in water), and the CODESSA program was used to obtain constitutional, electrostatic, geometrical, topological, and quantum descriptors. These values were correlated to toxicity data (EC50). The values of correlation coefficients were 0.9017 and 0.9209 for gas phase and water, respectively [38].

Alvarez-Guerra et al. proposed a new form to predict the ecotoxicity using 148 ionic liquids, 94 molecular descriptors, and the partial least square-discriminant method. The obtained a R^2 was 0.929 [50]. After this article, Das and Roy created other ecotoxicity predictive model using 157 ionic liquids, 12 molecular descriptors such as constitutional, topological, information indices, atom type E indices, molecular properties, connectivity, extended topochemical atom (ETA), and DRAGON software [51].

Cho et al. presented a paper where they generate a QSAR model using 160 ionic liquids based on different cations. They developed this method to predict the hydrophobicity and cytotoxicity toward IPC-81 (Cell line), *A. fischeri* and *Scenedesmus vacuolatus* (*S. vacuolatus*). To elaborate the model several linear free energy relationship descriptors were used such as excess molar refraction, dipolarity/polarizability, hydrogen bonding acidity, hydrogen bonding basicity, and McGowan volume. The correlation coefficient was 0.987, 0.778, 0.762, and 0.776 for hydrophobicity, IPC-81, *A. fischeri* and *S. vacuolatus* respectively [39].

Roy et al. described two QSAR models for the prediction of toxicity toward *Daphnia magna* (*D. magna*). In the first case, they used 62 ionic liquids and ETA indices, non-ETA (constitutional, ring descriptors, topological, connectivity, information, edge adjacency, atom type E-state indices) descriptors and thermodynamic descriptors (molecular properties and atom-centered fragments). They used DRAGON and PaDEL descriptors software. In the second case, 49 ionic liquids were used for developing the QSAR model; in this case, lipophilicity was included. In both cases, good correlation coefficients were obtained [40,41].

Yang et al. used 157 ionic liquids and 28 molecular descriptors to develop a QSAR model for predicting the toxicity toward *A. fischeri*, corresponding to 74 different cations and 22 anions. The R^2 obtained was 0.908 [52].

Wang et al. used 24 bromide ionic liquids in order to develop a QSAR model for predicting environmental toxicity of both *A. fischeri* and *D. magna*. The descriptors used were unoccupied molecular orbitals, dipole moment, total energy, and molecular volume. The correlation coefficients were 0.954 and 0.895 for *A. fischeri* and *D. magna*, respectively [42].

Narayan Das et al. where toxicity information was collected for the marine bacterium *A. fischeri* [43]. This work collected a significant number of results, important, with a training test of 302 ionic liquids, while the test set was 305 ionic mixtures. In total 6 mathematical models were developed, with a number of independent variables that varied from 3 to 7. When interpreting the results, all the independent variables could be related to lipophilicity, hydrogen bonding propensity, branching, aromaticity, and electronegativity of the ions that make up the ionic liquid. According to its authors, the models obtained offered a moderate or good prediction capacity, with R^2 values ranging from 0.61 to 0.88.

Jafari et al. used 187 ionic liquids to develop a QSAR method, which explains the correlation between Log EC50 and the number of carbon, hydrogen, nitrogen, and oxygen atoms in cation and in anion. Phosphorous atoms were not included because R^2 was not good. Additionally, they improved their results by including of some parameters, which increase and decrease the toxicity of ionic liquids by structural moieties. They could observe that the length of alkyl chain present an important influence in the toxicity [44].

Yan et al. developed a new norm QSTR for predicting the ecotoxicity of ionic liquids on IPC-81 cell line. They use 173 values of EC50 grouped in 8 kinds of ionic liquids. Additionally, a new family of descriptors (atomic distribution matrices) are used in this work in order to analyze the interactions that are form between anions and cations. This model presents a R^2 value of 0.954 [45].

It can be concluded that most of the QSAR methods described lead to good representations of the studied behavior. However, the casuistry is very different in each case, and no patterns are observed nor can a guide or recommendations be established when selecting the strategy or even the type of descriptors that best represent the behavior.

Deep eutectic solvents

Unlike what happens with ionic liquids, the composition of DES must necessarily be considered when developing QSAR models. This greatly hinders the process, and, perhaps for this reason, this problem has rarely been addressed in depth. As far as we know, there are really few studies on the matter. Contrary to expectations, three of these studies focus on the prediction of physicochemical properties (density, surface tension, and electrical conductivity), while the other two focus on the biological activity produced by eutectic solvents (ecotoxicity and cytotoxicity, specifically). It is also noteworthy that these studies are all very recent; it was not until 2018 that the study of the development of QSAR methods for eutectic systems began. Next, we review these studies, analyzing both the definition of eutectic substances and the descriptors used to develop the problem. The most important characteristics of the process and the results obtained are also considered. Table 30.2 shows a summary of the review carried out.

The work of Ahmadi et al. [56] was the first attempt to construct a QSAR model for eutectic mixtures in 2018. In this case, the dependent variable was toe toxic effect in form of IC_{50} in human embryonic kidney cells. A total of 28 DES were used to construct a QSAR model. The components of the eutectic mixtures were described thorough some topological, constitutional, and molecular properties. Furthermore, DRAGON software was used to obtain walk and path counts descriptors. It is said that the composition of the mixtures was taken into account in the form of the molar ratio of HBA to HBD. However, how the composition of the components of the mixtures was taken in was not made explicit. The final QSAR model was constructed using 3 parameters: the interaction of second power of carbon numbers with their corresponding molar ratio, the rotatable bond number, and the mean atomic van der Waals volume, with R^2 0.76. The small number of DES in the dataset could be a limitation to this study.

In 2020, there was another study that explored the prediction of the (eco)toxicity of DES toward the environmental biomodel *A. fischeri* [12]. In this case, the composition of the DES was considered explicitly via several mixing rules. Independent variables were molecular descriptors that meet some criteria related to cover the molecular characteristics of the components of the mixtures and the potential relationship with the toxic effect (bioaccumulation, bioavailability of molecular transport across membranes). The dataset was made up of 42 eutectic mixtures, and the final model ($R^2 = 0.86$) was constructed on the basis of two independent variables: topological surface area and dissociation constant (pKa), both related to the biological activity produced by the chemicals. These variables were also affected by a mixing rule that modulated their contribution. In this case, although the sample size was not excessively high either, the results obtained were significantly better than in the previous study, mainly due to the fact that the composition of the mixtures was at the time of determining the different descriptors.

The strategies carried out to model physicochemical properties of DES by means of QSAR are reviewed below. In the study of Lemaoui et al. the electrical conductivity is predicted [55]. Two models were developed, taking into account the binary interaction and without taking it into consideration, finding good results in both cases, but significantly better in the first case (interactions were considered) ($R^2 = 0.99$ vs. 0.81).

In the work of Halder et al. a QSAR model was developed for predicting the density of DES [54]. In this case, an important number of DES were studied (a total of 145) that led to the creation of a dataset with 1154 data, covering different experimental conditions of density. Composition was included in the model via a mixing rule, and several models were proposed. There were some of them that gave very good results (R^2 near 0.95). In addition, it has developed an open access QSARMx tool, based on the so-called Python ecosystem, which is specifically used to model mixtures by means of QSAR methods.

TABLE 30.2 Summary and main characteristics of QSAR studies for DES.

DES components	QSAR strategy	Endpoint	Independent variables	Composition	Author	References
Multiple combinations	Modeling or description of components individually	Surface tension	Molecular descriptors. DRAGON and temperature, absence/presence of halide anions	Molar ratio of HBA to HBD	Halder et al.	[53]
Multiple combinations	Modeling or description of components individually	Density	Molecular descriptors. DRAGON and temperature, absence/presence of halide anions	Molar ratio of HBA to HBD	Halder et al.	[54]
HBA: Benzyltriphenyl phosphonium chloride (BTPPCl), (ii) methyltriphenyl phosphonium bromide, choline chloride HBD: N,N-diethylethanol ammonium chloride ethylene glycol, glycerol, 2,2,2-trifluoroacetamide	Modeling or description of components individually and forming part or the eutectic mixture	Electrical conductivity	Molecular descriptors	Molar ratio of HBA to HBD	Lemaoui et al.	[55]
Choliniumdihydrogenocitrate, choliniumbutanoate, choliniumpropanoate, choliniumacetate, choliniumsalicylate, choliniumbitartrate, cholinium dihydrogenophosphate HBA: Choline chloride HBD: Acetic acid, lactic acid, glycolic acid, citric acid, ethlene glycol, glycerol, propionic acid, 2-propanediol, urea, 1-propanol	Modeling or description of components individually	Toxicity toward *A. fischeri* EC50	Molecular descriptors and properties	Mixing rules	Giner et al.	[12]
HBA: Choline chloride HBD: Ethylene glycol, glycerol, 1,2-propanediol, 1,3-B butanediol, 1,4-butanediol, sorbitol, xylitol, xylosa, frucotse, glucose, galactose, mannose, glucosamine, sucrose, maltose, rhamnose	Modeling or description of components individually	Cytotoxicity human embryonic kidney (HEK-293) IC50	Constitutional, topological and molecular properties. DRAGON descriptors	Molar ratio of HBA to HBD	Ahmadi et al.	[56]

This same model was later used to predict the surface tension of DES by Halder et al. [53]. In this case, a large number of data was used (113 binary DES with more than 600 data points collected from different data bases). The best model was obtained with a total of 7 independent variables, including temperature and R^2 close to 0.92, which is very high.

From the information provided, it is highlighted that most studies follow similar procedures to describe eutectic systems: the components are modeled first, and then the construction of each one of them is modulated, including the molar composition by means of a mixing rule. However, the work of Giner et al. [12] showed that the choice of the mixing rule is not obvious and does not have to be linear. In fact, different mixing rules can even be proposed for certain descriptors, being, not inferable which rule should be applied in each case. Therefore, at this stage of development, it is an eminently empirical method. As expected, the results obtained for predictive models of physicochemical properties lead to better results, probably because the starting dataset was larger. In addition, for the development of these models, the temperature was taken into account; temperature normally correlates in an almost linear way with the density and surface tension properties, at least in not very wide stretches of experimental conditions [57].

Oher complex systems

As mentioned before, complex systems are defined as substances of unknown or variable composition. In this category, we can find a large number of substances, such as polymeric systems of nonnatural origin, with different chain lengths of the same type, colloidal systems or even mixtures of various isomers. If we pay attention to the latter case, substances formed by stereoisomers pose a great challenge for QSAR methods. Although there are many of them, the molecular descriptors that are traditionally used are usually not sensitive enough to differentiate chiral centers. The simplex representation of molecular structure (SiRMS) method is a widely used possibility to address this problem [58]. According to this theory, it is possible to describe molecular stereochemistry by making use of fragments (simplexes) of the appropriate size. The chiral centers can be represented by the SiRMS method through five simplexes in which each atom is assigned a canonical number determined by algorithms. This representation corresponds to the R or S configuration of the chiral center. In addition to reproducing chirality, through this method, it is possible to describe the surroundings of molecules, even detailing the ability to establish hydrogen bonds that show some functional groups.

Polymers are very complex and different materials that, generally, are scared studied from the toxicological point of view since on many occasions it has been considered that the toxicological properties are not relevant because of the size of these substances is high. However, this has changed in recent years. It is necessary to use models that allow predicting the most relevant properties of these substances with respect to their toxicity. It is noticeable that there are not many published QSAR or QSPR studies, around 50, and this is mainly due to the complexity of the chemical structures. The main problem when these studies are carrying out is how to encode the chemical structure since; in most cases, the polymer has a long and polydisperse chain as its common chain, which makes its characterization and subsequent translation in molecular graphics very complex. To do this, some articles use approximations using monomeric entities that are repeated [59].

Other models consist of the new 2D and 3D hologram that is based on QSAR and hologram quantitative structure–activity relationship (HQSAR) fragments since it can determine properties to determine their toxicity. HQSAR is based on the use of fingerprint fragments (holograms) as an activity. It works by identifying patterns of relevant substructural fragments for that substance to generate biological activity in sets of several bioactive molecules. Produce a predictive relationship between substructural features in the data set and biological activity using Partial least squares (PLS) [60].

Conclusions or perspectives or future directions

In this work, the most common strategies used to approach the modeling of complex substances or mixtures of substances using the QSAR method have been reviewed.

Most of the studies carried out have focused on the modeling of ionic liquids to predict their toxic behavior against various organisms in biological systems. A large number of studies have also been aimed at determining the bactericidal capacity of ionic liquids. The description of ions has traditionally been done using quantum molecular descriptors and various molecular properties. The ions are taken into account separately in most cases. However, it is possible to model the complete ion pair, showing that good results are obtained using this approach.

An interesting case study is the modeling of eutectic systems, where the composition of the mixture may not be mole to mole and therefore must be considered when describing the system. Only five models have been described and in all of them, mixing rules were used to obtain the contribution of each component. In addition, in this case, it is highlighted that the models have been used to predict physicochemical properties to the detriment of biological properties, probably due to the availability of experimentally determined data.

Finally, the strategies used to model other complex systems including isomeric or polymeric substances, among others, have been reviewed. In these cases, the approximations depend largely on the nature of the system.

Although the number of studies carried out on DES is not very high, the trend followed with ionic liquids allows us to conclude that it would be necessary to carry out a global evaluation of all the knowledge that has been obtained up to now.

There does not seem to be a method or strategy of choice for each case or type of complex system to be analyzed, and the development of new QSAR methods seems to be opportunistic. The selection of descriptors may be conditioned by availability criteria rather than by utility or prior rational design. For this reason, a global study, covering a wide range of cases, could offer some guidelines that would allow the selection of descriptors and/or mixing rules, if applicable, in a more rational and guided way.

Acknowledgments

The GREENLIFE research group acknowledges financial support from Gobierno de Aragón and Fondo Social Europeo "Construyendo Europa desde Aragón" B58_23D. Furthermore, we thank EEE53 SL and the business groups Pinares de Venecia División Energética and Brial (ENATICA) for their support. Both business groups are committed to sustainable developments through environmental respect.

References

[1] Mutalib MIA, Ben Ghanem O. Ecotoxicity of ionic liquids towards Vibrio fischeri: experimental and QSAR studies. In: Progress and developments in ionic liquids; 2017. p. 429–49.
[2] Anonymous. Report of the OECD workshop on quantitative structure activity relationships (QSARs) in aquatic effects assessment ENV. 1992.
[3] Eriksson L, Jonsson J, Berglind R. External validation of a QSAR for the acute toxicity of halogenated aliphatic-hydrocarbons. Environ Toxicol Chem 1993;12:1185–91.
[4] Li M, Wei D, Du Y. Genotoxicity of quinolone antibiotics in chlorination disinfection treatment: formation and QSAR simulation. Environ Sci Pollut Control Ser 2016;23:20637–45.
[5] Jiang L, Lin ZF, Hu XL, Yin DQ. Toxicity prediction of antibiotics on luminescent bacteria, Photobacterium phosphoreum, based on their quantitative structure-activity relationship models. Bull Environ Contam Toxicol 2010;85:550–5.
[6] Chen C, Wang J. Correlating metal ionic characteristics with biosorption capacity using QSAR model. Chemosphere 2007;69:1610–6.
[7] Pathakoti K, Huang M-J, Watts JD, He X, Hwang H-M. Using experimental data of Escherichia coli to develop a QSAR model for predicting the photo-induced cytotoxicity of metal oxide nanoparticles. J Photochem Photobiol B-Biol 2014;130:234–40.
[8] Devillers J. QSAR modeling of pesticide toxicity to bees. In: Silico bees; 2014. p. 135–52.
[9] Devillers J, Flatin J. A general QSAR model for predicting the acute toxicity of pesticides to Oncorhynchus mykiss. SAR QSAR Environ Res 2000;11:25–43.
[10] Singh KP, Gupta S. Nano-QSAR modeling for predicting biological activity of diverse nanomaterials. Rsc Adv 2014;4:13215–30.
[11] Burello E, Worth AP. QSAR modeling of nanomaterials. Wiley Interdiscip Rev-Nanomed Nanobiotechnol 2011;3:298–306.
[12] Giner B, Lafuente C, Lapena D, Errazquin D, Lomba L. QSAR study for predicting the ecotoxicity of NADES towards Aliivibrio fischeri. Exploring the use of mixing rules. Ecotoxicol Environ Saf 2020;191.
[13] Tokuda H, Hayamizu K, Ishii K, Abu Bin Hasan Susan M, Watanabe M. Physicochemical properties and structures of room temperature ionic liquids. 1. Variation of anionic species. J Phys Chem B 2004;108:16593–600.
[14] Yoo CG, Pu Y, Ragauskas AJ. Ionic liquids. Promising green solvents for lignocellulosic biomass utilization. Curr Opin Green Sustain Chem 2017;5:5–11.
[15] Kuroda K. A simple overview of toxicity of ionic liquids and designs of biocompatible ionic liquids. New J Chem 2022;46(46):20047–52.
[16] Paiva A, Craveiro R, Aroso I, Martins M, Reis RL, Duarte ARC. Natural deep eutectic solvents—solvents for the 21st century. ACS Sustain Chem Eng 2014;2:1063–71.
[17] Juneidi I, Hayyan M, Hashim MA. Evaluation of toxicity and biodegradability for cholinium-based deep eutectic solvents. RSC Adv 2015;5: 83636–47.
[18] Liu Y, Friesen JB, McAlpine JB, Lankin DC, Chen SN, Pauli GF. Natural deep eutectic solvents. Properties, applications, and perspectives. J Nat Prod 2018;81:679–90.
[19] de Morais P, Goncalves F, Coutinho JAP, Ventura SPM. Ecotoxicity of cholinium-based deep eutectic solvents. ACS Sustain Chem Eng 2015;3: 3398–404.
[20] Lapena D, Errazquin D, Lomba L, Lafuente C, Giner B. Ecotoxicity and biodegradability of pure and aqueous mixtures of deep eutectic solvents: glyceline, ethaline, and reline. Environ Sci Pollut Control Ser 2021;28:8812–21.

[21] Abbott AP, Capper G, Davies DL, Rasheed RK, Tambyrajah V. Novel solvent properties of choline chloride/urea mixtures. Chem Commun 2003:70–1.
[22] Choi YH, van Spronsen J, Dai YT, Verberne M, Hollmann F, Arends I, et al. Are natural deep eutectic solvents the missing link in understanding cellular metabolism and physiology? Plant Physiol 2011;156:1701–5.
[23] Anonymous. Application of structure-activity relationships to the estimation of properties important in exposure assessment ENV. 1993.
[24] Anonymous. Guidance on grouping of chemicals series on testing and assessment. 2007.
[25] Das RN, Roy K, Popelier PLA. Exploring simple, transparent, interpretable and predictive QSAR models for classification and quantitative prediction of rat toxicity of ionic liquids using OECD recommended guidelines. Chemosphere 2015;139:163–73.
[26] Peng D, Picchioni F. Prediction of toxicity of ionic liquids based on GC-COSMO method. J Hazard Mater 2020;398.
[27] Hodyna D, Kovalishyn V, Rogalsky S, Blagodatnyi V, Petko K, Metelytsia L. Antibacterial activity of imidazolium-based ionic liquids investigated by QSAR modeling and experimental studies. Chem Biol Drug Des 2016;88:422–33.
[28] Cho C-W, Park J-S, Stolte S, Yun Y-S. Modelling for antimicrobial activities of ionic liquids towards Escherichia coli, Staphylococcus aureus and Candida albicans using linear free energy relationship descriptors. J Hazard Mater 2016;311:168–75.
[29] Lu L-Y, Zhang Y-J, Chen J-J, Tong Z-H. Toxicity of selected imidazolium-based ionic liquids on Caenorhabditis elegans. A quantitative structure-activity relationship study. Chin J Chem Phys 2017;30:423–8.
[30] Zhu P, Kang X, Zhao Y, Latif U, Zhang H. Predicting the toxicity of ionic liquids toward acetylcholinesterase enzymes using novel QSAR models. Int J Mol Sci 2019;20.
[31] Zanoni BV, Romao GB, Andrade RS, Barretto Cicarelli RM, Trovatti E, Chiari-Andreo BG, et al. Cytotoxic effect of protic ionic liquids in HepG2 and HaCat human cells. In vitro and in silico studies. Toxicol Res 2019;8:447–58.
[32] Rybinska-Fryca A, Sosnowska A, Puzyn T. Representation of the structure-a key point of building QSAR/QSPR models for ionic liquids. Materials 2020;13.
[33] Lotfi S, Ahmadi S, Zohrabi P. QSAR modeling of toxicities of ionic liquids toward Staphylococcus aureus using SMILES and graph invariants. Struct Chem 2020;31:2257–70.
[34] Semenyuta IV, Trush MM, Kovalishyn VV, Rogalsky SP, Hodyna DM, Karpov P, et al. Structure-activity relationship modeling and experimental validation of the imidazolium and pyridinium based ionic liquids as potential antibacterials of MDR Acinetobacter baumannii and Staphylococcus aureus. Int J Mol Sci 2021;22.
[35] Hodyna D, Kovalishyn V, Semenyuta I, Rogalsky S, Trokhimenko O, Gryniukova A, et al. Ester-functionalized imidazolium- and pyridinium-based ionic liquids. Design, synthesis and cytotoxicity evaluation. Biointerface Res Appl Chem 2022;12:2905–57.
[36] Chipofya M, Tayara H, Chong KT. Deep probabilistic learning model for prediction of ionic liquids toxicity. Int J Mol Sci 2022;23.
[37] Luis P, Garea A, Irabien A. Quantitative structure-activity relationships (QSARs) to estimate ionic liquids ecotoxicity EC50 (Vibrio fischeri). J Mol Liq 2010;152:28–33.
[38] Bruzzone S, Chiappe C, Focardi SE, Pretti C, Renzi M. Theoretical descriptor for the correlation of aquatic toxicity of ionic liquids by quantitative structure-toxicity relationships. Chem Eng J 2011;175:17–23.
[39] Cho CW, Ranke J, Arning J, Thoeming J, Preiss U, Jungnickel C, et al. In silico modelling for predicting the cationic hydrophobicity and cytotoxicity of ionic liquids towards the Leukemia rat cell line, Vibrio fischeri and Scenedesmus vacuolatus based on molecular interaction potentials of ions. SAR QSAR Environ Res 2013;24:863–82.
[40] Roy K, Das RN. QSTR with extended topochemical atom (ETA) indices. 16. Development of predictive classification and regression models for toxicity of ionic liquids towards Daphnia magna. J Hazard Mater 2013;254:166–78.
[41] Roy K, Das RN, Popelier PLA. Quantitative structure-activity relationship for toxicity of ionic liquids to Daphnia magna. Aromaticity vs. lipophilicity. Chemosphere 2014;112:120–7.
[42] Wang C, Wei Z, Wang L, Sun P, Wang Z. Assessment of bromide-based ionic liquid toxicity toward aquatic organisms and QSAR analysis. Ecotoxicol Environ Saf 2015;115:112–8.
[43] Das RN, Sintra TE, Coutinho JAP, Ventura SPM, Roy K, Popelier PLA. Development of predictive QSAR models for Vibrio fischeri toxicity of ionic liquids and their true external and experimental validation tests. Toxicol Res 2016;5:1388–99.
[44] Jafari M, Keshavarz MH, Salek H. A simple method for assessing chemical toxicity of ionic liquids on Vibrio fischeri through the structure of cations with specific anions. Ecotoxicol Environ Saf 2019;182.
[45] Yan F, Lan T, Yan X, Jia Q, Wang Q. Norm index-based QSTR model to predict the eco-toxicity of ionic liquids towards Leukemia rat cell line. Chemosphere 2019;234:116–22.
[46] Trush MM, Kovalishyn V, Ocheretniuk AD, Kalashnikova LE, Prokopenko VM, Holovchenko OV, et al. New 1,3-oxazolylphosphonium salts as potential biocides. QSAR study, synthesis, antibacterial activity and toxicity evaluation. Lett Drug Des Discov 2018;15:1259–67.
[47] Sosnowska A, Rybinska-Fryca A, Barycki M, Jagiello K, Puzyn T. Chemoinformatic approach to assess toxicity of ionic liquids. Comput Toxicol Methods Protoc 2018;1800:559–71.
[48] Couling DJ, Bernot RJ, Docherty KM, Dixon JK, Maginn EJ. Assessing the factors responsible for ionic liquid toxicity to aquatic organisms via quantitative structure-property relationship modeling. Green Chem 2006;8:82–90.
[49] Luis P, Ortiz I, Aldaco R, Irabien A. A novel group contribution method in the development of a QSAR for predicting the toxicity (Vibrio fischeri EC50) of ionic liquids. Ecotoxicol Environ Saf 2007;67:423–9.
[50] Alvarez-Guerra M, Irabien A. Design of ionic liquids. an ecotoxicity (Vibrio fischeri) discrimination approach. Green Chem 2011;13:1507–16.
[51] Das RN, Roy K. Development of classification and regression models for Vibrio fischeri toxicity of ionic liquids. Green solvents for the future. Toxicol Res 2012;1:186–95.
[52] Yan F, Shang Q, Xia S, Wang Q, Ma P. Topological study on the toxicity of ionic liquids on Vibrio fischeri by the quantitative structure-activity relationship method. J Hazard Mater 2015;286:410–5.
[53] Halder AK, Haghbakhsh R, Voroshylova IV, Duarte ARC, Cordeiro M. Predicting the surface tension of deep eutectic solvents. A step forward in the use of greener solvents. Molecules 2022;27.
[54] Halder AK, Haghbakhsh R, Voroshylova IV, Duarte ARC, Cordeiro M. Density of deep eutectic solvents. The path forward cheminformatics-driven reliable predictions for mixtures. Molecules 2021;26.

[55] Lemaoui T, Darwish AS, Hammoudi NE, Abu Hatab F, Attoui A, Alnashef IM, Benguerba Y. Prediction of electrical conductivity of deep eutectic solvents using COSMO-RS sigma profiles as molecular descriptors. A quantitative structure-property relationship study. Ind Eng Chem Res 2020;59:13343–54.
[56] Ahmadi R, Hemmateenejad B, Safavi A, Shojaeifard Z, Mohabbati M, Firuzi O. Assessment of cytotoxicity of choline chloride-based natural deep eutectic solvents against human HEK-293 cells. A QSAR analysis. Chemosphere 2018;209:831–8.
[57] Lapena D, Bergua F, Lomba L, Giner B, Lafuente C. A comprehensive study of the thermophysical properties of reline and hydrated reline. J Mol Liq 2020;303.
[58] Kuz'min V, Artemenko A, Ognichenko L, Hromov A, Kosinskaya A, Stelmakh S, et al. Simplex representation of molecular structure as universal QSAR/QSPR tool. Struct Chem 2021;32:1365–92.
[59] Rasulev B, Casanola-Martin G. QSAR/QSPR in polymers. Recent developments in property modeling. Int J Quant Struct-Prop Relatsh 2021.
[60] Sanderson H, Khan K, Hansen AMB, Connors K, Lam MW, Roy K, Belanger S. Environmental toxicity (Q)SARs for polymers as an emerging class of materials in regulatory frameworks, with a focus on challenges and possibilities regarding cationic polymers. Ecotoxicol QSARS 2020: 681–705.

C H A P T E R

31

Deploying QSAR to discriminate excess toxicity and identify the toxic mode of action of organic pollutants to aquatic organisms

Limin Su[1], Minghui He[1], Jiao Qu[1], Bingxin Gui[1], Jinjie Li[1,2], Rebecca Kusko[3], Huixiao Hong[4] and Yuanhui Zhao[1]

[1]Key Laboratory for Wetland Ecology and Vegetation Restoration of National Environmental Protection, School of Environment, Northeast Normal University, Changchun, Jilin, PR China [2]College of Marine Ecology and Environment, Shanghai Ocean University, Shanghai, PR China [3]Immuneering Corp, Cambridge, MA, United States [4]Division of Bioinformatics and Biostatistics, National Center for Toxicological Research, United States Food and Drug Administration, Jefferson, AR, United States

Abbreviations

AAR Average absolute residual
AR Average residual
BCF Bioconcentration factor
BCR Bioconcentration ratio
CBRs Critical body residuals
DM *Daphnia magna*
EC_{50} 50% effective concentration
F Fisher's criterion
IBC_{50} 50% bioluminescence inhibition concentration
IC_{50} 50% immobilization concentrations
IGC_{50} 50% growth inhibition concentration
LC_{50} 50% lethal concentration
MOA Mode of action
N Number of compounds
QSAR Quantitative structure–activity relationship
R^2 Coefficient of determination
S Standard error of the estimate
TP *Tetrahymena pyriformis*
TR Toxic ratio
VF *Vibrio fischeri*

Introduction

Scientific and technological developments have ushered in global industrialization while concomitantly promoting the progress of human society. The primary global industry has gradually transitioned from a traditional agricultural society to the modern industrial society, with science and technology as the driving productive force. To meet the driving imperatives of society, a large number of compounds have been synthesized with industrial chemical processes and other means. Some of these synthetic compounds eventually enter the environment in manifold ways. The American Chemical Society has registered more than 200 million chemicals as of 2022 [1], with the proportion of synthetic organic compounds being relatively large. Many synthetic compounds are toxic to humans and animals, with some substances displaying mutagenicity in lower organisms. Such chemicals being spread and amplified through the food chain thus directly or indirectly endanger human health as well as animal and plant safety. As a very important link in ecological risk assessment, clarifying the toxic effect(s) and mode(s) of action (MOA) of chemical compounds has increasingly attracted global attention. At present, many investigations have been performed on the toxicity of organic compounds to aquatic organisms. Studies have detailed that compound toxicity is mainly related to the following two factors. One is the ability of organic matter to be enriched in an

https://doi.org/10.1016/B978-0-443-15339-6.00017-5

organism through bioconcentration from the water phase, and the other is related to the ability of organic matter to interact with a given target in the organism [2]. Therefore, the bioconcentration and MOA of compound toxicity should be considered comprehensively. Accurate determination of compound toxicity mechanisms can help us establish toxicity prediction models based on different toxicity mechanisms of compounds, thus reducing the number of toxicity animal experiments [3]. Understanding a compound's toxic MOA is a prerequisite for organic pollutant safety evaluation. Around the globe, studies are underway to examine the mechanisms of toxicity of different types of organic pollutants on sensitive organisms and to establish methods for evaluating organic pollutant toxicity.

Mechanisms of toxicity of organic pollutants to aquatic organisms

Pinpointing toxicity mechanisms of chemicals helps the research community to further examine the toxic effects of organic pollutants on organisms and is the basis and premise of establishing quantitative structure–activity relationship (QSAR) models of toxicological effects. Based on the structures of compounds, Verhaar classified compounds into four toxic groups [4]: inert compounds that bring anesthetic toxic effects to organisms, weak inert compounds that bring polar anesthetic toxic effects to organisms, reactive compounds, and special reactive compounds that bring larger toxic effects to organisms.

Baseline compounds

Baseline compounds, or "nonpolar narcotics," are defined as chemicals that do not interact with specific receptors. Typically, such compounds are inert anesthesia, producing biological enrichment via noncovalent interaction in the cell membrane, and to a certain extent changing the structure and function of the cell membrane. This enriches organic matter, thus causing reversible biological toxicity [5]. All compounds have the ability to enter the organism, with the effect termed the background or baseline toxicity of the compound. The toxicity beyond the anesthetic toxicity is called residual toxicity. The toxicity of baseline compounds depends on their ability to enter the organism, so there is a good correlation between baseline toxicity and the octanol/water allocation coefficient [6–10].

Less inert compounds

Compounds that are not reactive but slightly more toxic than the hydrophobicity predicted baseline compounds are considered less inert compounds or "polar narcotics." Such compounds, such as phenols and anilines, typically possess hydrogen bond donor acidity. Less inert compounds also do not react with target receptors in the organism and do not display narcotic toxicity. However, unlike baseline compounds, the toxic effects of these compounds are more pronounced. Physiologically, it is not possible to distinguish between the two types of compounds, but if considering the molecular structure of the compounds, narcotic compounds do not have hydrogen bond donors, while polar narcotic compounds have hydrogen bond donors. Although the toxicity and hydrophobicity of both baseline and less inert compounds have a good correlation, there are certain differences in the toxicity mechanisms of these two types of compounds. Thus, QSAR modeling of the toxicity effect and hydrophobicity parameters also has certain differences, evident in slopes and intercepts [8].

Reactive compounds

Reactive compounds have special structures and properties that allow biochemical reactions with target sites in organisms (such as proteins, enzymes, peptides, and nucleophilic groups in nucleic acids). The toxic effects of such reactions are large, and commonly, reactions are electrophilic or nucleophilic, redox, or radical reactions between pollutants and biological macromolecules [11].

Special reactive compounds

Special reactive compounds include a variety of compounds such as dithiocarbates and pyrethroids. They have specific and complex interactions with receptor molecules after reaching their target through bioenrichment, resulting in greater toxic effects. For example, organophosphate inhibits the production of acetylcholinesterase, while DDT regulates sodium channels in ganglia to produce special toxicity or receptor toxicity [12].

Methods to distinguish the toxic action of organic compounds

A nonreactive mechanism is the noncovalent interaction between organic matter and the cell membrane, which reversibly changes the structure and function of the membrane and results in relatively low anesthetic toxicity. Importantly, this MOA is not the toxic mechanism of biochemical reactions between compounds and various targets in the organism.

The MOA is the toxic mechanism of biochemical reactions between organic compounds and various targets of organisms. To distinguish reactive and nonreactive compounds, there are two commonly used methods: structural warnings and toxicity ratios (TRs).

Structure warning method

Based on the toxicity data from commonly used aquatic model organisms such as bacteria, luminescent bacteria, pear-shaped tetrahymena, green algae, fish, and *Daphnia magna* (DM), many structural characteristics have been postulated to predict covalent binding to biological proteins in organisms. Hermens [13], Verhhar [4], Nendza Schultz [14,15], Von der Ohe [16], and Enoch [17] identified characteristics on the electrophilic structures of direct or indirect acting compounds, implicating the reactive mechanisms of residual toxicity.

TR method

In addition to the structural warning method, the concept of TRs is now widely employed to discriminate between reactive and nonreactive compounds.

TR refers to the ratio of baseline toxicity to the measured toxicity of a compound, as predicted by a QSAR model based on baseline toxicity [4,16,18,19]. The difference between the measured toxicity and the baseline toxicity is the excess toxicity of the compound. TR is used to describe excess toxicity and can be calculated using Eqs. (31.1) and (31.2):

$$TR = \frac{T_{pred\ (baseline)}}{T_{exp}} \tag{31.1}$$

$$\log(TR) = \log\frac{1}{T_{exp}} - \log\frac{1}{T_{pred\ (baseline)}} = \text{Residual} \tag{31.2}$$

where T_{pre}(baseline) is the toxicity value predicted by the QSAR model for baseline compounds, and T_{exp} is the experimental toxicity value of a compound. Using Eq. (31.2), TR can be converted into logarithmic space. Following this method, compounds can be distinguished as baseline or reactive/specifically acting toxicants by applying a $\log(TR) = 1$ cutoff. More details are in Table 31.1.

Factors affecting the classification of toxicity mechanisms of organic compounds

It is well known that different biological organisms have different adaptations or sensitivities to a particular chemical. Therefore, the differential sensitivity of a species to organic compounds affects the discrimination of toxic

TABLE 31.1 Interpretation of TR value [20,21].

Compound TR value	Interpretation
0–1	Toxicity of compound is baseline mode of toxic action
1–10	Nonreactive compound
>10	Significant excess toxicity from a reactive compound (chemical reaction occurred between organic matter and a biological target)
<10	Acting by a nonpolar narcosis mechanism
≥10	Toxicity >10× baseline toxicity, reactive, and MOA may drive excess toxicity

mechanisms of the organic pollutants. In addition, the bioconcentration and ionization rate of organic compounds affect the identification of toxicity mechanisms.

Species specific sensitivity

The toxicity of organic pollutants to organisms is a complex process. The difference in physiological structures and toxic targets in different organisms leads to different species sensitivities to various organic compounds. Therefore, investigating the sensitivity of organic pollutants in different organisms and the discrimination criteria of organic pollutant toxicity mechanisms helps the scientific community to have a more sophisticated understanding of organic pollutant toxicity mechanisms.

The sensitivity of species to compounds is influenced by the physiological structure of the organism, toxicity test standard and exposure time, toxicity effect parameters, and the sensitivity of organisms in different life cycles [22,23]. Different biological organisms have different adaptations or sensitivities to a chemical. Studies on the sensitivity of different types of organic pollutants found that the sensitivity of some types of organic matter was higher in *green algae* than in fish, DM, and other organisms [24], while toxicity sensitivity of some types of organic matter was higher in DM than in *green algae* and fish [25]. The toxicity sensitivity investigations on aniline and nitrile compounds to various aquatic organisms showed that sensitivity to aniline compounds was significantly different among aquatic organisms: DM and luminescent bacteria were more sensitive than *Tetrahymena pyriformis* (TP). In the contrast, the toxicity sensitivity to nitrile compounds was higher for fish and TP than for DM and luminaria [26,27]. Taken together, these results indicate that compounds with different properties may have substantial differences in sensitivity across organisms. On the one hand, the difference of physiological characteristics between different species leads to different biological enrichment abilities of organisms, indicating the toxicity mechanisms of organic matter for different organisms are not the same. Therefore, the reliability of results would be a question if the TR values of organic pollutants to fish are used to study the toxicity mechanisms of organic pollutants to other organisms with different sensitivity (Fig. 31.1).

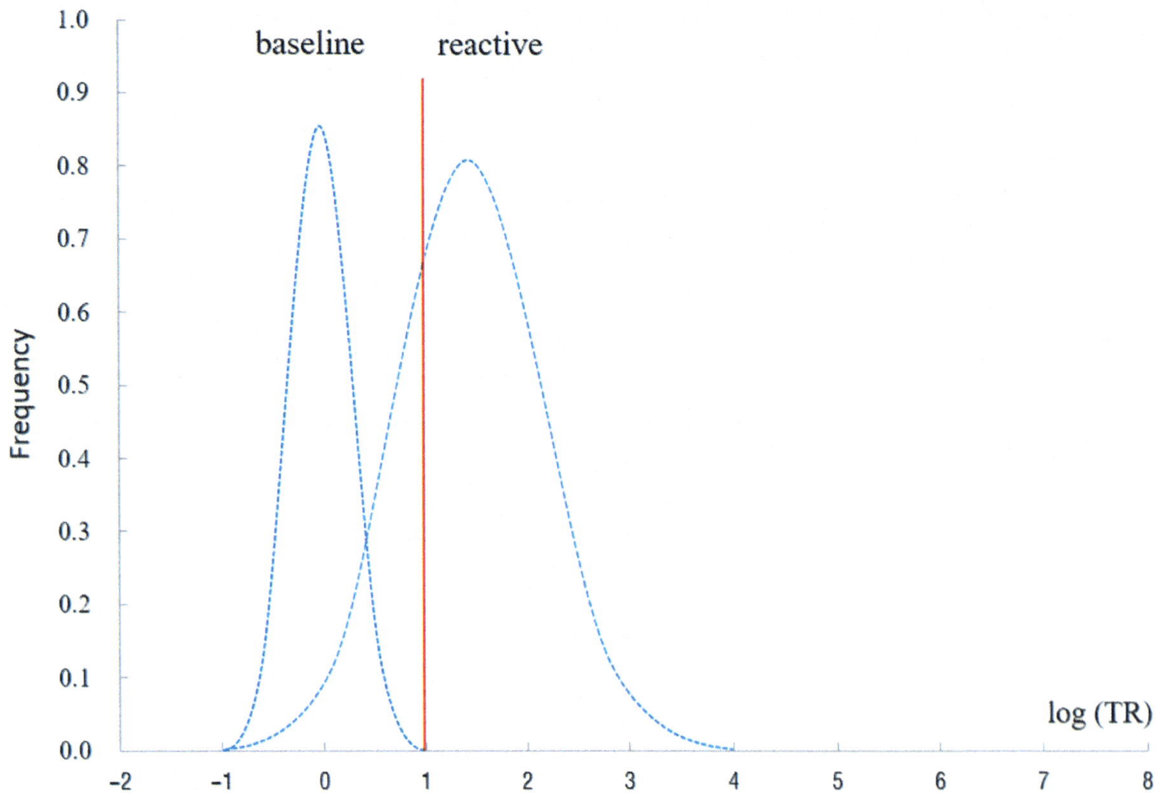

FIGURE 31.1 Normal distributions of log(TR) for baseline (*left*) and reactive (*right*) compounds for fish.

Bioconcentration

Bioconcentration is a phenomenon in which organic compounds accumulate from their surroundings (water, soil, and atmosphere) by nonphagocytic means so that their concentration in an organism exceeds the concentration in the surroundings. The species commonly employed for aquatic toxicity tests include *green algae*, DM, fish, TP, and VF. Fish are a higher aquatic organism with complex organelles and lipid-rich skin, DM are aquatic organisms, TP are single-celled eukaryotes, *green algae* are eukaryotes, and luminescent bacteria are prokaryotes. The physiological structures and bioconcentration ability are different among these organisms. The toxicity of benzoic acid compounds to fish, DM, and luminescent bacteria was found to be significantly different among various aquatic organisms, with luminescent bacteria displaying more toxicity and DM and fish displaying less toxicity. The correlation between the toxicity of each species was significantly increased by the introduction of $\log K_{ow}$ and E_{lumo}, which characterize the enrichment ability and reactivity with receptors of organic organisms, respectively [21]. The results suggested that the concomitant ability of the organism to bioconcentrate the compound and the mode of toxic interaction between the organism and the receptor together influence the toxic effects of the compounds. Therefore, studying the bioconcentration abilities of organic pollutants and clarifying the migration and transformation abilities of substances in the ecosystem are helpful to evaluate and distinguish the toxic mechanisms of compounds after entering the body.

Ionization

For ionizable organic pollutants, ionization rate is an important factor affecting their biological toxicity. Ionized organic pollutants such as weak organic acids and bases exist as nonionic and ionic forms in water, which have different toxicity to aquatic organisms. Moreover, the absorption rates of the two forms of ionizable compounds by aquatic organisms are distinct. Compared with the ionic state, the nonionic state has a higher absorption rate and is more toxic. Therefore, the ionization rate of organic pollutants influences compound aquatic toxicity, evidenced by the pH variation. For organic compounds with a higher pH, acidic compounds have higher ionization rates and lower toxicity, while alkaline organics have higher ionization rates and lower toxicity [28].

Toxic MOA of organic pollutants in species with different sensitivities based on TR: a case study

One of the most widely used methods in toxicity mechanism identification is TR ($\log(TR) > 1$ indicates excess toxicity), which has been thoroughly applied to investigate the toxic mechanisms of organic compounds on fish. However, there is a large deviation when TR is applied to study the toxic effects of organic compounds for other sensitive species. A total of 4995 toxicity datapoints for 3363 chemicals in fish (50% lethal concentration, LC_{50}), DM (50% immobilization concentrations, IC_{50}), TP (50% growth inhibition concentration, IGC_{50}), and *Vibrio fischeri* (VF) (50% bioluminescence inhibition concentration, IBC_{50}) were leveraged to investigate the effect of species sensitivity and bioconcentration on excess toxicity.

Interspecies correlations for toxicity spanning four different species

Toxicity relationships across species can be leveraged to further understand toxic MOA for the identical compounds [29]. Thus, a linear regression analysis interrogating toxicity data for identical compounds across fish, DM, VF and TP was executed.

Cross-species relationships measured by a coefficient of determination (R^2) are reported in Table 31.2. The impressive $R^2 = 0.72$ (models 1 and 2 in Table 31.2) indicated statistically meaningful correlations between fish, DM, and TP toxicity data. However, correlations for other species pairs are low with R^2 values ranging from 0.45 to 0.63 (models 3–6 in Table 31.2). The results suggested the striking differences between organic chemical toxicity in these four species.

Both the toxicant transport to the target site(s) and toxicant–target interactions can drive toxicity [31]. Comparing fish, DM, and TP toxicity data indicated that a substantial amount of compounds demonstrated correlation of biouptake and toxic metabolism. Thus, compounds could elicit identical MOA. Low correlations indicated that select compounds may have distinct toxicokinetics or MOAs. The cross-species differences could be due to metabolism differences (e.g., esters are metabolized to reactive compounds in fish but not in TP [32]). Alternatively, different species could drive different physiology, such as bacterium being more sensitive to antibiotics than fish. Notably, outliers (>1) have been noted within these correlations across species; clearly, in these cases, a compound's toxic effect was species-dependent [21,33].

TABLE 31.2 Interspecies relationships of toxicity for four species [30].

No.	Species A–B	Interspecies correlation	N	R^2
1	Fish–DM	$Log(1/IC_{50}) = 0.891 \log (1/LC_{50}) + 0.659$	467	0.72
2	Fish–TP	$Log(1/IGC_{50}) = 0.752 \log (1/LC_{50}) + 0.326$	478	0.72
3	Fish–VF	$Log(1/IBC_{50}) = 0.771 \log (1/LC_{50}) + 0.790$	304	0.54
4	DM–TP	$Log(1/IGC_{50}) = 0.703 \log (1/IC_{50}) + 0.361$	287	0.63
5	DM–VF	$Log(1/IBC_{50}) = 0.611 \log (1/IC_{50}) + 1.272$	294	0.45
6	TP–VF	$Log(1/IBC_{50}) = 0.947 \log (1/IGC_{50}) + 0.828$	556	0.63

N, number of compounds; R^2, coefficient of determination.

Species sensitivity

Table 31.2 lists the regression Eqs. (31.1–31.6) for the overall toxicities of compounds between species. Some slopes were significantly greater or less than one (e.g., <0.7 for Eq. 31.5), and several intercepts were greater or less than zero (e.g., >0.80 for Eqs. 31.5 and 31.6), pointing to the likely fact that fish, DM, TP, and VF may not have the same sensitivities to organic compounds. To further interrogate the sensitivity between different species, the average residuals (ARs) between compound toxicity values with relevant species data were tabulated (Table 31.3). Importantly, AR represents the difference between overall sensitivity across two species. DM was the most sensitive, and TP was the least sensitive of the four species, according to the AR values in Table 31.3. DM was the most sensitive, followed by fish, VF, and then TP. Higher AR values were associated with greater sensitivity differences between two species being compared. TP sensitivity was very different across the other three species (fish–TP =0 .64; DM–TP = 0.86; TP–VF = −0.65).

These data indicated that no species is sensitive across all tested compounds. Notably, species may be highly sensitive to one compound group but less sensitive to other compound groups [34–36]. Examining the toxicity data of classified compounds found that for secondary amino alcohols, anilines (N-alkyl) with fluoro and chloro groups, tertiary amino alcohols, anilides with hydroxy, and amino groups average $\log 1/EC_{50} > \log(1/LC_{50})$, but $<\log(1/LC_{50})$ for aldehydes and esters. Overall toxicity sensitivity of DM is greater than that of fish as more classes elicited higher toxicity in DM. A comparison of toxicities among other species produced similar findings.

To further investigate the toxicity classes (baseline, less inert, and reactive compounds) of classified compounds between four species, a statistical analysis was executed. The AR in toxicities across four species was calculated and shown in Fig. 31.2. Taken together, the AR values demonstrated similar toxicity sensitivity of DM, fish, and VF to baseline compounds. The ARs were 0.10 for DM–VF, −0.05 for fish–DM, and 0.13 for fish–VF. A similar trend was noted with DM, fish, and VF toxicity sensitivity for less inert compounds (fish–VF AR = −0.03; DM–VF AR = 0.04; fish–DM AR = −0.11). Importantly, TP sensitivity was very distinct from the three other species. Fish, DM, and VF are more sensitive for less inert and baseline compounds than TP. Reactive compound AR values

TABLE 31.3 Sensitivity of four species.

No.	Species A–B	AR	Sensitivity
1	Fish–DM	−0.19	Fish < DM
2	Fish–TP	0.64	Fish > TP
3	Fish–VF	0.11	Fish > VF
4	DM–TP	0.86	DM > TP
5	DM–VF	0.37	DM > VF
6	TP–VF	−0.65	VF > TP

$AR = \sum$ (Toxicity in species A − Toxicity in species B)/N.

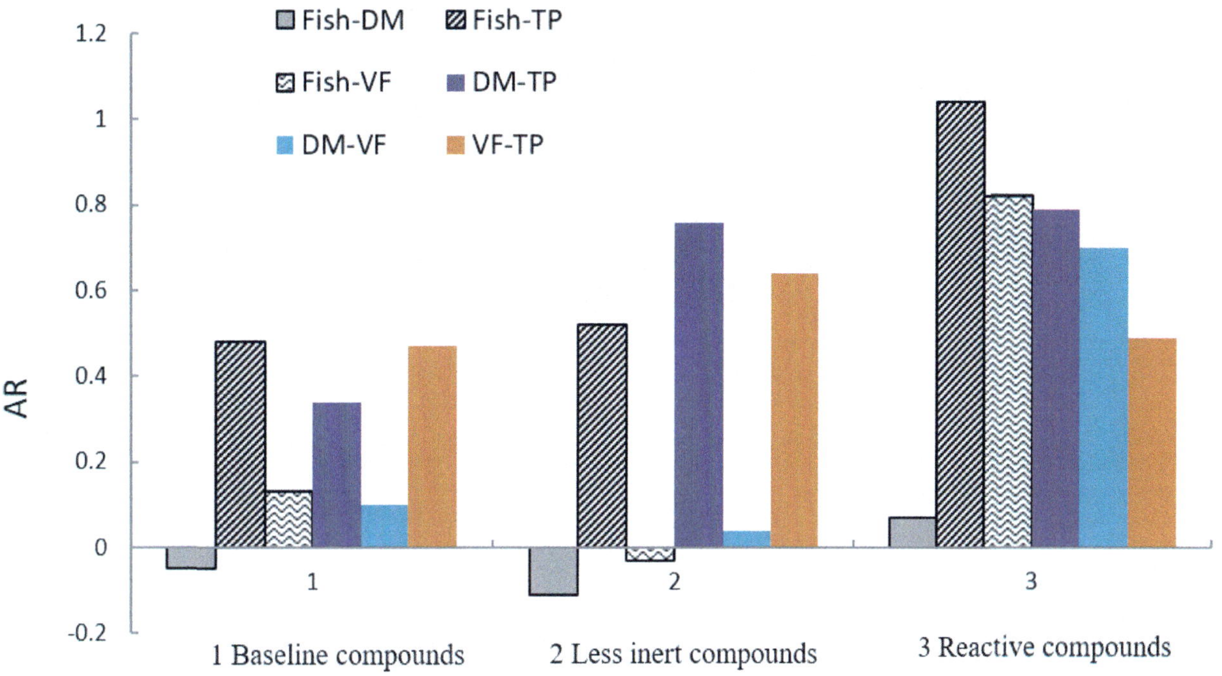

FIGURE 31.2 Average residuals between toxicities of any two of four species for baseline, less inert, and reactive compounds.

revealed that fish has similar toxicity sensitivity as DM (fish − DM = 0.07). Meaningful differences between toxicity sensitivity were noted for TP and VF versus DM or fish (Fig. 31.2). The toxicity sensitivity differences of a chemical between various organisms reflected specificity differences with the MOA, transport, or metabolic transformation site(s) [29,37,38].

Identification of toxic organic pollutant MOAs for differentially sensitive species based on TRs

The excess toxicity seen with fish, threshold log(TR) = 1, was based on the log(TR) normal distribution for reactive, less inert, and baseline compounds [4,21]. In principle, the thresholds of excess toxicity for DM, VF, and TP should be determined from the baseline, less inert, reactive, and specifically acting compound log(TR) distribution. Compounds pinpointed in fish as reactive or specifically acting, baseline, and less inert compounds may not be mirrored in other species. This is especially true for reactive or specifically acting compounds. Baseline compounds with high certainty of nonpolar narcotic action were selected and compared to obtain comparable thresholds for different species. These compounds include alcohols, alkanes, ethers, ketones, and benzenes with fluorine, alkyl, and chlorine groups. Select compounds pinpointed following the Verhaar scheme as baseline compounds but with log $K_{OW} > 7$ or <0 were not leveraged to develop the baseline model. The baseline models for VF, DM, and fish have similar slopes and intercepts (Table 31.4), indicating that the species sensitivity and MOAs are similar for the baseline compounds. This result is bolstered by AR analysis comparing baseline compound toxicity values across species (Fig. 31.1). Thus for DM and VF, the log(TR) = 1 threshold based on toxicity data distribution to fish can discriminate baseline narcotics from reactive or specifically acting compounds. Nearly all of the fish toxicity baseline compounds are DM and VF toxicity baseline compounds. Small differences in accuracy of prediction could be driven by experimental uncertainty in endpoints obtained from different species.

The baseline model's equation coefficient for TP was strikingly smaller than that for DM, FV, and fish (Table 31.4). This difference in TP sensitivity was observed comparing to VF, DM, and fish (Fig. 31.3). Thus, the log(TR) = 1 threshold obtained from fish toxicity is far from ideal for identifying excess toxicity to TP, and the threshold should be log(TR) < 1. To obtain a cutoff to separate narcotic level compounds from TP excess toxicity, the average absolute residual (AAR) for TP and fish was calculated for baseline compounds as the differences between observed and predicted toxicities. If minimal consideration is paid to toxicity endpoint experimental error for different species, the AAR value, reflecting the deviation from the baseline models, should have a positive relationship with sensitivity. Species with higher sensitivity should have more errors in fitting the baseline model. In this system, AAR should be tightly related to the log(TR) threshold between species and is expressed in Eq. (31.3):

TABLE 31.4 Baseline models for four species [39].

No.	Species	Models for baseline compounds	N	R^2	S	F
1	Fish	$\log (1/LC_{50}) = 0.858 \log K_{OW} + 1.22$	94	0.93	0.31	1268
2	DM	$\log 1/IC_{50} = 0.946 \log K_{OW} + 1.06$	63	0.90	0.42	543
3	TP	$\log 1/IGC_{50} = 0.775 \log K_{OW} + 0.96$	100	0.93	0.30	1369
4	VF	$\log 1/IBC_{50} = 0.995 \log K_{OW} + 0.92$	75	0.87	0.48	501

F, Fisher's criterion; *S*, standard error of the estimate.

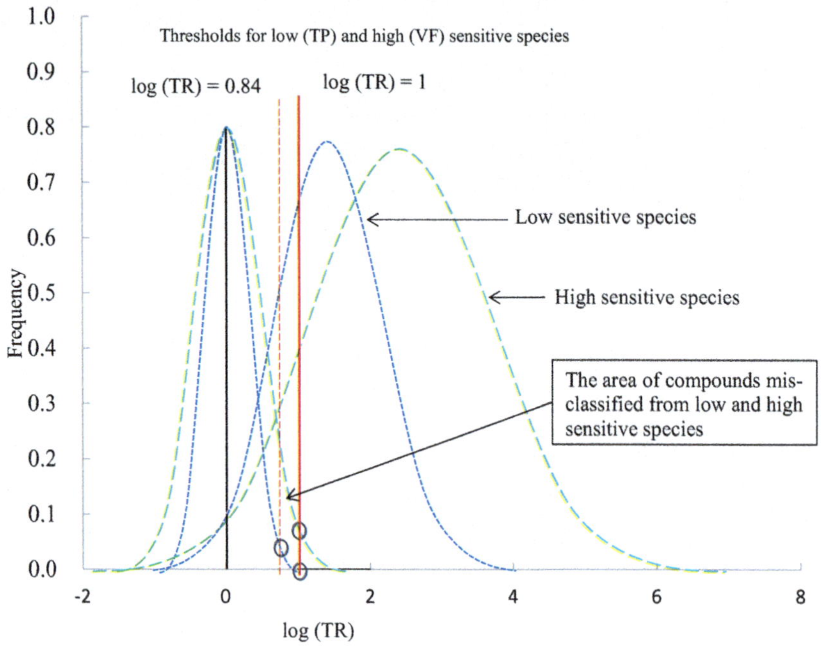

FIGURE 31.3 Log(TR) normal distributions for reactive (*right*) and baseline (*left*) compounds for two select species with varying toxicity sensitivity [30].

$$\frac{\log (TR_F)(=1.00)}{AAR_F(=0.25)} = \frac{\log(TR_{TP})(=0.84)}{AAR_{TP}(=0.21)} \tag{31.3}$$

where AAR_F (0.25) is AAR for fish, AAR_{TP} (0.21) represents AAR for TP, and $\log(TR_F)$ (1.00) indicates the threshold for fish. Therefore, the threshold for TP ($\log(TR_{TP})$) is calculated to be 0.84 by Eq. (31.3). Although this threshold was derived from fish data, the data points can be confirmed based on classification accuracy for these thresholds. Fig. 31.1 shows that the $\log(TR) = 0.84$ threshold can provide similar classification accuracy with $\log(TR) = 1$ for the TP toxicity of baseline compounds, which is almost identical to the $\log(TR) = 1$ fish toxicity threshold, suggesting that the TP $\log(TR) = 0.84$ threshold is comparable with the fish threshold for discriminating baseline compounds from reactive or specifically acting compounds. Using $\log(TR) = 0.84$, 142/760 (18%) more compounds were pinpointed as having potential excess toxicity than using $\log(TR) = 1$. For some classes, this has a striking influence. For example, of 31 substituted benzaldehyes, 6 were identified as having excess toxicity with $\log(TR) = 0.84$ as compared with 2 with $\log(TR) = 1$. Of note, not all specifically acting or reactive compounds had $\log(TR) > 1$. The interaction of reactive or specifically acting compounds with biological macromolecules may not have much more toxic contribution than that of baseline compounds, resulting in a log(TR) less than one [21,30]. Thus, not all reactive compounds exhibit excess toxicity with log(TR) > 1.

Toxic MOA identified using the same threshold

It is more difficult to compare differences in toxic mechanisms, despite the fact that analysis of toxicity sensitivity interspecies relationships suggests that there are differences among these four species. TR is a very handy tool to separate baseline toxicity from reactive or specifically acting compounds [16,19,20,40]. To calculate TR for each species in question, narcotic compounds need to be leveraged to develop baseline models. Interestingly, for baseline compounds for many species, log K_{OW} is linearly related to toxicity [9,41,42]. For fish, DM, TP, and VF, linear regression models have been established between log K_{OW} and the toxicity data. Importantly, the compounds deployed as input to baseline models were selected thoughtfully. Only compounds with $0 < \log K_{OW} < 7$ that were widely recognized as baseline (i.e., alcohols, ketones, alkanes, ethers, benzenes and their fluorine chlorine, and alkyl derivatives) were incorporated into the baseline models. Of note, compounds such as esters, alkenes, cycloalkanes, aliphatic acids, and their derivatives were not used to develop the baseline models. Compounds with residuals greater than 1 or less than −1 were also excluded from the development of baseline models.

The log(TR) ≥ 1 threshold indicated specifically acting or reactive compounds and log(TR) < 1 indicated baseline compounds. However, substantially more compounds have log(TR) lower than the baseline toxicity threshold for TP than that for VF, DM, and fish. Different excess toxicity for the same specifically acting or reactive compounds may be caused by different toxic mechanisms for different species. However, baseline compounds do not exhibit such behavior, because these compounds necessarily share the same cross-species toxic mechanism. The differing number of baseline compounds meeting the log(TR) = 1 threshold may be driven by toxicity sensitivity differences between different species.

For the species with different sensitivities, Fig. 31.3 details the toxicity distribution for reactive or specifically acting and baseline compounds. Theoretically, this should be a normal distribution. If an identical threshold such as log(TR) = 1 was leveraged to classify reactive or specifically acting compounds against a baseline level, the majority of predicted reactive or specifically acting and baseline compounds for one species could predict reactive or specifically acting and baseline compounds for other species. Notably, some compounds can suffer from incorrect classifications. As an example, certain compounds were predicted as baseline for a low sensitivity species but reactive or specifically acting compounds to high sensitivity species (Fig. 31.3). Therefore, the same threshold for species with differing sensitivities can drive incorrect classification for certain baseline and reactive or specifically acting compounds. In other words, at least for species considered in this chapter, the threshold is likely species dependent.

Identification of MOA with different thresholds for different species

Baseline compounds

The TRs calculated from baseline models (thresholds of log(TR) = 1 or 0.84) were calculated for all the compounds and species. Examination of compounds with log(TR) ≥ 0.84 or 1 found that most baseline compounds had log(TR) > 1 for DM, VF, and fish or 0.84 for TP. Only a small quantity of previously identified baseline compounds had log(TR) greater than the thresholds across the four species. Experimental error could drive these outliers. Most compounds have log(TR) under the thresholds, suggesting that these baseline compounds share a common MOA across the four species. They can all be described as neutral compounds acting via the "nonpolar narcosis" mechanism with unreactive and simple structures for the four species in question [41].

Less inert compounds

Taken together, the statistics demonstrated that many less inert compounds had log(TR) below the conventional thresholds. Importantly, some less inert compounds had log(TR) values greater than these thresholds for all four species as compared with baseline compounds. The less inert compounds are mainly "polar narcotics" and include substituted phenols and anilines [4,43]. Of note, less inert compounds were more toxic to the lower species (VF and TP) than to higher species (DM and fish). Additional compounds with log(TR) greater than the thresholds were observed to be less inert for VF and TP than for DM and fish. Differences in physiology between species may have resulted in different clearance rates and biouptake, thus driving different potency between species. For example, compounds sharing similar MOA between species have similar biouptake potential. Hydrophobicity was the main driver of neutral compound biouptake, with a linear relationship between hydrophobicity and the bioconcentration factor (BCF) for most hydrophobic compounds [44]. For hydrophilic compounds in lower species compared with higher species, there was a higher relative water volume fraction in the lower species. These hydrophilic compounds more readily permeate the cell membrane of lower species than the skin, scales, or other outer covering of higher organisms and thus have an increased observed BCF in lower species. This leads to more less inert compounds with log(TR) ≥ threshold 1 for TP and VF than for fish and DM species.

Reactive or specifically acting compounds

While certain compounds were clearly pinpointed as reactive with well-established MOAs, the MOAs of the remainder were less obvious. The number of log(TR) ≥ 1 compounds demonstrated that the majority of excess toxicity classes for one species also exhibit excess toxicity in other species. These select compounds were often reactive or specifically able to interact with target site biomacromolecules and thus showed excess toxicity specific to these species. However, there were a vanishing number of classes with excess toxicity to one species, but with no excess toxicity to other species. As an example, some β-halogenated alcohols displayed excess toxicity to DM, TP, and fish but not to VF. Nitrates showed robust toxicity to VF, but not to TP. Alkoxy-substituted benzenes and alcohol were relatively more toxic to VF than to DM, TP, and fish. These specific compounds could have a different MOA in each species.

Examination of log K_{OW} values pointed to the fact that hydrophobic compounds were relatively less toxic than hydrophilic compounds. Compounds with log $K_{OW} < 0$ had log(TR) >> 1. Two possible reasons exist for this excess toxicity:

1) They are specifically acting or reactive compounds with increased toxicity. The reactive mechanism consists of the formation of covalent bonds between a biological electron-rich (nucleophilic) target molecule and an electron-poor (electrophilic) substrate. This is especially true for biological macromolecules such as proteins and nucleic acids [17]. The hydrophilic chemicals considered in this chapter were intrinsically reactive and small, not specifically interacting with biomolecules or some receptors through acylation, Schiff base formation, aromatic nucleophilic substitution (S_NAr) reactions, and bimolecular nucleophilic substitution (S_N2) [13,45–47].
2) The observed or a bioconcentration potential of such compounds was underestimated from hydrophobicity [44]. The theoretical basis of discrimination of excess toxicity from baseline was the linear relationship between log(BCF) and log K_{OW} [33], causing a linear relationship between log K_{OW} and toxicities. The log K_{OW} was nonlinearly related with apparent log(BCF) for highly hydrophilic compounds. Thus, log K_{OW} could not be used as a surrogate for BCF for these highly hydrophilic compounds. Compared with the fish lipid phase, the aqueous-phase fraction was more of a driver for quantifying their concentration in an organism. When BCF is underestimated, it indicates that toxicity calculated from log K_{OW} from the baseline level was underestimated. For these hydrophilic compounds, this caused an overcalculation of log(TR) resulting from Eq. (31.2).

These statistics revealed that some ionizable compounds had relatively high toxicity with log(TR) greater than the threshold 1. As an example, many aliphatic diacids showed excess toxicity across all four species, but benzoic acids only to VF. In theory, excess toxicity presented in log(TR) should only apply to neutral compounds, and not to ionizable compounds. The baseline models were built upon the linear relationship between log K_{OW} and toxicities of the neutral species. Thus, for ionizable compounds, these log K_{OW} values were much overestimated [48]. The overestimation of log K_{OW} values demonstrated overestimation of baseline toxicity, and thus, ionizable compounds should indeed have lower log(TR) values than those resulting from Eq. (31.2). For these ionizable compounds, the excess toxicity in neutral species calculated by log K_{OW} indicated that they are indeed very reactive and toxic or specifically acting compounds for these species. Unsurprisingly, some of these compounds were used for antiseptic disinfection or were established antifungal in the food and pharmaceutical industries.

Effect of bioconcentration on excess toxicity and identification of MOA for organic chemicals

Based on the findings described so far, clearly TR is a valuable tool for differentiating baseline effect from excess toxicity. Although it was determined that critical body residual (CBR) or internal effect concentration should be calculated from BCF and used in the TR, the impact of BCF on excess toxicity discrimination from baseline effect has not been deeply described. In this chapter, acute toxicity data to fish (951 LC_{50}s) and BCFs (1088 LC_{50}s) were leveraged to investigate the relationship between BCT and TR.

Theoretical relationships between baseline toxicity and log K_{OW}

The narcosis model for baseline toxicity in aquatic toxicology is based on the earlier work of Ferguson, who suggested that narcosis is triggered when chemical thermodynamic activity achieves a threshold such that healthy physiological processes cannot continue normally. Chemicals inducing narcosis elicit their effect once a critical volume or concentration has been achieved within an organism's biophase site of action. The concentration at the target site is often challenging to directly obtain. As a surrogate, the total concentration of organisms that elicit a critical effect ("CBRs" or "internal effect concentrations") has been used in the toxicology field. When expressed on a molar basis, these so-called "critical body residues" were found to be equal for all nonreactive chemicals and were exposure time and concentration independent [40]. Theoretically, the relationship between CBR (internal effect concentration) and measured toxicity LC_{50} (external effect concentration) can be derived from the BCF resulting from Eqs. (31.4) and (31.5) [20,49].

$$\text{BCF} = \text{CBR}/\text{LC}_{50} \tag{31.4}$$

$$\text{Log}(1/\text{LC}_{50}) = \log(1/\text{CBR}) + \log(\text{BCF}) \tag{31.5}$$

The unit is L/kg for BCF, mmol/L (mM) for LC_{50}, and mmol/kg for CBR. Generally, it is accepted that hydrophobicity of organic compounds is a key bioconcentration driver. Linear relationships between log K_{OW} and log(BCF) have been noted [44,50] and employed to estimate BCF.

$$\text{Log (BCF)} = \text{a} \log\text{K}_{OW} + \text{b} \tag{31.6}$$

If Eq. (31.6) is introduced into Eq. (31.5), one can obtain a relationship between hydrophobicity and toxicity as log K_{OW}.

$$\text{Log}(1/\text{LC}_{50}) = \log(1/\text{CBR}) + \log(\text{BCF}) = \log(1/\text{CBR}) + \text{a} \log\text{K}_{OW} + \text{b} \tag{31.7}$$

CBRs were nearly a constant, and toxicity was linearly correlated with hydrophobicity, log K_{OW} for baseline narcotics. The reactive compounds had much higher CBRs than the baseline compounds, and log K_{OW} had a poor relationship with log(1/LC_{50}). An additional descriptor was required to detail the CBR difference between reactive and baseline compounds [23]. Of note, Eq. (31.7) is for discriminating baseline from excess toxicity. The log(BCF) underestimated or overestimated by log K_{OW} from baseline BCF model can potentially result in inaccurate predictions and classification between reactive, less inert, and baseline compounds.

The relationship between log K_{OW} and toxicity is delineated by Eqs. (31.4)–(31.7). Theoretically, the CBR should reach to a constant. Additionally, a linear relationship between log K_{OW} and log(BCF) should be noted for baseline compounds. For baseline compounds, Fig. 31.4 details measured log(BCF) against log K_{OW} and shows a linear relationship between log K_{OW} and log(BCF). The regression mode is given in Eq. (31.8).

$$\text{Log(BCF)} = 0.791 \log \text{K}_{OW} - 0.729 \tag{31.8}$$

$$S = 0.47 \quad N = 70 \quad R^2 = 0.83 \quad F = 334$$

To estimate log(BCF) for the baseline compounds and all other compounds, Eq. (31.8) was used. Compounds with log $K_{OW} < 0$ or > 7 were excluded from model development. It is important to note that the compounds used in Eq. (31.8) must be the same types of compounds used in the baseline compound model. It is used not only to estimate the log(BCF) for baseline compounds but also to compare the difference between experimental and estimated log(BCF) values.

Rather than the BCF, CBR can be calculated from the bioconcentration ratio (BCR) at 96 h in fish. However, BCR values are rare and challenging to measure for compounds with diverse structures. Thus, instead of BCR, BCF was leveraged in Eq. (31.4) or (31.5) to estimate CBR [20]. The literature reported BCFs were experimentally obtained in fish at a steady state, but not at 96 h. To achieve equilibrium, the uptake times of high K_{OW} chemicals from water can be greater than 96 h. The assumption of steady state in Eq. (31.4) is not necessarily valid for high hydrophobicity compounds, which tend to have lower bioconcentration kinetics [20]. If experimental measurements for highly hydrophobic compounds take a long time to achieve equilibrium, the CBRs may be overestimated using BCFs. To interrogate estimation of CBR from BCF, Eq. (31.4) was used to calculate the CBRs for baseline compounds. Fig. 31.4 shows the plot of log(1/CBR) calculated from Eqs. (31.4) and (31.8) against log K_{OW} for baseline compounds to fish. The log(1/CBR) values of baseline compounds approached a constant with AR of log(1/CBR) = −0.83. This

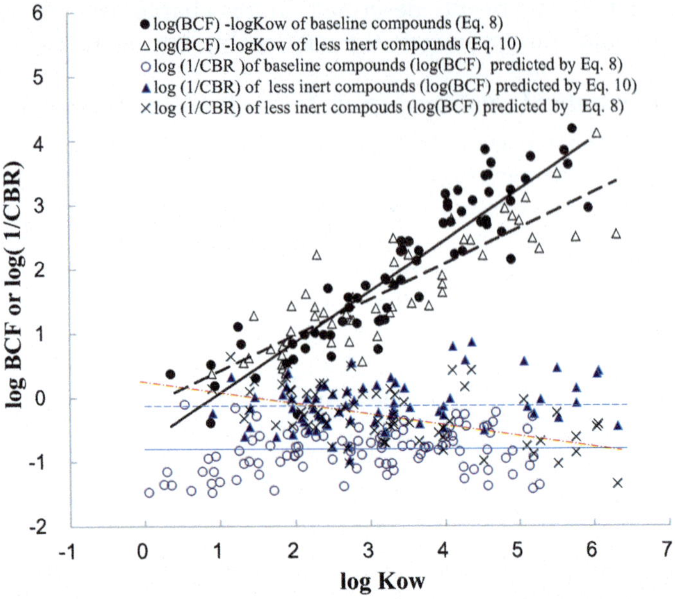

FIGURE 31.4 Comparing log(1/CBR) values calculated from log(BCF) and log K_{OW} for less inert and baseline compounds [51].

is within the expected bounds based on previous reports for baseline in the fathead minnow [20]. A subtle log(1/CBR) increase with log K_{OW} increasing demonstrated that for highly hydrophobic baseline compounds, CBRs estimated from BCF were slightly smaller than actual values. A long time was required for highly hydrophobic compounds to achieve equilibrium. These results are supported by the theory that CBRs are equal for all baseline compounds, when expressed on a molar basis [40]. It indicates that BCF can be employed to estimate CBR in fish. This finding explains the reason for a linear relationship between log K_{OW} and baseline toxicity.

Relationship between BCF and TR

The TR concept was deployed to separate out excess toxicity from baseline effect. The predicted baseline toxicity is divided by the experimental value to calculate excess toxicity using Eq. (31.1). The TR = 10 threshold was commonly employed for differentiating excess toxicity from baseline data [19,52]. However, this TR concept was building off a chemical's external effect concentration rather than the internal effect concentration. The actual excess toxicity for identifying reactive chemicals from baseline should be based on the internal critical concentration, rather than external critical concentration (i.e., LC_{50}). The relationship of TRs and internal effect concentration (Δlog1/CBR) can be derived from the definitions of excess toxicity in Eq. (31.2) and log(BCF) in Eq. (31.5) [20,21].

$$
\begin{aligned}
\text{Log(TR)} &= \log(1/LC_{50\text{exp}}) - \log(1/LC_{50\text{pred}})\ \text{(baseline)} \\
&= (\log(1/\text{CBR}) + \log(\text{BCF}))_{\text{exp}} - (\log(1/\text{CBR}) + \log(\text{BCF}))_{\text{pred}} \\
&= \Delta\log(1/\text{CBR}) + \log(\text{BCF}_{\text{exp}}) - \log(\text{BCF}_{\text{pred}})\Big) \\
&= \Delta\log(1/\text{CBR}) + \log(\text{BCF}_{\text{exp}}) - (a \log \text{Kow} + b)
\end{aligned}
\tag{31.9}
$$

If the BCF of a chemical was robustly estimated using log K_{OW} (the calculated BCF is very close to the experimental BCF), Δlog(BCF) (log(BCFexp) − log(BCFpred)) would approach zero. The external effect concentration difference would be similar to the internal effect concentration difference, and thus, the log(TR) would approach Δlog(1/CBR). Otherwise, log(TR) would not represent the internal effect concentration difference, described as Δlog(1/CBR). If not cautious, this could result in category misclassification from the log(TR) threshold.

BCF and outliers from log(TR) < −1

Deep diving into the residuals, defined as measured log(BCF) – predicted log(BCF), revealed that for some types of compounds the log(BCF) values were substantially overestimated (residual < −1) using Eq. (31.8), which expresses the relationship baseline log(BCF) with log K_{OW}. Notably, highly hydrophobic compounds with log

$K_{OW} > 7$ had impressively low solubility. This caused low bioavailability and lowering of BCF [53]. In addition, log(BCF) is also overestimated for long chain and polycyclic compounds with log $K_{OW} > 6$, and multibromo compounds with more than four bromo atoms. The structure of the epidermal tissue can restrict uptake of hydrophobic molecules of long chain or large cross-sectional area (>9.5 Å) by imposing a physical barrier to diffusion [54]. Third are the unstable and reactive compounds with three-membered rings. Fourth are the extremely ionizable compounds, organic acids. The log(BCF) values were overestimated for some acids. Generally, it is generally accepted that neutral species (unionized hydrophobic compounds) can be absorbed by organisms. Previous evidence indicates that ionized forms additionally contribute to bioconcentration within an organism. Moreover, log K_{OW} was calculated from the neutral form of compounds in this program. The log K_{OW} at a given pH for an acid can be substantially different from log K_{OW} in its unionized form.

In theory, baseline narcosis is the lowest level of toxicity demonstrated by a compound. Baseline compound TR values should approach one, but not dip significantly below one. Notably, $\log(TR) < -1$ outliers have been seen for certain compounds. Careful inspection of fish toxicity TRs revealed that 24 outlier compounds had $\log(TR) < -1$. These outliers were acids, polycyclic compounds, compounds with log $K_{OW} > 7$, and chemicals with long chain. Analysis of the relationship between log K_{OW} and log(BCF) pointed to the fact that these compound's log(BCF) values were substantially overestimated using Eq. (31.8) of the baseline log(BCF)−log K_{OW} relationship. Moving to Eq. (31.9), an overestimated log(BCF) points to an underestimation of log(TR). This then in turn drives a log(TR) of much less than one.

Less inert compounds and BCF

To interrogate the impact of BCF on less inert compounds, less inert compounds were used to develop the relationship between log K_{OW} and log(BCF). Similar to what was previously observed, the same types of less inert compounds should be used in this analysis. Fig. 31.4 shows the plot of log(BCF) versus log K_{OW} of the less inert compounds. The resulting regression model is given in Eq. (31.10).

$$\mathrm{Log(BCF)} = 0.555 \log K_{OW} - 0.145 \tag{31.10}$$

$$N = 54 \quad R^2 = 0.74 \quad S = 0.44 \quad F = 149$$

The result showed that although the estimated log(BCF) values for less inert compounds (brown squared in Fig. 31.4) are slightly different from the baseline log(BCF) values (blue diamonds in Fig. 31.4), their correlation lines with log K_{OW} are not parallel but with similar slopes.

When log(BCF) was estimated using Eq. (31.10), correlation lines were parallel for less inert and baseline compounds, suggesting that a $\log(LC_{50})$ shift toward baseline toxic effect with increasing log K_{OW} values was caused by the log(BCF) shifting toward baseline for less inert compounds. For phenols and highly hydrophobic anilines, toxic effect measured using log(1/CBR) should be higher than the observed toxicity, indicating that their toxicity might be caused through different mechanisms [55].

BCF and excess toxicity

Toxicity increase caused by specific or reactive toxicity is characterized as "excess toxicity." Importantly, reactive compounds not necessarily exhibit excess toxicity. The log(TR) values indicated that most reactive compounds identified by the literature exhibited excess toxicity. However, category misclassification by the log(TR) threshold could occur if the log(BCF) is underestimated from the baseline using Eq. (31.8). The residuals measured by log(BCF) − predicted log(BCF) showed that the log(BCF) values of highly hydrophilic compounds were significantly underestimated (residual >1) from the baseline using Eq. (31.8). These highly hydrophilic compounds had very polar functional groups. Notably, lipid tissue in fish is not a major storage site of these chemicals [56]. Uptake from other tissues/organs can play a much more important role than the lipid content, resulting larger measured BCF than predicted BCF for hydrophobic compounds. Reactive compounds in the literature included amino alcohols, polyamines, amides, diamines, and hydrazines. With most compounds having log $K_{OW} > 0$, their BCFs were substantially underrated from the baseline. The predicted residuals based on the log(TR) should be carefully compared against the residuals resulting from the relationship between log(BCF) and log K_{OW} for reactive compounds. Without taking care, misclassification occurs from the definition of log(TR). The benzenediamines, hydroxyphenols, and aminophenols displayed excess toxicity based on the log(TR) and were thus predicted to be reactive. These compounds are reported to be susceptible to quinine oxidation. The toxicity of quinones was greater than that of polar narcotics [57]. However, the predicted BCF values were significantly underestimated. The excess toxicity observed from log(TR) for these compounds may be due to the excess bioconcentration, rather than the excess toxicity inside body (Δlog 1/CBR).

Effect of pH on the discrimination of excess toxicity and identification of MOA for ionizable compounds

Toxic effects can be impacted by the pH of water because pH can drive the degree of ionization of ionizable compounds. Incorrect MOA classification could be caused by these apparent toxicities. The toxicity to DM across 61 compounds was measured at three different pH values and was employed to investigate the impact of pH on excess toxicity discrimination.

Discrimination of excess toxicity to DM

The compounds leveraged to create the DM toxicity baseline model were 23 compounds well recognized as baseline including chlorobenzenes, chloroalkanes, and alkyl. Our previous paper details the measured toxicities and the baseline model [28].

$$\mathrm{Log}(1/IC_{50}) = 0.809\ \log K_{OW} + 1.64 \tag{31.11}$$

$$N = 23\quad R^2 = 0.84\quad S = 0.30\quad F = 118$$

Log(TR) values were computed by Eqs. (31.2) and (31.11); the toxicity data measured at pH 7.8 for all compounds. Of the 61 studied compounds, 17 had $\log(TR) > 1$ showing excess toxicity to DM, including three hydroxyanilines, nine anilines, one dinitroaniline, two nitroanilines, one dinitrophenol, and one trihydroxybenzoic acid. Except trihydroxybenzoic acid, nearly all of the strongly ionizable compounds had $\log(TR) < 0$ and thus very low DM toxicity. The log(TR) values of the 10 benzoic acids were much less than −1, indicating substantially lower toxicity than the baseline level.

The majority of the weakly ionizable and neutral compounds are more toxic than the baseline compounds according to their log(TR) values. It has been reported that hydroxyanilines can be oxidized to form reactive compound electrophiles [57]. Being an electrophile, dinitroaniline can have noncovalent interaction with specific receptors to form irreversible covalent bonds [58] and thus displays a higher toxicity than the baseline level. Anilines and substituted phenols are reported across literature as less inert compounds [4]. However, only the substituted phenols, but not the anilines with bromo, alkyl, and chloro groups, showed toxicity slightly higher than the baseline, with $0 < \log(TR) < 1$. With a mean of 1.20 for their log(TR) values, the majority of the substituted anilines showed excess toxicity beyond baseline level. Interestingly, these substituted anilines were forecasted to be reactive compounds to DM but be less inert chemicals to other species. Excess toxicity to DM can be caused by these chemicals undergoing aromatic nucleophilic substitution [59]. Compared with the weakly ionizable and neutral compounds, much low toxicity was observed for the strongly ionizable compounds. Certain compounds were pinpointed as reactive compounds and did not show excess DM toxicity. As an example, pentachlorophenol was found to be a reactive compound associated specifically with weak respiratory uncoupling mechanism. This has been established to show toxicity in excess narcosis [57]. Its toxicity to DM was near to the baseline model predicted toxicity. With the averaged log(TR) of −0.74, the majority of these strongly ionizable compounds containing acidic group exhibited very low toxicity.

pH impact on excess toxicity

Following $\log(TR) < 0$, the substituted benzoic acids were predicted to be baseline compounds to DM. However, they may not act by the narcotic mechanism because the TRs were calculated following on IC_{50} or other external critical concentration, instead of the internal CBR. True excess toxicity is estimated using the internal rather than external critical concentration and can be used for distinguishing reactive chemicals from baseline compounds. Since BCF connects external critical concentration to internal critical concentration, the relationship of log(TR) and internal CBR can be represented by Eq. (31.13) based on the definition of excess toxicity in Eq. (31.2).

$$BCF = CBR/IC_{50} \tag{31.12}$$

$$\begin{aligned}\mathrm{Log(TR)} &= \log\left(1/\mathrm{IC}_{50}^{\mathrm{Exp}}\right) - \log\left(1/\mathrm{IC}_{50}^{\mathrm{Pred}}\right)_{\mathrm{(baseline)}} \\ &= (\log(1/\mathrm{CBR}) + \log(\mathrm{BCF}))^{\mathrm{Exp}} - (\log(1/\mathrm{CBR}) + \log(\mathrm{BCF}))^{\mathrm{Pred}}_{\mathrm{(baseline)}} \\ &= \Delta\log(1/\mathrm{CBR}) + \left(\log\left(\mathrm{BCF}^{\mathrm{Exp}}\right) - \log\left(\mathrm{BCF}^{\mathrm{Pred}}\right)_{\mathrm{(baseline)}}\right)\end{aligned} \tag{31.13}$$

In the case that the predicted log(BCF) approaches the log(BCF) experimentally determined, the calculated log(TR) will approach Δlog (1/CBR). The excess toxicity from baseline level as represented by log(TR) = Δlog(1/CBR) is reflected in log(TR). If a chemical's BCF cannot be readily estimated, the log(TR) will not reflect the difference between the baseline level and internal critical concentration. In this case, the mechanism of toxicity is not predicted well by log(TR). For the studied compounds, over- or underestimation of log(BCF) values will cause incorrect toxic mechanism classification using the log(TR) [51].

In water, an organic weak base and acid have two forms: unionized and less soluble form, and ionized and more soluble form. Unionized compounds have been noted to be absorbable by organisms. Eq. (31.13) predicted and experimentally measured log(BCF) values at a pH are apparent bioconcentration factors ($\mathrm{BCF^A}$).

$$\mathrm{Log}\left(\mathrm{BCF^A(acid)}\right) = \log\left(\frac{[\mathrm{AH}]_\mathrm{O}}{[\mathrm{AH}]_\mathrm{W} + [\mathrm{A^-}]_\mathrm{W}}\right) = \log\left(\frac{[\mathrm{AH}]_\mathrm{O}}{[\mathrm{AH}]_\mathrm{W}(1 + 10^{\mathrm{pH-pKa}})}\right) = \log(\mathrm{BCF}) + \log \mathrm{F_O} \tag{31.14}$$

$$\mathrm{Log}\left(\mathrm{BCF^A(base)}\right) = \log\left(\frac{[\mathrm{B}]_\mathrm{O}}{[\mathrm{BH^+}]_\mathrm{W} + [\mathrm{B}]_\mathrm{W}}\right) = \log\left(\frac{[\mathrm{B}]_\mathrm{O}}{[\mathrm{B}]_\mathrm{W}(1 + 10^{\mathrm{pKa-pH}})}\right) = \log(\mathrm{BCF}) + \log \mathrm{F_O} \tag{31.15}$$

$$\mathrm{F_O(acid)} = \frac{[\mathrm{AH}]_\mathrm{W}}{[\mathrm{AH}]_\mathrm{W} + [\mathrm{A^-}]_\mathrm{W}} = \frac{[\mathrm{AH}]_\mathrm{W}}{[\mathrm{AH}]_\mathrm{W}(1 + \mathrm{Ka}/[\mathrm{H^+}])} = \frac{1}{1 + 10^{\mathrm{pH-pKa}}} \tag{31.16}$$

$$\mathrm{F_O(base)} = \frac{[\mathrm{B}]_\mathrm{W}}{[\mathrm{AH^+}]_\mathrm{W} + [\mathrm{B}]_\mathrm{W}} = \frac{[\mathrm{B}]_\mathrm{W}}{[\mathrm{B}]_\mathrm{W}(1 + [\mathrm{H^+}]/\mathrm{pKa})} = \frac{1}{1 + 10^{\mathrm{pKa-pH}}} \tag{31.17}$$

Here, $\mathrm{F_O}$ represents the unionized fraction in water. For an ionizable compound, $\mathrm{F_O}$ can thus be treated as the degree of ionization correction factor. Of note, Eqs. (31.14) and (31.15) rely on the assumption that the organism has been treated as being in an organic phase. The ionized form in organic phase does not contribute to the $\mathrm{BCF^A}$. However, for an ionizable compound partitioning between organism and water, the BCF can be greater than the $\mathrm{BCF^A}$ calculated using Eq. (31.14) or (31.15). The log $\mathrm{BCF^A}$ decreased proportionally with the pH increase for an acid with pH > pKa. For a base with pH < pKa, log $\mathrm{BCF^A}$ was positively correlated with the pH increase. Importantly, EPI suite program predicted the log(BCF) only applies to the unionized compounds. Thus, the ionizable compound log(BCF) values were overestimated, especially for strongly ionizable compounds. This phenomenon resulted in log(TR) > 0 as calculated following Eq. (31.13). These findings indicated that pH could meaningfully impact the separation between baseline level and excess toxicity for ionizable compounds. If a TR is calculated from Eq. (31.2), it cannot fully reflect the internal critical concentration differences from baseline level. Based on this fact, determining whether baseline and ionizable compounds have the same MOA using TRs deserves further investigation.

pH impact on apparent toxicity

Interrogations into neutral compound toxicities showed that indeed pH did not impact toxicity. For neutral compounds such as nitrobenzene and methylbenzene, toxicity is at a similar level, while varying across a very narrow range. The toxicities slightly decreased for phenols and increased for anilines as pH increases for weakly ionizable compounds such as substituted phenols and anilines. At pH = 6, select anilines could be slightly ionized, and at pH = 9, some phenols could be partly ionized. For substituted benzoic acids and other strongly ionizable compounds, pH meaningfully impacted toxicity. For some compounds, a change of only one logarithmic unit in toxicity had been observed from pH = 6 to pH = 9. Taken together, these results point to pH having an impact on strongly ionizable compound toxicity.

Toxicity can be impacted in three ways. In the first way, the unionized form equally contributes to bioconcentration as the ionized form. In this scenario, toxicity is not impacted by pH. Toxicity values neither decrease nor increase with changing pH. In the second scenario, the ionized form does not contribute to bioconcentration, and pH impacts toxicity. $\mathrm{IC_{50}^{pH}}$ (the apparent toxicities) varies with pH. However, in principle, the internal CBR should not vary with pH in water. If BCF of unionized chemical is used, Eqs. (31.18) and (31.19) describe the relationship between pH and the apparent toxicity (IC_{50}^{pH}) values.

$$CBR = BCF \times IC_{50}{}^{U} = BCF \times IC_{50}{}^{pH} \times F_o \quad (31.18)$$

$$Log(1 / IC_{50}{}^{pH}) = log(1 / IC_{50}{}^{U}) + log\ F_O \quad (31.19)$$

where IC_{50}^{U} represents the contribution of a 100% unionized form to bioconcentration. This would mean pH >> pKa for a base and pH << pKa for an acid. $IC_{50}^{pH} \times F_O$ represents the unionized form contribution to bioconcentration at a certain pH. In the third scenario, pH indeed is a driver of toxicity, and the ionized form makes some contribution to bioconcentration. Thus, the correction factor F_O is included in Eq. (31.19).

$$Log\ 1 / IC_{50}{}^{pH} = log\ 1/IC_{50}{}^{U} + c\ log\ F_O \quad (31.20)$$

where $0 \leq c \leq 1$. In the case of $c = 0$, the ionized and unionized forms equally contribute to toxicity or bioconcentration. The ionized form does not contribute to bioconcentration or toxicity when $c = 1$. In the case of $0 < c < 1$, certain contribution to toxicity and bioconcentration are from the ionized form. The toxicity (described as $\log(1/IC_{50}^{U})$) and c value of a unionized form can be tabulated by Eq. (31.20) leveraging a linear regression analysis between fractions of unionized form (log F_O) and the apparent toxicities ($\log(1/IC_{50}^{pH})$) and at different pH values.

Fig. 31.5 shows the apparent toxicity plots varying with pH across three situations for 2-nitrophenol. This figure also includes the toxicities measured at pH = 6, 7.8, and 9. The c and $\log(1/IC_{50}^{U})$ values are shown in Table 31.2 for all ionizable compounds. The data showed that the calculated apparent toxicities fit well with the toxicities

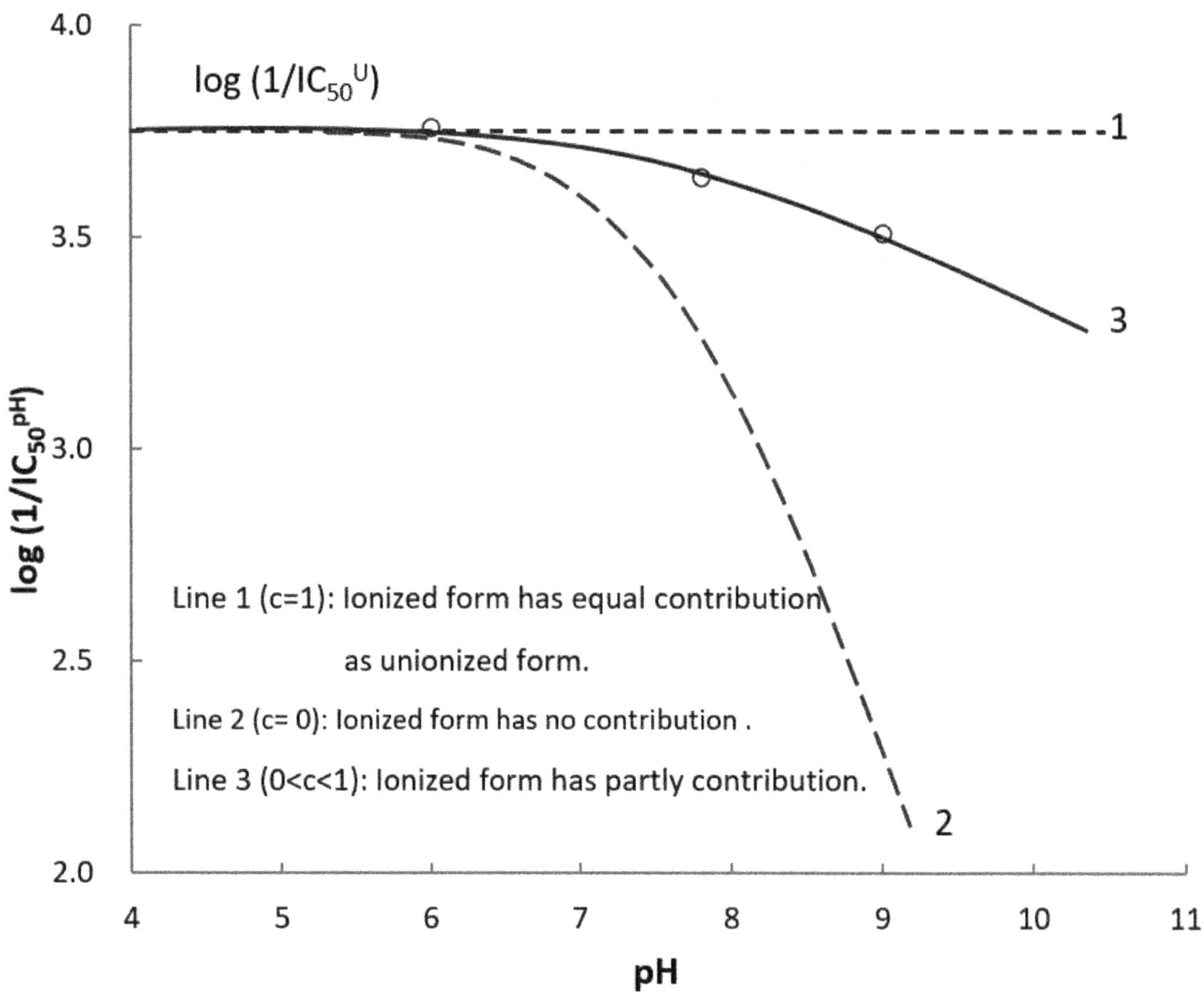

FIGURE 31.5 Relationships between pH and the apparent $\log(1/IC_{50}^{pH})$ 2-nitrophenol across three situations (o: Determined $\log(1/IC_{50}^{pH})$) [48].

calculated using Eq. (31.20). The relationship between toxicity and pH were nonlinear, but the relationship between toxicity and log F_O was linear. For $0 < c < 1$, the unionized form was key in driving toxicity, but the ionized form also contributed. The c values of strongly ionizable compounds are shown in Table 31.2, and the averaged $c = 0.34$ indicated that the ionized form contribution to toxicity was around 66%. There are two ways to explain why $0 < c < 1$. First, the ionized form can be absorbed by the organism, although the uptake was generally faster for the unionized form of weak electrolytes [60,61]. Second, only the neutral form is able to cross the organism's membranes. Once across, the deprotonated form is generated in the organism through an equilibrium between acid and base [62]. Both explanations suggest that the apparent toxicities occur at a pH, which was underestimated by Eq. (31.19). The adjusted Eq. (31.20) can better represent the relationship between pH and the apparent toxicities.

Ionizable compound toxic mechanisms

The DM toxicities of most strongly ionizable compounds were slightly above baseline level, according to their apparent toxicities. Theoretically, a compound exhibits baseline narcosis as the minimum toxicity. The predicted baseline toxicities should not be substantially higher than the measured toxicities. Of note, this finding was supported by the comparison of apparent toxicity with baseline toxicity data at pH 7.8, not the toxicities in compound unionized forms. As stated earlier, pH impacts the separation of baseline level from excess toxicity. Since the baseline model that was used for calculating the baseline toxicities was developed for the neutral species (EPI Suite was used for computing log K_{OW} for neutral species), the unionized form toxicity, not the apparent toxicities, was deployed in the comparison.

The $\log(1/IC_{50}^{U})$ values were not calculated for weak bases such as substituted anilines or acids such as certain substituted phenols since the ionization variation at pH from 6 to 9 was very small. Rigorous inspection of the strongly ionizable compound TRs revealed that when $\log(1/IC_{50}^{U})$ was employed for the comparison, nearly all compounds had higher or similar toxicity compared against baseline compounds. At pH = 7.8, this was very distinct from the TRs computed by the apparent toxicities. Impressively, only one compound has $\log(TR) > -1$. Select compounds (e.g., halogenated benzoic acids) had similar toxicities in comparison with baseline compounds, and some of them (e.g., chloro- and mono-nitro phenols) had slightly greater toxicity than baseline level. These important compounds could be considered either as less inert or baseline compounds to DM, and they act by the narcotic mechanism of action. There were multiple strongly ionizable compounds such as benzoic acids or phenols with dinitro, amino, and di- or trihydroxy groups exhibiting excess toxicity to DM. A reactive mechanism seemed to be the correct classification for these strongly ionizable compounds. At pH = 7.8, several strongly ionizable compounds did not exhibit excess toxicity to DM since the comparison was conducted using apparent toxicities.

Conclusions and perspectives

The TR was a very valuable tool to discriminate excess toxicity from baseline narcotic level. The excess toxicity indicates toxicity increase due to reactive or specific toxicity.

The baseline models for toxicity to fish, DM, and VF have similar sensitivity and MOA. As compared with the other three species, the sensitivity of baseline compounds to TP was different. Therefore, the log(TR) = 1 threshold based on the fish toxicity data distribution can distinguish baseline narcotics from specifically acting or reactive compounds for VF and DM. However, the threshold for TP was log(TR) = 0.84. The log(TR) summary statistics for TP, VF, DM, and fish showed that log(TR) for many baseline compounds was below these thresholds. More less inert compounds were found for VF and TP than for DM and fish and had log(TR) exceeding the thresholds. Physiological differences among species may have driven this effect, causing differences in bio-uptake. Consequently, this could lead to different toxic effects across species. The majority of specifically acting or reactive chemicals displaying excess toxicity to only one species showed excess toxicity to all other studied species. Small changes in toxic effects for certain compounds could possibly be attributed to differences in physiology and log K_{OW} over- or underestimation. These key compounds might not have the same MOA between these compared species. Hydrophobic compounds were relatively less toxic than hydrophilic compounds. In addition, hydrophilic compounds exhibited higher excess toxicity because of their intrinsic high reactivity and the bioconcentration potential underestimation from log K_{OW}. Strikingly, the majority of ionized compounds showed high toxicity; however, log(TRs) for some ionized compounds were lower as a result of the overestimated log K_{OW}. Certain noted toxicities were lower than what was predicted for different species from baseline models due to high hydrophobicity, experimental error, degree of ionization, and the size of the molecules. These findings from this study, taken together, posit that species-specific thresholds are good for classifying reactive/specifically acting, baseline, and less inert compounds.

Excess toxicity expressed as TR is related to both chemical reactivity and the bioconcentration–hydrophobicity relationship. A linear correlation between log K_{OW} and log(BCF) was the cornerstone in discriminating baseline effect from excess toxicity. Importantly, log K_{OW} is not linearly related to log(BCF) for all compounds. For compounds having log $K_{OW} > 7$ or log Know < 0, their log(BCF) values were either underestimated or overestimated by the linear model. A nonlinear relationship between log(BCF) and log K_{OW} was seen for compounds spanning a wide log K_{OW} range. The shift of log LC_{50} toward baseline toxic effect when log K_{OW} increases was caused by the increasing log(BCF) shift in the direction of baseline for less inert compounds. Parallel lines between calculated log CBR values of baseline and less inert compounds indicated that the two classes of compounds exerted their toxicity by different mechanisms of action [55]. The log(BCF) underestimated or overestimated by log K_{OW} arising from the baseline bioconcentration model could result in misprediction and misclassification among reactive, baseline, and less inert compounds. In the literature, polyamines, amino alcohols, and diamines were categorized as reactive compounds. Notably, these chemicals' log(BCF) values were greatly underestimated by log K_{OW} using the baseline bioconcentration model. This identical situation was seen for hydrazines and amide.

The MOA and toxic mechanism of ionic compounds were investigated based on excess toxicity and QSAR. The TRs pinpointed that weak bases were less inert compounds, more toxic than baseline level, and reactive compounds to DM. It is possible for them to undergo aromatic nucleophilic substitution, which causes excess toxicity. From pH = 6–9, their toxicities slightly varied. From baseline level with $\log(TR) < -1$, the majority of strongly ionizable compounds displayed impressively low toxicities. Toxicities can be impacted by pH for strongly ionizable compounds such as substituted benzoic acids, and their toxicities were observed to be lower with increasing pH. In contrast, for neutral compounds pH did not impact toxicity, and the toxicities approach a constant in the range of pH = 6–9. Based on the critical body residue, a relationship between pH and the apparent toxicities had been developed. The relationship of apparent toxicities to pH was nonlinear. However, the relationship was linearly related to unionized form. For ionizable compounds, the calculated toxic contributions showed that although the unionized form substantially contributed to toxicity or bioconcentration, the ionized form makes an additional contribution. The ionizable compound BCFs were substantially overestimated, which caused the log(TR) computed using the apparent toxicities to be <0. For ionizable compounds, pH can affect the discrimination between excess toxicity and baseline level. In unionized form, the toxicities showed that nearly all compounds had higher or similar toxicity compared against baseline level. These were very distinct from the apparent toxicities. Ionizable compounds could be either categorized as reactive, baseline, or less inert to DM based on reactive or narcotic mechanisms. At a certain pH, select ionizable compounds did not show excess toxicity not as a result of their poor target molecule reactivity, but due to their ionization in water. Thus, calculations arising from the apparent toxicity data underestimated their toxic ratios.

Disclaimer

This chapter reflects the views of the authors and does not necessarily reflect those of the US Food and Drug Administration.

References

[1] Chemical Abstracts Service. https://www.cas.org/support/documentation/chemical-substances. [Accessed 24 October 2022].
[2] Taylor K, Stengel W, Casalegno C, Andrew D. Experiences of the REACH testing proposals system to reduce animal testing. Altex-Altern Anim Ex 2014;31(2):107–28.
[3] Barron MG, Lilavois CR, Martin TM. MOAtox: a comprehensive mode of action and acute aquatic toxicity database for predictive model development. Aquat Toxicol 2015;161:102–7.
[4] Verhaar HJM, Vanleeuwen CJ, Hermens JLM. Classifying environmental pollutants. 1: structure-activity-relationships for prediction of aquatic toxicity. Chemosphere 1992;25(4):471–91.
[5] Schultz TW, Cronin MTD, Walker JD, Aptula AO. Quantitative structure-activity relationships (Qsars) in toxicology: a historical perspective. J Mol Struct 2003;622(1–2):1–22.
[6] Netzeva TI, Pavan M, Worth AP. Review of (quantitative) structure-activity relationships for acute aquatic toxicity. QSAR Comb Sci 2008; 27(1):77–90.
[7] Zhao YH, Qin WC, Su LM, Yuan X, Lu G, Michael AH. Toxicity of substituted benzenes and algae (*Scenedesmus obliquus*) with solvation equation. Chin Sci Bull 2009;54(1):1–7.
[8] Zhao YH, Zhang XJ, Wen Y, Sun FT, Guo Z, Qin WC, et al. Toxicity of organic chemicals to *Tetrahymena pyriformis*: effect of polarity and ionization on toxicity. Chemosphere 2010;79(1):72–7.
[9] Cronin MTD, Dearden JC. QSAR in toxicology 1. Prediction of aquatic toxicity. Quant Struct-Act Relat 1995;14(1):1–7.
[10] Su L, Fu L, He J, Qin W, Sheng L, Abraham MH, et al. Comparison of *Tetrahymena pyriformis* toxicity based on hydrophobicity, polarity, ionization and reactivity of class-based compounds. SAR QSAR Environ Res 2012;23(5–6):537–52.

[11] Roy DR, Parthasarathi R, Maiti B, Subramanian V, Chattaraj PK. Electrophilicity as a possible descriptor for toxicity prediction. Bioorg Med Chem 2005;13(10):3405–12.

[12] Escher BI, Ashauer R, Dyer S, Hermens JLM, Lee J-H, Leslie HA, et al. Crucial role of mechanisms and modes of toxic action for understanding tissue residue toxicity and internal effect concentrations of organic chemicals. Integrated Environ Assess Manag 2011;7(1):28–49.

[13] Hermens JLM. Electrophiles and acute toxicity to fish. Environ Health Perspect 1990;87:219–25.

[14] Nendza M, Wenzel A. Discriminating toxicant classes by mode of action - 1. (Eco)Toxicity profiles. Environ Sci Pollut Res 2006;13(3):192–203.

[15] Nendza M, Muller M. Discriminating toxicant classes by mode of action: 2. Physico-chemical descriptors. Quant Struct-Act Relat 2001;19(6): 581–98.

[16] von der Ohe PC, Kuhne R, Ebert RU, Altenburger R, Liess M, Schuurmann G. Structural alerts a new classification model to discriminate excess toxicity from narcotic effect levels of organic compounds in the acute daphnid assay. Chem Res Toxicol 2005;18(3):536–55.

[17] Enoch SJ, Ellison CM, Schultz TW, Cronin MTD. A review of the electrophilic reaction chemistry involved in covalent protein binding relevant to toxicity. Crit Rev Toxicol 2011;41(9):783–802.

[18] Neuwoehner J, Zilberman T, Fenner K, Escher BI. QSAR-analysis and mixture toxicity as diagnostic tools: influence of degradation on the toxicity and mode of action of diuron in algae and daphnids. Aquat Toxicol 2010;97(1):58–67.

[19] Schramm F, Mueller A, Hammer H, Paschke A, Schueuermann G. Epoxide and thiirane toxicity in vitro with the ciliates *Tetrahymena pyriformis*: structural alerts indicating excess toxicity. Environ Sci Technol 2011;45(13):5812–9.

[20] Maeder V, Escher BI, Scheringer M, Hungerbuhler K. Toxic ratio as an indicator of the intrinsic toxicity in the assessment of persistent, bioaccumulative, and toxic chemicals. Environ Sci Technol 2004;38(13):3659–66.

[21] Zhang X, Qin W, He J, Wen Y, Su LM, Sheng LX, et al. Discrimination of excess toxicity from narcotic effect: comparison of toxicity of class-based organic chemicals to *Daphnia magna* and *Tetrahymena pyriformis*. Chemosphere 2013;93(2):397–407.

[22] Roelofs W, Huijbregts MAJ, Jager T, Ragas AMJ. Prediction of ecological no-effect concentrations for initial risk assessment: combining substance-specific data and database information. Environ Toxicol Chem 2003;22(6):1387–93.

[23] Escher BI, Hermens JLM. Modes of action in ecotoxicology: their role in body burdens, species sensitivity, Qsars, and mixture effects. Environ Sci Technol 2002;36(20):4201–17.

[24] Weyers A, Sokull-Kluttgen B, Baraibar-Fentanes J, Vollmer G. Acute toxicity data: a comprehensive comparison of results of fish, Daphnia, and algae tests with new substances notified in the European union. Environ Toxicol Chem 2000;19(7):1931–3.

[25] Henegar A, Mombelli E, Pandard P, Pery ARR. What can be learnt from an ecotoxicity database in the framework of the reach regulation? Sci Total Environ 2011;409(3):489–94.

[26] Dom N, Knapen D, Benoot D, Nobels I, Blust R. Aquatic multi-species acute toxicity of (chlorinated) anilines: experimental versus predicted data. Chemosphere 2010;81(2):177–86.

[27] Huang C-P, Wang Y-J, Chen C-Y. Toxicity and Quant Struct-Act Relat of nitriles based on pseudokirchneriella subcapitata. Ecotoxicol Environ Saf 2007;67(3):439–46.

[28] Zhao YH, Ji GD, Cronin MTD, Dearden JC. QSAR study of the toxicity of benzoic acids to *Vibrio fischeri*, *Daphnia magna* and carp. Sci Total Environ 1998;216(3):205–15.

[29] Koleva YK, Cronin MTD, Madden JC, Schwöbel JAH. Modelling acute oral mammalian toxicity. 1. Definition of a quantifiable baseline effect. Toxicol Vitro 2011;25(7):1281–93.

[30] Li JJ, Zhang XJ, Yang Y, Huang T, Li C, Su LM, et al. Development of thresholds of excess toxicity for environmental species and their application to identification of modes of acute toxic action. Sci Total Environ 2018;616–617:491–9.

[31] Zhang XJ, Qin HW, Su LM, Qin WC, Zou MY, Sheng LX, et al. Interspecies correlations of toxicity to eight aquatic organisms: theoretical considerations. Sci Total Environ 2010;408(20):4549–55.

[32] Jaworska JS, Hunter RS, Schultz TW. Quantitative structure-toxicity relationships and volume fraction analyses for selected esters. Arch Environ Contam Toxicol 1995;29(1):86–93.

[33] Li JJ, Wang XH, Wang Y, Wen Y, Qin WC, Su LM, et al. Discrimination of excess toxicity from narcotic effect: influence of species sensitivity and bioconcentration on the classification of modes of action. Chemosphere 2015;120:660–73.

[34] Slooff W, Canton JH. Comparison of the susceptibility of 11 freshwater species to 8 chemical compounds. II. (Semi)Chronic toxicity tests. Aquat Toxicol 1983;4(3):271–82.

[35] Cairns Jr J, Weis JS. The myth of the most sensitive species; multispecies testing can provide valuable evidence for protecting the environment. (Includes related article on single species testing). Bioscience 1986;36:670.

[36] Vaal M, Van Der Wal JT, Hermens J, Hoekstra J. Pattern analysis of the variation in the sensitivity of aquatic species to toxicants. Chemosphere 1997;35(6):1291–309.

[37] Holmes E, Bonner FW, Nicholson JK. Comparative studies on the nephrotoxicity of 2-bromoethanamine hydrobromide in the fischer 344 rat and the multimammate desert mouse (*Mastomys natalensis*). Arch Toxicol 1995;70(2):89–95.

[38] Johnson JD, Ryan MJ, Toft JD, Graves SW, Hejtmancik MR, Cunningham ML, et al. Two-year toxicity and carcinogenicity study of methyleugenol in F344/N rats and B6C3F1 mice. J Agric Food Chem 2004;48(8):3620–32.

[39] Fan LY, Zhu D, Yang Y, Huang Y, Zhang SN, Yan LC, et al. Comparison of modes of action among different trophic levels of aquatic organisms for pesticides and medications based on interspecies correlations and excess toxicity: theoretical consideration. Ecotoxicol Environ Saf 2019; 177:25–31.

[40] Mccarty LS, Mackay D. Enhancing ecotoxicological modeling and assessment. Body residues and modes of toxic action. Environ Sci Technol 1993;27(9):1718–28.

[41] Raevsky OA, Grigor'ev VY, Weber EE, Dearden JC. Classification and quantification of the toxicity of chemicals to guppy, fathead minnow and rainbow trout: Part 1. Nonpolar narcosis mode of action. QSAR Comb Sci 2008;27(11–12):1274–81.

[42] Qin WC, Su LM, Zhang XJ, Qin HW, Wen Y, Guo Z, et al. Toxicity of organic pollutants to seven aquatic organisms: effect of polarity and ionization. SAR QSAR Environ Res 2010;21(5–6):389–401.

[43] Raevsky OA, Grigor'ev VY, Dearden JC, Weber EE. Classification and quantification of the toxicity of chemicals to guppy, fathead minnow, and rainbow trout. Part 2. Polar narcosis mode of action. QSAR Comb Sci 2009;28(2):163–74.

[44] Meylan WM, Howard PH, Boethling RS, Aronson D, Printup H, Gouchie S. Improved method for estimating bioconcentration/bioaccumulation factor from octanol/water partition coefficient. Environ Toxicol Chem 1999;18(4):664–72.
[45] Lipnick RL. Outliers: their origin and use in the classification of molecular mechanisms of toxicity. Sci Total Environ 1991;109–110(C):131–53.
[46] Aptula AO, Roberts DW. Mechanistic applicability domains for nonanimal-based prediction of toxicological end points: general principles and application to reactive toxicity. Chem Res Toxicol 2006;19(8):1097–105.
[47] Bohme A, Laqua A, Schuurmann G. Chemoavailability of organic electrophiles: impact of hydrophobicity and reactivity on their aquatic excess toxicity. Chem Res Toxicol 2016;29(6):952–62.
[48] Li JJ, Zhang XJ, Wang XH, Wang S, Yu Y, Qin WC, et al. Discrimination of excess toxicity from baseline level for ionizable compounds: effect of pH. Chemosphere 2016;147:382–8.
[49] Mccarty LS, Mackay D, Smith AD, Ozburn GW, Dixon DG. Interpreting aquatic toxicity Qsars: the significance of toxicant body residues at the pharmacologic endpoint. Sci Total Environ 1991;109–110(C):515–25.
[50] Pavan M, Netzeva TI, Worth AP(. Review of literature-based quantitative structure–activity relationship models for bioconcentration. QSAR Comb Sci 2008;27:21–31.
[51] Su LM, Liu X, Wang Y, Li JJ, Wang XH, Sheng LX, et al. The discrimination of excess toxicity from baseline effect: effect of bioconcentration. Sci Total Environ 2014;484(1):137–45.
[52] Russom CL, Bradbury SP, Broderius SJ, Hammermeister DE, Drummond RA. Predicting modes of toxic action from chemical structure: acute toxicity in the fathead minnow (*Pimephales promelas*). Environ Toxicol Chem 1997;16(5):948–67.
[53] Mackay D, Fraser A. Bioaccumulation of persistent organic chemicals: mechanisms and models. Environ Pollut (1987) 2000;110(3):375–91.
[54] Dimitrov SD, Dimitrova NC, Walker JD, Veith GD, Mekenyan OG. Predicting bioconcentration factors of highly hydrophobic chemicals. Effects of molecular size. Pure Appl Chem 2002;74(10):1823–30.
[55] Dearden JC, Cronin MTD, Zhao YH, Raevsky OA. QSAR studies of compounds acting by polar and non-polar narcosis: an examination of the role of polarisability and hydrogen bonding. Quant Struct-Act Relat 2000;19(1):3–9.
[56] Wen Y, He J, Liu X, Li J, Zhao Y. Linear and non-linear relationships between bioconcentration and hydrophobicity: theoretical consideration. Environ Toxicol Pharmacol 2012;34(2):200–8.
[57] Cronin MTD, Aptula AO, Duffy JC, Netzeva TI, Rowe PH, Valkova IV, et al. Comparative assessment of methods to develop Qsars for the prediction of the toxicity of phenols to *Tetrahymena pyriformis*. Chemosphere 2002;49(10):1201–21.
[58] Enoch SJ, Hewitt M, Cronin MTD, Azam S, Madden JC. Classification of chemicals according to mechanism of aquatic toxicity: an evaluation of the implementation of the verhaar scheme in toxtree. Chemosphere 2008;73(3):243–8.
[59] Schwobel JAH, Koleva YK, Enoch SJ, Bajot F, Hewitt M, Madden JC, et al. Measurement and estimation of electrophilic reactivity for predictive toxicology. Chem Rev 2011;111(4):2562–96.
[60] Barron MG. Bioconcentration. Environ Sci Technol 1990;24(11):1612–8.
[61] Blum DJW, Speece RE. Quant Struct-Act Relat for chemical toxicity to environmental bacteria. Ecotoxicol Environ Saf 1991;22(2):198–224.
[62] Escher BI, Hermens JL. Peer reviewed: internal exposure: linking bioavailability to effects. Environ Sci Technol 2004;38(23):455–62.

CHAPTER

32

Theoretical prediction for carrying capacity of microplastic toward organic pollutants

Xiaoxuan Wei, Bohao Li, Fang Xiao, Haiying Yu, Guangcai Ma and Xueyu Wang

College of Geography and Environmental Sciences, Zhejiang Normal University, Jinhua, China

Abbreviations

AdaBoost Adaptive boosting
BaP Benzo[a]pyrene
CBs Chlorinated benzenes
CPE Chlorinated polyethylene
DDT Dichlorodiphenyltrichloroethane
DFT Density functional theory
ELM Extreme learning machine
GA Genetic algorithm
GBDT Gradient boosting decision tree
HA Humic acid
HCHs Hexachlorocyclohexanes
kNN *k*-nearest neighbor
MD Molecular dynamics
MLP Multilayer perceptron
MLR Multiple linear regression
PA Polyamide
PAHs Polycyclic aromatic hydrocarbons
PCBs Polychlorinated biphenyls
PE Polyethylene
PP Polypropylene
PS Polystyrene
PSO Particle swarm optimization
PVC Polyvinyl chloride
QSPR Quantitative structure−property relationship
SVM Support vector machine
XGBoost Extreme gradient boosting

Introduction

As typical emerging pollutants, the environmental pollution and risk of microplastics have caused great concern [1,2]. Microplastics are defined as plastic particles or debris with a size less than 5 mm, which mainly come from plastic particles directly discharged into the environment (such as cleansing granules in personal care products and industrial polishing particles, etc.), as well as tiny debris formed by the long-term physical and chemical weathering of large plastic and textile wastes [3]. At present, microplastics have been continuously detected in seawater [4], freshwater [5], drinking water [6], sediments [7], soils [8], air [9], and living organisms [10] all over the world and even in polar regions [11]. As plastic waste, which is the "precursor" of microplastics, has accounted for three-fourths of marine trash [12], microplastics will certainly be a long-term pollution problem.

The long-term existence of microplastics in the environment may cause environmental risks. For example, microplastics floating on the surface of water can reduce the sunlight absorption of phytoplankton by reflecting sunlight and then affect their photosynthetic capacity [13]. Microplastics absorbed by the plant may block the cell wall and affect its normal growth [14,15]. Microplastics ingested by animals may also block and scratch their intestinal organs, damage the endocrine system, and even lead to the death [16,17]. Buteler et al. [18] found that the long-term existence of microplastics in bees may cause a fatal effect. Microplastics exposed to the environment will also affect the migration of microorganisms and diversity of communities [19].

In the combined pollution environment, microplastics can change the existing form, local concentration, environmental persistence, environmental behavior, and ecological risk of the coexisting pollutants through adsorption, due to its small particle size and large specific surface area [20−23]. For example, the adsorption of microplastics can

QSAR in Safety Evaluation and Risk Assessment
https://doi.org/10.1016/B978-0-443-15339-6.00031-X

inhibit the dispersion and transformation of phenanthrene and increase its bioaccumulation in *Daphnia magna* [24]. The bioaccessibility of sulfamethoxazole in the paddy soil is enhanced via the adsorption of polyethylene (PE) [25]. Zhang et al. [26] also found that PE can promote the accumulation of 9-nitroanthracene in zebrafish and affect its enzyme activity and intestinal microbial community composition. It can be concluded that studies on the adsorption between microplastics and coexisting pollutants are of great significance to fully and accurately clarify the environmental risks.

Previous experimental studies proved that microplastics can adsorb the coexisting pollutants, and the adsorption capacity may be affected by the properties of microplastics, coexisting pollutants, and environmental medium [20,21,23,27]. The adsorption capacity is often evaluated by the equilibrium partition coefficient (K_d) of organic pollutants in microplastics and aqueous phases (i.e., microplastic—aqueous sorption coefficient) [28]. This means that to obtain the basic adsorption data between different microplastics (composition, particle size, aging degree, surface roughness, etc.) and different organic pollutants under different environmental conditions, it is necessary to measure their K_d values one by one via experiments. It is bound to be a costly and time-consuming process. Therefore, it is necessary to establish a theoretical calculation method that can quickly and accurately predict the K_d values. Meanwhile, theoretical calculation can also help to clarify the complex adsorption mechanisms.

Considering that quantitative structure—property relationship (QSPR) models are often used to predict the partition equilibrium constant of organic compounds between two phases and show good predictive performance [29—31], it can also be used to predict the K_d value for the adsorption of microplastics and organic pollutants. However, as the reported experimental values of K_d are relatively insufficient and from different experimental conditions, it will undoubtedly increase the difficulty for constructing QSPR models. Therefore, only several QSPR models of K_d were reported so far. There are also a few studies that simulated the adsorption interaction between microplastics and organic pollutants by density functional theory (DFT) and molecular dynamics (MD) methods.

To construct more accurate and applicable prediction methods, the adsorption capacity of microplastics toward coexisting organic pollutants and its influencing factors will be first sorted out, and then the existing theoretical prediction methods on K_d and adsorption mechanisms will be summarized. Based on the analysis of the recent results, we hope to provide new ideas for the construction of fast prediction methods.

Adsorption between microplastics and organic pollutants

Previous studies proved that microplastics show a significant adsorption effect toward the coexisting organic pollutants. For example, Antunes et al. [32] detected polycyclic aromatic hydrocarbons (PAHs, 53—44,880 ng/g), polychlorinated biphenyls (PCBs, 2—223 ng/g), and dichlorodiphenyltrichloroethane (DDT, 1—41 ng/g) in all microplastics collected from 10 beaches along the Portuguese coast, and the concentrations of PAHs and PCBs on aged microplastics were higher. PAHs with concentrations of 282.4 ng/g, 364.2 ng/g, and 427.3 ng/g were also, respectively, detected on polypropylene (PP), PE, and expanded polystyrene collected from the Feilaixia Reservoir in the Beijiang River [33]. The concentration of PAHs on microplastics is much higher than that in water (2.61—18.9 ng/g). In addition, Camacho et al. [34] also found organochlorine pesticides, chemical sunscreens, and organic phosphorus flame retardants on the microplastics. Shi et al. [6] studied the correlation of concentration distribution between microplastics and antibiotics in groundwater, and the results proved that microplastics can adsorb the coexisting antibiotics and yield the strongest adsorption capacity to fluoroquinolones.

In addition to analyzing the adsorption concentration of organic pollutants on the surface of microplastics in the actual waters, adsorption studies are mostly carried out in the simulated water environment in the laboratory. Lee et al. [35] determined the log K_d value (2.04—7.87) of hydrophobic organic compounds, such as PAHs, hexachlorocyclohexanes (HCHs), and chlorinated benzenes (CBs), adsorbing on the surface of PE, PP, and polystyrene (PS) in seawater, and proved that microplastics have strong adsorption capacity for hydrophobic organic pollutants, which can significantly affect their local and global migration. Microplastics also can adsorb antibiotics and other dissociable organic pollutants, and its adsorption capacity will be affected by the dissociation state and charge population of antibiotics [36]. Wang et al. [37] proved that the adsorption capacity of microplastics toward tetracycline in soil is 2.5 times that of soil particles.

Based on the apparent experimental results, it is speculated that the possible adsorption mechanisms include hydrophobic interaction, electrostatic interaction, H-bond, $\pi-\pi$ interaction, etc. [20,27,38,39]. The reported studies on the adsorption of microplastics mainly focus on PE, PP, PS, polyamide (PA), and polyvinyl chloride (PVC), which are frequently detected in natural waters. The organic pollutants studied mainly include nondissociable persistent organic pollutants, such as PCBs, PAHs, CBs, HCHs, aromatic hydrocarbons, perfluorinated compounds, synthetic musks and aliphatic hydrocarbons, as well as emerging ionizable organic pollutants, e.g., antibiotics. The environmental medium mainly focuses on waters, such as seawaters, freshwaters, and pure water. Recently, attention has also been paid to the adsorption of microplastics in sediments and soil. In fact, the adsorption in soil is also the distribution of organic pollutants in soil solution and microplastics. It is of concern that the composition of soil solution may be more complex, and there is competitive adsorption to organic pollutants between soil and microplastics.

Influencing factors on the adsorption capacity

As shown in Fig. 32.1, the adsorption capacity of microplastics to organic pollutants may be affected by the properties of microplastics, coexisting pollutants, and environmental medium. Microplastics with different structures and properties may show different adsorption capacities for the coexisting organic pollutants. For example, the different functional groups contained in a microplastic structure can determine not only its hydrophobicity but also its surface charge [21,40,41]. It may also provide groups for special interactions such as H-bond and $\pi-\pi$ interaction. Generally, the decrease of particle size and the increase of surface roughness of microplastics can increase their specific surface area and exhibit stronger adsorption capacity to organic pollutants [20,42]. Microplastics entering the environment is subjected to physical fragmentation, chemical reactions, and biodegradation, resulting in varying degrees of aging. After aging, the physical and chemical properties of microplastics (e.g., structure of functional groups, particle size, surface roughness, etc.) may change, which further alter their specific surface area, hydrophilic and hydrophobic properties, and surface charge and affect the adsorption capacity of microplastics to organic pollutants [21,23,43,44].

Similarly, the structural difference in organic pollutants also makes them exhibit different hydrophobicity and surface charge population, resulting the same microplastics show different adsorption capacities for different organic pollutants. For example, Zhao et al. [45] proved that the order of K_d for five organic pollutants on the same microplastics positively correlated with their hydrophobicity. For the dissociable organic pollutants, such as antibiotics, they may present in either neutral molecular state, locally positive ionic state, or locally negative ionic state under different pH conditions, which could change their adsorption amount by affecting the hydrophobic and electrostatic interactions [20,38]. Of course, the structural differences of organic pollutants also affect the formation of special interactions between them and microplastics [27,45,46]. Tan et al. [33] found that the different concentrations of the coexisting PAHs affected their adsorption capacity on microplastics.

Environmental factors, such as coexisting components, pH, and temperature, may significantly affect the adsorption capacity as well. For example, Shen et al. [47] found that humic acid can reduce the adsorption affinity of microplastics to tetracycline by covering the surface of microplastics and changing their hydrophobicity and electrostatic repelling force as well as competing for the adsorption sites. The coexisting metal ions can decrease the adsorption capacity of microplastics toward organic pollutants through increasing the aggregation of microplastics [20] and competitive adsorption [39], while enhancing the adsorption via salting-out, which can increase the hydrophobicity of some organic pollutants [48]. The pH value mainly affects the electrostatic interaction by changing the surface charge of microplastics and the dissociation state of ionizable organic pollutants, thus affecting their adsorption [49,50]. In addition, pH can also alter the adsorption capacity of microplastics to organic pollutants by affecting

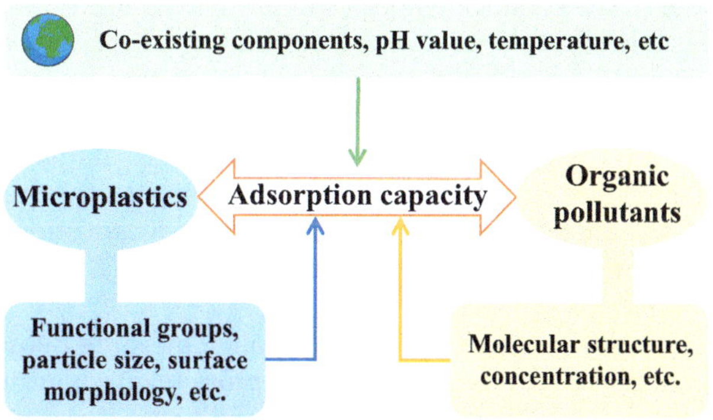

FIGURE 32.1 Influencing factors on the adsorption capacity of microplastics toward organic pollutants.

the agglomeration and corrosion of microplastics [51]. The hydrophobic interactions and van der Waals forces between microplastics and organics, surface tension of microplastics, and solubility of organic pollutants may be different under different temperature, which can cause distinct adsorption interaction. In general, the adsorption capacity increases first and then decreases with the increasing of temperature [48,52]. In summary, the comprehensive effect of environmental factors makes various adsorption capacity of the microplastic, even to the same organic pollutant [20,21].

Theoretical prediction models on K_d value

The K_d value is an important parameter for evaluating the adsorption capacity of microplastics toward coexisting organic pollutants in the environment. To overcome the limitations of experimental determination methods and provide more microscale basis for the analysis of adsorption mechanism, it is necessary to develop theoretical prediction methods for K_d values. QSPR models have been proved to be able to predict the partition of organic pollutants between two phases quickly and accurately [29–31]. However, due to the small number of experimental K_d values and the large differences in experimental values measured under different environmental conditions, constructing K_d prediction models is challenging. The currently reported prediction models on K_d can be divided into linear models and nonlinear models.

Linear prediction models

Initially, the purpose of constructing the QSPR model for K_d was not to predict, but to study the influence of sorbate properties on adsorption and to clarify the adsorption mechanism. For example, Velzeboer et al. [28] constructed the correlation between log K_d and log K_{ow} (n-octanol–water partition constant) based on the measured K_d values of 17 PCBs adsorbed by PE in seawater and freshwater. Hüffer et al. [40] established the correlation between log K_d and log K_{ow} (or hexadecane–water partitioning constant, log K_{hw}) based on the determined K_d values of 7 aliphatic and aromatic organic pollutants adsorbed on PA, PE, PVC, and PS in pure water. As the datasets are small and the purpose of model construction is to explain the adsorption mechanism, the datasets were not divided into training and testing chemicals and the application domain was not characterized in the aforementioned studies. As shown in Table 32.1, the fitting performance was only characterized by R^2 (0.66–0.94), which proves a strong relationship between the adsorption capacity of the microplastics and the hydrophobicity of the organic pollutants. Such results also indicate hydrophobic interactions as an important adsorption mechanism.

Abraham descriptors, including *E*, *S*, *A*, *B*, *V*, and *L*, have been frequently employed to establish prediction models of K_d. Here, *E* refers to excess molar refraction, *S* stands for dipolarity/polarizability parameter, *A* and *B* represent H-bond donating ability (acidity) and accepting ability (basicity), respectively, *V* is McGowan's molar volume with units of cubic centimeters per mole/100, and *L* is logarithmic hexadecane–air partitioning constant [58,59]. The developed models have been successfully applied to predict adsorption and helped to clarify the adsorption mechanisms. Each Abraham descriptor reflects the contribution from an individual interaction to the properties of organic pollutants and microplastics. Li et al. [53] and Wei et al. [30], respectively, employed stepwise regression and multiple linear regression (MLR) to screen descriptors and established prediction models for the adsorption of PE, PP, and PS in seawater, freshwater, and pure water. The statistical results (Table 32.1) and defined application domains indicate that the established models have good robustness and predictive ability, and the models can be used to predict the K_d values of PCBs, PAHs, HCHs, chlorobenzenes, antibiotics, aromatic hydrocarbons, aliphatic hydrocarbons, and perfluorinated compounds. The selected descriptors reflect that hydrophobic and π–π interaction are the main adsorption mechanism. H-bond basicity and cavity formation energy of compounds, crystallinity and aromaticity of microplastics, and salinity of aqueous medium exhibit important influences on the adsorption capacity. Hüffer et al. [54] used all Abraham descriptors (not screened) to construct the prediction model, to investigate the contribution of individual molecular interactions to overall sorption. In addition, some studies have established similar prediction models for the fitting parameters of Freundlich models (K_f and *n*) or K_d under different equilibrium concentrations of organic pollutants [60–62].

However, the absence of Abraham parameters for many nonpolar emerging pollutants may limit the construction of K_d model [30]. To expand the application range, more descriptors (e.g., quantum chemical descriptors, Dragon descriptors, and alvaDesc descriptors) have been employed to build the K_d prediction models. For example, for the adsorption of organic pollutants on PS in seawater, the K_d model employing log *D* (n-octanol/water distribution

TABLE 32.1 Linear prediction models on K_d values.

		Training set				Validation set				
Adsorption system	**Models**	n_{tra}	R^2	Q^2_{LOO}	$RMSE_{tra}$	n_{val}	R^2_{val}	Q^2_{val}	$RMSE_{val}$	**Refs.**
PE Freshwater	$\log K_d = (1.48 \pm 0.10) \times \log K_{ow} - (3.41 \pm 0.65)$	17	0.94							[28]
PE Seawater	$\log K_d = (1.18 \pm 0.18 \times \log K_{ow} - (0.88 \pm 1.17)$	17	0.75							
PE Pure water	$\log K_d = 1.34 \log K_{ow} - 0.73$	7	0.89							[40]
PE Pure water	$\log K_d = 0.89 \log K_{hw} - 0.05$	7	0.94							
PS Pure water	$\log K_d = 0.85 \log K_{ow} + 0.57$	7	0.76							
PA Pure water	$\log K_d = 0.58 \log K_{ow} + 1.79$	7	0.84							
PVC Pure water	$\log K_d = 0.75 \log K_{ow} + 1.06$	7	0.66							
PP Seawater	$\log K_d = 1.02\, E - 3.96\, B + 0.70\, V + 2.58$	27	0.903	0.827	0.317	6	0.971	0.954	0.317	[53]
PE Seawater	$\log K_d = -5.85\, B + 5.69\, V - 2.38$	25	0.794	0.763	0.985	6	0.955	0.950	0.460	
PE Freshwater	$\log K_d = -35.88\, B + 10.56$	14	0.869	0.863	0.320	3	0.886	0.825	0.200	
Aged PS Pure water	$\log K_d = 0.67\, E - 0.75\, S - 1.53\, A - 3.52\, B + 2.86\, V + 0.74$	21	0.949		0.167	6		0.903	0.231	[54]
PE Seawater	$\log K_d = (-3.822 \pm 0.222) \times B + (3.054 \pm 0.348) \times V + (1.293 \pm 0.173) \times E + (-1.410 \pm 0.684)$	26	0.907	0.907	0.721	10	0.928	0.923	0.583	[30]
PE Freshwater	$\log K_d = (-3.302 \pm 0.233) \times B + (5.594 \pm 0.887) \times V + (-3.960 \pm 1.691)$	16	0.897	0.897	0.651	7	0.934	0.932	0.563	
PE Pure water	$\log K_d = (-2.594 \pm 0.101) \times B + (2.940 \pm 0.121) \times V + (0.864 \pm 0.195)$	23	0.958	0.958	0.297	9	0.994	0.977	0.251	
PP Seawater	$\log K_d = (-3.357 \pm 0.219) \times B + (1.299 \pm 0.127) \times E + (3.108 \pm 0.287)$	25	0.914	0.914	0.471	10	0.937	0.896	0.463	
PS Seawater	$\log K_d = (4.141 \pm 0.371) \times \pi + (0.435 \pm 0.070) \times \log K_{ow} + (-3.050 \pm 0.585)$	20	0.880	0.880	0.655	8	0.832	0.812	0.981	
PE Seawater	$\log K_d = (0.725 \pm 0.\ 058) \times \log D + (-36.236 \pm 9.034) \times \varepsilon_\alpha + (-23.169 \pm 4.501) \times \varepsilon_\beta + (17.856 \pm 2.572)$	26	0.857	0.857	0.880	11	0.902	0.892	0.752	[29]
PE Freshwater	$\log K_d = (0.667 \pm 0.047) \times \log D + (1.714 \pm 0.302)$	17	0.896	0.896	0.732	7	0.947	0.910	0.661	

Continued

TABLE 32.1 Linear prediction models on K_d values.—cont'd

		Training set				Validation set				
Adsorption system	**Models**	n_{tra}	R^2	Q^2_{LOO}	$RMSE_{tra}$	n_{val}	R^2_{val}	Q^2_{val}	$RMSE_{val}$	**Refs.**
PE Pure water	$\log K_d = (0.486 \pm 0.035) \times \log D + (2.420 \pm 0.199)$	33	0.804	0.804	0.671	14	0.854	0.835	0.471	
PP Seawater	$\log K_d = (0.751 \pm 0.035) \times \log D + (-19.323 \pm 2.072) \times \varepsilon_\beta + (6.735 \pm 0.663)$	25	0.945	0.945	0.396	10	0.898	0.874	0.369	
PS Seawater	$\log K_d = (0.357 \pm 0.062) \times \log D + (3.766 \pm 0.384) \times \pi + (-2.080 \pm 0.540)$	20	0.829	0.829	0.853	8	0.859	0.843	0.714	
PE Freshwater	$\log K_d = 0.6612 \log K_{ow} + 0.6783\ E_R + 0.0498$	39	0.859	0.801	0.850	9	0.921	0.899	0.741	[55]
CPE Freshwater	$\log K_d = 0.9298 \log K_{ow} + 9.0085\ Q^-_{max,cl} + 0.5557$	9	0.930	0.749	0.236	4	0.993	0.913	0.361	
PP Pure water	$pK_d = 15.429 - 4.925$ Mor22e $- 0.531$ Mor02m $- 0.104$ nHAcc $+ 0.722$ nCb-	12	0.929		0.560					[56]
PP Pure water	$pK_d = 19.04 + 0.012\ E_{B3LYP} + 23.432\ E_{gap} + 0.047\ \mu$	12	0.700		1.511					
PE[a] Pure water	$pK_d = -26.443\ Mi - 3.852\ SpMAD\_B(m) - 0.161\ T - 3.035\ DLS\_cons + 0.526\ MLOGP + 0.377\ SpMax3\_Bh(s) + 41.076$	118	0.798	0.767	0.848	31	0.755	0.751	1.057	[57]
PE[a] Seawater	$pK_d = 3.570\ SpMax2\_Bh(m) - 7.914\ GATS2i + 4.789\ Mv + 2.531\ GATS4s - 0.080\ T - 4.639$	37	0.871	0.817	0.766	10	0.942	0.940	0.552	
PVC[a] Pure water	$pK_d = -3.116\ AVS\_B(s) + 11.439\ MATS1p + 0.019\ T - 2.665\ MATS2v + 17.803$	28	0.928	0.903	0.371	8	0.970	0.928	0.241	
PP[a] Seawater	$pK_d = -46.516\ Mi + 3.313\ SpMax3\_Bh(v) - 0.272\ T + 4.818\ GATS2s + 46.285$	30	0.929	0.857	0.485	9	0.959	0.885	0.634	
PE[b] Pure water	$pK_d = -25.080\ Mi - 2.720\ SpMAD\_B(m) - 0.152\ T - 2.990\ DLS\_cons + 0.574\ MLOGP + 0.382\ SpMax3\_Bh(s) + 37.349$	118	0.782	0.754	0.882	31	0.714	0.704	1.153	
PE[b] Seawater	$pK_d = 2.887\ SpMax2\_Bh(m) - 7.505\ GATS2i + 4.543\ Mv + 1.679\ GATS4s - 0.107\ T - 0.801$	37	0.863	0.758	0.788	10	0.945	0.935	0.573	

TABLE 32.1 Linear prediction models on K_d values.—cont'd

		Training set				Validation set				
Adsorption system	**Models**	n_{tra}	R^2	Q^2_{LOO}	$RMSE_{tra}$	n_{val}	R^2_{val}	Q^2_{val}	$RMSE_{val}$	**Refs.**
PVC[b] Pure water	$pK_d = -3.319$ $AVS\_B(s) + 0.006$ $MATS1p + 11.143\,T - 3.066$ $MATS2v + 19.282$	28	0.920	0.920	0.393	8	0.962	0.937	0.224	
PP[b] Seawater	$pK_d = -54.416Mi + 1.561$ $SpMax3\_Bh(v) -$ $0.198\,T + 3.278$ $GATS2s + 60.077$	30	0.907	0.870	0.554	9	0.898	0.726	0.978	

[a]MLR, *multiple linear regression.*
[b]SVM, *support vector machine.*
A, H-bond donating ability (acidity); *AVS_B(s)*, average vertex sum from Burden matrix weighted by I-State; *B*, H-bond accepting ability (basicity); *D*, n-octanol/water distribution coefficient at special pH value; *DLS_cons*, DRAGON consensus; *E*, excess molar refraction; E_{B3LYP}, total energy of the molecule; E_{gap}, gap energy of the lowest unoccupied molecular orbital and the highest occupied molecular orbital; E_R, excess molar refractive index; *GATS2s*, Geary autocorrelation of lag 2 weighted by I-state; *GATS4s*, Geary autocorrelation of lag 4 weighted by I-state; *GATS2i*, Geary autocorrelation of lag 2 weighted by ionization potential; K_{hw}, hexadecane-water partitioning constant; K_{ow}, n-octanol-water partition constant; *MATS1p*, Moran autocorrelation of lag 1 weighted by polarizability; *MATS2v*: Moran autocorrelation of lag 2 weighted by van der Waals volume; *Mi*, average first ionization potential (scaled on Carbon atom); *MLOGP*, Moriguchi octanol–water partition coeff. (logP); *Mor02m*, signal 2/weighted by atomic masses; *Mor22e*, signal 22/weighted by atomic Sanderson electronegativities; *Mv*, mean atomic van der Waals volume (scaled on Carbon atom); *nCb-*, number of substituted benzene *C(sp2)* atoms in the molecules; *nHAcc*, number of acceptor atoms for hydrogen bonding; $Q^-_{max,cl}$, most negative charge on Cl atom; *S*, dipolarity/polarizability parameter; *SpMAD_B(m)*, spectral mean absolute deviation from Burden matrix weighted by mass; *SpMax2_Bh(m)*, largest eigenvalue n. 2 of burden matrix weighted by mass; *SpMax3_Bh(v)*, largest eigenvalue n. 3 of burden matrix weighted by van der Waals volume; *SpMax3_Bh*(s), largest eigenvalue n. 3 of burden matrix weighted by I-state; *T*, temperature; *V*, McGowan's molar volume with units of cubic centimeters per mole/100; μ, dipole moment; π, the ratio of average molecular polarizability and molecular volume; ε_α, covalent acidity; ε_β, covalent basicity.

coefficient at special pH value) and quantum chemical descriptors shows better predictive ability than that with Abraham descriptors [29,30]. Yao et al. [56] constructed eight prediction models based on the measured K_d values of 12 pharmaceutical and personal care products adsorbed on PP in pure water using Dragon descriptors and 16 quantum chemical descriptors as initial parameters. Among them, the model with *Mor22e*, *Mor02m*, *nHAcc*, and *nCb* as descriptors showed the best prediction performance ($R^2 = 0.929$, $RMSE = 0.560$), which indicates that hydrophobic interaction and H-bond are the main adsorption mechanisms in this adsorption system. For the adsorption capacity of chlorinated polyethylene (CPE) in freshwater, a QSAR model with a satisfactory performance ($R^2 = 0.9302$) was established based on the log K_{ow} and 21 physicochemical parameters for 13 organic pollutants [55]. Zhu et al. [57] further expanded the dataset and constructed linear prediction models on K_d values for the adsorption systems of PE-pure water, PE-seawater, PVC-pure water, and PP-seawater based on 5290 alvaDesc descriptors and temperature via MLR and support vector machine (SVM) methods. For the first time, temperature was considered in the prediction models on K_d, and the results showed that increasing temperature can decrease the K_d values for PE-pure water, PE-seawater, and PP-seawater systems, while increasing that for PVC in pure water. In general, statistical results showed that the models for PVC in pure water and PP in seawater performed well, while PE in pure water and seawater obtained relatively poor performance. The analysis of residuals indicated that the MLR models attained better applicability than the SVM models.

Nonlinear models and neural network models

To further increase the prediction performance of K_d models and to reveal the complex relationships between K_d values and molecular descriptors, multiple nonlinear models have been developed. Zhu et al. [57] have made an outstanding advanced work in this research field. They established six nonlinear K_d models by four traditional nonlinear algorithms SVM, *k*-nearest neighbor (kNN), gradient boosting decision tree (GBDT), extreme gradient boosting (XGBoost), and two deep learning algorithms multilayer perceptron (MLP) and extreme learning machine (ELM) for the adsorption of PE in pure water and seawater. As shown in Table 32.2, the developed nonlinear models are acceptable with goodness-of-fit, robustness, and predictive ability. Compared with the statistical results in Table 32.1, all nonlinear models, except for the kNN model for the adsorption of PE in seawater, resulted better predictive ability than the corresponding linear models. However, all the deep learning models behaved insufficiently superior to other nonlinear models.

TABLE 32.2 Nonlinear prediction models on K_d values.[a]

		Training set (n_{tra} = 118)			Test set (n_{ext} = 31)		
Adsorption system	**Methods**	R^2_{adj}	Q^2_{LOO}	$RMSE_{tra}$	R^2_{ext}	Q^2_{ext}	$RMSE_{ext}$
PE Pure water	SVM	0.804	0.851	0.835	0.751	0.746	1.069
	kNN	0.944	0.83	0.446	0.804	0.799	0.951
	GBDT	0.994	0.85	0.142	0.889	0.87	0.765
	XGBoost	0.999	0.887	0.07	0.899	0.879	0.739
	MLP	0.886	0.833	0.639	0.851	0.844	0.837
	ELM	0.832	0.651	0.774	0.837	0.775	1.005
	MLP-GA	0.934	0.899	0.484	0.873	0.857	0.802
	MLP-PSO	0.925	0.855	0.518	0.9	0.874	0.753
	MLP-AdaBoost	0.901	0.844	0.594	0.884	0.861	0.79
	ELM-GA	0.879	0.81	0.656	0.913	0.892	0.695
	ELM-PSO	0.867	0.861	0.689	0.888	0.867	0.772
	ELM-AdaBoost	0.855	0.81	0.72	0.875	0.856	0.805
PE Seawater	SVM	0.988	0.805	0.235	0.911	0.896	0.727
	kNN	0.797	0.768	0.959	0.842	0.833	0.923
	GBDT	0.985	0.813	0.261	0.84	0.837	0.909
	XGBoost	0.973	0.841	0.352	0.848	0.842	0.895
	MLP	0.908	0.771	0.648	0.924	0.917	0.651
	ELM	0.888	0.606	0.712	0.866	0.844	0.892
	MLP-GA	0.987	0.787	0.245	0.968	0.962	0.437
	MLP-PSO	0.99	0.725	0.21	0.966	0.952	0.494
	MLP-AdaBoost	0.966	0.891	0.394	0.925	0.904	0.698
	ELM-GA	0.927	0.761	0.577	0.915	0.914	0.66
	ELM-PSO	0.926	0.681	0.579	0.882	0.873	0.803
	ELM-AdaBoost	0.922	0.875	0.596	0.95	0.947	0.52

[a]*Data from Ref. [57].*
AdaBoost, adaptive boosting; *ELM*, extreme learning machine; *GA*, genetic algorithm; *GBDT*, gradient boosting decision tree; *kNN*, *k*-nearest neighbor; *MLP*, multilayer perceptron; n_{tra} and n_{ext}, the number of chemicals in the training set and test set, respectively; *PSO*, particle swarm optimization; Q^2_{ext}, external explained variance; Q^2_{LOO}, leave one out cross-validated Q^2; R^2_{adj} and R^2_{ext}, the correlation coefficient square between observed and fitted values in training set and test set, respectively; *SVM*, support vector machine; *XGBoost*, extreme gradient boosting.

Furthermore, genetic algorithm (GA), particle swarm optimization (PSO), and adaptive boosting (AdaBoost) have been introduced into the optimization of the deep learning models, and hybrid intelligent models (optimized neural network models) were developed. The optimization models can significantly enhance the robustness and predictive performance of nonhybrid neural network models. Mechanistic analysis proved that the temperature, lipophilicity, ionization potential, molecular size, and molecular branching indices are main influencing factors on the adsorption capacity [57].

Other prediction methods on adsorption mechanism

In addition to the QSPR models, some other theoretical calculation methods, such as DFT and MD, can also be employed to simulate the adsorption interaction between microplastics and coexisting organic compounds. The main purpose of these studies is to provide microscopic evidence for the elucidation of adsorption mechanism and to explore the interaction strength based on the calculated binding energy (ΔE). For example, Bao et al. [63] and Mo et al. [64] took simple carbon chains (10 carbons) as microplastic models and simulated their interactions with organic pollutants based on DFT calculations. Adsorption mechanism was revealed by the calculated ΔE values and the analysis of structure and properties before and after adsorption, which can well explain the experimental results. Amorphous cluster-like structure (224 atoms) was also used as a nanoplastic model to study the adsorption of bisphenol A on the nano-PET [65]. The simulation results indicated that the contribution of dispersion and electrostatics effects to the adsorption mechanism are 38%−49% and 43%−50%, respectively. The formed complexes are also highly soluble, which may cause the high risk of nano-PET to adsorb and migrate pollutants in water. Most of the DFT calculations simulate the adsorption of one organic molecule on micro/nanoplastics.

MD method can simulate the adsorption process of multiple same or different organics on microplastics at a relatively large size. It can help to clarify the process of surface adsorption and internal diffusion, reveal the adsorption mechanism, and calculate the ΔE values [48,66,67]. By means of MD simulation, Feng et al. [68] clarified the adsorption mechanisms between benzo[a]pyrene (BaP) and five nanoplastics in the absence and presence of humic acid (HA). The formation of sandwiched π-stacking structures with benzene rings makes PP show the highest adsorption strength to BaP. HA can inhibit the adsorption of BaP to nanoplastics by competitive adsorption and encapsulation of BaP. Stable structure of the adsorption complex obtained after MD simulation can not only be used for further analyzing its surface property changes but also be directly used for simulating its environmental behavior and ecotoxicity [69].

Perspectives

This chapter analyzed the adsorption interaction between microplastics and coexisting organic pollutants in the environment and its influencing factors and reviewed the theoretical methods that can be used to predict the adsorption capacity or simulate the adsorption behavior. The existing theoretical prediction methods have been well used to explain the adsorption mechanism, but still need to improve in the prediction of adsorption capacity (K_d). From the perspective of application, if the constructed QSPR model can simultaneously predict the adsorption capacity of different microplastics to organic pollutants in different environments, instead of modeling specific microplastics and specific environments separately, it can better meet the requirements of K_d from researchers and decision-makers. Of course, this means that more environmental factors (such as salinity, pH, etc.) and structural properties of microplastics (functional group information, particle size, crystallinity, etc.) should be incorporated into the model construction. This assumption may be realized by nonlinear machine learning methods.

At the same time, it is worth pointing out that the binding energy (represents the adsorption strength) is not directly equivalent to the K_d value (represents the amount of an organic pollutant adsorbed by microplastics per unit mass). For example, the calculated ΔE value for the adsorption of PP to carbofuran is more negative than that for PE, but PP shows weaker adsorption capacity than PE [64]. Especially for the chemisorption, the calculated ΔE can reflect the involved interaction; however, the amount of the adsorbed organic pollutants is also limited by the active adsorption site on the exposed surface of microplastics, which may lead to a small K_d value. Therefore, the DFT and MD methods developed so far cannot be used to directly predict the K_d values. It is necessary to further establish the relationship between ΔE (or free energy change) and K_d, which can also expand the dataset for the development of QSPR model. Meanwhile, the effect of actual environmental conditions also needs to be considered in MD and DFT simulations.

To meet the demand for K_d values, it is necessary to develop theoretical prediction methods with better predictive accuracy and more comprehensive application domain. At present, there is still a very large development space in this research field.

Acknowledgments

We gratefully acknowledge the financial support from Natural Science Foundation of Zhejiang Provincial, China (LY22B070002) and National Natural Science Foundation of China (21806144, 22176177).

References

[1] Gibb BC. Plastics are forever. Nat Chem 2019;11:394–5.
[2] Rochman CM, Browne MA, Halpern BS, Hentschel BT, Hoh E, Karapanagioti HK, et al. Policy: classify plastic waste as hazardous. Nature 2013;494(7436):169–71.
[3] Thompson RC, Olsen Y, Mitchell RP, Davis A, Rowland SJ, John AWG, et al. Lost at sea: where is all the plastic? Science 2004;304(5672):838.
[4] Cui Y, Liu M, Selvam S, Ding Y, Wu Q, Pitchaimani VS, et al. Microplastics in the surface waters of the South China sea and the western Pacific Ocean: different size classes reflecting various sources and transport. Chemosphere 2022;299:134456.
[5] Liu Z, Huang QE, Wang H, Zhang SY. An enhanced risk assessment framework for microplastics occurring in the Westerscheldt estuary. Sci Total Environ 2022;817:153006.
[6] Shi JY, Dong YB, Shi YY, Yin TT, He W, An TY, et al. Groundwater antibiotics and microplastics in a drinking-water source area, northern China: occurrence, spatial distribution, risk assessment, and correlation. Environ Res 2022;210:112855.
[7] Li WF, Duo J, Wufuer R, Wang SZ, Pan XL. Characteristics and distribution of microplastics in shoreline sediments of the Yangtze River, main tributaries and lakes in China-from upper reaches to the estuary. Environ Sci Pollut Res 2022;29(32):48465.
[8] Zhang HX, Huang YM, An SS, Li HH, Deng XQ, Wang P, et al. Land-use patterns determine the distribution of soil microplastics in typical agricultural areas on the eastern Qinghai-Tibetan Plateau. J Hazard Mater 2022;426:127806.
[9] Liu K, Wu TN, Wang XH, Song ZY, Zong CX, Wei NA, et al. Consistent transport of terrestrial microplastics to the ocean through atmosphere. Environ Sci Technol 2019;53(18):10612–9.
[10] Wieczorek AM, Croot PL, Lombard F, Sheahan JN, Doyle TK. Microplastic ingestion by gelatinous zooplankton may lower efficiency of the biological pump. Environ Sci Technol 2019;53(9):5387–95.
[11] Pakhomova S, Berezina A, Lusher AL, Zhdanov I, Silvestrova K, Zavialov P, et al. Microplastic variability in subsurface water from the Arctic to Antarctica. Environ Pollut 2022;298:118808.
[12] Dias BFS. Marine debris: understanding, preventing and mitigating the significant adverse impacts on marine and coastal biodiversity. Montreal Secretariat of the Convention on Biological Diversity 2016;(83):78.
[13] Zhao SJ, Wang HY, Liu J. Influence of microplastics pollution on marine environment. Mar Sci 2009;33:84–6 [In Chinese)].
[14] Lian JP, Liu WT, Meng LZ, Wu JN, Zeb A, Cheng LP, et al. Effects of microplastics derived from polymer-coated fertilizer on maize growth, rhizosphere, and soil properties. J Clean Prod 2021;318:128571.
[15] Xu ZM, Zhang YX, Lin LP, Wang L, Sun WM, Liu CG, et al. Toxic effects of microplastics in plants depend more by their surface functional groups than just accumulation contents. Sci Total Environ 2022;833:155097.
[16] Tarasco M, Gavaia PJ, Bensimon-Brito A, Cordelieres FP, Santos T, Martins G, et al. Effects of pristine or contaminated polyethylene microplastics on zebrafish development. Chemosphere 2022;303(Pt 3):135198.
[17] Xiang XM, Zhou JJ, Lin SC, Zhang N, Abulipizi G, Chen GK, et al. Dual drive acute lethal toxicity of methylene blue to Daphnia magna by polystyrene microplastics and light. Sci Total Environ 2022;840:156681.
[18] Buteler M, Alma AM, Stadler T, Gingold AC, Manattini MC, Lozada M. Acute toxicity of microplastic fibers to honeybees and effects on foraging behavior. Sci Total Environ 2022;822:153320.
[19] Zhu MH, Yin H, Yuan YB, Liu H, Qi X, Ren Y, et al. Discrepancy strategies of sediment abundant and rare microbial communities in response to floating microplastic disturbances: study using a microcosmic experiment. Sci Total Environ 2022;835:155346.
[20] Fu LN, Li J, Wang GY, Luan YN, Dai W. Adsorption behavior of organic pollutants on microplastics. Ecotoxicol Environ Saf 2021;217:112207.
[21] Liu P, Zhan X, Wu XW, Li JL, Wang HY, Gao SX. Effect of weathering on environmental behavior of microplastics: properties, sorption and potential risks. Chemosphere 2020;242:125193.
[22] Wang JD, Tan Z, Peng JP, Qiu QX, Li MM. The behaviors of microplastics in the marine environment. Mar Environ Res 2016;113:7–17.
[23] Tourinho PS, Koci V, Loureiro S, van Gestel CAM. Partitioning of chemical contaminants to microplastics: sorption mechanisms, environmental distribution and effects on toxicity and bioaccumulation. Environ Pollut 2019;252:1246–56.
[24] Ma YN, Huang A, Cao SQ, Sun FF, Wang LH, Guo HY, et al. Effects of nanoplastics and microplastics on toxicity, bioaccumulation, and environmental fate of phenanthrene in fresh water. Environ Pollut 2016;219:166–73.
[25] Wang H, Yu B, Li BC, Zhao T, Cai Y, Luo YM, et al. A contrasting alteration of sulfamethoxazole bioaccessibility in two different soils amended with polyethylene microplastic: in-situ measurement using diffusive gradients in thin films. Sci Total Environ 2022;808:152187.
[26] Zhang JH, Meng H, Kong XC, Cheng XY, Ma T, He H, et al. Combined effects of polyethylene and organic contaminant on zebrafish (*Danio rerio*): accumulation of 9-Nitroanthracene, biomarkers and intestinal microbiota. Environ Pollut 2021;277:116767.
[27] Liu ZM, Qin QD, Hu ZX, Yan L, Ieong UI, Xu Y. Adsorption of chlorophenols on polyethylene terephthalate microplastics from aqueous environments: kinetics, mechanisms and influencing factors. Environ Pollut 2020;265:114926.
[28] Velzeboer I, Kwadijk C, Koelmans AA. Strong sorption of PCBs to nanoplastics, microplastics, carbon nanotubes, and fullerenes. Environ Sci Technol 2014;48(9):4869–76.
[29] Li M, Yu HY, Wang YF, Li JG, Ma GC, Wei XX. QSPR models for predicting the adsorption capacity for microplastics of polyethylene, polypropylene and polystyrene. Sci Rep 2020;10(1):14597.
[30] Wei XX, Li M, Wang YF, Jin LM, Ma GC, Yu HY. Developing predictive models for carrying ability of micro-plastics towards organic pollutants. Molecules 2019;24(9):1784.
[31] Ling YH, Klemes MJ, Steinschneider S, Dichtel WR, Helbling DE. QSARs to predict adsorption affinity of organic micropollutants for activated carbon and beta-cyclodextrin polymer adsorbents. Water Res 2019;154:217–26.
[32] Antunes JC, Frias JGL, Micaelo AC, Sobral P. Resin pellets from beaches of the Portuguese coast and adsorbed persistent organic pollutants. Estuar Coast Shelf Sci 2013;130:62–9.
[33] Tan XL, Yu XB, Cai LQ, Wang JD, Peng JP. Microplastics and associated PAHs in surface water from the Feilaixia Reservoir in the Beijiang River, China. Chemosphere 2019;221:834–40.
[34] Camacho M, Herrera A, Gomez M, Acosta-Dacal A, Martinez I, Henriquez-Hernandez LA, et al. Organic pollutants in marine plastic debris from Canary Islands beaches. Sci Total Environ 2019;662:22–31.
[35] Lee H, Shim WJ, Kwon JH. Sorption capacity of plastic debris for hydrophobic organic chemicals. Sci Total Environ 2014;470–471(2): 545–1552.

[36] Li J, Zhang KN, Zhang H. Adsorption of antibiotics on microplastics. Environ Pollut 2018;237:460–7.
[37] Wang Y, Wang XJ, Li Y, Liu YY, Sun Y, Hansen HCB, et al. Effects of struvite-loaded zeolite amendment on the fate of copper, tetracycline and antibiotic resistance genes in microplastic-contaminated soil. Chem Eng J 2022;430:130478.
[38] Zhang Y, Ni F, He JS, Shen F, Deng SH, Tian D, et al. Mechanistic insight into different adsorption of norfloxacin on microplastics in simulated natural water and real surface water. Environ Pollut 2021;284:117537.
[39] Guo X, Wang JL. Sorption of antibiotics onto aged microplastics in freshwater and seawater. Mar Pollut Bull 2019;149:110511.
[40] Hüffer T, Hofmann T. Sorption of non-polar organic compounds by micro-sized plastic particles in aqueous solution. Environ Pollut 2016;214: 194–201.
[41] Wang Z, Chen ML, Zhang LW, Wang K, Yu XB, Zheng ZM, et al. Sorption behaviors of phenanthrene on the microplastics identified in a mariculture farm in Xiangshan Bay, southeastern China. Sci Total Environ 2018;628–629:1617–26.
[42] Moura DS, Pestana CJ, Moffat CF, Hui JN, Irvine JTS, Edwards C, et al. Adsorption of cyanotoxins on polypropylene and polyethylene terephthalate: microplastics as vector of eight microcystin analogues. Environ Pollut 2022;303:119135.
[43] Sun Y, Yuan J, Zhou T, Zhao YC, Yu F, Ma J. Laboratory simulation of microplastics weathering and its adsorption behaviors in an aqueous environment: a systematic review. Environ Pollut 2020;265(Pt B):114864.
[44] Wang Y, Liu CQ, Wang FF, Sun QY. Behavior and mechanism of atrazine adsorption on pristine and aged microplastics in the aquatic environment: kinetic and thermodynamic studies. Chemosphere 2022;292:133425.
[45] Zhao LF, Rong LL, Xu JP, Lian JP, Wang L, Sun HW. Sorption of five organic compounds by polar and nonpolar microplastics. Chemosphere 2020;257:127206.
[46] Wang C, Liang SJ, Bai LH, Gu XY, Jin X, Xian ZY, et al. Structure dependent surface catalytic degradation of cephalosporin antibiotics on the aged polyvinyl chloride microplastics. Water Res 2021;206:117732.
[47] Shen XC, Li DC, Sima XF, Cheng HY, Jiang H. The effects of environmental conditions on the enrichment of antibiotics on microplastics in simulated natural water column. Environ Res 2018;166:377–83.
[48] Yang CH, Wu W, Zhou XT, Hao Q, Li TJ, Liu YZ. Comparing the sorption of pyrene and its derivatives onto polystyrene microplastics: insights from experimental and computational studies. Mar Pollut Bull 2021;173(Pt B):113086.
[49] Xu BL, Liu F, Brookes PC, Xu JM. Microplastics play a minor role in tetracycline sorption in the presence of dissolved organic matter. Environ Pollut 2018;240:87–94.
[50] Liu NN, Yu F, Wang YY, Ma J. Effects of environmental aging on the adsorption behavior of antibiotics from aqueous solutions in microplastic-graphene coexisting systems. Sci Total Environ 2022;806(Pt 4):150956.
[51] Wan T, Lu SH, Cheng W, Ren JH, Wang M, Hu BW, et al. A spectroscopic and theoretical investigation of interaction mechanisms of tetracycline and polystyrene nanospheres under different conditions. Environ Pollut 2019;249:398–405.
[52] Zhan ZW, Wang JD, Peng JP, Xie QL, Huang Y, Gao YF. Sorption of 3,3′,4,4′-tetrachlorobiphenyl by microplastics: a case study of polypropylene. Mar Pollut Bull 2016;110(1):559–63.
[53] Li GY, Chen JW, Li XH, Qiao XL. LSER models for several classes of organic pollutants partitioning between microplastics and water. J Ecotoxicol 2017;12(03):225–33 [in Chinese].
[54] Hüffer T, Weniger AK, Hofmann T. Sorption of organic compounds by aged polystyrene microplastic particles. Environ Pollut 2018;236: 218–25.
[55] Gui BX, Xu XT, Zhang SN, Wang Y, Li C, Zhang DM, et al. Prediction of organic compounds adsorbed by polyethylene and chlorinated polyethylene microplastics in freshwater using QSAR. Environ Res 2021;197:111001.
[56] Yao JJ, Wen JY, Li HP, Yang Y. Surface functional groups determine adsorption of pharmaceuticals and personal care products on polypropylene microplastics. J Hazard Mater 2022;423:127131.
[57] Zhu TY, Tao CC, Cheng HM, Cong HB. Versatile in silico modelling of microplastics adsorption capacity in aqueous environment based on molecular descriptor and machine learning. Sci Total Environ 2022;846:157455.
[58] Abraham MH, Ibrahim A, Zissimos AM. Determination of sets of solute descriptors from chromatographic measurements. J Chromatogr A 2004;1037:29–47.
[59] Abraham MH, Al-Hussaini AJM. Solvation parameters for the 209 PCBs: calculation of physicochemical properties. J Environ Monit 2005;7: 295–301.
[60] Uber TH, Huffer T, Planitz S, Schmidt TC. Characterization of sorption properties of high-density polyethylene using the poly-parameter linearfree-energy relationships. Environ Pollut 2019;248:312–9.
[61] Uber TH, Huffer T, Planitz S, Schmidt TC. Sorption of non-ionic organic compounds by polystyrene in water. Sci Total Environ 2019;682: 348–55.
[62] Xu JP, Wang L, Sun HW. Adsorption of neutral organic compounds on polar and nonpolar microplastics: prediction and insight into mechanisms based on pp-LFERs. J Hazard Mater 2021;408:124857.
[63] Bao ZZ, Chen ZF, Zhong YH, Wang GZ, Qi ZH, Cai ZW. Adsorption of phenanthrene and its monohydroxy derivatives on polyvinyl chloride microplastics in aqueous solution: model fitting and mechanism analysis. Sci Total Environ 2021;764:142889.
[64] Mo QM, Yang XJ, Wang JJ, Xu HJ, Li WY, Fan Q, et al. Adsorption mechanism of two pesticides on polyethylene and polypropylene microplastics: DFT calculations and particle size effects. Environ Pollut 2021;291:118120.
[65] Cortes-Arriagada D. Elucidating the co-transport of bisphenol A with polyethylene terephthalate (PET) nanoplastics: a theoretical study of the adsorption mechanism. Environ Pollut 2021;270:116192.
[66] Chen YJ, Li JN, Wang FH, Yang H, Liu L. Adsorption of tetracyclines onto polyethylene microplastics: a combined study of experiment and molecular dynamics simulation. Chemosphere 2021;265:129133.
[67] Li H, Wang FH, Li JN, Deng SP, Zhang ST. Adsorption of three pesticides on polyethylene microplastics in aqueous solutions: kinetics, isotherms, thermodynamics, and molecular dynamics simulation. Chemosphere 2021;264:128556.
[68] Feng H, Liu YJ, Xu Y, Li SX, Liu X, Dai YH, et al. Benzo[a]pyrene and heavy metal ion adsorption on nanoplastics regulated by humic acid: cooperation/competition mechanisms revealed by molecular dynamics simulations. J Hazard Mater 2022;424:127431.
[69] Li LZ, Xu Y, Li SX, Zhang XY, Feng H, Dai YH, et al. Molecular modeling of nanoplastic transformations in alveolar fluid and impacts on the lung surfactant film. J Hazard Mater 2022;427:127872.

C H A P T E R

33

QSAR models on degradation rate constants of atmospheric pollutants

Chao Li[1], *Hong-Bin Xie*[2] *and Jingwen Chen*[2]

[1]State Environmental Protection Key Laboratory of Wetland Ecology and Vegetation Restoration, Engineering Lab for Water Pollution Control and Resources Recovery, School of Environment, Northeast Normal University, Changchun, China [2]Key Laboratory of Industrial Ecology and Environmental Engineering (Ministry of Education), Dalian Key Laboratory on Chemicals Risk Control and Pollution Prevention Technology, School of Environmental Science and Technology, Dalian University of Technology, Dalian, China

Abbreviations

ADs Applicability domains
ANN Artificial neural network
BO_x Maximum bond order values in the molecule
·Cl Chlorine radical
C_X Substituent factors
DTB Decision tree boost
DTF Decision tree forest
E_{HOMO} Energy of the highest occupied molecular orbital
$f(-)_x$ The maximum Fukui index of electrophilic attack
$f(0)_n$ The minimum Fukui index of radical attack
GA Genetic algorithm
GCM Group contribution method
GRNN General regression neural network
GTIs Generalized topological indices
k_{basic} Rate constants of the basic structures
k_{Cl} Reaction rate constants of ·Cl with organics
k_{NO_3} Reaction rate constants of NO_3· with organics
k_{O_3} Reaction rate constants of O_3 with organics
k_{OH} Reaction rate constants of ·OH with organics
LR Linear regression
MLR Multiple linear regression
MRF Multivariate random forest
NO_3· Nitrate radical
O_3 Ozone
·OH Hydroxyl radical
OLS Ordinary least squares
PCBs Polychlorinated biphenyls
PCDD Polychlorinated dibenzopdioxin
PLS Partial least squares
PPR Projection pursuit regression
$q(C^-)_n$ The minimum negative partial net charge on a carbon atom
$q(C^-)_x$ The maximum negative partial net charge on a carbon atom
$q(CH^+)_x$ The maximum positive partial charge on a hydrogen atom linked with a carbon atom
QSAR Quantitative structure–activity relationship
RMSE Root mean square error
SAR Structure–activity relationships
SDEC Standard deviation error in calculation
SVM Support vector machine

Introduction

A vast number of organic pollutants are emitted into atmosphere by factories, automobile exhaust, and vegetation. Because of the complex compositions in the atmosphere, organic pollutants can undergo a variety of migration and transformation processes, resulting in their reduction in the atmosphere. Among the physical and chemical processes, the chemical reaction is the only way to essentially remove pollutants from the environment. Organic pollutants in the troposphere can be chemically transformed via photolysis by absorbing light >290 nm, reaction with atmospheric oxidants, hydrolysis, and heterogenous interfacial reactions on suspended particles and liquid particles (aerosols) [1].

QSAR in Safety Evaluation and Risk Assessment
https://doi.org/10.1016/B978-0-443-15339-6.00008-4

It should be recognized that the reaction with atmospheric oxidants can be considered as one of the most important removal processes for most gaseous pollutants. Hydroxyl radical (·OH), nitrate radicals (NO_3·), chlorine radical (·Cl), and ozone (O_3) are the most important atmospheric oxidants because of their high chemical potentials and abundance. Some studies have reported that these oxidants play important roles in the indoor chemistry of gaseous reactions of volatile organic compounds [2,3]. The kinetics is one of the key research areas in the chemical degradation process. The gaseous degradation kinetics of pollutants by oxidants follows second-order kinetics model. The second-order reaction rate constants (k_{O_3}, k_{OH}, k_{NO_3}, and k_{Cl}) can be used to measure the reactivity of pollutants toward oxidants. Based on these data, the lifetimes of pollutants can be obtained using $\tau = 1/(k_i \times [\text{oxidants}])$, where [oxidants] is the concentration of oxidants such as ·OH, NO_3·, ·Cl, and O_3 in the troposphere [4]. The kinetic rate constants (k_{O_3}, k_{OH}, k_{NO_3}, and k_{Cl}) are thus important parameters to reveal their degradation efficiency and to evaluate their fate and persistence in the atmosphere.

Up to now, although some efforts have been made to experimentally determine these rate constants by direct (flash photolysis, pulse radiolysis) and indirect (competition kinetics) methods, there is a huge gap in k_{O_3}, k_{OH}, k_{NO_3}, and k_{Cl} for existing chemicals [5]. Experiments usually suffer from some disadvantages such as high labor- and time consumption, high costs, and equipment dependence; thus, it is essential to develop reliable prediction methods to estimate the rate constants of diverse chemicals [6]. Quantitative structure–activity relationship (QSAR), which correlates "reactivity" with molecular structures, has been proven to be one of the important modeling methods for predicting reaction rate constants. Once QSAR models are developed, they can generate minimum computational costs for estimating rate constants. Some QSAR models have been established to estimate the k_{O_3}, k_{OH}, k_{NO_3}, and k_{Cl} values of various organic pollutants, which are of great help for risk assessments, management, and control of chemicals. In this chapter, the sources of ·OH, NO_3·, ·Cl, and O_3 in the atmosphere are introduced, and their corresponding QSAR models are reviewed and compared.

QSAR models for predicting reaction rate constants of pollutants with ·OH

Sources and levels of ·OH in the atmosphere

·OH is considered as the most important oxidant in the troposphere, while other oxidants especially NO_3·, ·Cl, and O_3 play important secondary roles [7]. ·OH determines the atmospheric lifetimes of pollutants in the troposphere and can be used as a measure of the self-cleaning capacity of local atmosphere. Therefore, ·OH is called "atmospheric cleaner." The concentration of ·OH in the troposphere can be used as an indicator of atmospheric oxidation capacity. The global mean ·OH level ([OH]) is ca. $(0.65–1.65) \times 10^6$ molecules cm^{-3} based on experimental determination [7,8]. The measured [OH] above clouds, inside a cloud, and in cloud free regions by Mauldin et al. are $(8–15) \times 10^6$, $(3–5) \times 10^6$, and 7.5×10^6 molecules cm^{-3}, indicating a direct loss of ·OH onto cloud droplet [9]. Additionally, [OH] in the atmosphere may change with time and space. Generally, high solar radiation intensity and high temperature conditions could enhance the generation of ·OH. Specifically, [OH] in summer is higher than that in winter; [OH] during the day is higher than that at night; [OH] in low-latitude regions is higher than that in high-latitude regions; and [OH] in the southern hemisphere is higher than that in the northern hemisphere. The previous study also found that [OH] in polluted areas is much higher than that in the clean area [10].

·OH can be generated via several pathways in the atmosphere (Table 33.1). Its main natural source is the reaction of H_2O with O atoms formed from the photolysis of O_3 after absorbing UV light of $\lambda < 320$ nm. In the polluted atmosphere, the photolysis of HNO_2, H_2O_2, NO_2, and HNO_3 can produce ·OH. The above ·OH formation pathways require sunlight, while only the ozonation of alkenes can generate ·OH in the absence of light and is an important pathway for indoor ·OH production. In some cases, the ozonation of alkenes is the main source of ·OH at night [11]. Another main source of the indoor ·OH is the direct photolysis of HNO_2, which is produced from fireplaces, gas stoves, and the combustion of indoor candles. Studies have shown that high concentrations of ·OH $(1.8–16) \times 10^6$ molecules cm^{-3}) have been detected in indoor environments [12,13]. Besides, HNO_2 in the soil can contribute to ·OH in the atmosphere [14].

k_{OH} QSAR models

There are relatively large number of researches on k_{OH} QSAR modeling compared with k_{O_3}, k_{Cl}, and k_{NO_3} modeling, as the k_{OH} values could determine the atmospheric persistence of pollutants to a great degree. The built k_{OH} QSAR models can be classified into two classes, which enable to predict k_{OH} for general organic compounds

TABLE 33.1 Main sources of ·OH in the atmosphere.

Compound	·OH formation pathway
O_3	$O_3 + hv$ (<320 nm) → $O + O_2$ $O + H_2O$ → 2·OH
	O_3 + alkenes → ·OH + products
HNO_3	$HNO_3 + hv$ → ·OH + NO_2
HNO_2	$HNO_2 + hv$ (<390 nm) → ·OH + NO
H_2O_2	$H_2O_2 + hv$ (<360 nm) → 2·OH
NO_2	$NO_2 + hv$ (250–650 nm) → ·NO_2
	$NO_2 + H_2O$ → OH + HNO_2

TABLE 33.2 Statistical parameters of QSAR models for predicting k_{OH} of general organic compounds.

					Training set		Validation set	
Model	Algorithm	*n*	*p*	*T* (K)	R^2_{tr}	Q^2_{cv}	R^2_{ext}	Q^2_{ext}
Gramatica et al. [15]	MLR	460	4	298	0.828	0.819	–	0.826
Öberg [16]	PLS	743	333	295–300	0.906	0.875	–	0.840
Wang et al. [17]	PLS	722	22	298	0.878	0.865	–	0.872
Li et al. [18]	MLR	1543	14	206–1364	0.873	0.871	0.838	0.835
		872	12	298	0.883	0.879	0.858	0.851
Gupta et al. [19]	DTB	1543	4	206–1364	0.920	–	0.920	–
	DTF	1543	4	206–1364	0.900	–	0.910	–
Gupta et al. [20]	DTB	127	3	298	0.948	–	0.946	–
Basant and Gupta [3]	MRF	61	5	~298	0.922	0.743	0.874	0.845–0.901
Liu et al. [21]	MLR	180	5	298	0.785	0.754	–	0.642
Sanches-Neto et al. [22]	XGBoost	903	1[a]	298	0.974	0.789	0.921	0.921
Shi et al. [23]	MLR	856	16	298	0.892	0.881	0.909	0.907

[a]*It refers to MACCS fingerprints;* n *represents the total number of chemical compounds in the dataset;* p *represents the number of molecular descriptors; MLR represents multiple linear regression; PLS represents partial least squares; DTB represents decision tree boost; DTF represents decision tree forest.*

[15–23] and for specific classes of organic compounds (e.g., polychlorinated dibenzo-p-dioxin [PCDD] [24], polychlorinated biphenyls [PCBs] [25,26], alkane [27], alkene [27–29], aromatic compounds [28]). Tables 33.2 and 33.3, respectively, summarize the statistical parameters of QSAR models for predicting k_{OH} of general organic compounds and specific classes of organic compounds.

The QSAR models of predicting k_{OH} of general organic compounds usually cover a large dataset (up to 1543 k_{OH} values) and are characterized by a high degree of structural diversity, implying that they have wide applicability domains (ADs). These k_{OH} QSAR models have been developed using different algorithms such as multiple linear regression (MLR), partial least squares (PLS), and other machine learning algorithms including decision tree boost (DTB), decision tree forest (DTF), and XGBoost. The models developed by nonlinear methods (DTB, DTF, XGBoost) usually possess better goodness-of-fit and predictability ($R^2_{tr} > 0.900$, $Q^2_{ext} > 0.900$) than those constructed by the linear methods such as MLR and PLS, as shown in Table 33.2. For instance, the QSAR models developed with DTF and DTB methods based on our dataset (1543 k_{OH} at different temperatures) by Gupta et al. [19] exhibit better performance than our MLR model [18]. Sanches-Neto et al. utilized the machine learning algorithm (XGBoost)

TABLE 33.3 Statistical parameters of k_{OH} QSAR models for specific classes of organic compounds.

				Training set		Validation set		
Model	Algorithm	n	p	R^2_{tr}	Q^2_{cv}	R^2_{ext}	Q^2_{ext}	Types of compounds
Pompe et al. [30]	MLR	58	6	0.878	–			Unsaturated hydrocarbons
Long and Niu [31]	PLS	14	10	–	0.728	–	–	Alkylnaphthalene
Fatemi et al. [32]	GA-ANN	98	5	–	0.86	–	–	Alkenes
Huang et al. [33]	GA-MLR	161	3	0.919	–	0.922, 0.910	0.920, 0.907	Alkanes
Xu et al. [29]	MLR	98	4	0.884	–	–	–	Alkenes
Chen et al. [28]	MLR	80	3	0.910	0.886	–	0.928	Aromatic compounds
		98	4	0.938	0.925	–	0.903	Alkenes
Li et al. [4]	MLR	31	3	0.943	–	0.960	0.856	Short-chain chlorinated paraffins
Yu et al. [34]	SVM	80	3	0.948	–	0.886, 0.885	–	Aromatics
		70	3	0.994	–	0.903, 0.929	–	Alkenes
Qi et al. [24]	MLR	75	2	0.865	0.852	–	0.806	PCDDs
Yang et al. [25]	LR	70[a]	1	0.900	–	–	0.700	PCBs
Luo et al. [26]	MLR	70[a]	3	0.789	–	–	–	PCBs

[a]70 k_{OH} data (294–300 K) for 26 PCB congeners; n represents the total number of compounds in the dataset; p represents the number of molecular descriptors; GA represents genetic algorithm; ANN represents artificial neural network; SVM represents support vector machine; LR represents linear regression.

combining molecular fingerprints to develop a k_{OH} QSAR model that performed the best so far [22]. Moreover, this model covers the largest k_{OH} dataset at 298 K. It is worth pointing out that Gupta et al. for the first time used inter-moieties reactivity correlations as a tool to develop QSAR models to predict k_{OH} and k_{O_3} of diverse chemicals by linking two reactants (·OH and O_3) [20]. These two models built by the DTB approach also exhibit better predictive performance compared with previous QSAR models. In addition, a robust multitarget quantitative structure–property relationship model was established to simultaneously predict multiple reaction rate constants (k_{O_3}, k_{OH}, and k_{NO_3}) of structurally diverse organic compounds with multivariate random forest (MRF) algorithm [3]. This model yielded high fitting ability ($R^2 = 0.904$–0.922) and predictive capacity ($R^2 = 0.874$–0.924) in simultaneously predicting k_{O_3}, k_{OH}, and k_{NO_3} values. This multitarget QSAR model would be of great help to reduce time and computational costs in assessing the ultimate fate of gaseous pollutants and their exposure risk.

Multiple classification models were developed specifically for predicting one certain class of organics such as alkanes, alkenes, aromatics, short-chain chlorinated paraffins (SCCPs), PCBs, and PCDD (Table 33.3). These models were developed using small datasets but had high performance. For example, we developed a QSAR model based on 9 SCCPs with k_{OH} values calculated by the quantum chemistry method and 22 polychlorinated alkanes having experimental k_{OH} values, and our model had high goodness-of-fit ($R^2_{tr} = 0.943$), robustness ($Q^2_{cv} = 0.917$), and predictivity ($R^2_{ext} = 0.960$) for estimating the k_{OH} of SCCPs [4]. The QSAR model built by Huang et al. for predicting k_{OH} of alkanes exhibits high fitting ability ($R^2_{tr} = 0.919$) and predictivity ($R^2_{ext} \geq 0.910$) [33].

In general, the most frequently used molecular descriptors include quantum chemical descriptors (such as the energy of the highest occupied molecular orbital (E_{HOMO}), Fukui index, polarizability (α), various atomic charge), 1D and 2D structural features, 3D molecular descriptors, molecular fingerprints, and norm descriptors. Among these molecular descriptors, the quantum chemical descriptors have clear physical–chemical meanings; thus, they could facilitate the mechanism interpretation of QSAR models. For example, Liu et al. used five quantum chemical descriptors (E_{HOMO}, the maximum negative partial net charge on a carbon atom ($q(C^-)_x$), maximum positive partial charge on a hydrogen atom linked with a carbon atom ($q(CH^+)_x$), maximum bond order values in the molecule (BO_x), the minimum Fukui index of ·OH radical attack ($f(0)_n$)) computed from Gaussian 09 and Material Studio 7.0 software to build a k_{OH} QSAR model [21]. However, the performance of this model is not high ($R^2 = 0.758$). We have employed a quantum chemical descriptor E_{HOMO} and multiple Dragon descriptors to develop two QSAR models (a temperature-dependent k_{OH} model and a room-temperature (298 K) k_{OH} model) for predicting k_{OH} of chemicals. Both models exhibit better performance than that built by Liu et al. [18]. In addition, some QSAR models developed with norm descriptors [23], a new variable distance-weighted zero-order connectivity index [27], or descriptors calculated from PaDEL [19,20] also exhibit better performance than that built based on the complete quantum chemical descriptors. It is worth pointing out that the developed model by Markelj et al. enables to distinguish different reaction pathways (addition vs. abstraction) on different reaction sites [27]. Furthermore, the calculation could be completed by their developed software Molecular Activity Characterization Indices (MACI).

It is found that most molecular descriptors used to construct QSAR models are calculated from neutral molecules. Some models are developed by combining the structural descriptors of ground states with the energy-rich transition states of alkenes [29]. For instance, Yu et al. used quantum chemical descriptors calculated from radical structures of alkenes and aromatics to predict k_{OH} [34]. Huang et al. also employed quantum chemical descriptors of free radicals produced in the H-abstraction pathway of alkanes to build a k_{OH} QSAR model, and the prediction results are satisfactory [33]. These studies provide insights into the screening and application of molecular descriptors for modeling the kinetics data of other oxidants.

QSAR models for predicting reaction rate constants of pollutants with O_3

Sources and levels of O_3 in the atmosphere

Ozone is a key constituent in the troposphere. At ordinary temperatures, ozone is a gas with light blue in color and pungent odor. As one of the important oxidants in the atmosphere, O_3 can also remove many natural and human-made pollutants, thus having an important influence on their transformation and fate [35]. Fig. 33.1 shows the resonance hybrid diagram of O_3. Its molecular structure represents triangle. The length of O–O bond is 1.272 Å, and the angle is 116.78° [36].

The average concentration of O_3 in the atmosphere is 2.71×10^{12} molecule cm^{-3} during the day, and 1.97×10^{12} molecule cm^{-3} at night [37]. In the cleaner troposphere, the O_3 concentration is ca. $(0.25-1.00) \times 10^{12}$ molecule cm^{-3} (i.e., 10–40 ppb) [38,39]. The concentration of surface O_3 usually increases with the altitude and also changes with latitude and seasons [40]. The sources of O_3 in the troposphere include in situ photochemical reactions and downward migration of the lower stratosphere. Studies show that the photochemical reactions of NO_2 are the main sources for O_3 [40]. NO_2 absorbs UV–visible light of ≤400 nm to produce NO and $O(^3P)$, and $O(^3P)$ reacts with O_2 to produce O_3. The reaction equations are shown as follows:

$$NO_2 + h\upsilon \rightarrow NO + O(^3P) \tag{33.1}$$

$$O(^3P) + O_2 + M \rightarrow O_3 + M \tag{33.2}$$

In addition, the downward migration of O_3 in the stratosphere can be expressed as follows [41]:

$$O_3 + h\upsilon\ (\lambda \leq 336\ \text{nm}) \rightarrow O(^1D) + O_2 \tag{33.3}$$

$$O(^1D) + N_2 \rightarrow O(^3P) + N_2 \tag{33.4}$$

$$O(^1D) + O_2 \rightarrow O(^3P) + O_2 \tag{33.5}$$

$$O(^3P) + O_2 + M \rightarrow O_3 + M \tag{33.6}$$

k_{O_3} QSAR models

Several literature report the applications of QSAR models to predict the k_{O_3} of organic compounds at 298 K and other temperatures. Some representative models are listed in Table 33.4. The modeling methods include linear (e.g., MLR, PLS) and nonlinear (e.g., ANN, DTB, DTF) methods. The dataset ranges from 95 to 379 k_{O_3} values. Pompe and

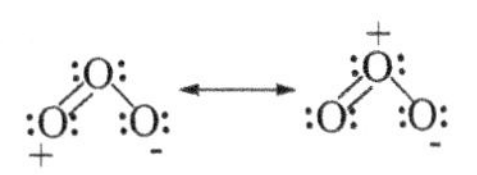

FIGURE 33.1 Resonance hybrid diagram of O_3.

TABLE 33.4 Summary of statistical parameters of k_{O_3} QSAR models.

Model	n	p	Algorithm	Training set R^2_{tr}	Q^2_{cv}	$RMSE_{tr}$	Validation set R^2_{ext}	Q^2_{ext}	$RMSE_{ext}$	Types of compounds
Pompe and Veber [42]	117	6	MLR	0.870	–	0.990	–	–	–	Diverse chemicals
Gramatica et al. [43]	125	6	MLR	0.881	0.847	–	–	0.904	0.770	Diverse chemicals
Fatemi [45]	120	6	ANN	0.980	–	0.357	0.962, 0.964	–	0.460, 0.481	Diverse chemicals
Ren et al. (2007) [46]	116	7	PPR	0.916	-	0.664	0.912	-	1.041	Diverse chemicals
Yu et al. [44]	139	3	MLR	0.839	–	0.827	0.814	–	0.981	Aliphatic compounds
			SVM	–	–	0.680	–	–	0.777, 0.709	
Li et al. [47]	379[a]	13	PLS	0.840	0.836	0.551	0.813	0.846	0.612	Diverse chemicals
Xu et al. [29]	95	3	MLR	0.857	–	0.540	–	–	0.540	Alkenes
			GRNN	–	–	0.265	–	–	0.448	
Gupta et al. [19]	379[a]	4	DTB	0.970	–	0.240	0.970	–	0.250	Diverse chemicals
			DTF	0.950	–	0.330	0.960	–	0.290	
Gupta et al. [20]	127	3	DTB	0.976	–	0.260	0.954	–	0.260	Diverse chemicals
Basant and Gupta [3]	61	5	MRF	0.904	0.834	0.570	0.901	0.838–0.846	0.620	Diverse chemicals
Liu et al. [48]	302[a]	4	MLR	0.830	0.820	0.490	–	0.720	–	Diverse chemicals

[a] *k_{O_3} values at different temperatures;* n *represents the total number of compounds in the dataset;* p *represents the number of molecular descriptors; PPR represents projection pursuit regression.*

Veber firstly established a six-descriptor MLR model based on 114 organic compounds, and the R^2 and root mean square error (*RMSE*) of the training set were 0.870 and 0.990 [42]. Gramatica et al. also built an MLR model using six descriptors based on 125 organic compounds [43], and this model showed slightly better performance than the one built by Pompe and Veber. Xu et al. employed stepwise MLR and general regression neural network (GRNN) algorithms to develop QSAR models for alkenes based on the molecular descriptors calculated from the ground-state alkenes and their transition-state structures [29]. The nonlinear GRNN model possessed better fitting ability and predictive power than the linear model. Yu et al. developed three-parameter MLR and SVM models to predict k_{O_3} of aliphatic compounds from radical structures, and the nonlinear method also shows better performance than the linear method [44]. The remaining QSAR models of predicting k_{O_3} of diverse organics at 298 K were developed by nonlinear methods such as PPR, DTB, and DTF, and all of them are very good in both fitting ($R^2_{tr} > 0.900$) and prediction ($R^2_{ext} > 0.900$) capacity [20,45,46].

In addition to these models of predicting k_{O_3} at 298 K, three studies have established QSAR models with large dataset (>300 k_{O_3} values) to estimate k_{O_3} values of diverse chemicals at different temperatures. Li et al. reported the first satisfactory temperature-dependent QSAR model, which is a 13-parameter PLS model built with 379 k_{O_3} values [47]. Subsequently, Gupta et al. built two nonlinear (DTB and DTF) QSAR models with the same dataset [19], and the performance is greatly improved compared with that of the PLS model. Recently, Liu et al. constructed a four-parameter MLR QSAR model using complete quantum chemical descriptors (e.g., E_{HOMO}, $f(0)_n$, the minimum negative partial net charge on a carbon atom ($q(C^-)_n$), BO_x) based on 302 k_{O_3} values [48]. This model has the advantages of fewer descriptors, satisfactory accuracy, and transparent mechanism, which is beneficial to applications.

QSAR models for predicting reaction rate constants of pollutants with $NO_3\cdot$

Sources and levels of $NO_3\cdot$ in the atmosphere

During the daytime, the $\cdot$OH-initiated oxidation is recognized as the most important chemical elimination pathway of organic pollutants in the atmosphere. In comparison, the reaction with $NO_3\cdot$ plays an important role in the removal of pollutants at night [49]. The formation pathways of $NO_3\cdot$ in the stratosphere and troposphere are shown as follows:

$$NO + O_3 \rightarrow O_2 + NO_2 \tag{33.7}$$

$$NO_2 + O_3 \rightarrow O_2 + NO_3\cdot \tag{33.8}$$

The concentration of $NO_3\cdot$ is low during the daylight, but it gradually increases during the nighttime. Its concentration could range from 5×10^7 to 1×10^{10} molecules cm^{-3} in the condition of low light and UV radiation [50], which is 25–5000 times higher than the average $\cdot OH$ concentration (2×10^6 molecules cm^{-3}). Although the $k_{NO_3\cdot}$ values of hydrocarbons are about 30 times lower than their k_{OH}, reaction with $NO_3\cdot$ could contribute greatly to the removal of organic pollutants from the atmosphere. It has been revealed that NO_3-initiated oxidation contributes 28% to the overall removal of organics in the continental boundary layer over Europe [51]. It is worth mentioning that the reactions of $NO_3\cdot$ with organics can form nitro-substituted compounds, which are usually very toxic and carcinogenic [52]. Thus, it is important to understand the degradation rates of organic contaminants to assess their environmental persistence and potential harm to humans.

k_{NO_3} QSAR models

Several QSAR models have been reported for predicting the gaseous k_{NO_3} values of diverse chemicals and special compound classes and are summarized in Table 33.5. These models include relatively small dataset compared with that of k_{OH} and k_{O_3} models. Most of the early k_{NO_3} models were established based on special compound classes and thus have limited ADs [35,53]. For example, Sabljic and Güsten developed two simple one-parameter (vertical ionization energy) prediction models of k_{NO_3}. The models covered aliphatics and benzene derivatives and exhibited good performance ($n = 62$, $R^2 = 0.899$; $n = 7$, $R^2 = 0.884$) [53]. Gramatica et al. utilized ordinary least squares (OLS) methods to build two k_{NO_3} models with 58 and 16 $\log k_{NO_3}$ values. The models were designed specifically

TABLE 33.5 Summary of statistical parameters of k_{NO_3} QSAR models.

				Training set			Validation set			
Model	n	p	Algorithm	R^2_{tr}	Q^2_{cv}	$RMSE_{tr}$	R^2_{ext}	Q^2_{ext}	$RMSE_{ext}$	Types of compounds
Sabljic and Güsten [53]	62	1	LR	0.899	–	–	–	–	–	Aliphatics, benzaldehyde, phenol and cresols
	7	1	LR	0.884	–	–	–	–	–	Benzene derivatives
Gramatica et al. [35]	58	5	OLS	0.840		0.792[b]	–	–	–	Aliphatic compounds
	16	3		0.978		0.281[b]	–	–	–	Aromatics
Gramatica et al. [54]	114	3	MLR	0.912	0.896	0.663[b]	–	0.957	–	Aliphatic and aromatic compounds
Long and Niu [31]	12	8	PLS	0.798	0.774	–	–	–	–	Alkylnaphthalenes
Papa and Gramatica [52]	160	3	MLR	0.899	–	0.670	–	0.746–0.894	–	Various organic contaminants
Gupta et al. [55]	174	4	DTF	0.939	–	0.510	0.944	–	0.450	Various organic contaminants
			DTB	0.981	–	0.290	0.974	–	0.310	
	335[a]	5	DTF	0.967	–	0.350	0.970	–	0.310	
			DTB	0.982	–	0.260	0.977	–	0.280	
Schindler et al. [56]	254	4	PLS	0.831	–	–	0.862	0.823	–	Various organic contaminants
Basant and Gupta [3]	61	5	MRF	0.915	0.818	0.660	0.924	0.853–0.876	0.620	Various organic contaminants
Luo et al. [50]	63	11	MLR	0.930	0.920	–	–	0.850	–	Various organic contaminants
Liu et al. [57]	189	2	MLR	0.880	0.872	0.623	–	0.861	–	Various organic contaminants

[a] *k_{NO_3} values at different temperatures;*
[b] *Standard deviation error in calculation (SDEC);*
n represents the total number of compounds in the dataset; p represents the number of molecular descriptors.

for aliphatic compounds and aromatics [35]. Both models possess good performance ($R^2 \geq 0.840$). Long and Niu developed a k_{NO_3} model for alkylnaphthalenes with PLS regression and quantum chemical descriptors ($n = 12$, $R^2 = 0.798$) [31]. However, the aforementioned models failed to perform external validations. Gramatica et al. subsequently built a new model based on a larger dataset ($n = 114$) and performed validation with a test set [54]. This model had high prediction ability, and thus, it could become a useful tool for generating reliable k_{NO_3} values for chemicals.

The modeled dataset and the ADs were further expanded by Papa and Gramatica ($n = 160$) [52], Schindler et al. ($n = 254$) [56], and Liu et al. ($n = 189$) [57] with linear methods (PLS or MLR). These models also exhibit satisfactory goodness-of-fit and prediction capacity. It is worth noting that Liu et al. for the first time introduced the Fukui index (the maximum Fukui index of electrophilic attack ($f(-)_x$)) for developing the k_{NO_3} model. The optimum QSAR model containing two descriptors (E_{HOMO} and $f(-)_x$) yielded adequate robustness and predictability ($R^2_{tr} = 0.880$, $Q^2 = 0.872$, $Q^2_{ext} = 0.861$). Basant and Gupta developed a robust multitarget QSAR model to simultaneously predict k_{OH}, k_{O_3}, and k_{NO_3} [3]. In addition, the quick and simple group contribution method (GCM) was employed by Luo et al. to construct a k_{NO_3} QSAR model based on 63 k_{NO_3} values and 11 molecular fragment descriptors [50]. The GCM model was validated internally and externally, exhibiting robust stability and good predictive power. All the aforementioned k_{NO_3} models are limited to predict k_{NO_3} at room temperature. Gupta et al. for the first time built a temperature-dependent k_{NO_3} QSAR model with machine learning algorithms (DTF and DTB), besides the room-temperature (298 K) model [55]. Both QSRR models are superior to the previously reported ones, and the temperature-dependent models offered much broader ADs. The developed models can be useful tools for predicting the reactivities of pollutants toward $NO_3\cdot$ in the atmosphere and for assessing their persistence and exposure risks.

QSAR models for predicting reaction rate constants of pollutants with ·Cl

Sources and levels of ·Cl in the atmosphere

In addition to ·OH, O_3, and $NO_3\cdot$ in the atmosphere, the ·Cl also possesses high oxidation capacity and is responsible for the removal of organic pollutants. ·Cl is traditionally considered to be formed mainly via heterogenous reaction cycles of sea salt [58], with the concertation of about 1%–10% of ·OH. Thus, the important role of ·Cl–initiated oxidation has been confined to the marine boundary layer at first. The photodissociation of natural or anthropogenic chlorinated compounds (e.g., Cl_2, $COCl_2$, HOCl, BrCl, HCl, $CHCl_3$) is also one of the most common pathways to global tropospheric ·Cl production [58]. Moreover, in recent years, researchers detected a very high concentration of $ClNO_2$ in the urban atmosphere and revealed $ClNO_2$ is a significant source for ·Cl in midcontinental areas (formulas (33.9) and (33.10)) [59]. The average global ·Cl concentration is about 1×10^3 atoms cm^{-3} in the troposphere [60]. Nevertheless, it would significantly increase to about 1×10^5 atoms cm^{-3} or more in some industrial areas and in the areas near/over the ocean in the early morning [61]. Additionally, the k_{Cl} values for organics are often at least 10 times higher than that of k_{OH} [62], suggesting that the ·Cl-initiated oxidation would play an important role in the elimination of pollutants and is even competitive with the hydroxylation of pollutants in coastal regions during early morning.

$$N_2O_{5(g)} + Cl^-_{(aq)} \rightarrow ClNO_{2(g)} + N_3O^-_{(aq)} \tag{33.9}$$

$$ClNO_{2(g)} + h\upsilon \rightarrow \cdot Cl + NO_2 \tag{33.10}$$

k_{Cl} prediction models

Some models have been developed for predicting k_{Cl} of atmospheric pollutants (Table 33.6). Two types of methods were used in these models: structure–activity relationships (SAR) and correlations with measured and calculated physicochemical parameters. Most of these models cover small datasets ($n < 65$) and are limited to predict certain classes of chemicals. These models are not valid QSAR models for practical uses according to the Organisation for Economic Co-operation and Development (OECD) guidelines for QSAR modeling since they have not been validated by external test sets and lack characterization of ADs.

SAR formulation assumes that the reactivity of oxidants toward organics could be estimated with GCM. Senkan and Quam established the k_{Cl} prediction method for (halogenated) hydrocarbon and oxygenated compounds with the SAR analysis based on the rate expression ($k = AT^n \exp(-E/RT)$) of the H-atom abstraction pathways from

TABLE 33.6 Summary of statistical parameters of k_{Cl} prediction models.

Model	*n*	*P*	Algorithm	R^2	Types of compounds
Senkan and Quam [63]	47	18	GCM	–	(Halogenated) hydrocarbon and oxygenated compounds
Teruel et al. [64]	14	14	GCM	–	Alkenes, unsaturated and oxygenated compounds
Blanco et al. [65]	11	8	GCM		Unsaturated esters and aldehydes
Aschmann and Atkinson [66]	6	4	GCM	–	Alkanes, acyclic and cyclic ketones
Ifang et al. [61]	18	3	GCM	–	Ester
Timerghazin and Ariya [67]	13	1	LR	0.790	Alkenes
McGillen et al. [62]	24	1	LR	0.74	C_2–C_{10} hydrocarbons
Estrada and Matamala [68]	24	1	LR	0.960	Alkanes and cycloalkanes
Long and Niu [31]	14	8	PLS	0.929	Alkylnaphthalenes
Blanco et al. [69]	13	1	LR	–	Unsaturated compounds
Poutsma [70]	64	4	MLR	>0.870	Various chemicals
Gallego-Iniesta et al. [71]	13	1	LR	–	α,β-Unsaturated esters

n represents the total number of compounds in the dataset; *p* represents the number of molecular descriptors.

primary, secondary, and tertiary carbon atoms by ·Cl [63]. The total rate constants can be calculated with formula (33.11). The group parameters (E_X) in this SAR are listed in Table 33.7. On average, this method allows k_{Cl} prediction within 50% of the experimental values.

$$k_{total} = \sum[k^0_{pri}\, F(X)] + \sum[k^0_{sec}\, F(X)\, F(Y)] + \sum[k^0_{ter}\, F(X)\, F(Y)\, F(Z)] \tag{33.11}$$

$$F(X) = \exp(-E_X/RT) \tag{33.12}$$

$$k^0_{pri} = 2.48 \times 10^7\, T^2 \exp(-760/RT)\ \mathrm{cm^3/(mol\ s)} \tag{33.13}$$

$$k^0_{sec} = 1.92 \times 10^7\, T^2 \exp(-69/RT)\ \mathrm{cm^3/(mol\ s)} \tag{33.14}$$

$$k^0_{ter} = 6.43 \times 10^6\, T^2 \exp(395/RT)\ \mathrm{cm^3/(mol\ s)} \tag{33.15}$$

TABLE 33.7 SAR parameters for H-atom abstraction of substituted hydrocarbon by ·Cl [63].

Group	E_X (cal/mol)	Group	E_X (cal/mol)
$-CH_3$	−960	−F	830
$-CH_2CH_3$	−2500	$-CH_2F$	210
$>CH_2$	−2020	$-CHF_2$	2470
>CH−	−2180	$-CF_3$	6350
>C<	−1950	−Br	390
−Cl	560	−OH	−2510
$-CH_2Cl$	160	$-OCH_3$	−3220
$-CHCl_2$	1470	$-OOCH_3$	−2950
$-CCl_3$	2050	−O	−1070

TABLE 33.8 Group rate constants (k_{basic}, $\times 10^{10}$ cm^3/molecule/s) and substituent factors (C_X) for ·Cl addition reactions to alkenes structural basic units at 298 K [64].

Basic structure	k_{basic}	Substituent X	C_X	Substituent X	C_X
CH_2=CHR	2.89	–F	0.64	$-C(O)OCH_2CH_3$	0.84
$CH_2=CR_2$	3.40	–Cl	0.57	$-C(O)O(CH_2)_3CH_3$	0.87
(Z)-RHC=CHR	3.63	$-CF_3$	0.31	$-OC(O)CH_3$	0.93
(E)-RHC=CHR	3.63	$-CH_2CH_3$	1.10	$-OC(O)CH_2CH_3$	0.73
$RHC=CR_2$	3.95	$-COCH_3$	0.73	$-CH_2OC(O)CH_3$	0.45
		$-COCH_2CH_3$	0.87	–C(O)H	0.87
		–C(O)OCH3	0.74	$-CH_2OH$	0.60

Teruel et al. extended the SAR approach to the application in the ·Cl addition reactions for alkenes and unsaturated and oxygenated compounds [64]. The rate constants of the basic structures (k_{basic}) and the substituent factors (C_X) are given in Table 33.8. The predicted k_{Cl} by this method showed a good correlation with the experimental ones. Blanco et al. also used the SAR approach to predict k_{Cl} of unsaturated esters and aldehydes [65]. The corresponding k_{basic} and C_X are listed in Table 33.9. Ifang et al. further revised the ester substituent factors based on their measured kinetics data and k_{Cl} from previous studies [61]. Aschmann and Atkinson have extended a previous SAR for the reactions of ·Cl with alkanes to include acyclic and cyclic ketones [66]. Gallego-Iniesta et al. estimated k_{Cl} of α,β-unsaturated esters using physicochemical correlations and SAR methods, and they concluded that the best method is the correlation of ln$k$ versus E_{HOMO} [71]. In addition, for the first time, they proposed the group reactivity factors f(–C(O)OR) for a variety of α,β-unsaturated esters, which would be beneficial to estimate k_{Cl} values of analogous compounds.

Some k_{Cl} prediction models were built based on physical–chemical parameters such as ionization potential, E_{HOMO}, and topological indices. McGillen et al. found that the k_{Cl} values of nonmethane hydrocarbons (C_2–C_{10}) correlate better with Randić index than ionization potential [62]. Estrada and Matamala improved the performance of the k_{Cl} model for alkanes and cycloalkanes by using the generalized topological indices (GTIs) ($R^2 = 0.960$) instead of Randić index ($R^2 = 0.860$) as descriptors [68]. Long and Niu developed a k_{Cl} model for alkylnaphthalenes with PLS method and eight quantum chemical descriptors [31]. This model has satisfactory goodness-of-fit ($R^2 = 0.929$) and robustness ($Q^2_{cum} = 0.896$), and it could facilitate mechanism interpretation. Blanco et al. [69] and Timerghazin et al. [67], respectively, built the linear relationship between k_{Cl} and E_{HOMO} for unsaturated compounds and alkenes. The good correlation allows these models to estimate k_{Cl} of chemicals for which the rate constants are unknown. It is worth pointing out that Poutsma developed SAR for predicting k_{Cl} of H-atom abstraction reactions by combining Evans–Polanyi and Hammett treatment based on a relatively large dataset ($n = 64$) [70]. The combination approach exhibits significantly better performance ($R^2 > 0.870$) than the individual one.

Perspectives

This chapter summarizes and compares the published k_{O_3}, k_{OH}, k_{NO_3}, and k_{Cl} prediction models. Overall, the modeling studies on predicting k_{O_3}, k_{OH}, and k_{NO_3} are relatively abundant. Some models are established based on

TABLE 33.9 Group rate constants (k_{basic}, $\times 10^{10}$ cm^3/molecule/s) and substituent factors (C_X) for the reactions of Cl with unsaturated esters and aldehydes at 298 K [65].

Basic structure	k_{basic}	Substituent X	C_X
CH_2=CHR	2.89	$-C(O)OCH_3$	0.74
$CH_2=CR_2$	3.40	$-C(O)OCH_2CH_3$	0.84
(Z/E)-RHC=CHR	3.63	$-C(O)O(CH_2)_3CH_3$	0.87
$RHC=CR_2$	3.95	–C(O)H	0.87

a large dataset that covers structurally diverse chemicals and could be used to predict rate constants at different temperatures. In contrast, the k_{Cl} modeling with a large dataset is scarce, implying narrow ADs. Considering the increasing importance of ·Cl in atmospheric chemistry, it is necessary to develop k_{Cl} QSAR models with broader ADs, and the temperature-dependent k_{Cl} models are also needed. Since gaseous pollutants react simultaneously with multiple oxidants in the atmosphere, it is necessary to obtain the reaction rate constants of pollutants with all the major oxidants. The multitarget model for simultaneously predicting multiple rate constants (k_{O_3}, k_{OH}, k_{NO_3}, and k_{Cl}) of the diverse chemicals is needed and would be helpful in reducing computational costs, time, and efforts.

It is found that combining different classes of descriptors to establish QSAR models could achieve better performance than a single class of descriptors. Moreover, the further combination of different classes of descriptors with various statistical algorithms including machine learning and deep learning algorithms (e.g., convolutional neural network) could generate much better predictive performance, especially for large datasets. However, QSAR models built by machine learning and deep learning algorithms are generally regarded as "black-box" models. Thus, it is better to uncover the "black-box" models and to identify the molecular features governing the predictive endpoints, which could facilitate the mechanistic understanding.

Defining ADs is important for regulatory acceptance of QSAR models. Nevertheless, some published models failed to characterize the ADs. The conventionally used characterization for MLR models may be not applicable for the new modeling scenarios such as machine learning algorithms. Therefore, it is necessary to develop characterization methods of ADs for QSAR models developed by machine learning algorithms.

Acknowledgments

This work is supported by the National Natural Science Foundation of China (21976027) and the Fundamental Research Funds for the Central Universities (2412019FZ019, 2412022ZD055).

References

[1] Finlayson-Pitts BJ, Pitts Jr JN. Chemistry of the upper and lower atmosphere: theory, experiments, and applications. Elsevier; 1999.
[2] Waring MS, Wells JR. Volatile organic compound conversion by ozone, hydroxyl radicals, and nitrate radicals in residential indoor air: magnitudes and impacts of oxidant sources. Atmos Environ 2015;106:382–91.
[3] Basant N, Gupta S. Multi-target QSPR modeling for simultaneous prediction of multiple gas-phase kinetic rate constants of diverse chemicals. Atmos Environ 2018;177:166–74.
[4] Li C, Xie H-B, Chen J, Yang X, Zhang Y, Qiao X. Predicting gaseous reaction rates of short chain chlorinated paraffins with ·OH: overcoming the difficulty in experimental determination. Environ Sci Technol 2014;48(23):13808–16.
[5] Atkinson R, Baulch DL, Cox RA, Crowley JN, Hampson RF, Hynes RG, et al. Evaluated kinetic and photochemical data for atmospheric chemistry: volume I—gas phase reactions of O_x, HO_x, NO_x and SO_x species. Atmos Chem Phys 2004;4:1461–738.
[6] Li C, Wei G, Chen J, Zhao Y, Zhang Y-N, Su L, Qin W. Aqueous OH radical reaction rate constants for organophosphorus flame retardants and plasticizers: experimental and modeling studies. Environ Sci Technol 2018;52(5):2790–9.
[7] Lawrence MG, Jöckel P, Von Kuhlmann R. What does the global mean OH concentration tell us? Atmos Chem Phys 2001;1(1):37–49.
[8] Prinn RG, Huang J, Weiss RF, Cunnold DM, Fraser PJ, Simmonds PG, et al. Evidence for substantial variations of atmospheric hydroxyl radicals in the past two decades. Science 2001;292(5523):1882–8.
[9] Mauldin RL, Madronich S, Flocke SJ, Eisele FL, Frost GJ, Prevot ASH. New insights on OH: measurements around and in clouds. Geophys Res Lett 1997;24(23):3033–6.
[10] Mount GH, Williams EJ. An overview of the tropospheric OH photochemistry experiment, Fritz Peak/Idaho Hill, Colorado, fall 1993. J Geophys Res-Atmos 1997;102(D5):6171–86.
[11] Siese M, Becker KH, Brockmann KJ, Geiger H, Hofzumahaus A, Holland F, et al. Direct measurement of OH radicals from ozonolysis of selected alkenes: a euphore simulation chamber study. Environ Sci Technol 2001;35(23):4660–7.
[12] Gómez Alvarez E, Amedro D, Afif C, Gligorovski S, Schoemaecker C, Fittschen C, et al. Unexpectedly high indoor hydroxyl radical concentrations associated with nitrous acid. Proc Natl Acad Sci 2013;110(33):13294–9.
[13] Bartolomei V, Gomez Alvarez E, Wittmer J, Tlili S, Strekowski R, Temime-Roussel B, et al. Combustion processes as a source of high levels of indoor hydroxyl radicals through the photolysis of nitrous acid. Environ Sci Technol 2015;49(11):6599–607.
[14] Su H, Cheng Y, Oswald R, Behrendt T, Trebs I, Meixner FX, et al. Soil nitrite as a source of atmospheric HONO and OH radicals. Science 2011; 333(6049):1616–8.
[15] Gramatica P, Pilutti P, Papa E. Validated QSAR prediction of OH tropospheric degradation of VOCs: splitting into training–test tets and consensus modeling. J Chem Inf Comput Sci 2004;44(5):1794–802.
[16] Öberg T. A QSAR for the hydroxyl radical reaction rate constant: validation, domain of application, and prediction. Atmos Environ 2005; 39(12):2189–200.
[17] Wang Y-N, Chen J, Li X, Wang B, Cai X, Huang L. Predicting rate constants of hydroxyl radical reactions with organic pollutants: algorithm, validation, applicability domain, and mechanistic interpretation. Atmos Environ 2009;43(5):1131–5.

[18] Li C, Yang X, Li X, Chen J, Qiao X. Development of a model for predicting hydroxyl radical reaction rate constants of organic chemicals at different temperatures. Chemosphere 2014;95:613–8.
[19] Gupta S, Basant N, Mohan D, Singh KP. Modeling the reactivities of hydroxyl radical and ozone towards atmospheric organic chemicals using quantitative structure-reactivity relationship approaches. Environ Sci Pollut Res 2016;23(14):14034–46.
[20] Gupta S, Basant N, Mohan D, Singh KP. Inter-moieties reactivity correlations: an approach to estimate the reactivity endpoints of major atmospheric reactants towards organic chemicals. RSC Adv 2016;6(55):50297–305.
[21] Liu Y, Cheng Z, Liu S, Tan Y, Yuan T, Yu X, Shen Z. Quantitative structure activity relationship (QSAR) modelling of the degradability rate constant of volatile organic compounds (VOCs) by OH radicals in atmosphere. Sci Total Environ 2020;729:138871.
[22] Sanches-Neto FO, Dias-Silva JR, De Oliveira VM, Aquilanti V, Carvalho-Silva VH. Evaluating and elucidating the reactivity of OH radicals with atmospheric organic pollutants: reaction kinetics and mechanisms by machine learning. Atmos Environ 2022;275:119019.
[23] Shi Y, Li J-J, Wang Q, Jia Q, Yan F, Luo Z-H, Zhou Y-N. Computer-aided estimation of kinetic rate constant for degradation of volatile organic compounds by hydroxyl radical: an improved model using quantum chemical and norm descriptors. Chem Eng Sci 2022;248:117244.
[24] Qi C, Zhang C, Sun X. Quantitative structure-activity relationships study on the rate constants of polychlorinated dibenzo-p-dioxins with OH radical. Int J Mol Sci 2015;16(8):18812–24.
[25] Yang Z, Luo S, Wei Z, Ye T, Spinney R, Chen D, Xiao R. Rate constants of hydroxyl radical oxidation of polychlorinated biphenyls in the gas phase: a single–descriptor based QSAR and DFT study. Environ Pollut 2016;211:157–64.
[26] Luo S, Wei Z, Spinney R, Yang Z, Chai L, Xiao R. A novel model to predict gas–phase hydroxyl radical oxidation kinetics of polychlorinated compounds. Chemosphere 2017;172:333–40.
[27] Markelj J, Pompe M. Modeling of atmospheric OH reaction rates using newly developed variable distance weighted zero order connectivity index. Atmos Environ 2016;131:418–23.
[28] Chen Z, Yu X, Huang X, Zhang S. Prediction of reaction rate constants of hydroxyl radical with organic compounds. J Chil Chem Soc 2014;59: 2252–9.
[29] Xu Y, Yu X, Zhang S. QSAR models of reaction rate constants of alkenes with ozone and hydroxyl radical. J Braz Chem Soc 2013;24:1781–8.
[30] Pompe M, Veber M, Randić M, Balaban AT. Using variable and fixed topological indices for the prediction of reaction rate constants of volatile unsaturated hydrocarbons with OH radicals. Molecules 2004;9(12):1160–76.
[31] Long X, Niu J. Estimation of gas-phase reaction rate constants of alkylnaphthalenes with chlorine, hydroxyl and nitrate radicals. Chemosphere 2007;67(10):2028–34.
[32] Fatemi MH, Baher E. Quantitative structure–property relationship modelling of the degradability rate constant of alkenes by OH radicals in atmosphere. SAR QSAR Environ Res 2009;20(1–2):77–90.
[33] Huang X, Yu X, Yi B, Zhang S. Prediction of rate constants for the reactions of alkanes with the hydroxyl radicals. J Atmos Chem 2012;69(3): 201–13.
[34] Yu X, Deng J, Yi B, Liu W. Predicting rate constants of hydroxyl radical reactions with alkenes and aromatics. J Atmos Chem 2015;72(2): 129–41.
[35] Gramatica P, Consonni V, Todeschini R. QSAR study on the tropospheric degradation of organic compounds. Chemosphere 1999;38(6): 1371–8.
[36] Tanaka T, Morino Y. Coriolis interaction and anharmonic potential function of ozone from the microwave spectra in the excited vibrational states. J Mol Spectrosc 1970;33(3):538–51.
[37] Percival C, Mcgillen M. Overview of structure-activity relationship methods for predicting gas-phase rate coefficients; proceedings of the simulation and assessment of chemical processes in a multiphase environment. Dordrecht: Springer Netherlands; 2008.
[38] Singh HB, Ludwig FL, Johnson WB. Tropospheric ozone: concentrations and variabilities in clean remote atmospheres. Atmos Environ 1978; 12(11):2185–96.
[39] Oltmans SJ. Surface ozone measurements in clean air. J Geophys Res Oceans 1981;86(C2):1174–80.
[40] Logan JA, Prather MJ, Wofsy SC, Mcelroy MB. Tropospheric chemistry: a global perspective. J Geophys Res Oceans 1981;86(C8):7210–54.
[41] Stockwell WR, Lawson CV, Saunders E, Goliff WS. A review of tropospheric atmospheric chemistry and gas-phase chemical mechanisms for air quality modeling. Atmosphere 2011;3(1):1–32.
[42] Pompe M, Veber M. Prediction of rate constants for the reaction of O_3 with different organic compounds. Atmos Environ 2001;35(22):3781–8.
[43] Gramatica P, Pilutti P, Papa E. QSAR prediction of ozone tropospheric degradation. QSAR Comb Sci 2003;22(3):364–73.
[44] Yu X, Yi B, Wang X, Chen J. Predicting reaction rate constants of ozone with organic compounds from radical structures. Atmos Environ 2012; 51:124–30.
[45] Fatemi MH. Prediction of ozone tropospheric degradation rate constant of organic compounds by using artificial neural networks. Anal Chim Acta 2006;556(2):355–63.
[46] Ren Y, Liu H, Yao X, Liu M. Prediction of ozone tropospheric degradation rate constants by projection pursuit regression. Anal Chim Acta 2007;589(1):150–8.
[47] Li X, Zhao W, Li J, Jiang J, Chen J, Chen J. Development of a model for predicting reaction rate constants of organic chemicals with ozone at different temperatures. Chemosphere 2013;92(8):1029–34.
[48] Liu Y, Liu S, Cheng Z, Tan Y, Gao X, Shen Z, Yuan T. Predicting the rate constants of volatile organic compounds (VOCs) with ozone reaction at different temperatures. Environ Pollut 2021;273:116502.
[49] Brown SS, Stutz J. Nighttime radical observations and chemistry. Chem Soc Rev 2012;41(19):6405–47.
[50] Luo S, Luo Q, Ye T, Zhang J, Xie X, Yang Y. Chemical structure based model to predict nitrate radical–mediated degradation of saturated organic contaminants in troposphere. Atmos Pollut Res 2022;13(2):101314.
[51] Geyer A, Alicke B, Konrad S, Schmitz T, Stutz J, Platt U. Chemistry and oxidation capacity of the nitrate radical in the continental boundary layer near Berlin. J Geophys Res Atmos 2001;106(D8):8013–25.
[52] Papa E, Gramatica P. Externally validated QSPR modelling of VOC tropospheric oxidation by NO_3 radicals. SAR QSAR Environ Res 2008; 19(7–8):655–68.
[53] Sabljic A, Güsten H. Predicting the night-time NO_3 radical reactivity in the troposphere. Atmos Environ Part A Gen Top 1990;24(1):73–8.

[54] Gramatica P, Pilutti P, Papa E. Predicting the NO_3 radical tropospheric degradability of organic pollutants by theoretical molecular descriptors. Atmos Environ 2003;37(22):3115–24.
[55] Gupta S, Basant N, Mohan D, Singh KP. Room-temperature and temperature-dependent QSRR modelling for predicting the nitrate radical reaction rate constants of organic chemicals using ensemble learning methods. SAR QSAR Environ Res 2016;27(7):539–58.
[56] Schindler M. A QSAR for the prediction of rate constants for the reaction of VOCs with nitrate radicals. Chemosphere 2016;154:23–33.
[57] Liu Y, Cheng Z, Liu S, Ren Y, Yuan T, Zhang X, et al. A quantitative structure activity relationship (QSAR) model for predicting the rate constant of the reaction between VOCs and NO_3 radicals. Chem Eng J 2022;448:136413.
[58] Faxon CB, Allen DT. Chlorine chemistry in urban atmospheres: a review. Environ Chem 2013;10(3):221–33.
[59] Thornton JA, Kercher JP, Riedel TP, Wagner NL, Cozic J, Holloway JS, et al. A large atomic chlorine source inferred from mid-continental reactive nitrogen chemistry. Nature 2010;464(7286):271–4.
[60] Wingenter OW, Blake DR, Blake NJ, Sive BC, Rowland FS, Atlas E, et al. Tropospheric hydroxyl and atomic chlorine concentrations, and mixing timescales determined from hydrocarbon and halocarbon measurements made over the Southern Ocean. J Geophys Res 1999;104(D17):21819–28.
[61] Ifang S, Benter T, Barnes I. Reactions of Cl atoms with alkyl esters: kinetic, mechanism and atmospheric implications. Environ Sci Pollut Res 2015;22(7):4820–32.
[62] Mcgillen MR, Percival CJ, Raventos-Duran T, Sanchez-Reyna G, Shallcross DE. Can topological indices be used to predict gas-phase rate coefficients of importance to tropospheric chemistry? Free radical abstraction reactions of alkanes. Atmos Environ 2006;40(14):2488–500.
[63] Senkan SM, Quam D. Correlation of reaction rate coefficients for the abstraction of hydrogen atoms from organic compounds by chlorine radical attack. J Phys Chem 1992;96(26):10837–42.
[64] Teruel MA, Achad M, Blanco MB. Kinetic study of the reactions of Cl atoms with α,β-unsaturated carbonyl compounds at atmospheric pressure and structure activity relations (SARs). Chem Phys Lett 2009;479(1):25–9.
[65] Blanco MB, Barnes I, Teruel MA. FTIR gas-phase kinetic study of the reactions of Cl atoms with $(CH_3)_2CCHC(O)H$ and $CH_3CHCHC(O)OCH_3$. Chem Phys Lett 2010;488(4):135–9.
[66] Aschmann SM, Atkinson R. Rate constants for the reactions of Cl atoms with a series of C6–C10 cycloalkanes and cycloketones at 297 ± 2 K. Int J Chem Kinet 2013;45(1):52–8.
[67] Timerghazin QK, Ariya PA. Kinetics of the gas-phase reaction of atomic chlorine with selected monoterpenes. Phys Chem Chem Phys 2001;3(18):3981–6.
[68] Estrada E, Matamala AR. Generalized topological indices. Modeling gas-phase rate coefficients of atmospheric relevance. J Chem Inf Model 2007;47(3):794–804.
[69] Blanco MB, Bejan I, Barnes I, Wiesen P, Teruel MA. Temperature-dependent rate coefficients for the reactions of Cl atoms with methyl methacrylate, methyl acrylate and butyl methacrylate at atmospheric pressure. Atmos Environ 2009;43(38):5996–6002.
[70] Poutsma ML. Evolution of structure–reactivity correlations for the hydrogen abstraction reaction by chlorine atom. J Phys Chem A 2013;117(4):687–703.
[71] Gallego-Iniesta MP, Cabañas B, Salgado S, Martínez E, Martín P. Estimation of gas-phase rate coefficients for the reactions of a series of α,β-unsaturated esters with OH, NO_3, O_3 and Cl. Atmos Environ 2014;90:133–45.

PART V

QSAR models in material science and other areas

CHAPTER

34

Significance of QSAR in cancer risk assessment of polycyclic aromatic compounds (PACs)

Gustav Gbeddy[1], *Prasanna Egodawatta*[2], *Ashantha Goonetilleke*[2], *Eric Glover*[1], *Eric Akortia*[1] *and Courage Egbi*[1]

[1]Ghana Atomic Energy Commission (GAEC), Legon-Accra, Ghana [2]Queensland University of Technology (QUT), Brisbane, QLD, Australia

Abbreviations

1MNAP 1-Methylnaphthalene
1MPHE 1-Methylphenanthrene
1NTD 1-Naphthaldehyde
2HBP 2-Hydroxybiphenyl
2MAQN 2-Methyl-9,10-anthraquinone
2MNAP 2-Methylnaphthalene
2NFLR 2-Nitrofluorene
3NFL 3-Nitrofluoranthene
4HBP 4-Hydroxybiphenyl
5NAC 5-Nitroacenaphthene
9FLN 9-Fluorenone
9NAN 9-Nitroanthracene
9NPHE 9-Nitrophenanthrene
ABS Dermal adsorption fraction
ACE Acenaphthene
ACNQ 1,2-Acenaphthenequinone
ACT Acenaphthylene
AD Applicability domain
AF Dermal adherence factor
ANT Anthracene
AQN 9,10-Anthraquinone
AT Averaging life span
BaAN Benz[a]anthracene
BaFN Benzo[a]fluoren-11-one
BAND Benzo[a]anthracene-7,12-dione
BANN 7H-benz[de]anthracen-7-one (benzanthrone)
BaP_{eq} Benzo[a]pyrene equivalent concentration
BcPH Benzo[c]phenanthrene
BghiPE Benzo[ghi]perylene
BP Boiling point
BPCX 2-Biphenylcarboxaldehyde
BPN Biphenyl
BW Body weight
CHR Chrysene
C_i Concentration of PAC
CPAHs Carbonyl-PAHs
CPPHE 4H-cyclopenta(def)phenanthrene
CPPHN 4H-cyclopenta(def)phenanthren-4-one
CRN Coronene
CSF_{dermal} Carcinogenic slope factor dermal uptake
$CSF_{ingestion}$ Carcinogenic slope factor ingestion
$CSF_{inhalation}$ Carcinogenic slope factor inhalation
DHAQ 1,8-Dihydroxyanthraquinone (Danthron)
DMBA 7,12-Dimethylbenz[a]anthracene
DMNA 1,3-Dimethylnaphthalene
DMPH 3,6-Dimethylphenanthrene
DNFL 2,7-Dinitrofluorene
ED Exposure duration
EF Exposure frequency
F Variance ratio
FA Factor analysis
FLR Fluorine
HMW Heavy molecular weight
HOPAHs Hydroxy-PAHs
HPAHs Halogenated-PAHs
HPYR 1-Hydroxypyrene
H_v Enthalpy of vaporization
IARC International Agency for Research on Cancer
IDN 1-Indanone
ILCR Incremental lifetime cancer risk
$IR_{ingestion}$ Dust ingestion rate
$IR_{inhalation}$ Inhalation rate
logBCF Bioaccumulation estimate from $logK_{ow}$
$logK_{oc}$ Soil adsorption coefficient
$logK_{ow}$ Octanol–water partitioning coefficient
LOO Leave-one-out
MAE Mean average error
MCHO 3-Methylcholanthrene
MLR Multiple linear regression
MP Melting point
mvol Molar volume
M_w Molecular weight
NAP Naphthalene
NNAP 1-Nitronaphthalene
NOAR Number of aromatic rings
NOR Number of rings
NPAHs Nitro-PAHs
NQN 1,4-Naphthaquinone

QSAR in Safety Evaluation and Risk Assessment
https://doi.org/10.1016/B978-0-443-15339-6.00025-4

NSO-PACs Nitrogen, sulfur, and oxygen PACs
NTCQ 5,12-Naphthacenequinone
OCPs Organochlorine pesticides
OPAHs Oxygenated-PAHs
PACs Polycyclic aromatic compounds
PAHs Polycyclic aromatic hydrocarbons
PCBs Polychlorinated biphenyls
PCDDs Polychlorinated dibenzo-p-dioxins
PCDFs Polychlorinated dibenzofurans
PEF Particle emission factor
PHE Phenanthrene
PIC Picene
Pl Polarizability
PLSR Partial least squares regression
POPs Persistent organic pollutants
PRESS Predicted residual sum of squares
PRL Perylene
PYR Pyrene
Q^2 Cross-validated determination coefficient
R^2 Determination coefficient
Ra^2 Adjusted determination coefficient
Rf Molar refractivity
RMSEC Root mean square error of calibration
RT Retene
s Standard error of estimate
SA Dermal exposure area
SDEP Standard deviation of error of prediction
St Surface tension
Sw Water solubility
TEF Toxicity equivalency factors
V_p Vapor pressure
ρ Density

Introduction

Polycyclic aromatic compounds (PACs) refer to a vast array of compounds entailing polycyclic aromatic hydrocarbons (PAHs) and their associated substituents or derivatives, as well as nitrogen, sulfur, and oxygen PACs (NSO-PACs). The substituted PAHs include alkylated-PAHs, nitro-PAHs (NPAHs), carbonyl-PAHs (CPAHs), cyano-PAHs, amino-PAHs, halogenated-PAHs (HPAHs), oxygenated-PAHs (OPAHs), and notably hydroxy-PAHs (HOPAHs) among others. The NSO-PACs contain either one or more nitrogen, sulfur, and oxygen atoms in the structure of the aromatic compound and can also be termed as NSO-heterocyclic PAHs (N/S/O-hPAHs) [1–4]. PACs are generally classified as persistent organic pollutants (POPs) due to their toxic, bioaccumulative, and persistent properties. A number of the pollutants in the PACs category of POPs have been identified as carcinogens and mutagens by the International Agency for Research on Cancer (IARC) [1,4–6]. In this regard, the assessment of the cancer risk posed by PACs is highly essential to protect human health and the environment.

The generic approach in estimating the cancer risk posed by PACs to living organisms especially humans is through the use of mathematical models that are based on the additive or cumulative effects of these pollutants. In this context, the concentrations and toxicity equivalency factors (TEFs) of the pollutants are vital parameters during the computation of the probable risk posed by these pollutants. In this regard, any limitation in the availability of these two parameters results in inaccurate estimation of the health risk posed by PACs in the environment. Therefore, the focus of this chapter is on how existing knowledge gaps with respect to the lack of TEF for PACs can be resolved via the application of quantitative structure–activity relationship (QSAR). Andersson and Achten [2] and Gbeddy et al. [6] have discussed the consequences of the limitations in determining the concentration of PACs on the output of potential toxic effects of these pollutants and how this can be resolved. Therefore, the concentration aspect of health risk assessment for PACs will not be discussed in this chapter.

Research studies conducted by Gbeddy et al. [5], Gbeddy et al. [7], and Gbeddy [3] have elucidated the consequences of the lack of TEF values for a large number of PACs especially in terms of cancer health risk under-estimation. For example, the application of QSAR by Gbeddy et al. [5] during an investigation of the cancer risk posed by PACs in road dust was found to lead to potential risk rather than no risk if the model was excluded. In this regard, appropriate remediation measure may be deployed by relevant authorities in curbing the toxic effects of these pollutants in the environment. Therefore, QSAR comes in handy to help address this challenge of under-estimation of health risk posed by PACs. Additionally, QSAR approach reduces the burden on animal toxicological test, and thus, the adoption of QSAR approach in cancer risk assessment of PACs offers great advantages over in vivo and in vitro approaches. As noted by Gbeddy et al. [5], QSAR offers an efficient, cost-effective, and less time-consuming approach in cancer health risk assessment from PACs exposure. This chapter presents a systematic QSAR approach in evaluating the cancer risk posed by PACs to living organisms.

QSAR methodology

QSAR refers to a mathematical association between pharmacological quantitatively expressed toxicities/activities of a series of pollutants to their relevant chemical properties. The development of QSAR involves five important

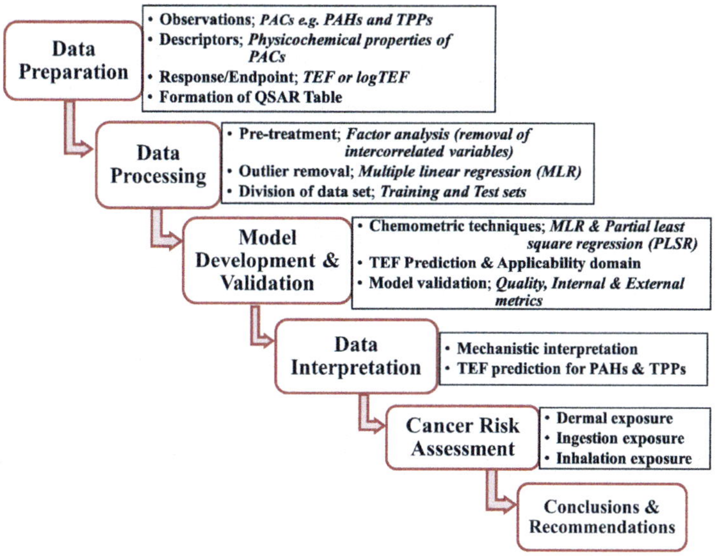

FIGURE 34.1 Sequence of steps for QSAR model development and application for PACs.

elements, namely (1) clearly defined endpoint; (2) model development using explicit algorithms; (3) unambiguous applicability domain (AD); (4) applicable statistical metrics for establishing the goodness-of-fit, robustness and prediction capability, and ultimately; (5) plausible mechanistic clarification [5,8–10]. The endpoint refers to the response or dependent variable (for instance, TEF) that has been derived from the predictor or independent variables such as physicochemical or structural properties of PACs (for example, molecular weight (M_w), vapor pressure (V_p), number of rings (NOR), boiling point (BP), enthalpy of vaporization (H_v), density (ρ), etc.). The developed QSAR model must be based on unambiguous mathematical algorithm, and thus the application of relevant chemometric techniques help address this requirement. The AD describes the boundaries of the physicochemical properties of PACs within which the QSAR model can produce reliable TEF results. The quality and dependability of the TEF output from the QSAR model that is produced must be hinged on sufficient and cogent statistical validation entailing internal and external validation techniques. Finally, the developed QSAR model may exhibit a potential mechanistic linkage between the predictors and the TEF endpoint. However, the absence of a mechanistic interpretation of the model does not denote lack of predictive power and reliability in the developed model [5,8].

Furthermore, the development of the QSAR model for assessing cancer health risk posed by PACs entails sequence of steps such as data preparation, data processing, model development and validation, data interpretation, cancer risk assessment, and conclusions and recommendations as outlined in Fig. 34.1 [8]. The execution of these steps ensures that all the stipulated five ingredients are duly incorporated into the model, thereby promoting the quality and reliability of the outcome. These steps are expatiated in subsequent sections of this chapter.

Data preparation and processing

The selection of suitable predictor variables especially from a large number of parameters constitutes a pivotal foundation on which a robust QSAR model can be developed. Kunal et al. [8] noted that such descriptors must possess some vital attributes such as reproducibility, produce varied values for compounds with dissimilar structure (minimal degeneracy), simplicity, applicability to a wide range of compounds, not dependent on experimental properties, good correlation with the endpoint but insignificant correlation with other descriptors, and have structural interpretation. In this context, the use of chemometric methods such as factor analysis (FA) and multiple linear regression (MLR) has been identified as essential in attaining the desired 1:5 ratio between predictor variables and observations (compounds). FA ensures that highly intercorrelated variables are removed, while MLR facilitates the identification and elimination of outlier observations or compounds prior to the model development. The choice of relevant chemometric software facilitates data processing, for instance, Minitab software is very useful in identifying and removing of outlier variables and compounds. From an initial number of 16 predictors comprising of density (ρ), octanol–water partitioning coefficient ($\log K_{ow}$), molar refractivity (Rf), vapor pressure (V_p), polarizability (Pl), melting point (MP), boiling point (BP), surface tension (St), water solubility (Sw), bioaccumulation estimate from $logK_{OW}$ (logBCF), molecular weight (M_W), soil adsorption coefficient ($logK_{OC}$), number of rings (NOR), molar volume (mvol), number of aromatic rings (NOAR), and enthalpy of vaporization (H_V), and 4 PAC predictors specifically BP, H_V, $logV_P$, and M_W were found to be relevant [5,10].

Furthermore, it has been reported by Kunal et al. [8] that the logarithmic transformation of the endpoint may facilitate better model output. However, this may not always be the case, and thus, the need to verify the appropriateness of both log-transformed endpoint and non–log-transformed endpoint. Moreover, the original dataset may be partitioned into training and test sets to facilitate the authentication process for the model in terms of its reliability. The training set is employed in developing the model, whereas the model's predictability is ascertained by utilizing the test set. However, as a result of the small number (less than 30) of PAHs and transformed PAH products (TPPs) in the dataset, the data was not divided since according to Hawkins et al. [11], such a partitioning will result in loss of information, thereby inhibiting the model's quality. In this regard, cross-validation was shown to be a suitable method for determining the predictability of the model. The formulated QSAR table for PAHs and TPPs is shown in Table 34.1.

Model development using chemometric technique and validation

A careful review of existing literature notably by Miller and Miller [12] shows that multiple linear regression (MLR) and partial least squares regression (PLSR) are two pertinent chemometric techniques in QSAR model development. PLSR method explains the variation in the predictors as a result of the linear combination of the underlying

TABLE 34.1 QSAR table for PAHs and TPPs.

		Selected predictor variables				Response variable	
PAHs	**ID**	**H_v**	**M_w**	**$logV_p$**	**BP**	**TEF**	**logTEF**
7H-benz[de]anthracen-7-one (benzanthrone)	BANN	69.3	230.3	−6.65560773	403.2	0.0039	−2.409
4H-cyclopenta(def) phenanthren-4-one	CPPHN	66.4	204.2	−5.57348874	375.8	0.005	−2.301
3-Nitrofluoranthene	3NFL	67.6	247.3	−14.8996295	586.7	0.0026	−2.585
9-Nitroanthracene	9NAN	62.9	223.2	−13.2932822	543.2	0.0032	−2.495
5-Nitroacenaphthene	5NAC	60.1	199.2	−10.9430951	478.3	0.01	−2.000
2-Nitrofluorene	2NFLR	59.8	211.2	−11.5767541	495.9	0.01	−2.000
Coronene	CRN	77	300.4	−9.35654732	493.5	0.01	−2.000
Perylene	PRL	70.2	252.3	−8.2798407	442.8	0.001	−3.000
1-Methylnaphthalene	1MNAP	46	142.2	−1.1739252	240	0.0025	−2.602
2-Methylnaphthalene	2MNAP	45.7	142.2	−1.25963731	241.1	0.001	−3.000
Pyrene	PYR	63	202.3	−5.34678749	404	0.001	−3.000
Phenanthrene	PHE	55.8	178.2	−3.91721463	340	0.0005	−3.301
Naphthalene	NAP	43.9	128.2	−1.07058107	217.9	0.001	−3.000
Fluorene	FLR	51.2	166.2	−3.22184875	295	0.0005	−3.301
Chrysene	CHR	67.9	228.3	−8.20551195	448	0.01	−2.000
Benzo[ghi]perylene	BghiPE	74.1	276.3	−9.05799195	486.3	0.01	−2.000
Benz[a]anthracene	BaAN	66.7	228.3	−6.67778071	437.6	0.005	−2.301
Anthracene	ANT	55.8	178.2	−5.18508682	339.9	0.0005	−3.301
Acenaphthylene	ACT	51.7	152.2	−2.17522354	280	0.001	−3.000
Acenaphthene	ACE	49.7	154.2	−2.66756154	279	0.001	−3.000
Range		43.9 to 77	128.2 to 300.4	−14.89963 to −1.07058107	217.9 to 586.7	0.0005 to 0.01	−3.301 to −2.000

Adopted from Gbeddy et al. [5].

TABLE 34.2 Comparison of statistical parameters for the developed QSAR model quality and validity.

	Quality statistics							Acceptability and reliability statistics		
Model	R^2	s	RMSEC	R_a^2	R	F	MAE	SDEP	Q^2	PRESS
a. logTEF response										
PLSR1	0.43	0.41	0.355	0.28	0.66	2.88	0.31	0.39	0.30	3.092
MLR	0.44	0.41	0.352	0.58	0.67	2.97	0.30	0.46	0.05	4.221
b. TEF response										
PLSR1	0.72	0.0034	0.00290	0.65	0.85	9.68	0.0024	0.0033	0.65	0.000212
MLR	0.73	0.0033	0.00288	0.65	0.85	9.91	0.0023	0.0038	0.51	0.000294

MAE, R, R^2, Ra^2, SDEP, F, RMSEC, s, Q^2, and PRESS represent the mean average error, regression correlation coefficient, determination coefficient, adjusted R^2, standard deviation of error of prediction, variance ratio, root mean square error of calibration, standard error of estimate, cross-validated determination coefficient, and predicted residual sum of squares, respectively.
Adopted from Gbeddy et al. [5].

predictors. PLSR is capable of analyzing a large dataset with missing, collinear, and noisy values. It can also simultaneously predict more than one response variable from a myriad of predictors by using PLSR1 or PLSR2 algorithms for one or more responses, respectively. On the other hand, MLR determines the regression equation for a dataset made up of a smaller number of predictor variables compared with the number of PAH and TPP compounds [5]. Gbeddy et al. [5] noted that Minitab statistical software is highly relevant for the QSAR model development due to its capability to identify potential outlier compounds during preprocessing for subsequent elimination prior to the actual model development. In addition to ease of use, Minitab also offers a comprehensive platform for performing both MLR and PLSR analysis, thereby facilitating the comparison of the results.

The quality of the developed QSAR model was evaluated using variance ratio (F), mean average error (MAE), adjusted determination coefficient (Ra^2), standard error of estimate (s), root mean square error of calibration (RMSEC), and determination coefficient (R^2). On the other hand, the validity and reliability of the developed model was assessed via leave-one-out (LOO) cross-validation due to the small data matrix of four predictors and 20 compounds [5,10]. In this regard, cross-validated determination coefficient (Q^2), standard deviation of error of prediction (SDEP), and predicted residual sum of squares (PRESS) LOO internal validation methods were used [5,8]. According to Hawkins et al. [11], external validation may not be necessary for a small data matrix since withholding of portions of the data during the model development process can lead to information deficit, thereby causing inaccurate model output. A comparative evaluation of these estimated statistical parameters as shown in Table 34.2 indicated that PLSR1 QSAR model for TEF response variable was the most appropriate model for estimating the TEFs of PAHs and TPPs. In particular, the PRESS value, which is an essential statistical parameter for assessing the model's predictability, was found to be comparatively smaller (that is closer to zero) for TEF response than that for logTEF response variable. The generated QSAR equation is shown in Eq. (34.1) [5].

$$TEF_{predicted} = (-5.99E-03) + (1.40E-05 * M_W) + (6.42E-05 * H_v) + (5.92E-06 * BP) - \left(1.48E-04 * logV_P\right) \tag{34.1}$$

PLSR1 QSAR model interpretation and TEF prediction

The developed model in Eq. (34.1) indicates that inverse correlation exist between only $logV_p$ and TEF response variable whereas H_v, BP, and M_w have positive relationship with TEF. The applicability domain using predictor bounding-box ranges for the four influential parameters, $logV_p$, BP, H_v, and M_w are −14.89963 – (−1.07058107) mmHg, 217.9–586.7°C, 43.9–77 kJ/mol, and 128.2–300.4 g/mol, respectively, are shown in Table 34.1 [5]. The model shows that TEF, in general, rises with ascending M_w PAHs and TPPs due to the larger octanol–water partitioning coefficient, higher lipophilicity, and bioaccumulative properties associated with pollutants [5,13]. Moreover, higher M_w PAHs and TPPs have high BP and H_v, thus the observed positive correlation between BP and H_v, and TEF. However, higher M_w PAHs and TPPs have lower vapor pressure with less volatility and, therefore, tend to be more bioaccumulative. This explains the inverse relationship between $logV_p$ and TEF [5,14].

TABLE 34.3 Estimated TEF values for PAHs and TPPs.

		Predictor variables				
PAHs	**ID**	**M_w**	**H_v**	**BP**	**$logV_p$**	**PLSR1_PredTEF**
Picene	PIC	278.4	76.2	519	−11.9914	0.008
7,12-Dimethylbenz[a]anthracene	DMBA	256.3	69.7	422.4	−6.59688	0.006
Benzo[c]phenanthrene	BcPH	228.3	66.7	448	−8.20551	0.005
3-Methylcholanthrene	MCHO	268.4	74.7	438.3	−7.36653	0.006
Biphenyl	BPN	154.2	47.6	257	−2.04915	0.001
Retene	RT	234.3	61.7	390	−5.5784	0.004
3,6-Dimethylphenanthrene	DMPH	206.3	58.5	351.49	−5.11976	0.003
Benzo[a]fluoren-11-one	BaFN	230.3	68.7	403.2	−6.41229	0.005
1,3-Dimethylnaphthalene	DMNA	156.2	48.1	263	−1.63451	0.001
9,10-Anthraquinone	AQN	208.2	62.5	377	−6.93554	0.004
1-Hydroxypyrene	HPYR	218.3	72.1	406.7	−8.31605	0.005
2-Hydroxybiphenyl	2HBP	170.2	54.2	286	−2.69897	0.002
1,8-Dihydroxyanthraquinone (Danthron)	DHAQ	240.2	73.9	433.3	−10.1192	0.006
1-Indanone	IDN	132.2	48.1	243	−1.57349	0.001
1-Nitronaphthalene	NNAP	173.2	52.3	472.3	−10.7282	0.004
1-Naphthaldehyde	1NTD	156.2	53.1	292	−2.60206	0.002
1,2-Acenaphthenequinone	ACNQ	182.2	60.9	350.3	−6.46852	0.003
2-Biphenylcarboxaldehyde	BPCX	182.2	57.5	324.5	−3.73283	0.003
2-Methyl-9,10-anthraquinone	2MAQN	222.2	66	375.2	−6.05899	0.004
5,12-Naphthacenequinone	NTCQ	258.3	73.5	434.5	−7.45593	0.006
Benzo[a]anthracene-7,12-dione	BAND	258.3	73.5	434.5	−7.41117	0.006
9-Nitrophenanthrene	9NPHE	223.2	64	543.2	−13.2933	0.006
4H-Cyclopenta(def)phenanthrene	CPPHE	190.2	57.4	353	−4.82974	0.003
2,7-Dinitrofluorene	DNFL	256.2	68.3	551.6	−13.6021	0.007
1-Methylphenanthrene	1MPHE	192.3	57.5	339.8	−4.82391	0.003
9-Fluorenone	9FLN	180.2	58.5	331.7		0.002
1,4-Naphthaquinone	NQN	158.2	53.8	301.2		0.001
4-Hydroxybiphenyl	4HBP	170.2	56.9	305		0.002

Adopted from Gbeddy et al. [5].

Based on the developed model and the relevant applicability domain, the unknown TEFs for 28 PAHs and TPPs were predicted as presented in Table 34.3. In this regard, the cancer risk posed by PAHs and TPPs can be assessed more holistically using these previously nonexisting TEF values in addition to the known TEFs for other analytes.

Cancer risk assessment

The estimation of the cancer risk posed by PAHs and TPPs exposure via ingestion, inhalation, and dermal routes using Eqs. (34.2)–(34.4) shows that the inclusion of the QSAR determined TEF values in Table 34.3 and originally existing TEFs for 10 [10] outlier compounds resulted in a more comprehensive and realistic incremental lifetime

TABLE 34.4 Parameters used in estimating the incremental lifetime cancer risk for PACs.

Parameters, units	Child	Adult
Dermal adherence factor (AF) [mg/cm]	0.2	0.07
Exposure duration (ED) [year]	6	24
Exposure frequency (EF) [day/year]	180	180
Dust ingestion rate ($IR_{ingestion}$) [mg/day]	200	100
Inhalation rate ($IR_{inhalation}$) [m^3/day]	10	20
Dermal adsorption fraction (ABS) [unitless]	0.13	0.13
Dermal exposure area (SA) [cm^2]	2800	5700
Carcinogenic slope factor ingestion ($CSF_{ingestion}$) of BaP [mg/kg/day]	7.3	7.3
Carcinogenic slope factor inhalation ($CSF_{inhalation}$) of BaP [mg/kg/day]	3.85	3.85
Carcinogenic slope factor dermal uptake (CSF_{dermal}) of BaP [mg/kg/day]	25	25
Averaging life span (AT) [years]	64.2	64.2
Body weight (BW) [kg]	15	70
Particle emission factor (PEF) [m^3/kg]	1.36E + 10	1.36E + 09

cancer risk (ILCR) estimates relative to reliance on only originally existing TEFs. The outlier compounds are predominantly heavy molecular weight (HMW) PAHs and TPPs having TEF magnitudes greater than one [5]. The ILCR model classifies cancer risk into three [3] ranges of $\leq 10^{-6}$, 10^{-4} to 10^{-6}, and $>10^{-4}$ corresponding to negligible risk, potential risk, and high risk, respectively [1,5,15].

$$ILCR_{ingestion} = \frac{\sum BaP_{eq} \times CSF_{ingestion} \times \left(\sqrt[3]{\frac{BW}{70}}\right) \times IR_{ingestion} \times EF \times ED}{BW \times AT \times 10^6} \quad (34.2)$$

$$ILCR_{inhalation} = \frac{\sum BaP_{eq} \times CSF_{inhalation} \times \left(\sqrt[3]{\frac{BW}{70}}\right) \times IR_{inhalation} \times EF \times ED}{BW \times AT \times PEF} \quad (34.3)$$

$$ILCR_{dermal} = \frac{\sum BaP_{eq} \times CSF_{dermal} \times \left(\sqrt[3]{\frac{BW}{70}}\right) \times SA \times AF \times ABS \times EF \times ED}{BW \times AT \times 10^6} \quad (34.4)$$

The benzo[a]pyrene equivalent concentration (BaP_{eq}) of PAHs and TPPs was assessed using $C_i \times TEF_i$ where C_i (g/g) and TEF_i represent the concentration and TEF, respectively, of each PAC. The meaning and values for all other terms in the equations are indicated in Table 34.4.

Based on the application of the QSAR model, a potential cancer risk to adults was observed, while dependence on only the originally existing TEFs would have underestimated the risk as negligible. However, in the case of children, regardless of the TEFs used, a negligible risk conclusion was obtained. It is postulated that the disparities in some parameters such as body weight, dermal adherence factor, dermal exposure area, exposure duration, dust ingestion rate, and inhalation rate between adults and children that were used in estimating the ILCR as shown in Eqs. (34.2)–(34.4) could account for the observed negligible risk in children.

Conclusions and future directions

The application of QSAR approach has helped to breach the knowledge gap with respect to toxicity equivalency factors (TEFs) for some PACs that hitherto existed. However, the results of the QSAR application for quantifying the cancer risk posed by PAHs and TPPs indicated the need to compliment modeling-based assessment with toxicological tests in particular for outlier PAHs and TPPs so as to establish a comprehensive TEF database for these hazardous compounds. The health effects of the PAHs and TPPs are often cumulative; thus the availability of such a comprehensive database will promote more accurate estimation of the associated risk, thereby enabling the formulation of well-informed mitigation measures. Considering the vast array of PACs notably the NSO-PACs that are prevalent in the environment, there is the need for future investigation into the applicability of QSAR in determining the ecological health risk posed by these compounds. Moreover, the relevance of QSAR in assessing the carcinogenic health risk posed by other classes of POPs such as PCBs, OCPs, PCDDs, and polychlorinated dibenzofurans requires future research.

Acknowledgments

The authors acknowledge Queensland University of Technology (QUT), Australia, for providing a postgraduate research award to Gustav Gbeddy to undertake the study, which culminated in this chapter. The study leave approved by the Ghana Atomic Energy Commission (GAEC) to Gustav Gbeddy to pursue this study is also gratefully acknowledged.

References

[1] Bandowe BA, Nkansah MA. Occurrence, distribution and health risk from polycyclic aromatic compounds (PAHs, oxygenated-PAHs and azaarenes) in street dust from a major West African Metropolis. Sci Total Environ 2016;553:439–49.

[2] Andersson JT, Achten C. Time to say goodbye to the 16 EPA PAHs? Toward an up-to-date use of PACs for environmental purposes. Polycycl Aromat Compd 2015;35(2–4):330–54.

[3] Gbeddy GK. Transformation and degradation of organic pollutants on urban road surfaces [Thesis by Publication]. Queensland, Australia: Queensland University of Technology; 2020.

[4] Idowu O, Semple KT, Ramadass K, O'Connor W, Hansbro P, Thavamani P. Beyond the obvious: environmental health implications of polar polycyclic aromatic hydrocarbons. Environ Int 2019;123:543–57.

[5] Gbeddy G, Egodawatta P, Goonetilleke A, Ayoko G, Chen L. Application of quantitative structure-activity relationship (QSAR) model in comprehensive human health risk assessment of PAHs, and alkyl-, nitro-, carbonyl-, and hydroxyl-PAHs laden in urban road dust. J Hazard Mater 2020;383:121154.

[6] Gbeddy G, Egodawatta P, Goonetilleke A, Ayoko G, Jayarathne A, Chen L, et al. Optimized simultaneous pressurized fluid extraction and in-cell clean-up, and analysis of polycyclic aromatic hydrocarbons (PAHs), and nitro-, carbonyl-, hydroxy-PAHs in solid particles. Anal Chim Acta 2020;1125:19–28.

[7] Gbeddy G, Goonetilleke A, Ayoko GA, Egodawatta P. Transformation and degradation of polycyclic aromatic hydrocarbons (PAHs) in urban road surfaces: influential factors, implications and recommendations. Environ Pollut 2019.

[8] Kunal R, Supratik K, Rudra ND. Understanding the basics of QSAR for applications in pharmaceutical sciences and risk assessment. Amsterdam; Boston: Academic Press, an imprint of Elsevier; 2015.

[9] Gadaleta D, Mangiatordi GF, Catto M, Carotti A, Nicolotti O. Applicability domain for QSAR models. Int J Quant Struct Relatsh 2016;1:45–63.

[10] Gbeddy G, Egodawatta P, Goonetilleke A, Ayoko G, Chen L. Dataset for the quantitative structure-activity relationship (QSAR) modeling of the toxicity equivalency factors (TEFs) of PAHs and transformed PAH products. Data Brief 2020;28:104821.

[11] Hawkins DM, Basak SC, Mills D. Assessing model fit by cross-validation. J Chem Inf Comput Sci 2003;43:579–86.

[12] Miller JN, Miller JC. Statistics and chemometrics for analytical chemistry. 6th ed. England: Pearson Education Limited; 2010.

[13] Forsgren AJ. Wastewater treatment: occurrence and fate of polycyclic aromatic hydrocarbons (PAHs). CRC Press Taylor & Francis Group; 2015 (A Series).

[14] Bates M, Bruno P, Caputi M, Caselli M, de Gennaro G, Tutino M. Analysis of polycyclic aromatic hydrocarbons (PAHs) in airborne particles by direct sample introduction thermal desorption GC/MS. Atmos Environ 2008;42:6144–51.

[15] Wei C, Bandowe BA, Han Y, Cao J, Zhan C, Wilcke W. Polycyclic aromatic hydrocarbons (PAHs) and their derivatives (alkyl-PAHs, oxygenated-PAHs, nitrated-PAHs and azaarenes) in urban road dusts from Xi'an, Central China. Chemosphere 2015;134:512–20.

CHAPTER

35

QSAR in risk assessment of nanomaterials

Ewelina Wyrzykowska[1], Maciej Gromelski[1], Krzesimir Ciura[1,2], Marta Swirog[1], Anita Sosnowska[1] and Tomasz Puzyn[1,3]

[1]QSAR Lab Ltd, Gdańsk, Poland [2]Department of Physical Chemistry, Faculty of Pharmacy, Medical University of Gdansk, Gdansk, Poland [3]Laboratory of Environmental Chemoinformatics, Faculty of Chemistry, University of Gdańsk, Gdańsk, Poland

Abbreviations

3R Replacement, reduction, and refinement
AD Applicability domain
ECHA European Chemicals Agency
ENMs Engineered nanomaterials
EPA United States Environmental Protection Agency
FDA Food and Drug Administration
OECD Organisation for Economic Co-operation and Development
QNAR Quantitative nanostructure–activity relationship
QSAR Quantitative structure–activity relationship
QSPR Quantitative structure–property relationship
REACH European Regulation on Registration, Evaluation, Authorisation and Restriction of Chemicals

Introduction

Considering the rapid development of nanotechnology, it is essential to comprehend how the engineered nanomaterials (ENMs) affect living organisms and the natural environment. Generally, the experimental procedures, both in vivo and in vitro, for ENMs are time-consuming and labor-intensive and frequently require much more effort than experimentation for classical chemicals. Hence, many researchers pay attention to the development of computational approaches for predicting the toxicological and ecotoxicological properties of ENMs. Among computational alternatives, especially important is nanoquantitative structure–activity relationships (nano-QSAR), also known as quantitative nanostructure–activity relationships (QNAR) [1]. Puzyn et al. initially proposed the nano-QSAR approach in 2011 to combine the cytotoxicity of *Escherichia coli* with the properties of metal–oxide nanoparticles [2]. From that time, nano-QSAR has evolved and found application in predicting many endpoints relevant to ENMs reaching more than 100 of articles published only in 2021. Currently, not only biologically relevant endpoints but also, important from the engineering point of view, physicochemical properties of ENMs are predicted. Such models are called quantitative structure–property relationships (QSPR).

Briefly, the QSAR/QSPR modeling is one of the nontesting methods applied to predict the biological activity, physicochemical properties, and (eco)toxicological or environmental fate properties of substances. Using various mathematical approaches, QSAR/QSPR models link the relationships between chemical structures and experimentally measured properties of substances. Information about chemical structures is represented numerically by so-called "descriptors." In the case of nano-QSAR, ENMs are noticeably more complex and complicated than simple chemical molecules; therefore, describing ENMs is still a demanding task [3]. Hence, both theoretical- and experimental-driven descriptors are used to describe ENMs. Another significant challenge is the influence of exposure medium, which is usually rich in macromolecules and can interact with EMNs. Consequently, ENMs can change over time, including physical changes such as aggregation or sedimentation and chemical changes such as oxidation.

QSAR in Safety Evaluation and Risk Assessment
https://doi.org/10.1016/B978-0-443-15339-6.00021-7

The reliability of QSAR/QSPR is proven with a series of statistical parameters, including internal and external validation based on OECD recommendations [4]. At the same time, the credibility of predictions for new substances depends on their presence in the model's applicability domain (AD). Furthermore, unambiguous identification and characterization as well as harmonized chemical nomenclature regarding the nanomaterials and nanoforms is strongly suggested. Consequently, each nano-QSAR/QSPR modeling and analysis should be preceded by comprehensive documentation regarding analyzed substances.

In toxicological research, nano-QSAR/QSPR plays an essential role going beyond the prediction of bioactivity properties of newly designed ENMs. The nano-QSAR/QSPR models can be used to fill wide data gaps for substances for which a modeled parameter is not available, which is especially common in the case of ENMs. Furthermore, nano-QSAR/QSPR allows insights into the molecular mechanism of toxicity of ENMs. This is an additional advantage since there are still many gaps in how ENMs affect living organisms.

This chapter aims at providing support to the application of QSAR/QSPR approach for nanomaterials and nanoforms (nano-QSAR/QSPR). Here we present some recommendations on how to bridge data gaps on specific nanoforms (or set of nanoforms) through the observation of relationships between physicochemical characteristics of nanoforms of the same substance or different substances with the relevant endpoint.

Special attention was placed on covering the development and validation of nano-QSAR/QSPR models as well as the adequacy of modeling results, documentation, and justification for scientific, manufacturing, and regulatory purposes. Only the fulfillment of the criteria developed for traditional QSAR/QSPR modeling approach dedicated for classic compounds (OECD principles) [4,5], adapted to the specificity of nanomaterials, can ensure a plausible usage of nano-QSAR/QSPR models under the 3Rs' idea of replacement, reduction, and refinement for providing more humane animal research proposed by Russell and Burch [6]. The standardized model development procedure based on reliable experimental data and adequate and documented methodology consistent with the FAIR data concept (i.e., findable, accessible, interoperable, reusable data) is a prerequisite for conducting research of environmental, health, and safety potential of nanomaterials, including toxicity assessment, life cycle, and environmental fate.

Critical aspects in nano-QSAR/QSPR modeling

The nano-QSAR/QSPR modeling has a great potential to be used as a tool for filling the data gaps in the developed datasets obtained through years in the whole of nanomaterials science. However, the development of new experimental and computational approaches dedicated to nanomaterials that would ensure credibility of the specifying nanomaterials properties is required. The predictions of ENMs properties are strictly dependent on the appropriate nanoparticles' structure representation. However, representing an ENM structure is challenging for a number of reasons. The most critical aspects related to the representation of ENM structures are described in detail in the following. However, due to the consistent identification of new issues and concepts of ENM characterization, the presented aspects may not be exhaustive in this respect, and developers of nano-QSAR/QSPR models are advised to approach the subject critically and objectively [7].

Nanomaterial and nanoform

Depending on the regulatory institution, the "nanomaterial" definition encloses various aspects of nanostructure and properties of material, and due to the lack of consensus, the definition is under constant revision.

In the United States, the Food and Drug Administration (FDA) has not established working regulatory definition of "nanomaterial" or other related "nano-" terms. However, in the context of nanotechnology products, FDA requires information on (1) *"whether a material or end product is engineered to have at least one external dimension, or an internal or surface structure, in the nanoscale range (approximately 1–100 nm)"*; or (2) *"whether a material or end product is engineered to exhibit properties or phenomena, including physical or chemical properties or biological effects, which are attributable to its dimension(s), even if these dimensions fall outside the nanoscale range, up to 1 μm (1000 nm)"* [8,9].

Apart from the FDA, the organization that plays a key role in regulating chemicals, pesticides, food, cosmetics, and drugs is the US Environmental Protection Agency (EPA), which also has no formal regulatory definition of nanomaterial. However, they proposed guidelines to distinguish nanoscale materials from chemicals. According to the EPA's working guidance, nanomaterials can be defined as *"a solid at 25°C and standard atmospheric pressure, which is manufactured or processed in a form where any particles, including aggregates and agglomerates, are in the size range of 1–100 nm in at least one dimension, and which is intentionally manufactured or processed to exhibit unique and novel properties because of its size. A reportable chemical substance does not include a chemical substance that is manufactured or processed in a form where less than 1% of any particles, including aggregates and agglomerates measured by weight, are in the size range of 1–100 nm in at least one dimension*" [10,11].

In turn, Health Canada is based on the working definition of nanomaterial, which considers "*any manufactured substance or product and any component material, ingredient, device, or structure to be a nanomaterial if (1) it is at or within the nanoscale in at least one external dimension, or has internal or surface structure at the nanoscale; or (2) it is smaller or larger than the nanoscale in all dimensions and exhibits one or more nanoscale properties/phenomena*" [12].

The Australian Government Department of Health's National Industrial Chemicals Notification and Assessment Scheme (NICNAS), now called the Australian Industrial Chemicals Introduction Scheme (AICIS), regulates industrial chemicals and has developed a working definition of industrial NMs. It states that "*industrial materials are intentionally manufactured or engineered to have unique properties or specific compositions at the nanoscale, which is in the size range typically 1–100 nm, and are either a nanoobject (i.e., that is limited to one, two, or three dimensions at the nanoscale) or is nanostructured (i.e., having an internal or surface structure at the nanoscale)*" [13]. A chemical is considered to be nano if it is introduced in solid or dispersed form and consists of particles in an unbound state or as an aggregate or agglomerate where at least 50% (by number size distribution) of the particles have at least one nanoscale external dimension.

According to the European Commission [14], the term "nanomaterial" refers to "*a natural, incidental or manufactured material containing particles, in an unbound state or as an aggregate or as an agglomerate and where, for 50% or more of the particles in the number size distribution, one or more external dimensions is in the size range 1–100 nm, including also by derogation fullerenes, graphene flakes, and single-walled carbon nanotubes with one or more external dimensions below 1 nm*" [14]. However, the recent update of the nanomaterial definition states that "*nanomaterial means a natural, incidental, or manufactured material consisting of solid particles that are present, either on their own or as identifiable constituent particles in aggregates or agglomerates, and where 50% or more of these particles in the number-based size distribution fulfill at least one of the following conditions: (1) one or more external dimensions of the particle are in the size range 1–100 nm; (2) the particle has an elongated shape, such as a rod, fiber or tube, where two external dimensions are smaller than 1 nm and the other dimension is larger than 100 nm; (3) the particle has a plate-like shape, where one external dimension is smaller than 1 nm and the other dimensions are larger than 100 nm.*" However, materials are not considered as nanomaterials when in the number-based analysis of size distribution, "particles with at least two orthogonal external dimensions (are) larger than 100 μm", or their specific surface area by volume is lower than 6 m^2/cm^3 [15].

Moreover, it is important to notice that the revised Annex VI of REACH on December 3, 2018 [16] introduced the concept of "nanoform of the substance" into the regulation, which addresses the information requirements concerning nanoforms of chemical substances. Within the REACH Regulation, nanomaterials are considered as special forms of a chemical substance, whereas term "nanoform" is used to distinguish different nanomaterials of the same registered substance [17]. In the context of chemical substances registration, the revised Annex VI establishes principles of registering nanoforms of particular substance. This results in a requirement that all the nanoforms of a substance are identified and characterized in the registration dossiers.

A nanoform has to be characterized in accordance with Annex VI Section 2.4 of REACH [16,18], which states that a substance may comprise of one or more different nanoforms, depending on the differences in the following parameters:

- particle size distribution and number fraction of constituent particles,
- shape, aspect ratio, and other morphological characterization (e.g., crystallinity),
- surface treatment and functionalization,
- specific surface area of the particles (specific surface area by volume, specific surface area by mass or both).

Any variation in one or several of these characteristics will result in a different nanoform (unless it is a batch-to-batch variability).

Mono- and multicomponent engineered nanomaterials

Nanomaterials require detailed description of their chemical composition and the physicochemical properties to identify the appropriate nanoform of the substance. The issues of describing the physicochemical properties of an ENM are common to all nanomaterials regardless of the complexity of their chemical composition (these are described in detail in the following sections). Still, the proper description of the ENMs chemical composition is a complex issue.

The ENMs could be identified as a mono- or multicomponent materials, which properties are affected with additives (i.e., polymers for stabilizing effect) and impurities (i.e., surface impurities formed during synthesis). Another critical issue is related to ENMs construction. Even ENMs composed of one chemical component can have various properties depending on crystalline forms such as titanium dioxide (where rutile, anatase, and mixed forms are available) or nanomaterials structures such as single- and multiwalled carbon base nanotubes. Unless the description of single-component ENMs requires detailed characterization of used component and the influences of additives and impurities, the multicomponent ENMs require also details concerning components proportions and their mutual relations. The multicomponent ENMs may have the form of nanocomposites (i.e., bimetallic or multimetallic cores, dopped ENMs, etc.) or surface-modified ENMs (i.e., through coatings or shells, attached surface organic/inorganic functionalization, etc.) (Fig. 35.1). However, the appropriate identification of actual quantity of individual components is problematic due to limitations of the experimental characterization (i.e., the identification of actual amount of doped metal on the ENMs, verification of uniform thickness of shell on the ENMs surface, etc.). Furthermore, the multicomponent ENMs require the identification of the combined effects of the relevant components in terms of physicochemical properties or (eco)toxicological responses. A positive additive effect where the sum of the effects of individual components is observed, synergistic effect where an enhanced effect of individual component is observed, or an antagonistic effect where the effect of the entire ENM is weaker than it would appear from the sum of the effects of the individual components should be recognized. Regarding the toxicological aspect, the observed effect is strictly linked to the mechanism of action of individual components, so the multifaceted properties of nanocomposites may be observed.

The characterization of nanoforms' chemical composition in the nano-QSAR/QSPR models should be as much detailed as possible to avoid any ambiguities because the presence and amount of a proper component could substantially modify the ENMs application potential and (eco)toxicological features.

System-dependent properties of engineered nanomaterials

The ENMs structure is changing during their life cycle. ENMs and ENM dispersion systems in particular show susceptibility to structure changes caused by various factors. The reliability of ENMs characterization in powder form leaves little doubt. However, the characterization of the ENMs dispersions is affected with relatively higher

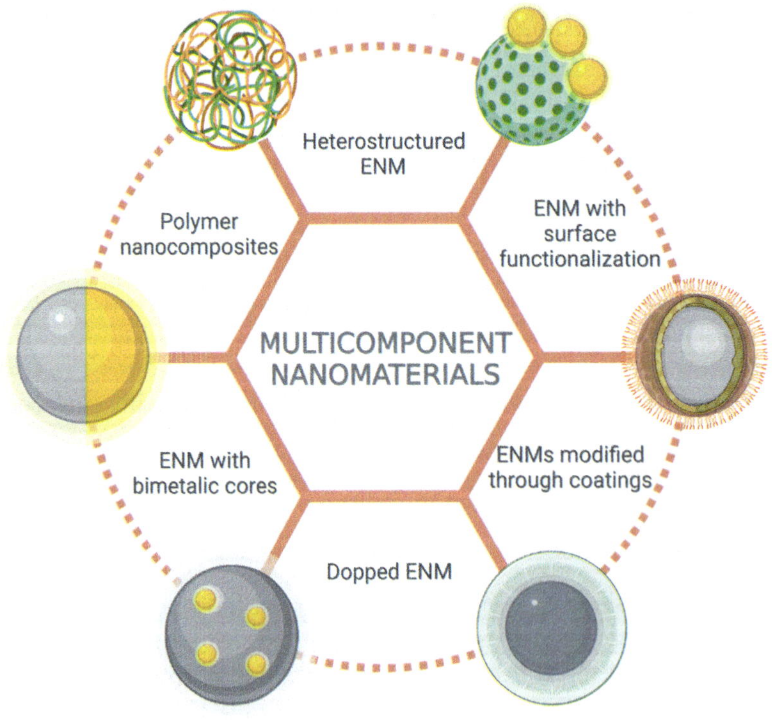

FIGURE 35.1 The spectrum of multicomponent ENMs.

uncertainty because of the state of dynamic equilibrium resulted from a series of overlapping processes. The external conditions may influence the ENMs properties in an extraordinary manner, i.e., dispersion stability, aggregation/agglomeration processes, ions release and dissolution, transformation of the ENMs structure through the surface modification by adsorption of surrounding ions or molecules such as organic, inorganic, biomolecules, and surfactants as well as modification of the surface activity through externally (i.e., with UV light) induced generation of radical oxygen species (Fig. 35.2). Moreover, the physical conditions such as pressure, electronic or magnetic field, and temperature can also induce the structural changes of ENMs structure. As a result of the changing structure, the ENMs properties are also changed [19,20]. The external conditions determine the ENMs properties and could be responsible, in the case of activity, for toxicity or safety and, in the case of application, for their enhanced functionality or the lack of it. Thus, the prediction of actual ENMs properties should take into consideration of the external conditions. Although the nano-QSAR/QSPR modeling based on the description of ENMs structure in well-established and unchanging conditions allows to recognize the relationships between the ENMs structure in these strict conditions and predict a specific effect (activity or property), it is nearly impossible to estimate the further fate of the investigated ENMs, i.e., when the conditions change over life span of an ENM. For instance, the ENMs could exhibit different levels of toxicity in the human body depending on the biological conditions related to different sections of the digestive system and, similarly, in different conditions in the environment. Thus, the improvement of nano-QSAR/QSPR modeling requires rather flexible approaches for the quantification of the system-dependent properties of ENMs and how it relates to the changes in their structures.

Characterization of the structures and properties of ENMs requires the use of standardized protocols supported by metadata detailing the experimental measurement conditions to ensure that the unique fingerprint properties of the tested ENMs are identified and, at the same time, that the ENM effects are reproducible (i.e., activity, (eco)toxicological activity, physicochemical properties, potential application, etc.) and result from the appropriate features of the ENM. Where possible, the nanoform descriptors in the nano-QSAR/QSPR models should reflect the features of the nanoform under the appropriate measurement conditions.

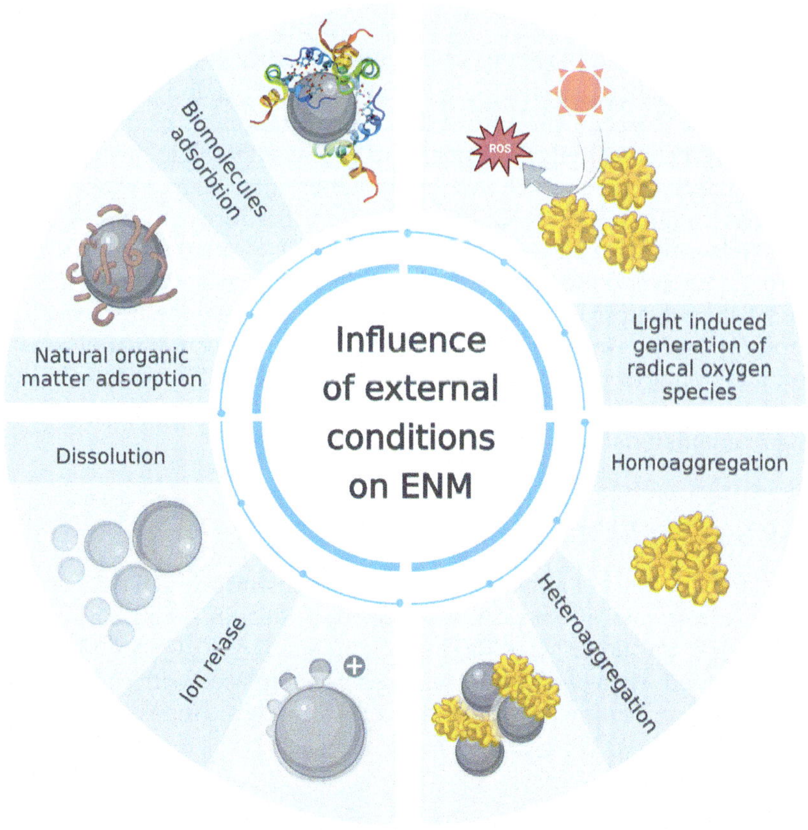

FIGURE 35.2 The processes of nanostructure changes induced by external conditions.

Heterogenicity of ENMs

Vast majority of ENMs' physicochemical characteristics are specified as median values supported with a standard deviation. However, ENMs are produced more or less not as homogenic but as heterogenic materials where individual particles differ from each other in size, crystallinity structure, surface charges, coating thickness, surface modifiers position, etc. Therefore, the median value does not reflect the actual ENMs features but is a generalized approximation. Through using the median value only, some information about the ENMs' structure is lost, including information related to the polydispersity of ENMs, i.e., whole range of ENMs size and diversified amount of particular fraction (the smallest, median, and the largest particles). The analysis of a size distribution is also notably important because a particular fraction of the tested ENMs may or may not be responsible for the occurrence of a specific process or mechanism, and the median value would provide misleading information [21]. In the context of (eco) toxicity, even small fraction of the lowest size of ENMs can induce toxic effect, while the median size of ENMs would not suggest the possible toxic effect [17]. Currently, the ECHA recommends reflecting the size distribution through the values of d_{10} that is the size for which 10% of the whole dataset of nanoparticles has lower size; d_{50} that is the median size of particles, and d_{90} that is the size for which 90% of nanoparticles have lower size. Undoubtedly, the increase of the credibility of the nano-QSAR/QSPR modeling requires further development of techniques for ENMs characterization and new descriptors for ENMs structure representation covering their heterogenicity.

Nano-QSAR/QSPR development

Nano-QSAR/QSPR concept

In general, QSAR/QSPR approaches allow to generate nontesting data based on the similarity principle, i.e., hypothesis that structurally similar compounds should exhibit similar properties and biological activity. QSAR/QSPR approach assumes development of qualitative (SAR/SPR) and quantitative (QSAR/QSPR) theoretical models to predict defined endpoints related to physicochemical properties, biological activity, or environmental fate properties, based on the structural characteristic of compounds. The structural characteristic of compounds is expressed in a numerical manner, so-called (molecular) descriptors. The process of selection of a set of descriptors that "code" relevant structural features of molecules and are subsequently employed in the developed models to predict desired response is the aftermath of (1) application of advanced algorithms for feature selection (often including statistical and machine learning methods, called "statistical approach") and/or (2) expert decision (known as "knowledge-based approach" or "mechanistic approach"). Identification of the quantitative relationships between substantial descriptors and predicted activity or property of ENMs is possible through the application of machine learning algorithms. High quality of the prediction is provided by internal and external validation of obtained results (Fig. 35.3). Detailed description of QSAR/QSPR modeling procedures are available elsewhere [1,2,22–25].

Nano-QSAR/QSPR models constitute the adaptation of QSAR/QSPR methodology for nanomaterials, where the models may be applied for nanoforms of the same substance or nanoforms of different substances [7]. The specificity of nanomaterials requires a special approach in a few aspects: (1) appropriate characterization of pristine nanoforms and in the test conditions, (2) descriptors especially enabling the distinguish of nanoforms of the same substances, and (3) appropriate definition of applicability domain of nano-QSAR/QSPR models. In the following, we present some suggestions on how to develop appropriate scientifically validated nano-QSAR/QSPR models for the development of safe-by-design nanomaterials as well as regulatory purposes.

Regulatory framework for QSAR/QSPR modeling

Despite the differences in the context of application, regulatory institutes such as EPA, ECHA, or Health Canada [5,7,26,27] recommend the application of QSAR/QSPR approach to fulfill the data gaps whenever possible. For instance, in Europe, according to the Annex XI of REACH regulation [16,18], QSAR/QSPR approach and other nontesting methods are recommended when (vertebrate) testing can be replaced with computational methods to achieve the same level of information. Especially with regard to (Q)SAR models, Annex XI indicates:

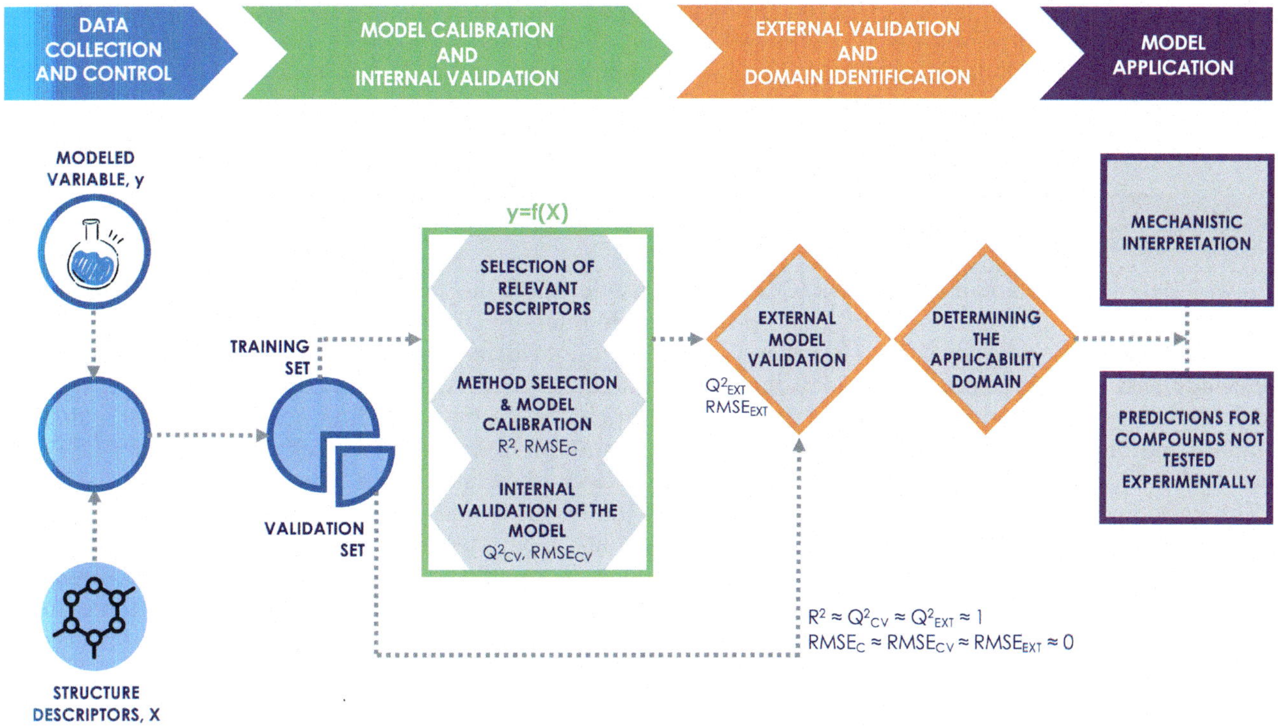

FIGURE 35.3 The scheme of QSAR/QSPR modeling procedure.

"Results obtained from valid qualitative or quantitative structure-activity relationship models ((Q)SARs) may indicate the presence or absence of a certain dangerous property. Results of (Q)SARs may be used instead of testing when the following conditions are met:

- results are derived from a (Q)SAR model whose scientific validity has been established,
- the substance falls within the applicability domain of the (Q)SAR model,
- results are adequate for the purpose of classification and labeling and/or risk assessment, and
- adequate and reliable documentation of the applied method is provided."

The aforementioned excerpt explains that computational models can be used in place of experimental testing when a number of conditions are met. These conditions ensure obtaining relevant, reliable, and adequate results that are well documented in an appropriate manner. For the registration purposes, the modeling results can be supported with a weight of evidence approach to complete the information not contained in the developed model.

However, the correctness, applicability, and acceptance of QSAR/QSPR models in the context of REACH requirements as well as other recommendations [5,7,26,27] are reinforced with the internationally agreed OECD principles, adopted in March 2007 by the OECD Member Countries and the Commission [4]. The OECD principles are not formal criteria but concern crucial issues during the development of QSAR/QSPR models and constitute as conceptual framework for the assessment of models for regulatory purposes. OECD principles are as follows:

"to facilitate the consideration of a (Q)SAR model for regulatory purposes, it should be associated with the following information:

1. a defined endpoint;
2. an unambiguous algorithm;
3. a defined domain of applicability;
4. appropriate measures of goodness-of-fit, robustness and predictivity;
5. a mechanistic interpretation, if possible."

The original wording of the OECD principles is included in the *ECHA Guidance on information requirements and chemical safety assessment, Chapter R.6: QSARs and grouping of chemicals* [5], whereas this chapter presents the suggested adaptation of well-established principles for nanomaterials.

The adaptation of OECD principles for nano-QSAR/QSPR modeling

A defined endpoint

A well-defined endpoint in nano-QSAR/QSPR model is relevant to the valuable application and is strictly dependent on the purpose of its development. The endpoint should refer directly to *"any physicochemical property, biological effect or environmental parameter"* [4]. However, while there are available protocols for physicochemical measurements adjusted to nanomaterials specificity, there are limited (eco)toxicological and environmental properties protocols dedicated to nanomaterials. More precise descriptions of the nano-specific procedures (test guides and standardized operating procedures) for (eco)toxicological and environmental measurements are currently developed within Horizon 2020 project RiskGONE and other projects associated in the NanoSafety Cluster [28,29].

The endpoint should be clearly defined according to the experimental protocol and test conditions used. The nano-QSAR/QSPR modeling should be based on high-quality experimental data that were obtained according to standardized operating procedures. It is mandatory to include any changes of procedure in the documentation, especially when dealing with the adaptation of protocols dedicated for classic compounds and its acceptance for nanomaterials was not confirmed yet. A clearly defined endpoint should be evaluated according to the criteria of (1) scientific intended purpose, (2) experimental protocol of measurement/assay, experimental conditions, and sources of experimental errors and uncertainty, (3) defined concentration/dose-related units of measurements, (4) consistency of experimental procedures for the whole dataset used in the model development, and (5) where registration is the purpose of modeling, an endpoint response to a specific regulatory need is recommended [4].

The terminology used should be precise enough to avoid any ambiguity in the interpretation of the output and the modeled effect or physicochemical properties. For instance, the cellular "cytotoxicity effect" should be properly expressed as reduced viability, apoptosis and necrosis, or sublethal effect. To ensure the transparency and coherence of terminology, it is recommended to base on BioAssay Ontology (BAO), which indicates different assays related to different aspects of cytotoxicity [30].

An unambiguous algorithm

A nano-QSAR/QSPR model can be built using statistical or machine learning methods. The method should be properly selected by taking into account the number and the type of data. All available and recognized methods can be used, especially those dedicated to small datasets, which are often generated for nanomaterials tested under the same experimental protocol. Accurate and transparent description of applied algorithm, including all algorithm parameters, the so-called hyperparameters, should be a matter of public information. The details of the training and validation data are obligatory and increase reproducibility of the model.

Descriptors, or in other words, explanatory variables, in nano-QSAR/QSPR models should reflect key features of nanomaterials and their corresponding nanoforms such as size and size distribution, shape and aspect ratio, surface chemistry and functionality (regarding core, shell, coatings, surface modifiers, etc.), and intrinsic and/or extrinsic properties. The molecular descriptors can base solely on experimental measurements or theoretically obtained characteristics. Those numerical values may relate to the chemical structure of nanomaterial's constituents, i.e., core, shells, surface functionalization, and impurities, as well as directly to the physical structure of nanoform. It is noteworthy that one can distinguish "coding" and "noncoding" descriptors. The first kind allows to recreate the structure of nanomaterial through the knowledge about chemical structure of constituents and physical aspects of nanoparticles. The second kind provides information about general properties of constituents and nanomaterial, which could be the same for diverse nanomaterials and preclude the recreation of the nanostructure. However, regarding the intended purpose of modeling, different kinds of descriptors are recommended. For the cognitive purposes of nano-QSAR/QSPR modeling, both kinds of descriptors are approved. In turn, for predictive purposes in the safe-by-design approach for new nanomaterials, there are required descriptors that directly "code" the structure and composition of nanoform. The procedure of obtaining descriptors should be precisely described pointing out required and available software for potential stakeholders to ensure the reproducibility of the model.

Variables selection aims to reduce the quantity of considered descriptors (that directly refers to the quality of produced model) and indicates the most relevant features for the analyzed case: (1) nano-(Q)SAR model—mechanistic interpretation of nanomaterials activity and adverse effects or (2) nano-(Q)SPR model—mechanistic interpretation of physicochemical processes. Similar to the classic QSAR/QSPR approach, a variable selection should be performed based on the highest correlation of variables with the modeled endpoint. Methods used for ranking of the features relevance are usually specific for the machine learning technique applied in nano-QSAR/QSPR model development. Machine learning methods used in QSAR/QSPR modeling approach include among others (multi)linear and logistic regression, *k*-nearest neighbors, decision trees and random forests, XG boost, naive Bayes, support vector machine and support vector regression, neural networks, etc.

Reproducibility of nano-QSAR/QSPR models mostly depends on the availability and transparency of model development details and its metadata. The nano-QSAR/QSPR model reporting format should be consistent with the recommendations of the European Commission's Joint Research Centre (JRC)—(Q)SAR Model Reporting Format (QMRF) [31]. Technical details and hyperparameters should be reported in a machine-readable fashion, independently on the applied software, e.g., QsarDB format [32], which uses Predictive Model Markup Language [33] and uploaded in QSAR/QSPR repositories, e.g., QsarDB [32] and JaqPot Quattro [34].

A defined applicability domain

The reliability of predictions should be confirmed through applicability domain analysis. In the case of nanomaterials, applicability domain of QSAR/QSPR models can be defined in various aspects. The basis is to examine the extent of applied descriptor values through statistical algorithms. The reliability assessment should be proved with statistical parameters adequate to regression/classification methods applied for nano-QSAR/QSPR model development regarding variance, confidence, sensitivity, local errors, divergence probability distributions, conformal predictions, and additional metrics to confirm credibility of predictions.

Applicability domain of a nano-QSAR/QSPR model should be limited to the type of nanomaterials used during the model development. The type of nanomaterials can be recognized on diversified levels, regarding

- the complexity of nanomaterial composition, i.e., one-component (nonmodified or with coatings, shells, surface modifiers, functionalization, etc.) or multicomponent (alloys, doping, copolymers, etc.);
- the main constituent, i.e., metals, metal oxides, carbon derivatives, etc.;
- the size/shape/aspect ratio, i.e., spheres, fibers, sheets, nanotubes, fullerenes, etc.

or could be reduced to

- nanoforms of the same substance, which provides the maximal credibility of prediction and is strongly recommended.

Moreover, in the case of nano-(Q)SAR models based on omics data, the applicability domain should remain within similar mode of action through the same molecular initiating events or key events.

Appropriate measures of goodness-of-fit, robustness, and predictivity

OECD principles state the need to perform statistical analysis to provide measures defining the quality of a model. These measures, obtained through internal and external validation, cover the models' *goodness-of-fit*, *robustness*, and *predictivity* and are adequate for vast majority of the regression of classification algorithms.

The *goodness-of-fit* is a measure that defines the statistical significance of a developed model. It can be used to assess how well the variance of the response (modeled variable) in the training set is explained by the model itself. Consequently, a model of poor *goodness-of-fit* will be statistically insignificant, and its predictions will be burdened with a larger error. In the regression approaches, the *goodness-of-fit* is reflected through the multiple correlation coefficient R^2 and error of the estimate. Both measures can be calculated in diversified manners; however, the value of the first measure is expected to tend to 1, while the second one should be as low as possible, however, not be lower than the experimental error of the data used in the model development. In the case of classification methods, the Cooper statistics are used to confirm the *goodness-of-fit* through specificity, sensitivity, and concordance, i.e., true positive and negative rates as well as false positive and negative rates, and the judgment of the nano-QSAR/QSPR models quality should be based on them [5].

The measure of *robustness* describes the models' stability and is based on the internal validation using observations from the training set. Among the internal validation, the permutation tests and cross-validation techniques are the most common, i.e., leave-one-out (LOO), leave-many-out (LMO), bootstrapping, and randomization approaches [5]. Usually, the models' *robustness* is assessed based on several iterative procedures of removing and restoring observations within the training set and is expressed with a cross-validated regression coefficient (internal) Q^2, which should be close to 1. Due to the diversity of available techniques, there is required the detailed and transparent description of the applied method in the QMRF documentation.

The *predictivity* measure verifies the predictive ability of the model through the external validation and is crucial for the model application. In contrary to the *goodness-of-fit* and the *robustness*, the direct evaluation of the *predictivity* is based on the model application for the validation set (the data that were not used during the model development but are still in the applicability domain of the model). The predictive ability is expressed with the external Q^2, and analogically to R^2 and internal Q^2, it should tend to 1. The crucial thing is the representativeness of the objects in the validation set according to the applicability domain of the developed model. Also, here, the external Q^2 can be calculated using diversified techniques, and no method is imposed in advance. Nevertheless, the unsatisfactory results of measures reflecting *predictivity* imply that the model is overfitted to the training dataset and has low predictive ability for other objects.

Mechanistic interpretation

OECD suggests that a mechanistic interpretation, whenever possible, should be associated with the developed QSAR/QSPR model [5]. In principle, the mechanistic interpretation interlinks the observed phenomena (either biological activity or observed physicochemical property) with the structural features of analyzed chemicals and is usually supported by extensive evidence from the literature. Due to the specificity of the nanomaterials, the mechanistic interpretation of the nano-QSAR/QSPR models has to be considered in the context of the possible mechanisms of action (nano-(Q)SAR) or physical processes (nano-(Q)SPR). The mechanistic interpretation should be strictly related to applicability domain of the nano-QSAR/QSPR models, especially, in the context of ENMs types and nanoforms of the substance(s). In this manner, the mechanistic interpretation can support defining the domain of applicability of the nano-QSAR/QSPR model.

The mechanistic interpretation inherently boosts the model's confidence and reduces the possibility of bias related to chance correlations. From a scientific point of view, a mechanistic interpretation of a given model is not always possible, for example, due to not yet fully understood biological mechanism or physicochemical phenomena. Nevertheless, a model without mechanistic interpretation may still be useful in a regulatory context, or it may serve as a tool to support the understanding of the underlying mechanism.

Conclusions

Computational approaches such as QSAR/QSPR methods are very attractive alternatives to time-consuming experimental research methodology. They are also recommended by EU regulations to support experimental data or to prioritize chemicals when the experimental data for regulatory purpose is not available. However, as in case of in vitro methods, to use currently accepted tests for classic chemicals, it is necessary to adapt them to the specificity of nanomaterials. The very important aspect regarding nano-QSAR/QSPR model development is the description of the nanoforms; here, characterization of nanoforms' chemical composition should be as much detailed as possible, and descriptors applied for modeling should reflect the structural features of the nanoform under the appropriate measurement conditions. There is a pressing need to develop new ENMs descriptors reflecting their size distribution and other heterogenicity aspects, e.g., irregularly distributed surface charges or varied coating thickness. Special attention should be paid to the adaptation of universal OECD principles for QSAR/QSPR modeling to nanomaterials specificity. Consideration of detailed nanostructure characteristics in the models will allow an increase in understanding of the mechanisms of action of the observed adverse effects, as well as better identification of applicability domains and practical usage in the design of safer nanomaterials. Going beyond the classic paradigm of descriptors reflecting only the chemical structure of substances, nano-QSAR/QSPR models will give great opportunities for the reliable usage of in silico methods complying 3Rs' principles in the manufactory, regulatory, and academic purposes.

Acknowledgments

The authors acknowledge the funding received from the European Union Horizon 2020 projects: NanoSolveIT (grant agreement No. 814572) and RiskGONE (grant agreement No. 814425). Graphics in the figures were created with BioRender.com.

References

[1] Muratov EN, Bajorath J, Sheridan RP, Tetko IV, Filimonov D, Poroikov V, et al. QSAR without borders. Chem Soc Rev 2020;49:3525–64.

[2] Puzyn T, Rasulev B, Gajewicz A, Hu X, Dasari TP, Michalkova A, et al. Using nano-QSAR to predict the cytotoxicity of metal oxide nanoparticles. Nat Nanotechnol 2011;6:175–8.

[3] Wyrzykowska E, Mikolajczyk A, Lynch I, Jeliazkova N, Kochev N, Sarimveis H, et al. Representing and describing nanomaterials in predictive nanoinformatics. Nat Nanotechnol 2022;17:924–32.

[4] OECD. Guidance document on the validation of (quantitative) structure-activity relationship (Q)SARs models, ENV/JM/MONO(2007)2. 2007. Paris, France.

[5] ECHA. Guidance on information requirements and chemical safety assessment. Chapter R.6: QSARs and grouping of chemicals. 2008.

[6] Russell WMS, Burch RL. The principles of humane experimental technique. London: Methuen & Co; 1959.

[7] ECHA. Guidance on information requirements and chemical safety assessment: Appendix R.6-1 for nanomaterials applicable to the guidance on QSARs and grouping of chemicals. Version 20. 2019. https://doi.org/10.2823/273911.

[8] FDA. Guidance for industry. Considering whether an FDA-regulated product involves the application of nanotechnology. 2014.

[9] FDA. Guidance for industry. Safety of nanomaterials in cosmetic products. 2014.

[10] US EPA. Technical fact sheet—nanomaterials. 2017. https://www.epa.gov/sites/default/files/2014-03/documents/ffrrofactsheet_emergingcontaminant_nanomaterials_jan2014_final.pdf.

[11] US EPA. Working guidance on EPA's section 8(a) information gathering rule on nanomaterials in commerce. 2017.

[12] Policy statement on health Canada's working definition for nanomaterial; 2011. https://www.canada.ca/en/health-canada/services/science-research/reports-publications/nanomaterial/policy-statement-health-canada-working-definition.html.

[13] Ferraris C, Rimicci C, Garelli S, Ugazio E, Battaglia L. Nanosystems in cosmetic products: a brief overview of functional, market, regulatory and safety concerns. Pharmaceutics 2021. https://doi.org/10.3390/pharmaceutics13091408.

[14] EU. Commission recommendation of 18 October 2011 on the definition of nanomaterial (2011/696/EU). Brussels, Belgium: European Commission (EC); 2011.

[15] EC. Commission recommendation of 10.6.2022 on the definition of nanomaterial. 2022. https://ec.europa.eu/environment/chemicals/nanotech/pdf/C_2022_3689_1_EN_ACT_part1_v6.pdf.

[16] EC. Commission Regulation (EU) 2018/1881 of 3 December 2018 amending Regulation (EC) No 1907/2006 of the European Parliament and of the Council on the Registration, Evaluation, Authorisation and Restriction of Chemicals (REACH) as regards Annexes I, III, VI, V. 2018.

[17] ECHA. Appendix for nanoforms applicable to the guidance on registration and substance identification. 2019. https://doi.org/10.2823/832485.

[18] EC. Regulation (EC) No 1907/2006 of the European Parliament and of the Council of 18 December 2006. 2006.

[19] Lynch I, Afantitis A, Leonis G, Melagraki G, Valsami-Jones E. Strategy for identification of nanomaterials' critical properties linked to biological impacts: interlinking of experimental and computational approaches. Challenges Adv Comput Chem Phys 2017.

[20] Lynch I, Lee RG. In: Support of the inclusion of data on nanomaterials transformations and environmental interactions into existing regulatory frameworks; 2016.

[21] Chetwynd AJ, Zhang W, Thorn JA, Lynch I, Ramautar R. The nanomaterial metabolite corona determined using a quantitative metabolomics approach: a pilot study. Small 2020.

[22] Roy K, Kar S, Das RN. A primer on QSAR/QSPR modeling: fundamental concepts. Springer; 2015.

[23] Roy K, Kar S, Das RN. Background of QSAR and historical developments. Underst Basics QSAR Appl Pharm Sci Risk Assess 2015. https://doi.org/10.1016/b978-0-12-801505-6.00001-6.

[24] Singh KP, Gupta S. Nano-QSAR modeling for predicting biological activity of diverse nanomaterials. RSC Adv 2014;4:13215–30.

[25] Puzyn T, Leszczynski J, Cronin MT, Puzyn T, Leszczynski J, Cronin MT. Recent advances in QSAR studies: methods and applications. Challenges Adv Comput Chem Phys 2010. https://doi.org/10.1007/978-1-4020-9783-6.

[26] U.S. EPA: using predictive methods to assess hazard under TSCA. https://www.epa.gov/tsca-screening-tools/using-predictive-methods-assess-hazard-under-tsca.

[27] Survey of the use of (quantitative) structure activity relationship [(Q) SAR] tools and approaches at health Canada and partner departments; 2012.

[28] RiskGONE project. https://riskgone.wp.nilu.no/.

[29] NanoSafety Cluster. https://www.nanosafetycluster.eu/eu-nanosafety-cluster-projects/seventh-framework-programme-projects/enpra.html.

[30] Visser U, Abeyruwan S, Vempati U, Smith RP, Lemmon V, Schürer SC. BioAssay Ontology (BAO): a semantic description of bioassays and high-throughput screening results. BMC Bioinf 2011. https://doi.org/10.1186/1471-2105-12-257.

[31] The European Commission's Science and Knowledge Service. JRC QSAR Model Database. EURL ECVAM DataBase service on alternative methods to animal experimentation. 2017.

[32] QSAR DataBase. https://qsardb.org.

[33] Predictive Model Markup Language. http://dmg.org.

[34] JaqPot Quattro. http://www.enanomapper.net/library/jaqpot-quattro-ui.

CHAPTER

36

In silico and in vivo ecotoxicity—QSAR-based predictions and experimental assays for the aquatic environment

Melanie Voigt and Martin Jaeger

Niederrhein University of Applied Sciences, Department of Chemistry and ILOC, Krefeld, Germany

Abbreviations

AOPs Advanced oxidation processes
BDD Boron-doped diamond
CAS Chemical abstract service
ChV Chronic value
DEMETRA Development of environmental modules for evaluation of toxicity of pesticide residues in agriculture
ECHA European Chemicals Agency
ECOSAR Ecological Structure–Activity Relationship
ECOTOX ECOTOXicology Knowledgebase
EPISuit Estimation Programs Interface Suite
EU European Union
GC Glassy carbon
LANUV Office for Nature, Environment and Consumer Protection (Landesamt für Natur, Umwelt und Verbraucherschutz)
LC Lethal concentration
LOEC Lowest observed effect concentration
NOEC No observed effect concentration
OECD Organisation for Economic Co-operation and Development
PFAS Per- and polyfluoroalkyl substances
PFOA Perfluorooctanoic acid
PFOS Perfluorooctanesulfonic acid
PLS Partial least squares
PPCPs Pharmaceutical personal care products
QSAR Quantitative structure–activity relationship
R^2 Coefficient of determination
REACH Registration, Evaluation, Authorisation and Restriction of Chemicals
RMSE Root mean square error
SMILES Simplified molecular-input line-entry system
SSE Sum square error
SVR Support vector regression
T.E.S.T. Toxicity estimation software tool
US EPA United States Environmental Protection Agency
UV Ultraviolet
VOS Visualization of similarities

Introduction

In recent years, more and more anthropogenic micropollutants, such as pharmaceuticals, pesticides, personal care products, and plasticizers, have been observed in numerous water bodies worldwide [1–3]. These compounds and their transformation products (TPs) of chemical, photochemical, and biological origins represent a serious environmental hazard often expressed in terms of acute and chronic toxicity to aquatic organisms. Knowledge of the toxicity is hence of paramount importance. Depending on the type of water, reported concentrations of micropollutants varied from several pg/L up to several hundred of μg/L [1–3]. While the rapid development of analytical instruments provides sufficient sensitivity to allow to access picogram-per-liter ranges, an ever-increasing number of new substances are being produced in industry, which may eventually penetrate into the aquatic environment [4–6]. Over 196 million organic and inorganic substances have been registered by the chemical abstract service (CAS), and every day, thousands of new substances are added [7]. Unsurprisingly, millions of tons of chemicals are released into the environment every year [8]. Many of these substances are highly persistent and poorly degradable through biodegradation by naturally occurring bacteria, through photodegradation by sunlight, or through chemical degradation

QSAR in Safety Evaluation and Risk Assessment
https://doi.org/10.1016/B978-0-443-15339-6.00018-7

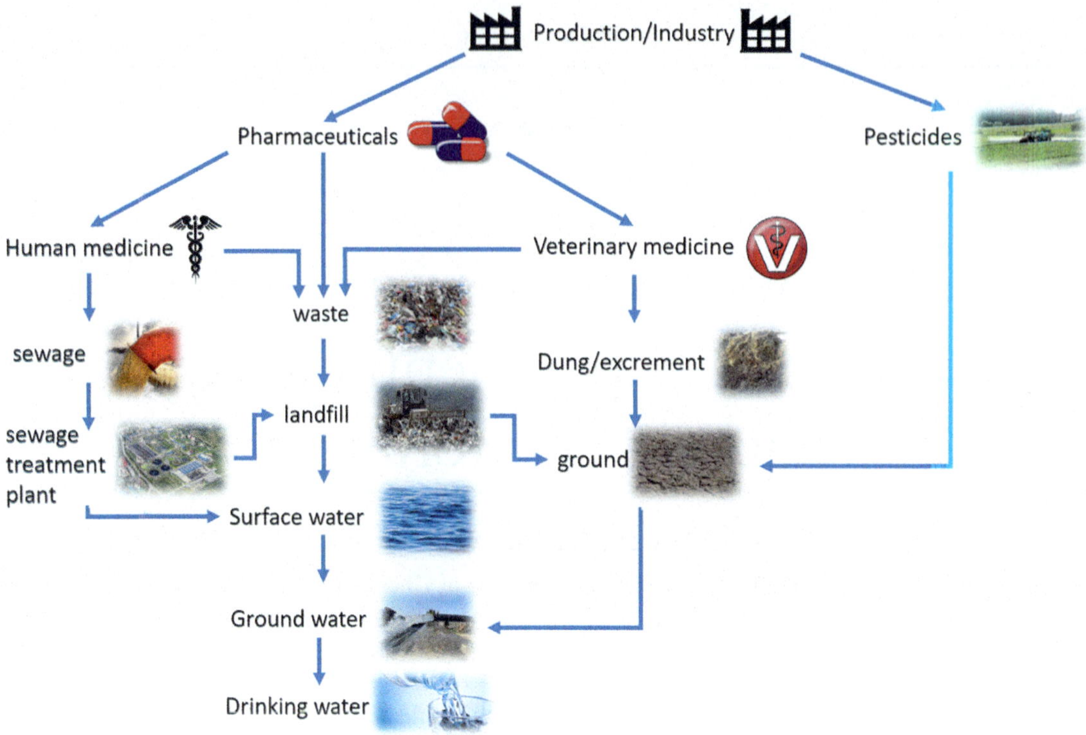

FIGURE 36.1 Entry routes of anthropogenic micropollutants into the aquatic environment.

[8]. The entry routes for these micropollutants are known [9–11]. Those reported most frequently are shown in Fig. 36.1.

Anthropogenic micropollutants of pharmaceutical origin enter the municipal sewage system through household and industrial wastewater. Conventional sewage treatment plants equipped with the usual three-step mechanical, biological, and chemical purification are often not able to completely remove the anthropogenic micropollutants from the water cycle [12,13]. Thus, an unknown number of these contaminants enter the aquatic environment. Another route of entry exists for the use of pharmaceuticals in animal husbandry [14]. Pharmaceuticals are applied to animals, from which they are excreted and enter agriculture via the slurry. The soil, which is then contaminated with the pharmaceuticals, may be extracted by rainwater, and the pharmaceuticals find their way into surface waters. Analogously, pesticides after application may be washed off the ground and transferred directly into surface waters [15,16].

To eliminate pharmaceutical, pesticidal, and cosmetic micropollutants from effluents, research has been carried out for several years on advanced purification stages for sewage treatment plants [17]. The research methods comprise advanced oxidation processes (AOPs). The common feature of AOPs is the formation of hydroxyl radicals, which is the essential mode of action potentially accompanied by other mechanisms [18–20]. With an oxidation potential of 2.80 V, hydroxyl radicals are strong oxidizing agents, which can attack and deconstruct organic substances [21]. Hydroxyl radicals are formed on exposure to ozone, electrochemistry, ultraviolet (UV) radiation, or sonolysis [17,22]. Fig. 36.2 shows an overview of the common AOPs.

Several degradation mechanisms may occur during AOPs. Indirect degradation is seen as the predominant pathway. It is caused by the hydroxyl radicals formed from hydrolysis. The direct degradation occurs through the interaction of the AOP agent with the chemical substance itself. When it is exposed to radiation, the compound may absorb light of suitable wavelength, enters an excited stated, undergoes chemical reactions, and is destroyed as a result [23–25]. Other direct pathways may be initiated through electron transfer, sound-induced excitation, and chemical reactions. However, both indirect AOP-induced and direct pathways may lead to transformation products that may be even more ecotoxic than the initial drug or pesticide [26–28]. Depending on the ecotoxicity test organism and on the substances, increase of ecotoxicity, decrease, and increase–decrease have been observed on AOP treatment. In their review article, Wang et al. show an overview of previous studies on ecotoxicity when using AOPs in wastewater. Yet, no significant correlation between AOP and ecotoxicity could be concluded [28].

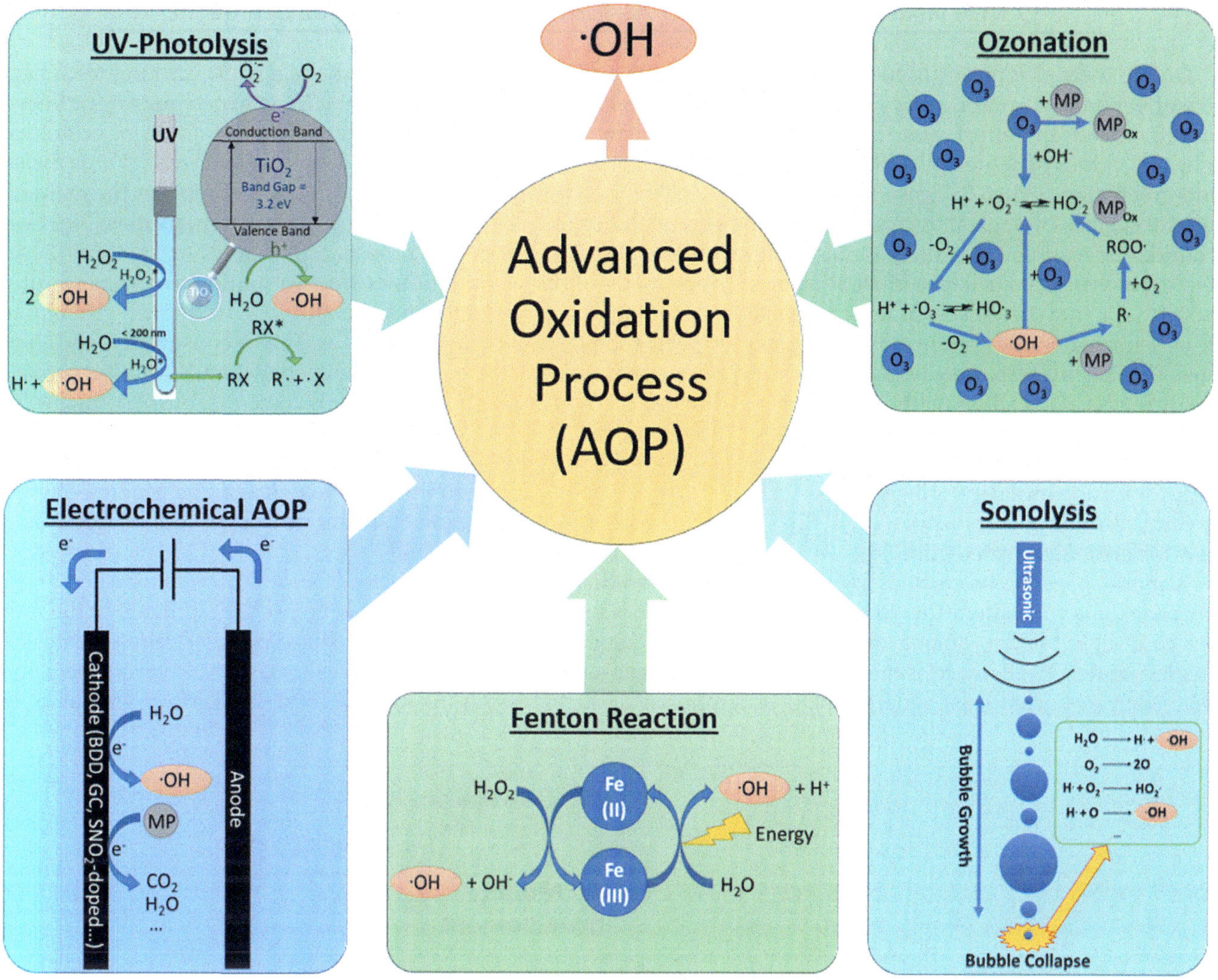

FIGURE 36.2 Overview of common AOPs. *BDD*, boron-doped diamond; *GC*, glassy carbon [17].

Ecological hazard or ecotoxicity is expressed as the chronic or acute toxicity to aquatic organisms such as bacteria, algae, or fish. Suitable data are usually obtained from experimental assays and often made available in public domain databases. In many cases, however, sufficient data on the toxicity of micropollutants to aquatic organisms are not available or unknown. In particular, toxicity data of transformation products stemming from AOPs but even from sewage treatment metabolization were determined as part of scientific research and remain as singletons. Toxicity determination of TPs depends on the knowledge of the chemical structure. Hence, isolation and structure elucidation are required. When sufficient quantities cannot be obtained for experimental in vitro or in vivo testing, e.g., due to inaccessibility through chemical synthesis, in silico methods have established themselves as a remedy. Among them, quantitative structure–activity relationship (QSAR) analysis has become the prominent means to assess ecotoxicity quickly and easily [24,29–32].

This chapter provides an overview of QSAR with the most common databases containing experimental ecotoxicity data. Their features will be compared. Focus is laid on the ECOTOXicology Knowledgebase (ECOTOX) as the largest collection of data. ETOX and the European Chemicals Agency (ECHA) Registration, Evaluation, Authorisation and Restriction of Chemicals (REACH) are presented as well. In addition, software tools for in silico ecotoxicity prediction are reviewed. In particular, the Ecological Structure–Activity Relationship (ECOSAR) tool is discussed in detail, as its reported use is steadily increasing. For illustration purposes, in silico and in vivo ecotoxicity values will be compared for diclofenac, oxadiazon, prochloraz, sulfamethoxazole, tebuconazole, tetraconazole, thiacloprid, and triallate.

QSAR modeling for ecotoxicity assessment in the aquatic environment

Due to their ubiquitous occurrence, micropollutants and their TPs most often have a short- or long-term negative effect on the aquatic ecosystems. A comprehensive knowledge of their ecotoxicity is hence important to evaluate the need, the effort, and the means to prevent their entry into surface waters. The definition of ecotoxicity relies on values of acute and chronic toxicity to aquatic organisms such as *Daphnia magna*, *Danio rerio*, *Vibrio fischeri*, and green algae [33–37]. These have often been determined for the active agent and less frequent for the main metabolites, since biological assays depend on the availability of pure compounds in sufficient quantities [38]. Compounds for testing are usually obtained from chemical or biochemical synthesis and repetitive isolation [39,40]. Yet, their chemical structure must be known. Since many of the degradation or transformation products resulting from AOPs may be inaccessible or too expensive for chemical synthesis, in vitro and in vivo assays with aquatic organisms cannot be conducted. In the same respect, such tests may be laborious and time-consuming, and therefore, ecotoxicity tests may be avoided for economic reasons [8,41–43]. As an inexpensive and fast alternative to in vitro assays, in silico methods such as read-across procedures or QSAR models have established themselves as promising tools [42,43]. Aquatic organism and chemical compounds need not be used. Ecotoxicity is predicted using multivariate data analysis or statistical mathematical models based on molecular structure, molecular descriptors, and similarity principles [8,41,42,44]. With the help of these models, ecotoxicity can be predicted within several seconds and at low cost [8,41,42]. Fig. 36.3 depicts a typical scheme of the mode of operation of QSAR analysis for ecotoxicity prediction [8].

A model needs to be created first for the adequate prediction of the ecotoxicity of an unknown substance. To this purpose, data from literature, databases, or directly from in vitro experiments may be used. Fragments or the number of double bonds—among others—serve as molecular descriptors. Based on these data, a training dataset is created, and a QSAR model is built. The model represents a mathematical equation as used in multivariate data analysis, such as partial least squares (PLS) or support vector regression (SVR). The model is further validated; the

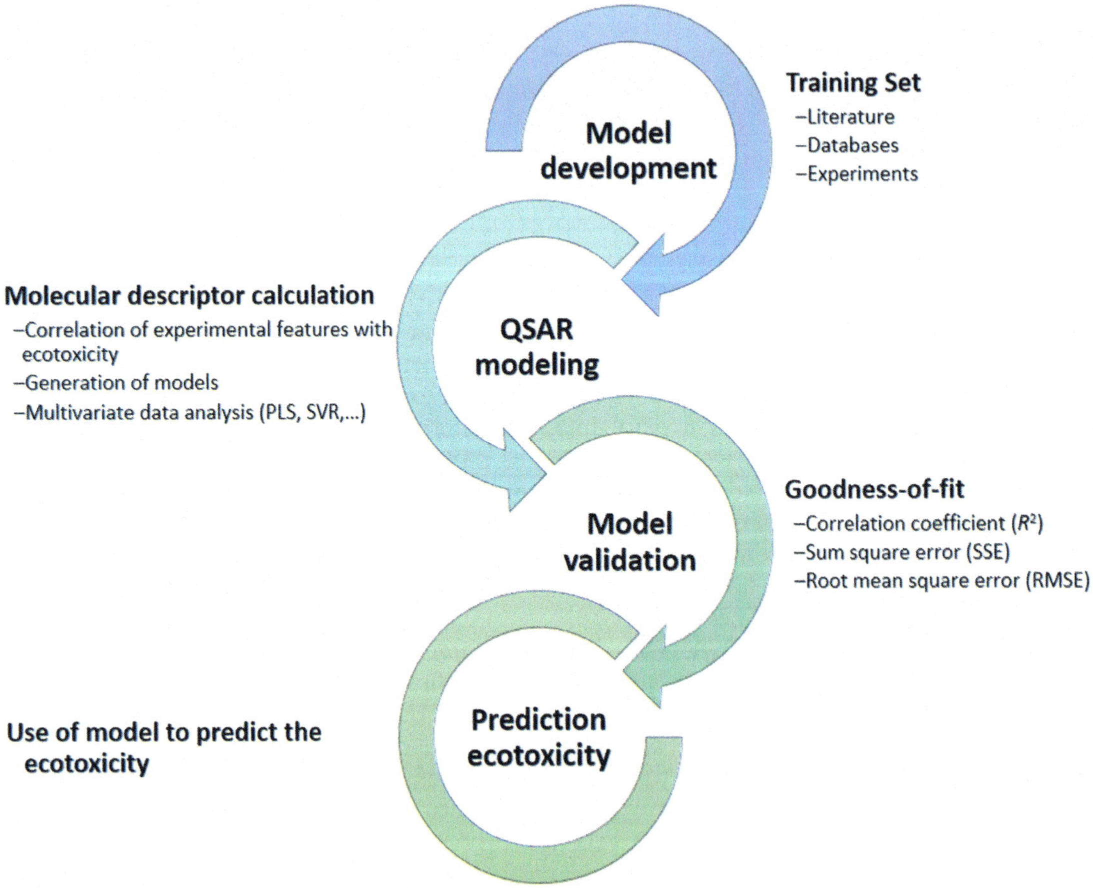

FIGURE 36.3 Mode of operation of QSAR modeling for ecotoxicity prediction.

goodness-of-fit is estimated through the coefficient of determination (R^2), sum square error (SSE), or the root mean square error (RMSE). After successful validation, the model can be applied to unknown substances to predict their ecotoxicity. As common in statistical analysis, the higher the number of available data as training set, the more precise and accurate the ecotoxicity prediction will be [41,45].

Several organizations have defined good application practices for the use of QSAR analysis, especially for use in a regulatory context. Five principles for the validation of QSAR models were introduced in November 2004 by the Organisation for Economic Co-operation and Development (OECD) [46]. A QSAR model for regulatory purposes is required to have

(1) a defined endpoint,
(2) an unambiguous algorithm,
(3) a defined domain of applicability,
(4) an appropriate measure of goodness-of-fit, robustness, and predictivity,
(5) a mechanistic interpretation, if possible.

The first principle, defined endpoint, is important to identify the experimental system, on which the QSAR model is built. As an example, the lethal concentration (LC_{50}) value against an organism may represent the defined endpoint. The second principle, unambiguous algorithm, is of high importance. The model algorithm must be transparent and comprehensible to the user. A defined applicability domain should demonstrate where the model predictions are reliable. The fourth principle requires model information on their internal performance in terms of robustness and goodness-of-fit as well as on the predictive power. Yet, the predictability of a model cannot be described in absolute terms, since values may vary with the statistical methods and parameters used. This information depends on the training datasets used. Wherever possible, mechanistic interpretation should be possible according to principle 5. Since this is not always the case, all connections between the descriptors should be documented [46]. The correlation between the main features of molecules, represented by the molecular descriptors, and the bioactivities represent an example for a mechanistic interpretation [47].

The United States Environmental Protection Agency (US EPA) does not recommend prediction for inorganic and organometallic chemicals, polymers, nanoparticles, and chemicals with a molecular weight greater than 1000 g/mol. They emphasize that due to their unique properties, nanoparticles may have unaccounted toxicity mechanisms [48]. The molecular weight limitation is due to the decreased ability of chemicals to absorb through the respiratory membranes of organisms. Respiratory absorption is hampered at molecular weights larger than 600 Da. At molecular weight larger than 1000 Da, chemicals are assumed no longer absorbable and hence nontoxic. Exceptions to this assumption are surface-active compounds, such as cationic polymers, since their toxic effect depends on nonrespiratory absorption mechanisms [48].

The REACH regulation came into force in the European Union on June 1, 2007. REACH regulates access of chemical substances to the market. Only previously, chemicals registered within the scope may be sold. When registering under REACH, manufacturers and importers are obliged to disclose the chemicals' hazardous potential to humans and the environment. The scope of the required data depends on the annual production or import quantity, so that no or an incomplete risk assessment may need to be carried out for small quantities of chemicals [49]. Annex XI of the regulation defines criteria for the application of QSAR models for ecotoxicity assessment [49]:

- The results were obtained using a scientifically validated (Q)SAR model.
- The substance falls within the scope of the (Q)SAR model.
- The results are sufficient to classify and label the substance and assess its risk.
- The method used is adequately and meaningfully documented.

The criteria compare to the OECD principles. Scope and documentation requirements are similar. Scientific validation, if taken seriously, is in agreement with principles 1 and 2. Yet, the requirements of a defined domain and quality parameters seem more specific than a result sufficient for substance classification and labeling together with risk assessment. Nevertheless, regulatory initiatives have included and defined the application of QSAR models for in silico ecotoxicity assessment.

The use of in silico methods to predict ecotoxicity has been growing steadily for two decades. The bibliometric visualization of similarities (VOS) analysis from the perspective QSAR and aquatic ecotoxicity assessment reveals the relevant organisms (blue area), the fields of QSAR applications (red area), the modeling methods (green area), the modeling-related subjects (purple area), and QSAR-related topics for aquatic ecotoxicity determination (yellow area). Findings are of interest for future research as research trends and activities and aquatic ecotoxicity assessment will be relevant for QSAR analyses.

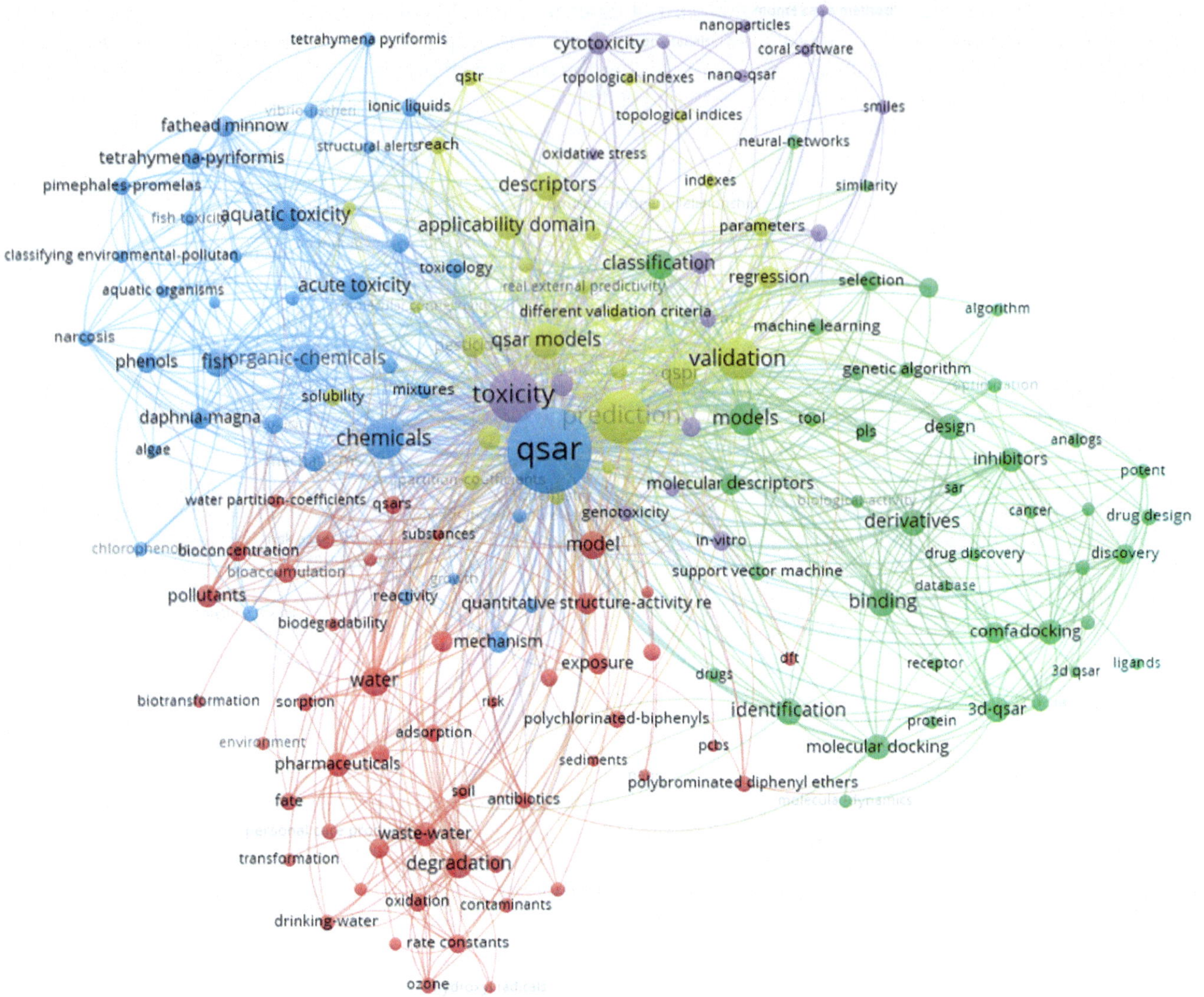

FIGURE 36.4 VOS of QSAR analysis in environmental science created with VOSviewer version 1.6.18 (Leiden University, Centre for Science and Technology Studies (CWTS), Netherlands).

The VOS scheme in Fig. 36.4 demonstrates how QSAR analysis has spread in environmental science. When entering "QSAR" and "Environmental Science" in "Web of Science," a total of 2887 scientific publications were found. A full keyword search with at least 25 entries was selected for the creation of the VOS scheme.

A total of five clusters were formed. The blue area primarily describes the organisms for the ecotoxicity determination. These are *D. magna*, fish, fathead minnow, *Tetrahymena pyriformis*, and *Pimephales promeas*. Areas of application such as pharmaceuticals, contaminants, wastewater, pollutants, or antibiotics are shown in the red area. The green area color-codes the type of QSAR analysis and the methods used for modeling, e.g., support vector machines, PLS, neural networks, database, and machine learning. Subjects, algorithms, and subordinated methods related to modeling, such as validation or prediction, are spread in the purple area. Less specific or very general topics are found in the yellow part.

Databases for experimentally determined aquatic ecotoxicity values

The commonly used databases for aquatic ecotoxicity will be presented in this section. The most widespread and hence indicated by search engines on the keywords "database aquatic ecotoxicology" is the data collection ECOTOX (https://cfpub.epa.gov/ecotox/) as abbreviation for ECOTOXicology Knowledgebase. The database is maintained by the US EPA and is the largest one with curated ecotoxicity data [50]. About 12,500 chemicals, 13,700 species, and 53,000 references comprising over 1.1 million results are currently contained. It thus provides the most of the freely available data on aquatic ecotoxicity. Pharmaceutical personal care products (PPCPs), per- and polyfluoroalkyl substances (PFAS), neonicotinoides, conazoles, and phthalate esters belong to the 10 anthropogenic micropollutants with the highest number of records in the database ECOTOX.

The German Environment Agency (The UBA) have established an information system of ecotoxicology and environmental quality objectives called ETOX (http://webetox.uba.de/webETOX/). It contains about 55,000 entries on the adverse effects on aquatic organisms, about 5000 entries on the adverse effects on terrestrial organisms, and about 4500 entries on environmental quality targets for the compartments water and soil. A third source of data is provided by the ECHA as ECHA REACH (https://echa.europa.eu/). The database stems from registered chemicals and substances that had to be registered and whose risk potential had to be assessed. More than 21,500 substances have been registered since 2008. For several substances, ecotoxicity value entries for fish, aquatic invertebrates, and algae are available. A distinction was made between short- and long-term toxicity.

The extent to which these and other ecotoxicity databases may be useful with respect to future research will be discussed in more detail in the following. Further to experimental determination, tools will be presented how ecotoxicity data can be obtained through nonexperimental means and how they are predicted.

In silico tools

Over the past two decades, several software packages and tools have been designed and introduced into the field of ecotoxicity. In these applications, models for aquatic ecotoxicity prediction have been implemented. The user only needs to have knowledge of the chemical structure, from which the ecotoxicity is readily predicted [51,52]. Due to the computational prediction, these methods are referred to as in silico, although the models rely on experimental data [53]. In analogy to computational modeling, the term semiempirical methods might seem appropriate as well. Table 36.1 gives an overview of actual software tools for predicting ecotoxicity [43].

Hou et al. and Melkinov et al. investigated the applicability and performance of aquatic ecotoxicity software tools [43,56]. The best performing algorithm in combination with the available database was found within ECOSAR, followed by T.E.S.T [43,56]. The tool VEGA was not examined in their study.

Another study compared seven in silico tools for the ecotoxicity prediction for daphnia and fish. VEGA showed the best performance, whereas ECOSAR and T.E.S.T performed similarly, which was slightly inferior to VEGA [57]. For this purpose, the R^2, RMSE, and percentage of accuracy between measured and the predicted values were compared [57]. When testing several prediction models, the State Office of Nature, Environment and Consumer Protection (LANUV) of North Rhine-Westphalia concluded that the prediction of ecotoxicity was usually in the range of the experimental data allowing a variation by a factor of 10 [58]. This level of accuracy would suffice many requirements for an initial toxicity assessment when no experimental data are available. When the toxicity of a substance was based on a particular mode of action, the prediction tools could only deliver accurate results in cases, where this specific mode of action had been integrated in the prediction tool. To reduce the probability of false positives and false negatives, LANUV recommends using several prediction models in parallel, especially when data are sparse. In addition, the results of the toxicity assessments should be checked by a toxicological expert [58]. Yet, the need for an expert opinion may not always be met. Hence, expert knowledge should be included in the software, e.g., as a reliability indicator taking into account number of entries, relevant models, and structural similarity.

TABLE 36.1 Overview of free available software tools for predicting the aquatic ecotoxicity.

Software	Link	References
ECOSAR (Ecological Structure–Activity Relationship)	https://www.epa.gov/tsca-screening-tools/ecological-structure-activity-relationships-ecosar-predictive-model	[48,54]
OECD QSAR Toolbox	https://www.oecd.org/chemicalsafety/risk-assessment/oecd-qsar-toolbox.htm	
T.E.S.T. (Toxicity Estimation Software Tool)	https://oasis-lmc.org/	[55]
VEGA	https://www.vegahub.eu/portfolio-item/vega-qsar/ https://www.vega-qsar.eu/	

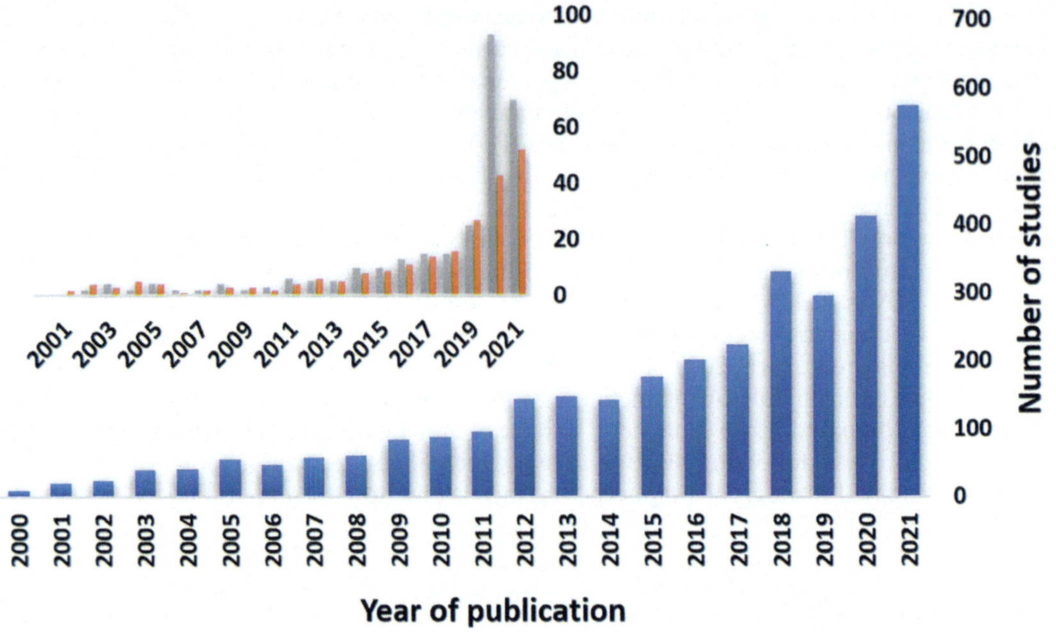

FIGURE 36.5 Number of studies per year as found by Google Scholar (*blue*), SciFinder (*orange*), and Web of Science (*gray*) in search of the keyword *ECOSAR*.

The most widely used predictive tool is the ECOSAR software tool box. Study reports making use of ECOSAR increased exponentially during the past decade (see Fig. 36.5). ECOSAR was developed by the US EPA. It can be obtained through download as stand-alone software, but is also implemented in the OECD QSAR Toolbox. The actual free-ware version is ECOSAR Application 2.2. Chemical structures can be input as Chemical Abstract Service (CAS) number or as simplified molecular-input line-entry system (SMILES). The chemical structure may also be drawn within ECOSAR or input as chemical table files, such as .mol file, SDF Multiple Mol file (.sdf, .sd). In the early 1980s, data evaluation by the US EPA and others revealed that the octanol—water distribution coefficient or K_{OW} was a compound's property exhibiting the best correlation between structure and ecotoxic effects [59,60]. In the early 1990s, the first version of ECOSAR was released, and the focus was laid on K_{OW} prediction for neutral organic substances using the relationship of log K_{OW} values and the median toxicity values such as LC_{50} and EC_{50} [48,54]. In the meantime, more and new toxicity data were implemented in the program allowing to predict acute median toxicity and chronic value (ChV) of fish, daphnids, and algae. Other organisms such as earthworms or sea urchins can also be included. ChV was calculated from the geometric mean of the no observed effect concentration (NOEC) and the lowest observed effect concentration (LOEC), according to Eq. (36.1) [61].

$$ChV = 10^{[\log(LOEC \times NOEC)]/2} \tag{36.1}$$

One hundred and eleven chemical classes were implemented in ECOSAR version 1.11 such that 704 QSAR values for numerous endpoints and organisms were computed [48,54]. ECOSAR training datasets were derived from ECOTOX database entries originating from ecotoxicity assays with US EPA compliance [57]. The selected and applied data fulfill all required criteria: The experimental values were generated at pH 7, which corresponds best to standard environmental conditions. As further requirements, the total content of organic carbon was not to exceed 2 mg/L, and the temporary water hardness was about 150 mg/L $CaCO_3$ [48]. The latest version of ECOSAR predicts log K_{OW} values of chemical substances using the KOWWIN software from the Estimation Programs Interface Suite (EPISuit) program [48,57,62].

As mentioned before, compounds with a molecular weight of larger than 1000 Da are assumed nontoxic. Since molecular mass restrictions are not implemented into ECOSAR, the results need checking for adequacy [48].

In general, the ECOSAR prediction model leads to reliable toxicity value prediction provided two requirements are fulfilled. Firstly, the scope of application must be respected. Acute toxicity values for fish and daphnia require log K_{OW} values less or equal to 5.0, while green algae need values of less or equal to 6.4. Outside this range, substances are considered insufficiently soluble in water and hence unable to cause aquatic toxicity. For chronic toxicity values, the application domain extends to log K_{OW} values less or equal to 8.0 [48]. Water solubility is an important criterion to consider when assessing the toxicity of chemicals. A meaningful toxicity value must not exceed water solubility. When dealing with solid materials at room temperature, the melting point is required and needs consideration when using ECOSAR for toxicity assessment, since water solubility depends inversely proportional on the melting point for constant K_{OW} [48]. Secondly, the interpretation of chemical similarity needs to be sound. Its realization relies on the performance of the SMILES code. At last, the amount of data entries for model building must be sufficiently large. This criterion is not satisfied occasionally.

For a comprehensive risk assessment, chronic toxicity values are of great importance, since some hazards only arise on long-term exposure. However, there are no explicit models for chronic values. The models ECOSAR provides for chronic toxicity assessments are often extrapolations from acute toxicity levels using certain factors, e.g., Eq. (36.1) [58,61]. In general, many prediction models perform poorly on estimating the toxicity of ionizable substances such as diclofenac, since ions have hydrophobicities changing with the pH of the environment [58].

Further to the restrictions mentioned, it is being discussed whether pharmaceuticals fall within the application scope of ECOSAR [63]. In the process of toxicity assessment, ECOSAR divides chemicals into structural classes. Since many pharmaceuticals possess structures that lead to an ambiguous interpretation as a consequence of the SMILES code, unequivocal classification is sometimes impossible, so that many chemicals are assigned to several classes [63]. In addition, drugs usually have special modes of action that need to be known and stored in the program [64]. In 2009, a comparative study assessed the toxicity of 364 pharmaceuticals using ECOSAR. Madden et al. investigated whether the compounds were assigned to reasonable chemical classes in a representative manner. The predicted values were compared with available experimental toxicity values. The authors concluded that the 364 pharmaceuticals had been accurately assigned to the chemical classes that correspond to the application domains of ECOSAR [65].

In a study by Bu et al., prediction models, which were specifically developed to estimate the toxicity values of pharmaceuticals, were implemented into ECOSAR and tested for their performance. The results were evaluated using drugs that had not been included in the training datasets for any of the models. ECOSAR was found to tend to underestimate the ecotoxicity of the drugs [63]. Yet, other studies using ECOSAR reported results in favor of the applicability of ECOSAR to pharmaceuticals [66]. In the past decade, the data availability has greatly improved. Ecotoxicity data for some new drugs have been made publicly available by the regulatory agencies and pharmaceutical companies. Developers have also enhanced the applicability of the prediction model. Yamada et al. reported that they rarely observed significantly underestimated toxicity predictions [67]. It should be noted that the application scopes for the QSAR models complied with the corresponding criteria [32]. When the log K_{OW} value is outside the application range and when the predicted value exceeds the solubility of the substance, the ECOSAR prediction engine issues warnings [57]. As mentioned before, experts may further check the scope of the training datasets.

The OECD QSAR Toolbox, current version 4.4.1, is freeware as well. It was first released in 2008 after initiation by the OECD and development in the Mathematical Chemistry laboratory of the Prof. As. Zlatarov University at Bourgas, Bulgaria. Over time, other contributors added to the development. This toolbox is by far the most extensive. In addition to the QSAR prediction, trend analyses and read-across can be carried out. Read-across refers to interpolation of qualitative endpoints but may also be applied to quantitative endpoints with sparse analogues. Structures could be input as CAS number, systematic name, drawing, SMILES notation, or International Chemical Identifier (InChI) or imported as file. Supported file formats are SMILES, .mol, .sdf, or ascii files. Profiling may be carried out to determine which models are applicable for the chemical structure under investigation. The QSAR models originate from ECOSAR, as mentioned before. Further available databases are the Danish QSAR database (https://qsar.food.dtu.dk/) and the ones from Veit et al. and Pavan et al. [68,69]. As many as 683 QSAR models are implemented in this tool, where 669 stem from ECOSAR, 9 from the Danish QSAR Database, and 4 from Veit et al. and Pavan et al. The software can retrieve experimental data from databases such as ECOSAR, ECHA REACH, or Aquatic Oasis.

The Toxicity Estimation Software Tool (T.E.S.T.) was developed by the US EPA and is freely available. The software predicts the acute ecotoxicity of fathead minnow after 96 h, *D. magna* after 48 h and *T. pyriformis* after 96 h as endpoints. Structure drawing tools, .mol file input, searching through CAS numbers, SMILES notation, systematic name, InChI, InChIKey, which is a compact chemical identifier derived from InChI and is only 27-characters long, or DSSTox substance identifier (DTXSID), which is used in the US EPA CompTox Dashboard, are implemented. The software toolbox provides different methods for predicting ecotoxicity: the hierarchical method, the single model method, the group contribution method, the nearest neighbor method, and the consensus method. The hierarchical method predicts the ecotoxicity using a weighted average of several cluster models. The single model and group contribution methods apply a multilinear regression model. The nearest neighbor method takes the average of three training set compounds, which are structurally most similar to the compound under consideration. The consensus method is assumed to yield the most accurate prediction, since the predicted ecotoxicity is estimated as the average of all the aforementioned methods [55]. T.E.S.T also uses the ECOTOX database for model training [57].

VEGA also comes as freeware. Version 1.2.0 BETA was released in 2021. Again, SMILES notation input and import as .sdf or .sd file are allowed. Aquatic acute and chronic toxicity models have been implemented. Five models can predict acute fish toxicity, two models toxicity to fathead minnow, four to *D. magna*, one to guppy, and three to algae. For aquatic chronic toxicity, one model each are available for fish, *D. magna*, and algae. A helpful feature in VEGA is that the ecotoxicity prediction is evaluated using a traffic light system and that the reliability of the prediction is indicated with stars. Explanations for the assessment are output as well.

VEGA's traffic light system comprises four colors representing the following comments:

Red: : Compound is highly toxic, LC_{50} or EC_{50} value is less than 1 mg/L.
Orange:: Compound is toxic, LC_{50} or EC_{50} value is less than 10 mg/L and more than 1 mg/L.
Yellow: : Compound is toxic, LC_{50} or EC_{50} value is less than 100 mg/L and more than 10 mg/L.
Green: : Compound is nontoxic, LC_{50} or EC_{50} value is more than 100 mg/L.

This classification is similar to the Passino–Smith classification system [70]. This scheme uses the following indicators: supertoxic, i.e., 0.01–0.1 mg/L; highly toxic, i.e., 0.1–1 mg/L toxic, i.e., 1–10 mg/L; slightly toxic, i.e., 10–100 mg/L; practical nontoxic, i.e., 100–1000 mg/L, and relatively harmless, i.e., >1000 mg/L. Furthermore, VEGA discloses all substances that have been used in the training set of the applied model.

Comparison of experimental versus predicted data

To monitor the occurrence and evaluate the impact of anthropogenic micropollutants in more detail, the EU introduced a watch list, which is updated every 3 years [71–73]. The first watch list of the relevant substances for EU-wide monitoring was introduced within Decision 2015/495/EU, the second within 2018/840/EU, and the third in 2021. The list includes substances that have the potential to exceed the proposed environmental quality standard, as assessed by the Predicted No Effect Concentration (PNEC) value. Yet, Europe-wide monitoring data were insufficient or data were unavailable with a determination limit below the proposed environmental quality standard, which would allow Europe-wide environmental quality standards to be set. A total of 36 substances have been listed on the three version today. For only eight of them, experimental ecotoxicity values in form of LC_{50} after 48 h were found against *D. magna* in the database ECOTOX (see Table 36.2). The databases ETOX and ECHA REACH do not contain values of 50% mortality of *D. magna* after 48 h for any of these substances. For illustrative purposes, QSAR analysis of diclofenac, oxadiazon, prochloraz, sulfamethoxazole, tebuconazole, tetraconazole, thiacloprid, and triallate was carried out for illustration purposes for this chapter, using various QSAR tools with the results shown in Table 36.2.

In the ECOTOX database, the mortality values of *D. magna* after 48 h exposure (LC_{50} 48 h) were found for the eight substances listed in Table 36.2. It is also interesting that seven different values were experimentally determined for tebuconazole. These spread from 0.75 to 40.1 mg/L, which ranges from highly to moderately toxic according to the Passino–Smith classification. Different assay conditions, such as the purity of a substance or application frequencies, have a major impact on the resulting ecotoxicity values. Comparison with the prediction revealed that most QSAR models did not match the experimental data. In most cases, predicted ecotoxicity indicated a larger hazard than did the experimental data. This finding is in agreement with the previously mentioned LANUV study, which found a deviation by a factor of 10 and a higher predicted ecotoxicity [58]. From an environmentally friendly perspective, such false positives might be acceptable, since a potential hazard would not be underestimated and measures would be more than sufficient for environmental protection.

The ECOSAR ecotoxicity prediction yielded several different values for some substances. This occurred when ECOSAR assigned the compound to more than one chemical class. As a consequence, different models were used for the prediction according to the individual class. Furthermore, ECOSAR neglects species such that prediction is made for the genus *Daphnid*, whereas experimental data originate mostly from the species *D. magna*. One substance in particular catches the eye: sulfamethoxazole, which was classified as *amides* and as *anilines (unhindered)*. The values predicted for both classes show large differences. This finding can be traced back to the training set data for each model. The training set for the ECOSAR model Daphnid 48 h LC_{50} *amides* contains 11 compounds, whereas the model *anilines (unhindered)* comprises 18 substances. Test data were not provided for either model, so that the quality of the prediction cannot be evaluated in more detail. Previous studies suggested that as in all statistical methods,

TABLE 36.2 Selected substances of the EU watch lists, experimental data from ECOTOX database and VEGA and QSAR analyses with toxicity organisms using the predictive tools ECOSAR, T.E.S.T., and VEGA; colors according to Passino–Smith classification, cf. above [70]: dark red: supertoxic 0.01–0.1 mg/L; red: highly toxic 0.1–1 mg/L; orange: moderately toxic 1–10 mg/L; yellow: slightly toxic 10–100 mg/L; green: practically nontoxic 100–1000 mg/L; dark green: relatively harmless >1000 mg/L.

EU-Watch-list	Structure, Substance, CAS-number, SMILES	ECOTOX *Daphnia magna* LC_{50} 48h mg/L	ECOSAR DAPHNID LC_{50} 48h		T.E.S.T. LC_{50} 48h mg/L	VEGA *Daphnia magna* LC_{50} 48h (EPA) 1.0.8 mg/L			VEGA *Daphnia magna* LC_{50} 48h (DEMETRA) 1.0.4 mg/L			VEGA Experimental value LC_{50} 48h mg/L
			mg/L	Class			Prediction	Reliability		Prediction*	Reliability	
1st	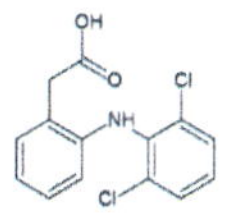 Diclofenac, 15307-86-5, OC(=O)Cc1ccccc1Nc1c(Cl)cccc1Cl	80.1 94.1	No prediction	-	3.50	0.9859	Red	1 star	17.19	Green	1 star	-
1st	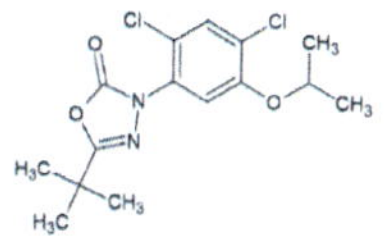 Oxadiazon, 19666-30-9, CC(C)Oc1cc(N2N=C(OC2=O)C(C)(C)C)c(Cl)cc1Cl	<10	0.374 0.741	Carbamate Esters Hydrazines	0.0386	0.1042	Red	1 star	2.44	Red		0.5333
3rd	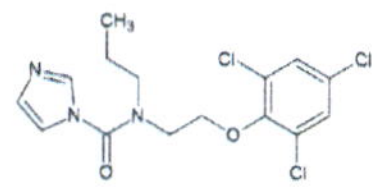 Prochloraz, 67747-09-5, CCCN(CCOc1c(Cl)cc(Cl)cc1Cl)C(=O)n1ccnc1	2.98 3.54 4.94	0.733	Imidazoles	9.92	7.73	Yellow	1 star	0.7033	Red	1 star	-
3rd	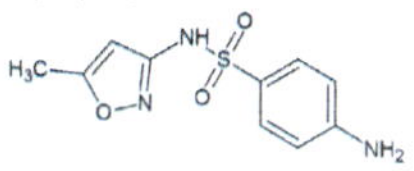 Sulfamethoxazole, 723-46-6, Cc1cc(NS(=O)(=O)c2ccc(N)cc2)no1	234	1.87 2050	Amides Anilines (Unhindered)	No prediction	37.24	Yellow	1 star	4.7	Orange	1 star	-
3rd	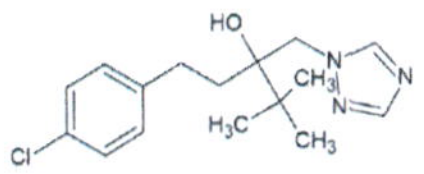 Tebuconazole, 107534-96-3, CC(C)(C)C(O)(CCc1ccc(Cl)cc1)Cn1cncn1	0.75 3.63 3.83 9.19 10.3 40.1	2.78	Triazoles (Non-Fused)	9.63	49.03	Yellow	1 star	2.64	Orange		1.17
3rd	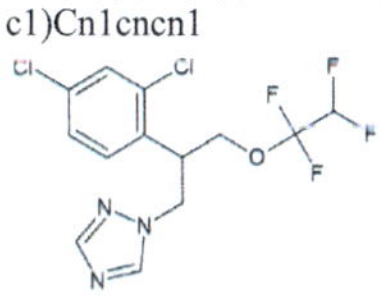 Tetraconazole, 112281-77-3, FC(F)C(F)(F)OCC(Cn1cncn1)c1ccc(Cl)cc1Cl	7.2	2.4	Triazoles (Non-Fused)	9.18	18.34	Yellow	1 star	0.0733	Red	1 star	-
1st, 2nd	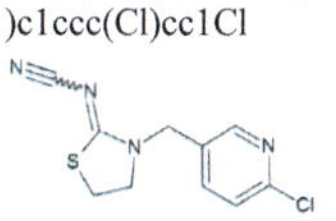 Thiacloprid, 111988-49-9, Clc1ccc(CN2CCSC2=NC#N)cn1	88.0	3.65 1.39 12.2	Aliphatic Amines Halopyrdines Neonicotinoids	3.99	0.1493	Red	1 star	1.03	Orange	1 star	-
1st	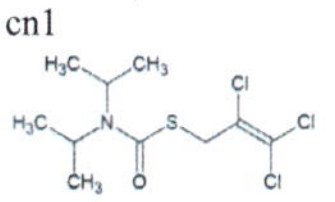 Triallate, 2303-17-5, CC(C)N(C(C)C)C(=O)SCC(Cl)=C(Cl)Cl	0.057	0.449 0.262	Thiocarbamates, Mono Vinyl/Allyl Halides	0.44	2.49	red		0.9612	Red	1 star	0.0566

*Colors are from the traffic-light system from VEGA

prediction reliability increases with increasing numbers [41,45]. Nevertheless, this tool can be used well for a first estimation of ecotoxicity.

The application T.E.S.T. also associates lower values and thus higher toxicity than their experimental counterparts. The predicted values result from the consensus method, which was considered best performing [55]. Yet, the substances used for test or training sets showed little structural similarity to the substance for which ecotoxicity was predicted. The coefficient of determination R^2 for the consensus model amounted to 0.642 and the RMSE to 1.042 [74].

VEGA has the advantageous feature that reliability of the prediction is reported. The star rating enhanced by explanations allows the user to evaluate the results by quick inspection. For the eight substances under consideration, VEGA found the prediction mostly nonreliable. This evaluation was confirmed by comparison with the experimental values. Again, in most cases, the substances were predicted more ecotoxic. Two models served for computing LC_{50} values against *D. magna*. One model stemmed from the US EPA, which is a reimplementation of the original model developed by Todd Martin inside T.E.S.T. [55], while the other one originated from the EU-funded project "Development of Environmental Modules for Evaluation of Toxicity of pesticide Residues in Agriculture" (DEMETRA). For both models, more than 220 substances were contained in the training set and over 40 served as test set, which are significantly higher numbers than used for the ECOSAR models. The US EPA model yielded $R^2 = 0.71$ for the training set and 0.49 for the test set, and RMSE of 0.96 for the training set and 1.02 and for the test set. Quality statistics for the DEMETRA training set were obtained as $R^2 = 0.75$ and RMSE = 1.37. For the test set, $R^2 = 0.68$ and RMSE = 1.49 were determined, indicating a better performance by the US EPA model. Yet, the quality parameters of the models show that there is still room for improvement. More experimental data are required to improve the performance of the models. The variety of compounds would need to be increased, leading to a higher probability for similarity that would be as advantageous as an increased number itself.

Among the studies comparing predicted to experimental ecotoxicity, Xin et al. examined the aquatic toxicity of pharmaceuticals and personal care products to algae [75]. In vitro experiments with different algal species, the toxicity of a mixture of substances, and QSAR analyses were discussed in this study. Especially in vitro assays with *chlorophyta* and *diatoms* were found more sensitive to these substances. In contrast, QSAR has not frequently been applied to mixtures of substances for predicting the ecotoxicity to algae. Model compounds were exemplarily compared with antiinflammatory drugs. Their acute effects on algae were considered relatively low [75]. The authors did not elaborate on the QSAR model, hence leaving the model quality unknown.

Massarsky et al. compared experimental data with ECOSAR-predicted values for per- and polyfluoroalkyl substances (PFAS), especially perfluorooctanoic acid (PFOA) and perfluorooctanesulfonic acid (PFOS), which are also observed in the aquatic environment [61]. Reported acute ecotoxicity values of PFOA were 2- to 108-fold higher than the predicted ones, i.e., the experimental ecotoxicity was lower than the predicted one, which is in line with previous observations. PFOS experimental values were reported to be 1- to 3-fold higher than predicted ones [61]. Situation changed for chronic values. Most of the ChVs were predicted in the range of the reported values. A clear trend as was the case for LC_{50} values was not recognized.

In their study, Melnikov et al. examined 83 structurally diverse chemicals and compared 5 in silico tools including ECOSAR and T.E.S.T. They used data from the US EPA ECOTOX database [56]. They found that ECOSAR prognoses were more accurate than T.E.S.T predictions. Success estimation was based on total accuracy, i.e., predicted and measured LC_{50} fell within the same regulatory category. Success was reported for 61% of the compounds when using ECOSAR versus 48% when using T.E.S.T [56]. Lower values were predicted for 27 substances. For 54 chemicals, LC_{50} values were predicted between 1.02 and 1.53×10^4-fold higher, which contradicts previous observations [56,61]. Yet, an older version of ECOSAR was employed in the study by Melnikov et al. Hence, the previous findings that predicted LC_{50} values lower than the experimental data may not apply to this study due to older models.

As a summary, the performance of in silico ecotoxicity prediction depends on statistical parameters. The number of database entries, structural similarity, descriptors, mode of action consistency, model suitability, and endpoint choice contribute to the goodness of prediction. It is obvious that in the absence or inaccessibility of the chemical compound such as a transformation product or a rare metabolite, QSAR analysis for ecotoxicity is the mean of choice. Computational cost might also be lower than experimental testing and chemical synthesis or isolation of TPs and metabolites. The majority of researchers found that predicted ecotoxicity was higher. Hence, the environmental hazard would be overestimated. Further advantages of in silico methods are the avoidance of laborious assays and extermination of living organisms such as fish. For accurate results, careful choice of tool and model are prerequisites. The best overall performance was reported for ECOSAR, when experimental data for model building stemmed from ECOTOX. Also, rules for good application practice and quality checks have been established and should be followed to evaluate the validity of the obtained results. Experimental data can and should further be recorded and integrated into databases and models. Yet in agreement with the guiding rules, special care should be taken to identify if sufficient experimental data for model training and appropriate models were applied.

Conclusions, perspectives, and future directions

Determining ecotoxicity and aquatic ecotoxicity and assessing environmental hazard represent a major challenge. Synthetically difficult to access transformation products from AOPs, their large number and metabolites even increase the endeavor. This chapter reviewed the application of in silico QSAR methodology as a remedy to these problems. After an overview of the major databases for ecotoxicities and their experimental entries, the working principle of QSAR software tools was sketched. Again, the most prominent representatives were introduced. Several sets of established rules for evaluating the reliability of the prediction were presented. Using an example of ecotoxicity prediction of eight micropollutants, the quality of the software tools was discussed by comparing the predicted with experimental values. Application examples for the aquatic environment were presented from the literature. Both experimental and in silico methods will be relevant for ecotoxicity assessment for the aquatic environment. There is still more need for experimental data to integrate into suitable QSAR models. For a rapid or preliminary assessment of aquatic ecotoxicity, e.g., in pharmaceutical drug discovery, or in the absence of reference substance e.g., after AOP-induced transformation, QSAR analysis is a valuable tool. For an accurate aquatic ecotoxicity analysis, experimental assays with relevant organisms are still preferable.

Acknowledgment

The authors would like to thank B. Sc. Prisca Sandmann for her support.

References

[1] Spindola Vilela CL, Damasceno TL, Thomas T, Peixoto RS. Global qualitative and quantitative distribution of micropollutants in the deep sea. Environ Pollut 2022;307:119414.
[2] Kumar NM, Sudha MC, Damodharam T, Varjani S. Chapter 3—micro-pollutants in surface water: impacts on the aquatic environment and treatment technologies. In: Varjani S, Pandey A, Tyagi RD, Ngo HH, Larroche C, editors. Current developments in biotechnology and bioengineering. Amsterdam: Elsevier; 2020. p. 41–62.
[3] Rogowska J, Cieszynska-Semenowicz M, Ratajczyk W, Wolska L. Micropollutants in treated wastewater. Ambio 2020;49(2):487–503.
[4] Chang H, Wu F, Jin F, Feng C, Zhao X, Liao H. Picogram per liter level determination of hydroxylated polybrominated diphenyl ethers in water by liquid chromatography-electrospray tandem mass spectrometry. J Chromatogr A 2012;1223:131–5.
[5] Kaziur W, Salemi A, Jochmann MA, Schmidt TC. Automated determination of picogram-per-liter level of water taste and odor compounds using solid-phase microextraction arrow coupled with gas chromatography-mass spectrometry. Anal Bioanal Chem 2019;411(12):2653–62.
[6] Rösch A, Beck B, Hollender J, Singer H. Picogram per liter quantification of pyrethroid and organophosphate insecticides in surface waters: a result of large enrichment with liquid–liquid extraction and gas chromatography coupled to mass spectrometry using atmospheric pressure chemical ionizati. Anal Bioanal Chem 2019;411(14):3151–64.
[7] CAS. CAS-Registry. 2022. https://www.cas.org/cas-data/cas-registry.
[8] Satpathy R. Quantitative structure–activity relationship methods for the prediction of the toxicity of pollutants. Environ Chem Lett 2019;17:123–8.
[9] Antakyali D, Morgenschweis C, Kort D, Sasse R, Schulz J, Herbst H. Micropollutants in the aquatic environment and their removal in wastewater treatment works. In: 9th European waste water management conference, Manchester UK; 2015.
[10] Gerbersdorf SU, Cimatoribus C, Class H, Engesser KH, Helbich S, Hollert H, et al. Anthropogenic Trace Compounds (ATCs) in aquatic habitats—research needs on sources, fate, detection and toxicity to ensure timely elimination strategies and risk management. Environ Int 2015;79:85–105.
[11] Pathak N, Tran VH, Merenda A, Johir MAH, Phuntsho S, Shon H. Removal of organic micro-pollutants by conventional membrane bioreactors and high-retention membrane bioreactors. Appl Sci 2020;10(8):2969.
[12] Barbosa MO, Moreira NFF, Ribeiro AR, Pereira MFR, Silva AMT. Occurrence and removal of organic micropollutants: an overview of the watch list of EU Decision 2015/495. Water Res 2016;94:257–79.
[13] Sousa JCG, Ribeiro AR, Barbosa MO, Pereira MFR, Silva AMT. A review on environmental monitoring of water organic pollutants identified by EU guidelines. J Hazard Mater 2018;344:146–62.
[14] Kovalakova P, Cizmas L, McDonald TJ, Marsalek B, Feng M, Sharma VK. Occurrence and toxicity of antibiotics in the aquatic environment: a review. Chemosphere 2020;251:126351.
[15] Rajmohan KS, Chandrasekaran R, Varjani S. A review on occurrence of pesticides in environment and current technologies for their remediation and management. Indian J Microbiol 2020;60(2):125–38.
[16] Goulson D. An overview of the environmental risks posed by neonicotinoid insecticides. J Appl Ecol 2013;50(4):977–87.
[17] Voigt M, Wirtz A, Hoffmann-Jacobsen K, Jaeger M. Prior art for the development of a fourth purification stage in wastewater treatment plant for the elimination of anthropogenic micropollutants-a short-review. AIMS Environ Sci 2020;7(1):69–98.

[18] Coha M, Farinelli G, Tiraferri A, Minella M, Vione D. Advanced oxidation processes in the removal of organic substances from produced water: potential, configurations, and research needs. Chem Eng J 2021;414:128668.
[19] Mazivila SJ, Ricardo IA, Leitão JMM, Esteves da Silva JCG. A review on advanced oxidation processes: from classical to new perspectives coupled to two- and multi-way calibration strategies to monitor degradation of contaminants in environmental samples. Trends Environ Anal Chem 2019;24:1–10.
[20] Deng Y, Zhao R. Advanced oxidation processes (AOPs) in wastewater treatment. Curr Pollut Reports 2015;1(3):167–76.
[21] Sillanpää M. Advanced water treatment—advance oxidation process. Amsterdam: Elsevier; 2020.
[22] Oppenländer T. Photochemical purification of water and air: advanced oxidation processes (AOPs): principles, reaction mechanisms, reactor concepts (chemistry). Weinheim: WILEY-VCH Verlag; 2003.
[23] Voigt M, Jaeger M. Structure and QSAR analysis of photoinduced transformation products of neonicotinoids from EU watchlist for ecotoxicological assessment. Sci Total Environ 2021;751:141634.
[24] Voigt M, Langerbein V, Jaeger M. In silico ecotoxicity assessment of photoinduced imidacloprid degradation using HPLC–HRMS, QSAR and ecotoxicity equivalents. Environ Sci Eur 2022;34(1):47.
[25] Zhang Y, Zhang J, Xiao Y, Chang VWC, Lim TT. Direct and indirect photodegradation pathways of cytostatic drugs under UV germicidal irradiation: process kinetics and influences of water matrix species and oxidant dosing. J Hazard Mater 2017;324:481–8.
[26] Fatta-Kassinos D, Vasquez MI, Kümmerer K. Transformation products of pharmaceuticals in surface waters and wastewater formed during photolysis and advanced oxidation processes—degradation, elucidation of byproducts and assessment of their biological potency. Chemosphere 2011;85(5):693–709.
[27] Vasconcelos TG, Henriques DM, König A, Martins AF, Kümmerer K. Photo-degradation of the antimicrobial ciprofloxacin at high pH: identification and biodegradability assessment of the primary by-products. Chemosphere 2009;76(4):487–93.
[28] Wang J, Wang S. Toxicity changes of wastewater during various advanced oxidation processes treatment: an overview. J Clean Prod 2021;315: 128202.
[29] Voigt M, Hentschel B, Theiss N, Savelsberg C, Bartels B, Nickisch-Hartfiel A, et al. Lomefloxacin—occurrence in the German River Erft, its photo-induced elimination, and assessment of ecotoxicity. Clean Technol 2020;2(1):74–90.
[30] Voigt M, Bartels I, Schmiemann D, Votel L, Hoffmann-Jacobsen K, Jaeger M. Metoprolol and its degradation and transformation products using AOPS-assessment of aquatic ecotoxicity using QSAR. Molecules 2021;26(11):3102.
[31] Villaverde JJ, Sevilla-Morán B, López-Goti C, Calvo L, Alonso-Prados JL, Sandín-España P. Photolysis of clethodim herbicide and a formulation in aquatic environments: fate and ecotoxicity assessment of photoproducts by QSAR models. Sci Total Environ 2018;615:643–51.
[32] Holmquist H, Lexén J, Rahmberg M, Sahlin U, Palm JG, Rydberg T. The potential to use QSAR to populate ecotoxicity characterisation factors for simplified LCIA and chemical prioritisation. Int J Life Cycle Assess 2018;23(11):2208–16.
[33] ISO. ISO 15088:2007 water quality—determination of the acute toxicity of waste water to zebrafish eggs. 2007.
[34] OECD. Test no. 203: fish, acute toxicity test. Paris: OECD Publishing; 2019.
[35] OECD. Test no. 202: *Daphnia* sp. acute immobilisation test. Paris: OECD Publishing; 2004.
[36] OECD. Test no. 201: alga, growth inhibition test. Paris: OECD Publishing; 2006.
[37] ISO. ISO 11348-3:2007 water quality—determination of the inhibitory effect of water samples on the light emission of Vibrio fischeri (Luminescent bacteria test)—part 3: method using freeze-dried bacteria. 2008.
[38] Parolini M. Toxicity of the Non-Steroidal Anti-Inflammatory Drugs (NSAIDs) acetylsalicylic acid, paracetamol, diclofenac, ibuprofen and naproxen towards freshwater invertebrates: a review. Sci Total Environ 2020;740:140043.
[39] Brusotti G, Cesari I, Dentamaro A, Caccialanza G, Massolini G. Isolation and characterization of bioactive compounds from plant resources: the role of analysis in the ethnopharmacological approach. J Pharm Biomed Anal 2014;87:218–28.
[40] Mikulášová M. Resurgence of the interest in microorganisms and plants as sources of secondary metabolites for treating staphylococcal infections. Nat Prod J 2019;9(1):14–25.
[41] Sakkiah S, Guo W, Pan B, Kusko R, Tong W, Hong H. Computational prediction models for assessing endocrine disrupting potential of chemicals. J Environ Sci Heal Part C Environ Carcinog Ecotoxicol Rev 2018;36(4):192–218.
[42] Tang W, Chen J, Wang Z, Xie H, Hong H. Deep learning for predicting toxicity of chemicals: a mini review. J Environ Sci Heal Part C Environ Carcinog Ecotoxicol Rev 2018;36(4):252–71.
[43] Hou P, Jolliet O, Zhu J, Xu M. Estimate ecotoxicity characterization factors for chemicals in life cycle assessment using machine learning models. Environ Int 2020;135:105393.
[44] Schneider M, Pons JL, Labesse G, Bourguet W. In silico predictions of endocrine disruptors properties. Endocrinology 2019;160(11):2709–16.
[45] Zhang L, Sedykh A, Tripathi A, Zhu H, Afantitis A, Mouchlis VD, et al. Identification of putative estrogen receptor-mediated endocrine disrupting chemicals using QSAR- and structure-based virtual screening approaches. Toxicol Appl Pharmacol 2013;272(1):67–76.
[46] OECD. OECD principles for the validation, for regulatory purposes, of (quantitative) structure-activity relationship models. In: 37th joint meeting of the chemicals committee and working party on chemicals, pesticides and biotechnology, Paris; 2004.
[47] Zaki MEA, Al-hussain SA, Masand VH, Sabnani MK, Samad A. Mechanistic and predictive QSAR analysis of diverse molecules to capture salient and hidden pharmacophores for anti-thrombotic activity. Int J Mol Sci 2021;22(15):8352.
[48] Mayo-Bean K, Moran K, Meylan B, Ranslow P. ECOSAR methodology document. US Environmental Protection Agency; 2012.
[49] European Commission. Regulation (EC) No 1907/2006 of the European Parliament and of the council. Off J Eur Union 2006;49.
[50] Olker JH, Elonen CM, Pilli A, Anderson A, Kinziger B, Erickson S, et al. The ECOTOXicology Knowledgebase: a curated database of ecologically relevant toxicity tests to support environmental research and risk assessment. Environ Toxicol Chem 2022;41(6):1520–39.
[51] Roy K. Ecotoxicological QSARs. New York: Springer US; 2020.
[52] Kar S, Sanderson H, Roy K, Benfenati E, Leszczynski J. Ecotoxicological assessment of pharmaceuticals and personal care products using predictive toxicology approaches. Green Chem 2020;22(5):1458–516.
[53] Raies AB, Bajic VB. In silico toxicology: computational methods for the prediction of chemical toxicity. Wiley Interdiscip Rev Comput Mol Sci 2016;6(2):147–72.
[54] Wright RT, Fay K, Kennedy A, Mayo-Bean K, Moran-Bruce K, Meylan W, et al. ECOlogical structure-activity relationship model (ECOSAR) class program. US Environmental Protection Agency; 2022.

[55] Martin T, Harten P, Venkatapathy R, Young D. T.E.S.T (toxicity estimation software tool). 2016.
[56] Melnikov F, Kostal J, Voutchkova-Kostal A, Zimmerman JB, Anastas PT. Assessment of predictive models for estimating the acute aquatic toxicity of organic chemicals. Green Chem 2016;18(16):4432–45.
[57] Zhou L, Fan D, Yin W, Gu W, Wang Z, Liu J, et al. Comparison of seven in silico tools for evaluating of daphnia and fish acute toxicity: case study on Chinese Priority Controlled Chemicals and new chemicals. BMC Bioinf 2021;22(1):1–31.
[58] Rhiem S. Anwendung von Stoffpriorisierungsverfahren für Oberflächengewässer Datenrecherche und Einsatz modellierter Daten. LANUV-Fachberichte 72. 2016.
[59] Hermens J, Leeuwangh P, Musch A. Quantitative structure-activity relationships and mixture toxicity studies of chloro- and alkylanilines at an acute lethal toxicity level to the guppy (Poecilia reticulata). Ecotoxicol Environ Saf 1984;8(4):388–94.
[60] Könemann H. Quantitative structure-activity relationships in fish toxicity studies—part 1: relationship for 50 industrial pollutants. Toxicology 1981;19(3):209–21.
[61] Massarsky A, Donnell MT, de Gandiaga E, Kozal JS, Garnick L, Kubitz JA, et al. Critical evaluation of ECOSAR and E-FAST platforms to predict ecological risks of PFAS. Environ Adv 2022;8:100221.
[62] Mayo-Bean K, Moran-Bruce K, Nabholz JV, Meylan WM, Howard PH, Cassidy L. Operation manual for the ecological structure-activity relationship model (ECOSAR) class program estimating toxicity of industrial chemicals to aquatic organisms using the ECOSAR (ecological structure activity relationship) class program MS-Windows Version: 41. 2017.
[63] Bu Q, Li Q, Liu Y, Cai C. Performance comparison between the specific and baseline prediction models of ecotoxicity for pharmaceuticals: is a specific QSAR model inevitable? J Chem 2021;2021:1–8.
[64] Sanderson H, Johnson DJ, Wilson CJ, Brain RA, Solomon KR. Probabilistic hazard assessment of environmentally occurring pharmaceuticals toxicity to fish, daphnids and algae by ECOSAR screening. Toxicol Lett 2003;144(3):383–95.
[65] Madden JC, Enoch SJ, Hewitt M, Cronin MTD. Pharmaceuticals in the environment: good practice in predicting acute ecotoxicological effects. Toxicol Lett 2009;185(2):85–101.
[66] Sanderson H, Thomsen M. Comparative analysis of pharmaceuticals versus industrial chemicals acute aquatic toxicity classification according to the United Nations classification system for chemicals. Assessment of the (Q)SAR predictability of pharmaceuticals acute aquatic toxicity and their predominant acute toxic mode-of-action. Toxicol Lett 2009;187(2):84–93.
[67] Yamada T, Kawamura T, Maruyama T, Kurimoto M, Yamamoto H, Katsutani N, et al. Quantitative structure-activity relationship and a category approach to support algal toxicity assessment of human pharmaceuticals. Fundam Toxicol Sci 2021;8(7):195–204.
[68] Veith GD, Mekenyan OG. A QSAR approach for estimating the aquatic toxicity of soft electrophiles [QSAR for soft electrophiles]. Quant Struct Relatsh 1993;12(4):349–56.
[69] Pavan M, Worth A, Netzeva T. Comparative assessment of QSAR models for aquatic toxicity. 2005.
[70] Passino DRM, Smith SB. Acute bioassays and hazard evaluation of representative contaminants detected in great lakes fish. Environ Toxicol Chem 1987;6(11):901–7.
[71] European Commission. Commission implementing decision (EU) 2015/495. Off J Eur Union 2015;58.
[72] European Commission. Commission implementing regulation (EU) 2018/837. Off J Eur Union 2018;61.
[73] European Commission. Decision (EU) 2020/1161. Off J Eur Union 2020;257:32–5.
[74] Environmetnal Protection Agency, Todd. User's guide for T. E. S. T. (toxicity estimation software tool version 5.1). 2020.
[75] Xin X, Huang G, Zhang B. Review of aquatic toxicity of pharmaceuticals and personal care products to algae. J Hazard Mater 2021;410:124619.

C H A P T E R

37

In vitro to in vivo extrapolation methods in chemical hazard identification and risk assessment

Yuxuan Zhang, Minghua Zhu, Wenjia Liu, Zhiqiang Fu, Xuehua Li and Jingwen Chen

Key Laboratory of Industrial Ecology and Environmental Engineering (Ministry of Education), Dalian Key Laboratory on Chemicals Risk Control and Pollution Prevention Technology, School of Environmental Science and Technology, Dalian University of Technology, Dalian, China

Abbreviations

AC_{10} Activity concentration at 10% of maximal activity
AC_{50} Activity concentration at 50% of maximal activity
ADME Absorption, distribution, metabolism, and excretion
AED Administrated equivalent dose
AOP Adverse outcome pathway
A_{ROC} Average area under the receiver operating characteristic curve
BCF Biological enrichment factor
CL_H Hepatic clearance
CL_{int} Intrinsic clearance rate
DL Deep learning
ENM Engineering nanomaterial
EPA The US Environmental Protection Agency
HTTK High-throughput toxicokinetics
IVIVE In vitro to in vivo extrapolation
IVIVE-ADME IVIVE for ADME parameters
IVIVE-Dose IVIVE for exposure dose
KEGG Kyoto Gene and Genome Encyclopedia
ML Machine learning
MPPD Multiple-path particle dosimetry model
PBPK Physiologically based pharmacokinetic
PBTK Physiologically based toxicokinetic
QSAR Quantitative structure–activity relationship
RT-S9 *Rainbow trout* liver subcellular fractions

Introduction

More than 350,000 commercial chemicals and their mixtures have been registered for production and use in the global market [1]. Although chemicals can bring convenience to daily human life, they can also enter the environment via various pathways throughout their lifecycles, from raw material extraction, chemical synthesis, product manufacturing, and use to end of life, causing adverse effects on ecosystems and human health [2]. Therefore, it is crucial to obtain exposure and hazard data on chemicals, including toxicokinetics, persistence, bioaccumulability, toxicity, and other parameters, to support chemical risk assessment and management [3]. Nevertheless, many testing methods for these parameters, especially bioaccumulability and toxicity tests, are time-consuming, laborious, expensive, and requiring use of many animals, which raises concerns on ethical issues. Therefore, it is not practical to generate exposure and hazard data (such as bioaccumulability and toxicity) for multitudinous chemicals only through traditional in vivo testing.

As an alternative, high-throughput screening, in vitro assays such as those in Tox21 and ToxCast programs have been developed and applied to screen a broad range of chemicals and have generated large amounts of reliable in vitro activity data [4,5]. Despite the fact that the in vitro tests can identify potential hazards of chemicals to some extent, nominal in vitro test concentrations of chemicals do not directly represent equivalent in vivo external

QSAR in Safety Evaluation and Risk Assessment
https://doi.org/10.1016/B978-0-443-15339-6.00034-5

dose [6]. Because the in vitro assays use cells or tissues outside the biological environment to test the effects of chemicals on biological processes, they are short of absorption, distribution, metabolism, and excretion (ADME) processes of chemicals in organisms. Given that in vivo data remain the preferred choice for policy development by regulatory agencies, in vitro to in vivo extrapolation (IVIVE) methods that can qualitatively or quantitatively convert in vitro data to in vivo ones have been proposed.

Currently, the IVIVE methods have been recognized as a promising tool for chemical risk assessment and have been validated with some chemicals [7,8]. Nevertheless, these studies have not yet been fully summarized. In addition, the concept of IVIVE is different in disparate application scenarios. Accordingly, this chapter aims to (1) introduce two concepts and workflows of IVIVE, (2) summarize the application status of IVIVE, and (3) discuss the challenges and perspectives of IVIVE in chemical risk assessment.

Concept and workflow of IVIVE

According to different applications, IVIVE methods can be divided into IVIVE for ADME parameters (IVIVE-ADME) and IVIVE for exposure doses (IVIVE-Dose). The flowcharts of the two IVIVE methods are depicted in Fig. 37.1.

IVIVE for ADME parameters

IVIVE-ADME estimates in vivo ADME parameters by scaling the parameters measured in vitro such as hepatic metabolism and plasma protein binding fraction [9]. The estimated or measured in vivo ADME parameters can be used to construct toxicokinetics or physiologically based toxicokinetics (PBTK) models, which are referred to as pharmacokinetics or physiologically based pharmacokinetics (PBPK) in pharmacological research. In addition, other

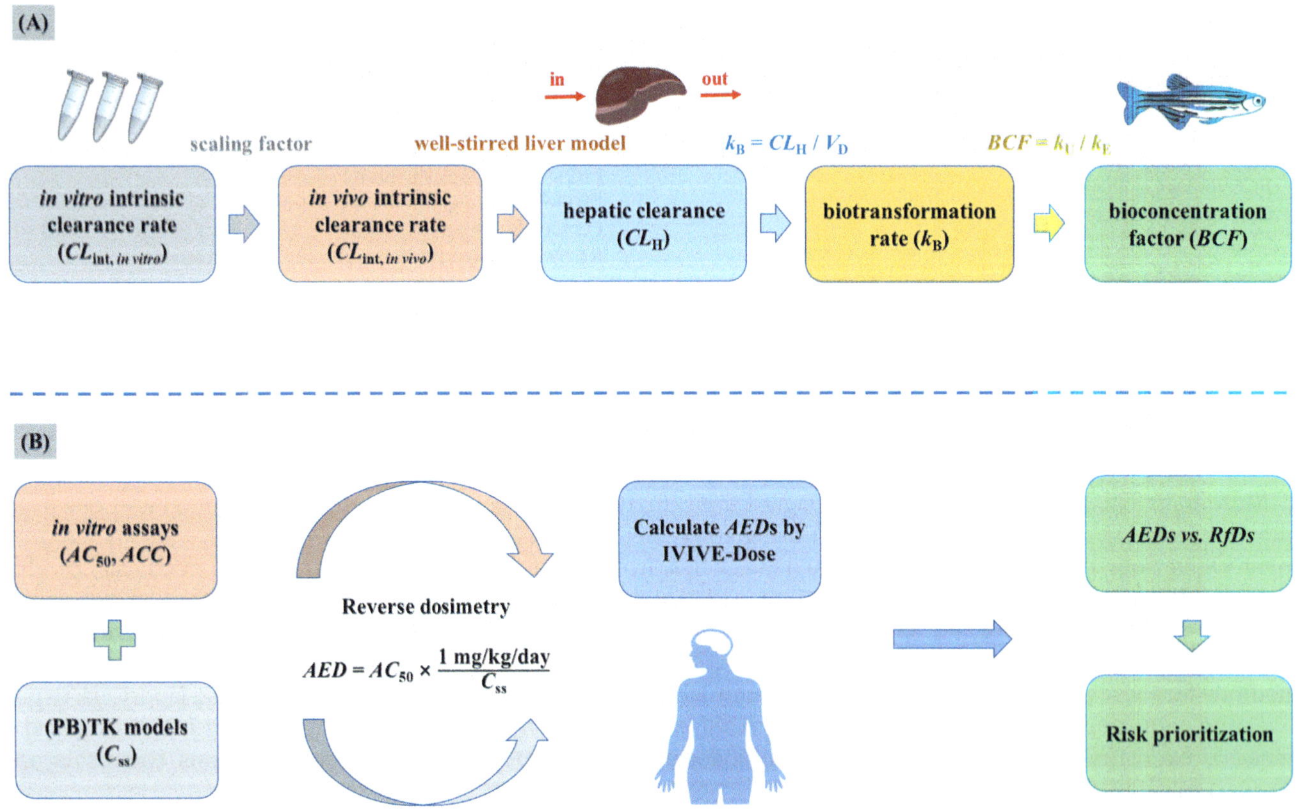

FIGURE 37.1 Flowcharts of IVIVE for ADME parameters (A) and IVIVE for exposure doses (B). V_D is apparent volume for distribution of xenobiotics, k_U is uptake rate constant, k_E is total elimination rate constant, AC_{50} stands for activity concentration at 50% of maximal activity, ACC stands for activity concentration at cutoff, C_{ss} stands for steady-state concentration in blood, IVIVE-Dose stands for IVIVE for exposure doses, AED stands for administrated equivalent dose, and RfD stands for reference dose.

ADME parameters such as blood–brain barrier and tissue–plasma partition coefficients have also been successfully extrapolated from in vitro to in vivo levels [10,11].

Liver is a primary organ of metabolism and clearance of xenobiotics from the organism. Many studies have combined in vitro biotransformation assays and the IVIVE-ADME method to extrapolate in vitro clearance rates to in vivo values that were further used to calculate biotransformation rates and bioconcentration factor (*BCF*) [12,13], which are important metrics for bioaccumulation studies within risk assessment.

Here, procedures of extrapolating the clearance rate and *BCF* are taken as an example to describe the general steps of IVIVE-ADME:

(1) Obtain in vitro intrinsic clearance rate ($CL_{\text{int}, \textit{in vitro}}$; mL/h/mg protein or mL/h/10^6 cells) through in vitro assays (such as *Rainbow trout* liver subcellular fractions (RT-S9) or hepatocyte systems):

$$CL_{\text{int},\textit{in vitro}} = \frac{k_{\text{DEP}}}{C_{\text{S9}} \text{ or } C_{\text{HEP}}}$$

where k_{DEP} (1/h) represents in vitro first-order depletion rate constant; C_{S9} (mg/mL) and C_{HEP} (cells/mL) represent the S9 protein and hepatocytes concentrations, respectively.

(2) Convert $CL_{\text{int}, \textit{in vitro}}$ into in vivo intrinsic clearance ($CL_{\text{int}, \textit{in vivo}}$; mL/h/g liver) with a scaling factor:

$$CL_{\text{int},\textit{in vivo}} = CL_{\text{int},\textit{in vitro}} \times \text{scaling factor}$$

where scaling factor represents S9 concentrations or hepatocytes in liver (mg/g liver or cells/g liver). $CL_{\text{int}, \textit{in vivo}}$ can be further converted to the whole-body one (mL/h/g body weight) by combining liver weight fraction of the whole-body weight.

(3) Calculate hepatic clearance (CL_{H}; mL/h/g liver or mL/h/g body weight) with a well-stirred liver model, which is also known as the venous balance model. This model assumes that the flow rate of liver blood limits the distribution of chemicals into liver, the concentration of chemicals in each part of liver is the same, and the concentration of chemicals in the venous blood of the tissue reaches a dynamic equilibrium with the concentration of chemicals in the liver tissue instantly [14,15].

$$CL_{\text{H}} = Q_{\text{H}} \times \frac{f_{\text{u}} \times CL_{\text{int},\textit{in vivo}}}{Q_{\text{H}} + f_{\text{u}} \times CL_{\text{int},\textit{in vivo}}}$$

where Q_{H} (mL/h/g liver) is blood flow rate to the liver; f_{u} (unitless) is the unbound fraction of xenobiotics and can be calculated as the ratio of the unbound fraction in blood plasma ($f_{\text{u,b}}$; unitless) and that in the in vitro assays ($f_{\text{u,assay}}$; unitless) [16].

(4) Calculate biotransformation rate constant (k_{B}; 1/h) by CL_{H} [17]:

$$k_{\text{B}} = \frac{CL_{\text{H}}}{V_{\text{D}}}$$

where V_{D} (L/kg) is xenobiotic's apparent volume of distribution.

(5) Calculate *BCF* (L/kg) based on uptake rate constant (k_{U}; L/kg/h) and total elimination rate constant (k_{E}; 1/h) [18]:

$$BCF = \frac{k_{\text{U}}}{k_{\text{E}}}$$

where k_{E} is summary of k_{B}, growth rate constant, fecal egestion rate constant, and others.

IVIVE for exposure dose

IVIVE-Dose transforms in vitro bioactivity concentrations into equivalent exposure doses that may induce in vivo bioactivity, which is based on the assumption that xenobiotics in an in vitro system behave the same way they behave in blood or tissue in organisms [19,20].

IVIVE-Dose analysis requires two essential constituents:

(1) Available and reliable in vitro bioactivity data such as activity concentration at 50% of maximal activity (AC_{50}), activity concentration at 10% of maximal activity (AC_{10}), lowest effective concentration, activity concentration at baseline, and activity concentration at cutoff [21].

(2) Models that relate in vitro bioactivity concentrations to in vivo doses that produce equivalent blood or target tissue concentrations, such as PBTK models.

PBTK models describe the ADME processes that control fate and transport of xenobiotics in organisms between interconnected compartments through mathematical equations [22]. Conventionally, PBTK models can calculate blood or tissue concentrations based on exposure doses or concentrations [23]. In comparison, PBTK models can also be used inversely, referred to as "reverse dosimetry" [24], and applied for two scenarios: (1) estimation of human exposure levels based on biomonitoring data (such as levels of xenobiotics in human fluids) [25]; (2) prediction of toxicity dose levels based on in vitro activity data [26].

With the second scenario, administrated equivalent doses (*AEDs*) required for the in vivo steady-state concentrations equivalent to in vitro results generated at the cellular or subcellular level (such as AC_{50}) can be estimated by using the "reverse dosimetry" method. Meanwhile, *AEDs* can be considered as dose–response thresholds and theoretically used as points of departure [27]. Subsequently, *AEDs* derived from in vitro data could be compared with the observed effect values in human or animal assays to establish confidence of IVIVE-Dose in chemicals risk assessment. The calculation formula for *AEDs* is as follows [19]:

$$AEDs\ (\mathrm{mg/kg/day}) = AC_{50}\ (\mu\mathrm{M}) \times \frac{1\ \mathrm{mg/kg/day}}{C_{ss}\ (\mu\mathrm{M})}$$

where AC_{50} can be replaced by AC_{10}, lowest effective concentration, activity concentration at baseline, activity concentration at cutoff, or other parameters, according to different application scenarios (such as extrapolating different endpoints); C_{ss} indicates steady-state concentration in blood and is calculated by (PB)TK models [15].

Given that it is impractical to construct PBTK models for each of the thousands of chemicals by experimentally measuring the relevant toxicokinetic parameters, a high-throughput toxicokinetics (HTTK) approach has been developed [28], which is also configured on the Integrated Chemical Environment (https://ice.ntp.niehs.nih.gov/) [29]. HTTK uses a generic toxicokinetic model that allows for rapid parameterization of large amounts of chemicals, adopting in vitro measurements and/or in silico predictions of chemical-specific toxicokinetic parameters such as CL_{int} and $f_{u,b}$, enhancing the efficiency of IVIVE-Dose analysis.

Although IVIVE-Dose provides a method to translate in vitro concentration–response relationships to in vivo dose–response relationships, it mainly considers ADME processes in organisms and relies on the development of PBTK models applicable to different chemicals, resulting in its inability to predict in vivo levels of multitudinous chemicals in a high-throughput and low-cost manner. Recently, in combination with machine learning (ML) algorithms commonly used in artificial intelligence to process "big data," data-driven models represented by quantitative structure–activity relationships (QSARs) in environmental toxicology have been rapidly developed [30]. Therefore, in vitro data can also be used with additional information, such as chemical structure and omics data, to construct data-driven IVIVE models for predicting interactions of chemicals with biological systems in vivo (Fig. 37.2).

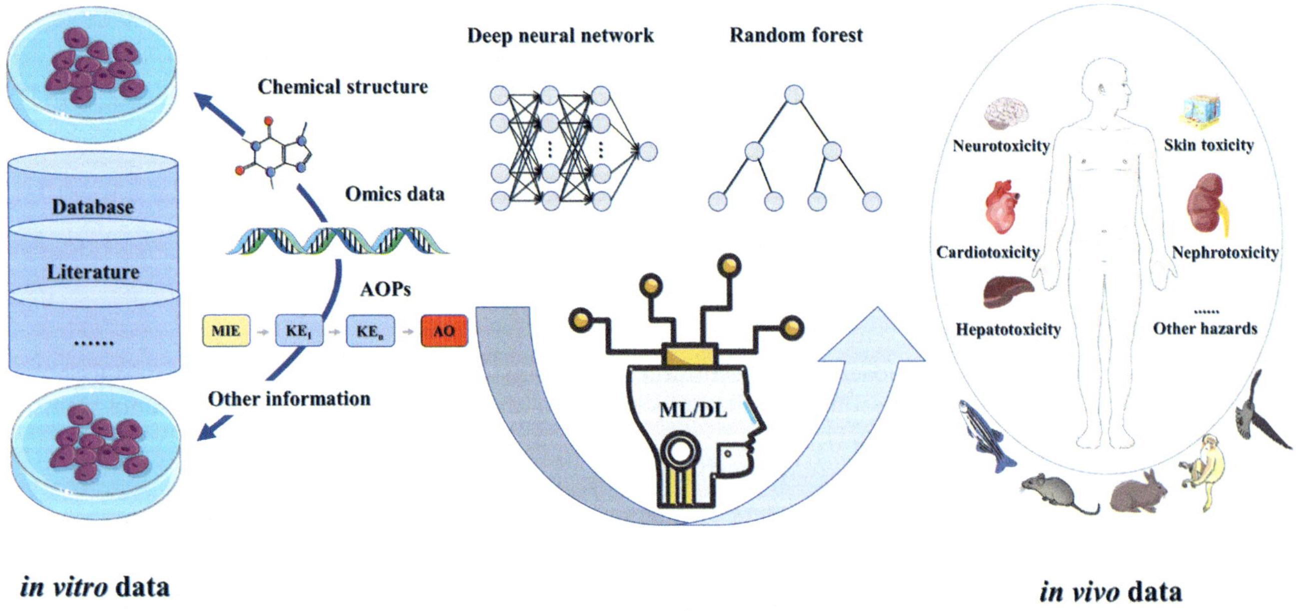

FIGURE 37.2 Schematic diagram of data-driven IVIVE models. AOPs stand for adverse outcome pathways, MIE stands for molecular initiating event, KE stands for key event, AO stands for adverse outcome, ML is machine learning, DL is deep learning.

Application status of IVIVE

The terms "in vitro to in vivo extrapolation" or "in vitro–in vivo extrapolation" or "IVIVE" were used to search for literature during 2017–22 on the Web of Science Core Collection Database in November 2022. The number and classification of IVIVE publications during 2017–22 are shown in Fig. 37.3.

The literature dataset was narrowed down to 255 articles (research type) of direct relevance to this chapter via screening titles and abstracts (Fig. 37.3A). One hundred sixty-four studies in environmental toxicology field were screened by terms "Toxicology," "Environmental Science," and "Engineering Environmental" categories. Some non-research types such as perspective articles were excluded. Finally, 98 articles were included and analyzed in this chapter (Fig. 37.3B).

Many tools used for the IVIVE were initially developed for drugs, leading to less research on nondrug chemicals (such as environmental chemicals) over the years. It was reported that nearly 1000 environmental chemicals had been detected in human body [31], some of which can cause specific health hazards. Therefore, IVIVE studies on environmental chemicals have gradually increased in recent years. In view of this, the following section focuses on progress of IVIVE application in nondrug chemical risk assessment.

Application of IVIVE-ADME

As a promising alternative, IVIVE-ADME has been widely used to extrapolate in vitro biotransformation of well-biotransformed chemicals into in vivo levels, improving the performance of *BCF* prediction models. *Rainbow trout* hepatocellular and RT-S9 are typically used to measure in vitro biotransformation rate data. In addition, the inhibitory capability of chemicals toward metabolic enzymes such as sulfotransferases, UDP-glucuronosyltransferases, and cytochrome P450 enzymes can also be extrapolated with IVIVE-ADME, which can be found in other literature [32–34] and are not described here.

Existing studies concerning biotransformation rate extrapolation and *BCF* prediction have focused on a relatively narrow chemical space. For instance, Laue et al. [35] tested in vitro biotransformation rates of 30 chemicals (mostly fragrance ingredients) in RT-S9 and converted them to in vivo ones that can predict *BCF* using the IVIVE-ADME method. They also investigated uncertainties involved in application of IVIVE-ADME. Ribbenstedt et al. [36] compared in vitro (RT-S9) and in vivo (*Rainbow trout*) hepatic clearance of six anionic surfactants with a range of alkyl chain lengths and different polar head groups. Their results demonstrated that these parameters are in good agreement, enhancing confidence in subsequent use of in vitro RT-S9 assays and IVIVE-ADME to predict *BCF*.

Some studies extended IVIVE-ADME to a broad chemical space and further investigated uncertainties affecting application of IVIVE-ADME. Schultz et al. [13] tested hepatic clearances of five chemicals in an isolated perfused liver model and compared the measured clearance with published in vitro clearance predictions to validate IVIVE-ADME models. In addition to two previously investigated compounds (pyrene and phenanthrene), three

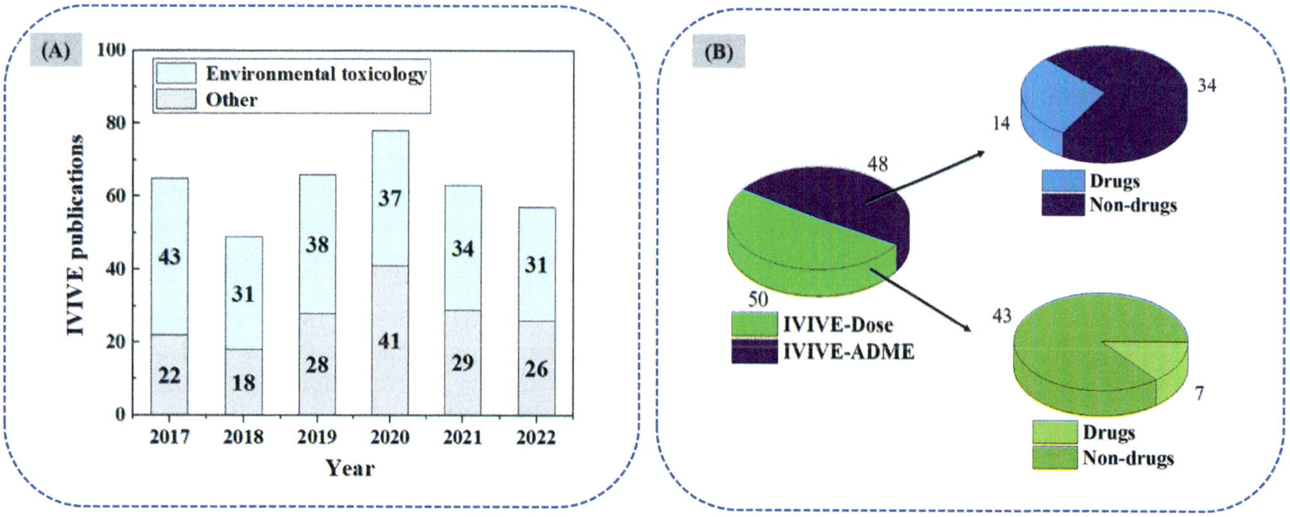

FIGURE 37.3 Number (A) and classification (B) of IVIVE publications during 2017–22.

chemicals (4-nonyl phenol, deltamethrin, and methoxychlor) with relative chemical diversities were chosen to expand the chemical space. They also investigated effects of protein binding on free chemical concentrations, which may significantly affect hepatic clearance, using three different concentrations of bovine serum albumin. The results indicated that the current IVIVE-ADME approach could reliably predict in vivo clearance rates. The protein binding levels did not significantly affect predicted hepatic clearance and *BCF*s for the aforementioned chemicals.

In addition to bioconcentration prediction in aquatic organisms such as fish, a recent study applied the IVIVE-ADME method to predict bioaccumulation of xenobiotics in humans. Due to the COVID-19 epidemic, various disinfectants such as quaternary ammonium compounds are heavily used. Zheng et al. [37] employed IVIVE-ADME to estimate bioaccumulation potential of 18 quaternary ammonium compounds with alkyl chain lengths of C8 ~ C18 based on results of in vitro human liver metabolism and serum protein binding assays. These results can deepen understanding of internal exposure of quaternary ammonium compounds and guide future epidemiological and risk assessment studies.

Overall, it is necessary to apply the IVIVE-ADME method to a broad chemical space for biotransformation and *BCF*s prediction to reduce animal experiments. In addition, factors such as nonuniform *BCF* tests, extrahepatic biotransformation, and changes in enzyme activity that may cause discrepancies between measured and estimated *BCF*s should be considered in future IVIVE-ADME studies.

Application of IVIVE-Dose

In vitro bioactivity data (or transcriptomic data) for different targets can be associated with many health hazards. Based on the in vitro data, IVIVE-Dose coupled with PBTK models has been used for hazard and risk assessment of several groups of environmental chemicals such as endocrine-disrupting chemicals, per- and polyfluoroalkyl substances, and nanomaterials to reduce animal testing and assist regulatory agencies in proposing reasonable hazard-based reference dose.

IVIVE-Dose in health hazards

IVIVE-Dose has been applied and validated in different health risk assessments, covering endpoints such as endocrine disruption effects, hepatotoxicity, cardiotoxicity, development and neurotoxicity, nephrotoxicity, inhalation toxicity, and genotoxicity [38–45]. Given that IVIVE-Dose studies on endocrine disruption effects are more comprehensive relative to other health hazards, the application of IVIVE-Dose in health hazards, represented by the endocrine-disrupting effects, is summarized here.

Endocrine-disrupting effects. Endocrine-disrupting chemicals can interfere with the production, secretion, transport, metabolism, binding, and elimination of natural hormones, disturbing physiological homeostasis balance and reproductive development of organisms, thereby leading to adverse effects [46]. IVIVE-Dose studies on endocrine-disrupting effects mainly focused on estrogenic and antiandrogenic potencies of chemicals. For instance, coupled with PBTK models and IVIVE-Dose, in vitro endocrine potency data (estrogenic and antiandrogenic activity) of bisphenol A and its analogs were extrapolated to *AED*s for comparing their risk priorities [47]. Scholze et al. [48] constructed a PBTK model for gestational exposure in rats and subsequently correlated in vitro antiandrogenic activity data of seven pesticides with in vivo data that manifested as a shortened anogenital distance in male rats, in combination with the IVIVE-Dose method.

Because xenobiotics may bind to proteins and lipids in cell culture media, bind to plastic containers, and undergo evaporation or degradation processes, the actual exposure concentrations in in vitro systems are commonly less than the nominal concentrations [49]. Casey et al. [50] combined IVIVE-Dose and three pharmacokinetic models, including one-compartment, three-compartment, and multiple compartments models, for extrapolating in vitro estrogen receptor activities (activity concentration at cutoff values) of 29 environmental chemicals to estrogenic effects measured in rodent uterotrophic tests (in vivo). In their extrapolation, mass balance models were also used to correct the actual exposure concentrations of chemicals in in vitro systems, improving prediction performance of the models and strengthening confidence of IVIVE-Dose in chemicals risk assessment.

Even though IVIVE-Dose can be used to derive points of departure, single in vitro assays may not adequately characterize in vivo effects. Adverse outcome pathways (AOPs) describe the link between molecular initiating events such as molecular interactions between a xenobiotic and a specific biomolecule and tissue and organism-level (in vivo) adverse outcomes [51]. Therefore, identifying AOP-relevant in vitro bioassays may facilitate recognizing biologically plausible and related sets of key events induced by xenobiotics that ultimately manifest as adverse outcomes at the tissue or organism level.

A recent study used an AOP framework to select molecular initiating events, key events, and relevant in vitro bioactivity assays, focusing on adverse outcomes of reduced thyroid hormone synthesis [52]. On this basis, AC_{50} values from thyroid-based bioactivity in vitro assays of 45 pesticides were converted into *AED*s by using IVIVE-Dose and HTTK models. Meanwhile, the in vitro-derived *AED*s were compared with in vivo-derived points of departure that serve as the basis for regulatory guidance values for pesticides in support of hazard decision-making. Effects of HTTK model parameterization and dosing regimen on *AED*s estimation were also evaluated.

Other health hazards. Similar to endocrine-disrupting effects, in vitro bioactivities of multiple environmental chemicals based on different targets that correspond to different health hazards were translated into *AED*s by integrating IVIVE-Dose and PBTK models, including nephrotoxicity [39], inhalation toxicity [40], hepatotoxicity [41], cardiotoxicity [42], development toxicity [43], neurotoxicity [44], and genotoxicity [45]. Detailed information on the chemicals and in vitro data involved in these models is shown in Table 37.1. Notably, Paini et al. [53] implemented PBTK models into Konstanz Information Miner, a user-friendly graphical platform for data analysis processes such as data access, merge, transformation, visualization, and management (https://www.knime.com/), which can be used to automatically perform IVIVE-Dose in support of chemicals risk assessment.

IVIVE-Dose on engineering nanomaterials

The rapid development of nanotechnology is accompanied by the intentional or unintentional introduction of many engineering nanomaterials (ENMs) into the environment, becoming an important risk source to human health [54]. In recent years, researchers have increasingly applied in vitro methods to identify and characterize hazards of nanomaterials, leading to an urgent need to translate in vitro dose metrics for ENMs to relevant in vivo human exposures.

Notably, kinetic parameters that influence fate of ENMs are primarily different from those of conventional chemicals, leading to the reported PBTK models for conventional chemicals, may not be suitable for ENMs. Thus, Smith and Skinner [55] employed the multiple-path particle dosimetry (MPPD) model to predict deposition of silver nanoparticles from inhalation exposure. Subsequently, they combined the MPPD model with available lung lining fluid and nanoparticle dissolution data to predict regional dose metrics for nanoparticles exposure. Finally, by comparing these dose metrics with toxicity thresholds determined from in vitro systems, the dose metrics for nanoparticles exposure were successfully translated from cellular systems to humans.

TABLE 37.1 Representative IVIVE-Dose studies in different health hazards.

Health hazards	Chemicals	Models	In vitro data (unit)	Reference
Nephrotoxicity	Cadmium	PBTK	*BMDL* (μM)	[39]
Inhalation toxicity	Chlorothalonil	CFPD	*BMDL* (mg/cm^2)	[40]
Hepatotoxicity	Organophosphate esters	HTTK	AC_{50} (μM)	[41]
Cardiotoxicity	82 environmental chemicals	HTTK	EC_{10} (μM)	[42]
Development toxicity	1,2,4-Triazole derivatives, ethers, phthalates	PBTK	*BMC* (μM), IC_{50} (μM)	[43]
Neurotoxicity	(2-Pyridylthio)-N-oxide, dibutyltin dichloride	TK	AC_{50} (μM)	[44]
Genotoxicity	292 environmental chemicals	HTTK	*BMC* (μM)	[45]

AC_{50}, activity concentration at 50% of maximal activity; *BMC*, benchmark concentration; *BMDL*, benchmark doses lower limits; *CFPD*, computational fluid-particle dynamics; EC_{10}, 10% effect concentration; *HTTK*, high-throughput toxicokinetics; IC_{50}, half-inhibition concentration; *(PB)TK*, (physiologically based) toxicokinetics.

Additional information can be found in a recent review, which summarized efforts made thus far to apply the IVIVE-Dose method for the human risk assessment of ENMs [56].

IVIVE-Dose with toxicogenomics data

From the view point of protecting human health, governmental agencies have issued health-based guidance values for synthetic chemicals such as reference doses from the US Environmental Protection Agency (EPA). Nevertheless, the reference dose values were usually derived based on a single critical effect. For example, in 2009 and 2016, EPA published reference dose values for perfluorooctane sulfonate based on the decreased serum T3 levels in cynomolgus monkeys (77 ng/kg/day) [57] and the developmental effect of decreased pup body weight in rats (20 ng/kg/day) [58].

Toxicogenomic dose–response data involve multiple endpoints at levels of gene expression, which in turn are associated with multiple toxicity pathways. Recently, perfluorooctane sulfonate was adopted in a case study [59], and in vitro dose–response data of differentially expressed genes were converted to pathway- and disease-based *AED*s by integrating IVIVE-Dose and PBPK modeling. The differentially expressed genes were analyzed using the Kyoto Gene and Genome Encyclopedia (KEGG) and Disease Ontology Enrichment Analysis to identify significantly enriched pathways and diseases. The pathway- and disease-based AEDs were extrapolated to relevant reference doses, which are similar to the 2016 EPA's recommended reference doses on perfluorooctane sulfonate. Therefore, compared with traditional in vitro activity data, incorporating transcriptomic dose–response data into the IVIVE-Dose approach takes into account multiple endpoints, provides insights into mechanisms of toxicity, and is independent of the apical toxicity endpoints.

Data-driven IVIVE models in toxicity prediction

Data-driven IVIVE models employ in vitro activity data based on different targets as input features for chemicals and analyze association between these data and adverse effects of the organism (in vivo endpoints).

The Tox21 project tested over 10,000 molecules on 70 toxicity endpoints, which provides an important source of in vitro activity data for building data-driven IVIVE models. Huang et al. [60] adopted both in vitro activity data obtained from human-derived cell lines in the Tox21 project and structural information of chemicals as input features to construct a QSAR model for predicting 72 in vivo toxicity endpoints such as acute toxicity, hepatotoxicity, reproductive toxicity, and carcinogenicity for different species. The model exhibited robust prediction performance with an average area under the receiver operating characteristic curve (A_{ROC}) of 0.84. For in vivo endpoints prediction, QSAR models based on chemical structures and in vitro activities showed a better performance than those based solely on chemical structures ($A_{ROC} = 0.78$) or in vitro data ($A_{ROC} = 0.64$). Notably, there was no significant difference in the performance of the toxicity prediction models built for different species (animal: $A_{ROC} = 0.84$; human: $A_{ROC} = 0.86$). Furthermore, a recent study further confirmed the reliability of in vitro data when applied to in vivo endpoint prediction by comparing relevant chemical toxicity rankings obtained using benchmark doses and benchmark doses lower limits derived from in vivo and in vitro assays [61].

Identifying molecular targets and biological pathways involved in chemical-induced in vivo toxicity improves understanding of toxicity mechanisms. Compound–molecular target interaction annotations collected from DrugBank, KEGG, ChEMBL, and Comparative Toxicogenomics Database were used simultaneously with in vitro data as input features to construct data-driven IVIVE models for eight human organ level (in vivo) toxicity endpoints [62]. Toxicity-related molecular targets obtained from predictive models and literature mining processes were subsequently employed to identify significant pathways for eight in vivo toxicity endpoints, which contribute to understanding biological mechanisms of toxicity and serve as a helpful complement to AOPs.

In addition to using in vitro activities to construct data-driven IVIVE models and directly predict in vivo toxicity endpoints, researchers are beginning to utilize the in vitro data to develop quantitative AOP models. By integrating information (in vitro) on the androgen receptor and estrogen receptor mediated AOPs and structural information on more than 8000 chemicals, single-target models were constructed for molecular initiating events, key events, and adverse outcomes in AOPs [63]. After that, an AOP-based consensus model (RepTox) for human reproductive toxicity prediction was constructed using the prediction results of single-target models. The RepTox model showed robust prediction performances with an accuracy of 96.3%. The reliability of using in vitro data to predict in vivo toxicity endpoint was also verified.

To further explore the role of the AOP framework in correlating toxicity data in vitro and in vivo, Ciallella et al. [64] developed a knowledge-based deep neural network approach to organize in vitro nuclear estrogen receptor α and β (ERα and ERβ) binding potentials that drive the target rodent uterotrophic bioactivity (in vivo) of compounds by encoding in vitro endpoints (ERα and ERβ binding) layers into neurons in hidden layers of a deep neural network. The constructed model controls the neuronal messaging based on bioactivities between chemical and in vitro endpoints, predicting in vivo uterotrophic bioactivities of chemicals on rodents (in vivo). The results demonstrated that this approach can successfully derive key relationships between in vitro and in vivo tests, shown in the network as weights of 6521 edges between 1071 neurons.

In short, data-driven IVIVE models can directly predict in vivo toxicity endpoints from in vitro data and overcome limitations faced by IVIVE-Dose, including differences in individual biological specificity and lack of toxicity mechanisms, which can be considered as a promising tool in chemicals risk assessment.

Challenges and perspectives

Although the IVIVE method has been successfully applied in risk assessments of some chemicals in different scenarios, many efforts are still needed to improve its credibility and scope of applications.

(1) Nominal concentrations are conventionally employed as exposure indicators in in vitro assays. However, in most cases, the nominal concentrations are not fully representative of the effective concentrations in in vitro systems. Despite the fact that several mass balance models for predicting in vitro actual exposure concentrations have been developed [65], there is still a paucity of data on several important partition coefficients such as bovine serum albumin–water distribution ratios and phospholipid–water distribution ratios, limiting the use of corrected effective concentrations of chemicals in the IVIVE analysis.

In silico methods, represented by ML-based QSAR models, can rapidly and effectively predict the parameters for numerous chemicals. For example, a recent study proposed an algebraic graph-assisted bidirectional transformer framework by fusing representations generated by algebraic graphs and bidirectional transformers, which shows good prediction performance in endpoints with a small size of dataset (about 300 records) [66]. This method makes it possible to predict bovine serum albumin–water distribution ratios and phospholipid–water distribution ratios for thousands of chemicals to support the use of corrected effective concentrations in IVIVE analysis.

(2) An essential step in IVIVE-ADME analysis is the conversion of $CL_{\text{int},\, in\ vitro}$ into CL_H by the liver model, including the well-stirred liver model, parallel tube model, and dispersion model. Even though the well-stirred liver model is the most commonly used one because of its simplicity in mathematical calculations, it is the least physiologically compatible among the three models [67]. Recent studies have questioned the assumptions of the well-stirred liver model, suggesting that the model leads to an underestimation of CL_H [68]. Therefore, it is necessary to further investigate uncertainty in applying the well-stirred model and, simultaneously, to use the other liver models closer to the physiological state in the IVIVE-ADME analysis as appropriate.

(3) Generally, a limited number of model species are used to obtain hazard and exposure data such as *BCF*, while these surrogate species are not representative of the full diversity of species. For example, due to the lack of various types of kinetic parameters, the number of fully parameterized PBTK models, which can be used to predict bioaccumulation of chemicals in the organism, has been restricted to a few species.

On the one hand, the IVIVE-ADME method can be used to extrapolate in vitro toxicokinetic parameters to in vivo levels in a forward way. Moreover, the IVIVE-ADME method can also be used in reverse, extrapolating the whole-body liver clearance predicted by QSAR models to CL_{int}, which can be input into a mass-balance equation for liver tissue during the construction of the PBTK models. Consequently, the two usage scenarios for the IVIVE-ADME method make it possible to develop PBTK models applicable to multiple-specie or across-specie [69].

(4) In general, anthropogenic chemicals in the environment exist in a mixture rather than individual chemicals, resulting in human exposure to chemical mixtures from the environment. Nevertheless, most IVIVE-Dose studies have focused on individual chemicals, and little information has been available regarding chemical mixtures. As an alternative, the classical concentration addition and independent action prediction models may provide basic information, such as toxicity prediction data for IVIVE on mixtures, and promote the study on environmental risk of chemical mixtures extrapolated from IVIVE [70].

(5) Omics technology has been used in the environmental toxicology field to elucidate the mechanisms of chemical action. Therefore, integrating omics technology into IVIVE analysis can help improve the mechanistic understanding of IVIVE [59]. Meanwhile, omics data can also facilitate identification of molecular changes, contribute to developing AOP frameworks, and provide more comprehensive physiological information for cross-species extrapolation of IVIVE.

Acknowledgments

This study was supported by the National Natural Science Foundation of China (22136001) and the National Key R&D Program of China (2022YFC3902100). The authors would like to acknowledge SciDraw (https://www.scidraw.io/) and Servier Medical Art (https://smart.servier.com/) for providing free image elements that were used to create Figs. 37.1 and 37.2 of this chapter.

References

[1] Wang ZY, Walker GW, Muir DCG, Nagatani-Yoshida K. Toward a global understanding of chemical pollution: a first comprehensive analysis of national and regional chemical inventories. Environ Sci Technol 2020;54(5):2575–84.
[2] Landrigan PJ, Fuller R, Acosta NJR, Adeyi O, Arnold R, Basu NN, et al. The lancet commission on pollution and health. Lancet 2018; 391(10119):462–512.
[3] Jaworska JS, Comber M, Auer C, Van Leeuwen CJ. Summary of a workshop on regulatory acceptance of (Q)SARS for human health and environmental endpoints. Environ Health Perspect 2003;111(10):1358–60.
[4] Richard AM, Judson RS, Houck KA, Grulke CM, Volarath P, Thillainadarajah I, et al. Toxcast chemical landscape: paving the road to 21st century toxicology. Chem Res Toxicol 2016;29(8):1225–51.
[5] Attene-Ramos MS, Miller N, Huang R, Michael S, Itkin M, Kavlock RJ, et al. The Tox21 robotic platform for the assessment of environmental chemicals-from vision to reality. Drug Discov Today 2013;18(15):716–23.
[6] Blaauboer BJ. Biokinetic modeling and *in vitro-in vivo* extrapolations. J Toxicol Environ Health B 2010;13(2–4):242–52.
[7] Bell SM, Chang XQ, Wambaugh JF, Allen DG, Bartels M, Brouwer KLR, et al. *In vitro* to *in vivo* extrapolation for high throughput prioritization and decision making. Toxicol Vitro 2018;47:213–27.
[8] Nicolas CI, Linakis MW, Minto MS, Mansouri K, Clewell RA, Yoon M, et al. Estimating provisional margins of exposure for data-poor chemicals using high-throughput computational methods. Front Pharmacol 2022;13:980747.
[9] Bowman CM, Benet LZ. *In vitro-In vivo* extrapolation and hepatic clearance-dependent underprediction. J Pharmaceut Sci 2019;108(7):2500–4.
[10] Ball K, Bouzom F, Scherrmann J, Walther B, Declèves X. Physiologically based pharmacokinetic modelling of drug penetration across the blood-brain barrier—towards a mechanistic IVIVE-based approach. AAPS J 2013;15(4):913–32.
[11] Holt K, Ye M, Nagar S, Korzekwa K. Prediction of tissue-plasma partition coefficients using microsomal partitioning: incorporation into physiologically based pharmacokinetic models and steady-state volume of distribution predictions. Drug Metab Dispos 2019;47(10):1050–60.
[12] Nichols JW, Huggett DB, Arnot JA, Fitzsimmons PN, Cowan-Ellsberry CE. Toward improved models for predicting bioconcentration of well-metabolized compounds by Rainbow trout using measured rates of *in vitro* intrinsic clearance. Environ Toxicol Chem 2013;32(7):1611–22.
[13] Schultz M, Krause S, Brinkmann M. Validation of methods for *in vitro-in vivo* extrapolation using hepatic clearance measurements in isolated perfused fish livers. Environ Sci Technol 2022;56(17):12416–23.
[14] Rowland M, Benet LZ, Graham GG. Clearance concepts in pharmacokinetics. J Pharmacokinet Biopharm 1973;1(2):123–36.
[15] Wilkinson GR, Shand DG. A physiological approach to hepatic drug clearance. Clin Pharmacol Ther 1975;18(4):377–90.
[16] Obach RS. The importance of nonspecific binding in *in vitro* matrices, its impact on enzyme kinetic studies of drug metabolism reactions, and implications for *in vitro-in vivo* correlations. Drug Metab Dispos 1996;24(10):1047–9.
[17] Nichols JW, Schultz IR, Fitzsimmons PN. *In vitro-In vivo* extrapolation of quantitative hepatic biotransformation data for fish: I. A review of methods, and strategies for incorporating intrinsic clearance estimates into chemical kinetic models. Aquat Toxicol 2006;78(1):74–90.
[18] OECD. OECD guidelines for the testing of chemicals, Test No. 305: bioaccumulation in fish: aqueous and dietary exposure. 2012.
[19] Rotroff DM, Wetmore BA, Dix DJ, Ferguson SS, Clewell HJ, Houck KA, et al. Incorporating human dosimetry and exposure into high-throughput *in vitro* toxicity screening. Toxicol Sci 2010;117(2):348–58.
[20] Wetmore BA, Wambaugh JF, Ferguson SS, Sochaski MA, Rotroff DM, Freeman K, et al. Integration of dosimetry, exposure, and high-throughput screening data in chemical toxicity assessment. Toxicol Sci 2012;125(1):157–74.
[21] Filer DL, Kothiya P, Setzer RW, Judson RS, Martin MT. Tcpl: the toxcast pipeline for high-throughput screening data. Bioinformatics 2017; 33(4):618–20.
[22] Paini A, Leonard JA, Joossens E, Bessems JGM, Desalegn A, Dorne JL, et al. Next generation physiologically based kinetic (NG-PBK) models in support of regulatory decision making. Comput Toxicol 2019;9:61–72.
[23] Caldwell JC, Evans MV, Krishnan K, Ng J. Cutting edge PBPK models and analyses: providing the basis for future modeling efforts and bridges to emerging toxicology paradigms. J Toxicol 2012;2012:852384.
[24] Tan Y, Liao KH, Clewell HJ. Reverse dosimetry: interpreting trihalomethanes biomonitoring data using physiologically based pharmacokinetic modeling. J Expo Sci Environ Epidemiol 2007;17(7):591–603.
[25] Lyons MA, Yang RSH, Mayeno AN, Reisfeld B. Computational toxicology of chloroform: reverse dosimetry using Bayesian inference, Markov chain Monte Carlo simulation, and human biomonitoring data. Environ Health Perspect 2008;116(8):1040–6.
[26] Louisse J, Beekmann K, Rietjens IMCM. Use of physiologically based kinetic modeling-based reverse dosimetry to predict *in vivo* toxicity from *in vitro* data. Chem Res Toxicol 2017;30(1):114–25.
[27] Paul Friedman K, Gagne M, Loo L, Karamertzanis P, Netzeva T, Sobanski T, et al. Utility of *in vitro* bioactivity as a lower bound estimate of *in vivo* adverse effect levels and in risk-based prioritization. Toxicol Sci 2020;173(1):202–25.

[28] Pearce RG, Setzer RW, Strope CL, Sipes NS, Wambaugh JF. Httk: R package for high-throughput toxicokinetics. J Stat Software 2017;79(4): 1–26.

[29] Abedini J, Cook B, Bell S, Chang XQ, Choksi N, Daniel AB, et al. Application of new approach methodologies: ICE tools to support chemical evaluations. Comput Toxicol 2021;20:100184.

[30] Hemmerich J, Ecker GF. *In silico* toxicology: from structure-activity relationships towards deep learning and adverse outcome pathways. Wires Comput Mol Sci 2020;10(4):e1475.

[31] Plassmann MM, Fischer S, Benskin JP. Nontarget time trend screening in human blood. Environ Sci Technol Lett 2018;5(6):335–40.

[32] Jia RY, Zhang ZP, Qin G, Zhang WQ, Yang K, Liu YZ, et al. Inhibition of hydroxylated polychlorinated biphenyls (OH-PCBs) on sulfotransferases (SULTs). Environ Pollut 2021;291:118214.

[33] Habenschus MD, Carrão DB, de Albuquerque NCP, Perovani IS, Moreira Da Silva R, Nardini V, et al. *In vitro* enantioselective inhibition of the main human CYP450 enzymes involved in drug metabolism by the chiral pesticide tebuconazole. Toxicol Lett 2021;351:1–9.

[34] Yang QY, Bai Y, Qin GQ, Jia RY, Zhu WH, Zhang DF, et al. Inhibition of UDP-glucuronosyltransferases (UGT) by polycyclic aromatic hydrocarbons (PAHs) and hydroxy-PAHs (OH-PAHs). Environ Pollut 2020;263:114521.

[35] Laue H, Hostettler L, Badertscher RP, Jenner KJ, Sanders G, Arnot JA, et al. Examining uncertainty in *in vitro-in vivo* extrapolation applied in fish bioconcentration models. Environ Sci Technol 2020;54(15):9483–94.

[36] Ribbenstedt A, Armitage JM, Günther F, Arnot JA, Droge STJ, Mclachlan MS. *In vivo* bioconcentration of 10 anionic surfactants in Rainbow trout explained by *in vitro* data on partitioning and S9 clearance. Environ Sci Technol 2022;56(10):6305–14.

[37] Zheng GM, Webster TF, Salamova A. Quaternary ammonium compounds: bioaccumulation potentials in humans and levels in blood before and during the covid-19 pandemic. Environ Sci Technol 2021;55(21):14689–98.

[38] Lin YJ, Lin ZM. *In vitro-In silico*-based probabilistic risk assessment of combined exposure to bisphenol a and its analogues by integrating toxcast high-throughput *in vitro* assays with *in vitro* to *in vivo* extrapolation (IVIVE) via physiologically based pharmacokinetic (PBPK) modeling. J Hazard Mater 2020;399:122856.

[39] Zhang YC, Liu ZQ, Wang ZW, Gao H, Wang YQ, Cui MX, et al. Health risk assessment of cadmium exposure by integration of an *in silico* physiologically based toxicokinetic model and *in vitro* tests. J Hazard Mater 2023;443:130191.

[40] Ramanarayanan T, Szarka A, Flack S, Hinderliter P, Corley R, Charlton A, et al. Application of a new approach method (NAM) for inhalation risk assessment. Regul Toxicol Pharmacol 2022;133:105216.

[41] Ding JQ, Liu WX, Zhang H, Zhu LY, Zhu L, Feng JF. Liver-based probabilistic risk assessment of exposure to organophosphate esters via dust ingestion using a physiologically based toxicokinetic (PBTK) model. Int J Environ Res Publ Health 2021;18(23):12469.

[42] Blanchette AD, Burnett SD, Grimm FA, Rusyn I, Chiu WA. A bayesian method for population-wide cardiotoxicity hazard and risk characterization using an *in vitro* human model. Toxicol Sci 2020;178(2):391–403.

[43] Fragki S, Piersma AH, Rorije E, Zeilmaker MJ. *In vitro* to *in vivo* extrapolation of effective dosimetry in developmental toxicity testing: application of a generic PBK modelling approach. Toxicol Appl Pharmacol 2017;332:109–20.

[44] Turley AE, Isaacs KK, Wetmore BA, Karmaus AL, Embry MR, Krishan M. Incorporating new approach methodologies in toxicity testing and exposure assessment for tiered risk assessment using the Risk21 approach: case studies on food contact chemicals. Food Chem Toxicol 2019; 134:110819.

[45] Kuo B, Beal MA, Wills JW, White PA, Marchetti F, Nong A, et al. Comprehensive interpretation of *in vitro* micronucleus test results for 292 chemicals: from hazard identification to risk assessment application. Arch Toxicol 2022;96(7):2067–85.

[46] Diamanti-Kandarakis E, Bourguignon J, Giudice LC, Hauser R, Prins GS, Soto AM, et al. Endocrine-disrupting chemicals: an endocrine society scientific statement. Endocr Rev 2009;30(4):293–342.

[47] Punt A, Aartse A, Bovee TFH, Gerssen A, van Leeuwen SPJ, Hoogenboom RLAP, et al. Quantitative *in vitro*-to-*in vivo* extrapolation (QIVIVE) of estrogenic and anti-androgenic potencies of BPA and BADGE analogues. Arch Toxicol 2019;93(7):1941–53.

[48] Scholze M, Taxvig C, Kortenkamp A, Boberg J, Christiansen S, Svingen T, et al. Quantitative *in vitro* to *in vivo* extrapolation (QIVIVE) for predicting reduced anogenital distance produced by anti-androgenic pesticides in a rodent model for male reproductive disorders. Environ Health Perspect 2020;128(11):117005.

[49] Henneberger L, Mühlenbrink M, Heinrich DJ, Teixeira A, Nicol B, Escher BI. Experimental validation of mass balance models for *in vitro* cell-based bioassays. Environ Sci Technol 2020;54(2):1120–7.

[50] Casey WM, Chang XQ, Allen DG, Ceger PC, Choksi NY, Hsieh J, et al. Evaluation and optimization of pharmacokinetic models for *in vitro* to *in vivo* extrapolation of estrogenic activity for environmental chemicals. Environ Health Perspect 2018;126(9):97001.

[51] Ankley GT, Bennett RS, Erickson RJ, Hoff DJ, Hornung MW, Johnson RD, et al. Adverse outcome pathways: a conceptual framework to support ecotoxicology research and risk assessment. Environ Toxicol Chem 2010;29(3):730–41.

[52] Carlson JM, Janulewicz PA, Kleinstreuer NC, Heiger-Bernays W. Impact of high-throughput model parameterization and data uncertainty on thyroid-based toxicological estimates for pesticide chemicals. Environ Sci Technol 2022;56(9):5620–31.

[53] Paini A, Sala Benito JV, Bessems J, Worth AP. From *in vitro* to *in vivo*: integration of the virtual cell based assay with physiologically based kinetic modelling. Toxicol Vitro 2017;45:241–8.

[54] Hochella MF, Mogk DW, Ranville J, Allen IC, Luther GW, Marr LC, et al. Natural, incidental, and engineered nanomaterials and their impacts on the earth system. Science 2019;363(6434):eaau8299.

[55] Smith JN, Skinner AW. Translating nanoparticle dosimetry from conventional *in vitro* systems to occupational inhalation exposures. J Aerosol Sci 2021;155:105771.

[56] Jagiello K, Ciura K. *In vitro* to *in vivo* extrapolation to support the development of the next generation risk assessment (NGRA) strategy for nanomaterials. Nanoscale 2022;14(18):6735–42.

[57] Seacat AM, Thomford PJ, Hansen KJ, Olsen GW, Case MT, Butenhoff JL. Subchronic toxicity studies on perfluorooctanesulfonate potassium salt in cynomolgus monkeys. Toxicol Sci 2002;68(1):249–64.

[58] Luebker DJ, Case MT, York RG, Moore JA, Hansen KJ, Butenhoff JL. Two-generation reproduction and cross-foster studies of perfluorooctanesulfonate (PFOS) in rats. Toxicology 2005;215(1):126–48.

[59] Chen QR, Chou WC, Lin ZM. Integration of Toxicogenomics and physiologically based pharmacokinetic modeling in human health risk assessment of perfluorooctane sulfonate. Environ Sci Technol 2022;56(6):3623–33.

[60] Huang RL, Xia MH, Sakamuru S, Zhao JH, Shahane SA, Attene-Ramos M, et al. Modelling the Tox21 10 K chemical profiles for *in vivo* toxicity prediction and mechanism characterization. Nat Commun 2016;7(1):10425.

[61] Phifer A, Gray G, Kratchman J, Attene-Ramos MS. Assessing how *in vitro* assay types predict *in vivo* toxicology data. J Toxicol Environ Health Part A 2021;84(17):710–28.

[62] Xu T, Wu LH, Xia M, Simeonov A, Huang RL. Systematic identification of molecular targets and pathways related to human organ level toxicity. Chem Res Toxicol 2021;34(2):412–21.

[63] Tan HY, Wu JQ, Zhang R, Zhang C, Li W, Chen QC, et al. Development, validation, and application of a human reproductive toxicity prediction model based on adverse outcome pathway. Environ Sci Technol 2022;56(17):12391–403.

[64] Ciallella HL, Russo DP, Aleksunes LM, Grimm FA, Zhu H. Revealing adverse outcome pathways from public high-throughput screening data to evaluate new toxicants by a knowledge-based deep neural network approach. Environ Sci Technol 2021;55(15):10875–87.

[65] Dimitrijevic D, Fabian E, Nicol B, Funk-Weyer D, Landsiedel R. Toward realistic dosimetry *in vitro*: determining effective concentrations of test substances in cell culture and their prediction by an *in silico* mass balance model. Chem Res Toxicol 2022;35(11):1962–73.

[66] Chen D, Gao KF, Nguyen DD, Chen X, Jiang Y, Wei GW, et al. Algebraic graph-assisted bidirectional transformers for molecular property prediction. Nat Commun 2021;12(1):3521.

[67] Poulin P, Hop CECA, Ho Q, Halladay JS, Haddad S, Kenny JR. Comparative assessment of *in vitro-in vivo* extrapolation methods used for predicting hepatic metabolic clearance of drugs. J Pharmaceut Sci 2012;101(11):4308–26.

[68] Chang TY, Wang HJ, Hsu SH, Chang ML, Kao LT, Pao LH. Evidence of the need for modified well-stirred model *in vitro* to *in vivo* extrapolation. Eur J Pharmaceut Sci 2022;177:106268.

[69] Mangold-Döring A, Grimard C, Green D, Petersen S, Nichols JW, Hogan N, et al. A novel multispecies toxicokinetic modeling approach in support of chemical risk assessment. Environ Sci Technol 2021;55(13):9109–18.

[70] Hsieh N, Chen ZW, Rusyn I, Chiu WA. Risk characterization and probabilistic concentration-response modeling of complex environmental mixtures using new approach methodologies (NAMs) data from organotypic *in vitro* human stem cell assays. Environ Health Perspect 2021;129(1):17004.

CHAPTER

38

QSAR models in marine ecotoxicology and risk assessment

Fei Li[1], *Xiaoqing Wang*[1,2] *and Yuefa Teng*[1,2]

[1]CAS Key Laboratory of Coastal Environmental Processes and Ecological Remediation, Yantai Institute of Coastal Zone Research (YIC), Chinese Academy of Sciences (CAS), Shandong Key Laboratory of Coastal Environmental Processes, YICCAS, Yantai, PR China [2]University of Chinese Academy of Sciences, Beijing, PR China

Abbreviations

EC_{50} Median effect concentration
ISC QSAAR Interspecies correlation quantitative structure activity −activity relationship
K_{OA} Octanol−air distribution coefficient
K_{OC} Sediment−water distribution coefficient
K_{OW} Octanol−water distribution coefficient
LC_{50} Lethal concentration 50
NOEC No observed effect concentration
PNECs Predicted no-effect concentrations
QSAR Quantitative structure−activity relationship
QSTR Quantitative structure−toxicity relationship
SSD Species sensitivity distribution

Introduction

Marine ecosystems are under increasing pressures to eutrophication, climate change, and chemical pollution [1−3]. Tens of thousands of anthropogenic chemicals and wastes enter the marine environment each year due to the ever-increasing anthropogenic activities and demographic growth of the human population, which is majorly concentrated along coastal areas [3]. Therefore, the recognition of serious and harmful effects of these compounds indicates an absolute and urgent need to assess the potential risks arising from exposure to these compounds.

The ultimate use of toxicity and exposure data is for risk assessment to determine the probability that a recipient will suffer harmful effects from exposure to environmental contaminants. According to the Registration, Evaluation and Authorization of Chemicals (REACH) statistics, there are more than 140,000 chemicals commonly used in our daily market, and the rate of new chemicals entering the market is about 500−1000 per year. If only experimental methods are used for risk assessment of chemicals, it would consume a lot of material and financial resources. The presence of hundreds of contaminants at a single site and the absence of data for many contaminants lead to the use of predictive models [4−9].

In recent years, computational toxicology can effectively simulate and predict chemical environmental exposure, hazard, and risk by constructing computational models, which greatly improves the speed of chemical risk assessment [10−13]. Quantitative structure−activity relationship (QSAR) models can be used to establish the relationship between molecular structure descriptors of compounds and biological activity end points and to predict quantitative or qualitative end points, which can fill the data gap faced by chemical hazard assessment and green chemistry [14−16]. Currently, QSAR has made important contributions to marine ecotoxicology and risk assessment [17−19]. This approach can not only supplement ecotoxicological data but also infer privileged structural fragments and warm high-risk emerging chemicals, providing new insights into the mechanism of toxicity even at low environmental doses [16,20].

QSAR in Safety Evaluation and Risk Assessment
https://doi.org/10.1016/B978-0-443-15339-6.00027-8

Development characteristics of QSAR model in marine ecotoxicology

Through the literature survey, publications on the application of QSAR model in marine ecotoxicology research were identified. All 188 candidate publications retrieved from Web of Science Core Collection (WOSCC) database and divided into six types (article, review, proceedings, paper, book chapter, editorial, material, and early access) (Fig. 38.1). Majority of the papers were articles (91.49%), followed by proceeding paper (9.57%) and review (4.79%). Dynamic change in numbers of publications is an important index of the development status of a research topic and may also reveal the potential prospects of its future research [21,22]. These articles and reviews could greatly reflect the development trends and changes of QSAR models in marine ecotoxicology. The number of publications increased from 2 in 1991 to 18 in 2021 (Fig. 38.1).

Furthermore, the first 20 research fields were ranked by the number of articles published. Among these periodicals, the Environmental Sciences (97), Toxicology (53), Chemistry Multidisciplinary (37), Engineering Environmental (26), and Chemistry Medicinal (23) appeared as the major subject area with the most papers (Fig. 38.1). Besides, the analysts have also found the number of published papers in the fields of SAR and QSAR in Environmental Research, Chemosphere, Environmental Toxicology and Chemistry, Science of the Total Environment, and Journal of Hazardous Materials continued to increase through the evolution of the scientific sources of the research fields (Fig. 38.1).

Word clouds further visualize the high-frequency keywords and gather deep information of QSAR application in marine ecotoxicology. The trend of high-frequency keywords presented an increasing trend along with time. The top 10 keywords with the highest density were QSAR (occurrences: 45), toxicity (occurrences: 20), norm index (occurrences: 10), ionic liquids (occurrences: 8), mixture toxicity (occurrences: 8), acute toxicity (occurrences: 7), molecular docking (occurrences: 6), 3D-QSAR (occurrences: 5), aromatic hydrocarbons (occurrences: 5), and docking (occurrences: 5) (Fig. 38.2A). Furthermore, the Pearson correlation coefficient was used to explore the data analysis, which had been widely applied to interpret the correlation analysis among important keywords. The correlation coefficient (Q) between 0.50 and 0.79 was considered a moderate correlation among QSAR, toxicity, molecular docking, and norm index (Fig. 38.2B).

An overlay visualization map showed that the frequency of co-occurrence keywords along with time periods (Fig. 38.3). Before 2010, QSAR models mainly focused on toxicological joint effects, phototoxicity, genotoxic effects, cyanogenic toxicant, and toxicity prediction of traditional marine organism, such as *photobacterium phosphoreum*, Rainbow trout, and mussel. The cluster from 2010 to 2015 showed that QSAR models mainly applied to predict

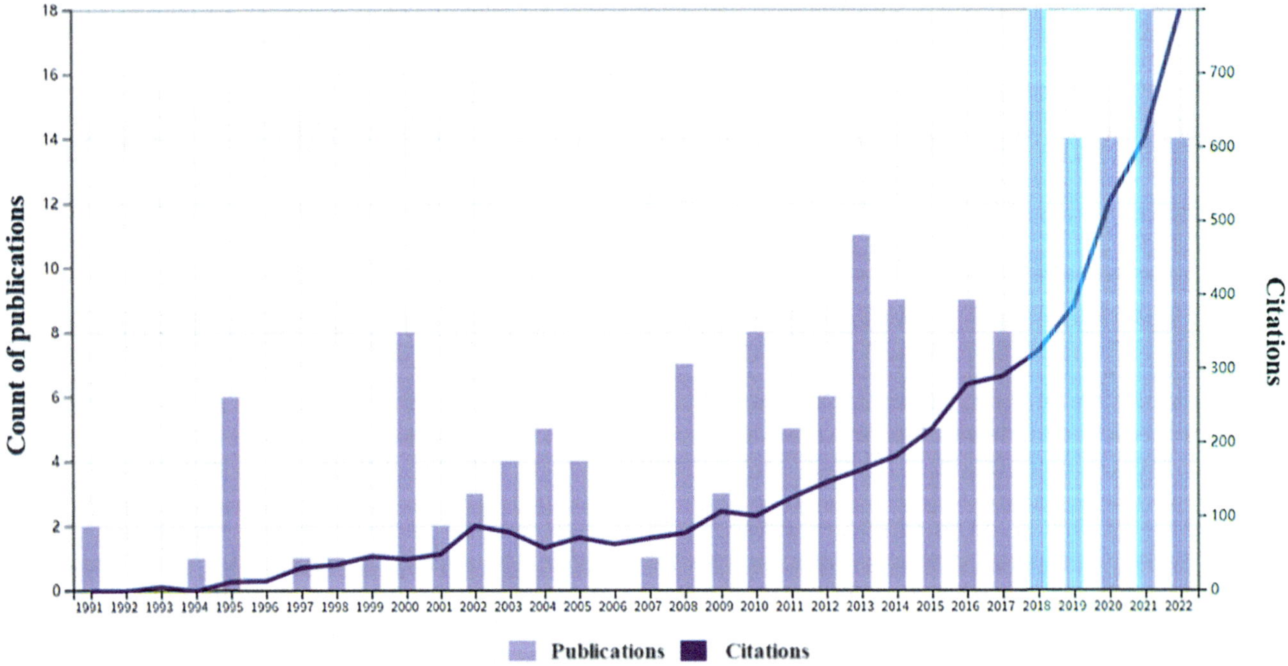

FIGURE 38.1 The annual and cumulative quantity of publications and citations of QSAR models in marine ecotoxicology from 1991 to June 2022.

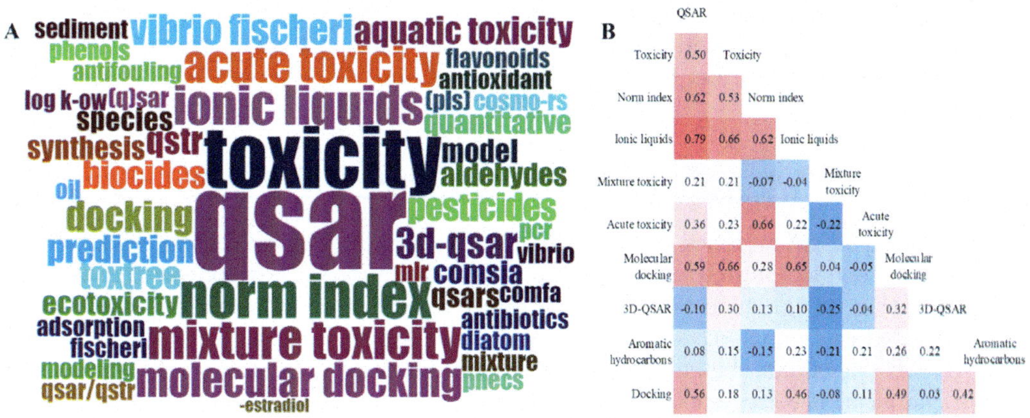

FIGURE 38.2 (A) Word-cloud of most frequent words extracted from the author keywords based on total link strength and (B) their correlation based on Pearson correlation coefficient from January 1991 to 2022.

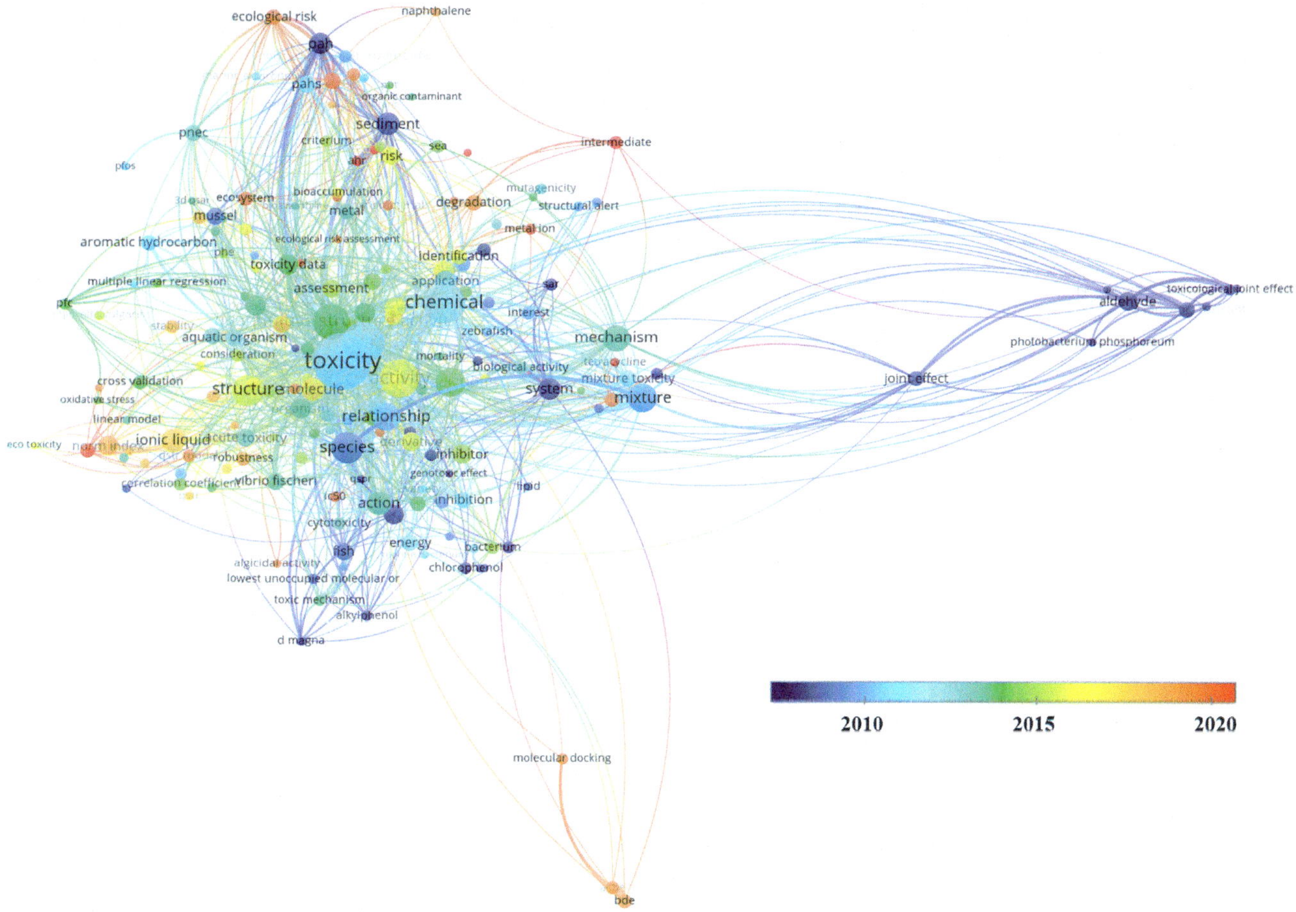

FIGURE 38.3 The network visualization map for the QSAR in marine ecotoxicology from 1991 to June 2022. The colors (from blue to red) represented the different clusters. The red dots reflected more recent the study.

mixture toxicity, biological activity, mutagenicity, and cytotoxicity. In addition, it was also applied to risk assessment and structural alter of metals, polycyclic aromatic hydrocarbon (PAHs), and others in the marine environment. In 2015–2020, most studies investigated the bioaccumulation, bioavailability, carcinogenicity, degradation of organic contaminant (pesticide), nanomaterial, metal ions, and their ecological assessment. The toxic mechanisms of developmental toxicity for marine organisms such as marine bacterium and molecular docking were also investigated. In the last 2 years, the keywords with the highest frequency were ecotoxicity of persistent organic pollutants, intermediates, and aryl hydrocarbon receptor (Fig. 38.3).

Framework of marine ecological risk assessment

Marine ecological risk assessment relies on a large amount of data. Marine ecological risk assessment includes a systematic assessment of chemical source release, including concentrations of environmental media, target exposure doses, and effect doses (Fig. 38.4), which can be carried out from following three perspectives. Firstly, the environmental exposure model of pollutants is constructed based on the physical–chemical and environmental behavior parameters of chemicals to evaluate the environmental distribution and trend of chemicals. Secondly, the internal exposure model of chemicals is constructed according to the physiological and biochemical reactions of chemicals in vivo to evaluate the biological process of pollutants. Thirdly, the toxicity exposure model is developed based on physiological dose–response data to evaluate the harmful effects of chemicals.

Application of QSAR model in charactering environmental distribution and trend of chemicals

Physicochemical and environmental behavioral parameters of chemicals include distribution coefficient, which has certain effects on the adsorption and degradation of chemicals in the environment [23,24]. Therefore, it is important to establish relationships between chemical structural characteristics and environmental behavior parameters for the chemical risk assessment. QSAR models can be used to determine the structural factors that play a decisive role in the properties of compounds, and it is also helpful to understand some environmental action mechanisms of compounds [25].

The distribution coefficients commonly used in environmental risk assessment include octanol–water distribution coefficient (K_{OW}), octanol–air distribution coefficient (K_{OA}), and sediment–water distribution coefficient (K_{OC}), which determine the migration and transformation of chemicals in the environment and the exchange behavior between different layers. For example, Zhu et al. [26] developed the K_{OW} QSAR model of liquid crystal monomer (LCM) and found that LCM tends to accumulate in the adipose tissue of aquatic organisms. Toropov et al. [27] developed the K_{OW} QSAR models for different types of vitamins based on SMILES codes, and the correlation coefficient (R^2) of verification set was as high as 0.902, providing important basis for the prediction of vitamin K_{OW}. Wang et al. [28] established the QSAR models of K_{OW} and K_{OA} for polychlorinated biphenyls (PCB) series by using molecular hologram and partial least squares regression, with R^2 of all models above 0.900, providing a highly

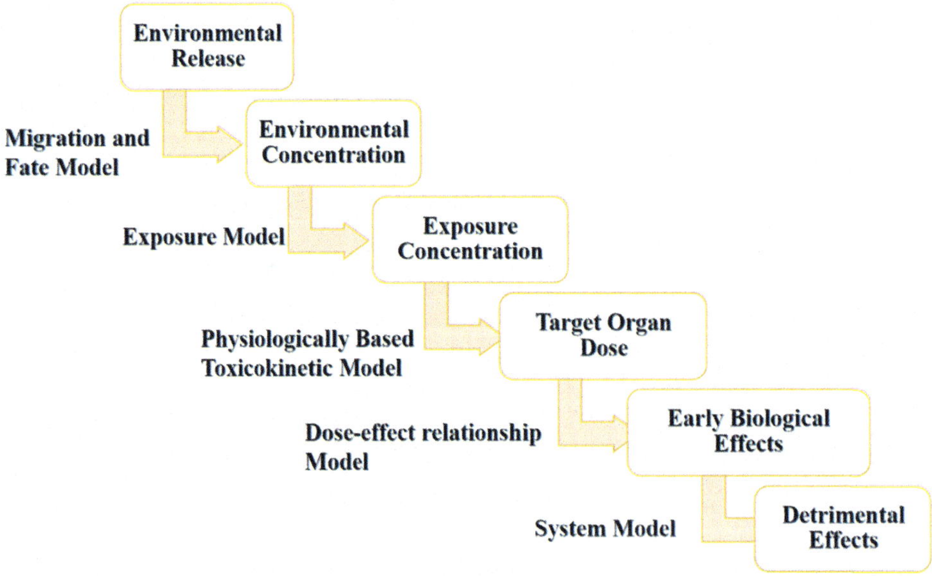

FIGURE 38.4 The process of marine ecological risk assessment.

effective method for the prediction of important physical and chemical properties of PCB. Meylan et al. [29] established a K_{OA} prediction model for 310 chemicals by using K_{OW} and Henry's constant, and the correlation coefficient of the model reached 0.972, greatly improving the prediction accuracy of K_{OA}.

Application of QSAR model in assessing biological process of pollutants

Some chemicals can be transmitted through the food chain and accumulated in organisms, thus reaching higher concentrations than their environmental concentrations [30]. Bioenrichment coefficient (BCF) reflects the accumulation of chemicals in living organisms. Bioamplification at each trophic level exposes top predators to high concentrations of exogenous substances, the effects of which may manifest later in life or several generations later [31]. Therefore, the establishment of QSAR model for chemical BCF is helpful to evaluate its ecological effects. For example, Francesca et al. [32] developed an expert system to predict the BCF of chemicals, and compared with the benchmark BCF, this model has a huge improvement in prediction accuracy and transparency. Ding et al. [33] established five single models and 11 integrated models for predicting fish BCF by using a variety of machine learning algorithms based on the measured BCF values of more than 1300 organic compounds. The results show that compared with the single model, the integrated model has better fitting ability, robustness, prediction accuracy, and wider application domains. Zheng et al. [34] constructed a QSAR model of BCF in fish using 12 descriptors and multiple linear regression method. The R^2 of the QSAR model is 0.809, and the leave-one-out cross-validation coefficient is 0.803, indicating that the model has good goodness of fit, robustness, and prediction ability. Meylan et al. [35] constructed the QSAR prediction model of BCF for 694 chemicals, which solved the problem of large error of other models in predicting organic-metal compounds, nonionic compounds with long alkyl chains, and aromatic azo compounds.

Application of QSAR model in toxicity assessment for marine organisms

The Ecotoxicology Database (ECOTOX, https://cfpub.epa.gov/ecotox/) contains 1,018,565 test records that cover 13,266 aquatic and terrestrial species and 12,223 chemicals, compiled from more than 50,000 references. This database provides information on adverse effects of single chemical stressors to ecologically relevant aquatic and terrestrial species. However, only 53 compounds' acute toxicity data met the requirements of the minimum datasets for the construction of species sensitivity distribution (SSD) models (8 species in five families at three trophic levels) (4.34‰), while the amount of data on toxicity of compounds to organisms (>10 species) was only 43 (3.24‰), and especially importantly, marine species accounted for only 20% (Fig. 38.5). Notably, the REACH regulation encourages the use of QSAR to supplement aquatic toxicity data to provide information to assess hazards of increasing emerging chemicals [36,37]. QSARs have been considered as a rapid and low-cost option because acute short-term and chronic aquatic toxicity data for many chemicals are still scarce [17,38]. This study reviewed in silico prediction of aquatic toxicity for a range of chemicals to marine organisms to point research progress and foreground (Fig. 38.6).

QSAR models of acute toxicity for marine organisms

QSAR models for predicting acute toxicity end points lethal concentration 50 (LC_{50}) or median effect concentration (EC_{50}) could be used for detecting early biological effects of contaminants [36,39,40]. For example, Almeida et al. [41] used the acute toxicity of 19 synthetic chemicals to mussel larvae and developed predictive QSAR models to predict the toxicity effect of chemicals. The inhibition activity of PAHs to *Chlorella vulgaris* larvae was evaluated

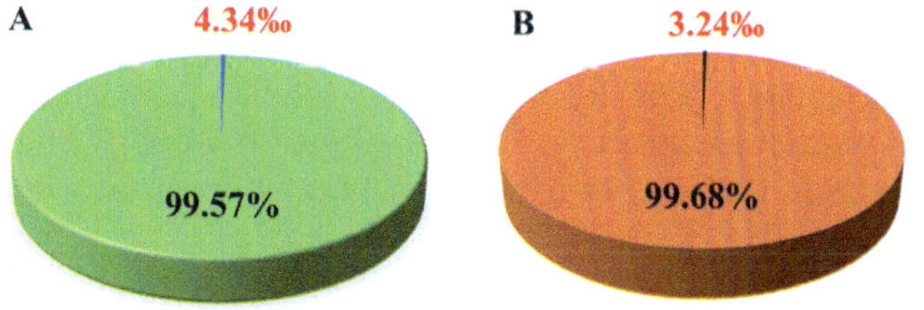

FIGURE 38.5 The summary of acute toxicity data used to develop the SSD model from ECOTOX.

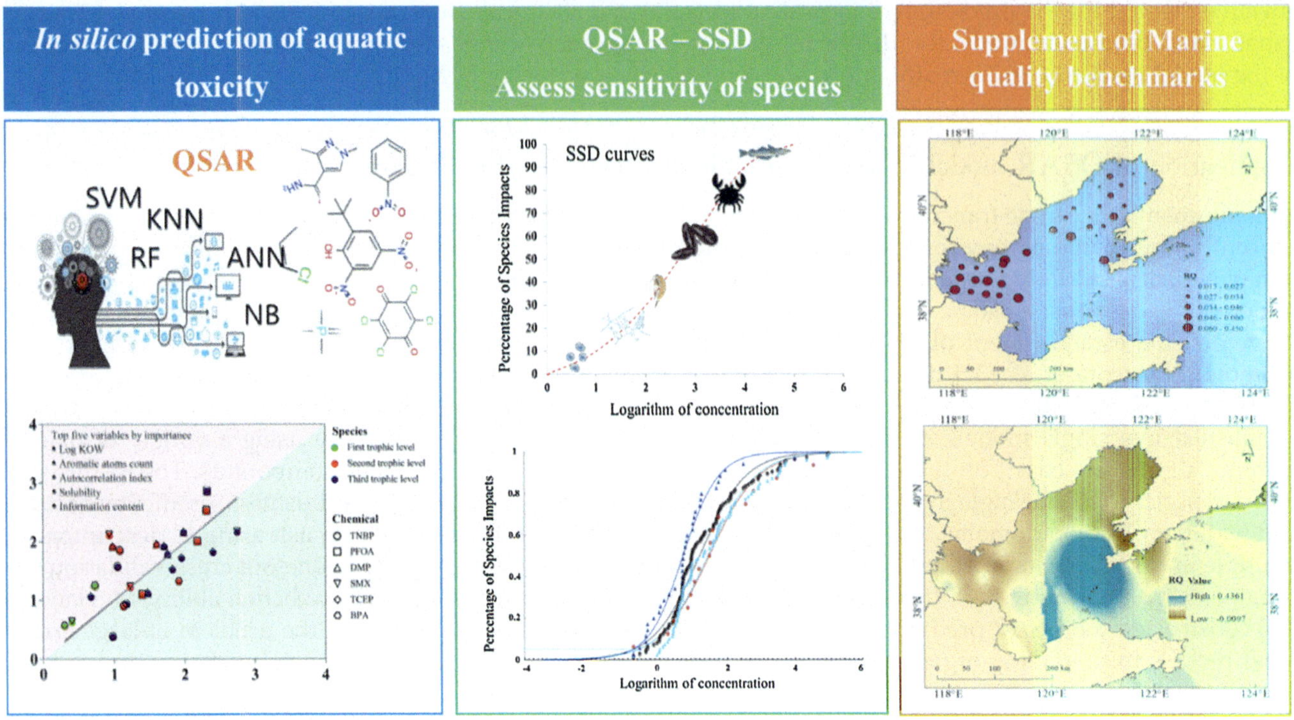

FIGURE 38.6 Application of QSAR–SSD to supply marine quality benchmarks.

through the QSAR model developed by Liu et al. [42], which provided the scientific data for predicting the risk of PAHs to marine biological safety. Zhang et al. [43] constructed QSAR models for the acute toxicity of four aquatic organisms through a multiple linear regression algorithm, which provided data to support the ecological risk assessment of per- and polyfluoroalkyl substances (PFASs) in the marine environment. Basant et al. [44] established local and global quantitative structure–toxicity relationship (QSTR) and interspecies correlation quantitative structure activity–activity relationship (ISC QSAAR) models to predict the toxicity of pesticides using the toxicity data of crustacean (*Daphnia magna*, *Americamysis bahia*, *Gammarus fasciatus*, and *Penaeus duorarum*) and fish (*Oncorhynchus mykiss* and *Lepomis macrochirus*) species. The constructed local, global, and interspecies QSAAR models yielded high correlations (R^2) of >0.941; >0.943 and >0.826, respectively, between the measured and model predicted end point toxicity values in the test data. The developed QSTR/QSAAR models in this study could not only reliably predict the aquatic toxicity of structurally diverse pesticides in multiple test species but also screen and prioritize the new pesticides [44]. Wang et al. [45] adopted density functional theory (DFT) descriptors to develop optimal QSAR model using freely dissolved concentration to predict the toxicity data of alkylated and parent aromatic hydrocarbons to *Vibrio fischeri* (R^2 values of 0.96 and 0.94). The results demonstrated that the average molecular polarizability and several selected thermodynamic parameters reflecting the intermolecular interactions played important roles in the partition of aromatic hydrocarbons between the water phase and biomembrane. Energy of the highest occupied molecular orbital was the most influential descriptor which dominated the toxicity of aromatic hydrocarbons through the electron-transfer reaction with biomolecules [45]. Liu et al. [46] developed four QSAR models based on norm descriptors for acute toxicity of different exposure times toward zebrafish embryo of organic compounds. Their work proved that the effect of different exposure time on the acute toxicity of organic compounds to zebrafish embryos was crucial, and the developed models could predict the potential toxicity of various compounds to zebrafish embryo at four kinds of exposure times, which was valuable for environmental risk assessment [46]. Jia et al. [47] used two norm index formulas to establish QSAR acute toxicity (median lethal concentration) toward rainbow trout of various pesticides with high prediction accuracy. They proposed that QSAR models might be efficacious in predicting the environmental hazards of various pesticides toward rainbow trout, which was of great significance for ecological risk assessment [47].

QSAR models of chronic toxicity for marine organisms

Most of the QSAR models focus on acute aquatic toxicity, and only a few involve chronic aquatic toxicity. Toma et al. [17] developed predictive models for *Raphidocelis subcapitata*, *Daphnia magna*, and fish chronic toxicities on growth rate, immobilization, reproduction, and mortality. The toxicity predictions yielded good performance for the training set ($R^2_{train} = 0.94 - 0.96$) and test set ($R^2_{validation} = 0.56 - 0.78$) [17]. Wang et al. [48] adopted QSAR method to achieve the predicted no-effect concentrations (PNECs) of 16 polycyclic aromatic hydrocarbons in surface seawater and 10 parent- and alkyl-polycyclic aromatic hydrocarbons (alkyl-PAHs) in seawater and marine sediment of Liaodong Bay, China. The model found the individual high ecological risk of benzo[a]pyrene, benzo[b]fluoranthene, and benz[a]anthracene [48]. Donkin et al. [49] developed QSARs by using the physiological responses of mussels to assess the sublethal responses of marine *Mytilus edulis*. Claeys et al. [18] and Austin and Eadsforth [19] developed models for chronic narcosis toxicity and chronic nonpolar narcosis toxicity in fish, respectively. Wang et al. [38] selected the no observed effect concentration (NOEC) for population abundance of marine rotifers as end points to develop 48 binary models for the prediction of chemical marine chronic toxicity. The optimal binary model (1D and 2D molecular descriptors—random trees model) identified the primary factors (*SpMAD_Dzp, AMW, MATS2v*) and filtered out three high alerting substructures [c1ccc(Cl)cc1, CNCO, CCOP(=S)(OCC)O]. These results showed that the compounds with low molecular volume, high polarity, and large molecular weight could contribute to adverse effects on rotifer's population, which facilitated the deeper understanding of chronic toxicity mechanisms [38]. Fan et al. [50] developed a local model that was based on 48 substituted benzenes to predict the chronic toxicity of chemicals toward *Daphnia magna*. Ding and colleagues [51] developed classification models to predict the chronic toxicity of chemicals toward *Daphnia magna* and the green algae *Pseudokirchneriella subcapitata*. In addition, the chronic toxicity of four simple alkylphenols with butyl and propyl substitutions was evaluated with aquatic microbes, invertebrates, and fish [52]. The QSAR models revealed that marine bacterium *Vibrio fischeri* was the most susceptible to the alkylphenols, and the energy of the lowest unoccupied molecular orbital explained the observed toxicity of the individual alkylphenols to *V. fischeri* ($R^2 = 0.92$, $P < 0.05$) [52].

Application of QSAR model in development of marine quality benchmarks

The values of toxicological dose descriptors (EC_{50} or LC_{50}) can only be used to evaluate the harmful effects of chemicals from an individual perspective and single species. However, ecological risk assessment mainly focuses on the extrapolation of signal toxicity end points of multiple species to biomes or even the highest levels (Fig. 38.6). Therefore, the SSD method was developed, which could describe the sensitivity of different species to a stressor in a complex ecosystem [53].This method could randomly select toxicological data of different species, which could represent the community to some extent and assess the conservation level based on the proportion of affected species [54]. Therefore, SSDs play an important role in ecological risk assessment [55]. Estimating SSDs requires toxicity data for many species. However, reports on saltwater species are often limited compared to freshwater species. The QSAR-based SSDs can accurately predict the SSD parameters of data-poor chemicals and supports screening-level evaluations of the potential ecotoxicity of compounds for which data are lacking [56]. Currently, SSD method has been used as a standard method for deriving water quality benchmarks for pollutants by many countries and organizations such as Canada, the European Union (including the Netherlands), Australia, and New Zealand. For example, Wu and Mu et al. used metal QSAR model with the SSD method to predict a series of metal water quality benchmark to protect the marine aquatic organisms. When compared with the benchmark values recommended by the United States Environmental Protection Agency (US EPA), the model achieved good performances [57–59].

Song et al. [60] established QSAR models for eight species using artificial neural networks (ANN) to predict the LC_{50} values and to fit these values of 8424 compounds using SSD curves to evaluate their potential ecological risks. Mu et al. [61] predicted the ecological risks of heavy metals in coastal areas of different countries by constructing QSAR-SSD models. Mu et al. [62] used the SSD method to assess the water quality benchmarks of cadmium (Cd) in the ocean and made a preliminary evaluation of the ecological risks of Cd in seawater by combining the distribution characteristics of Cd in seawater in China. Jung et al. [63] used the water quality benchmarks of nonspherical microplastics via SSD method and evaluated the ecological risk of microplastics in coastal, continental shelf, and deep-sea areas of Korea. Garner et al. [64] evaluated the water quality benchmarks of 10 specific engineered nanomaterials (ENMs) using the SSD method and found that the size and solubility differences of ENMs were the most important reasons for the differences in water quality benchmarks. Lu et al. [65] evaluated the water quality benchmarks of five types of brominated flame retardants in freshwater, and the results showed that the toxicity classes of the five brominated flame retardants were tetrabromobisphenol A > pentabromodiphenyl ether - > octabromodiphenyl ether > decabromodiphenyl ether > hexabromocyclododecane. In addition, QSAR models were constructed to predict the toxicity values of compounds among different species, thus complementing the missing data from the construction of SSD models. Zhang et al. [43] constructed QSAR models for four aquatic species by multiple linear regression algorithm to provide data support for water ecological risk assessment of PFASs and constructed SSD models to assess the pollution status of PFASs around electronics plants. Huang et al. [66] successfully derived the chronic toxicity of 35 pesticides by constructing QSAR-SSD models, and the difference between the model predicted and actual values was within one order of magnitude, providing a reliable method, technique, and data support for the water ecological risk assessment of pesticides. Mu et al. [61] developed a QSAR model for predicting heavy metal toxicity for different species and constructed SSD curves at different temperatures and salinities to assess the ecological risk of heavy metals in coastal areas of different countries.

Although different QSAR-SSD models have been developed for assessing water ecological risks, there are still some problems. First, the trophic levels involved in species selection are not comprehensive enough to represent the community structure of the ecosystem, which may result in "over-protection" or "under-protection" when deriving water quality benchmarks. Second, the differences in geographical and national conditions are not taken into account, and the water quality benchmarks extrapolated through their models may not truly and completely reflect the objective laws and requirements of water ecological protection in different countries and regions. Third, the dataset used for modeling is too small and cannot be well extrapolated to other compounds. Therefore, to improve the QSAR prediction method of existing toxicity data and establish a scientific methodological system of water environmental quality benchmarks suitable for social, economic, and human characteristics are the most urgent taskd in ecological risk assessment research.

Perspectives

Due to the wide acceptance of QSAR models in the environmental field and the application of EPI Suite and other software, QSAR prediction of physical–chemical properties and toxicity effects have become a necessary step in the assessment of chemical ecological risks [67–69]. Therefore, the predicted QSARs models will become a routine means of marine risk assessment in the future. The goal of this review is, therefore, not to provide an exhaustive coverage of modes of toxic action and their underlying biochemical mechanisms but rather to discuss the application of this knowledge in ecotoxicological risk assessment. Knowing the mechanism or, at least, the mode of toxic action is indispensable for developing descriptive and predictive models in ecotoxicology. This review seeks to show the crucial role of target sites, interactions with the target site(s), and mechanisms for an adequate and efficient ecotoxicological risk assessment. Emphasis in the discussion is on target effect concentrations (or target occupancy), species selectivity and species sensitivity, time perspective of effect studies, QSAR, and mixture toxicity. A particular focus of this review is on multiple mechanisms. Although the illustrative examples were mainly taken from studies in aquatic ecotoxicology, the proposed conceptual approach is also in principle applicable and even particularly useful for soil and sediment systems. Recommendations for further research and developments include the use of internal effect concentrations, target site concentrations in site-specific risk assessment, and mixture toxicity parameter as well as general considerations for the derivation of mechanistically meaningful QSAR and other predictive models.

References

[1] Oberdorster E, Cheek AO. Gender benders at the beach: endocrine disruption in marine and estuarine organisms. Environ Toxicol Chem 2001;20(1):23–36.

[2] Liu L, Yang H, Cai Y, Cao Q, Sun L, Wang Z, et al. In silico prediction of chemical aquatic toxicity for marine crustaceans via machine learning. Toxicol Res 2019;8(3):341–52.

[3] Connon RE, Geist J, Werner I. Effect-based tools for monitoring and predicting the ecotoxicological effects of chemicals in the aquatic environment. Sensors 2012;12(9):12741–71.

[4] Gourmelon A, Ahtiainen J, Organization for Economic Co, Development. Developing test guidelines on invertebrate development and reproduction for the assessment of chemicals, including potential endocrine active substances-the OECD perspective. Ecotoxicology 2007;16(1): 161–7.

[5] Cao Q, Liu L, Yang H, Cai Y, Li W, Liu G, et al. *In silico* estimation of chemical aquatic toxicity on crustaceans using chemical category methods. Environ Sci Process Impacts 2018;20(9):1234–43.

[6] Sears ME, Genuis SJ. Environmental determinants of chronic disease and medical approaches: recognition, avoidance, supportive therapy, and detoxification. J Environ Public Health 2012;2012:356798.

[7] Zhang X, Xia P, Wang P, Yang J, Baird DJ. Omics advances in ecotoxicology. Environ Sci Technol 2018;52(7):3842–51.
[8] Xue Q, Liu X, Liu XC, Pan WX, Fu JJ, Zhang AQ. The effect of structural diversity on ligand specificity and resulting signaling differences of estrogen receptor alpha. Chem Res Toxicol 2019;32(6):1002–13.
[9] Kostal J, Voutchkova-Kostal A, Anastas PT, Zimmerman JB. Identifying and designing chemicals with minimal acute aquatic toxicity. Proc Natl Acad Sci USA 2015;112(20):6289–94.
[10] Wang Z, Chen J, Qiao X, Li H, Xie H, Cai X. Computational toxicology: oriented for chemicals risk assessment. Sci China Chem 2016;46(02): 222–40.
[11] Yang H, Lou C, Li W, Liu G, Tang Y. Computational approaches to identify structural alerts and their applications in environmental toxicology and drug discovery. Chem Res Toxicol 2020;33(6):1312–22.
[12] Luechtefeld T, Rowlands C, Hartung T. Big-data and machine learning to revamp computational toxicology and its use in risk assessment. Toxicol Res 2018;7(5):732–44.
[13] Ciallella HL, Zhu H. Advancing computational toxicology in the big data era by artificial intelligence: data-driven and mechanism-driven modeling for chemical toxicity. Chem Res Toxicol 2019;32(4):536–47.
[14] Li J, Li KJ, Zhang C, Peng QP, Tang Y. Computational systems toxicology: emergence, development and application. Sci Bull 2015;60(19): 1751–61.
[15] Wang Z, Chen J, Hong H. Developing QSAR models with defined applicability domains on PPARgamma binding affinity using large data sets and machine learning algorithms. Environ Sci Technol 2021;55(10):6857–66.
[16] Chatterjee M, Roy K. Prediction of aquatic toxicity of chemical mixtures by the QSAR approach using 2D structural descriptors. J Hazard Mater 2021;408:124936.
[17] Toma C, Cappelli CI, Manganaro A, Lombardo A, Arning J, Benfenati E. New models to predict the acute and chronic toxicities of representative species of the main trophic levels of aquatic environments. Molecules 2021;26(22):6983.
[18] Claeys L, Iaccino F, Janssen CR, Van Sprang P, Verdonck F. Development and validation of a quantitative structure-activity relationship for chronic narcosis to fish. Environ Toxicol Chem 2013;32(10):2217–25.
[19] Austin TJ, Eadsforth CV. Development of a chronic fish toxicity model for predicting sub-lethal NOEC values for non-polar narcotics. SAR QSAR Environ Res 2014;25(2):147–60.
[20] Judson R, Richard A, Dix DJ, Houck K, Martin M, Kavlock R, et al. The toxicity data landscape for environmental chemicals. Environ Health Perspect 2009;117(5):685–95.
[21] Wu M, Jiang Y, Kwong RWM, Brar SK, Zhong H, Ji R. How do humans recognize and face challenges of microplastic pollution in marine environments? A bibliometric analysis. Environ Pollut 2021;280:116959.
[22] Gao W, Chen Y, Liu Y, Guo HC. Scientometric analysis of phosphorus research in eutrophic lakes. Scientometrics 2015;102(3):1951–64.
[23] Wang T, Zhai Y, Zhu Y, Li C, Zeng G. A review of the hydrothermal carbonization of biomass waste for hydrochar formation: process conditions, fundamentals, and physicochemical properties. Renew Sustain Energy Rev 2018;90:223–47.
[24] Ngoc Han T, Reinhard M, Gin KY-H. Occurrence and fate of emerging contaminants in municipal wastewater treatment plants from different geographical regions—a review. Water Res 2018;133:182–207.
[25] Lu J. Application of SAR/QSAR methods in predictions of pysicohemical properties of huge organic compounds data sets. Lanzhou University; 2009.
[26] Zhu M, Su H, Bao Y, Li J, Su G. Experimental determination of octanol-water partition coefficient (K_{OW}) of 39 liquid crystal monomers (LCMs) by use of the shake-flask method. Chemosphere 2022;287:132407.
[27] Toropov AA, Toropova AP, Raska I. QSPR modeling of octanol/water partition coefficient for vitamins by optimal descriptors calculated with SMILES. Eur J Med Chem 2008;43(4):714–40.
[28] Wang X, Tang S, Liu S, Cui S, Wang L. Molecular hologram derived quantitative structure–property relationships to predict physico-chemical properties of polychlorinated biphenyls. Chemosphere 2003;51(7):617–32.
[29] Meylan WM, Howard PH. Estimating octanol–air partition coefficients with octanol water partition coefficients and Henry's law constants. Chemosphere 2005;61(5):640–4.
[30] Gobas FAPC, Morrison HA. Bioconcentration and biomagnification in the aquatic environment. 2000.
[31] van der Oost R, Beyer J, Vermeulen NPE. Fish bioaccumulation and biomarkers in environmental risk assessment: a review. Environ Toxicol Pharmacol 2003;13(2):57 149.
[32] Grisoni F, Consonni V, Vighi M, Villa S, Todeschini R. Expert QSAR system for predicting the bioconcentration factor under the REACH regulation. Environ Res 2016;148:507–12.
[33] Ding R, Chen J, Yu Y, Lin J, Wang Z, Tang W, et al. Using ensemble learning algorithms to develop QSAR models on bioconcentration factors of organic chemicals in multispecies fish. Environ Chem 2021;40(5):1295–304.
[34] Zheng Y, Qiao X, Yu Y, Lin J, Ding Q. Quantitative structure-activity relationship model for bioconcentration factors of organic chemicals. Asian J Androl 2019;14(2):214–21.
[35] Meylan WM, Howard PH, Boethling RS, Aronson D, Printup H, Gouchie S. Improved method for estimating bioconcentration/bioaccumulation factor from octanol/water partition coefficient. Environ Toxicol Chem 1999;18(4):664–72.
[36] Singh KP, Gupta S, Kumar A, Mohan D. Multispecies QSAR modeling for predicting the aquatic toxicity of diverse organic chemicals for regulatory toxicology. Chem Res Toxicol 2014;27(5):741–53.
[37] Cassotti M. QSAR study of aquatic toxicity by chemometrics methods in the framework of REACH regulation. 2015. https://boa.unimib.it/retrieve/handle/10281/77501/114693/phd_unimib_066014.pdf.
[38] Wang X, Li F, Chen J, Teng Y, Ji C, Wu H. Critical features identification for chemical chronic toxicity based on mechanistic forecast models. Environ Pollut 2022;307:119584.
[39] Basant N, Gupta S, Singh KP. Predicting aquatic toxicities of chemical pesticides in multiple test species using nonlinear QSTR modeling approaches. Chemosphere 2015;139:246–55.
[40] Singh KP, Gupta S, Rai P. Predicting acute aquatic toxicity of structurally diverse chemicals in fish using artificial intelligence approaches. Ecotoxicol Environ Saf 2013;95:221–33.

[41] Almeida JR, Palmeira A, Campos A, Cunha I, Freitas M, Felpeto AB, et al. Structure-antifouling activity relationship and molecular targets of bio-inspired(thio)xanthones. Biomolecules 2020;10(8):1126.
[42] Liu S, Liu H, Wei H, Zhao X, Song X. Acute toxicity of four pahs to larva of patinopecten yessoensis. Asian J Ecotoxicol 2020;15(5):352–60.
[43] Zhang J, Zhang M, Tao H, Qi G, Guo W, Ge H, et al. A QSAR-ICE-SSD model prediction of the pnecs for per- and polyfluoroalkyl substances and their ecological risks in an area of electroplating factories. Molecules 2021;26(21):6574.
[44] Basant N, Gupta S, Singh KP. Modeling the toxicity of chemical pesticides in multiple test species using local and global QSTR approaches. Toxicol Res 2016;5(1):340–53.
[45] Wang Y, Yang X, Wang J, Cong Y, Mu J, Jin F. A DFT-based toxicity QSAR study of aromatic hydrocarbons to Vibrio fischeri: consideration of aqueous freely dissolved concentration. J Hazard Mater 2016;308:149–56.
[46] Liu T, Yan F, Jia Q, Wang Q. Norm index-based QSAR models for acute toxicity of organic compounds toward zebrafish embryo. Ecotoxicol Environ Saf 2020;203:110946.
[47] Jia Q, Liu T, Yan F, Wang Q. Norm index-based QSAR model for acute toxicity of pesticides toward rainbow trout. Environ Toxicol Chem 2020;39(2):352–8.
[48] Wang Y, Wang J, Mu J, Wang Z, Cong Y, Yao Z, et al. Aquatic predicted no-effect concentrations of 16 polycyclic aromatic hydrocarbons and their ecological risks in surface seawater of Liaodong Bay, China. Environ Toxicol Chem 2016;35(6):1587–93.
[49] Donkin P, Widdows J, Evans SV, Brinsley MD. QSARs for the sublethal responses of marine mussels (Mytilus edulis). Sci Total Environ 1991; 109–110:461–76.
[50] Fan D, Liu J, Wang L, Yang X, Zhang S, Zhang Y, et al. Development of quantitative structure-activity relationship models for predicting chronic toxicity of substituted benzenes to daphnia magna. Bull Environ Contam Toxicol 2016;96(5):664–70.
[51] Ding F, Wang Z, Yang X, Shi L, Liu J, Chen G. Development of classification models for predicting chronic toxicity of chemicals to Daphnia magna and Pseudokirchneriella subcapitata. SAR QSAR Environ Res 2019;30(1):39–50.
[52] Choi K, Sweet LI, Meier PG, Kim PG. Aquatic toxicity of four alkylphenols (3-tert-butylphenol, 2-isopropylphenol, 3-isopropylphenol, and 4-isopropylphenol) and their binary mixtures to microbes, invertebrates, and fish. Environ Toxicol 2004;19(1):45–50.
[53] Vighi M, Finizio A, Villa S. The evolution of the environmental quality concept: from the us epa red book to the European water framework directive. Environ Sci Pollut Res 2006;13(1):9–14.
[54] Zhang R, Wu F, Li H, Feng C, Guo G. Deriving aquatic water quality criteria for inorganic mercury in China by species sensitivity distributions. Acta Sci Circumstantiae 2012;32(02):440–9.
[55] Aldenberg T, Jaworska J, Traas TP. Normal species sensitivity distributions and probabilistic ecological risk assessment. 2001.
[56] Yanagihara M, Hiki K, Iwasaki Y. Can chemical toxicity in saltwater be predicted from toxicity in freshwater? A comprehensive evaluation using species sensitivity distributions. Environ Toxicol Chem 2022;41(8):2021–7.
[57] Mu YS, Wu FC, Chen C, Liu YD, Zhao XL, Liao HQ, et al. Predicting criteria continuous concentrations of 34 metals or metalloids by use of quantitative ion character-activity relationships-species sensitivity distributions (QICAR-SSD) model. Environ Pollut 2014;188:50–5.
[58] Wu F, Mu Y, Chang H, Zhao X, Giesy JP, Wu KB. Predicting water quality criteria for protecting aquatic life from physicochemical properties of metals or metalloids. Environ Sci Technol 2013;47(1):446–53.
[59] Chen C, Mu Y, Wu F, Zhang R, Su H, Giesy JP. Derivation of marine water quality criteria for metals based on a novel QICAR-SSD model. Environ Sci Pollut Res Int 2015;22(6):4297–304.
[60] Song R, Li D, Chang A, Tao M, Qin Y, Keller AA, et al. Accelerating the pace of ecotoxicological assessment using artificial intelligence. Ambio 2022;51(3):598–610.
[61] Mu Y, Wang Z, Wu F, Zhong B, Yang M, Sun F, et al. Model for predicting toxicities of metals and metalloids in coastal marine environments worldwide. Environ Sci Technol 2018;52(7):4199–206.
[62] Mu J, Wang J, Hong M. Methods of deriving marine water quality criterion and proposal for establishment of national marine water quality criterion in China. Asian J Ecotoxicol 2010;5(6):761–8.
[63] Jung J-W, Park J-W, Eo S, Choi J, Song YK, Cho Y, et al. Ecological risk assessment of microplastics in coastal, shelf, and deep sea waters with a consideration of environmentally relevant size and shape. Environ Pollut 2021;270:116217.
[64] Garner KL, Suh S, Lenihan HS, Keller AA. Species sensitivity distributions for engineered nanomaterials. Environ Sci Technol 2015;49(9): 5753–9.
[65] Lu C, Yang S, Yan Z, Ling J, Jiao L, He H, et al. Deriving aquatic life criteria for PBDEs in China and comparison of species sensitivity distribution with TBBPA and HBCD. Sci Total Environ 2018;640:1279–85.
[66] Huang P, Liu S-S, Wang Z-J, Ding T-T, Xu Y-Q. Deriving the predicted no effect concentrations of 35 pesticides by the QSAR-SSD method. Chemosphere 2022;298:134303.
[67] Li J, Su G, Letcher RJ, Xu W, Yang M, Zhang Y. Liquid crystal monomers (LCMs): a new generation of persistent bioaccumulative and toxic (PBT) compounds? Environ Sci Technol 2018;52(9):5005–6.
[68] Conder JM, Hoke RA, De Wolf W, Russell MH, Buck RC. Are PFCAs bioaccumulative? A critical review and comparison with regulatory lipophilic compounds. Environ Sci Technol 2008;42(4):995–1003.
[69] Reemtsma T, Benito Quintana J, Rodil R, Garcia-Lopez M, Rodriguez I. Organophosphorus flame retardants and plasticizers in water and air I. Occurrence and fate. Trac-Trends Anal Chem 2008;27(9):727–37.

Index

'*Note:* Page numbers followed by "f" indicate figures and "t" indicate tables.'

D

E

T

U

V

W

X

Z

Printed and bound by CPI Group (UK) Ltd, Croydon, CR0 4YY
14/09/2026
02216388-0012